Machine Learning in MRI

Advances in Magnetic Resonance Technology and Applications Series

Titles published:
Volume 1 – Quantitative Magnetic Resonance Imaging – Edited by Nicole Seiberlich and Vikas Gulani
Volume 2 – Handbook of Pediatric Brain Imaging – Edited by Hao Huang and Timothy Roberts
Volume 3 – Hyperpolarized Carbon-13 Magnetic Resonance Imaging and Spectroscopy – Edited by Peder Larson
Volume 4 – Advanced Neuro MR Techniques and Applications – Edited by In-Young Choi and Peter Jezzard
Volume 5 – Breast MRI – Edited by Katja Pinker, Ritse Mann and Savannah Partridge
Volume 6 – Motion Correction in MR – Edited by Andr_e Van Der Kouwe and Jalal B. Andre
Volume 7 – Magnetic Resonance Image Reconstruction – Edited by Mehmet Akçakaya, Mariya Doneva and Claudia Prieto
Volume 8 – MR Linac Radiotherapy – Edited by Enis Ozyar, Cem Onal and Sara L. Hackett
Volume 9 – Imaging Neuroinflammation – Edited by Cornelia Laule and John D. Port
Volume 10 – Ultra-High Field Neuro MRI – Edited by Karin Markenroth Bloch, Maxime Guye and Benedikt A. Poser
Volume 11 – Cheng and Strijkers: *Quantitative Perfusion MRI*, ISBN: 9780323952095 August 2023
Volume 12 – The Chemistry of Hyperpolarized Magnetic Resonance Probes – Edited by Eul Hyun Suh and Zoltan Kovacs
Visit the Series webpage at https://www.elsevier.com/books/book-series/advances-in-magneticresonance-technology-and-applications

Advances in Magnetic Resonance Technology and Applications

Machine Learning in MRI: From Methods to Clinical Translation

Volume 13

Edited by

Thomas Küstner
Medical Imaging and Data Analysis (MIDAS.lab), Diagnostic and Interventional Radiology, University Hospital of Tüebingen, Tübingen, Germany

Hao Huang
Radiology in the Perelman School of Medicine, University of Pennsylvania, Philadelphia, PA, United States
Small Animal Imaging Facility, Children's Hospital of Philadelphia, Philadelphia, PA, United States

Christian F. Baumgartner
Machine Learning for Medical Image Analysis Group, University of Tübingen, Tübingen, Germany

Sam Payabvash
Yale University, New Haven, CT, United States

ELSEVIER
ACADEMIC PRESS
An imprint of Elsevier

Academic Press is an imprint of Elsevier
125 London Wall, London EC2Y 5AS, United Kingdom
50 Hampshire Street, 5th Floor, Cambridge, MA 02139, United States

ISBN: 978-0-443-14109-6
ISSN: 2666-9099

For information on all Academic Press publications visit our website at https://www.elsevier.com/books-and-journals

Publisher: Mara Conner
Acquisitions Editor: Saniya Puri
Editorial Project Manager: Ritika Dagar
Production Project Manager: Fizza Fathima
Cover Designer: Mark Rogers

Typeset by TNQ Technologies

Contents

PART 5 Generalization and fairness

Contributors

Ben Adcock
Simon Fraser University, Burnaby, BC, Canada

Pouria Yazdian Anari
Radiology Department, Creighton University/CHI, Omaha, NE, United States; Radiology Department, Clinical Center, NIH, Bethesda, MD, United States

Vegard Antun
University of Oslo, Oslo, Norway

Jessica A.M. Bastiaansen
Department of Diagnostic, Interventional and Pediatric Radiology (DIPR), Inselspital, Bern University Hospital, University of Bern, Bern, Switzerland; Translational Imaging Center (TIC), Swiss Institute for Translational and Entrepreneurial Medicine, Bern, Switzerland

Christian Baumgartner
University of Tübingen, Cluster of Excellence – Machine Learning for Science, Tübingen, Germany

Wolfgang Bogner
High-Field MR Center, Department of Biomedical Imaging and Image-Guided Therapy, Medical University of Vienna, Vienna, Austria; Comprehensive Center for AI in Medicine (CAIM), Medical University of Vienna, Vienna, Austria; Christian Doppler Laboratory for MR Imaging Biomarkers (BIOMAK), Department of Biomedical Imaging and Image-Guided Therapy, Medical University of Vienna, Vienna, Austria

Rene M. Botnar
King's College London, London, United Kingdom; Pontificia Universidad Católica de Chile, Santiago de Chile, Chile

Matthan W.A. Caan
Amsterdam UMC Location University of Amsterdam, Biomedical Engineering and Physics, Amsterdam, The Netherlands

Gustavo Carneiro
Centre for Vision, Speech and Signal Processing (CVSSP), University of Surrey, Surrey, United Kingdom

Akshay S. Chaudhari
Department of Radiology, Stanford University, Stanford, CA, United States; Department of Biomedical Data Science, Stanford University, Stanford, CA, United States

Chen (Cherise) Chen
School of Computer Science, Insigneo Institute, University of Sheffield, Sheffield, United Kingdom

Mingshi Chen
Amsterdam UMC Location University of Amsterdam, Biomedical Engineering and Physics, Amsterdam, The Netherlands

Tolga Çukur
Bilkent University, Department of Electrical and Electronics Engineering, Ankara, Turkey

Onat Dalmaz
Stanford University, Department of Electrical Engineering, Stanford, CA, United States

O. Burak Demirel
Department of Radiology, Mayo Clinic Rochester, Rochester, MN, United States; Philips Healthcare, MR Clinical Science, Philips North America, Cambridge, MA, United States

Mariya Doneva
Philips Innovative Technologies, Hamburg, Germany

Maryam Etesami
University of Toronto Temerty Medicine, Department of Medical Imaging, Toronto, ON, Canada

Qiuyun Fan
Academy of Medical Engineering and Translational Medicine, Medical College, Tianjin University, Tianjin, China

Li Feng
Center for Advanced Imaging Innovation and Research (CAI2R), New York University Grossman School of Medicine, New York, NY, United States

Aasa Feragen
DTU Compute, Technical University of Denmark, Lyngby, Denmark

Melanie Ganz-Benjaminsen
Department of Computer Science, University of Copenhagen, Copenhagen, Denmark; Neurobiology Research Unit, Rigshospitalet, Copenhagen University Hospital, Copenhagen, Denmark

Sergios Gatidis
Department of Radiology, Stanford University, Stanford, CA, United States

Nina M. Gottschling
Deutsches Zentrum für Luft- und Raumfahrt (DLR), Köln, Germany

Kerstin Hammernik
Technical University of Munich, School of Computation, Information and Technology, Munich, Germany

Anders C. Hansen
University of Cambridge, Cambridge, United Kingdom

Reinhard Heckel
Department of Computer Engineering, Technical University of Munich, Munich, Germany

Kelly K. Horst
Department of Radiology, Mayo Clinic Rochester, Rochester, MN, United States

Mathews Jacob
Department of Electrical and Computer Engineering, University of Lowa, Lowa, IA, United States

Sergey Kastryulin
Yandex Research, Moscow, Russia

Eric Kerfoot
School of Biomedical Engineering & Imaging Sciences, King's College London, London, United Kingdom

Iman Khodarahmi
New York University School of Medicine, New York, NY, United States

Andreas Kofler
Physikalisch-Technische Bundesanstalt (PTB), Braunschweig and Berlin, Germany

Christoph Kolbitsch
Physikalisch-Technische Bundesanstalt (PTB), Braunschweig and Berlin, Germany

Thomas Küstner
University Hospital Tübingen, Medical Image and Data Analysis, Department of Radiology, Tübingen, Germany

Taro Langner
Antaros Medical, BioVenture Hub, Mölndal, Sweden

Tim Leiner
Department of Radiology, Mayo Clinic Rochester, Rochester, MN, United States

Haoxiang Li
School of Biomedical Engineering, Tsinghua University, Beijing, China

Zeju Li
FMRIB Centre, Wellcome Centre for Integrative Neuroimaging, University of Oxford, Oxford, United Kingdom

Zihan Li
School of Biomedical Engineering, Tsinghua University, Beijing, China

Ziyu Li
School of Biomedical Engineering, Tsinghua University, Beijing, China; Wellcome Centre for Integrative Neuroimaging, FMRIB, Nuffield Department of Clinical Neurosciences, University of Oxford, Oxford, United Kingdom

Fang Liu
Athinoula A. Martinos Center for Biomedical Imaging, Massachusetts General Hospital, Charlestown, MA, United States; Harvard Medical School, Boston, MA, United States

Adèle L.C. Mackowiak
Department of Medical Radiology, Lausanne University Hospital (CHUV-UNIL), Lausanne, Switzerland

Ashkan A. Malayeri
Radiology Department, Clinical Center, NIH, Bethesda, MD, United States

Karla L. Miller
Wellcome Centre for Integrative Neuroimaging, FMRIB, Nuffield Department of Clinical Neurosciences, University of Oxford, Oxford, United Kingdom

Marc Modat
School of Biomedical Engineering & Imaging Sciences, King's College London, London, United Kingdom

Stanislav Motyka
High-Field MR Center, Department of Biomedical Imaging and Image-Guided Therapy, Medical University of Vienna, Vienna, Austria; Comprehensive Center for AI in Medicine (CAIM), Medical University of Vienna, Vienna, Austria; Christian Doppler Laboratory for MR Imaging Biomarkers (BIOMAK), Department of Biomedical Imaging and Image-Guided Therapy, Medical University of Vienna, Vienna, Austria

Sara Mozayan
UCLA Health (University of California), Los Angeles, CA, United States

Benjamin Murray
School of Biomedical Engineering & Imaging Sciences, King's College London, London, United Kingdom

Jovitha Nelson
Interventional Radiology Department, Case Western University Hospitals Cleveland Medical Center, Cleveland, OH, United States

Sebastien Ourselin
School of Biomedical Engineering & Imaging Sciences, King's College London, London, United Kingdom

Haoyang Pei
Center for Advanced Imaging Innovation and Research (CAI2R), New York University Grossman School of Medicine, New York, NY, United States; Department of Electrical and Computer Engineering, NYU Tandon School of Engineering, New York, NY, United States

Eva S. Peper
Department of Diagnostic, Interventional and Pediatric Radiology (DIPR), Inselspital, Bern University Hospital, University of Bern, Bern, Switzerland; Translational Imaging Center (TIC), Swiss Institute for Translational and Entrepreneurial Medicine, Bern, Switzerland

Fernando Pérez-García
Microsoft Research Cambridge, Cambridge, United Kingdom

Eike Petersen
DTU Compute, Technical University of Denmark, Lyngby, Denmark

Nicola Pezzotti
Eindhoven University of Technology, Eindhoven, The Netherlands; Philips Healthcare, Eindhoven, The Netherlands

Claudia Prieto
King's College London, London, United Kingdom; Pontificia Universidad Católica de Chile, Santiago de Chile, Chile

Haikun Qi
School of Biomedical Engineering, ShanghaiTech University, Shanghai, China

Chen Qin
I-X and Department of Electrical and Electronic Engineering, Imperial College London, London, United Kingdom

Ravinder R. Regatte
NYU Grossman School of Medicine, New York, NY, United States

Annika Reinke
German Cancer Research Center (DKFZ) Heidelberg, Division of Intelligent Medical Systems, Heidelberg, Germany; German Cancer Research Center (DKFZ) Heidelberg, HI Helmholtz Imaging, Heidelberg, Germany

David Schote
Physikalisch-Technische Bundesanstalt (PTB), Braunschweig and Berlin, Germany

Delaram Shakoor
Hospital for Special Surgery, New York, NY, United States

Efrat Shimron
Department of Electrical and Computer Engineering, Technion-Israel Institute of Technology, Haifa, Israel; Department of Biomedical Engineering, Technion-Israel Institute of Technology, Haifa, Israel

Dongyue Si
King's College London, London, United Kingdom

Kyle Tegtmeyer
Department of Radiology and Biomedical Imaging, Yale University School of Medicine, New Haven, CT, United States

Qiyuan Tian
School of Biomedical Engineering, Tsinghua University, Beijing, China

Minu D. Tizabi
German Cancer Research Center (DKFZ) Heidelberg, Division of Intelligent Medical Systems, Heidelberg, Germany; National Center for Tumor Diseases (NCT), NCT Heidelberg, a partnership between DKFZ and University Medical Center, Heidelberg, Germany

Long Tu
Department of Radiology and Biomedical Imaging, Yale University School of Medicine, New Haven, CT, United States

Shanshan Wang
Paul C. Lauterbur Research Center for Biomedical Imaging, Shenzhen Institutes of Advanced Technology, Chinese Academy of Sciences, Shenzhen, China

Ruoyou Wu
Paul C. Lauterbur Research Center for Biomedical Imaging, Shenzhen Institutes of Advanced Technology, Chinese Academy of Sciences, Shenzhen, China; University of Chinese Academy of Sciences, Beijing, China

Wenchuan Wu
Wellcome Centre for Integrative Neuroimaging, FMRIB, Nuffield Department of Clinical Neurosciences, University of Oxford, Oxford, United Kingdom

Bin Yang
University of Stuttgart, Department of Electrical Engineering and Information Technology, Stuttgart, Germany

Marcelo V.W. Zibetti
Center for Biomedical Imaging, Department of Radiology, NYU Grossman School of Medicine, New York University, New York, NY, United States

Preface

We can all agree that artificial intelligence (AI), machine learning (ML), and deep learning (DL) are currently hot topics. AI serves as an umbrella term to solve any task that usually requires human intelligence. ML is a subbranch of AI that describes any data-driven learning model to predict the output of a particular task and summarizes (statistical) methods that enable machines to improve with experience. DL is a subbranch of ML that enables automated feature extraction and comprises learnable networks. In the following, we will use the term ML as the wholesome collection of statistical and learnable models with an emphasis on the model and learning concept.

With the rise of large language and vision models, these technologies have become more accessible than ever. At the same time, an increasing number of products are entering the market, making it difficult to navigate and identify the most suitable solutions. This growing landscape underscores the importance of understanding the potentials, strengths, and limitations of ML for magnetic resonance imaging (MRI). To achieve this, we need to dive deep into the respective methods and solutions. This book is therefore named *Machine Learning in MRI: From Methods to Clinical Translation* and is part of a book series on *Advances in Magnetic Resonance Technology and Applications*. The book is targeted to be a reference for medical imaging researchers, industry scientists and engineers, advanced undergraduate and graduate students, and clinicians.

Given the complexity of MRI as an imaging modality, we structured this book into distinct parts, each focusing on the different stages of the MR processing chain—from acquisition over reconstruction to postprocessing and analysis. Accordingly, we explore how ML solutions influence these various stages, presenting both well-established and cutting-edge approaches in MRI. The book highlights how ML-driven advancements can shape the future of patient treatment and planning, with a special emphasis on the applications and anatomical regions where ML has already made, or is about to make, significant contributions. As we strive toward more generalizable and scalable models, this book explores ideas and concepts of data harmonization, standardized processing, and how this can contribute to a reproducible research environment. The book also outlines the role of human expertise in this evolving landscape and explores how we can develop objective and equitable ML solutions that integrate seamlessly into clinical workflows. To this end, each chapter addresses the potential challenges and current limitations, offering insights into how they can be overcome to better meet the clinical needs.

In "Basics of machine learning and magnetic resonance imaging" (Part I), the main ingredients for understanding the proposed methods and concepts of the following chapters are presented, covering the statistical basics (Chapter 1), ML concepts and networks (Chapter 2), and the physics of MR imaging (Chapter 3). A comprehensive review of all the basics in ML and MRI would be too extensive and fill one (or multiple) books by itself. We, therefore, encourage the reader to check out further material as referenced in these chapters, or relate to the other books of this book series.

In “MR image acquisition” (Part II), the existing and potential contributions of ML to the acquisition side are discussed. Chapter 4 discusses the advancements in MRI data acquisition, moving from traditional human-designed sampling patterns to ML-driven approaches that optimize k-space trajectories and reconstruction for enhanced efficiency. Chapter 5 focuses on the resurgence of low-field and ultralow-field MRI, addressing their advantages, challenges, and how ML-based techniques like image reconstruction and denoising can enhance their clinical utility and accessibility. Chapter 6 examines the growing role of ML in MR spectroscopy, highlighting its applications in data acquisition, processing, and analysis.

In “MR image reconstruction” (Part III), the chapters describe the role of ML in MRI reconstruction, resolution enhancement, motion correction, quantitative imaging, and its trustworthiness. Chapter 7 explores the techniques for solving inverse problems in MRI reconstruction, covering classical and ML methods, including GANs and diffusion models, while also addressing challenges like robustness and generalization. Chapter 8 focuses on ML approaches to enhance spatial resolution, examining their clinical and neuroscientific impact, as well as the challenges of reliability and feasibility. Chapter 9 tackles motion correction in MRI, discussing traditional and ML-based methods for reducing motion artifacts and estimating motion in both the image and k-space. Chapter 10 provides an overview of ML in quantitative MRI, outlining its integration into the workflow and potential for clinical impact. Finally, Chapter 11 highlights critical concerns about the reliability of ML-based MRI reconstruction, detailing risks, like hallucinations, instabilities, and poor generalization, and emphasizing the need for trustworthy ML in medical imaging.

In “MR image postprocessing” (Part IV), the chapters explore the key advancements in ML for MRI analysis, covering segmentation, quality assessment, and image translation. Chapter 12 examines ML-driven semantic segmentation, detailing architectures, training methodologies, and challenges such as data scarcity and bias, with applications in brain and cardiac imaging. Chapter 13 focuses on automated image quality assessment, discussing objective evaluation metrics and the challenges of aligning automated methods with radiologists’ assessments to improve diagnostic reliability. Chapter 14 explores ML in MRI analysis, covering radiomics, DL, and their clinical applications in diseases such as cancer and stroke, while also highlighting emerging areas like normative modeling and integration with language models for personalized medicine. Chapter 15 describes image-to-image translation, where techniques like transformers and diffusion models facilitate the reconstruction of missing contrasts from acquired MRI data, reducing scan times and improving efficiency in multimodal imaging.

The chapters in “Generalization and Fairness” (Part V) cover the aspects of how we can scale-up the processing to different domains while integrating the human expert into learning, and while considering the causal relationships between data and ML model outcomes. Chapter 16 discusses the limitations of ML models in MRI, highlighting the issues of generalizability, domain shifts, bias, and shortcut learning, while also presenting potential solutions such as model invariance and causal modeling. Chapter 17 examines the use of large-scale MRI datasets, addressing the challenges related to distribution shifts and exploring techniques like transfer learning, federated learning, and representation learning to enhance ML model training across diverse data sources. Chapter 18 introduces human-in-the-loop approaches, emphasizing the role of clinicians in ML model development to mitigate biases, improve reliability, and ensure clinical relevance through techniques like active learning and reinforcement learning.

The translation of developed models—in the research and product stage—is presented in "Clinical Applications" (Part VI) for a selected range of MR-imaged body regions. Chapter 19 covers ML applications in neuroradiology, including the brain, neck, and spine MRI for conditions like cancer, dementia, and stroke, while addressing the challenges in clinical adoption. Chapter 20 focuses on ML in cardiovascular MRI, detailing its role in automating acquisition, reconstruction, and biomarker extraction to improve diagnostic precision, despite the challenges in standardization and interpretability. Chapter 21 examines ML contributions to body MRI, particularly in abdominal, renal, pelvic, and gastrointestinal imaging, where learning improves disease detection and segmentation. Chapter 22 highlights the ML-driven advancements in breast MRI, including lesion detection, tumor classification, and treatment response prediction, while acknowledging barriers like model generalizability and data standardization. Chapter 23 explores the role of ML in musculoskeletal MRI, improving the diagnosis of injuries, tumors, and infections, although clinical validation remains a challenge.

To foster future research and collaborative efforts, the chapters in "Reproducibility" (Part VII) present the current initiatives and good practice guidelines. Chapter 24 discusses the role of open-source software, datasets, and ML models in advancing research and commercial applications in medical imaging. It highlights the benefits of reproducibility, security, and interdisciplinary collaboration while addressing the challenges and future directions in the open-source landscape. Chapter 25 focuses on the role of benchmarking challenges in evaluating ML algorithms for MRI image analysis. It examines the best practices in challenge design, including metric selection, ranking methodologies, and transparent reporting, while also exploring the recent developments and open research questions in MRI-related competitions.

In the last part "Conclusion" (Part VIII), we—the editors—draw a summary of the discussed limitations, pitfalls, challenges, and outlooks in each of the chapters, to initiate a discussion on the potential future research directions. With this book, we would like to highlight the potentials of ML in MRI, but also raise awareness for the easy-to-miss pitfalls. We hope these discussions initiate further open-source and collaborative research.

We are very grateful for all the contributions and time invested by the chapter authors. We thank the Elsevier team, especially Emily Thomson (our Editorial Project Manager at Elsevier) for her dedication and support in guiding this book to completion and Tim Pitts for initiating the discussions that laid the foundation for this book.

Editors
Thomas Küstner
Medical Image and Data Analysis laboratory (MIDAS.lab),
Department of Radiology
University Hospital of Tübingen
Hao Huang
Department of Radiology
University of Pennsylvania
Christian Baumgartner
Machine Learning for Medical Image Analysis Group
University of Tübingen
Sam Payabvash
Department of Radiology
Yale University

PART

Basics of machine learning and magnetic resonance imaging

1

CHAPTER

The statistics behind machine learning

1

Bin Yang

University of Stuttgart, Department of Electrical Engineering and Information Technology, Stuttgart, Germany

Corresponding author: E-mail: bin.yang@iss.uni-stuttgart.de

1.1 Introduction

To understand machine learning (ML), one needs mainly three mathematical tools: vector and matrix calculation, probability theory, and optimization. This chapter gives an introduction into the first two areas. We give a concise summary of the mathematical concepts and tools, which are necessary for an understanding of the next chapters. A detailed derivation of these concepts and a discussion of all properties is beyond the scope of this chapter.

Optimization is not addressed in this chapter. The whole area of optimization including all variants such as single-objective versus multiobjective, deterministic versus stochastic, continuous versus discrete, smooth versus nonsmooth, linear versus nonlinear, unconstrained versus constrained, convex versus nonconvex, and local versus global optimization is too broad for one chapter. Luckily, machine learning requires only one pretty narrow field of optimization. In most cases, the training of a machine learning model with continuous-valued parameters (weight, bias, kernel) leads to a single-objective, continuous, and mostly nonconvex minimization problem, optionally subject to some regularization constraints. In particular, when a deep neural network (DNN) with a large number of parameters is used as a machine learning model, only the simplest family of optimization algorithms, i.e., gradient descent and its various modifications, is applicable due to complexity reason. Therefore, only relatively shallow knowledge about optimization is required for machine learning. Two recommended text books about optimization for machine learning are *Convex optimization* and *Optimization for machine learning* [1,2].

1.2 Vector and matrix calculation

1.2.1 Basic notations

We use the following notations throughout this book. $\mathbb{N}, \mathbb{Z}, \mathbb{R}$, and $\mathbb{C}$ denote the set of natural, integer, real, and complex numbers, respectively. x denotes a scalar number. $\mathbf{x}$ is a column vector with the i-th element x_i or $[\mathbf{x}]_i$. $\mathbf{X}$ denotes a matrix

Machine Learning in MRI. https://doi.org/10.1016/B978-0-443-14109-6.00001-8

with the (i,j)-th element x_{ij} or $[\mathbf{X}]_{ij}$. The transpose of a matrix $\mathbf{X}$ is denoted by $\mathbf{X}^T$, while $\mathbf{X}^*$ and $\mathbf{X}^H = (\mathbf{X}^*)^T$ represent the complex conjugate and Hermitian transpose of $\mathbf{X}$, respectively. Given two matrices $\mathbf{X}$ and $\mathbf{Y}$, $\mathbf{X} \circ \mathbf{Y}$ is the Hadamard (elementwise) product while $\mathbf{X} \otimes \mathbf{Y}$ denotes the Kronecker product of $\mathbf{X}$ and $\mathbf{Y}$.

Vector and matrix are one- or two-dimensional arrays of numbers. An extension to high-dimensional arrays is called tensor. 3D, 4D, and even higher-dimensional data and kernel tensors will be required later in convolutional layers of neural networks or in any operations involving feature maps. They are also denoted by a boldface capital letter $\mathbf{X}$. The notation $\text{vec}(\mathbf{X})$ is used to vectorize a matrix or tensor $\mathbf{X}$, i. e., stacking all elements of $\mathbf{X}$ into a single long column vector.

Given a square matrix $\mathbf{X}$, the determinant $\det(\mathbf{X})$ and trace $\text{tr}(\mathbf{X})$ are two scalar functions of $\mathbf{X}$. The trace is defined as the sum of all main diagonal elements of $\mathbf{X}$. Given any matrix $\mathbf{X}$, $\text{rank}(\mathbf{X})$ denotes the rank of $\mathbf{X}$, i.e., the number of linearly independent columns (or rows) of $\mathbf{X}$.

1.2.2 Vector and matrix norms

The norm $\|\mathbf{x}\| \geq 0$ of a vector $\mathbf{x} \in \mathbb{C}^M$ is a nonnegative scalar function of $\mathbf{x}$. It reflects the "length" of $\mathbf{x}$ in a specific sense. A popular family of vector norms are the p-norms

$$\|\mathbf{x}\|_p = \left(\sum_{i=1}^{M} |x_i|^p \right)^{1/p}, \quad p \geq 0. \tag{1.1}$$

Among them, the L_2, L_1, and L_0 norms are mostly used in machine learning. The L_2 norm $\|\mathbf{x}\|_2 = \sqrt{\sum_{i=1}^{M} |x_i|^2}$ is also known as the Euclidean norm. The squared L_2 norm $\|\mathbf{x}\|_2^2$ reflects the vector energy of $\mathbf{x}$. The L_1 norm $\|\mathbf{x}\|_1 = \sum_{i=1}^{M} |x_i|$ is the sum of magnitude of the elements of $\mathbf{x}$. The L_0 norm $\|\mathbf{x}\|_0$ counts the number of nonzero elements and is a nondifferentiable measure for the sparsity of $\mathbf{x}$. Different vector norms lead to different shapes of the unit-norm contour line $\{\mathbf{x} \big| \|\mathbf{x}\|_p = 1\}$, see Fig. 1.1. They are all used in machine learning to define a suitable loss function and different regularization terms.

There are also different matrix norms $\|\mathbf{X}\| \geq 0$. One popular family of matrix norms are the p-norms, induced by the corresponding vector norms:

$$\|\mathbf{X}\|_p = \max_{\|\mathbf{a}\|_p = 1} \|\mathbf{X}\mathbf{a}\|_p, \quad p \geq 1. \tag{1.2}$$

The L_2 matrix norm $\|\mathbf{X}\|_2 = \sigma_1$ is the largest singular value σ_1 of $\mathbf{X}$ and is also called the spectral norm. The L_1 matrix norm $\|\mathbf{X}\|_1 = \max_j \sum_i |x_{ij}|$ is the largest L_1 norm of the columns of $\mathbf{X}$, while the L_∞ matrix norm $\|\mathbf{X}\|_\infty = \max_i \sum_j |x_{ij}|$ is the largest L_1 norm of the rows of $\mathbf{X}$ [3].

Two other matrix norms are widely used in machine learning. The Frobenius norm $\|\mathbf{X}\| = \|\text{vec}(\mathbf{X})\|_2 = \sqrt{\sum_i \sum_j |x_{ij}|^2}$ is a straightforward extension of the

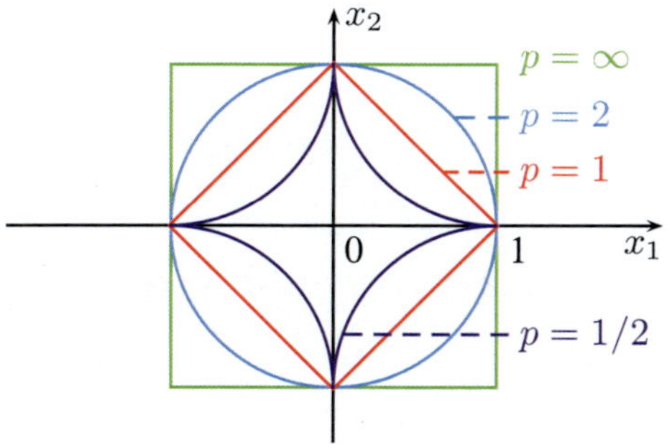

FIGURE 1.1

Unit-norm contour lines for different vector norms.

Euclidean vector norm to matrices and reflects the square root of the matrix energy. The nuclear norm $\|\mathbf{X}\|_* = \sum_i \sigma_i$ is the sum of all singular values σ_i of $\mathbf{X}$.

1.2.3 Vector derivative

In machine learning, one has often to minimize a loss function of a parameter vector (or matrix or tensor), which is a function of another parameter vector (or matrix or tensor). In this case, vector derivative becomes useful.

Let $\mathbf{y} \in \mathbb{R}^M$ be a differentiable vector function of $\mathbf{x} \in \mathbb{R}^N$. The derivative of $\mathbf{y}(\mathbf{x})$ with respect to $\mathbf{x}$ is defined as

$$\mathbf{J} = \frac{\partial \mathbf{y}}{\partial \mathbf{x}} = \left[\frac{\partial y_i}{\partial x_j}\right]_{ij} = \begin{bmatrix} \frac{\partial y_1}{\partial x_1} & \cdots & \frac{\partial y_1}{\partial x_N} \\ \vdots & \ddots & \vdots \\ \frac{\partial y_M}{\partial x_1} & \cdots & \frac{\partial y_M}{\partial x_N} \end{bmatrix} \in \mathbb{R}^{M \times N} \tag{1.3}$$

and is called the Jacobi matrix. Clearly, the derivative of a vector $\mathbf{y}(x) \in \mathbb{R}^M$ with respect to a scalar number x is a column vector

$$\frac{\partial \mathbf{y}}{\partial x} = \begin{bmatrix} \frac{\partial y_1}{\partial x} \\ \vdots \\ \frac{\partial y_M}{\partial x} \end{bmatrix}. \tag{1.4}$$

But the derivative of a scalar function $y(\mathbf{x})$ with respect to a vector $\mathbf{x} \in \mathbb{R}^N$ is a row vector $\frac{\partial y}{\partial \mathbf{x}} = \left[\frac{\partial y}{\partial x_1}, \ldots, \frac{\partial y}{\partial x_N}\right]$. It is the transpose of the gradient vector of $y(\mathbf{x})$

$$\nabla y = \begin{bmatrix} \frac{\partial y}{\partial x_1} \\ \vdots \\ \frac{\partial y}{\partial x_N} \end{bmatrix} = \left(\frac{\partial y}{\partial \mathbf{x}}\right)^T. \tag{1.5}$$

As an example, let $L(\mathbf{x}(\mathbf{y}(\theta))) \in \mathbb{R}$ be a cascaded function of the scalar parameter $\theta \in \mathbb{R}$ where $\mathbf{x}$ and $\mathbf{y}$ are two intermediate vectors. Then the chain rule in calculus tells us

$$\frac{\partial L}{\partial \theta} = \frac{\partial L}{\partial \mathbf{x}} \cdot \frac{\partial \mathbf{x}}{\partial \mathbf{y}} \cdot \frac{\partial \mathbf{y}}{\partial \theta}. \tag{1.6}$$

This chain rule is, for example, heavily used in backpropagation when training a neural network.

1.3 Random variable and probability distribution

A random variable (RV) is a mathematical formalization of a quantity whose possible values depend on the occurrence of random events. There are basically two types of random variables, discrete and continuous. When we study an MRI dataset, the input MR image for a machine learning model is represented by a random vector $\mathbf{x}$ (or matrix/tensor $\mathbf{X}$). In classification applications, the output of the machine learning model is a discrete-valued random vector $\mathbf{y}$ containing, e.g., estimated subject gender, health state (cancer or not), etc. In regression tasks, the output of the machine learning model is a continuous-valued random vector $\mathbf{y}$ (or matrix/tensor $\mathbf{Y}$), which may contain estimated subject age, body mass index (BMI), or estimated images (e.g., image reconstruction, image superresolution, image-to-image translation). Machine learning for MRI aims to draw conclusions about one random vector $\mathbf{y}$ from another one $\mathbf{x}$.

1.3.1 One random vector

Without loss of generality, we first consider a single real-valued random vector $\mathbf{x} = [x_1, \ldots, x_N]^T \in \mathbb{R}^N$. Any random matrix or tensor $\mathbf{X}$ can be reshaped to a random vector by vectorization $\text{vec}(\mathbf{X})$. Also a complex-valued vector $\mathbf{z}$, which occurs, for example, in the k-space of MRI, can be reformulated to a real-valued one

$\mathbf{x} = \begin{bmatrix} \text{Re}(\mathbf{z}) \\ \text{Im}(\mathbf{z}) \end{bmatrix}$ where $\text{Re}(\mathbf{z})$ and $\text{Im}(\mathbf{z})$ represent the real and imaginary part of $\mathbf{z}$, respectively.

The statistical properties of a random vector $\mathbf{x}$ are described by the distribution. It can be the probability mass function (PMF) $P()$ or the probability density function (PDF) $p()$. When $\mathbf{x}$ is discrete-valued, i.e., $\mathbf{x} \in \{\mathbf{x}_1, \mathbf{x}_2, \ldots\}$, one prefers to use the PMF

$$P(\mathbf{x} = \mathbf{x}_i) = P(\mathbf{x}_i) = p_i \geq 0, \sum_i p_i = 1 \tag{1.7}$$

which denotes the probability of $\mathbf{x}$ taking the value $\mathbf{x}_i$. When $\mathbf{x}$ is continuous-valued, the use of the PDF $p(\mathbf{x})$ with $\int p(\mathbf{x})d\mathbf{x} = 1$, i.e., probability per unit volume, is more popular.

The simplest distribution is the uniform distribution. A continuous in $[a, b]$ uniformly distributed random variable x has the PDF

$$p(x) = \begin{cases} \frac{1}{b-a} & \text{for} \quad a \le x \le b \\ 0 & \text{for} \quad \text{otherwise} \end{cases}. \tag{1.8}$$

A more important distribution in machine learning is the normal or Gaussian distribution. A normal distributed random variable $x \sim N(\mu, \sigma^2)$ with the mean μ and variance σ^2 has the PDF

$$p(x) = \frac{1}{\sqrt{2\pi}\sigma} e^{-\frac{(x-\mu)^2}{2\sigma^2}}. \tag{1.9}$$

As a generalization, a length-d normal distributed random vector $\mathbf{x} \sim N(\boldsymbol{\mu}, \boldsymbol{\Sigma})$ with the mean vector $\boldsymbol{\mu} \in \mathbb{R}^d$ and covariance matrix $\boldsymbol{\Sigma} \in \mathbb{R}^{d \times d}$ has the PDF

$$p(\mathbf{x}) = \frac{1}{(2\pi)^{d/2} \det(\boldsymbol{\Sigma})^{1/2}} e^{-\frac{1}{2}(\mathbf{x}-\boldsymbol{\mu})^T \boldsymbol{\Sigma}^{-1} (\mathbf{x}-\boldsymbol{\mu})}. \tag{1.10}$$

$\mathbf{x}$ is called standard normal distributed when $\boldsymbol{\mu} = \mathbf{0}$ and $\boldsymbol{\Sigma}$ is equal the identity matrix $\mathbf{I}$. A random variable is Laplace distributed, in notation $x \sim \text{Laplace}(\mu, b)$, if

$$p(x) = \frac{1}{2b} e^{-\frac{|x-\mu|}{b}}. \tag{1.11}$$

In comparison with the normal distribution, the exponent is proportional to $|x - \mu|$ instead of $(x - \mu)^2$. This gives rise later to the L_1 and L_2 loss, respectively.

Among the discrete random variables, the Bernoulli distribution describes a binary random variable $x \in \{0, 1\}$. Its PMF is $P(x = 0) = 1 - p$ and $P(x = 1) = p$ with $0 \le p \le 1$. As a generalization, the categorical distribution characterizes a discrete random variable $x \in \{1, 2, \ldots, c\}$ with $c \ge 2$ possible values. Its PMF is

$$P(x = i) = p_i \ge 0, \sum_{i=1}^{c} p_i = 1. \tag{1.12}$$

Instead of this quite usual integer coding $x = i \in \mathbb{N}$ for a discrete random variable, the so-called one-hot coding is often employed in machine learning. In this case, the random variable x is rewritten as a random vector $\mathbf{x} = [x_1, \ldots, x_N]^T \in \{\mathbf{e}_1, \ldots, \mathbf{e}_c\}$ whose c possible values are the c unit vectors $\mathbf{e}_i$. Alle elements of $\mathbf{x}$ are 0 except for a single element of 1, hence the name one-hot coding. The position of 1 in $\mathbf{x}$ indicates the class index i. The advantage of the one-hot coding is a compact expression of the PMF

$$P(\mathbf{x}) = \begin{cases} p_1 & \text{if } \mathbf{x} = \mathbf{e}_1 \quad \text{or} \quad x_1 = 1 \\ \vdots & \\ p_c & \text{if } \mathbf{x} = \mathbf{e}_c \quad \text{or} \quad x_c = 1 \end{cases} = p_1^{x_1} \ldots p_c^{x_c} = \prod_{i=1}^{c} p_i^{x_i}. \tag{1.13}$$

This expression is useful later for an efficient calculation of the categorical cross-entropy loss.

The distribution, either PMF or PDF, is a complete characterization of the statistical properties of a random vector **x**. It is a function of **x**. Sometimes the distribution is not available or inconvenient to handle. In this case, the calculation of the moments of **x** is easier. They are expectation E(), i.e., statistical average over all random realizations of **x**, of some power functions $g(\mathbf{x})$ of **x**:

$$\mathrm{E}(g(\mathbf{x})) = \int g(\mathbf{x})p(\mathbf{x})\mathrm{d}\mathbf{x} \overset{\text{discreteRV}}{=} \sum_i g(\mathbf{x}_i)P(\mathbf{x}_i) \tag{1.14}$$

The first integral expression is used for a continuous random vector with the PDF $p(\mathbf{x})$ while the second sum expression is easier for a discrete random vector with the PMF $P(\mathbf{x}_i)$. Moments are just numbers. They are easier to handle than distribution, but represent often a partial description of the statistical properties. Moments can always be calculated from the distribution, but not vice versa in general.

The mean of **x** is the first-order moment

$$\boldsymbol{\mu} = \mathrm{E}(\mathbf{x}) = \int \mathbf{x}p(\mathbf{x})\mathrm{d}\mathbf{x} \overset{\text{discreteRV}}{=} \sum_i \mathbf{x}_i P(\mathbf{x}_i) \in \mathbb{R}^d. \tag{1.15}$$

The correlation matrix of **x** is a second-order moment

$$\mathbf{R} = \mathrm{E}(\mathbf{x}\mathbf{x}^T) = \int \mathbf{x}\mathbf{x}^T p(\mathbf{x})\mathrm{d}\mathbf{x} \overset{\text{discreteRV}}{=} \sum_i \mathbf{x}_i\mathbf{x}_i^T P(\mathbf{x}_i) \in \mathbb{R}^{d\times d}. \tag{1.16}$$

The covariance matrix is a second-order central moment of **x** after first removal of $\boldsymbol{\mu}$ from **x**

$$\boldsymbol{\Sigma} = \mathrm{E}\left((\mathbf{x}-\boldsymbol{\mu})(\mathbf{x}-\boldsymbol{\mu})^T\right). \tag{1.17}$$

1.3.2 Multiple random vectors

In machine learning, the case of two random vectors **x** and **y** is much more relevant than one random vector because we aim to draw conclusions about the hidden but desired state **y** from the observation **x**.

Again we assume two real-valued random vectors **x** and **y**. For simplicity, we consider only PDF for continuous random vectors. The joint statistical properties of the pair $(\mathbf{x}, \mathbf{y})$ are described by the joint PDF

$$p(\mathbf{x}, \mathbf{y}) = p\left(\begin{bmatrix}\mathbf{x}\\ \mathbf{y}\end{bmatrix}\right). \tag{1.18}$$

In comparison, the distributions of **x** or **y** alone regardless of **y** or **x** are called the marginal distributions. They can be calculated from the joint distribution

$$p(\mathbf{x}) = \int p(\mathbf{x}, \mathbf{y})\mathrm{d}\mathbf{y}, p(\mathbf{y}) = \int p(\mathbf{x}, \mathbf{y})\mathrm{d}\mathbf{x}. \tag{1.19}$$

In addition, the conditional distributions are defined as the ratio between the joint and marginal distributions

$$p(\mathbf{x}|\mathbf{y}) = \frac{p(\mathbf{x},\mathbf{y})}{p(\mathbf{y})}, \quad p(\mathbf{y}|\mathbf{x}) = \frac{p(\mathbf{x},\mathbf{y})}{p(\mathbf{x})}. \tag{1.20}$$

They describe the distribution of $\mathbf{x}$ or $\mathbf{y}$ assuming we know the value of $\mathbf{y}$ or $\mathbf{x}$. $\mathbf{x}$ and $\mathbf{y}$ are said to be (statistically) independent if

$$p(\mathbf{x}|\mathbf{y}) = p(\mathbf{x}) \quad \text{or} \quad p(\mathbf{y}|\mathbf{x}) = p(\mathbf{y}) \quad \text{or} \quad p(\mathbf{x},\mathbf{y}) = p(\mathbf{x})p(\mathbf{y}), \tag{1.21}$$

i.e., the value taken by $\mathbf{y}$ or $\mathbf{x}$ has no influence on the distribution of $\mathbf{x}$ or $\mathbf{y}$.

From the definition of the conditional distributions, the product rule of probability

$$p(\mathbf{x},\mathbf{y}) = p(\mathbf{x}|\mathbf{y})p(\mathbf{y}) = p(\mathbf{y}|\mathbf{x})p(\mathbf{x}) \tag{1.22}$$

easily follows. Another immediate result is the famous Bayes rule

$$p(\mathbf{y}|\mathbf{x}) = p(\mathbf{x}|\mathbf{y})\frac{p(\mathbf{y})}{p(\mathbf{x})}. \tag{1.23}$$

It states that one can calculate the conditional distribution $p(\mathbf{y}|\mathbf{x})$ from the other one $p(\mathbf{x}|\mathbf{y})$. In Section 1.5, we will see that this Bayes rule lays the statistical fundament for all machine learning models.

The case of two random vectors can be easily extended to multiple random vectors $\mathbf{x}_1, \ldots, \mathbf{x}_N$ with the joint distribution $p(\mathbf{x}_1, \ldots, \mathbf{x}_N)$. The aforementioned product rule of probability then generalizes to the chain rule of probability

$$p(\mathbf{x}_1, \ldots, \mathbf{x}_N) = p(\mathbf{x}_N|\mathbf{x}_{N-1}, \ldots, \mathbf{x}_1)p(\mathbf{x}_{N-1}|\mathbf{x}_{N-2}, \ldots, \mathbf{x}_1)\ldots p(\mathbf{x}_2|\mathbf{x}_1)p(\mathbf{x}_1). \tag{1.24}$$

The joint distribution is determined by a sequence of conditional distributions and one final marginal distribution. $\mathbf{x}_1, \ldots, \mathbf{x}_N$ are said to be independent and identically distributed (i.i.d.) if all random vectors are mutually independent with

$$p(\mathbf{x}_1, \ldots, \mathbf{x}_N) = p_1(\mathbf{x}_1)\ldots p_N(\mathbf{x}_N) \tag{1.25}$$

and all marginal distributions are identical with

$$p_i(\mathbf{x}_i) = p(\mathbf{x}_i) \quad \forall i. \tag{1.26}$$

This implies

$$p(\mathbf{x}_1, \ldots, \mathbf{x}_N) = \prod_i p(\mathbf{x}_i) \tag{1.27}$$

1.3.3 Kernel-based density estimation

Machine learning follows a probabilistic framework, which assumes to know the distribution of the underlying quantities. This is an academic assumption. Except for a few cases (e.g., normal distribution due to the central limit theorem), the precise distribution is unknown in practice. Instead, we only measure some samples of random variables. An MRI dataset contains samples of MR image for samples of subject collected on samples of MR scanner. One step to close this gap between theory and practice is to estimate the required distribution from samples.

For simplicity, we consider a continuous real-valued random vector $\mathbf{x}$ whose PDF $p(\mathbf{x})$ has to be estimated from N i.i.d. samples $\mathbf{x}_1, \ldots, \mathbf{x}_N$. There are different approaches for density estimation such as parametric and nonparametric methods. To train a machine learning model using a distribution-based loss function, typically the nonparametric kernel-based approach is used for density estimation

$$\hat{p}(\mathbf{x}) = \frac{1}{N} \sum_{n=1}^{N} k(\mathbf{x} - \mathbf{x}_n). \tag{1.28}$$

$k(\mathbf{x})$ is the so called kernel function. It satisfies all properties of a PDF: nonnegativity $k(\mathbf{x}) \geq 0$ and normalization $\int k(\mathbf{x})d\mathbf{x} = 1$. Therefore, the estimated PDF $\hat{p}(\mathbf{x})$ satisfies these two properties as well. Often a smooth Gaussian kernel $k(\mathbf{x}) \sim N(\mathbf{0}, \sigma^2\mathbf{I})$ is used if a differentiable PDF estimate is desired. σ is the so-called kernel width.

In Section 1.5, the Dirac kernel $k(\mathbf{x}) = \delta(\mathbf{x})$ will be used where $\delta(\mathbf{x})$ is the multidimensional Dirac function satisfying the sampling property

$$\int \delta(\mathbf{x} - \mathbf{x}_0) f(\mathbf{x}) d\mathbf{x} = f(\mathbf{x}_0). \tag{1.29}$$

The resulting PDF estimate is known as the empirical distribution. For two random vectors $\mathbf{x}$ and $\mathbf{y}$, the joint distribution $p(\mathbf{x}, \mathbf{y})$ can be estimated in a similar way.

$$\hat{p}(\mathbf{x}) = \frac{1}{N} \sum_{n=1}^{N} \delta(\mathbf{x} - \mathbf{x}_n) \tag{1.30}$$

1.4 Kullback-Leibler divergence and cross entropy

1.4.1 Kullback-Leibler divergence

Given a random vector $\mathbf{x}$ with the true distribution, say PDF $p(\mathbf{x})$. As we will see in the next section, a machine learning model is a parametric computational model, which approximates (learns) $p(\mathbf{x})$ by another PDF $q(\mathbf{x})$. The goal is to adjust the machine learning model such $q(\mathbf{x})$ best approximates $p(\mathbf{x})$. Just like comparing two numbers or two vectors, the comparison between two PDFs also requires a similarity or dissimilarity measure.

One popular dissimilarity measure between two PDFs $p(\mathbf{x})$ and $q(\mathbf{x})$ is the Kullback-Leiber divergence (KLD) [4]

$$D_{\mathrm{KL}}(p \| q) = \mathrm{E}_p \ln\left(\frac{p(\mathbf{x})}{q(\mathbf{x})}\right) = \int p(\mathbf{x}) \ln\left(\frac{p(\mathbf{x})}{q(\mathbf{x})}\right) d\mathbf{x}. \tag{1.31}$$

The subscript p in E_p indicates an expectation over $p(\mathbf{x})$. Is $\mathbf{x} \in \{\mathbf{x}_1, \mathbf{x}_2, \ldots\}$ discrete with the true PMF $P(\mathbf{x}_i)$ and an approximating PMF $Q(\mathbf{x}_i)$, the KLD is defined in a similar way:

$$D_{KL}(P\|Q) = E_P \ln\left(\frac{P(\mathbf{x})}{Q(\mathbf{x})}\right) = \sum_i P(\mathbf{x}_i) \ln\left(\frac{P(\mathbf{x}_i)}{Q(\mathbf{x}_i)}\right). \tag{1.32}$$

The KLD has the following properties:

- Nonnegativity: $D_{KL}(p\|q) \geq 0 \quad \forall p, q$
- Equality: $\forall x D_{KL}(p\|q) = 0 \quad \text{iff} \quad p(\mathbf{x}) = q(\mathbf{x})$ for all x
- Nonsymmetry: $D_{KL}(p\|q) \neq D_{KL}(q\|p)$.

The first two properties demonstrate that KLD is indeed a suitable dissimilarity measure between $p(\mathbf{x})$ and $q(\mathbf{x})$. Minimizing $D_{KL}(p\|q)$ will force $q(\mathbf{x})$ to approximate $p(\mathbf{x})$ as well as possible. Hence the minimization of KLD is one general machine learning principle. The third property says that $D_{KL}(p\|q)$ is not a symmetric measure. The forward KLD $D_{KL}(p\|q)$ is in general different from the backward KLD $D_{KL}(q\|p)$. This is the reason why the word "divergence" is used instead of "distance," which is typically used for symmetric dissimilarity measures. If one needs a symmetric dissimilarity measure, the Jensen-Shannon divergence can be used.

$$D_{JS}(p\|q) = \frac{1}{2}\left[D_{KL}\left(p\|\frac{p+q}{2}\right) + D_{KL}\left(q\|\frac{p+q}{2}\right)\right] = D_{JS}(q\|p) \tag{1.33}$$

1.4.2 Cross-entropy

In information theory, the entropy of a random variable is a measure for its average information content in the sense of uncertainty. The larger the entropy, the larger the information content. Let $\mathbf{x} \in \{\mathbf{x}_1, \ldots, \mathbf{x}_c\}$ be a discrete random vector with c possible values $\mathbf{x}_i$ and the PMF $P(\mathbf{x} = \mathbf{x}_i) = p_i$. Its entropy is defined as

$$H(\mathbf{x}) = H(P) = -E(\ln(P(\mathbf{x}))) = -\sum_{i=1}^{c} p_i \ln(p_i) \geq 0. \tag{1.34}$$

If $\mathbf{x} \sim p(\mathbf{x})$ is a continuous random vector with the PDF $p(\mathbf{x})$, its (differential) entropy is

$$H(\mathbf{x}) = H(p) = -E(\ln(p(\mathbf{x}))) = -\int p(\mathbf{x}) \ln(p(\mathbf{x}))d\mathbf{x} \geq 0. \tag{1.35}$$

For two random vectors $\mathbf{x} \sim p(\mathbf{x})$ and $\mathbf{y} \sim q(\mathbf{y})$, the cross-entropy (CE) between them is defined as

$$H(\mathbf{x}, \mathbf{y}) = H(p, q) = -E_p \ln(q(\mathbf{x})) = -\int p(\mathbf{x}) \ln(q(\mathbf{x}))d\mathbf{x} \geq 0. \tag{1.36}$$

There is a simple relationship between KLD and CE

$$
\begin{aligned}
D_{\mathrm{KL}}(p\|q) &= \int p(\mathbf{x}) \ln\left(\frac{p(\mathbf{x})}{q(\mathbf{x})}\right) \mathrm{d}\mathbf{x} \\
&= \int p(\mathbf{x}) \ln p(\mathbf{x}) \mathrm{d}\mathbf{x} - \int p(\mathbf{x}) \ln q(\mathbf{x}) \mathrm{d}\mathbf{x} \\
&= -H(p) + H(p,q).
\end{aligned} \tag{1.37}
$$

The forward KLD $D_{\mathrm{KL}}(p\|q)$ differs from the CE $H(p,q)$ in the additive term $-H(p)$. Since in machine learning $p(\mathbf{x})$ is the given fixed distribution, $-H(p)$ is fixed as well, i.e., independent of the approximating distribution $q(\mathbf{x})$ to be learned. The minimization of $D_{\mathrm{KL}}(p\|q)$ is thus equivalent to the minimization of $H(p,q)$. This will be used in the next section.

1.5 Probabilistic framework of machine learning

1.5.1 Real world and machine learning

Fig. 1.2 illustrates a machine learning model, which aims to learn the behavior of a real world from data. The real world may be an MR scanner. The input to the MR scanner may be the tumor state $\mathbf{y}$ in an organ of a patient. The output of the MR scanner is an observation, an MR image $\mathbf{x}$. Clearly, there is a statistical relationship between $\mathbf{x}$ and $\mathbf{y}$, described by the conditional PDF $p(\mathbf{x}|\mathbf{y})$. It describes the impact of $\mathbf{y}$ on the MR image $\mathbf{x}$.

The machine learning model aims to predict the hidden state vector $\mathbf{y}$ from the observation $\mathbf{x}$. It processes the input $\mathbf{x}$ according to the machine learning model (e.g., a neural network) and returns a prediction $\hat{\mathbf{y}}$ for $\mathbf{y}$. This step to draw conclusion about $\mathbf{y}$ from $\mathbf{x}$ is called inference. For this purpose, we need to learn $p(\mathbf{y}|\mathbf{x})$, which describes which values of $\mathbf{y}$ are more likely for a given observation $\mathbf{x}$.

Both the real word and machine learning model are described in a probabilistic way. Given two random vectors $\mathbf{x}$ and $\mathbf{y}$, there exist five different distributions:

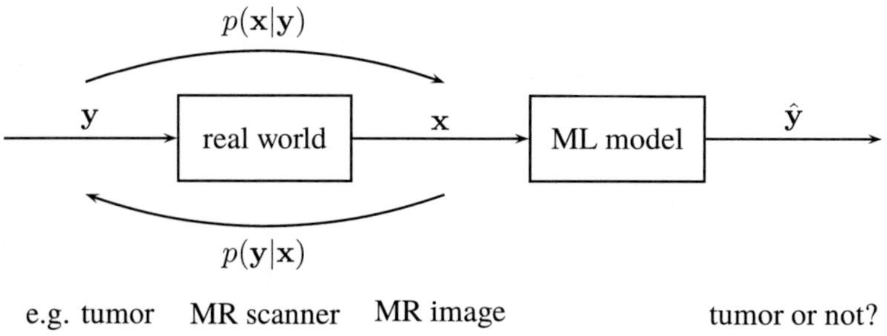

FIGURE 1.2

Real world and machine learning.

- The joint distribution $p(\mathbf{x}, \mathbf{y})$ is also called the data generating distribution. It describes the joint occurrence of $\mathbf{x}$ and $\mathbf{y}$. In practice, $p(\mathbf{x}, \mathbf{y})$ is analytically unknown. Only paired samples of $(\mathbf{x}, \mathbf{y})$ (in supervised learning) or samples of $\mathbf{x}$ only (in unsupervised learning) are collected.
- The marginal distribution $p(\mathbf{y})$ is called the prior of $\mathbf{y}$. It describes the distribution of $\mathbf{y}$ regardless of $\mathbf{x}$.
- The conditional distribution $p(\mathbf{x}|\mathbf{y})$ is called the likelihood. It describes the real world behavior, e.g., how does the MR scanner acquire an MR image $\mathbf{x}$ for a given $\mathbf{y}$. Learning the likelihood from data means learning about the imaging process.
- The marginal distribution $p(\mathbf{x})$ is called the evidence. Among the five distributions, it has the lowest relevance for machine learning because it is not related to $\mathbf{y}$.
- The second conditional distribution $p(\mathbf{y}|\mathbf{x})$ is called the posterior of $\mathbf{y}$. In contrast to the prior $p(\mathbf{y})$, it describes which values of $\mathbf{y}$ are more likely given a particular observation $\mathbf{x}$. The goal of a machine learning model is to learn $p(\mathbf{y}|\mathbf{x})$ from data to draw conclusion about $\mathbf{y}$ from $\mathbf{x}$.

1.5.2 Bayes rule

As already introduced in Section 1.3, the Bayes rule relates the posterior to the likelihood

$$p(\mathbf{y}|\mathbf{x}) = p(\mathbf{x}|\mathbf{y})\frac{p(\mathbf{y})}{p(\mathbf{x})} \quad \text{or} \quad \text{posterior} = \text{likelihood}\frac{\text{prior}}{\text{evidence}}. \tag{1.38}$$

Clearly, the posterior $p(\mathbf{y}|\mathbf{x})$ contains information from both the likelihood $p(\mathbf{x}|\mathbf{y})$ and the prior $p(\mathbf{y})$. The training of a machine learning model implies the learning of $p(\mathbf{y}|\mathbf{x})$ using the Bayes rule from samples. The next inference step draws a conclusion about $\mathbf{y}$ for a new sample $\mathbf{x}$ based on the learned posterior $p(\mathbf{y}|\mathbf{x})$. This is the basic idea of machine learning, also known as Bayesian learning.

1.5.3 Machine learning tasks

There exist different machine learning tasks depending on the type of the desired output $\mathbf{y}$:

- If $\mathbf{y}$ is discrete-valued like a class label (e.g., tumor or not), the task is called classification.
- If $\mathbf{y}$ is continuous-valued like the age of a subject or the volume of an organ, the task is referred as regression.
- Given an input MR image $\mathbf{x}$, semantic segmentation aims to partition $\mathbf{x}$ into nonoverlapping semantic regions corresponding to different body parts (e.g., bones, organs, tissues). This is actually a voxelwise classification problem where each voxel of the MR image $\mathbf{x}$ is assigned to a class label.

- Object detection aims to localize objects of interest (e.g., organs, tumors) in the MR image $\mathbf{x}$ in terms of bounding boxes and classify the objects. This is a mixture of regression and classification.
- An even larger family of machine learning tasks is image enhancement or more general image translation. The goal is to improve the quality of distorted MR images or translate a medical image from one modality (e.g., CT, MRT, PET) to another one. The concrete machine learning task can be denoising, motion artifact correction, superresolution, inpainting in MR image or PET-to-CT translation, etc. In this case, the machine learning model translates an input image $\mathbf{x}$ to an output image $\widehat{\mathbf{y}}$ based on the learned posterior $p(\mathbf{y}|\mathbf{x})$.

1.5.4 Supervision of machine learning

In addition to the different machine learning tasks, there also exist different supervision schemes of machine learning:

- In supervised learning, there is one labeled dataset $D = \{(\mathbf{x}_n, \mathbf{y}_n)\}$ where each input sample $\mathbf{x}_n$ is assigned to a unique corresponding ground truth $\mathbf{y}_n$ as the desired output. The pairs $(\mathbf{x}_n, \mathbf{y}_n)$ specify the real-world behavior in terms of examples and guide the learning of this behavior. Classification, regression, semantic segmentation, and object detection are all supervised learning tasks. Supervised learning is highly effective because the ground truth $\mathbf{y}_n$ guides the learning process, but the collection of labeled data can be very time-consuming, expensive, or even impractical in some situations.
- Unsupervised learning requires only an unlabeled dataset $D = \{\mathbf{x}_n\}$, which is much easier to collect. Due to the lack of specification of the desired output $\mathbf{y}_n$ for a given input sample $\mathbf{x}_n$, unsupervised learning is unable to learn a mapping rule $\mathbf{x} \rightarrow \mathbf{y}$ by definition. Instead, it is used for clustering, i.e., group input samples $\mathbf{x}_n$ into different clusters based on their similarity, or for representation learning, i.e., learn an efficient hidden space representation of the input $\mathbf{x}$.
- An important special case of unsupervised learning is the self-supervised representation learning. The goal is to leverage highly effective supervised learning methods to learn a useful representation of the input from an unlabeled dataset over modern pretext task or contrastive learning techniques. The learned representation contains common discriminative features of the input, which can be used later to solve different downstream tasks. Starting with the learned representation, the fine-tuning of this pretrained model to a supervised downstream task requires only a pretty small labeled dataset. This is much more data efficient than training a model for the downstream task from scratch using a large labeled dataset. The combination of the unsupervised (self-supervised) representation learning using a large unlabeled dataset with a supervised downstream task learning using a small labeled dataset is also known as semisupervised learning.

1.5.5 Loss function

In machine learning, the loss function specifies the goal of learning and guides the training. Since supervised learning is also the core of self-supervised representation learning, we focus now on the loss function in supervised learning. It is in general a scalar real-valued function $L(\boldsymbol{\theta}; D)$ depending on the parameter vector $\boldsymbol{\theta}$ of the underlying machine learning model, calculated on a labeled dataset $D = \{(\mathbf{x}_n, \mathbf{y}_n) | 1 \leq n \leq N\}$ containing N training samples.

1.5.5.1 Distribution-based loss

There are different families of loss functions. The first one is distribution-based, i.e., the loss is derived from the aforementioned probabilistic framework of machine learning.

Let $p(\mathbf{x}, \mathbf{y})$ be the joint true PDF of $(\mathbf{x},\mathbf{y})$. A machine learning model tries to learn $p(\mathbf{x}, \mathbf{y})$ by an approximating distribution $q(\mathbf{x}, \mathbf{y}; \boldsymbol{\theta})$ parametrized by the parameter vector $\boldsymbol{\theta}$. The function $q()$ is determined by the architecture of the chosen machine learning model (e.g., a neural network). As introduced in Section 1.4, the aim of machine learning is the minimization of the forward KLD or cross-entropy between $p(\mathbf{x}, \mathbf{y})$ and $q(\mathbf{x}, \mathbf{y}; \boldsymbol{\theta})$

$$D_{\mathrm{KL}}(p(\mathbf{x}, \mathbf{y}) \| q(\mathbf{x}, \mathbf{y}; \boldsymbol{\theta})) = H(p(\mathbf{x}, \mathbf{y}), q(\mathbf{x}, \mathbf{y}; \boldsymbol{\theta})) + \text{const.} \tag{1.39}$$

Note $q(\mathbf{x}, \mathbf{y}; \boldsymbol{\theta}) = q(\mathbf{y}|\mathbf{x}; \boldsymbol{\theta}) q(\mathbf{x})$. Since only the first term $q(\mathbf{y}|\mathbf{x}; \boldsymbol{\theta})$ describes the machine learning model and depends on $\boldsymbol{\theta}$, the following simplification yields

$$\begin{aligned} H(p(\mathbf{x}, \mathbf{y}), q(\mathbf{x}, \mathbf{y}; \boldsymbol{\theta})) &= -\int p(\mathbf{x}, \mathbf{y}) \ln (q(\mathbf{x}, \mathbf{y}; \boldsymbol{\theta})) \mathbf{dxdy} \\ &= -\int p(\mathbf{x}, \mathbf{y}) \ln (q(\mathbf{y}|\mathbf{x}; \boldsymbol{\theta})) \mathbf{dxdy} \\ &\quad \underbrace{-\int p(\mathbf{x}, \mathbf{y}) \ln (q(\mathbf{x})) \mathbf{dxdy}}_{\text{const}} \qquad (1.40) \\ &= \int p(\mathbf{x}, \mathbf{y}) [-\ln (q(\mathbf{y}|\mathbf{x}; \boldsymbol{\theta}))] \mathbf{dxdy} + \text{const.} \end{aligned}$$

In practice, $p(\mathbf{x}, \mathbf{y})$ is unknown and is estimated by the empirical distribution (see Section 1.3)

$$\widehat{p}(\mathbf{x}, \mathbf{y}) = \frac{1}{N} \sum_{n=1}^{N} \delta(\mathbf{x} - \mathbf{x}_n, \mathbf{y} - \mathbf{y}_n). \tag{1.41}$$

By using the sampling property (1.29) of the Dirac function $\delta(\mathbf{x}, \mathbf{y})$ and ignoring all constant terms, the final cost function to be minimized is

$$L(\boldsymbol{\theta}; D) = \frac{1}{N} \sum_{n=1}^{N} l(\mathbf{x}_n, \mathbf{y}_n; \boldsymbol{\theta}). \tag{1.42}$$

It is the average of the cross-entropy loss or negative log-likelihood (NLL) loss.

$$l(\mathbf{x}, \mathbf{y}; \boldsymbol{\theta}) = -\ln(q(\mathbf{y}|\mathbf{x}; \boldsymbol{\theta})) \tag{1.43}$$

Over all N sample pairs $(\mathbf{x}_n, \mathbf{y}_n)$.

We can further simplify the loss $l(\mathbf{x}, \mathbf{y}; \boldsymbol{\theta})$ depending on the type of the machine learning task. In regression tasks, the ground truth $\mathbf{y}$ is continuous-valued. It is often assumed as normal distributed $\mathbf{y} \sim N(\widehat{\mathbf{y}}(\mathbf{x}; \boldsymbol{\theta}), \sigma^2\mathbf{I})$. This means that the mean of the desired hidden state vector $\mathbf{y}$ is well estimated by the output $\widehat{\mathbf{y}}(\mathbf{x}; \boldsymbol{\theta})$ of the machine learning model for the input $\mathbf{x}$. In this case, the resulting loss

$$l(\mathbf{x}, \mathbf{y}; \boldsymbol{\theta}) = -\ln(q(\mathbf{y}|\mathbf{x}; \boldsymbol{\theta})) = \text{const} + \frac{1}{2\sigma^2}\|\mathbf{y} - \widehat{\mathbf{y}}(\mathbf{x}; \boldsymbol{\theta})\|^2 \tag{1.44}$$

is called the L_2 loss due to the underlying L_2 vector norm. The final cost function

$$L(\boldsymbol{\theta}) = \frac{1}{N}\sum_{n=1}^{N}\|\mathbf{y}_n - \widehat{\mathbf{y}}(\mathbf{x}_n; \boldsymbol{\theta})\|^2 \tag{1.45}$$

after ignoring all constant terms is the mean squared error (MSE) between $\mathbf{y}_n$ and $\widehat{\mathbf{y}}(\mathbf{x}_n; \boldsymbol{\theta})$.

By assuming independent elements y_i of $\mathbf{y}$ and Laplace $(\widehat{y}_i(\mathbf{x}; \boldsymbol{\theta}), b)$ distribution for each element y_i, the L_1 loss yields due to the involved L_1 vector norm

$$l(\mathbf{x}, \mathbf{y}; \boldsymbol{\theta}) = \sum_{i}|y_i - \widehat{y}_i(\mathbf{x}; \boldsymbol{\theta})| = \|\mathbf{y} - \widehat{\mathbf{y}}(\mathbf{x}; \boldsymbol{\theta})\|_1. \tag{1.46}$$

This leads to the minimization of the mean absolute error (MAE)

$$L(\boldsymbol{\theta}) = \frac{1}{N}\sum_{n=1}^{N}\|\mathbf{y}_n - \widehat{\mathbf{y}}(\mathbf{x}_n; \boldsymbol{\theta})\|_1. \tag{1.47}$$

In a c-class classification task, we use the one-hot coding $\mathbf{y} \in \{\mathbf{e}_1, \ldots, \mathbf{e}_c\}$ from Section 1.3 for the class label $\mathbf{y}$. Let $p_i = P(\mathbf{y} = \mathbf{e}_i|\mathbf{x})$ be the true but unknown posterior class probability for $\mathbf{x}$ belonging to class i and

$$P(\mathbf{y}|\mathbf{x}) = \prod_{i=1}^{c} p_i^{y_i} \tag{1.48}$$

be the posterior PMF. In this case, we design the machine learning model such that its output $\widehat{\mathbf{y}}(\mathbf{x}; \boldsymbol{\theta}) \in \mathbb{R}^c$ is a length-c vector containing class probability estimates $\widehat{y}_i$ for p_i. This can be easily done by using the so-called softmax activation function in the output layer of the neural network. Hence the approximating PMF is $Q(\mathbf{y}|\mathbf{x}; \boldsymbol{\theta}) = \prod_{i=1} \widehat{y}_i^{y_i}$, and the NLL loss becomes

$$l(\mathbf{x}, \mathbf{y}; \boldsymbol{\theta}) = -\ln(Q(\mathbf{y}|\mathbf{x}; \boldsymbol{\theta})) = -\sum_{i=1}^{c} y_i \ln(\widehat{y}_i(\mathbf{x}; \boldsymbol{\theta})) = -\mathbf{y}^T \ln(\widehat{\mathbf{y}}(\mathbf{x}; \boldsymbol{\theta})). \tag{1.49}$$

The last expression is a scalar product of two vectors, and ln() is applied elementwise to the column vector $\widehat{\mathbf{y}}$. This is the so-called categorical cross-entropy loss. The final cost function to train a classification model is

$$L(\boldsymbol{\theta}) = \frac{1}{N}\sum_{n=1}^{N} -\mathbf{y}_n^T \ln(\widehat{\mathbf{y}}(\mathbf{x}_n; \boldsymbol{\theta})). \tag{1.50}$$

1.5.5.2 Region-based loss

Conceptually, the aforementioned categorical cross-entropy loss can also be applied to the voxelwise classification in semantic image segmentation. The cost function in a 2D segmentation becomes then

$$L(\boldsymbol{\theta}) = \frac{1}{HW} \sum_{h=1}^{H} \sum_{w=1}^{W} -\mathbf{y}_{hw}^{T} \ln(\widehat{\mathbf{y}}(\mathbf{x}_{hw};\boldsymbol{\theta})) \tag{1.51}$$

where $H \times W$ represents the size of the 2D input MRI slice.[1] This cost function, however, is sensitive to class imbalance in MRI. There are major classes with a large number of voxels (image background, large organs) and minor classes with a small number of voxels (small organs, rare tumors). Due to the unweighted sum over all voxels in Eq. (1.51), the major classes and in particular the large image background contribute much more to the cost function than the minor classes. As a consequence, the machine learning model is learned to fit to the major classes, and the minor classes will have a poor segmentation accuracy. This class imbalance can be partially compensated by using a weighted cross-entropy loss. The NLL loss in Eq. (1.51) for each voxel is weighted by a factor, which is large for minor and small for major classes.

Another better solution is the use of so-called region-based loss instead of distribution-based loss as illustrated in Fig. 1.3 for a binary segmentation: object versus background. A represents the ground truth object mask (a subset of pixels), B is the estimated object mask, and the two solid- and dashed-line rectangles symbolize two different sizes of the image background. Two popular region-overlap metrics are used to measure the overlap of the ground truth and estimated mask. The Jaccard index or intersection-over-union (IoU) is defined as

$$0 \leq J = \frac{|A \cap B|}{|A \cup B|} = \frac{|A \cap B|}{|A| + |B| - |A \cap B|} \leq 1. \tag{1.52}$$

$A \cap B$ and $A \cup B$ denote the intersection and union of A and B, respectively. $|A| \in \mathbb{N}$ is the cardinality of A, i.e., the number of elements in A. The Dice coefficient has a slightly different definition

$$0 \leq D = \frac{2|A \cap B|}{|A| + |B|} \leq 1. \tag{1.53}$$

Both metrics are related by $J = \frac{D}{2-D} \leq D$ and can be converted bijective into each other. If $A \cap B = \emptyset$, $J = D = 0$. If $A = B$, $J = D = 1$. Obviously, J and D only quantify the ratio of overlapping and are fully independent of the size of the background as indicated by the dashed-line rectangle in Fig. 1.3.

The previous Jaccard index and Dice coefficient are defined for one object class only. In a multiclass segmentation, they can be calculated for each of the c object

---

[1]In a 3D semantic segmentation, the cost function is calculated by a triple sum over all voxels in a volume scan

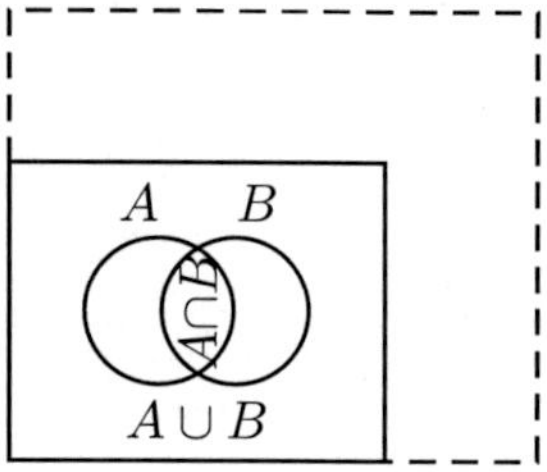

FIGURE 1.3

Overlap of two regions *A* and *A*.

classes. Let A_i and B_i denote the ground truth and estimated masks for class $1 \leq i \leq c$ (e.g., different organs), respectively. Then

$$0 \leq J_i = \frac{|A_i \cap B_i|}{|A_i \cup B_i|} \leq 1 \tag{1.54}$$

is the Jaccard index for class i and

$$J = \frac{1}{c} \sum_{i=1}^{c} J_i \tag{1.55}$$

is the mean Jaccard index or mean IoU. The mean Dice coefficient is defined in a similar way.

The aforementioned metrics are used for the evaluation of a semantic segmentation, i.e., comparison of the ground truth and estimated masks. They are not suitable for training because they count pixels and are not differentiable. For the training, differentiable soft Jaccard and Dice loss are employed.

Let $\mathbf{y}_{hw} \in \{\mathbf{e}_1, \ldots, \mathbf{e}_c\}$ be the ground truth one-hot coding of c classes and $\widehat{\mathbf{y}}_{hw} \in \mathbb{R}^c$ be the c-class softmax output for the voxel (h, w), respectively. The soft Jaccard loss used for training is

$$\begin{aligned}
\alpha &= \sum_{h=1}^{H} \sum_{w=1}^{W} \mathbf{y}_{hw} \odot \widehat{\mathbf{y}}_{hw} = [\alpha_i] \in \mathbb{R}^c, \\
\beta &= \sum_{h=1}^{H} \sum_{w=1}^{W} (\mathbf{y}_{hw} + \widehat{\mathbf{y}}_{hw}) = [\beta_i] \in \mathbb{R}^c, \\
J_i &= \frac{\alpha_i + \epsilon}{\beta_i - \alpha_i + \epsilon}, \\
l(\boldsymbol{\theta}) &= 1 - \frac{1}{c} \sum_{i=1}^{c} J_i
\end{aligned} \tag{1.56}$$

$\odot$ is the elementwise multiplication. α_i and β_i represent arithmetic equivalents of $|A_i \cap B_i|$ and $|A_i| + |B_i|$, respectively. J_i is the soft Jaccard index for class i where $\epsilon > 0$ is a suitable small number to avoid $0/0$ if $\alpha_i = \beta_i = 0$. $l(\boldsymbol{\theta})$ is the soft Jaccard loss, which is differentiable w.r.t. $\boldsymbol{\theta}$. The soft Dice loss is defined in a similar way.

In machine learning for MRI, sometimes other more specialized losses than the previously presented ones are used.

1.6 Estimators

One subarea of statistical signal processing is parameter estimation. Given an observation vector modeled by the random vector $\mathbf{x} \in \mathbb{R}^N$. In contrast to machine learning, we assume to know the analytical expression of the PDF $p(\mathbf{x};\boldsymbol{\theta})$ of $\mathbf{x}$ parameterized by a deterministic parameter vector $\boldsymbol{\theta} \in \mathbb{R}^M$. This probabilistic signal model provides an explicit description of the relationship between $\mathbf{x}$ and $\boldsymbol{\theta}$. This kind of a prior domain knowledge may be available in some simple problems with a good human understanding of the underlying physical and biomedical phenomenon. The same function $p()$ is known under two different names. For a given $\boldsymbol{\theta}$, $p(\mathbf{x};\boldsymbol{\theta})$ as a function of $\mathbf{x}$ is known as the PDF in probability theory. Given an observation vector $\mathbf{x}$, $p(\mathbf{x};\boldsymbol{\theta})$ as a function of $\boldsymbol{\theta}$ is called the likelihood function in estimation theory.

The goal of parameter estimation is to estimate $\boldsymbol{\theta}$ from $\mathbf{x}$. We denote the estimator for $\boldsymbol{\theta}$, a function of $\mathbf{x}$, by $\widehat{\boldsymbol{\theta}}(\mathbf{x})$ or simply $\widehat{\boldsymbol{\theta}}$. By intuition, the stronger the relationship between $\mathbf{x}$ and $\boldsymbol{\theta}$, the higher the estimation accuracy will be. The relationship is described by the likelihood function.

1.6.1 Accuracy of an estimator

Typically, we use the bias and covariance matrix to characterize the accuracy of the estimator $\widehat{\boldsymbol{\theta}}$. The bias

$$\mathbf{b}_{\widehat{\boldsymbol{\theta}}} = \mathrm{E}(\widehat{\boldsymbol{\theta}}) - \boldsymbol{\theta} \in \mathbb{R}^M \tag{1.57}$$

measures the systematic deviation of $\widehat{\boldsymbol{\theta}}$ from $\boldsymbol{\theta}$. An estimator is said to be unbiased if $\mathbf{b}_{\widehat{\boldsymbol{\theta}}} = \mathbf{0}$. The covariance matrix

$$\mathbf{C}_{\widehat{\boldsymbol{\theta}}} = \mathrm{E}\left[\left(\widehat{\boldsymbol{\theta}} - \mathrm{E}(\widehat{\boldsymbol{\theta}})\right)\left(\widehat{\boldsymbol{\theta}} - \mathrm{E}(\widehat{\boldsymbol{\theta}})\right)^T\right] \in \mathbb{R}^{M\times M} \tag{1.58}$$

of $\widehat{\boldsymbol{\theta}}$ describes its random variation around the mean $\mathrm{E}(\widehat{\boldsymbol{\theta}})$. One favorite optimum estimator is the minimum variance unbiased estimator (MVUE): In the family of all unbiased estimators, we look for the best one with the smallest possible covariance matrix. Unfortunately, there are no systematic approaches to derive the MVUE in all cases. Cramer-Rao bound and maximum likelihood are two popular approaches to find the MVUE under certain conditions or to find a reasonable estimator.

1.6.2 Cramer-Rao bound

The Cramer-Rao bound (CRB) is a lower bound for the covariance matrix of all unbiased estimators. Under certain regularity conditions,

$$\mathbf{C}_{\widehat{\boldsymbol{\theta}}} \geq \mathbf{J}^{-1}(\boldsymbol{\theta}) \tag{1.59}$$

holds for each unbiased estimator $\widehat{\boldsymbol{\theta}}$ with $\mathrm{E}(\widehat{\boldsymbol{\theta}}) = \boldsymbol{\theta}$ [5].

$$\mathbf{J}(\boldsymbol{\theta}) = \mathrm{E}\left[(\nabla \ln p(\mathbf{x};\boldsymbol{\theta}))(\nabla \ln p(\mathbf{x};\boldsymbol{\theta}))^T\right] = -\mathrm{E}\left[\nabla\nabla^T \ln p(\mathbf{x};\boldsymbol{\theta})\right] \tag{1.60}$$

is the Fisher information matrix (FIM). $\ln p(\mathbf{x};\boldsymbol{\theta})$ is the log-likelihood function. $\nabla\left[\frac{\partial}{\partial\theta_1},\ldots,\frac{\partial}{\partial\theta_M}\right]^T$ is the gradient vector containing all first-order derivatives as defined in Section 1.2.3. $\nabla\nabla^T = \left[\frac{\partial^2}{\partial\theta_i\partial\theta_j}\right]_{ij}$ is the Hessian matrix containing all second-order derivatives. The FIM $\mathbf{J}(\boldsymbol{\theta})$ represents the information amount contained in $\mathbf{x}$ about $\boldsymbol{\theta}$. It only depends on the probabilistic model $p(\mathbf{x};\boldsymbol{\theta})$ and can be calculated in advance before developing any estimator. The stronger the relationship between $\mathbf{x}$ and $\boldsymbol{\theta}$, the higher the information amount, the larger the FIM $\mathbf{J}(\boldsymbol{\theta})$, the smaller the CRB $\mathbf{J}^{-1}(\boldsymbol{\theta})$, and the more accurate the parameter estimation will be.

In Eq. (1.60), $\mathbf{J}^{-1}(\boldsymbol{\theta})$ is the lower bound for $\mathbf{C}_{\widehat{\boldsymbol{\theta}}}$. The matrix inequality means that the difference $\mathbf{C}_{\widehat{\boldsymbol{\theta}}} - \mathbf{J}^{-1}(\boldsymbol{\theta})$ is a nonnegative definite matrix. If the covariance matrix achieves the lower bound, i.e., $\mathbf{C}_{\widehat{\boldsymbol{\theta}}} = \mathbf{J}^{-1}(\boldsymbol{\theta})$, $\widehat{\boldsymbol{\theta}}$ is called an efficient estimator. By definition, an efficient estimator is always MVUE because there is no other unbiased estimator with a smaller covariance matrix. But the opposite is not true: An MVUE is not always efficient because the CRB may be not tight enough.

Luckily, there exists one necessary and sufficient condition for the existence of an efficient estimator, namely the existence of two functions $\mathbf{g}(\mathbf{x})$ and $\mathbf{J}(\boldsymbol{\theta})$ such that

$$\nabla \ln p(\mathbf{x};\boldsymbol{\theta}) = \mathbf{J}(\boldsymbol{\theta})[\mathbf{g}(\mathbf{x}) - \boldsymbol{\theta}]. \tag{1.61}$$

If this factorization of $\nabla \ln p(\mathbf{x};\boldsymbol{\theta})$ is possible, $\widehat{\boldsymbol{\theta}} = \mathbf{g}(\mathbf{x})$ is the unbiased efficient estimator (and MVUE) with $\mathbf{C}_{\widehat{\boldsymbol{\theta}}} = \mathbf{J}^{-1}(\boldsymbol{\theta})$. Hence the aforementioned equation provides an easy way to find the MVUE provided that the factorization in Eq. (1.61) is satisfied. Unfortunately, this condition is not satisfied for most signal models. In this case, we still have a valid CRB in Eq. (1.60), but we don't have an optimum estimator. If this happens, the maximum likelihood approach is the next choice.

1.6.3 Maximum likelihood estimation

Maximum likelihood (ML)[2] is a general principle in signal processing and machine learning. Given the likelihood function $p(\mathbf{x};\boldsymbol{\theta})$ or the log-likelihood function $\ln p(\mathbf{x};\boldsymbol{\theta})$, the ML estimator of $\boldsymbol{\theta}$ for a given observation $\mathbf{x}$ is given by

$$\widehat{\boldsymbol{\theta}}_{\mathrm{ML}}(\mathbf{x}) = \underset{\boldsymbol{\theta}}{\mathrm{argmax}}\, p(\mathbf{x};\boldsymbol{\theta}) = \underset{\boldsymbol{\theta}}{\mathrm{argmax}} \ln p(\mathbf{x};\boldsymbol{\theta}). \tag{1.62}$$

It makes the observation $\mathbf{x}$ the most likely one. Conceptually, this is a simple unconstrained maximization problem and can be solved analytically or numerically by considering the necessary condition $\nabla \ln p(\mathbf{x};\boldsymbol{\theta}) = \mathbf{0}$.

[2]Unfortunately, the same abbreviation ML is widely used for two different terms: machine learning and maximum likelihood. The correct meaning of ML is mostly visible from the context.

The relationship between MVUE, efficient estimator, and ML estimator is [5]

$$\text{MUVE} \quad \underset{\nrightarrow}{\leftarrow} \quad \text{efficient estimator} \quad \underset{\nleftarrow}{\rightarrow} \quad \text{ML estimator} \tag{1.63}$$

This means that an efficient estimator (condition (1.61) is satisfied) is automatically MVUE and ML, but not vice versa. While MVUE represents "what we want to find," efficient estimator means "what we hope to find" and ML estimator is "what we can find."

For a finite number N of samples in $\mathbf{x}$, an ML estimator is in general not optimum. It can be biased, not efficient, and has an unknown distribution. But an ML estimator has, under certain regularity conditions and for i.i.d. samples, nice asymptotic properties. As $N \to \infty$, $\widehat{\boldsymbol{\theta}}_{\text{ML}}$ is asymptotically unbiased, asymptotically Gaussian distributed, and asymptotically efficient [5].

1.7 Markov and hidden Markov process

A discrete-time stochastic process $x(n), n \in \mathbb{Z}$ is a sequence of random variables. At time n, the joint PDF $p(x(n), \ldots, x(1))$ for continuous-valued $x(n)$ or the joint PMF $P(x(n), \ldots, x(1))$ for discrete-valued $x(n)$ describes the joint distribution of all observed random variables $x(n), \ldots, x(1)$. This joint distribution can be quite complex in general. A simplification is the Markov process, a family of stochastic processes with arbitrary distribution but local dependency. In the following, we focus on discrete-valued (finite-state) Markov processes with c states, i.e., $x(n) \in \{1, \ldots, c\}$. Such a Markov process is also called a Markov chain.

A Markov chain of order $m \in \mathbb{N}$ is characterized by [6]

$$P(x(n)|x(n-1), \ldots, x(1)) = P(x(n)|x(n-1), \ldots, x(n-m)). \tag{1.64}$$

Conditioned on $x(n-1), \ldots, x(n-m)$, the current sample $x(n)$ is independent of the even older samples $x(n-m-1), \ldots, x(1)$. The simplest case of a Markov chain is that of first order with

$$P(x(n)|x(n-1), \ldots, x(1)) = P(x(n)|x(n-1)). \tag{1.65}$$

In this case, the joint PMF $P(x(n), \ldots, x(1))$ is completely described by a sequence of first-order transition probabilities $P(x(i)|x(i-1)), 2 \le i \le n$ and one marginal probability $P(x(1))$:

$$P(x(n), \ldots, x(1)) = \left[\sum_{i=2}^{n} P(x(i)|x(i-1))\right] P(x(1)). \tag{1.66}$$

This follows immediately from the chain rule of probability Eq. (1.24) and the Markov property (1.65).

Let $\pi_i(n) = P(x(n) = i)$ be the probability of $x(n)$ taking the state value i. Let $\boldsymbol{\pi}(n) = [\pi_1(n), \ldots, \pi_c(n)]^T$ be the probability vector of $x(n)$ for all c states. It describes fully the Markov chain at time n. Let $\phi_{ij}(n) = P(x(n) = i|x(n-1) = j)$ be the transition probability from the old state $x(n-1) = j$ to the new one $x(n) = i$. All $c \times c$ transition probabilities can be summarized in one transition probability matrix $\phi(n) = [\phi_{ij}(n)]_{1 \le i,j \le c} \in \mathbb{R}^{c \times c}$. According to the law of total probability,

$$P(x(n)=i)=\sum_{j=1}^{c}P(x(n)=i|x(n-1)=j)P(x(n-1)=j)$$

or equivalently

$$\pi_i(n)=\sum_{j=1}^{c}\phi_{ij}(n)\pi_j(n-1). \quad (1.67)$$

In vector notation,

$$\boldsymbol{\pi}(n)=\phi(n)\boldsymbol{\pi}(n-1)=\ldots=\phi(n)\ldots\phi(2)\boldsymbol{\pi}(1). \quad (1.68)$$

As previously mentioned, the distribution of the current sample $x(n)$ is governed by a sequence of transition probability matrices $\phi(n),\ldots,\phi(2)$ and the initial state distribution $\boldsymbol{\pi}(1)$. This greatly simplifies the probabilistic description of a stochastic process.

In natural language processing, examples of Markov process are as follows:

Examples of Markov chain in natural language processing

(a) Word	= a sequence of letters with local dependency
(b) Sentence	= a sequence of words with local dependency
(c) Text	= a sequence of sentences with local dependency

Using methods of machine learning, the transition probabilities of the Markov chains are learned from data, i.e., (a) dictionary, (b) grammar, and (c) semantics. They are then used for inference.

Unfortunately, a Markov chain $x(n)$ is often not directly observable in practice. It is a hidden Markov process (HMM). Instead, each hidden state $x(n)$ of the Markov chain induces a corresponding noisy observation $\mathbf{y}(n)$ with the likelihood $p(\mathbf{y}(n)|x(n))$. One frequent task in machine learning is the estimation of the hidden Markov sequence $x(n)$ from the sequence of noisy observations $\mathbf{y}(n)$ by leveraging the likelihood $p(\mathbf{y}(n)|x(n))$ and the state transition probabilities. Instead of first-order Markov chains, also high-order Markov chains can be applied leading to a better performance but higher complexity.

References

[1] Boyd S. Convex optimization. Cambridge University Press; 2004.
[2] Sra S, Nowozin S, Wright SJ. Optimization for machine learning. MIT Press; 2011.
[3] Golub GH, Van Loan CF. Matrix computations. Johns Hopkins Press; 1983.
[4] Bishop C. Pattern recognition and machine learning. Springer; 2006.
[5] Kay SM. Fundamentals of statistical signal processing, volume 1: Estimation theory. Prentice-Hall; 1993.
[6] Papoulis A. Probability, random variables and stochastic process. McGraw-Hill; 1965.

CHAPTER

The ingredients for machine learning

2

Thomas Küstner[1,*], Kerstin Hammernik[3], Christian Baumgartner[2]
[1]*University Hospital Tübingen, Medical Image and Data Analysis, Department of Radiology, Tübingen, Germany;* [2]*University of Tübingen, Cluster of Excellence – Machine Learning for Science, Tübingen, Germany;* [3]*Technical University of Munich, School of Computation, Information and Technology, Munich, Germany*
[*]*Corresponding author: E-mail: thomas.kuestner@med.uni-tuebingen.de*

2.1 Introduction

Artificial intelligence (AI) and machine learning (ML) have evolved tremendously during the last decade. In the field of medicine, including in the field of medical imaging, AI and ML have emerged as techniques that are likely to fundamentally transform clinical practice in the coming years. In particular in medical imaging, AI and ML have the potential to impact all stages of the imaging pipeline, from (1) image acquisition and reconstruction to (2) image analysis and interpretation and (3) diagnosis and prognosis. The terms AI, ML, and more recently deep learning (DL) are often used as a synonyms; however, there are substantial differences between them.

2.1.1 Artificial Intelligence

In 1956, the term *artificial intelligence (AI)* was defined by John McCarthy as "the science and engineering of making intelligent machines" at the 1956 Dartmouth Conference [1]. Since then, AI serves as an umbrella term to solve any task that usually requires human intelligence.

2.1.2 Machine learning

ML is a subbranch of AI that describes any *data-driven* learning model to predict the output of a particular task and summarizes (statistical) methods that enable machines to improve with experience. Inspired by the biological neuron, Frank Rosenblatt introduced the perceptron [2], a single artificial neuron. After the AI winter, ML started to flourish in the 1980s with developments on Boltzman machines [3], recurrent neural networks [4–7], multilayer perceptron and backpropagation [8]. The neocognitron [9] reported in 1980 can be seen as the origin of convolutional

Machine Learning in MRI. https://doi.org/10.1016/B978-0-443-14109-6.00002-X

neural networks (CNNs) [10]. All these publications formed the basis for later developments in the field of DL.

The basic working principle of ML is summarized in Fig. 2.1. The ML algorithm receives as input *features* or a reduced/combined subset of features that are extracted from input data. These input data include acquired sensor signals (e.g., raw MR), (reconstructed) images, clinical and anthropometric metaparameters, and many more. Feature extraction requires hand-crafting and manual fine-tuning of features according to the underlying task and application. Hence, the quality of the predicted output depends not only on the ML algorithm itself but also on the extracted features.

2.1.3 Deep learning

DL [11] is a subbranch of ML that enables automated feature extraction and comprise deep neural networks consisting of multiple layers. The basics of these methods were established in the late 1980s, but only received widespread attention with the remarkable performance boost of AlexNet [12] in the ImageNet challenge 2012 for object recognition. The increasing availability and performance of graphic processing units (GPUs) as well as annotated large-scale databases helped in the development of novel neural network building blocks (see Section 2.7) and deep architectures (see Section 2.8).

2.2 Organization of this chapter

In the following, we will briefly introduce the basic concepts of machine/deep learning such as the types of learning (see Section 2.4), their association to the network parameter optimization (see Section 2.5), and how the data is split for training, testing, and validation (see Section 2.5). We will then show some

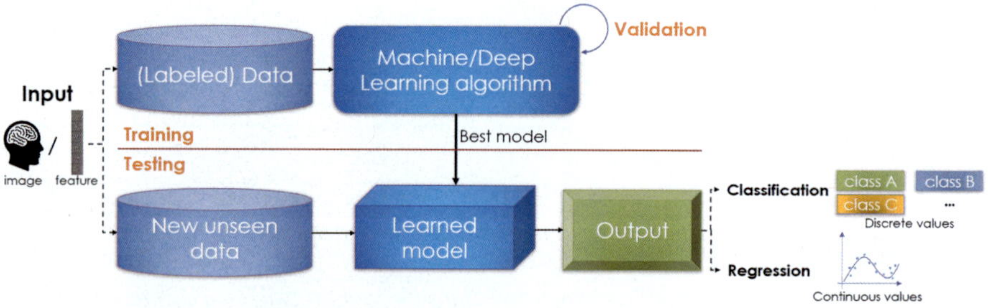

FIGURE 2.1 Processing steps in machine learning (ML).

Processing steps of ML algorithms with corresponding image or feature input and task-specific output of classification or regression. Algorithms are learned on labeled (supervised) or unlabeled (unsupervised) data in a training step. Validation with independent datasets allows for tracking and optimization of learning. Testing (if not unsupervised learning) is performed on new unseen data for the best trained model.

commonly used network building blocks (see Section 2.7) and highlight well-known established architectures (see Section 2.8). We conclude with an overview of generative models (see Section 2.9).

2.3 Learning models

The learning model represents any (non)linear and parametric model

$$\mathbf{x}_{\mathrm{NN}} = f_{\boldsymbol{\theta}}(\mathbf{x}) \tag{2.1}$$

which maps the inputs $\mathbf{x}$ to the neural network outputs $\mathbf{x}_{\mathrm{NN}}$. The mapping function $f_{\boldsymbol{\theta}}$ represents a neural network whose parameters $\boldsymbol{\theta}$ can be optimized under some given cost function $\mathscr{L}$ (see Section 2.5). During a training process, the model sees different data samples such that the parameters $\boldsymbol{\theta}$ of the neural network can be learned to predict a reasonable output $\mathbf{x}_{\mathrm{NN}}$ for any new input $\mathbf{x}$. During testing, the model is fixed and the trained parameters generate the output from new unseen test data. The models perform a sequence of operations on the inputs to yield a task-specific output. The most commonly used learning models and their basic building blocks are summarized in Section 2.7. The reader is referred to textbooks [13–15] for more details on learning models.

2.4 Types of learning

Types of learning can be differentiated based on the availability of label information and the type of label integration during training. In principle, this relates to how closely a human observer is involved during training (Fig. 2.2).

Supervised learning assumes that all data samples have associated task-specific labels and that all samples as well as their labels are available in the database. This allows us to directly learn the parametric model in Eq.2.1. However, labeling is very time- and cost-intensive and often requires human interaction, ranging from data sorting over curating to annotating structures within the image. For example, in undersampled MR image reconstruction, a labeled ground-truth corresponds to a fully sampled dataset or a dataset acquired with mild undersampling. In segmentation, the ground-truth refers to the manually annotated organ or tissue of interest.

In *unsupervised learning*, no labeled data or any type of a priori information about the data is available that could be leveraged to guide the training. The network learns to identify patterns in the data. Common approaches are clustering [13] (e.g., k-means or Gaussian mixture models), principal component analysis (PCA), (variational) autoencoders (VAEs) [16,17], deep belief networks [18], or generative adversarial networks (GANs) [19].

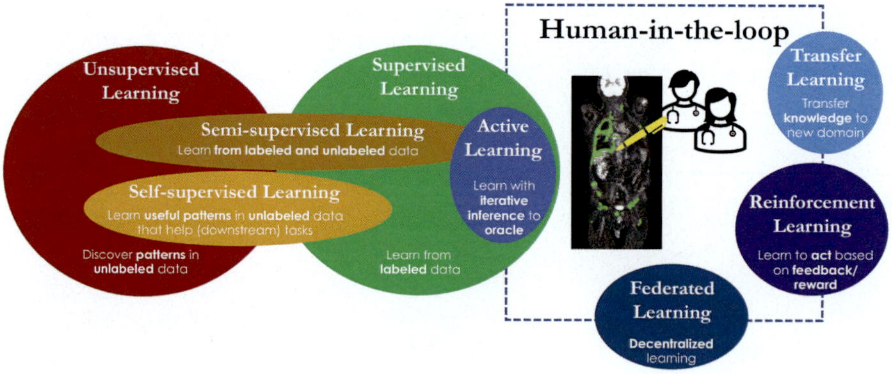

FIGURE 2.2 Types of learning.

Types of learning machine/deep learning models. Various types can be distinguished based on the availability and usage of labeled data. Fundamentally, this concerns the extent to which a human is actively engaged during training, known as human-in-the-loop.

In *semisupervised* learning, both labelled and unlabelled data are included in the learning process. The unlabelled data provides additional information about the underlying data distribution.

Self-supervised learning circumvents the problem of external labels. The input data itself is used to guide the learning.

Active learning integrates an oracle into the training procedure, which is periodically queried to either label or select the next most meaningful samples for training. The oracle is in most cases a human observer, but can also be another algorithm [20].

In *reinforcement learning*, the model receives feedback from an environment in the form of rewards or penalties, based on its current prediction, which subsequently drives the training procedure.

Transfer learning investigates the possibility to transfer knowledge between models or tasks. Information might be transferred from simpler to more complex tasks, or from a source domain to another (but similar) target domain [21–23].

Federated learning [24–26] trains a model across multiple decentralized devices, where each device holds their own set of training data and only the model weights are shared across devices. This allows for training across highly heterogenous datasets. Furthermore, federated learning has a vast potential in medicine, where sharing of data across multiple centers is challenging due to data protection and data privacy [27].

2.5 Cost function, optimization, and backpropagation

2.5.1 Cost function

The *cost function* (*error*, *loss*) $\mathscr{L}$ is a quantitative measure that describes how well the neural network $f_\theta(\mathbf{x})$ predicts the output $\mathbf{x}_{\mathrm{NN}}$ from the input $\mathbf{x}$. In supervised learning, the aim is to minimize the loss between prediction and label/ground-truth. In general, the ML optimization task can be formulated as

$$\min_{\boldsymbol{\theta}} \mathscr{L}(\boldsymbol{\theta}) = \min_{\boldsymbol{\theta}} \sum_{j} \lambda_j \mathscr{L}_j(\boldsymbol{\theta}) \tag{2.2}$$

which is composed of several λ_j weighted sublosses $\mathscr{L}_j(\boldsymbol{\theta})$. This can include, e.g., regularization of weights or penalties on intermediate model outputs. Note that the result of a loss function is a real-valued scalar, even if the input data $\mathbf{x}$ and model output $\mathbf{x}_{\mathrm{NN}}$ are in the complex domain as often encountered in MRI.

Depending on the chosen architecture (see Section 2.8) and the desired task (regression, classification), the appropriate losses are selected. Common cost functions in regression problems are the mean-squared-error (MSE, L_2 distance)

$$\mathscr{L}_{MSE}(\boldsymbol{\theta}) = \frac{1}{2N_s} \sum_{i=1}^{N_s} \| f_{\boldsymbol{\theta}}(\mathbf{x}_i) - \mathbf{x}_{\mathrm{GT}}, i \|_2^2 \tag{2.3}$$

or the mean-absolute error (MAE, L_1 distance)

$$\mathscr{L}_{MAE}(\boldsymbol{\theta}) = \frac{1}{N_s} \sum_{i=1}^{N_s} \| f_{\boldsymbol{\theta}}(\mathbf{x}_i) - \mathbf{x}_{\mathrm{GT},i} \|_1, \tag{2.4}$$

to minimize the error between predicted output $f_{\boldsymbol{\theta}}(\mathbf{x})$ and ground-truth $\mathbf{x}_{\mathrm{GT}}$ for all N_s data samples. Differences between commonly used loss functions, including MSE, MAE, or structural similarity index (SSIM) [28], can be found in Ref. [29]. Practically, any piece-wise differential function can be used as a loss function.

In classification, particularly function is the cross-entropy loss, which can be stated as

$$\mathscr{L}_{CE}(\boldsymbol{\theta}) = -\sum_{c=1}^{C} y_{\mathrm{GT},i} \log f_{\boldsymbol{\theta}}(\mathbf{x}_i) \tag{2.5}$$

for a predicted probability output $f_{\boldsymbol{\theta}}(\mathbf{x}_i)$ and binary indicator $y_{\mathrm{GT},i} \in \{0, 1\}$ that the sample i belongs to any of C classes. Cross-entropy measures the difference between the predicted probability and the actual label and penalizes the model more when being confident about an incorrect prediction.

Perceptual or style transfer losses are often derived from feature maps of pre-trained networks and try to mimic the human visual system by capturing differences between high-level image feature representations [30].

A number of losses are aimed at reducing the distance between two probability distributions. For example, the Kullback-Leibler divergence is used in VAEs [31] to constrain the latent embeddings to follow a standard normal distribution. The adversarial loss used in GANs [19] can be shown to minimize the Jenson-Shannon divergence between the real image distribution and the generated samples. In later work, adversarial losses were developed, which implicitly minimize other distances between distributions, such as the Wasserstein distance in Wasserstein GANs [32] or general f-divergences in f-GANs [33]. Generative models will be discussed in greater detail in Section 2.9 (Box 2.1).

2.5.2 Optimization

Once a cost function is defined, the model parameters $\boldsymbol{\theta}$ need to be found which minimize the cost function $\mathscr{L}$. Gradient descent algorithms are commonly used to perform this optimization. While the negative gradient of the loss function $-\frac{\partial \mathscr{L}}{\partial \boldsymbol{\theta}}$ points toward the local minimum, optimal parameters $\boldsymbol{\theta}$ can be found by iteratively going along this direction with a predefined step size τ, known as learning rate. A simple gradient descent algorithm iteratively solves for

$$\boldsymbol{\theta}^{[it+1]} = \boldsymbol{\theta}^{[it]} - \tau \frac{\partial \mathscr{L}}{\partial \boldsymbol{\theta}}. \tag{2.6}$$

Key challenges for gradient-based optimizations are the proper selection of the learning rate τ to reach the (global) minimum and to avoid getting trapped in numerous suboptimal local minima for the commonly nonconvex optimization problem. Many other *optimizers* exist to toggle these challenges in various ways. They differ in the way how much data is incorporated to update the parameters, if gradient momentum is included and/or if the learning rate is adapted, known as learning rate scheduling [34].

The most common optimizers in ML are stochastic gradient descent (SGD) [35], adaptive moment estimation (ADAM) [36], RMSProp [37], Nesterov accelerated gradient (NAG) [38], Adagrad [39], Adadelta [40], AdaMax [36], or Nesterov-accelerated adaptive moment estimation (NADAM) [41]. An overview and comparison of these optimizers can be found in Ref. [42] (Box 2.2).

2.5.3 Backpropagation

In general, the network $f_{\boldsymbol{\theta}}$ consists of L connected layers

$$\mathbf{x}_{\text{NN}} = f_{\boldsymbol{\theta}}(\mathbf{x}) = f_L \circ f_{L-1} \circ \ldots \circ f_1(\mathbf{x}) = \mathbf{x}_L. \tag{2.7}$$

> **Box 2.1 Implementation of losses in ML frameworks**
>
> Common loss functions are provided in `tf.keras.losses` or `torch.nn`. A loss function needs to be defined to compile a model. Additional losses on, e.g., intermediate outputs of the model can be defined in the model using the `add_loss()` method in the `call` method of a layer or subclassed model. Weight regularization applies penalty on the network parameters. Weight regularization is defined in `tf.keras.regularizers` and can be added when creating a `tf.keras.layer`. The Keras API collects the defined losses in the model. In case of PyTorch, this is handled in self-defined layers or as a parameter `weight_decay` in the optimizer for an L_2 regularization.

Box 2.2 Implementation of optimizers in ML frameworks

Common optimizers are provided in `tf.keras.optimizers` or `torch.optim`. An optimizer needs to be defined to compile a model. A learning rate scheduler (see `tf.keras.optimizers.schedules` or `torch.optim.lr_scheduler`) can be added to the optimizer call.

The output of the l-th layer is denoted as $\mathbf{x}_l = f_l(\cdot), 1 \leq l \leq L$. In training, the parameters $\boldsymbol{\theta}$ of the network $f_{\boldsymbol{\theta}}$ are learned to minimize the loss function Eq. (2.2). This involves two steps as depicted in Fig. 2.3. First a feedforward prediction from the input $\mathbf{x}$ to the output $\mathbf{x}_{\mathrm{NN}} = f_{\boldsymbol{\theta}}(\mathbf{x})$ is required to evaluate the loss $\mathscr{L}$ for the current set of parameters $\boldsymbol{\theta}$. Second, the error between the predicted output $f_{\boldsymbol{\theta}}(\mathbf{x})$ and the ground-truth labels $\mathbf{x}_{\mathrm{GT}}$ is backpropagated to adapt the parameters $\boldsymbol{\theta}$. Backpropagation [43] can be realized by applying the chain rule of differentiation:

$$\frac{\partial \mathscr{L}}{\partial \boldsymbol{\theta}} = \frac{\partial \mathscr{L}}{\partial \mathbf{x}_{\mathrm{NN}}} \frac{\partial \mathbf{x}_{\mathrm{NN}}}{\partial \boldsymbol{\theta}}. \tag{2.8}$$

More specifically, the gradient needs to be backpropagated from the output through the single layers to update the layer parameters $\boldsymbol{\theta}_l$. Together with Eq. (2.6), the update rule for the parameters $\boldsymbol{\theta}_l$ of layer f_l in the $[it]$-th training iteration becomes

$$\begin{aligned} \boldsymbol{\theta}_l^{[it+1]} &= \boldsymbol{\theta}_l^{[it]} - \tau \frac{\partial \mathscr{L}}{\partial \boldsymbol{\theta}_l} \\ &= \boldsymbol{\theta}_l^{[it]} - \tau \frac{\partial \mathscr{L}}{\partial \mathbf{x}_L} \frac{\partial \mathbf{x}_L}{\partial \mathbf{x}_{L-1}} \cdots \frac{\partial \mathbf{x}_l}{\partial \boldsymbol{\theta}_l}. \end{aligned} \tag{2.9}$$

If complex-valued layers and network outputs $\mathbf{x}_{\mathrm{NN}}$ are involved, complex backpropagation following Wirtinger calculus [44–47] has to be considered (Box 2.3).

2.5.4 Training, validation, and testing

Training describes the process of updating the model parameters $\boldsymbol{\theta}$ in the ML model $f_{\boldsymbol{\theta}}(\mathbf{x})$ optimizing a cost function. The training database is the composition of N_{train}

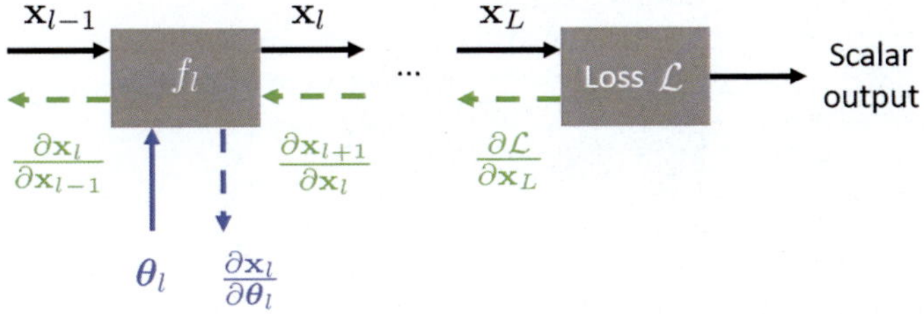

FIGURE 2.3 Backpropagation.

The network consists of L connected layers with $f_l = \mathrm{x}_l$ being the output of the l-th layer, $1 \leq l \leq L$. The output prediction $\mathrm{x}_{\mathrm{NN}} = \mathrm{x}_L$ is passed to a loss function $\mathscr{L}$. This gives an error that is backpropagated using the chain rule of differentiation to update the network parameters $\boldsymbol{\theta}_l$.

Box 2.3 Backpropagation in practice

Automatic differentiation is supported in ML frameworks. Complex backpropagation according to Wirtinger calculus is supported in Tensorflow/Keras and PyTorch. Since version 1.6 PyTorch starts to implement support for complex-valued processing.

training samples. In case of supervised learning, the cost function involves the labels/reference output $\mathbf{x}_{GT}$; hence, the training database contains both training samples and their associated annotated labels. The gradient descent in Eq. (2.9) would require all training samples for the update; however, this requires a huge computing effort especially for large datasets. In ML, stochastic optimization is used to overcome this computational burden. SGD (stochastic version of GD) computes a single gradient update only on a subset of training samples, provided in a single batch. The *batch size* N_b is set a priori. For N_{train} training samples and a batch size N_b, we obtain $\lfloor N_{\text{train}}/N_b \rfloor$ iterations per epoch. An epoch refers to one pass over the full training set. The amount of epochs determines the training duration. Careful consideration needs to be taken in terms of amount of training data, model capacity, and problem complexity to avoid under- or overfitting, which can be observed by comparing discrepancies between the *training* and *validation loss* [15]. During training, dropout layers (see Section 2.7.7) can be included in neural networks, weights can be regularized, or early stopping can be deployed to reduce the risk of overfitting. The model behavior can be observed by the training loss. However, it is advisable to monitor the training progress with a separate validation set.

Validation is performed during model training to track the model performance or to optimize model hyperparameters. However, during validation, the model parameters $\boldsymbol{\theta}$ themselves are not updated. Therefore, a separate validation set N_{val} data samples with annotated labels (in case of supervised learning) is used to evaluate the performance of the currently learned model following a predetermined validation cycle. Evaluation metrics are on one hand the loss function itself (validation loss), but can also include other quantitative measures (MSE, MAE, SSIM) [29].

In *testing/inference/prediction*, the trained model is tested with N_{test} new unseen data samples. Test data is usually not accompanied by annotated labels (in case of supervised learning). In case annotated labels or ground-truth references are available, nonblinded similarity measures (e.g., MSE) can be applied to assess the model performance.

2.6 Database splitting

Training, validation, and test set are disjoint. Available data samples N_s can be split up into training N_{train}, validation N_{val}, and test set N_{test} according to different inclusion/exclusion criteria or completely random. Database creation and curation is an important step toward a reliable ML method. The outline and organization (file structure, data types, etc.) of the database specifies the database pipeline needed to load the data into the ML model.

Selection of representative training samples N_{train} is crucial to provide enough diversity for training, but with sufficient consistency to converge. The validation set size N_{val} is usually much smaller than N_{train}, e.g., 10%–20% of the number of training samples, depending on the underlying application, task, and hyperparameter optimization. Test samples N_{test} are never seen by the method during training and ideally (but not necessarily) reflect the same underlying trained application, which enable to infer performance, robustness, and/or generalization of the model.

2.6.1 Random splitting

While randomly splitting data samples into training, validation, and test set (or between training and validation set) is feasible and justified for some applications, it is strongly discouraged in the field of medical imaging. A random splitting ignores the information from which patient the data sample is derived and hence samples from a specific patient can be assigned to training, validation, and test set, which at the end potentially boosts performance, but definitively biases results.

2.6.2 Patient leave-out approach

Therefore, to avoid biases and confounding via the selection of the database, common splitting methods consider a patient leave-out approach. Only unique patients are fed through the model in training, validation, and testing. This avoids any bias of the model toward specific anatomies or pathologies. A database of P patients with each patient having S data samples (e.g., slices in a volume) consists of $N_s = P \cdot S$ samples:

$$\left\{ \underbrace{x_{1,1}, \ldots, x_{1,S}}_{S}, \underbrace{x_{2,1}, \ldots, x_{2,S}}_{S}, \underbrace{x_{3,1}, \ldots, x_{3,S}}_{S}, \ldots, \underbrace{x_{P,1}, \ldots, x_{P,S}}_{S} \right\}. \tag{2.10}$$

In a patient leave-out approach, unique patients $\{x_{i,1}, \ldots, x_{i,S}\}$ are assigned for the training, validation, and test set as illustrated in Fig. 2.4. In a similar fashion, one could also define other and/or additional means to sort and split the data along S, e.g., imaging contrast, multiparametric information, trajectory, and so on, depending on the specific reconstruction problem.

2.6.3 Cross-validation

For smaller databases, it is advisable to perform a cross-validation to reduce bias toward selected samples and investigate the method's robustness. In a k-fold cross-

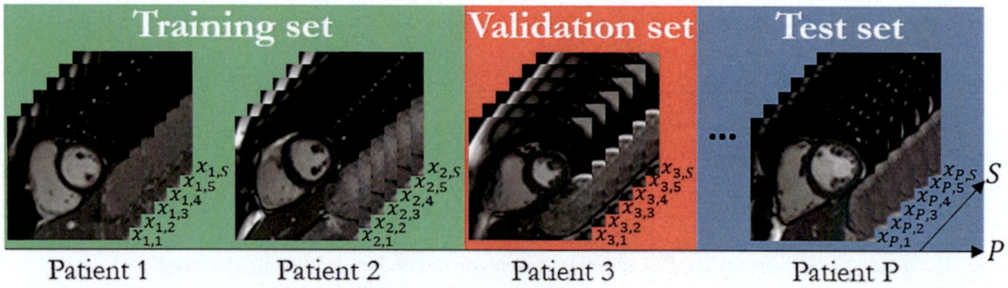

FIGURE 2.4 Database splitting with patient leave-out.

Splitting of available data in training, validation, and test set. Each of the P patients has S samples (slices in a volume, imaging contrasts, ...). A patient leave-out approach is chosen, i.e., unique patients are assigned for each set.

validation, the available database of N_s samples is split into k folds—considering unique patients are sorted into each fold—with each having N_s/k samples. Assignment of samples to these folds can either be random (after patient consideration) or follow specific criteria. In total, k networks are then trained, one for each of the folds. Finally a mean performance across all folds can be reported. Deviation in evaluation metrics between folds gives an indicator on the method's robustness.

2.7 Fundamental network elements

Neural networks are composed of several layers. Each layer operates on received input variables and outputs output variables, which are denoted by $\widehat{\mathbf{x}}$ in the following. A layer may have trainable and nontrainable parameters. Different types of layers are commonly used in ML-based MR reconstruction, including convolution, normalization, activation, and so on. These layers are further described hereafter in the order in which they typically appear in a neural network.

2.7.1 Convolution layer

Convolutions are linear operations that extract *features* of an image. While shallow convolution layers extract low-level features such as edges, deeper layers extract more abstract, high-level features such as shapes or objects. A convolution layer maps $N_{f,\text{in}}$ input feature channels to $N_{f,\text{out}}$ output feature channels

The discrete convolution of an image $\mathbf{x}$ with filter kernels $\mathbf{k}_{i,j}$ reads as

$$\widehat{\mathbf{x}}_j = \sum_{i=1}^{N_{f,\text{in}}} \mathbf{x}_i * \mathbf{k}_{i,j} = \mathbf{K}_{i,j}\mathbf{x}_i, j = 1, \ldots, N_{f,\text{out}}, \tag{2.11}$$

where the subscripts denote the feature channels. This produces $N_{f,\text{out}}$ feature maps in $\mathbf{x}_{\text{NN}}$. Note that convolutional layers can be considered a special case of fully connected layers (see Section 2.7.4) with a sparse connectivity and weight sharing. Therefore, the convolution with a filter kernel $\mathbf{k}_{i,j}$ can also be represented as a matrix-vector multiplication, where $\mathbf{K}_{i,j} : \mathbb{K}^N \rightarrow \mathbb{K}^N$ is a sparse matrix. Note that convolution filters are local feature extractors and invariant to translation. The convolution operation is depicted in Fig. 2.5.

For real-valued convolutions, it is $\mathbb{K} = \mathbb{R}$. For complex convolutions, $\mathbb{K} = \mathbb{C}$, the convolution operation is extended to

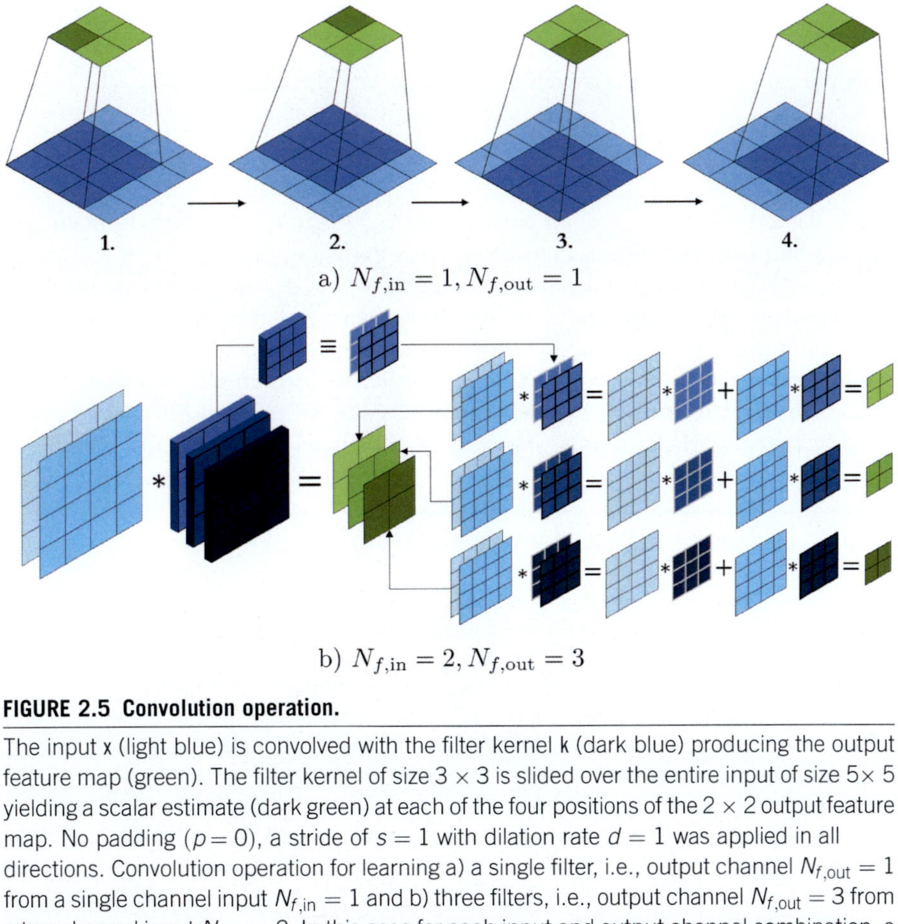

FIGURE 2.5 Convolution operation.

The input $\mathbf{x}$ (light blue) is convolved with the filter kernel $\mathbf{k}$ (dark blue) producing the output feature map (green). The filter kernel of size 3×3 is slided over the entire input of size 5×5 yielding a scalar estimate (dark green) at each of the four positions of the 2×2 output feature map. No padding ($p = 0$), a stride of $s = 1$ with dilation rate $d = 1$ was applied in all directions. Convolution operation for learning a) a single filter, i.e., output channel $N_{f,\text{out}} = 1$ from a single channel input $N_{f,\text{in}} = 1$ and b) three filters, i.e., output channel $N_{f,\text{out}} = 3$ from a two channel input $N_{f,\text{in}} = 2$. In this case for each input and output channel combination, a kernel is learned, i.e., $2 \cdot 3 = 6$ kernels of size 3×3 are trained. Input channels are convolved with the respective kernel and summed up to form the 2×2 output for each of the three output channels.

$$\begin{aligned}\mathbf{x}_i * \mathbf{k}_{i,j} &= \left(Re(\mathbf{x}_i) * Re\left(\mathbf{k}_{i,j}\right) - Im(\mathbf{x}_i) * Im\left(\mathbf{k}_{i,j}\right)\right) \\ &\quad + \mathrm{i}\cdot\left(Im(\mathbf{x}_i) * Re\left(\mathbf{k}_{i,j}\right) + Re(\mathbf{x}_i) * Im\left(\mathbf{k}_{i,j}\right)\right) \\ &= \mathbf{K}_{i,j}\mathbf{x}_i \\ &= \left(Re\left(\mathbf{K}_{i,j}\right)\cdot Re(\mathbf{x}_i) - Im\left(\mathbf{K}_{i,j}\right)\cdot Im(\mathbf{x}_i)\right) \\ &\quad + \mathrm{i}\cdot\left(Im\left(\mathbf{K}_{i,j}\right)\cdot Re(\mathbf{x}_i) + Re\left(\mathbf{K}_{i,j}\right)\cdot Im(\mathbf{x}_i)\right).\end{aligned} \tag{2.12}$$

If real and imaginary parts are represented as a vector, the following notation can be used:

$$\begin{bmatrix} Re\left(\mathbf{K}_{i,j}\mathbf{x}_i\right) \\ Im\left(\mathbf{K}_{i,j}\mathbf{x}_i\right) \end{bmatrix} = \begin{bmatrix} Re\left(\mathbf{K}_{i,j}\right) & -Im\left(\mathbf{K}_{i,j}\right) \\ Im\left(\mathbf{K}_{i,j}\right) & Re\left(\mathbf{K}_{i,j}\right) \end{bmatrix} \begin{bmatrix} Re(\mathbf{x}_i) \\ Im(\mathbf{x}_i) \end{bmatrix}. \tag{2.13}$$

The amount of trainable parameters is determined by the kernel size k along each direction, and the amount of input and output features $N_{f,\mathrm{in}}$ and $N_{f,\mathrm{out}}$. The amount of trainable parameters in a convolution layer can be calculated as

$$\#\text{parameters} = N_{f,\mathrm{out}}\cdot\left(1 + N_{f,\mathrm{in}}\cdot\prod_p k_p\right), \tag{2.14}$$

where the kernel sizes k_p along each direction $p \in \{x, y, z, t\}$ are multiplied and where a bias term per filter is considered (Box 2.4).

If complex-valued input data are considered, two processing modes are possible: (1) real/imaginary are considered in two input channels; hence, the output for each convolution filter is real-valued. However, the complex relationship between real and imaginary parts is lost. (2) Complex-valued convolutions as stated in Eq. (2.12) can be performed, where the output for each convolution filter is complex-valued. The drawback of this approach is that twice the amount of trainable

Box 2.4 Implementation of a convolution layer

A convolution layer is characterized by the following items:

- `in_channels` $N_{f,\mathrm{in}}$: Number of channels in the input image
- `out_channels` $N_{f,\mathrm{out}}$: Number of output channels, respectively, feature maps in the result of the convolution, equals the number of filters to be trained
- `kernel_size` k: Size of the convolution kernel, integer or tuple
- `padding` p: Boundary handling, impacts the image size, string, integer or tuple
- `strides` s (optional): Spacing between pixel elements considered in the convolution, integer or tuple
- `dilation_rate` d (optional): Spacing between kernel elements, integer or tuple
- `bias` (optional): Adds a trainable offset to the output of the convolution, bool

A simple Keras 2D convolution layer can be created using

```
tf.keras.layers.Conv2D(out_channels, kernel_size, use_bias=True)
```

The input channels are automatically estimated in a separate `build` function.

All parameters need to be *initialized*. Commonly used initializers for filter kernels follow a Glorot Uniform [48] or an He normal distribution [49]. The initializer can be passed directly to the Keras `Conv2D` by passing, e.g., `kernel_initializer="glorot_uniform"`.

parameters is required as a real and imaginary filter kernel needs to be learned. Additionally, the number of convolution operations doubles compared to (1).

2.7.1.1 Padding

Padding appends values at the boundary of the input, which for the majority of frameworks, is a zero-padding, i.e., a constant value of zero, but in general can follow periodic or symmetric padding as well. Three special cases for padding are of importance: (1) no or *valid padding* uses the input without padding ($p = 0$), (2) half or *same padding* appends zeroes so that the output has the same size as the input, and (3) *full padding* appends zeroes such that every pixel in the input contributes equally to the output. Full padding may be useful in applications where important information is located at the edges of the image, as it allows us to treat edge pixels the same as other pixels. However, full padding results in an increased output size. The amount of padding p in these cases depends on the kernel size k and stride s. Fig. 2.6 illustrates these three special cases.

2.7.1.2 Stride

The *stride* describes the step size with which the filter kernel is moved over the input. A stride can be used to perform a downsampling operation similar to a pooling layer (see Section 2.7.5). Fig. 2.7 shows a convolution operation with stride.

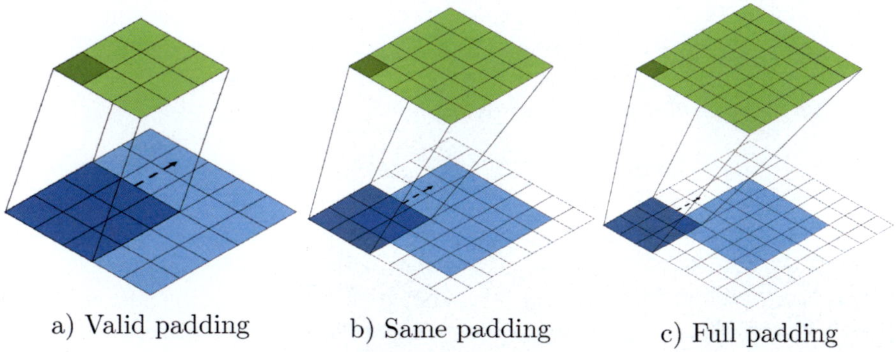

FIGURE 2.6 Convolution operation: Padding.

Padding operations of the input. The input x (light blue) is padded with zeros (white) and then convolved with the filter kernel k (dark blue) producing the output feature map (green). The filter kernel of size 3×3 is slided over the entire input of size 5×5 yielding a scalar estimate (dark green) at each position of the output feature map of size a) 3×3 for valid padding, i.e., $p = 0$, b) 5×5 for same padding with $p = 1$ in each direction, i.e., 1×1 padding so that output size is equal to input size, or c) 7×7 for full padding with $p = 2$ in each direction, i.e., 2×2 padding to enlarge output size. A stride of $s = 1$ and dilation rate $d = 1$ was used in all cases and all directions. Only a single filter is learned, i.e., output channel $N_{f,\text{out}} = 1$ from a single input channel $N_{f,\text{in}} = 1$.

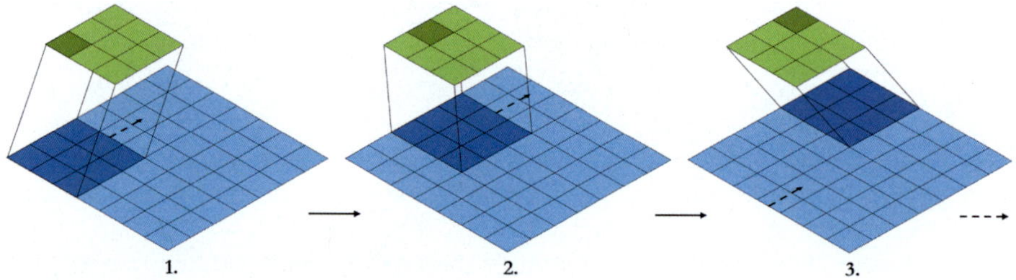

FIGURE 2.7 Convolution operation: stride.

Stride of the filer kernel moving over the input. The input x (light blue) is convolved with the filter kernel k (dark blue) producing the output feature map (green). The filter kernel of size 3×3 is slided over the entire input of size 7×7 with a stride of $s = 2$ yielding a scalar estimate (dark green) at each of the output feature map of size 3×3. No padding ($p = 0$) and a dilation rate $d = 1$ was applied in all directions. Only a single filter is learned, i.e., output channel $N_{f,\text{out}} = 1$ from a single input channel $N_{f,\text{in}} = 1$.

2.7.1.3 Receptive field

The receptive field describes the spatial extent of the input that a particular feature map within the network observes or is affected by. It is determined by the `kernel_size` k, `padding` p, `stride` s, and `dilation_rate` d (see Section 2.7.1.1). These values are prescribed for each input direction and consequently influence the output size and receptive field, respectively. The output size along each dimension is for an input of size R_{in} given by

$$R_{\text{out}} = \left\lfloor \frac{R_{\text{in}} + 2p - d \cdot (k-1) - 1)}{s} \right\rfloor + 1. \tag{2.15}$$

For N-dimensional filters, these values can be stated as $R_{x,\text{out}} \times R_{y,\text{out}} \times \ldots$. For an input $R_{x,\text{in}} \times R_{y,\text{in}} \times \ldots \times N_{f,\text{in}}$, the convolution with $N_{f,\text{out}}$ filters of size $k_x \times k_y \times \ldots$ (with padding $p_x \times p_y \times \ldots$ and stride $s_x \times s_y \times \ldots$) yields an output of size $R_{x,\text{out}} \times R_{y,\text{out}} \times \ldots \times N_{f,\text{out}}$. For a single layer, the receptive field is the same as the size of the output from this layer. If several layers are applied sequentially, the receptive field can be retrieved by consecutive calculation of Eq. (2.15).

2.7.1.3.1 Dilated convolution

Dilated convolution, also known as atrous convolution, introduce a dilation rate along each direction to spread out the filter kernel. With this concept, the same receptive field can be achieved but with less trainable parameters. Fig. 2.8 depicts an example for a convolutional layer with dilation rate of $d = 1$, i.e., no dilation, and $d = 2$. To achieve the same receptive field, a filter kernel of 5×5 for $d = 1$ and of 3×3 for $d = 2$ can be used, yielding 25 ($d = 1$) versus 9 ($d = 2$) trainable parameters.

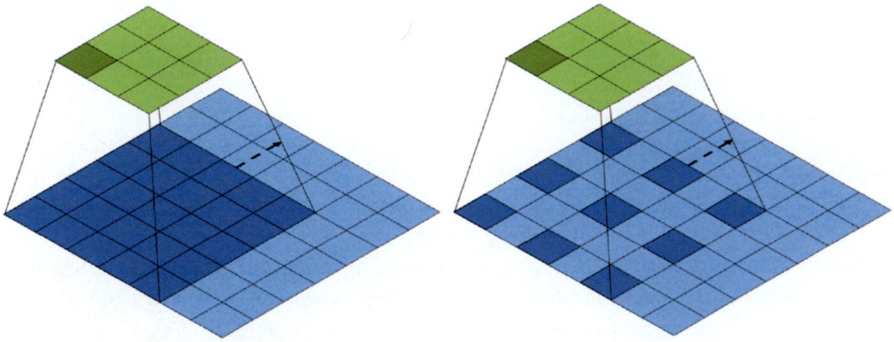

a) Dilation rate $d = 1$ b) Dilation rate $d = 2$

FIGURE 2.8 Dilated convolution.

Dilated convolutions on the input x (light blue) which is convolved with the filter kernel k (dark blue) producing the output feature map (green). The filter kernel of size (a) 5×5 for dilation rate of $d = 1$, i.e., no dilation, and (b) 3×3 for a dilation rate of $d = 2$, i.e., dilation of 2×2, is slided over the entire input of size 7×7 yielding a scalar estimate (dark green) at each position of the output feature map of size 3×3. A stride of $s = 1$ and valid padding $p = 0$ was used in all cases. Only a single filter is learned, i.e., output channel $N_{f,\mathrm{out}} = 1$ from a single input channel $N_{f,\mathrm{in}} = 1$.

2.7.1.3.2 Separable convolution

While 2D and 3D implementations exist in machine learning frameworks, spatio-temporal and higher-dimensional convolutions might be designed in a different way. Separable convolution splits the N-dimensional kernel operation into multiple steps. *Spatial separable convolution* splits the convolution along the spatial direction if they can be factored into independent directions. While N-dimensional convolutions can be used, various work have shown that separable convolutions can perform superior to N-dimensional convolutions [50,51]. Furthermore, less trainable parameters are required.

For example in a separable 2D + t convolution, the 3D convolution kernel $\mathbf{k}$ along x,y, and t can be separated into a 2D spatial and 1D temporal convolution filter if $\mathbf{k} = \mathbf{k}_{x,y} \otimes \mathbf{k}_t$ applies. Various other variants exist to further improve the performance [52] or can be extended toward higher dimensions as shown in Refs. [51,53].

A *depthwise separable convolution* splits the spatial component from the depth/output channel component. An example calculation for the reduction in trainable parameters is shown in Appendix A.1.

2.7.1.3.3 Transposed convolution

Transposed convolutions generally arise from the desire to use a transformation going in the opposite direction of a normal convolution, i.e., from something that has the shape of the output of some convolution to something that has the shape of its input while maintaining a connectivity pattern that is compatible with convolutions.

Transposed convolutions can thus be used for upsampling operations (see Section 2.7.6). Following Eq. (2.11), this yields the transposed filter $\mathbf{K}^T$ for $\widehat{\mathbf{x}} = \mathbf{K}^T\mathbf{x}$.

2.7.2 Normalization layer

Adding normalization layers directly after convolution layers is a common way to enable faster and more stable training of neural networks. Statistics are estimated from the input and used to reparametrize the input. Hence, the subsequent layers are less tolerant to changes in previous layers. The selection of the normalization layer is task dependent, and there is no general recipe which layer should be selected. Although normalization layers are often important to make the neural network train, they might lead to unwanted artifacts for image restoration tasks [54].

In this section, we define normalization layers n for the 1D case of a feature vector $\mathbf{x}$ with the dimensions $[N_b, N_x, N_f]$. We introduce the notation $E_{d=\cdot}[\mathbf{x}]$ and $\mathrm{Var}_{d=\cdot}[\mathbf{x}]$ to compute the mean and variance values over the dimension $d \in \{b, x, f\}$. Common normalization methods are depicted in Fig. 2.9.

The normalization layer is defined as

$$\widehat{\mathbf{x}} = \frac{\mathbf{x} - E_d[\mathbf{x}]}{\sqrt{Var_d[\mathbf{x}] + \epsilon}}\alpha + \beta \tag{2.16}$$

where $E[\mathbf{x}]$ and $Var[\mathbf{x}]$ are the estimated mean and variances of the input $\mathbf{x}$. The scaling parameter $\alpha \in \mathbb{R}$ and bias $\beta \in \mathbb{R}$ are optional parameters that can be learned.

Trabelsi et al. [55] introduced complex-valued batch normalization that ensures a circular distribution, i.e., same variance for real and imaginary part. The scaling terms are replaced by covariance matrices; hence, complex batch-normalization reads as

$$\widehat{\mathbf{x}} = \boldsymbol{A}\boldsymbol{V}^{-\frac{1}{2}}(\mathbf{x} - E_d[\mathbf{x}]) + \beta. \tag{2.17}$$

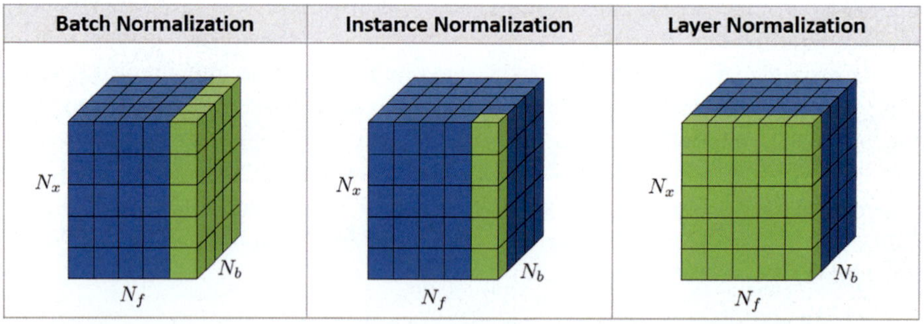

FIGURE 2.9 Normalization techniques.

Visualization of common normalization techniques. The green area denotes the dimensions that are used to calculate the normalization statistics, i.e., mean and standard deviation, over the input batch with dimensions $[N_b, N_x, N_f]$.

The matrix $\boldsymbol{V}$ defines the covariance matrix with real-valued components V_{rr}, V_{ii}, and V_{ri} following

$$\boldsymbol{V} = \begin{bmatrix} E_d\left(Re(\mathbf{x})^2\right) & E(Re(\mathbf{x})Im(\mathbf{x})) \\ E_d(Im(\mathbf{x})Re(\mathbf{x})) & E_d\left(Im(\mathbf{x})^2\right) \end{bmatrix} = \begin{bmatrix} V_{rr} & V_{ri} \\ V_{ri} & V_{ii} \end{bmatrix}. \tag{2.18}$$

The scaling operator A contains (trainable) real-valued parameters α_{rr}, α_{ii}, and α_{ri} following

$$\mathbf{A} = \begin{bmatrix} \alpha_{rr} & \alpha_{ri} \\ \alpha_{ri} & \alpha_{ii} \end{bmatrix}. \tag{2.19}$$

The bias $\beta \in \mathbb{C}$ is complex-valued.

2.7.2.2 Batch normalization

Ioffe et al. [56] introduced batch normalization to reduce the internal covariance shift. Statistics are computed over a mini-batch $d = b, x$, with batch size $N_b > 1$. During training, mini-batch statistics are accumulated over time. The accumulated running means $E[\overline{\mathbf{x}}]$ and variances $Var[\overline{\mathbf{x}}]$ are computed using a running average scheme with momentum ν

$$\begin{aligned} E[\overline{\mathbf{x}}] \quad &= \nu \cdot E[\mathbf{x}] + (1-\nu) \cdot E[\overline{\mathbf{x}}] \\ Var[\overline{\mathbf{x}}] \quad &= \nu \cdot Var[\mathbf{x}] + (1-\nu) \cdot Var[\overline{\mathbf{x}}] \end{aligned}$$

where $E[\mathbf{x}]$ and $Var[\mathbf{x}]$ are the estimated mean and variance of the current mini-batch $\mathbf{x}$. It follows that in training mode the running means and averages are only updated, while the real batch statistics are used to compute Eq. (2.16). In inference mode, the running means and averages are used to compute Eq. (2.16).

2.7.2.3 Instance normalization

Initially introduced for style transfer, instance normalization is another widely used normalization technique for neural networks [57]. Here, $E[\mathbf{x}]$ and $\mathrm{Var}[\mathbf{x}]$ are computed during training and testing individually for each sample and feature independently, i.e., $d = x$. The normalization is computed in training and inference mode directly from the batch; hence, no moving average and moving variance have to be tracked during training. It is optional to learn additional scaling α and bias β parameters. The complex instance normalization follows Eq. (2.17).

2.7.2.4 Layer and group normalization

In contrast to batch normalization, *layer normalization* [58] computes the statistics over a whole layer, i.e., $d = x,f$. This is a common way to stabilize training of recurrent neural networks. *Group normalization* [59] divides the feature channels N_f into groups, and statistics are calculated within each individual group.

2.7.3 Activation layer

Nonlinear activation layers are typically added after the main layer operation (e. g., convolution). This is a crucial component allowing neural networks to learn nonlinear functions. This is realized by applying a nonlinear activation function in an element-wise way. The activation layer can also be seen as a decision on whether a neuron should be "fired," in analogy to action potentials in biological neurons.

Rectified linear units (ReLUs) are one of the most commonly used activation functions in ML. These piece-wise linear functions $\phi_{\text{ReLU}} : [-\infty, \infty] \rightarrow [0, \infty]$ set negative values to 0 and keep positive values following

$$\phi_{\text{ReLU}}(\mathbf{x}) = \max(\mathbf{x}, 0).$$

Although ReLUs are not differentiable at $\mathbf{x} = 0$, which might cause issues in gradient-based learning, gradient-based optimizers use the subgradient for backpropagation and have shown to be efficient for a variety of ML tasks [14].

A common problem in training networks are *dying ReLUs*, which output all zeros independent of the input values, making parts of the network inactive. To avoid this problem, *leaky ReLUs* are used, which apply a small gradient for negative values scaled by $\alpha < 0$, i.e.,

$$\phi_{\text{Leaky ReLU}}(\mathbf{x}) = \begin{cases} \mathbf{x} & \text{if} \quad \mathbf{x} \geq 0 \\ \alpha\mathbf{x} & \text{if} \quad \mathbf{x} < 0 \end{cases}.$$

This extends the output range $\phi_{\text{Leaky ReLU}} : [-\infty, \infty] \rightarrow [-\infty, \infty]$. A further extension is *parametrized ReLUs (PReLUs)*, which allow the scale α to be learned for the individual feature channels.

When applying nonlinear activation functions ϕ to complex values, several things have to be considered. One possibility is to apply the activation function to the real and imaginary part separately; however, the natural correlation between real and imaginary channels is not considered in this case. Furthermore, the phase information is mapped to the first quadrant, i.e., the interval $\left[0, \frac{\pi}{2}\right]$. Another variant is to keep the phase information fixed and only alter the magnitude information. An example is the *ModReLU* [60]:

$$\phi_{\text{ModReLU}}(\mathbf{x}) = \max(0, |\mathbf{x}| + \beta)\frac{\mathbf{x}}{|\mathbf{x}|}$$

where β is the bias that is trainable.

Virtue et al. [46] proposed a new complex activation function called *cardioid*:

$$\phi_{\text{Cardioid}}(\mathbf{x}) = \frac{1}{2}(1 + \cos(\angle\mathbf{x} + \beta))\mathbf{x}$$

The complex cardioid can be seen as a generalization of ReLU activation functions to the complex plane. Compared with other complex activation functions, the complex cardioid acts on the input phase rather than the input magnitude. A bias β can be additionally learned.

2.7.4 Fully connected layer

Fully connected (FC) (or dense) layers connect every input unit to every output unit and are formulated as dense matrix–vector multiplication:

$$\widehat{\mathbf{x}} = \mathbf{W}\mathbf{x} + \beta, \tag{2.20}$$

with an additional bias β. Trainable parameters are the dense matrix $\mathbf{W}$ and the bias β. FC layers are global feature extractors and are core components in multilayer perceptrons (MLPs), object recognition, classification, and global transform learning [61]. In contrast to convolution layers, the number of trainable parameters depends on the amount of input nodes N_i and output nodes N_o (including bias β), yielding # parameters $= N_i \cdot N_o + N_o$. This becomes inefficient to compute when learning global transforms of large image sizes.

2.7.5 Downsampling layer

Downsampling layers reduce the spatial resolution and make neural networks more efficient to train as the number of operations is reduced substantially in subsequent layers. *Pooling layers* are used as downsampling operation to reduce the spatial resolution in the image and to introduce approximate invariance to small translations [14]. Small patches of size k are analyzed in the individual features maps to keep important information about extracted features. Common pooling layers are *average pooling* and *max-pooling*. For complex-valued images, the maximum operation does not exist. Instead, the pooling layer is modified such that it keeps values with, e.g., the maximum magnitude response. A pooling layer does not have any trainable parameters.

Convolutional downsampling layers are convolution layers (see Section 2.7.1) with a stride greater than 1, resulting in an output with decreased spatial resolution. Downsampling with convolution layers can be seen as learning the pooling operation. However, the number of trainable parameters is increased when using convolutional downsampling.

2.7.5.1 Upsampling layer

Upsampling layers increase the spatial resolution and are important to get back from coarser scales to the original image resolution in image-to-image networks. *Convolutional upsampling* with transposed convolutions (see Section 2.7.1) allows to recover the original input size, for strides $s > 1$. However, this kind of upsampling is known to introduce checkerboard artifacts [62]. The number of trainable parameters increases with convolutional upsampling. To overcome checkerboard artifacts in convolutional upsampling, nearest neighbor or bilinear *interpolation* provides an alternative. Interpolation is commonly followed by a convolution layer for further computation. Bilinear interpolation in the complex domain can be conducted by either interpolating real/imaginary part or magnitude/phase separately.

2.7.6 Dropout layer

Dropout [63] is a regularization technique during training to prevent overfitting of the network. At each training iteration, a random selection of nodes and connections to and from the layer is temporarily set to zero with probability $1 - p$ or kept with probability p. The result of dropout is scaled with a factor $1/p$ to maintain the scale of the output layer. The complex dropout is realized the same way to the complex number, but not to real and imaginary parts separately.

2.7.7 Merging layers

Concatenation stacks tensors together along a certain dimension, mostly the feature dimension. The aim is to increase feature diversity or to gather feature information from different parts of the network. It is also possible to merge further inputs into the network at deeper layers. Note that the spatial dimensions have to match when using merging layers at various scales. If concatenation is followed by a convolution layer, this can be interpreted as weighted combination of different inputs. In a similar sense, other feature map combinations such as *addition*, *subtraction*, *multiplication*, *averaging*, *minimum*, or *maximum* can be performed.

2.7.8 Recursive layer

Recursive layers enable processing of spatiotemporal sequences of arbitrary length. The temporal frames are processed sequentially, but the temporal features are captured and stored in hidden units, which act as memory. Later frames have access to this memory. A basic recursive layer reads as

$$\begin{aligned} \mathbf{h}_t &= \phi_h(\mathbf{W}_h \boldsymbol{h}_{t-1} + \mathbf{W}_x \mathbf{x}_t + b_x) \\ \widehat{\mathbf{x}}_t &= \phi_{\widehat{x}}(\mathbf{W}_{\widehat{x}} \boldsymbol{h}_t + b_{\widehat{x}}) \end{aligned} \tag{2.21}$$

where $\mathbf{W}_h$, $\mathbf{W}_x$, and $\mathbf{W}_{\widehat{x}}$ are weight matrices, b_x and $b_{\widehat{x}}$ are the bias, ϕ_h and $\phi_{\widehat{x}}$ are the activation functions, and $\mathbf{h}$ are the hidden units receiving input $\mathbf{x}$ and output $\widehat{\mathbf{x}}$ for the processed temporal frame t. Fig. 2.10 depicts the recursive layer together with its unrolled version in which the temporally sequential processing can be appreciated. The formulation in Eq. (2.21) might result in vanishing gradients during backpropagation as each step recurrently depends on all previous steps. This might be avoided by introducing additional *gates* such as in long short-term memory (LSTMs) [6] and gated recurrent units (GRUs) [64]. Note that the weight matrices are defined by convolutions in the case of image reconstruction tasks. While Eq. (2.21) can be straightforwardly translated to complex-valued operations, LSTMs and GRUs require real-valued gating functions, as defined in, e.g., Ref. [65].

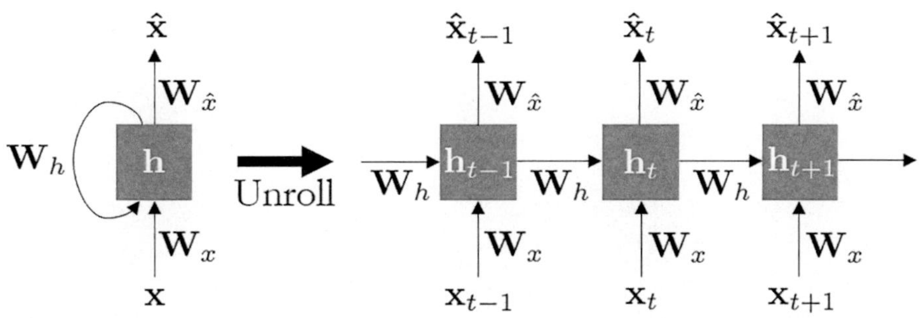

FIGURE 2.10 Recursive layer.

Recursive layer with input x, hidden unit *h*, and output $\hat{x}$ to process a sequence of time frames *t*. Inputs and outputs are weighted by learned filters W_x and $W_{\hat{x}}$, respectively, as well as the recursive feedback W_h. Unrolling the recursive unit reveals its temporally sequential processing.

2.7.9 Attention mechanisms

Attention mechanisms originated from the field of natural language processing help to learn dependencies between source and target, which by that focus the network's attention to the most significant information. The methods differ in the used alignment scores (content [66], additive [67], location [68], or scaled dot-product [69]) and information sharing (self-attention [70], global/soft [71], local/hard [68,71]). Spatial/channel squeezing/excitation, known as *squeeze-and-excite* [72] or *merge-and-run mapping* [73], is another form to learn a weighting of the significant contributing regions in the feature map and thus to focus the network's attention. Attention layers were predominantly used in recent years in transformer architectures [69] (Section 2.8.6) including large language models and vision transformers [74,75], and in generative models (Section 2.9) including diffusion models (Section 2.9.4). More details about the attention mechanisms are described in Section 2.8.6.

2.7.10 Building blocks

Several layers can be bundled to form larger building blocks as shown in Fig. 2.11. *Convolution blocks* in neural networks commonly consist of a sequence of convolution, normalization (optional), and activation. They bundle together to a CNN. *Residual blocks* [76] learn a residual, which is added to the input of the block. The direct identity connections are also known as skip or residual connections. Furthermore, this avoids the vanishing gradient problem and makes the network easier to train. Depending on the kernel sizes and channels, these residual blocks can be also referred to as *bottleneck residual* (including 1*x*1 convolutions) or *wide residuals* (increased channel depth) [77]. *Dense blocks* [78] follow a similar principle, but concatenate feature maps from several previous layers, determined by the growth rate *g*, to build a collective knowledge. It is also possible to train a convolution filter

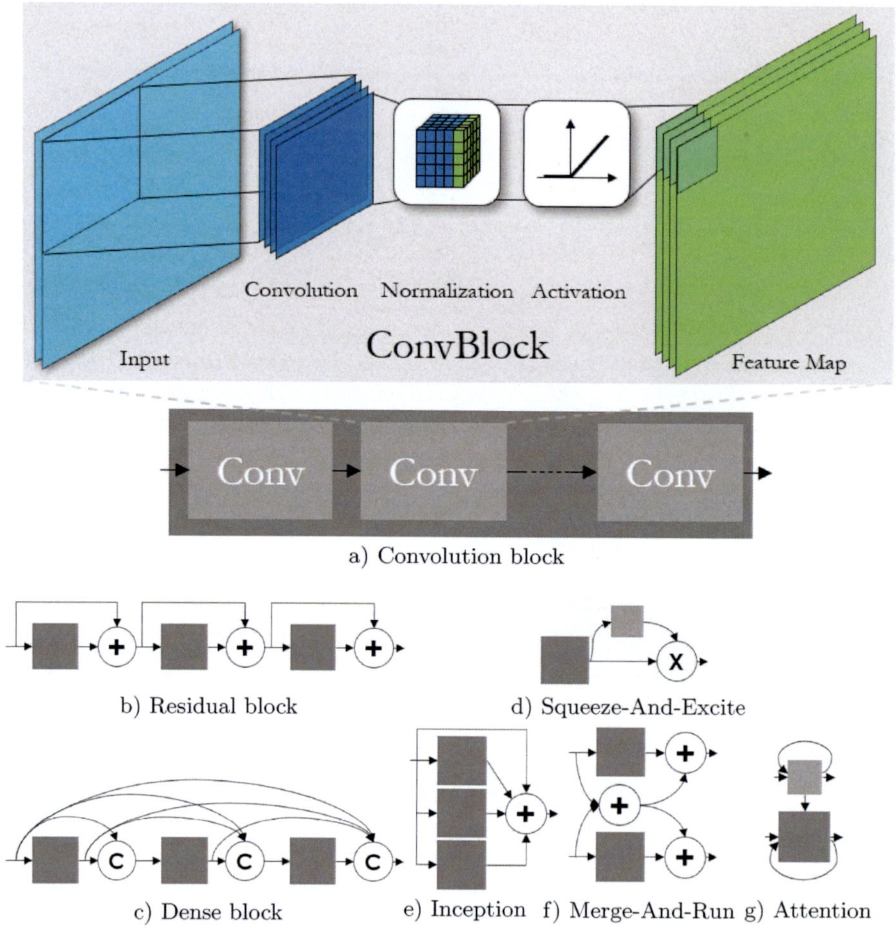

FIGURE 2.11 Building blocks.

Common building blocks in neural networks. (a) Convolution blocks are used for processing image input via convolution layer, followed by normalization (optional) and activation function to form the feature map output. Several consecutive convolution blocks are usually taken inside other building blocks. (b) Residual block with residual/skip connection and addition of feature map at output. (c) Dense block with skip connections over several stages and concatenation of feature maps. (d) Squeeze-and-excitation by learning a spatial/channel weighting to scale the output feature map. (e) Inception with multiple parallel convolution blocks of various kernel sizes to allow a multiresolution approach. (f) Merge-and-run mapping for merging information between parallel branches. (g) Attention mechanisms via recursive layers to focus the layer output.

for the residual/dense path, which acts then as a weighted combination or attention focusing depending on if acting on spatial dimension, also known as *fire blocks* [79], or spatial/channel squeezing/excitation, known as *squeeze-and-excite* [72]. Several convolution blocks acting in parallel to a residual connection form an *assembly block*. An alternative assembling can be achieved with *merge-and-run mapping* [73]: average the inputs of the residual branches (merge) and add the average to the output of each residual branch as the input of the subsequent residual branch (run). A multiresolutional approach is achieved if convolution layers/blocks of different kernel sizes are operating in parallel, known as *inception blocks* [80]. Several versions of this have been proposed [81,82]. Within the concept of attentions or sequential data processing, *recurrent neural networks (RNNs)* (see Section 2.7.9) are used to extract the sequence of information based on their relative spatial/temporal spreading. RNNs thus form another type of building blocks that are especially advantageous to process spatiotemporal sequences of arbitrary length. The temporal frames are processed sequentially with temporal information stored in hidden units to which later frames have access to.

2.8 Fundamental architectures

A multitude of neural network architectures for diverse tasks can be obtained by simply stacking the fundamental network elements and building blocks described before. Since all the operations are differentiable, all network parameters can be optimized end-to-end using the backpropagation algorithm (c.f. Section 2.5.3). We will now described a number of popular architectures for image classification, segmentation, and image-to-image regression.

2.8.1 VGG

A significant landmark was the publication of a range of deep network architectures by the Visual Geometry Group (VGG) of the University of Oxford in 2015 [83]. Those architectures are composed of groups of 3×3 convolutions followed by a max-pooling step and a concurrent doubling of the channels. After a number of these blocks the convolutional output is transformed by three fully connected layers followed by a softmax operation to obtain the final class prediction. The authors proposed four variations of their architecture with 11, 13, 16, and 19 layers, respectively. Due to their simplicity and state-of-the-art performance on the 2014 ImageNet classification challenge, in particular the latter two, often referred to as VGG16 and VGG19, were highly influential.

One of the main insights of this work is that increasing the depth of neural network architectures is often beneficial for performance. However, depth cannot be increased indefinitely due to the gradients becoming small to effectively train the earlier layers of the network.

The publication of the VGG16 and VGG19 also marked a shift from developing application-specific architectures for each problem to building on established architecture backbones. VGG networks with pretrained weights were used as starting point for many medical and nonmedical applications (Refs. [84–86]). We note that any classification architecture including VGG16 and VGG19 can be also used for regression tasks by removing the final softmax operation. Moreover, the VGG architectures are commonly used to calculate a perceptual loss [87,88].

2.8.2 ResNet

He et al. [49] addressed the difficulty of training very deep networks by introducing residual blocks, which were already briefly introduced in Section 2.7.11. The proposed architectures, which are typically referred to as ResNets, are reminiscent of the VGG architectures in their use of blocks of fixed sized 3×3 convolutions. The most important difference is that all convolution operations are embedded as residual blocks with a skip connection. The skip connections facilitate the gradients flowing back to the earlier layers of the network. Each residual block further uses—the then newly proposed—batch normalization operation [56] to further facilitate the training. In contrast to the VGG architectures, the last convolution of a block is performed with a stride of 2 rather than a max-pooling operation. Training the network required periodic reduction of the learning rate.

The deeper architectures use a "bottleneck" in the residual layer, which employs a 1×1 convolution layer to reduce the number of channels for the 3×3 convolution operation, and another 1×1 convolution to increase the number of channels back to the original input.

For the final classification, the ResNet employs an average pooling step over all spatial dimensions of the final convolutional feature maps and uses this as input to a single fully connected classification layer. This is computationally more efficient than using three fully connected layers like in the VGG architectures, and became the de facto standard for all subsequent classification architectures.

Using these innovations, the authors explored architectures from 18 to 152 layers deep (i.e., ResNet-18 to ResNet-152). Remarkably, the ResNet-152 still had lower complexity compared with the VGG19 network due to the bottleneck connections.

While the deepest ResNet-152 performed the best, and set a new high-score for the ImageNet challenge, in practice the lower complexity ResNet-18, ResNet-34, and ResNet-50 gained widespread popularity. Due to their simplicity and easy availability of pretrained weights, they are to this day one of the most popular choices for medical and nonmedical image classification and regression tasks.

2.8.3 DenseNet

Huang et al. [78] introduced the dense convolutional network (DenseNet) architecture, which extends the ideas of the ResNet architectures. DenseNets are built on the concept of ensuring maximum information flow between layers in the network.

Instead of residual blocks, the architecture employs dense blocks, for which each layer receives inputs from all preceding layers and passes its own feature maps to all subsequent layers. Similar, to ResNets, DenseNets employ 3×3 convolutions and batch normalization as well as bottlenecks based on 1×1 convolutions to temporarily reduce the number of feature map channels, thereby improving computational efficiency.

Unlike the ResNets, DenseNets use feature map concatenation rather than summation to combine and implement the skip connections. This concatenation leads to feature reuse, enhancing feature propagation and reducing the vanishing gradient problem. Furthermore, in contrast to the ResNet and other preceding architectures, the DenseNet can have very "narrow" layers with few channels. The growth of the number of channels with the depth of the network is defined by a growth rate. Hence, in this architecture, we see a clear emphasis on depth at the cost of the width of the network.

The standard configurations of DenseNet are defined by their depth and growth rate, leading to variants such as DenseNet-121, DenseNet-169, DenseNet-201, and DenseNet-264.

DenseNets have demonstrated significant improvements in image recognition tasks, offering competitive performance with fewer parameters and less computational complexity compared with deeper architectures. In practice, they are widely used in both medical and nonmedical image analysis tasks, thanks to their ability to manage feature redundancy and efficiency in parameter usage.

2.8.4 U-Net

Our discussion so far has focused on networks for image classification and regression, which take an image as input and, through a number of neural network elements, reduce it to a one- or few-dimensional number (e.g., a classification). We will now discuss the U-Net architecture [89] where the output has the same (or similar) dimensions as the input. This image-to-image architecture has its origin medical image segmentation, but now forms the building blocks for a vast range of applications such as image-to-image regression, or denoising diffusion probabilistic models (DDPMs).

The U-Net was originally proposed for 2D segmentation of neuronal structures in electron microscopic stacks as well as cell tracking on transmitted light microscopy images [89]. Similar to the VGG architecture, it consists of a downsampling path consisting of blocks of 3×3 convolutions followed by a max-pooling operation. The downsampling path is followed by an upsampling path, which increases the size of the convolutional feature maps again via transposed convolutions (c.f. Section 2.7.1) followed, again, by blocks of 3×3 convolutions. The downsampling and upsampling paths are almost identical with the only difference that in the upsampling path the max-pooling operations are replaced by transposed convolutions. The symmetric nature of this arrangement gives it the appearances of the letter "U" when drawn diagrammatically.

A key feature of the U-Net are skip connections linking the final convolutional features of each block of the downsampling path with its corresponding block from the upsampling path. The features are simply concatenated with the output of the transposed convolutions and given as input to the next convolutional block of the upsampling path. The final output (e.g., the segmentation) is obtained by an additional convolution layer followed by an appropriate nonlinearity (e.g., the softmax function). The 2D U-Net architecture was later also extended to a 3D version [90], which follows a similar paradigm but uses $3 \times 3 \times 3$ convolutions instead.

A detail of the U-Net architecture is that nonpadded convolutions were used, which reduces the image size by 1 pixel on each side after each convolution. While an enormous amount of work builds on the U-Net architecture, specific implementations do not always follow the exact implementation originally proposed in Ref. [89]. For example, in practice, "same" padding is often used for simplicity, and batch normalization is added to the convolution operations for faster training [91,92]. Newer implementations also sometimes replace the max-pooling operation by average pooling and the transposed convolutions by bilinear upsampling (see Ref. [91]).

The U-Net also has seen many extensions with ideas from the popular classification architectures discussed before. For example, Zhang et al. [93] used residual connections in each of the convolution blocks of the up- and downsampling pathways, and Oktay et al. [94] used an attention mechanism in the upsampling path.

While newer segmentation architectures exist (e.g., DeepLab [95]), the U-Net is still a popular choice to this day. Interestingly, it was recently shown that a simple U-Net architecture within a robust image processing framework was able to win a large number of medical image segmentation challenges [96].

U-Nets are also prominently used in image-to-image regression tasks, such as MR image reconstruction (see Chapter 7), image superresolution (see Chapter 8), motion correction (see Chapter 9), or image translations (see Chapter 15).

2.8.5 Neural radiance fields

Neural radiance fields (NeRF) enable the creation of highly detailed 3D reconstructions of scenes from 2D images [97]. NeRF have demonstrated superior performance compared with traditional 3D reconstruction methods, particularly in capturing fine details and handling challenging scenes with complex lighting and geometry. NeRF leverages neural networks to represent both the geometry and appearance of a scene.

The key idea behind NeRF is to model a scene as a continuous 3D volume of radiance, where every point in space has an associated color and density. This is achieved by training a neural network to predict the radiance (color) and density at any given 3D point, conditioned on its spatial coordinates and the corresponding 2D image observations.

During training, NeRF learns from a set of images and corresponding camera poses, optimizing the neural network parameters to minimize the discrepancy between predicted and observed pixel colors. The input to the network forms the pixel coordinates. By iteratively refining the network, NeRF effectively learns to reconstruct complex scenes with high fidelity, capturing intricate details such as lighting effects, reflections, and occlusions.

Once trained, NeRF can generate novel views of the scene from any viewpoint, enabling realistic rendering from arbitrary camera angles. Additionally, NeRF can be extended to handle dynamic scenes by incorporating temporal information, enabling applications such as interactive virtual reality experiences, augmented reality, image registration (Chapter 9), and image reconstruction (Chapter 7) including neural implicit networks [98].

2.8.6 Transformer

Transformer model architectures are widely used across language, vision, and reinforcement learning tasks [99]. Transformers are primarily used in three ways: (1) encoder-only (e.g., for classification), (2) decoder-only (e.g., for language modeling), and (3) encoder–decoder (e.g., for machine translation). A fundamental component are the attention layers. The intention is to focus the attention toward most similar patterns, structures, textures, etc. within/between the input data/tensor. Images are therefore often passed through multiple attention layers after previous feature extraction or after being patched in image domain, or chunked apart by any other transformation to be embedded into a richer feature representation. To aid the transformer in spatially relating the input tensor parts, a *positional embedding* can be added by projecting the spatial location via, e.g., cosine and sine functions [69]. The most common types of attention layers are discussed in the following. *Self-attention*: Self-attention embeds information about its input's context. The first step involves the creation of key $K \in \mathbb{R}^{n \times d_K}$, query $Q \in \mathbb{R}^{n \times d_Q}$, and value $V \in \mathbb{R}^{n \times d_V}$ tensors from the input tensor $X \in \mathbb{R}^{n \times d}$. The input tensor is multiplied with a learnable key/query/value matrix $W_K/W_Q/W_V$ to create these projections, following:

$$Q = W_Q X \tag{2.22}$$

$$K = W_K X \tag{2.23}$$

$$V = W_V X. \tag{2.24}$$

In a second step, a score is calculated between the key K and query Q by the dot product:

$$S = QK^T \tag{2.25}$$

reflecting the "self-similarity" within the projections. To stabilize gradients during training, the score is normalized by the square root of the size of the key tensor d_K, followed by a softmax operation on each row of the score. The output is called the attention/softmax score $A = \text{softmax}\left(\frac{QK^T}{\sqrt{d_K}}\right) \in \mathbb{R}^{n \times n}$. The attention/softmax score is then multiplied with the value tensor to keep only relevant information, i.e., dropping irrelevant projections. This yields the summed up and weighted value scores along the rows of V to form the output of the layer:

$$\mathrm{Attention}(\mathrm{K}, \mathrm{Q}, \mathrm{V}) = \mathrm{softmax}\left(\frac{\mathrm{QK}^{\mathrm{T}}}{\sqrt{d_K}}\right)\mathrm{V} \in \mathbb{R}^{n \times d_V} \tag{2.26}$$

Multihead attention: Instead of only paying attention to each other in one dimension, transformers use the concept of multihead attention to strengthen the embedding by paying attention to various aspects within the input tensor in relation to its surroundings. It is thus a combination of individual self-attention layers with n parallel heads.

Cross-attention: Instead of purely focusing on the same input tensor, cross-attention mixes or combines to different input tensors $X_1 \in \mathbb{R}^{n \times d}$ and $X_2 \in \mathbb{R}^{m \times d}$, which can be of different dimensions. While queries Q are only extracted from X_1, keys K and values V are obtained from X_2:

$$Q = W_Q X_1 \tag{2.27}$$

$$K = W_K X_2 \tag{2.28}$$

$$V = W_V X_2. \tag{2.29}$$

yielding thus the cross-attention following Eq. (2.26), and with the attention score being $A \in \mathbb{R}^{n \times m}$. For $X_1 = X_2$, we obtain again the self-attention.

2.9 Generative models

In the broadest sense, generative models are statistical models that can generate all values for a phenomenon, both observed and unobserved. Generative models can often be understood as approximating the probability density of our data $p(\mathbf{x})$ from some samples within that distribution. This contrasts with discriminative models such as classifiers or regression models, which focus solely on the relationship between observed variables $\mathbf{x}$ and a target variable y. Note that generative models can also be used to solve discriminative problems by modeling class conditional distributions $p(\mathbf{x}|y)$ and applying Bayes theorem to obtain $p(y|\mathbf{x})$.

Before the advent of deep learning, generative modeling was already a crucial part of the machine learning landscape. However, those techniques frequently relied on explicit assumptions about the data distribution. With advances in neural network technology, generative modeling has seen a shift toward models such as GANs, VAEs, and DDPMs, which can automatically learn complex, high-dimensional data distributions without many of the assumptions required by earlier models.

In recent years, generative models have received substantial attention for their ability to generate highly realistic images and text samples. These advances are epitomized by tools such as DALL-E [100] and ChatGPT [101].

However, generative modeling also has many other interesting applications that have received less of the spotlight. Examples include outlier detection [102], unsupervised anomaly detection [103–107], and modeling conditional densities for probabilistic predictions [91,92,108,109]. As an example of the last, some recent MR reconstruction techniques use DDPMs to directly model the conditional distribution $p(\mathbf{x}|\mathbf{y})$ of the reconstructed image $\mathbf{x}$ given the k-space measurements $\mathbf{y}$ [110–113].

In the following, we will briefly introduce some of the currently most influential neural network–based generative models: autoencoders, VAEs, GANs, and DDPMs. The mathematical derivations of these models are often involved; however, the resulting algorithms are simple to understand and to implement. We will focus on a short intuitive sketch of the derivations and refer the readers to the original publications or any of the many online tutorials for a more in-depth introduction.

2.9.1 Autoencoders

An autoencoder [114] is a neural network that encodes data into a low-dimensional latent representation $\mathbf{z}$, and then decodes this representation again to reconstruct the original input image. Most commonly, a simple similarity-based reconstruction loss such as the L1 or L2-distance is used as the training objective. After training, one can randomly draw a $\mathbf{z}$ and decode it to obtain ideally a novel data point from the original input distribution. In practice, however, simple autoencoders tend to memorize the training examples and do not interpolate well in the latent space, i.e., similar high-dimensional samples are not embedded close to each other in latent space.

2.9.2 Variational Autoencoders

On a purely mechanistic level, VAEs [31] are very similar to autoencoders with the sole difference that there is an additional loss to regularize the latent representations. However, the mathematical basis of VAEs actually has surprisingly little to do with regular autoencoders.

We first make the assumption of the data generating process of the autoencoder more rigorous. Specifically, we assume that there is a low dimensional $\mathbf{z}$ that is distributed according to some simple probability density $p(\mathbf{z})$ that is exclusively responsible for generating the high-dimensional data $\mathbf{x}$. That is,

$$p(\mathbf{x}) = \int p(\mathbf{z}) p_\theta(\mathbf{x}|\mathbf{z}) d\mathbf{z}.$$

Typically, $p(\mathbf{z})$ is defined as an axis-aligned Gaussian distribution. We have access to training samples from the high-dimensional distribution $p(\mathbf{x})$; however, the latent embedding $\mathbf{z}$ is unknown, i.e., hidden to us. The distribution $p_\theta(\mathbf{x}|\mathbf{z})$ is typically also parametrized as a Gaussian with a fixed variance, but a mean that is predicted by a neural network with weights θ. While directly maximizing the likelihood of $p(\mathbf{x})$ by randomly sampling $\mathbf{z}$'s is mathematically feasible, it is, however, not practical. In reality, $p(\mathbf{x}|\mathbf{z})$ will be nearly zero for most values of $\mathbf{z}$.

The key innovation of VAEs is to approximate $p(\mathbf{z})$ with a learnable variational distribution q. To make sampling of the $\mathbf{z}$'s more efficient, q is chosen such that it depends on $\mathbf{x}$, i.e., $q_\phi(\mathbf{z}|\mathbf{x})$. This variational approximation is typically also chosen as a Gaussian distribution with its mean and variance parameterized by a neural network.

While the exact derivation is beyond the scope of this chapter, this variational approximation can be shown to produce the following lower bound on the true probability distribution:

$$p(\mathbf{x}) \geq \mathbb{E}_{q_\phi(\mathbf{z}|\mathbf{x})}[p_\theta(\mathbf{x}|\mathbf{z})] + \mathrm{KL}\left[q_\phi(\mathbf{z}|\mathbf{x}) \middle\| p(\mathbf{z})\right].$$

In practice, we optimize ϕ and θ to maximize this lower bound and get as close to the actual data distribution as possible. Under the Gaussian assumption, the first loss term simply becomes an L2-loss. The KL-loss in the second term ensures that the embedding is well-behaved.

In practice, the network q_ϕ is typically implemented as an encoder that predicts a mean and a variance. Subsequently, a random variable $\mathbf{z}$ from which mean and variance is sampled, is passed on to the decoder network $p_\theta(\mathbf{x}|\mathbf{z})$. It can thus be seen that this results in an autoencoder architecture. After training, we can sample from the simple distribution $p(\mathbf{z})$, and decode them, to get samples from the approximated data distribution. In practice, VAEs often produce blurry samples due to their bottleneck architecture. More recently, newer techniques such as GANs and DDPMs have received more attention. However, it is important to note that in contrast to those techniques, VAEs produce a potentially interpretable latent space, which may be desirable for some applications. It should also be noted that hierarchical VAEs have been shown to produce highly competitive results on image generation [115].

2.9.3 Generative adversarial networks

GANs [19] are comprised of two networks competing with each other: a generator network g_θ that aims to generate realistic data, and a discriminator network d_ϕ that tries to distinguish the generated data from real data. The two networks are trained simultaneously with an adversarial loss function, whereby the generator tries to produce better and better samples to "cheat" the discriminator, while the discriminator tries to get better and better at distinguishing generated samples from real samples. More specifically, the generator typically takes a randomly sampled low-dimensional variable $\mathbf{z}$ as input and upsamples it into a generated data point $\widehat{\mathbf{x}} = g_\theta(\mathbf{z})$. The discriminator is implemented as a classifier $d_\phi(\widehat{\mathbf{x}})$ that sees both real and generated images and tries to classify them accordingly. The networks are trained with the following min-max objective:

$$\min_\theta \max_\phi \mathbb{E}_{\mathbf{x}\sim p_{data}(\mathbf{x})}\left[\log d_\phi(\mathbf{x})\right] + \mathbb{E}_{\mathbf{z}\sim p_\mathbf{z}(\mathbf{z})}\left[\log\left(1 - d_\phi(g_\theta(\mathbf{z}))\right)\right].$$

The aforementioned objective can be shown to minimize the Jenson-Shannon divergence between the real data distribution and the generated data distribution.

While GANs are capable of generating sample images of far superior quality to VAEs, they are notoriously hard to train, as training requires careful balancing of the generator and discriminator. Many extensions of the basic GAN idea have been proposed that make training more stable (Refs. [116,117]), and improve image quality (Refs. [118,119]). However, GANs have recently been eclipsed by Denoising Diffusion Probabilistic Models (DDPMs) in both easy of training and image quality.

2.9.4 Denoising probabilistic diffusion models

DDPMs, introduced by Ho et al. [120], effectively learn the data distribution $p(\mathbf{x})$ by progressively corrupting the training data with noise and subsequently learning to reverse this noise addition. During the training phase, Gaussian noise is iteratively applied to the input images $\mathbf{x}$ following a predetermined sequence of noise levels, often referred to as a noise schedule. This procedure continues until the images are wholly obscured, resembling pure Gaussian noise with zero mean and unit variance. Typically, this involves a fixed number of steps, commonly set at $T = 1000$, with each increment referred to as a time step. The original, uncorrupted image is usually denoted as $\mathbf{x_0}$, and the fully noised image as $\mathbf{x_T}$. This method of adding noise mirrors the natural diffusion process observed in physics, where particles spread from regions of higher concentration to lower concentration over time.

After the noising process, a neural network is then trained to reverse the noising process at each step. To this end typically a U-Net-like architecture is used with an image $\mathbf{x}_t$ as input, the less noisy image $\mathbf{x}_{t-1}$ as target, and the L2-distance as loss function. Since each denoising step satisfies the Markov property and only depends on the time step immediately before it, random time steps (as well as random images) can be sampled during training.

To produce a new data sample after training, a new noise sample $\mathbf{x}_T$ is drawn from a Gaussian distribution with zero mean and unit variance. $\mathbf{x}_T$ is then repeatedly denoised through the whole chain until we arrive at a novel sample $\mathbf{x}_0$.

Similar to VAEs, DDPMs minimize a variational lower bound on the true probability $p(\mathbf{x})$. Alternatively, via Tweedie's formula, DDPMs can also be interpreted to be matching the score function $\Delta \log p(\mathbf{x})$ of the original data distribution. An excellent discussion on the relationship between VAEs and DDPMs, as well as score-based generative models, can be found in the tutorial by Calvin Luo [121].

DDPMs have become a popular tool for generating novel images from a data distribution and power popular tools for natural image generation. A particularly popular tool are latent diffusion models often also called StableDiffusion [122], which performs the noising and denoising process in the latent space of a powerful pretrained autoencoder. DDPMs have also been shown to be powerful tools for solving inverse problems such as superresolution [123], or inpainting [124]. Moreover, DDPMs have been used successfully for solving the inverse problem of MR reconstruction, where they have been shown to outperform other neural network–based reconstruction approaches [110,112,125].

Despite their success, DDPMs have a major drawback. The technique requires very long inference times, with each image generation requiring a high number of passes through a U-Net to perform the whole denoising process (typically $T = 1000$). Therefore, accelerating the inference time of DDPMs is an active area of research (see Ref. [126]).

2.9.5 Modeling conditional probability distributions

All of the generative models discussed before can also be used to model conditional probability distributions $p(\mathbf{y}|\mathbf{x})$ of a target variable $\mathbf{y}$ conditioned on some input $\mathbf{x}$. Conditional VAEs [127] can be obtained using some simple modifications in the original derivation. They can be a powerful tool to explore prediction uncertainty for many applications such as segmentation [91,92], or MR reconstruction [109].

GANs can also be easily extended to the conditional setting. They have been extensively used to model image-to-image translation problems, such as synthesizing CT images from MR images [128], MR motion correction [129–131], or image superresolution [132–134]. An interesting property of GANs is that this translation can also be achieved without access to paired data during training, using cyclic GANs [135].

DDPMs are relatively straightforward to condition on arbitrary inputs [136,137] by simply inputting the condition to the denoising network along with the noised image in a process known as classifier-free guidance [136]. This approach has been used for conditional density modeling, for instance, in the context of brain MR segmentation [108].

2.10 Summary

This chapter introduced the basic concepts of machine learning that underpins much of the advances in medical image analysis and reconstruction in the past decade. First, the fundamentals of ML were discussed, including types of learning, the role of cost functions, optimization, backpropagation, as well as training/validation/test set composition and their usage. We then covered various approaches for splitting databases for ML applications.

The chapter also introduced several fundamental network elements that form the basis for the majority of neural network architectures, which are widely used today. While some of the building blocks such as fully connected layers and convolutional layers have been at the core of neural networks for decades, the past years have seen many key innovations that allow to build powerful deep architectures and to efficiently optimize them. Prominent examples include the residual unit, batch normalization, and attention mechanisms.

Following the fundamental network elements, the chapter discusses how those elements were combined into four popular neural network architectures that played pivotal roles in machine learning. Many of those architectures such as the ResNet or the U-Net have themselves become basic machine learning components that are now often used as black-box building blocks for various applications.

While advances in neural network technology were initially driven by discriminative tasks such as segmentation, regression, or classification, recently they have also had a substantial impact on the field of generative modeling. We devoted the last part of this chapter a discussion of such generative modeling approaches. Beyond simply creating novel data points as seen in tools such as DALL-E, generative modeling has emerged as a powerful tool to model the probability distribution of data. This makes it a key component for integrating probabilistic knowledge about the data into various frameworks for image analysis and image reconstruction.

While a number of topics covered in this chapter are subject to constant advancements such as neural network architectures, other topics are at the core of machine learning regardless of new developments. Topics relating to splitting of the data for training and model evaluation or topics relating to optimization are essential to every machine learning project. While the fundamental network elements keep evolving, it should be noted that many innovations build on the basic building blocks discussed in this chapter.

As machine learning is gaining in importance, also the availability of tools and resources is growing exponentially. Training a neural network classifier used to be only possible for select group of ML researchers. Now, getting started with machine learning is easier than ever thanks to several easy-to-use Python frameworks such as Keras or PyTorch, a wealth of online resources, as well as easily available compute resources via services such as Google Colab or Amazon Web Services (AWS). However, despite this newfound ease in initiating and executing ML projects, it remains crucial for practitioners to not overlook the foundational principles and theories discussed in this chapter.

2.11 Further resources and tutorials

In the following, a list of resources for further reading is provided. This (incomplete) list includes online resources, tutorials, books, book chapters, and survey papers.

- Keras: https://keras.io/
- Tensorflow tutorials: https://www.tensorflow.org/tutorials
- Tensorflow playground: https://playground.tensorflow.org
- PyTorch tutorials: https://pytorch.org/tutorials
- Peter Chang's DL tutorials: https://github.com/peterchang77/dl_tutor
- Neural network zoo: https://www.asimovinstitute.org/neural-network-zoo/
- Interactive deep learning tutorial: https://d2l.ai/
- Udacity AI courses: https://www.udacity.com/courses/school-of-ai
- Bishop *Pattern Recognition and Machine Learning* [13]
- Vapnik *The Nature of Statistical Learning Theory* [138]

- Murphy *Machine Learning: A Probabilistic Perspective* [15]
- Goodfellow et al. *Deep Learning* [14]
- Survey papers [139–142]
- Google Colab: https://colab.research.google.com

Appendix

A.1 Trainable parameters of separable convolutions

Depthwise separable convolution is best illustrated in an example. For a normal convolution, let us assume the input of size 32×32 with three input channels is convolved with $f = 64$ kernels of size 7×7 (no padding $p = 0$ and stride of $s = 1$) yielding an output of size 26×26 with 64 feature maps. The operation can be split up by first depth convolving each input channel independently with a 7×7 convolutional filter ($p = 0, s = 1, f = 1$), i.e., requiring in total 3 2D filters. The output feature maps are stacked along the channel direction giving a tensor of size $26 \times 26 \times 3$ (in channels last notation), which is then convolved pointwise with 64 filters of size $1 \times 1 \times 3$ ($p = 0, s = 1$) producing the final output tensor of size $26 \times 26 \times 64$. The advantage is that less multiplications are required in a depthwise separable convolution (3 kernels with $7 \times 7 \times 1$ sliding 32×32 times followed by 64 kernels with $1 \times 1 \times 3$ sliding 32×32 times, yielding $3 \cdot 7 \cdot 7 \cdot 1 \cdot 32 \cdot 32 + 64 \cdot 1 \cdot 1 \cdot 3 \cdot 32 \cdot 32 = 216,064$ multiplications) than in a normal convolution (64 kernels with $7 \times 7 \times 3$ sliding 32×32 times, yielding $64 \cdot 7 \cdot 7 \cdot 3 \cdot 32 \cdot 32 = 9,633,792$ multiplications).

References

[1] McCarthy J, Minsky ML, Rochester N, Shannon CE. A proposal for the Dartmouth summer research project on artificial intelligence. 2006.

[2] Rosenblatt F. The perceptron: a probabilistic model for information storage and organization in the brain. Tech Rep 1958;6. https://doi.org/10.1037/h0042519.

[3] Ackley DH, Hinton GE, Sejnowski J. A learning algorithm for Boltzmann machines. In: Proceedings of the IEEE conference on decision and control. 169. IEEE; 1985. p. 147–69. https://doi.org/10.1207/s15516709cog0901{_}7.

[4] Hopfield JJ. Neural networks and physical systems with emergent collective computational abilities. Proc Natl Acad Sci USA 1982;79(8):2554–8. https://doi.org/10.1073/pnas.79.8.2554.

[5] Jordan MI. Attractor dynamics and parallelism in a connectionist sequential machine. In: Proceedings of the eighth annual conference of the cognitive science society; 1986. p. 531–46.

[6] Hochreiter S, Schmidhuber J. Long short-term memory. Neural Comput 1997;9(8): 1735–80. https://doi.org/10.1162/neco.1997.9.8.1735.

[7] Williams RJ, Zipser D. A learning algorithm for continually running fully recurrent neural networks. Neural Comput 1989;1(2):270–80. https://doi.org/10.1162/neco.1989.1.2.270.

[8] Rumelhart DE, Hinton GE, Williams RJ. Learning representations by back-propagating errors. Nature 1986;323(6088):533–6. https://doi.org/10.1038/323533a0.

[9] Fukushima K. Neocognitron: a self-organizing neural network model for a mechanism of pattern recognition unaffected by shift in position. Biol Cybern 1980;36(4):193–202. https://doi.org/10.1007/BF00344251.

[10] LeCun Y, Haffner P, Bottou L, Bengio Y. Object recognition with gradient-based learning. Lect Notes Comput Sci 1999;1681:319–45. https://doi.org/10.1007/3-540-46805-6{_}19.

[11] Lecun Y, Bengio Y, Hinton G. Deep learning. Nature 2015;521(7553):436–44. https://doi.org/10.1038/nature14539.

[12] Krizhevsky A, Sutskever I, Geoffrey EH, Hinton GE. ImageNet classification with deep convolutional neural networks. In: Advances in neural information processing systems; 2012. p. 1097–105. http://code.google.com/p/cuda-convnet/.

[13] Bishop C. Pattern recognition and machine learning, vol 9. Springer; 2006.

[14] Goodfellow I, Bengio Y, Courville A. Deep learning. MIT Press; 2016.

[15] Murphy KP. Machine learning: A probabilistic perspective. The MIT Press; 2012.

[16] Kramer MA. Nonlinear principal component analysis using autoassociative neural networks. AIChE J 1991;37(2):233–43. https://doi.org/10.1002/aic.690370209.

[17] Kingma DP, Welling M. In: An introduction to variational autoencoders; 2019. https://doi.org/10.1561/2200000056.

[18] Hinton GE, Salakhutdinov RR. Reducing the dimensionality of data with neural networks. Science 2006;313(5786):504–7. https://doi.org/10.1126/science.1127647.

[19] Goodfellow IJ, Pouget-Abadie J, Mirza M, Xu B, Warde-Farley D, Ozair S, et al. Generative adversarial nets. Adv Neural Inf Process Syst 2014;3:2672–80.

[20] Burr S. Active learning literature survey. In: Computer Sciences Technical Report; 2009. https://minds.wisconsin.edu/bitstream/handle/1793/60660/TR1648.pdf?sequence=1&isAllowed=y.

[21] Pan SJ, Yang Q. In: A survey on transfer learning; 2010. https://doi.org/10.1109/TKDE.2009.191.

[22] Tan C, Sun F, Kong T, Zhang W, Yang C, Liu C. A survey on deep transfer learning. Lect Notes Comput Sci 2018;11141:270–9. https://doi.org/10.1007/978-3-030-01424-7{_}27.

[23] Raghu M, Zhang C, Kleinberg J, Bengio S. Transfusion: understanding transfer learning for medical imaging. Proceedings of the 33rd International Conference on Neural Information Processing Systems 2019 2019:3347–57. Article No.: 301.

[24] Brendan McMahan H, Moore E, Ramage D, Hampson S, Agüera y Arcas B. Communication-efficient learning of deep networks from decentralized data. Proceedings of the International Conference on Neural Information Processing Systems 2017 2017:1273–82.

[25] Yang Q, Liu Y, Chen T, Tong Y. Federated machine learning: concept and applications. ACM Trans Intell Syst Technol 2019;10(2):1–19. https://doi.org/10.1145/3298981.

[26] Li T, Sahu AK, Talwalkar A, Smith V. Federated learning: challenges, methods, and future directions. IEEE Signal Process Mag 2020;37(3):50–60. https://doi.org/10.1109/MSP.2020.2975749.

[27] Rieke N, Hancox J, Li W, Milletarì F, Roth HR, Albarqouni S, et al. The future of digital health with federated learning. Npj Digit Med 2020;3(1):1–7. https://doi.org/10.1038/s41746-020-00323-1.

[28] Wang Z, Bovik AC, Sheikh HR, Simoncelli EP. Image quality assessment: from error visibility to structural similarity. IEEE Trans Image Process 2004;13(4):600–12.

[29] Zhao H, Gallo O, Frosio I, Kautz J. Loss functions for image restoration with neural networks. IEEE Trans Comput Imaging 2016;3(1):47–57.

[30] Johnson J, Alahi A, Fei-Fei L, Assari SM, Idrees H, Shah M, et al. Perceptual losses for real-time style transfer and super-resolution. In: Proceedings of the European conference on computer vision. vol 9906. Springer Verlag; 2016. p. 694–711. https://doi.org/10.1007/978-3-319-46475-6{_}43.

[31] Kingma DP, Welling M. Auto-encoding variational bayes. arXiv preprint arXiv: 13126114 2013.

[32] Arjovsky M, Chintala S, Bottou L. Wasserstein generative adversarial networks. In: Proceedings of the international conference on machine learning; 2017. p. 214–23.

[33] Nowozin S, Cseke B, Tomioka R. f-gan: training generative neural samplers using variational divergence minimization. Adv Neural Inf Process Syst 2016;29.

[34] Robbins H, Monro S. A stochastic approximation method. Ann Math Stat 1951;22(3): 400–7. https://doi.org/10.1214/aoms/1177729586.

[35] Richard SS. Two problems with backpropagation and other steepest-descent learning. procedures for networks. In: Proc 8th annual conf cognitive science society. Erlbaum; 1986. p. 823–31.

[36] Kingma DP, Ba J. Adam: a method for stochastic optimization. arXiv preprint arXiv: 14126980 2014. https://doi.org/10.1063/1.4902458.

[37] Hinton GE, Srivastava NN, Swersky K. Neural networks for machine learning lecture 6a overview of mini-batch gradient descent. Tech Rep 2012;14(8):1–4. 241.

[38] Nesterov Y. A method for unconstrained convex minimization problem with the rate of convergence o(1/k̂2). Doklady AN USSR 1983;269:543–7.

[39] Duchi J, Hazan E, Singer Y. Adaptive subgradient methods for online learning and stochastic optimization. Tech Rep 2010. https://doi.org/10.5555/1953048.2021068.

[40] Zeiler MD. ADADELTA: an adaptive learning rate method. arXiv Prepr arXiv: 12125701 2012.

[41] Dozat T. Workshop track-ICLR 2016 incorporating Nesterov momentum into adam. Tech Rep 2016.

[42] Ruder S, An overview of gradient descent optimization algorithms. arXiv Prepr arxiv: 160904747 2016; .

[43] LeCun Y, Bottou L, Orr G, Müller KR. Efficient BackProp. Lect Notes Comput Sci 1998;1524:5–50.

[44] Kreutz-Delgado K. The complex gradient operator and the CR-calculus. arXiv preprint arXiv:09064835 2009.

[45] Wirtinger W. Zur formalen Theorie der Funktionen von mehr komplexen Veränderlichen. Math Ann 1927;97(1):357–75. https://doi.org/10.1007/BF01447872.

[46] Virtue P, Yu SX, Lustig M. Better than real: complex-valued neural nets for MRI fingerprinting. In: IEEE international conference on image processing; 2018. p. 3953–7.

[47] Hammernik K, Küstner T, Rueckert D. Chapter 11 - machine learning for mri reconstruction. In: Akçakaya M, Doneva M, Prieto C, editors. Magnetic resonance image

reconstruction. vol. 7. Academic Press; 2022. p. 281–323. https://doi.org/10.1016/B978-0-12-822726-8.00021-X.

[48] Glorot X, Bengio Y. Understanding the difficulty of training deep feedforward neural networks. Tech Rep JMLR Workshop and Conference Proceedings 2010:249–56.

[49] He K, Zhang X, Ren S, Sun J. Delving deep into rectifiers: surpassing human-level performance on imagenet classification. Proceedings of the IEEE International Conference on Computer Vision 2015;2015:1026–34. https://doi.org/10.1109/ICCV.2015.123.

[50] Sandino CM, Cheng JY, Chen F, Mardani M, Pauly JM, Vasanawala SS, Compressed S. From research to clinical practice with deep neural networks: shortening scan times for magnetic resonance imaging. IEEE Signal Process Mag 2020; 37(1):117–27. https://doi.org/10.1109/MSP.2019.2950433.

[51] Küstner T, Fuin N, Hammernik K, Bustin A, Qi H, Hajhosseiny R, et al. CINENet: deep learning-based 3D cardiac CINE MRI reconstruction with multi-coil complex-valued 4D spatio-temporal convolutions. Sci Rep 2020;10(1):1–13. https://doi.org/10.1038/s41598-020-70551-8.

[52] Stergiou A, Poppe R. Spatio-temporal FAST 3D convolutions for human action recognition. In: Proceedings - 18th IEEE international conference on machine learning and applications, ICMLA 2019. Institute of Electrical and Electronics Engineers Inc.; 2019. p. 183–90. https://doi.org/10.1109/ICMLA.2019.00036.

[53] Choy C, Gwak J, Savarese S. 4D spatio-temporal convnets: minkowski convolutional neural networks. Proceedings of the IEEE computer society conference on computer vision and pattern recognition. 2019. p. 3070–9. https://doi.org/10.1109/CVPR.2019.00319.

[54] Wang Y, Song Y, Xie H, Li W, Hu B, Yang G. Reduction of Gibbs artifacts in magnetic resonance imaging based on Convolutional Neural Network. In: Proceedings - 2017 10th international congress on image and signal processing; 2018.

[55] Trabelsi C, Bilaniuk O, Zhang Y, Serdyuk D, Subramanian S, Santos JF, et al. Deep complex networks. In: International conference on learning representations; 2018. https://github.com/ChihebTrabelsi/deep_complex_networks.

[56] Ioffe S, Szegedy C. Batch normalization: accelerating deep network training by reducing internal covariate shift. In: Proceedings of the international conference on machine learning; 2015. p. 448–56.

[57] Ulyanov D, Vedaldi A, Lempitsky V. Instance normalization: The missing ingredient for fast stylization. arXiv preprint arXiv:160708022 2016.

[58] Ba JL, Kiros JR, Hinton GE, Layer normalization. arXiv Prepr arXiv:160706450 2016; .

[59] Wu Y, He K. Group normalization. Tech Rep 2020;3. https://doi.org/10.1007/s11263-019-01198-w.

[60] Arjovsky M, Shah A, Bengio Y. Unitary evolution recurrent neural networks. In: Proceedings of the international conference on machine learning (ICML); 2016. p. 1120–8.

[61] Zhu B, Liu JZ, Cauley SF, Rosen BR, Rosen MS. Image reconstruction by domain-transform manifold learning. Nature 2018;555(7697):487–92.

[62] Odena A, Dumoulin V, Olah C. Deconvolution and checkerboard artifacts. Distill 2016. https://doi.org/10.23915/distill.00003.

[63] Srivastava N, Hinton G, Krizhevsky A, Sutskever I, Salakhutdinov R. Dropout: a simple way to prevent neural networks from overfitting. Tech Rep JMLR 2014;15(56): 1929–58.

[64] Cho K, Van Merriënboer B, Gulcehre C, Bahdanau D, Bougares F, Schwenk H, et al. Learning phrase representations using RNN encoder-decoder for statistical machine translation. In: EMNLP 2014–2014 conference on empirical methods in natural language processing, proceedings of the conference. Association for Computational Linguistics (ACL); 2014. p. 1724–34. https://doi.org/10.3115/v1/d14-1179.

[65] Wolter M, Yao A. Complex gated recurrent neural networks. In: Advances in neural information processing systems; 2018. p. 10536–46. https://github.com/v0lta/Complex-gated-recurrent-neural-networks.

[66] Graves A, Wayne G, Danihelka I. Neural turing machines. arXiv Prepr arXiv: 14105401 2014.

[67] Bahdanau D, Cho KH, Bengio Y. Neural machine translation by jointly learning to align and translate. In: 3rd international conference on learning representations, ICLR 2015 - conference track proceedings, international conference on learning representations, ICLR; 2015. https://arxiv.org/abs/1409.0473v7.

[68] Luong MT, Pham H, Manning CD. Effective approaches to attention-based neural machine translation. In: Conference proceedings - emnlp 2015: conference on empirical methods in natural language processing. Association for Computational Linguistics (ACL); 2015. p. 1412–21. https://doi.org/10.18653/v1/d15-1166.

[69] Vaswani A, Shazeer N, Parmar N, Uszkoreit J, Jones L, Gomez AN, et al. Attention is all you need. Tech Rep Advances in Neural Information Processing Systems 2017;30.

[70] Cheng J, Dong L, Lapata M. Long short-term memory-networks for machine reading. In: EMNLP 2016 - conference on empirical methods in natural language processing, Proceedings. 2; 2016. p. 551–61. https://doi.org/10.18653/v1/d16-1053. 3.

[71] Xu K, Ba JL, Kiros R, Cho K, Courville A, Salakhutdinov R, et al. Show, attend and tell: neural image caption generation with visual attention. In: 32nd International conference on machine learning, ICML. 3. International Machine Learning Society (IMLS); 2015. p. 2048–57.

[72] Hu J, Shen L, Albanie S, Sun G, Wu E. Squeeze-and-Excitation networks. IEEE Trans Pattern Anal Mach Intell 2020;42(8):2011–23. https://doi.org/10.1109/TPAMI.2019.2913372.

[73] Zhao L, Li M, Meng D, Li X, Zhang Z, Zhuang Y, et al. Deep convolutional neural networks with merge-and-run mappings. In: IJCAI international joint conference on artificial intelligence. vol 2018; 2018. p. 3170–6. https://doi.org/10.24963/ijcai.2018/440.

[74] Zhang J, Huang J, Jin S, Lu S. Vision-language models for vision tasks: a survey. IEEE Trans Pattern Anal Mach Intell 2024;46(8):5625–44. https://doi.org/10.1109/TPAMI.2024.3369699.

[75] Minaee S, Mikolov T, Nikzad N, Chenaghlu M, Socher R, Amatriain X, et al. Large language models: a survey. arXiv preprint arXiv:240206196 2024.

[76] He K, Zhang X, Ren S, Sun J. Deep residual learning for image recognition. In: IEEE conference on computer vision and pattern recognition; 2016. p. 770–8.

[77] Zagoruyko S, Komodakis N. Wide residual networks. In: British machine vision conference 2016. BMVC; 2016. p. 1–87. https://doi.org/10.5244/C.30.87.

[78] Huang G, Liu Z, Van Der Maaten L, Weinberger KQ. Densely connected convolutional networks. In: Proceedings - 30th IEEE conference on computer vision and pattern recognition, CVPR. 2017; 2017. p. 2261–9. https://doi.org/10.1109/CVPR.2017.243.

[79] Iandola FN, Han S, Moskewicz MW, Ashraf K, Dally WJ, Keutzer K. SqueezeNet: AlexNet-level accuracy with 50x fewer parameters and <0.5MB model size. arXiv Prepr arXiv:160207360 2016.

[80] Szegedy C, Liu W, Jia Y, Sermanet P, Reed S, Anguelov D, et al. Going deeper with convolutions. In: Proceedings of the IEEE computer society conference on computer vision and pattern recognition. vol 7. IEEE Computer Society; 2015. p. 1–9. https://doi.org/10.1109/CVPR.2015.7298594.

[81] Szegedy C, Vanhoucke V, Ioffe S, Shlens J, Wojna Z. Rethinking the inception architecture for computer vision. In: Proceedings of the IEEE computer society conference on computer vision and pattern recognition. vol 2016. IEEE Computer Society; 2016. p. 2818–26. https://doi.org/10.1109/CVPR.2016.308.

[82] Szegedy C, Ioffe S, Vanhoucke V, Alemi AA. Inception-v4, inception-ResNet and the impact of residual connections on learning. In: 31st AAAI conference on artificial intelligence, AAAI 2017. AAAI press; 2017. p. 4278–84.

[83] Simonyan K, Zisserman A. Very deep convolutional networks for large-scale image recognition. In: International conference for learning representations; 2015. p. 1–14.

[84] Kaur T, Gandhi TK. Automated brain image classification based on vgg-16 and transfer learning. In: 2019 international conference on information technology (ICIT), IEEE; 2019. p. 94–8.

[85] Majib MS, Rahman MM, Sazzad TS, Khan NI, Dey SK. Vgg-scnet: a vgg net-based deep learning framework for brain tumor detection on mri images. IEEE Access 2021;9:116942–52.

[86] Toğaçar M, Cömert Z, Ergen B. Classification of brain mri using hyper column technique with convolutional neural network and feature selection method. Expert Syst Appl 2020;149:113274.

[87] Pham CH, Tor-Díez C, Meunier H, Bednarek N, Fablet R, Passat N, et al. Multiscale brain mri super-resolution using deep 3d convolutional networks. Comput Med Imag Graph 2019;77:101647.

[88] Küstner T, Munoz C, Psenicny A, Bustin A, Fuin N, Qi H, et al. Deep-learning based super-resolution for 3d isotropic coronary mr angiography in less than a minute. Magn Reson Med 2021;86(5):2837–52.

[89] Ronneberger O, Fischer P, Brox T. U-Net: convolutional networks for biomedical image segmentation. In: International conference on medical image computing and computer assisted intervention; 2015. p. 234–41.

[90] Çiçek Ö, Abdulkadir A, Lienkamp SS, Brox T, Ronneberger O. 3d u-net: learning dense volumetric segmentation from sparse annotation. In: Medical image computing and computer-assisted intervention–MICCAI 2016: 19th international conference, Athens, Greece, October 17–21, 2016, Proceedings, Part II 19. Springer; 2016. p. 424–32.

[91] Kohl S, Romera-Paredes B, Meyer C, De Fauw J, Ledsam JR, Maier-Hein K, et al. A probabilistic u-net for segmentation of ambiguous images. Adv Neural Inf Process Syst 2018;31.

[92] Baumgartner CF, Koch LM, Pollefeys M, Konukoglu E. An exploration of 2d and 3d deep learning techniques for cardiac mr image segmentation. In: Statistical atlases and computational models of the heart. ACDC and MMWHS challenges: 8th international workshop, STACOM 2017, held in conjunction with MICCAI 2017, Quebec city, Canada, September 10–14, 2017, Revised selected papers 8. Springer; 2018. p. 111–9.

[93] Zhang Z, Liu Q, Wang Y. Road extraction by deep residual u-net. In: IEEE geoscience and remote sensing letters. 15; 2018. p. 749–53.

[94] Oktay O, Schlemper J, Folgoc LL, Lee M, Heinrich M, Misawa K, et al. Attention u-net: learning where to look for the pancreas. arXiv preprint arXiv:180403999 2018.

[95] Chen LC, Papandreou G, Kokkinos I, Murphy K, Yuille AL. Deeplab: semantic image segmentation with deep convolutional nets, atrous convolution, and fully connected crfs. IEEE Trans Pattern Anal Mach Intell 2017;40(4):834–48.

[96] Isensee F, Jaeger PF, Kohl SA, Petersen J, Maier-Hein KH. nnu-net: a self-configuring method for deep learning-based biomedical image segmentation. Nat Methods 2021; 18(2):203–11.

[97] Mildenhall B, Srinivasan PP, Tancik M, Barron JT, Ramamoorthi R, Ng R. Nerf: representing scenes as neural radiance fields for view synthesis. Commun ACM 2021;65 (1):99–106.

[98] Huang W, Li HB, Pan J, Cruz G, Rueckert D, Hammernik K. Neural implicit k-space for binning-free non-cartesian cardiac mr imaging. In: Frangi A, de Bruijne M, Wassermann D, Navab N, editors. Information processing in medical imaging. Cham: Springer Nature Switzerland; 2023. p. 548–60.

[99] Catania F, Spitale M, Garzotto F. Conversational agents in therapeutic interventions for neurodevelopmental disorders: a survey. ACM Comput Surv 2023;55(10):1–34.

[100] Ramesh A, Dhariwal P, Nichol A, Chu C, Chen M. Hierarchical text-conditional image generation with clip latents. arXiv preprint arXiv:220406125 2022;1(2):3.

[101] Ray PP. Chatgpt: a comprehensive review on background, applications, key challenges, bias, ethics, limitations and future scope. Internet of Things and Cyber-Phys Syst 2023;3:121–54. https://doi.org/10.1016/j.iotcps.2023.04.003.

[102] Lee K, Lee K, Lee H, Shin J. A simple unified framework for detecting out-of-distribution samples and adversarial attacks. Adv Neural Inf Process Syst 2018;31.

[103] Schlegl T, Seeböck P, Waldstein SM, Langs G, Schmidt-Erfurth U. f-anogan: fast unsupervised anomaly detection with generative adversarial networks. Med Image Anal 2019;54:30–44.

[104] Kascenas A, Pugeault N, O'Neil AQ. Denoising autoencoders for unsupervised anomaly detection in brain mri. In: International conference on medical imaging with deep learning. PMLR; 2022. p. 653–64.

[105] Wyatt J, Leach A, Schmon SM, Willcocks CG. Anoddpm: anomaly detection with denoising diffusion probabilistic models using simplex noise. In: Proceedings of the IEEE/CVF conference on computer vision and pattern recognition; 2022. p. 650–6.

[106] Bercea CI, Neumayr M, Rueckert D, Schnabel JA. Mask, stitch, and re-sample: enhancing robustness and generalizability in anomaly detection through automatic diffusion models. arXiv preprint arXiv:230519643 2023.

[107] Frotscher A, Kapoor J, Wolfers T, Baumgartner CF. Unsupervised anomaly detection using aggregated normative diffusion. arXiv preprint arXiv:231201904 2023.

[108] Wolleb J, Sandkühler R, Bieder F, Valmaggia P, Cattin PC. Diffusion models for implicit image segmentation ensembles. In: International conference on medical imaging with deep learning. PMLR; 2022. p. 1336–48.

[109] Fischer P, Thomas K, Baumgartner CF. Uncertainty estimation and propagation in accelerated mri reconstruction. In: International workshop on uncertainty for safe utilization of machine learning in medical imaging. Springer; 2023. p. 84–94.

[110] Jalal A, Arvinte M, Daras G, Price E, Dimakis AG, Tamir J. Robust compressed sensing mri with deep generative priors. Adv Neural Inf Process Syst 2021;34:14938–54.

[111] Peng C, Guo P, Zhou SK, Patel VM, Chellappa R. Towards performant and reliable undersampled mr reconstruction via diffusion model sampling. In: International conference on medical image computing and computer-assisted intervention. Springer; 2022. p. 623–33.

[112] Chung H, Ye JC. Score-based diffusion models for accelerated mri. Med Image Anal 2022;80:102479.

[113] Xie Y, Li Q. Measurement-conditioned denoising diffusion probabilistic model for under-sampled medical image reconstruction. In: International conference on medical image computing and computer-assisted intervention. Springer; 2022. p. 655–64.

[114] Rumelhart DE, Hinton GE, Williams RJ, et al. In: Learning internal representations by error propagation; 1985.

[115] Vahdat A, Kautz J. Nvae: a deep hierarchical variational autoencoder. Adv Neural Inf Process Syst 2020;33:19667–79.

[116] Arjovsky M, Chintala S, Bottou L. Wasserstein generative adversarial networks. In: International conference on machine learning. PMLR; 2017. p. 214–23.

[117] Miyato T, Kataoka T, Koyama M, Yoshida Y. Spectral normalization for generative adversarial networks. arXiv preprint arXiv:180205957 2018.

[118] Karras T, Laine S, Aila T. A style-based generator architecture for generative adversarial networks. In: Proceedings of the IEEE/CVF conference on computer vision and pattern recognition; 2019. p. 4401–10.

[119] Brock A, Donahue J, Simonyan K. Large scale gan training for high fidelity natural image synthesis. arXiv preprint arXiv:180911096 2018.

[120] Ho J, Jain A, Abbeel P. Denoising diffusion probabilistic models. Adv Neural Inf Process Syst 2020;33:6840–51.

[121] Luo C. Understanding diffusion models: a unified perspective. arXiv preprint arXiv: 220811970 2022.

[122] Rombach R, Blattmann A, Lorenz D, Esser P, Ommer B. High-resolution image synthesis with latent diffusion models. In: Proceedings of the IEEE/CVF conference on computer vision and pattern recognition; 2022. p. 10684–95.

[123] Choi J, Kim S, Jeong Y, Gwon Y, Yoon S. Ilvr: conditioning method for denoising diffusion probabilistic models. arXiv preprint arXiv:210802938 2021.

[124] Song Y, Ermon S. Generative modeling by estimating gradients of the data distribution. Adv Neural Inf Process Syst 2019;32.

[125] Song Y, Shen L, Xing L, Ermon S. Solving inverse problems in medical imaging with score-based generative models. arXiv preprint arXiv:211108005 2021.

[126] Meng C, Rombach R, Gao R, Kingma D, Ermon S, Ho J, et al. On distillation of guided diffusion models. In: Proceedings of the IEEE/CVF conference on computer vision and pattern recognition; 2023. p. 14297–306.

[127] Sohn K, Lee H, Yan X. Learning structured output representation using deep conditional generative models. Adv Neural Inf Process Syst 2015;28.

[128] Wolterink JM, Dinkla AM, Savenije MH, Seevinck PR, van den Berg CA, Išgum I. Deep mr to ct synthesis using unpaired data. In: Simulation and synthesis in medical imaging: second international workshop, SASHIMI 2017, held in conjunction with MICCAI 2017, Québec City, QC, Canada, September 10, 2017, Proceedings 2. Springer; 2017. p. 14–23.

[129] Küstner T, Armanious K, Yang J, Yang B, Schick F, Gatidis S. Retrospective correction of motion-affected mr images using deep learning frameworks. Magn Reson Med 2019;82(4):1527–40. https://doi.org/10.1002/mrm.27783.

[130] Armanious K, Jiang C, Fischer M, Küstner T, Hepp T, Nikolaou K, et al. Medgan: medical image translation using gans. Comput Med Imag Graph 2020;79:101684. https://doi.org/10.1016/j.compmedimag.2019.101684.

[131] Johnson PM, Drangova M. Conditional generative adversarial network for 3d rigid-body motion correction in mri. Magn Reson Med 2019;82(3):901–10. https://doi.org/10.1002/mrm.27772.

[132] Küstner T, Munoz C, Psenicny A, Bustin A, Fuin N, Qi H, et al. Deep-learning based super-resolution for 3d isotropic coronary mr angiography in less than a minute. Magn Reson Med 2021;86(5):2837–52. https://doi.org/10.1002/mrm.28911.

[133] Chen Y, Christodoulou AG, Zhou Z, Shi F, Xie Y, Li D. Mri super-resolution with gan and 3d multi-level densenet: smaller, faster, and better. arXiv preprint arXiv: 200301217 2020.

[134] Zhao M, Liu X, Liu H, Wong KK. Super-resolution of cardiac magnetic resonance images using laplacian pyramid based on generative adversarial networks. Comput Med Imag Graph 2020;80:101698.

[135] Zhu JY, Park T, Isola P, Efros AA. Unpaired image-to-image translation using cycle-consistent adversarial networks. In: Proceedings of the IEEE international conference on computer vision; 2017. p. 2223–32.

[136] Ho J, Salimans T. Classifier-free diffusion guidance. arXiv preprint arXiv:220712598 2022.

[137] Ramesh A, Pavlov M, Goh G, Gray S, Voss C, Radford A, et al. Zero-shot text-to-image generation. In: International conference on machine learning. PMLR; 2021. p. 8821–31.

[138] Vapnik VN. The nature of statistical learning theory, vo. 4. Springer Science and Business Media; 2013. p. 409.

[139] Lundervold AS, Lundervold A. In: An overview of deep learning in medical imaging focusing on MRI; 2019. https://doi.org/10.1016/j.zemedi.2018.11.002.

[140] Knoll F, Hammernik K, Zhang C, Moeller S, Pock T, Sodickson DK, et al. Deep-learning methods for parallel magnetic resonance imaging reconstruction: a survey of the current approaches, trends, and issues. IEEE Signal Process Mag 2020;37(1): 128–40. https://doi.org/10.1109/MSP.2019.2950640.

[141] Wang G, Ye JC, Mueller K, Fessler JA. Image reconstruction is a new frontier of machine learning. IEEE Trans Med Imag 2018;37(6):1289–96.

[142] Fessler JA. Optimization methods for MR image reconstruction (long version). arXiv Prepr arXiv:190303510 2019.

CHAPTER

Introduction to the physics behind MR

3

Eva S. Peper[1, 2, *], Adèle L.C. Mackowiak[3] and Jessica A.M. Bastiaansen[1, 2]

[1]Department of Diagnostic, Interventional and Pediatric Radiology (DIPR), Inselspital, Bern University Hospital, University of Bern, Bern, Switzerland; [2]Translational Imaging Center (TIC), Swiss Institute for Translational and Entrepreneurial Medicine, Bern, Switzerland; [3]Department of Medical Radiology, Lausanne University Hospital (CHUV-UNIL), Lausanne, Switzerland

[*]Corresponding author: E-mail: evaspeper@gmail.com

3.1 Introduction

Magnetic resonance imaging (MRI) has revolutionized medical diagnostics and research, offering a window into the human body. MRI remains unparalleled in its clarity and detail as an imaging technique that does not use ionizing radiation. A brain MRI scan can distinguish between gray and white matter, detect abnormalities such as tumors, and reveal inflammation or demyelination in conditions such as multiple sclerosis. Similarly, cardiac MRI leverages MR physics to visualize heart structures and function, aiding in diagnosing and managing cardiovascular diseases. Understanding the underlying physics of magnetic resonance (MR) is a significant academic endeavor that continues to push the boundaries in medical science. To this day, new contrast mechanisms are being discovered by manipulating the magnetic properties of biological tissues within a magnetic field. This chapter introduces the fundamental principles of MR physics, covering magnetization, excitation, relaxation, signal and image formation, and essential MRI sequences.

The phenomenon that specific atomic nuclei can absorb and emit electromagnetic energy was first observed in 1946 by Purcell [1] and Bloch [2], earning them the Nobel Prize in Physics. In 1950, Proctor [3] and Dickinson [4] further revealed that nuclei within the same molecule absorb energy at different frequencies, influenced by the applied magnetic field and their chemical environment. This insight broadened the application of MR to studying molecular structures and their surroundings. In the following decades, MR was applied to living organisms, driven by a growing interest in understanding water properties in biological tissues. Researchers observed that the MR properties of healthy and cancerous tissues differ significantly [5]. In the mid-1970s, Lauterbur [6], Mansfield [7], and their teams demonstrated that spatial information could be reconstructed using position-

Machine Learning in MRI. https://doi.org/10.1016/B978-0-443-14109-6.00003-1

dependent magnetic fields, laying the foundation for medical imaging. This breakthrough laid the groundwork for MRI and was later recognized with the 2003 Nobel Prize in Medicine. Over the past decades, the field of MRI has undergone remarkable advancements, becoming an essential tool for studying tissue structure, function, composition, and dynamic metabolic processes in living subjects.

3.2 Magnetization formation and spin dynamics

Magnetic resonance is based on the collective behavior of nuclear spins, particularly those of hydrogen atoms in water, the most abundant molecule in the human body. When exposed to an external magnetic field, these spins partially align, producing a macroscopic magnetization. Their interaction with additional perturbations, such as radiofrequency pulses, follows spin dynamics, which include precession and relaxation processes. The Bloch equations describe the effects of both the static magnetic field and the applied oscillating field. A thorough understanding of these mechanisms is crucial for optimizing MR techniques and improving signal contrast.

3.2.1 Atomic nuclei in a magnetic field

The interaction between atomic nuclei possessing a nuclear spin and an external static magnetic field [8] is governed by the Larmor equation:

$$\omega_0 = \gamma B_0 \tag{3.1}$$

where ω_0 is the Larmor frequency, representing the precession frequency of the nuclear spin around the magnetic field B_0. The gyromagnetic ratio $\gamma = 2\pi \cdot 42.576 \cdot 10^6$ rad/sT is a unique constant for each isotope with a nuclear spin.

More specifically, atomic nuclei with an odd number of protons, neutrons, or both possess a quantum mechanical property known as nuclear spin $\boldsymbol{I}$, characterized by the spin quantum number I. When placed in a magnetic field B_0, the energy levels of these nuclei split into $2I + 1$ distinct states due to the Zeeman effect (Fig. 3.1).

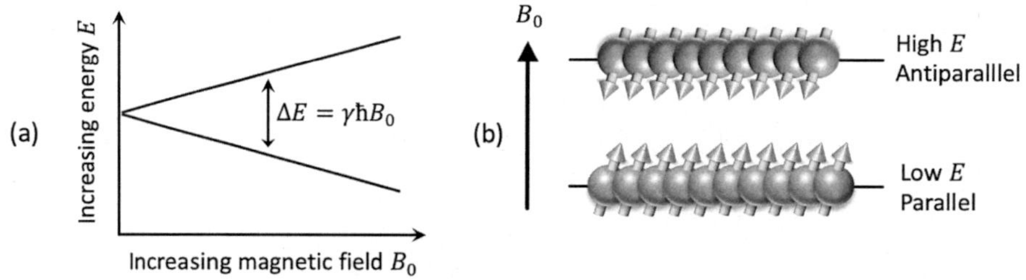

FIGURE 3.1

(a) Representation of the energy levels for nuclear spins with $I = 1/2$ as a function of the magnetic field strength B_0. (b) The lower energy level corresponds to magnetic moments aligned parallel to the magnetic field B_0, and the higher energy level represents magnetic moments with an antiparallel alignment relative to B_0.

The separation between these energy levels depends on the strength of the magnetic field and the magnetic properties of the nucleus. The Larmor frequency ω_0 corresponds to the frequency of the electromagnetic radiation required to induce transitions between these distinct nuclear spin energy levels.

Hydrogen (^{1}H) is the most abundant atom in the human body. It has a nuclear spin quantum number of $I = 1/2$. As a result, when placed in an external magnetic field, hydrogen nuclei exhibit two distinct energy levels. The nuclear spin vector can align either parallel or antiparallel to the field direction, corresponding to the lower and higher energy states, respectively (Fig. 3.1b).

Resulting from its nuclear spin, each atom can have a nuclear magnetic moment $\boldsymbol{\mu}$ with the relation

$$\boldsymbol{\mu} = \gamma\hbar\boldsymbol{I} \tag{3.2}$$

with the reduced Planck's constant $\hbar = 6.626 \cdot 10^{-34}$ Js. The orientation of the nuclear magnetic moment is random in the absence of an external magnetic field. However, within a magnetic field, the nuclear magnetic moments are aligned or antialigned with $\boldsymbol{B_0}$. In semiclassical descriptions, the nuclear magnetic moment rotates on a cone of quantized μ values around the main magnetic field direction with a resonance frequency ω_0 (Fig. 3.2a). When considering many nuclear magnetic moments of, e.g., all hydrogen atoms of a tissue sample, this alignment characterizes the net magnetization direction $\boldsymbol{M}$ of the sample, which will be covered in the subsequent section and is crucial to MRI (Fig. 3.2b and c).

The energy levels for atomic nuclei with $I = 1/2$ in a magnetic field are equal to

$$E = \pm\mu B_0 = \pm\frac{1}{2}\gamma\hbar B_0 \tag{3.3}$$

which indicates the low and high energy levels induced by the Zeeman effect relative to the magnetic field orientation. The energy difference required for a transition between these two levels is given by

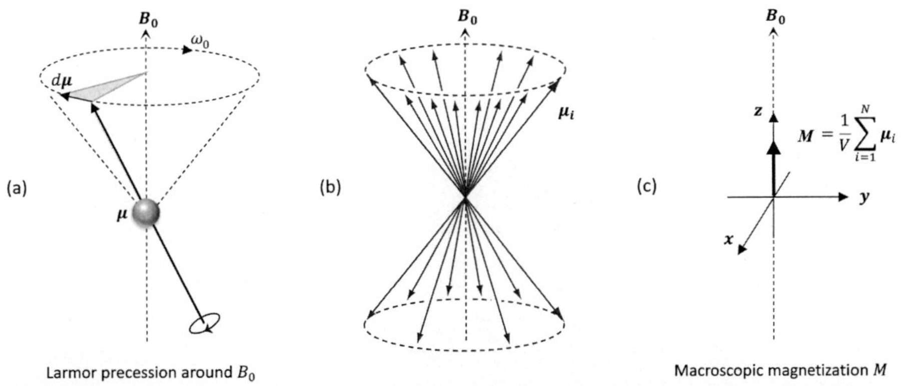

FIGURE 3.2

(a) A classical depiction of the Larmor precession of a single nuclear magnetic moment μ with $I = 1/2$ around a magnetic field B_0. (b) The collective sum of many individual nuclear magnetic moments μ within a sample leads to the formation of the (c) net magnetization vector M.

$$\Delta E = \gamma \hbar B_0 = \hbar \omega_0 \tag{3.4}$$

This equation emphasizes the fundamental principle behind MRI: the energy difference between two states corresponds to the energy associated with the Larmor frequency, thus facilitating the absorption and emission of energy by the nuclei. In subsequent sections, we explore how MRI leverages the concepts of excitation and relaxation to manipulate the energy states.

3.2.2 Macroscopic magnetization

Although the phenomenon of magnetic resonance is purely a quantum mechanical process, its macroscopic manifestation in the net magnetization $\boldsymbol{M}$ and its interaction with additional external magnetic fields is, under most circumstances, well described by classical physics.

In MRI, in a macroscopic view, each imaging volume element (voxel) contains many atomic nuclei with nuclear magnetic moments distributed on a cone (Fig. 3.2b). The energy distribution of all nuclear magnetic moments of this sample at thermal equilibrium can be calculated using Boltzmann statistics:

$$\frac{n_+}{n_-} = e^{\Delta E / k_B T} \tag{3.5}$$

where n_+ and n_- are the population of both, the low and high energy state, $k_B = 1.3806 \cdot 10^{-23}$ J/K is the Boltzmann constant, and T is the absolute temperature. Although the population difference between the two energy states of hydrogen is relatively small at room temperature, about one in a million, it generates a macroscopic net magnetization within the imaging voxel (Fig. 3.2c). The net magnetization $\boldsymbol{M}$ is the vector sum of all magnetic moments $\boldsymbol{\mu}$ within the voxel with sample volume V (Fig. 3.2c). This net magnetization vector is finite, and observable, and when the system is in a state of equilibrium, this vector is parallel to the main magnetic field:

$$\boldsymbol{M} = \frac{1}{V} \sum_{i=1}^{N} \boldsymbol{\mu}_i \tag{3.6}$$

$N = n_+ + n_-$ is the total number of atomic nuclei within the imaging voxel. The absolute magnetization for a given spin density N/V can then be expressed as follows:

$$M_0 = \frac{N}{V} \frac{(\gamma \hbar)^2 B_0 I(I+1)}{3 k_B T} \tag{3.7}$$

If exposed to not a static, but a time-varying external magnetic field $\boldsymbol{B}(t)$, each nuclear magnetic moment $\boldsymbol{\mu}$ experiences a torque and precesses with ω_0 around the axis of the direction of $\boldsymbol{B}(t)$.The precession is described by the following equation of motion:

$$\frac{d\boldsymbol{\mu}(t)}{dt} = \gamma\,(\boldsymbol{\mu}(t) \times \boldsymbol{B}(t)) \tag{3.8}$$

A sum over the equation of motion for all individual nuclear magnetic moments leads to the following description for the macroscopic magnetization vector $\boldsymbol{M}$ in a magnetic field $\boldsymbol{B}(t)$

$$\frac{d\boldsymbol{M}(t)}{dt} = \gamma\,(\boldsymbol{M}(t) \times \boldsymbol{B}(t)), \tag{3.9}$$

which is known as the Bloch Equation [2,9]. These equations were developed by Felix Bloch in 1946 and include the effects of the external static magnetic field and any applied oscillating magnetic field.

The interaction between atomic nuclei and magnetic fields produces a net magnetization, guided by the Larmor equation. In the next section, to understand the process of creating images from magnetic signals, we delve into how magnetization is observed and manipulated.

3.3 MR signal and relaxation

Generating a measurable MR signal requires perturbing the net magnetization vector from its equilibrium state, as discussed in this chapter. Following the application of an RF excitation pulse, the magnetization gradually returns to thermal equilibrium through two simultaneous relaxation processes. The Bloch equations are extended to incorporate these relaxation mechanisms, which play a crucial role in shaping MR signal characteristics and are fundamental to MRI sequence design.

3.3.1 Excitation of the macroscopic magnetization

At thermal equilibrium, all net magnetization is parallel to $\boldsymbol{B}_0$; therefore, a transverse component of the magnetization vector $\boldsymbol{M}$, defined as $M_{xy} = M_x + iM_y$, is nonexistent ($M_{xy} = 0$), and the longitudinal magnetization $M_z = M_0$ is constant (Fig. 3.2c):

$$\frac{dM_z(t)}{dt} = 0 \tag{3.10}$$

This equation shows that, without external influences, the longitudinal magnetization remains steady over time.

The required perturbation of the magnetization is achieved by applying an additional, time-varying magnetic field, $\boldsymbol{B}_1$, oriented perpendicular to the static magnetic field $\boldsymbol{B}_0$. The $\boldsymbol{B}_1$ field is a radiofrequency (RF) field that oscillates at the Larmor frequency w_0 of the nuclei of interest (Fig. 3.3a):

$$\boldsymbol{B}_1(t) = B_1\,cos(\omega_0 t) + B_1\,sin(\omega_0 t) \tag{3.11}$$

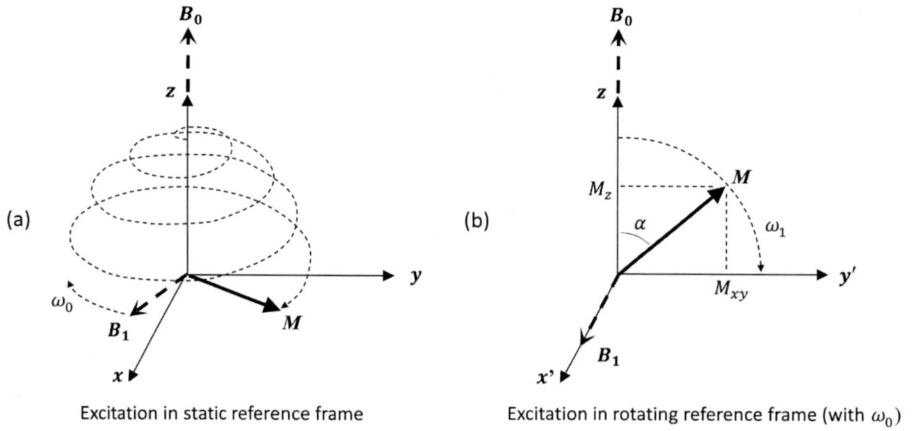

FIGURE 3.3

(a) Illustration of the precessional motion of magnetization in the presence of an oscillating magnetic field B_1 applied orthogonally to the static magnetic field B_0 within a static reference frame. (b) Illustration of the same precessional motion within a rotating reference frame, where the x' and y' axes rotate about the z'-axis at the Larmor frequency $\boldsymbol{\omega}_0$. This frame simplifies the depiction of magnetization during RF excitation with B_1.

The $\boldsymbol{B}_1$ field causes the nuclear spins to absorb energy and transition from their lower energy state to a higher one, called excitation. On a macroscopic level, the magnetization $\boldsymbol{M}$ experiences a torque and is tipped away from the $\boldsymbol{B}_0$ axis toward the transverse plane. The magnetization starts to precess around the $\boldsymbol{B}_0$ axis with ω_0 and around the $\boldsymbol{B}_1$ axis with $\omega_1 = \gamma B_1$ (Fig. 3.3a). This creates an oscillating transverse magnetization M_{xy}, inducing an observable signal in the MRI receiver coils.

In a rotating frame of reference (Fig. 3.3b), the magnetization vector experiences $\boldsymbol{B}_1$ as a static magnetic field when $\boldsymbol{B}_1$ is oscillating with ω_0. The degree to which $\boldsymbol{M}$ is tipped away from the $\boldsymbol{B}_0$ axis is determined by the amplitude and duration of the RF field and is described by the RF excitation angle α.

$$\alpha = \gamma B_1 t \tag{3.12}$$

Since the time needed to flip the magnetization is typically very short, ranging from 100 μs to a few ms, the applied $\boldsymbol{B}_1$ field is known as an RF excitation pulse. For instance, $\alpha = 90°$ RF excitation pulse rotates $\boldsymbol{M}$ into the transverse plane, maximizing M_{xy}, while $\alpha = 180°$ pulse inverts $\boldsymbol{M}$, aligning it opposite to the direction of $\boldsymbol{B}_0$. These manipulations control the MR signal characteristics and are fundamental to MRI pulse sequence design.

3.3.2 Relaxation processes

After applying the RF excitation pulse, the magnetization returns to thermal equilibrium through relaxation (Fig. 3.4a). Two simultaneous relaxation processes occur:

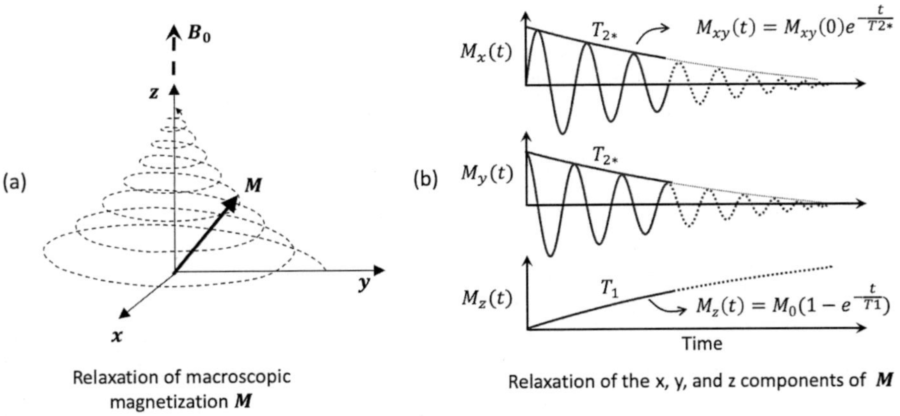

FIGURE 3.4

(a) Precessional motion of the net magnetization vector M after a 90° RF pulse in a static reference frame. Following RF excitation, M relaxes back toward thermal equilibrium. (b) Relaxation of the individual components of M along the x, y, and z axes. The transverse components (M_x and M_y) diminish according to the T_{2*} relaxation time. In comparison, the longitudinal component (M_z) recovers to equilibrium under the T_1 relaxation time constant.

- Longitudinal relaxation (T_1 recovery) involves the return of the longitudinal component M_z to its equilibrium state M_0, a process driven by energy exchange between the nuclear spins and their environment. The T_1 relaxation time is a measure of how quickly energy is exchanged, allowing M_z to realign with $\boldsymbol{B}_0$.
- Transverse relaxation (T_2 decay) refers to the dephasing and decaying of the M_{xy} component due to spin–spin interactions, reducing the coherent transverse magnetization. The T_2 relaxation time reflects how quickly the nuclear spins lose phase coherence in the plane perpendicular to $\boldsymbol{B}_0$, diminishing the MR signal.

The Bloch equations [9] can be extended to include these two relaxation processes:

$$\frac{dM_{xy}(t)}{dt} = \gamma\,(\boldsymbol{M}(t) \times \boldsymbol{B}_1(t))_{xy} - \frac{M_{xy}(t)}{T_2} \tag{3.13}$$

and

$$\frac{dM_z(t)}{dt} = \gamma\,(\boldsymbol{M}(t) \times \boldsymbol{B}_1(t))_z - \frac{M_0 - M_z(t)}{T_1} \tag{3.14}$$

The rate of recovery of $M_z(0)$ to the thermal equilibrium value M_0 can then be described as follows:

$$M_z(t) = M_z(0)\exp\left(-\frac{t}{T1}\right) + M_0\left(1 - \exp\left(-\frac{t}{T1}\right)\right) \tag{3.15}$$

where $M_z(0) = 0$, when $\boldsymbol{M}$ is rotated by a 90° excitation pulse (Fig. 3.4b), called a saturation recovery experiment, and $M_z(0) = -M_0$, when $\boldsymbol{M}$ is rotated by 180°, called an inversion recovery experiment. Changes in T_1 between different tissue types, leading to changes in the image contrast when using MRI sequences which are sensitive to variations in T_1, are referred to as T_1-weighted sequences. The inverse of the longitudinal relaxation time, $\frac{1}{\mathrm{T1}}$, is referred to as the longitudinal relaxation rate, R_1.

The decay of M_{xy} (Fig. 3.4b) is a result of local field inhomogeneities on a microscopic scale. The field variations are introduced by various shielding effects at the molecular level and macroscopic field inhomogeneities due to variations in the local susceptibility. This leads to different local precession frequencies and the dephasing of M_{xy}. The decay time of M_{xy} is described by T_2 as follows:

$$M_{xy}(t) = M_{xy}(0)\exp\left(-\frac{\mathrm{t}}{\mathrm{T2}}\right) \tag{3.16}$$

In practice, an additional dephasing of the magnetization is observed. Within the tissue sample, local $\boldsymbol{B}_0$ inhomogeneities create a distribution of Larmor frequencies. This introduces an acceleration of transverse magnetization loss directly after excitation. This additional loss can be described with the relaxation time constant T_2^*:

$$\frac{1}{T_2^*} = \frac{1}{T_2} + \gamma\Delta B \tag{3.17}$$

where $\gamma\Delta B$ indicates the effect of static external field inhomogeneities. The additional loss of transverse magnetization is recoverable with specific MRI sequences [10] (see Section 3.5).

In a biological context, the relaxation dynamics described by T_1, T_2, and T_{2*} reflect different mechanisms within the tissue environment and can be used to differentiate tissue types. They are fundamental descriptors of how nuclear magnetization returns to equilibrium, but they each describe different physical processes. T_1, or longitudinal relaxation time, is typically shorter than T_2, the transverse relaxation time because T_1 processes are driven by interactions between the spin system and the surrounding lattice, which allows energy exchange (spin–lattice relaxation). These interactions are relatively strong and frequent compared with the spin–spin interactions that govern T_2, where energy is not exchanged with the environment but is redistributed among the spins themselves. Furthermore, T_{2*} is always the shortest relaxation time in an inhomogenous magnetic field because it includes all the dephasing effects present in T_2, as well as additional dephasing due to inhomogeneities in the static magnetic field. In essence, T_{2*} represents the observed decay rate of the transverse magnetization and accounts for the intrinsic spin–spin relaxation and external magnetic field inhomogeneities.

In summary, relaxation times follow the relationship $T_1 > T_2 > T_{2*}$, which reflects how internal and external factors within biological tissues affect relaxation processes. Relaxometry, a branch of MRI parametric mapping, aims to quantify these relaxation times to identify tissue biomarkers. Tissue relaxation times reflect the local molecular environment (size, interactions, motion), the state of which can provide information on pathology [11,12].

3.3.3 MR signal and contrast

In the MR experiment, the sample is exposed to a static homogenous magnetic field $\boldsymbol{B}_0$ induced by a superconducting coil or permanent magnet. Transmitting radiofrequency coils are arranged to irradiate the radiofrequency field $\boldsymbol{B}_1(t)$ perpendicular to the main magnetic field. During the relaxation phase, MRI receiver coils detect the oscillating transverse magnetization M_{xy} as an electromotive force. The detected MR signal decays exponentially due to relaxation processes and is termed the free induction decay (FID). The measured complex signal can be described as follows:

$$S(t) \propto \int M_{xy}(\boldsymbol{r},t)\, e^{-i\,\omega_{RF}\,t}\, dV \tag{3.18}$$

where $S(t)$ denotes the MR signal strength, and $M_{xy}(\boldsymbol{r},t)$ represents the spatial distribution of the transverse magnetization. The measured signal directly reflects the transverse magnetization immediately following the RF excitation pulse. Its decay rate is determined by both T_2 relaxation processes, which differ for each tissue type, and additional dephasing effects due to magnetic field inhomogeneities.

The next extension of the MRI experiment is the application of several consecutive RF excitation pulses. By carefully designing the timing between consecutive RF excitations, termed repetition time (TR), and the timing between the RF excitation and the signal acquisition, termed the echo time (TE), MRI can exploit differences in relaxation times across different tissues to generate images with different types of contrast (Fig. 3.5). The following section focuses on how measured signals translate to spatial information and how magnetic resonance images are formed.

3.3.4 Magnetic field strength

The magnetic field strength $\boldsymbol{B}_0$ plays a crucial role in MR imaging and influences the choice of the scanner, as the signal increases approximately quadratically with field strength. An increase in $\boldsymbol{B}_0$ leads to a higher net magnetization, as described by Equation [7]. This results in a stronger signal and improved signal-to-noise ratio (SNR) (Fig. 3.6), enabling the use of smaller voxel sizes and allowing for high-resolution imaging. However, increasing $\boldsymbol{B}_0$ not only enhances the signal but also alters tissue relaxation properties. At high magnetic field strengths (>1T), noise increases linearly with $\boldsymbol{B}_0$. Despite this, the net increase in SNR and enhanced tissue contrast (Fig. 3.5b) make 1.5 and 3.0T scanners the standard in clinical practice. Ultrahigh magnetic fields (7.0T and above) are primarily used in research settings, as they can produce highly detailed images but require strict safety protocols and may introduce additional artifacts. Meanwhile, low-field (0.2–0.5T) and mid-field (0.5–1.0T) scanners are gaining interest due to their cost-effectiveness and reduced magnetic field inhomogeneities despite their lower SNR. The choice of magnetic

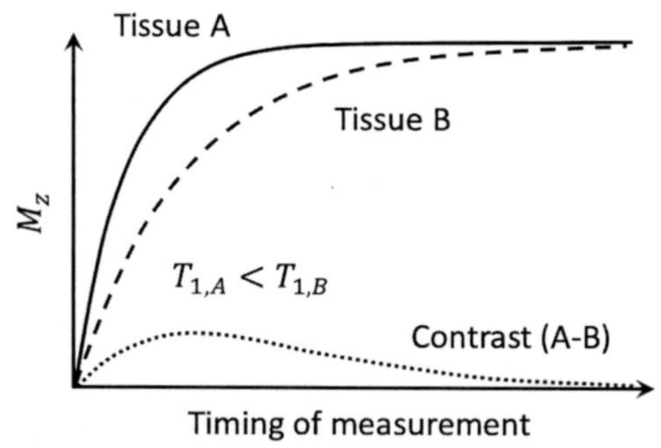

FIGURE 3.5

Recovery of longitudinal magnetization over time for two different tissue types, A and B. Tissue A has a shorter T_1 relaxation time than tissue B. The contrast between the two tissue types can be maximized by carefully timing the MR experiment [13]: manipulating the repetition time (TR) between excitations induces a change in the magnetization recovery curves, while changing the time of echo formation (echo time TE) allows to sample data when the difference in signal intensity is the largest (see also Section 3.5. Pulse Sequences).

field strength seeks to balance scanner costs, patient accessibility, SNR, scan time, and image quality.

3.4 Image formation and encoding in k-space

Transforming a magnetic resonance signal into an MR image involves two key steps: applying gradient fields to encode magnetization in the spatial frequency domain and performing a Fourier transform on the acquired data to reconstruct the spatially resolved image. Various sampling trajectories can be employed in this process, each with distinct advantages and limitations, which are discussed in this chapter.

3.4.1 Gradients and spatial encoding

MRI signal detection alone does not directly produce a spatially resolved image, as it lacks inherent spatial localization. In other words, the detected signal does not inherently contain the positional information needed to form an image composed of pixels (in 2D) or voxels (in 3D) with specific spatial resolution. To achieve spatial encoding, gradient coils are integrated into the MRI system. These coils generate controlled magnetic field gradients that can be switched on or off during the scan. Positioned around the bore of the magnet (Fig. 3.6a), they introduce spatial variations in the magnetic field, allowing the MRI signal to be encoded with location-specific information.

Magnetic gradients can be switched on or off to generate a controlled variation in the magnetic field along any spatial direction. An MRI sequence diagram defines the

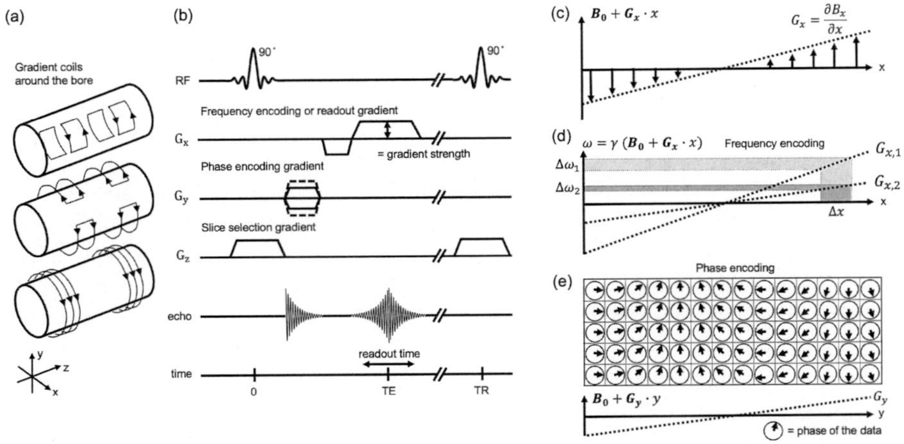

FIGURE 3.6

(a) Gradient coils are positioned around the bore of the magnet, generating three additional magnetic fields that vary linearly along the x-, y-, and z-axes of the scanner. (b) Pulse sequence diagram: The top row shows the RF pulse and flip angle, while the second to fourth rows display the activation timing of the G_x, G_y, and G_z gradients, respectively. The fifth row indicates the time point of echo formation. The height of each gradient waveform represents its relative strength. The scheme spans one repetition time (TR) and is repeated for multiple phase-encoding steps, with the phase-encoding gradient (G_y) varying in strength across repetitions, as indicated by the dotted line in the G_y gradient scheme. (c) Gradient strength and spatial encoding: When a gradient field is applied, the magnetic field strength along that axis varies linearly. This gradient is superimposed on the main magnetic field (B_0) and alters the local magnetic field along its direction. The gradient strength refers to the proportionality constant governing this change in field strength. (d) Frequency-encoding: During frequency-encoding, different positions along the x-axis (in this example) experience distinct local magnetic field strengths, causing spins at various locations to resonate at different frequencies. Applying an inverse Fourier transform (IFT) along this direction reconstructs spatial information into the final image. (e) Phase-encoding: During phase-encoding, a phase shift is induced in the signal due to the application of varying G_y gradients. These phase-encoding steps must be repeated multiple times, with each step corresponding to a unique point in k-space.

timing, order, and axis of gradient application (Fig. 3.6b). These additional magnetic fields, $\boldsymbol{B}_x$, $\boldsymbol{B}_y$, and $\boldsymbol{B}_z$ are superimposed on the main magnetic field $\boldsymbol{B}_0$ (Fig. 3.6c).

Each gradient encodes spatial information along a specific direction $\boldsymbol{r} = (x, y, z)$ by applying a magnetic field that varies linearly with the position along that axis. The proportionality constants defining the strength of these field variations are given by G_x, G_y, and G_z:

$$G_x = \frac{\partial B_x}{\partial x}, G_y = \frac{\partial B_y}{\partial y}, G_z = \frac{\partial B_z}{\partial z} \tag{3.19}$$

The introduction of additional gradient fields causes the precession frequencies of the nuclear spins to vary across space (Fig. 3.6d). This establishes a relationship between frequency and spatial location, expressed as follows:

$$\omega(\boldsymbol{r}) = \frac{\gamma}{2\pi} \cdot (\boldsymbol{B_0} + \boldsymbol{G_r} \cdot \boldsymbol{r}) \tag{3.20}$$

where $\boldsymbol{G_r} = (G_x, G_y, G_z)$ represents the gradient vector. This principle enables two essential steps in MRI data acquisition. Frequency-encoding (Fig. 3.6d) exploits differences in precession frequencies along a specific spatial direction to localize signal. Since frequency-encoding can only be applied along one axis at a time, phase encoding (Fig. 3.6e) is introduced to further encode spatial information within the phase of the MRI signal. For three-dimensional imaging, both frequency-encoding and phase-encoding gradients are applied in all three spatial directions, ensuring full spatial localization, using the following:

- **Slice-selection gradients**, which refer to a frequency-encoding gradient typically applied along the z-axis. This is accomplished by simultaneously applying a gradient field (G_z) and an RF excitation with a frequency matching that of the spins in the desired location. In 3D scans, where a volume of data is acquired instead of just a single slice, slice selection can incorporate both frequency-encoding and phase-encoding techniques.
- **Phase-encoding gradients**, which are applied between RF excitation and echo formation, usually along the y-axis. By activating the G_y gradient for a specific duration, spin dephasing occurs with a linear dependency along the y-axis, enabling the distinction of signal sources based on their spatial location.
- **Frequency-encoding gradients**, which are also known as the readout gradient and are applied during signal reception (readout) using the G_x gradient.

While 2D imaging is generally faster and allows a higher in-plane resolution, 3D MRI sequences provide valuable volumetric information and flexible reconstruction options that aid in accurate diagnosis and surgical planning.

3.4.2 K-space formalism

Magnetic field gradients modulate the MR signal during MRI data acquisition, mapping spatial information into k-space. The receiver coils capture time-varying voltages induced by the free induction decay (FID) and echo signals as complex-valued data throughout the readout period. K-space represents the spatial frequency domain in which the sampled MR signals are organized during the data acquisition. It is mathematically analogous to Fourier space and is typically visualized as a 2D or 3D Cartesian grid, with each point corresponding to a unique spatial frequency component (Fig. 3.7a).

The trajectory in k-space $k_{\boldsymbol{r}}(t)$ along an arbitrary direction $\boldsymbol{r}$ is defined by integrating over the time-dependent gradient field $G(\boldsymbol{r}, t)$:

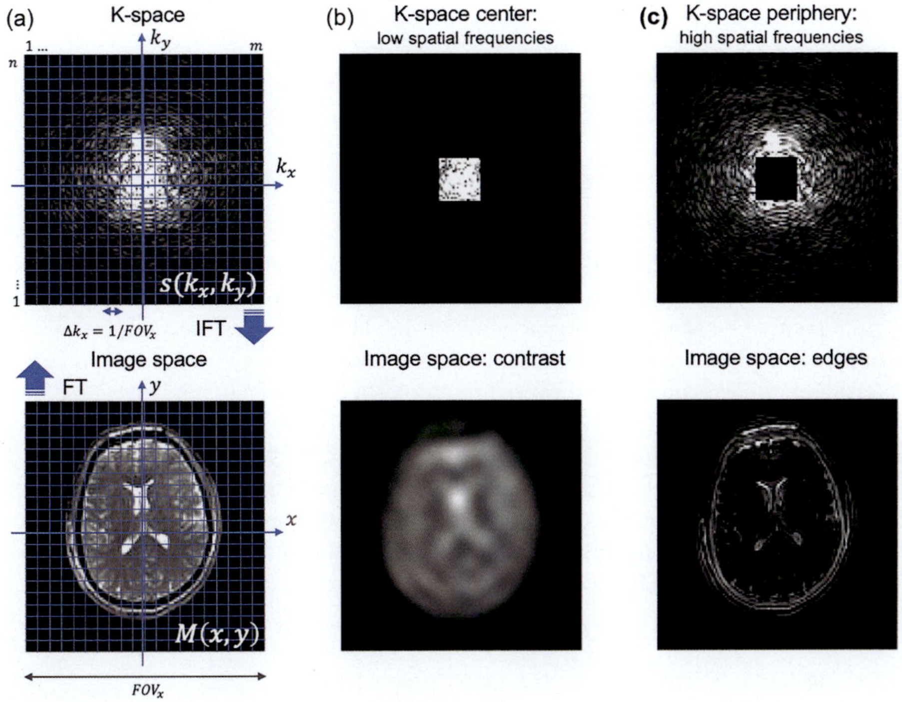

FIGURE 3.7

(a) The MRI image (bottom row) is obtained by applying the inverse Fourier transform (IFT) to the raw k-space data (top row). In k-space, the frequency-encoding steps during the sampling time Δt are represented by $k_x = \gamma G_x\, m\Delta t$, where m denotes the number of frequency-encoding steps. Similarly, the phase-encoding steps for a gradient duration t are given by $k_y = \gamma n \Delta G_y t$, where n represents the number of phase-encoding steps. (b) The central region of k-space contains low spatial frequencies, which, after the IFT, determine the overall image contrast and coarse structural features. (c) The outer regions of k-space contain high spatial frequencies, encoding fine details and sharp edges in the final image.

$$k_r(t) = \frac{\gamma}{2\pi} \int_0^t G(\mathbf{r}, t)\, dt. \tag{3.21}$$

In Cartesian sampling, k-space is typically described in spatial frequency coordinates k_x and k_y, corresponding to the readout and phase-encoding directions, respectively. The **acquired MRI data is collected in k-space as** the signal $s(k_x, k_y)$.

K-space is related to image space through the forward and inverse Fourier transform. The inverse Fourier transform (IFT) on $s(k_x, k_y)$ converts the acquired k-space data into the spatial image $M(x, y)$:

$$M(x,y) = \int s(k_x,k_y)e^{i2\pi(k_x x + k_y y)}dk_x dk_y \tag{3.22}$$

The **FT** converts the spatial image $M(x,y)$ (or the magnetization distribution) into its frequency (k-space) representation $s(k_x,k_y)$:

$$s(k_x,k_y) = \int M(x,y)e^{-i2\pi(k_x x + k_y y)}dxdy \tag{3.23}$$

These operations are typically performed using fast Fourier transform (FFT) algorithms [14], which efficiently process discrete k-space data. The signal intensity at each voxel is then mapped to a grayscale value to generate the final image.

A feature of k-space is the organization of spatial frequencies. Readouts acquired near the center of k-space contain low spatial frequencies (Fig. 3.7b), primarily determining image contrast and overall intensity. In comparison, readouts at the periphery contain high spatial frequencies (Fig. 3.7c), which encode fine details and sharp edges.

When multiple receiver coils are used, the coil sensitivity operator $C(k_x,k_y)$ must be included in the reconstruction process. This operator accounts for variations in the sensitivity of different coils that affect how the signal is received at different locations in k-space. The **IFT** with coil sensitivity is expressed as follows:

$$M(x,y) = \int C(k_x,k_y)s(k_x,k_y)e^{i2\pi(k_x x + k_y y)}dk_x dk_y \tag{3.24}$$

Coil sensitivity, in combination with k-space sampling, enables various undersampling and reconstruction techniques, including parallel imaging [15,16] and compressed sensing [17]. Further details on image reconstruction will be discussed in Chapter 7.

3.4.3 K-space trajectories and artifacts

The dynamic modulation of the gradient fields progressively fills the k-space with data. However, the scan must be repeated for the number of phase-encoding steps, with a repetition time TR. Because raw image data is stored in k-space, where a single point contains information from all the points in image space, the order of the sampling data points can be arbitrary, providing multiple trajectory possibilities. The choice of k-space sampling trajectory significantly impacts image quality and scan efficiency, motivating the design of trajectories tailored to specific imaging needs.

Cartesian trajectories (Fig. 3.8a and b) sample k-space along an orthogonal grid of k-space. Each point along the phase-encoding direction k_y (and k_z in 3D) is sampled within a TR. Cartesian sampling patterns are straightforward to implement and reconstruct, making them the standard for most MRI sequences. However, Cartesian patterns can suffer from motion and flow-induced errors [19], leading to blurring or ghosting artifacts if motion is not handled effectively.

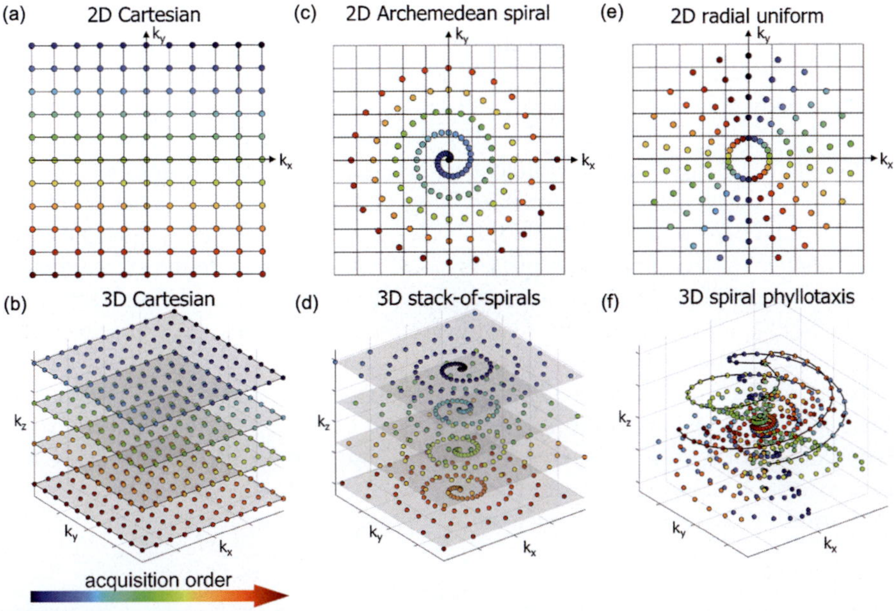

FIGURE 3.8

(a) Cartesian sampling trajectories are the most commonly used MRI acquisition schemes since they are acquired on an orthogonal grid. In 2D imaging, k-space is typically sampled line by line, with one row (frequency-encoding direction k_X) acquired per TR. (b) 3D Cartesian acquisitions repeat the 2D trajectory, either through slice excitation or by using a second phase-encoding gradient k_Z. (c) 2D spiral sampling of k-space is achieved within one TR. The center of the grid is sampled more densely than the periphery. The k-space samples must be matched to a Cartesian grid during reconstruction, accounting for this variable sampling density. (d) 3D stack-of-spirals similarly repeat the spiral trajectory for each slice but interleave the pattern by adding an in-plane rotation from one slice to the next. (e) In 2D radial sampling, radial lines go from one periphery of k-space to the other, always crossing the k-space center. (f) With the 3D phyllotaxis pattern [18], radial segments make 3D spirals (black lines) interleaved with a rotation of the golden angle $\theta = 137.51°$. The color-coding arrow on the bottom left indicates a conventional temporal sampling order for each trajectory type.

Spiral [20] and radial [21] trajectories (Fig. 3.8c–f) do not have frequency-encoding or phase-encoding steps along any of the spatial axes. Their readout follows a spiral or radial trajectory, respectively. This characteristic leads to motion artifacts manifesting as diffuse noise across the image, which is less disruptive compared with Cartesian imaging. In addition, the repeated sampling of the k-space center can be used for additional corrections during image reconstruction. 3D radial trajectories (Fig. 3.8f) offer the possibility of high-resolution isotropic imaging. Continuous, uniform data sampling over time can be useful for time-resolved (e.g., cardiac) imaging and advanced image reconstruction pipelines [21]. However, non-Cartesian trajectories exhibit reduced scanning efficiency and increased susceptibility to hardware imperfections, such as gradient delays, trajectory infidelities [22], and eddy currents [23].

Reconstruction of non-Cartesian MRI data requires additional steps, such as applying a density compensation function to account for variable sampling density sampling – i.e., to prevent the overweighting of low frequencies at the center of k-space – and performing a convolution with a windowing kernel for interpolation onto a regular Cartesian grid (in a process called *gridding*) prior to the inverse Fourier transform (FT). Several nonuniform FFT (nuFFT) algorithms have been proposed [24,25], often complemented with postprocessing steps to ensure accurate image reconstruction.

Image artifacts can occur in all forms of 2D, 3D, Cartesian, or non-Cartesian imaging and are distortions or erroneous features in reconstructed images. They arise from a variety of sources, including intrinsic properties of the imaged object, acquisition errors, or reconstruction inaccuracies [26–28]. Understanding and mitigating these artifacts is crucial for the accurate interpretation of MRI data.

3.5 Pulse sequences

Image formation does not solely rely on free induction decays. Robust signals can also be generated by realigning dephased spins, a process known as echo formation. This can be achieved in two ways: (1) applying an additional refocusing RF excitation pulse to generate a spin echo (SE) or (2) superimposing magnetic field gradients onto the B_0 and B_1 fields to produce a gradient echo (GRE).

An MRI pulse sequence is a programmed set of instructions that dictate RF and gradient field changes. Sequence diagrams display the most basic block of imaging instructions, which is repeated multiple times to obtain the full image. The timing parameters, notably the repetition time TR (interval between two subsequent RF pulses) and the echo time TE (time elapsed between RF excitation and echo formation), can be manipulated to vary the final image contrast, to emphasize either on T_1, T_2, or proton density differences between adjacent tissues. MRI sequences broadly fall into two categories, SE and GRE sequences, based on the mechanism of how the echo is formed (Fig. 3.9a and b). In clinical examinations, different sequences are subsequently used to visualize a wide range of anatomical and functional aspects, forming a comprehensive MRI protocol.

3.5.1 Spin echo sequences

SE sequences are widely used to mitigate the effects of magnetic field inhomogeneities ΔB. Applying a 180° refocusing pulse following an initial 90° RF excitation pulse, SE sequences correct for phase discrepancies among nuclear spins

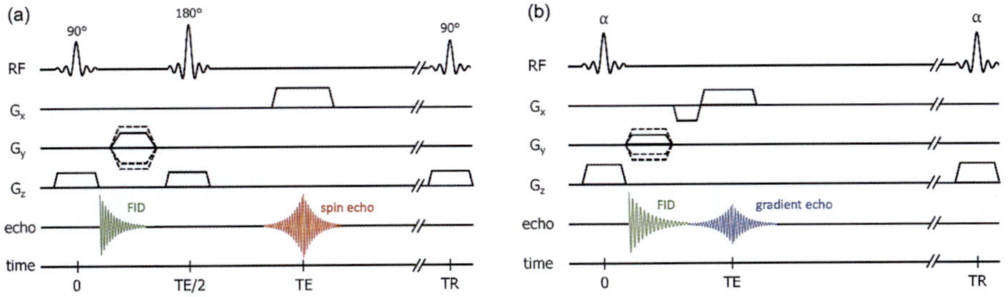

FIGURE 3.9

(a) In a 2D spin echo (SE) sequence diagram, a 90° RF pulse is followed by a 180° refocusing pulse. The slice-select gradient (G_Z) is turned on while the RF pulses are sent. The phase-encoding gradient (G_y) is applied in an orthogonal direction to the slice-select gradient and is turned on with a different amplitude (dashed lines) at each repetition time (TR). The spin echo is formed at t = TE, during which the readout gradient (G_X) is turned on. (b) In a basic (unspoiled) gradient echo (GRE) sequence, a single RF pulse of flip angle α is played each TR, allowing to reach shorter TE than with SE sequences. Typically, the flip angles used in GRE are lower than 90°, which can result in moderate SNR and decreased T_1 contrast compared with SE. Note that the term "GRE sequence" is interchangeably used to refer to the subclass of spoiled GRE sequences: they employ a combination of RF spoiling (RF phase changes at each TR) and gradient spoiling (addition of a G_Z gradient just before the next RF pulse) to eliminate the residual transverse magnetization. While GRE sequences are typically faster than SE sequences, they are more sensitive to susceptibility changes.

(Fig. 3.10). This refocusing pulse is applied midway between the initial excitation and the SE signal acquisition, effectively reversing the phase dispersion caused by magnetic field inhomogeneities. The initial FID decays with T_2^*, but the resulting echo at $t = TE$ provides a signal predominantly influenced by the T_2 relaxation properties of the tissue, free from the additional loss in transverse magnetization induced by ΔB (Eq. 3.17).

3.5.2 Gradient echo sequences

GRE sequences diverge from the SE approach by utilizing gradient fields instead of a refocusing RF excitation pulse to generate echoes (Fig. 3.9). By altering the gradient direction, GRE sequences induce dephasing and subsequent rephasing of the spins, creating a GRE signal. Unlike SE sequences, GRE sequences do not compensate for magnetic field inhomogeneities, making them more susceptible to variations in magnetic susceptibility across different tissues. However, GRE sequences can achieve faster imaging times and have greater flexibility in adjusting image contrast, mainly through manipulating the RF excitation angle and echo time. GRE sequences are characterized by their sensitivity to T_2^* decay, which includes intrinsic spin–spin relaxation and dephasing due to field inhomogeneities.

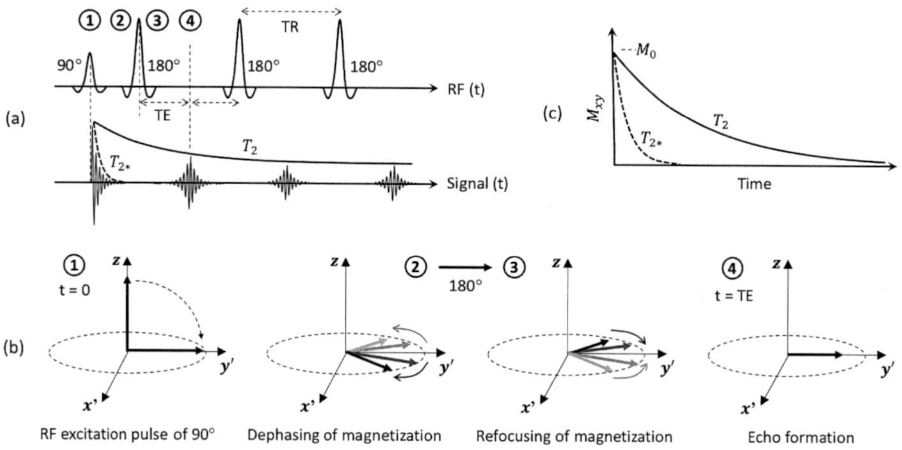

FIGURE 3.10

(a) A typical spin echo sequence involves the application of a 90° RF excitation pulse followed by several 180 degrees RF excitation pulses. (b) After the initial pulse, the transverse magnetization decays with T_2^* due to the dephasing of individual nuclear spins caused by both static and dynamic magnetic field inhomogeneities . The 180° refocusing pulse rotates all magnetization vectors, which leads to the rephasing of the nuclear spins in the second half of the echo time, counteracting the effects of static magnetic field inhomogeneities and frequency offsets. At the echo time $t = TE$ is reached, all nuclear spins have rephased, forming a spin echo signal.

3.5.3 Balanced steady-state free precession

Balanced steady-state free precession (bSSFP) sequences are rapid GRE sequences characterized by a high signal-to-noise ratio. The distinguishing feature of bSSFP is its capacity to preserve the transverse magnetization M_{xy} from one repetition time TR to the next, contrarily to other GRE sequences in which the transverse component is nulled using either gradient or RF spoilers. When a short TR is chosen (i.e., TR $\ll T_2 < T_1$), magnetization dynamics exhibit complex oscillatory patterns for a finite duration, known as the transient state [29], before eventually reaching a steady state. The steady state corresponds to a constant transverse magnetization formed of both the pre-RF excitation echo reformation and the postexcitation ("FID-like") signals. Fully balanced gradients used along all three axes refocus both signal components at the midpoint of the TR interval. The bSSFP sequence is T_1/T_2-weighted. This mixed contrast causes fat to have high signal intensity, which accentuates the visibility of adipose tissue in the resulting images. Additionally, this T_1/T_2-weighting also offers high blood-to-myocardium contrast, which, combined with the high signal-to-noise and speed of bSSFP compared with other GRE sequences, makes the sequence highly suitable for cardiovascular applications, e.g., for cine MRI images.

bSSFP sequences are, however, sensitive to off-resonance, which leads to banding artifacts.

3.5.4 Inversion recovery

Inversion recovery (IR) sequences enhance tissue contrast or highlight specific tissue types and pathologies by selectively suppressing certain signals. These sequences begin with a 180° RF pulse, inverting the longitudinal magnetization (M_z), followed by a second RF pulse after a delay, known as the inversion time (TI). Since different tissue types recover their longitudinal magnetization at varying rates (due to distinct T_1 relaxation times), an appropriately chosen TI can selectively suppress specific tissues. For example, fat or fluids can be suppressed, improving the visibility of other tissues. This technique is particularly beneficial for brain and vascular imaging.

3.5.5 Signal weighting

In MRI, signal weighting derives from the intrinsic properties of tissue relaxation times and proton densities. It allows the production of discernible contrasts, facilitating the differentiation of anatomical structures and pathological conditions. T_1-weighted images are characterized by heightened signal intensity from tissues with short T_1 relaxation times, such as adipose-rich regions. Muscles appear with intermediate signal intensities, while fluids have low intensities. T_1-weighted imaging offers good anatomical resolution and lesion conspicuity. Conversely, T_2-weighted images accentuate tissues with long T_2 relaxation times, typically resulting in hyperintense signals from fluid-filled spaces, which aid in detecting edema and tissue abnormalities. Proton density weighting refers to contrast based on tissue hydrogen proton density (i.e., the number of nuclei in a given imaging region) rather than displaying magnetic properties. It is less commonly used but can provide valuable insights into tissue composition. MR pulse sequence parameters can be tuned to achieve the desired signal weighting. The relevant parameters are the repetition time TR and echo time TE. While TR determines the T_1 weighting, TE determines the T_2 weighting (Fig. 3.11):

- **T_1-weighted imaging** requires a short TR value (and therefore a short TE value, too) so that the longitudinal magnetization recovery of different tissue types is minimized, and maximal contrast between tissues with long and short T_1 values can be obtained.
- **T_2-weighted imaging** requires a long TE setting to accentuate differences in transverse magnetization decays.
- **Proton density (PD) weighted imaging** is achieved by setting a long TR to minimize T_1 differences between tissue types and a relatively short TE to minimize T_2 differences. This ensures a balanced representation of proton density across various tissues.

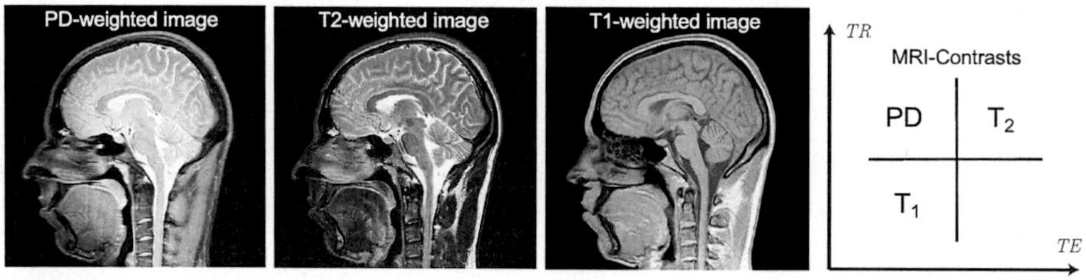

FIGURE 3.11

Signal weighted using a spin echo sequence for brain imaging.

3.5.6 Specialized MRI sequences

MRI sequences can be broadly categorized into qualitative and quantitative sequences. Qualitative sequences are primarily used to provide tissue contrast for anatomical imaging, but the values of the imaging voxels do not correspond to reproducible physical measurements. In contrast, quantitative MRI addresses this limitation by using specialized sequences to encode physical properties, such as T_1 and T_2 relaxation times (in milliseconds), blood flow velocity (in cm/second), diffusion coefficients (in cm^2/second), and magnetic susceptibility (in ppm). This approach enables reproducible and quantifiable measurements of tissue properties. From the broad categories of MRI pulse sequences discussed earlier, several variants have emerged. Besides classical contrast mechanisms based on relaxation time differences, MRI can also explore other biophysical and microstructural properties of biological tissues.

3.5.6.1 Relaxometry

Relaxometry includes MRI sequences designed to quantitatively measure T_1 and T_2 relaxation times [30]. These measurements are crucial for assessing tissue characteristics and detecting pathological changes, such as edema or fibrosis, which can alter relaxation times.

3.5.6.2 Diffusion sequences

Diffusion sequences measure the random Brownian motion of water molecules in tissue [31]. Diffusion-weighted imaging (DWI) measures diffusivity and shows areas of restricted diffusion, useful for stroke imaging, while diffusion tensor imaging (DTI) encodes the directionality of diffusion, enabling fiber tractography in brain imaging. Both sequences rely on diffusion-induced transverse relaxation, which causes irreversible signal loss. For diffusion imaging, a pulsed SE sequence with two additional diffusion gradients $\mathbf{G}_{\text{diff}}$ of duration δ is typically used. The sequence consists of a diffusion gradient applied after SE excitation to induce spatially dependent phase shifts. Then a 180° RF pulse reverses the phase shifts and a second diffusion gradient is applied after a diffusion time Δ, introducing phase offsets proportional to the molecular motion along the gradient direction.

3.5.6.3 Flow and perfusion sequences

Flow sequences [32] encode the coherent motion of flowing blood into the MRI signal's phase. Motion-sensitive bipolar gradients create a net phase accumulation of zero for stationary spins, whereas spins moving along the gradient direction accumulate phase proportional to their velocity. These gradients can be applied along each gradient axis, known as 4D flow MRI [33] enabling the reconstruction and visualization of blood flow velocity as 3D streamlines. This method is crucial for assessing blood flow patterns or angiograms in patients with congenital heart disease, valvular diseases, and vascular disorders. Perfusion MRI [34] can investigate the blood supply to tissues, helping to identify perfusion deficits in conditions such as stroke, myocardial ischemia, and tumor vascularity. Arterial spin labeling uses magnetically labeled arterial blood as a tracer for perfusion without the need of contrast agents.

3.5.6.4 Quantitative susceptibility mapping

Quantitative susceptibility mapping (QSM) [35,36] is an MRI technique that utilizes the phase of the MRI signal to measure tissue magnetic susceptibility, providing valuable insights into tissue composition, such as iron content and structural changes. This technique is beneficial for assessing regions with varying magnetic susceptibility, such as deep brain structures, and can be applied to study conditions involving abnormal iron accumulation, such as neurodegenerative diseases. Quantitative susceptibility-weighted imaging (SWI), a related approach, generates susceptibility-weighted magnitude images emphasizing differences in tissue magnetic susceptibility.

3.5.6.5 Elastography

MR elastography [37] assesses tissue elasticity by applying a harmonic mechanical excitation to the body, creating waves that propagate through the tissue, which are detected by changes in the phase of the MRI signal. This technique is particularly valuable in liver studies and cancer diagnosis, where tissue stiffness can serve as a biomarker for disease progression.

3.5.6.6 Functional MRI

Functional MRI (fMRI) [38] detects changes in blood oxygenation and flow, which occur in response to neural activity, allowing for the mapping of functional brain areas involved in movement, language, and vision. This is primarily achieved through blood oxygen level–dependent (BOLD) contrast [39].

3.5.6.7 Magnetization transfer

Magnetization transfer (MT) [40] exploits the exchange of magnetization between protons in free water and protons bound to macromolecules, such as proteins and lipids. This exchange process can provide insight into tissue composition and integrity. MT imaging is particularly useful for detecting changes in brain tissue, such as myelin loss and altered macromolecular content, making it valuable in conditions such as multiple sclerosis and other neurodegenerative diseases.

3.5.6.8 Chemical exchange saturation transfer

Chemical exchange saturation transfer (CEST) [41] detects proton exchange between water and other molecules, such as creatine, allowing for the imaging of metabolic processes and offering insights into tissue biochemistry. It can be used to detect pH and glucose changes, and cancer biomarkers.

3.5.6.9 Spectroscopic imaging and metabolic imaging

Spectroscopic imaging [42] provides a chemical spectrum from each voxel, offering insights into tissue metabolic composition. This technique is particularly useful in neurological and oncological imaging to detect alterations in metabolite concentrations associated with diseases. Metabolic imaging [43] is an MRI technique that evaluates tissue metabolism, providing insights into biochemical processes within the body, with applications in cancer, neurology, and exercise physiology.

3.5.6.10 Thermometry

Thermometry [44] enables noninvasive temperature mapping of tissues using MRI, which is useful in monitoring therapeutic interventions.

The continuous evolution of MRI techniques, driven by advances in hardware and software, has significantly expanded MRI's clinical applications. From detailed anatomical imaging to functional and metabolic assessments, MRI now plays an indispensable role in diagnosis, treatment planning, and research. As technology progresses, more sophisticated sequences will emerge, providing deeper insights into human anatomy and physiology, further revolutionizing medical diagnostics and therapeutic monitoring.

References

[1] Purcell EM, Torrey HC, Pound RV. Resonance absorption by nuclear magnetic moments in a solid. Phys Rev 1946;69:37.
[2] Bloch F. The principle of nuclear induction. Science 1979;1953(118):425–30.
[3] Proctor WG, Yu FC. The dependence of a nuclear magnetic resonance frequency upon chemical compound. Phys Rev 1950;77:717.
[4] Dickinson WC. Dependence of the F 19 nuclear resonance position on chemical compound. Phys Rev 1950;77:736.
[5] Damadian R. Tumor detection by nuclear magnetic resonance. Science 1979;1971 (171):1151–3.
[6] Lauterbur PC. Image formation by induced local interactions: examples employing nuclear magnetic resonance. Nature 1973;242:190–1.
[7] Mansfield P, Grannell PK. NMR'diffraction'in solids? J Phys C Solid State Phys 1973; 6:L422.

[8] Haacke EM, Brown RW, Thompson MR, Venkatesan R. Haacke—magnetic resonance imaging - physical principles and sequence Design.pdf. J Appl Phys 1999:914.
[9] Bloch F. Nuclear induction. Phys Rev 1946;70:460.
[10] Hahn EL. Spin echoes. Phys Rev 1950;80:580–94.
[11] Messroghli DR, Moon JC, Ferreira VM, Grosse-Wortmann L, He T, Kellman P, et al. Clinical recommendations for cardiovascular magnetic resonance mapping of T1, T2, T2* and extracellular volume: a consensus statement by the Society for Cardiovascular Magnetic Resonance (SCMR) endorsed by the European Association for Cardiovascular Imaging (EACVI). J Cardiovasc Magn Reson 2016;19:75.
[12] Dekkers IA, Lamb HJ. Clinical application and technical considerations of T 1 \& T 2 (*) mapping in cardiac, liver, and renal imaging. Br J Radiol 2018;91:20170825.
[13] Perman WH, Hilal SK, Simon HE, Maudsley AA. Contrast manipulation in NMR imaging. Magn Reson Imaging 1984;2:23–32.
[14] Cooley JW, Tukey JW. An algorithm for the machine calculation of complex Fourier series. Math Comput 1965;19:297–301.
[15] Pruessmann KP, Weiger M, Scheidegger MB, Boesiger P. SENSE: sensitivity encoding for fast MRI. Magn Reson Med 1999;42:952–62.
[16] Griswold MA, Jakob PM, Heidemann RM, Nittka M, Jellus V, Wang J, et al. Generalized autocalibrating partially parallel acquisitions (GRAPPA). Magn Reson Med 2002; 47:1202–10.
[17] Lustig M, Donoho D, Pauly JM. Sparse MRI: the application of compressed sensing for rapid MR imaging. Magn Reson Med 2007;58:1182–95.
[18] Bianchi V, Yerly J, Chaptinel J, Yvan M, Prsa M, Alamo L, et al. Small structures, big challenges: fetal cardiac magnetic resonance imaging. Adv Clin Imaging Technol 2019: 66–70.
[19] Zaitsev M, Maclaren J, Herbst M. Motion artifacts in MRI: a complex problem with many partial solutions. J Magn Reson Imag 2015;42:887–901.
[20] Block KT, Frahm J. Spiral imaging: a critical appraisal. J Magn Reson Imag 2005;21: 657–68.
[21] Feng L. Golden-angle radial MRI: basics, advances, and applications. J Magn Reson Imag 2022;56:45–62.
[22] Tan H, Meyer CH. Estimation of k -space trajectories in spiral MRI. Magn Reson Med 2009;61:1396–404.
[23] Moussavi A, Untenberger M, Uecker M, Frahm J. Correction of gradient-induced phase errors in radial MRI. Magn Reson Med 2014;71:308–12.
[24] Dutt A, Rokhlin V. Fast fourier transforms for nonequispaced data. SIAM J Sci Comput 1993;14:1368–93.
[25] Liu QH, Nguyen N. An accurate algorithm for nonuniform fast Fourier transforms (NUFFT's). IEEE Microw Guid Wave Lett 1998;8:18–20.
[26] Smith TB, Nayak KS. MRI artifacts and correction strategies. Imag Med 2010;2: 445–57.
[27] Bekiesińska-Figatowska M. Artifacts in magnetic resonance imaging. Pol J Radiol 2015;80:93–106.
[28] Zhuo J, Gullapalli RP. MR artifacts, safety, and quality control. Radiographics 2006;26: 275–97.
[29] Scheffler K. On the transient phase of balanced SSFP sequences. Magn Reson Med 2003;49:781–3.

[30] Damadian R. Tumor detection by nuclear magnetic resonance. Science (1979) 1971; 171:1151–3.
[31] Stejskal EO, Tanner JE. Spin diffusion measurements: spin echoes in the presence of a time-dependent field gradient. J Chem Phys 1965;42:288–92.
[32] Bryant DJ, Payne JA, Firmin DN, Longmore DB. Measurement of flow with NMR imaging using a gradient pulse and phase difference technique. J Comput Assist Tomogr 1984;8:588–93.
[33] Markl M, Frydrychowicz A, Kozerke S, Hope M, Wieben O. 4D flow MRI. J Magn Reson Imag 2012;36:1015–36.
[34] Detre JA, Leigh JS, Williams DS, Koretsky AP. Perfusion imaging. Magn Reson Med 1992;23:37–45.
[35] Marques JP, Bowtell R. Application of a Fourier-based method for rapid calculation of field inhomogeneity due to spatial variation of magnetic susceptibility. Concepts Magn Reson B Magn Reson Eng 2005;25B:65–78.
[36] de Rochefort L, Liu T, Kressler B, Liu J, Spincemaille P, Lebon V, et al. Quantitative susceptibility map reconstruction from MR phase data using bayesian regularization: validation and application to brain imaging. Magn Reson Med 2010;63:194–206.
[37] Muthupillai R, Lomas DJ, Rossman PJ, Greenleaf JF, Manduca A, Ehman RL. Magnetic resonance elastography by direct visualization of propagating acoustic strain waves. Science (1979) 1995;269:1854–7.
[38] Ogawa S, Lee T, Nayak AS, Glynn P. Oxygenation-sensitive contrast in magnetic resonance image of rodent brain at high magnetic fields. Magn Reson Med 1990;14:68–78.
[39] Ogawa S, Lee TM, Kay AR, Tank DW. Brain magnetic resonance imaging with contrast dependent on blood oxygenation. In: Proceedings of the National Academy of Sciences. vol 87; 1990. p. 9868–72.
[40] Wolff SD, Balaban RS. Magnetization transfer contrast (MTC) and tissue water proton relaxation *in vivo*. Magn Reson Med 1989;10:135–44.
[41] Ward KM, Aletras AH, Balaban RS. A new class of contrast agents for MRI based on proton chemical exchange dependent saturation transfer (CEST). J Magn Reson 2000; 143:79–87.
[42] Bottomley PA, Foster TH, Argersinger RE, Pfeifer LM. A review of normal tissue hydrogen NMR relaxation times and relaxation mechanisms from 1–100 MHz: dependence on tissue type, NMR frequency, temperature, species, excision, and age. Med Phys 1984;11:425–48.
[43] Chance B, Eleff S, Leigh JS, Sokolow D, Sapega A. Mitochondrial regulation of phosphocreatine/inorganic phosphate ratios in exercising human muscle: a gated 31P NMR study. Proc Natl Acad Sci 1981;78:6714–8.
[44] Depoorter J, Dewagter C, Dedeene Y, Thomsen C, Stahlberg F, Achten E. The proton-resonance-frequency-shift method compared with molecular diffusion for quantitative measurement of two-dimensional time-dependent temperature distribution in a phantom. J Magn Reson B 1994;103:234–41.

PART

2

MR image acquisition

CHAPTER

4 Adjust to your imaging scenario: Learning and optimizing MRI data sampling

Marcelo V. W. Zibetti[1],* and Ravinder R. Regatte[2]

[1]*Center for Biomedical Imaging, Department of Radiology, NYU Grossman School of Medicine, New York University, New York, NY, United States;* [2]*NYU Grossman School of Medicine, New York, NY, United States*

**Corresponding author: E-mail: Marcelo.WustZibetti@nyulangone.org*

4.1 MRI data acquisition: Pulse sequences, k-space trajectories, and data sampling

4.1.1 Pulse sequences: Excitation, encoding, and sampling

The MRI data acquisition process is a sequential process controlled by a series of signals, called a pulse sequence [1], as illustrated in Fig. 4.1a. The pulse sequence contains all the controlling signals for the radio-frequency (RF) system and magnetic field gradient systems necessary for data acquisition. These signals are used to (1) prepare the magnetization of the object, (2) excite a 3D volume or a 2D slice that will resonate emitting a measurable response, and (3) apply the appropriate encoding scheme to the response signals to allow for later 2D or 3D image reconstruction.

As seen in Fig. 4.1a, part 1 of the pulse sequence is the magnetization-preparation step, which is responsible for generating the necessary contrast in the images. This part depends on the specific type of MRI exam; for example, the inverse recovery preparation used to produce T_1-weighted images [1]. The specific signals used and the duration of this part in the pulse sequence depend on the type of preparation. In Part 2 of the pulse sequence, the RF pulse excites spins of a volume or slice, in a physical setup as seen in Fig. 4.1b. In this example, the G_z gradient is used to change the local precessional frequency so that the RF pulse will excite only the frequencies corresponding to a particular slice. The excited spins resonate, emitting signals that are measured by the RF coils. These measurements are averages of the signals coming from all excited voxels in the object. To separate them later, MRI scanners use a spatial data encoding scheme employing the spatially linear magnetic gradient system, which is controlled by Part 3 of the pulse sequence.

Machine Learning in MRI. https://doi.org/10.1016/B978-0-443-14109-6.00004-3

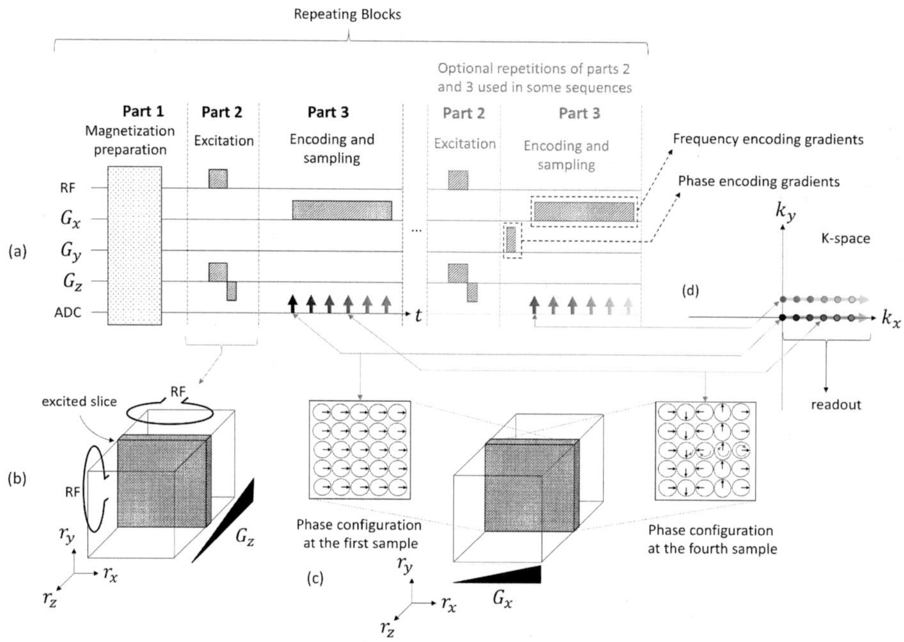

FIGURE 4.1

(a) A pulse sequence controlling MRI data acquisition is composed of three main parts: (1) magnetization preparation, (2) volume or slice excitation, and (3) spatial encoding and data sampling. In some sequences, parts 2 and 3 may be repeated before going back again to the first part. (b) An illustration of the physical setup, showing the RF coils and the magnetic gradient G_z in relation to the object and the excited slice, as used in Part 2 of the pulse sequence. (c) Illustration of the change in the phase in the voxels of the excited slice along the r_x axis, controlled by the activation of the gradient G_x. (d) The corresponding sampling positions in the k-space when the analog-to-digital converter (ADC) is activated.

The gradients change the local precession frequency, continually changing the phase of the spins in different spatial locations over time, as seen in Fig. 4.1c. As data are acquired, a different spatial phase configuration is applied, creating an encoding system that allows for image reconstruction.

The spatial encoding is parametrized by a frequency parameter $\boldsymbol{k}$, leading to the notion of k-space, which corresponds to the Fourier transform [2] of the signals coming from the excited object:

$$s(\boldsymbol{k}) = \int_{object} \rho(\boldsymbol{r}) e^{-i2\pi \boldsymbol{k}\cdot\boldsymbol{r}} d\boldsymbol{r}, \tag{4.1}$$

where $s(\boldsymbol{k})$ is the measured signal at 3D k-space position $\boldsymbol{k} = [k_x\ k_y\ k_z]$. The measured signal is the sum of the signals from all excited voxels of the object $\rho(\boldsymbol{r})$, with 3D spatial position $\boldsymbol{r} = [r_x\ r_y\ r_z]^T$, where superscript T means transpose.

At each moment in time, during encoding and sampling, the activated magnetic gradients change the spatial phase distribution [3], which corresponds to moving around in the k-space, as seen in Fig. 4.1d, following a k-space trajectory $\boldsymbol{k}(t)$, given by:

$$\boldsymbol{k}(t) = \frac{\gamma}{2\pi} \int_0^t \boldsymbol{G}(\tau) d\tau, \tag{4.2}$$

where $\boldsymbol{G}(t) = [G_x(t)\ G_y(t)\ G_z(t)]$ is the magnetic gradients of the pulse sequence as a function of time.

Using this encoding system, once all k-space data are collected, an inverse Fourier transform can be used to retrieve the image of the object $\rho(\boldsymbol{r})$ from each spatial location.

4.1.2 Requirements for sampling the k-space

The sampling scheme requirements begin with the definition of the voxel size $\boldsymbol{\Delta r} = [\Delta r_x\ \Delta r_y\ \Delta r_z]^T$ and field of view (FOV) in all three axes of the volume, where $\text{FOV} = [\text{FOV}_x\ \text{FOV}_y\ \text{FOV}_z]^T$. The number of voxels within the volume is $N = N_x \times N_y \times N_z$, as seen in Fig. 4.2. The smaller the voxel size, the larger the k-space frequency that needs to be acquired, which must satisfy $k_{xmax} \geq 1/2\Delta r_x$ for the x-axis, for example. In addition, the larger the FOV, the smaller the sampling spacing required to avoid aliasing, or superposition, of the object, which must satisfy $\Delta k_x \leq 1/FOV_x$ for the x-axis, for example. The sampling spacing is defined by $\boldsymbol{\Delta k} = [\Delta k_x\ \Delta k_y\ \Delta k_z]$. See more details in Ref. [2].

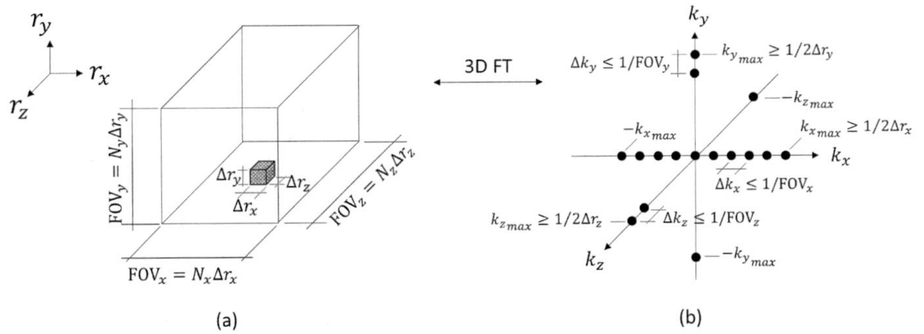

FIGURE 4.2

Discrete 3D volume and 3D k-space sampling requirements. In (a), the imaged volume is given by the FOV, and the voxel size is given by $\boldsymbol{\Delta r}$. In (b), the 3D k-space, with a maximum frequency k_{max} and sampling space $\boldsymbol{\Delta k}$.

4.1.3 Gradient system, k-space trajectories, and data collection

Ideally, one needs all the k-space measurements to completely separate the signals coming from every voxel. However, after each excitation, only a small amount of the k-space samples can be collected before the integrity of the MR signal and the k-space encoding are severely affected by relaxation, undesired dephasing (other than the one controlled by the gradients), and other practical issues. A simple solution is to keep repeating the pulse sequence, changing the signals used in the gradient system with each repetition, to collect different k-space positions in different trajectories until all necessary k-space samples are collected. Fig. 4.1 shows a repetition of parts 2 and 3 of the pulse sequence elements to sample a second line in the k-space.

In general, there is a lot of freedom in how we can collect the k-space data. Usually, we classify the acquisition as Cartesian and non-Cartesian, as exemplified by Figs. 4.3 and 4.4. Cartesian acquisition typically uses linear trajectories with a uniform sampling scheme, as shown in Fig. 4.3b. Non-linear trajectories can also be used as long as k-space is sampled exactly when the trajectory crosses the Cartesian grid, as in echo-planar imaging (EPI) seen in Fig. 4.3c [4].

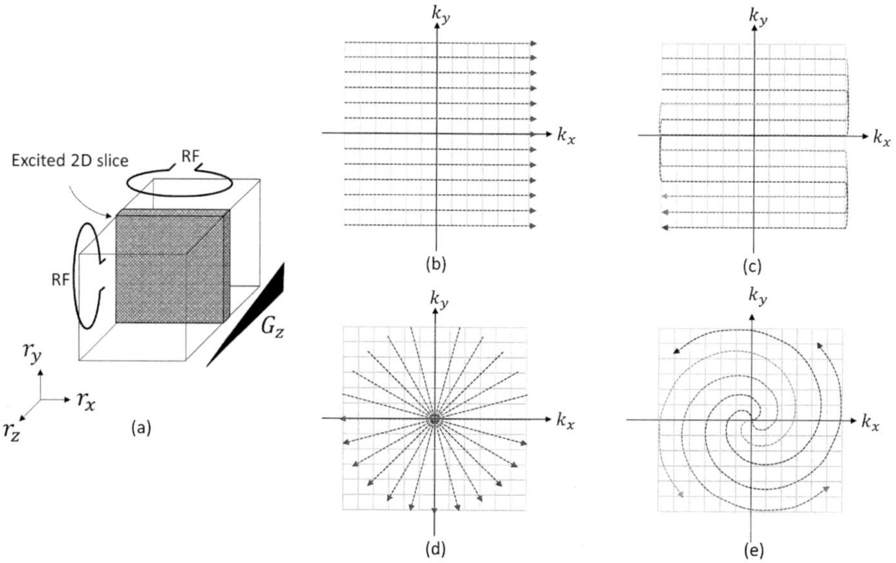

FIGURE 4.3

Representative examples of 2D Cartesian and non-Cartesian acquisitions. In (a), the setup used to excite a 2D slice is shown. In (b), Cartesian acquisition with several short linear trajectories. In (c), Cartesian acquisition with few long echo-planar imaging (EPI) trajectories. In (d), non-Cartesian acquisition with several short radial (linear) trajectories. In (e), non-Cartesian acquisition with few long spiral-like trajectories. Cartesian acquisitions always sample in the Cartesian grid, illustrated as a light gray grid behind the readout lines.

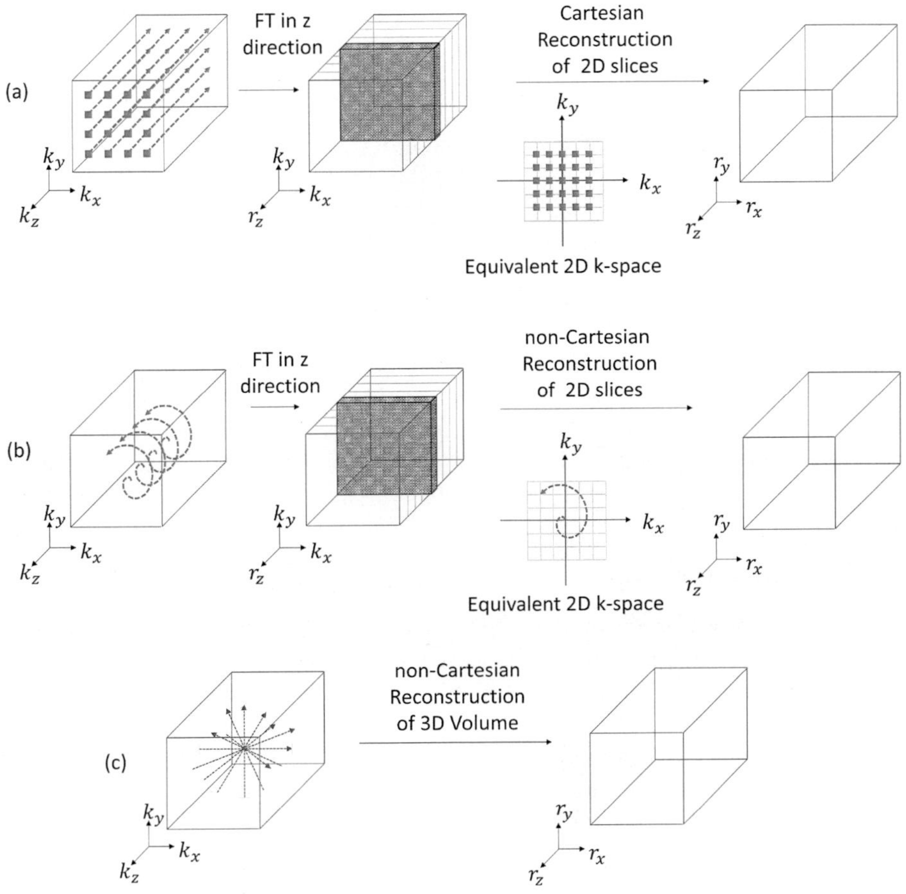

FIGURE 4.4

In (a), a 3D Cartesian acquisition with linear trajectories is shown. The 3D data is easily separable for 2D reconstructions using Fourier transform in the readout direction. In (b), an acquisition composed of a stack of non-Cartesian trajectories is shown. The 3D data is also separable in the stack direction because the trajectories are repeated. In (c), a non-separable 3D acquisition is composed of 3D non-Cartesian trajectories.

Each segment in which data samples are collected across the k-space after an excitation is called a *readout*. Note that in the simple configuration shown as an example in Fig. 4.3b, the encoding created by the gradients used before sampling starts is called *phase encoding*, which is used to move the starting position of the trajectory from the center of the k-space to a different position, moving in the k_y direction in this example. The encoding created during the readout, when the trajectories are linear and parallel is called *frequency encoding*, which corresponds to k_x direction in this example. The terms phase and frequency encoding are not used in non-Cartesian acquisitions.

Non-Cartesian acquisition [5] is often used in MRI due to several advantages such as faster and more efficient data collection, higher SNR, and reduced motion artifacts. Radial trajectories, such as the one shown in Fig. 4.3d, were the first used in MRI [6]. Over the years, different non-Cartesian acquisition schemes have been proposed, such as the spiral, illustrated in Fig. 4.3e.

In many MRI applications, we are interested in imaging the entire 3D volume, not just one 2D slice. While one can simply repeat the 2D acquisition for other slices, there are advantages in acquiring the entire 3D volume at once, such as improved SNR per unit of time, thinner slices, and potentially faster scans when undersampling is used, as will be discussed later. When acquiring the entire volume, one can either use 3D Cartesian, a Cartesian stack of non-Cartesian 2D trajectories, or 3D non-Cartesian trajectories, as illustrated in Fig. 4.4. The first two types use a Cartesian configuration in the k_z direction, and if all required samples in this direction are acquired, the 3D k-space data can be easily separated into multiple 2D k-space data for each slice by using a one-dimensional inverse Fourier transform in the z-axis. The 2D reconstruction of each slice can be performed individually.

4.1.4 Cartesian versus non-Cartesian acquisition

The Cartesian acquisition is simpler to understand and reconstruct, provides uniform k-space coverage, requires only simple piecewise constant gradient signals, and uses the *Fast Fourier Transform* (FFT), leading to fast reconstruction algorithms. The non-Cartesian acquisition is more flexible, has more freedom in covering k-space, and can potentially be better when combined with different undersampling schemes. However, it requires the use of *Non-Uniform Discrete Fourier Transforms* (NUDFT), which is very computationally costly. To reduce the computational cost, the NUDFT is approximated by a *Non-Uniform FFT* (NUFFT) [7], and reconstruction usually needs some type of density compensation, ultimately leading to slower reconstructions compared to Cartesian cases. Some approaches for parallel MRI replace the NUFFT with faster operators such as the *GRAPPA operator gridding* (GROG) [8], that approximate the k-space values in the nearest position in the grid. However, they incur a loss of quality since they are not as accurate as the NUFFT in approximating the NUDFT.

4.2 Faster MRI data acquisition

The overall MRI scan time is determined by the duration of the pulse sequence that controls the data acquisition. There are essentially three ways to reduce the overall scan time: (1) modifying the time of each step of the sequence and its components, including the magnetization-preparation step, to make it faster; (2) using undersampling schemes to acquire less k-space data, which reduces the number of repetition of the pulse sequence blocks, as seen in Fig. 4.1; and (3) improving k-space trajectory to cover the k-space faster, acquiring more k-space data per block (increasing the size of part 3 in Fig. 4.1), while reducing the number of repetitions.

The first method, improving each step of the sequence, requires rethinking the structure and components of the sequence, and the shape of the pulses, as all these factors affect the evolution of the MRI signal. This kind of MRI pulse sequence work will not be discussed here, but some examples can be found in Refs. [1,9–11]. The second method, undersampling k-space, is usually easier to implement on the scanner. However, it leads to more sophisticated reconstruction methods because capturing less k-space data leads to undersampling artifacts in images recovered by inverse Fourier transform. The third method has led to the development of a variety of k-space trajectories, and most importantly, to the development of long trajectories, such as EPI and spiral, as shown in Fig. 4.3. Later in this chapter, we will discuss how current machine learning methods are improving the design of non-Cartesian k-space trajectories. Note that one can accelerate MRI acquisition using only one or any combination of all three approaches.

4.2.1 Accelerated MRI data acquisition by undersampling k-space

Undersampling is the easiest way to accelerate MRI acquisition [2]. The scan time is reduced because one does not need to repeat the pulse sequence as many times as required for a full k-space acquisition. The most important point here is: which k-space elements are essential for imaging and which can be left out of the acquisition. To answer this question, we need to learn what is redundant in the k-space data (that can be left out because it is easily predicted). This leads to an important question: Does the redundancy of the k-space depend on how we compute the images (reconstruct) or on the k-space data alone?

In the past, researchers and MRI experts noticed that several redundancies occurred with k-space data (under certain circumstances), leading to classical undersampling k-space methods. The first one, named Partial Fourier [2,12], comes from the observation of one of the Fourier transform properties on real-valued signals and images [13]. If the signal is real-valued, then the k-space has complex-conjugated symmetry, that is, $s(\boldsymbol{k}) = s(-\boldsymbol{k})^*$, where $*$ denote complex-conjugated. Essentially, the k-space value at a particular frequency $\boldsymbol{k}$ is the complex conjugate of its value in its negative frequency $-\boldsymbol{k}$. In practice, one acquires only half of the k-space and fills the gaps in the negative frequencies by applying complex conjugation on the acquired data, as seen in Fig. 4.5a. This, of course, only works if the image is real-valued. However, if the phase is spatially smooth, minor modifications such as capturing a little bit more than half of the k-space can be used [2]. This point led to the development of the phase-constrained reconstructions [14–16]. Note that this already answers our previous question: The redundancy depends on the data, but can only be exploited using the right reconstruction.

Keyhole [17] is another approach that exploits temporal redundancy of images when the object is not moving, such as those used for quantitative relaxation mapping or multiple contrast imaging. In this approach, the k-space data of the first

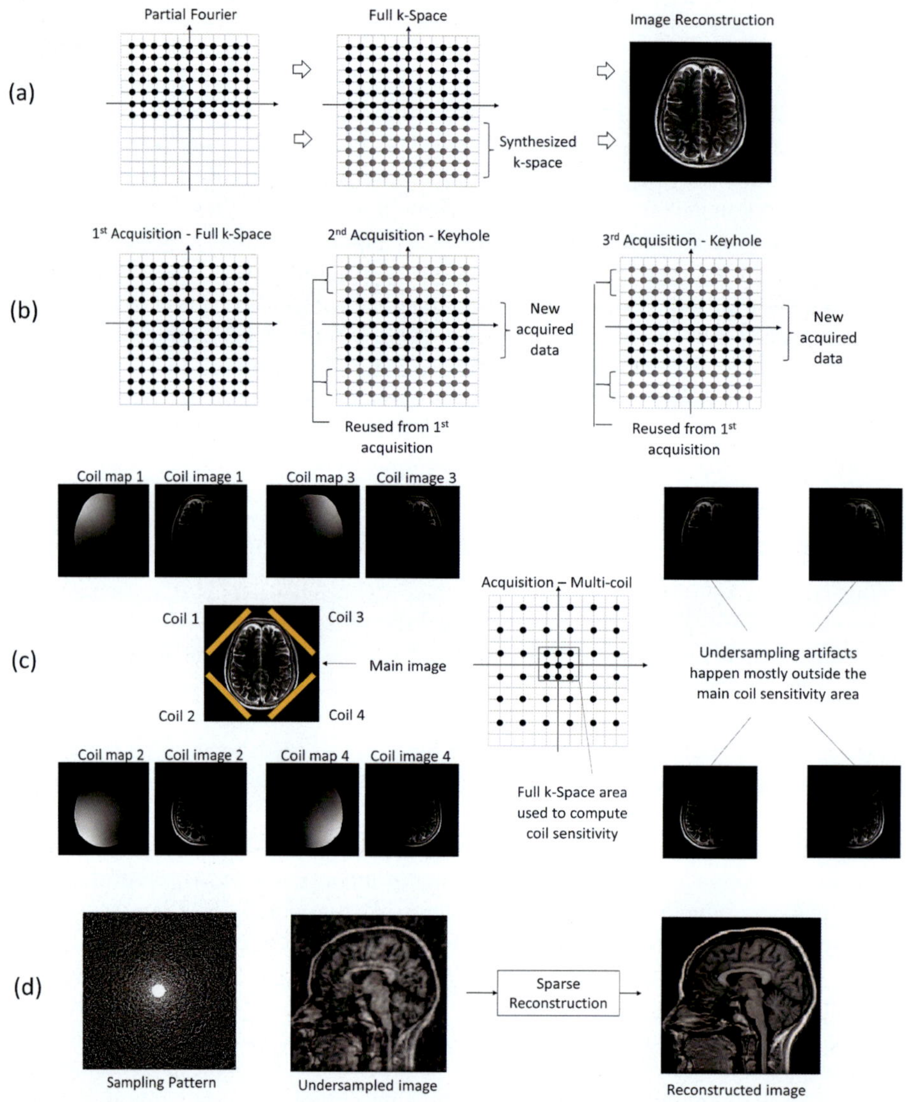

FIGURE 4.5

Illustration of different classical types of undersampling approaches for fast Cartesian MRI acquisition. In (a), Partial Fourier acquisitions. In (b), Keyhole acquisitions. In (c) parallel MRI acquisitions. In (d), compressed sensing acquisition and reconstruction.

image is fully acquired. In the following images; however, only the central area of the k-space is acquired. Because the high frequencies are very similar, one can simply copy the high-frequencies from the first image to the gaps of the following images, as shown in Fig. 4.5b.

More recently, the development of multi-coil acquisitions, also called parallel MRI [18,19], has led to important improvements in data acquisition quality and acceleration. By using a coil array, with multiple RF receive coils, more data are acquired simultaneously. Essentially, each coil measures the k-space of a different version of the original image, weighted by the spatial sensitivity function of the coil. Because an individual coil in the coil array is usually much smaller than the object and located in very close proximity to the object, the sensitivity is spatially local, and the FOV of the coil-sensitivity-weighted image is much smaller than that of the entire object. This way, one can accelerate the acquisition by increasing the space between samples Δk (see subsection "4.1.2. Requirements to Sample the k-Space"). Aliasing may occur, but the replicas due to aliasing will happen in areas where the coil is not sensing.

Another way to understand this effect is using Fourier properties. The point-by-point multiplication of the image with the coil sensitivity function leads to a convolution of the original k-space data with a point-spread function that corresponds to the Fourier transform of the coil sensitivity function, which usually resembles a smoothing filter. This creates a more correlated or smoothed version of the k-space. In the same way, one can increase the space between samples and still have a good acquisition of these smoother versions of the k-space. Because of this, uniform k-space undersampling is typically used in parallel MRI, as shown in Fig. 4.5c.

However, there are various approaches to reconstruct the images in parallel MRI. The two dominant approaches are: (1) *Sensitivity Encoding* (SENSE)-based reconstructions [20,21], that use coils sensitivities and formulate the problem as an inverse problem; and (2) frequency-domain approaches that recover the undersampled k-space data using linear or nonlinear methods, usually some kind of filtering or interpolation in the k-space, taking advantage of the smoother k-space, and later recombine the images from different coils [22,23].

SENSE reconstructions of each slice start with the image to k-space model, given by:

$$\mathbf{s} = \mathbf{E}_\Omega \boldsymbol{\rho} + \boldsymbol{\eta}, \tag{4.3}$$

where $\boldsymbol{\rho}$ is a $N \times 1$ vector that represents the $N = N_x \times N_y$ image to be reconstructed (in 3D case $N = N_x \times N_y \times N_z$). $\mathbf{s} = [\mathbf{s}_1^H \ldots \mathbf{s}_{Nc}^H]^H$, is $MN_c \times 1$ vector that corresponds to the acquired multi-coil k-space data, where $\mathbf{s}_c$ is the data from coil c, of a total of N_c coils. The $MN_c \times N$ forward encoding matrix $\mathbf{E}_\Omega = \text{diag}(\mathbf{E}_1, \ldots, \mathbf{E}_{N_c})$ is composed of forward encoding matrices of each coil, $\mathbf{E}_\text{c} = \mathbf{F}_\Omega \mathbf{C}_\text{c}$, which include coil sensitivities transform $\mathbf{C}_\text{c}$, which is a $N \times N$ diagonal matrix with coil sensitivity map, and $\mathbf{F}_\Omega$ is the $M \times N$ combined Fourier transform and undersampling. Note that the undersampling factor (UF), also called acceleration factor (AF) when only undersampling is used for acceleration, is given by $UF = N/M$. In non-Cartesian acquisition $\mathbf{F}_\Omega$ uses NUFFT for non-Cartesian k-space positions Ω. For Cartesian acquisitions, $\mathbf{F}_\Omega = \mathbf{B}_\Omega \mathbf{F}$, composed of a 2D Fourier transform $\mathbf{F}$, and undersampling matrix $\mathbf{B}_\Omega$. We emphasize here the use of Ω to represent the collection of acquired k-space positions, given by a sampling pattern or a k-space trajectory. We may use the representation $\bar{s} = E_\Gamma \rho$ to denote the fully sampled k-space data, where $\Omega \subset \Gamma$. Undersampled data is also produced from fully sampled data by $s = B_\Omega \bar{s}$. However, for simplicity of notation, only the symbol E will be used without Ω or Γ when the definition of the sampling pattern is clear.

SENSE reconstructions solve the reconstruction problem by:

$$\hat{\rho} = \arg\min_{\rho} \|s - E\rho\|_2^2 \cong E^\dagger s, \tag{4.4}$$

where $E^\dagger$ is the pseudo inverse of E [24]. If the problem is not underdetermined, then $E^\dagger = (E^H E)^{-1} E^H$, where E^H is the adjoint (or Hermitian, that is the transposed and complex-conjugated) of E.

Another important breakthrough in undersampling is the theory of *Compressed Sensing* (CS) [25–28]. CS states that if the image is sparse or compressible in some domain, and the sampling space is incoherent with the sparse domain, one can recover the sparse values using undersampled acquisitions. The number of samples can be much smaller than the one proposed by sampling theories [29], but still larger than the number of sparse elements of the representation [26,30]. Since CS first appeared, MRI has always been a good imaging application for its use [25], mainly because the wavelets domain was a good sparsifying transform, and because incoherence could be obtained with certain kinds of random undersampling patterns [27], as seen in Fig. 4.5d.

In CS, images are reconstructed with nonlinear methods, usually formulated as an optimization problem, such as:

$$\hat{\rho} = \arg\min_{\rho} \|s - E\rho\|_2^2 + \lambda R(\rho), \tag{4.5}$$

where $R(\rho)$ is the regularizing penalty, also known as Prior, that enforces sparsity in some transformed domain, while λ is the regularization parameter. The data-discrepancy term, $\|s - E\rho\|_2^2$, is the same used in parallel MRI based on SENSE reconstruction.

The flexibility of CS and many parallel MRI methods in choosing the sampling pattern (SP) has led to extensive research on how to better undersample the k-space. This is especially true for problems that combine CS and parallel MRI [31–34]. Some commonly known SPs for CS and CS-parallel MRI are Random [25], *Variable Density* (VD) [35–37], *Poisson-Disc* (PD) [33,38], and combined VD and PD [39,40], as shown in Fig. 4.6. Note that the 1D-SPs in Fig. 4.6a and b (even though the sampling mask is 2D) are used in 2D Cartesian acquisitions as shown in Fig. 4.3b or c, usually exploiting undersampling in the phase-encoding direction since there is no advantage in undersampling the readout. On the other hand, the 2D-SPs in Fig. 4.6c–f are used in 3D Cartesian acquisitions as shown in Fig. 4.4a, since these acquisitions have two phase-encoding directions to be exploited.

In the examples shown in Fig. 4.6, note that in most SPs there is a fully sampled central area, used to compute the coil sensitivities [41] necessary in CS-parallel MRI.

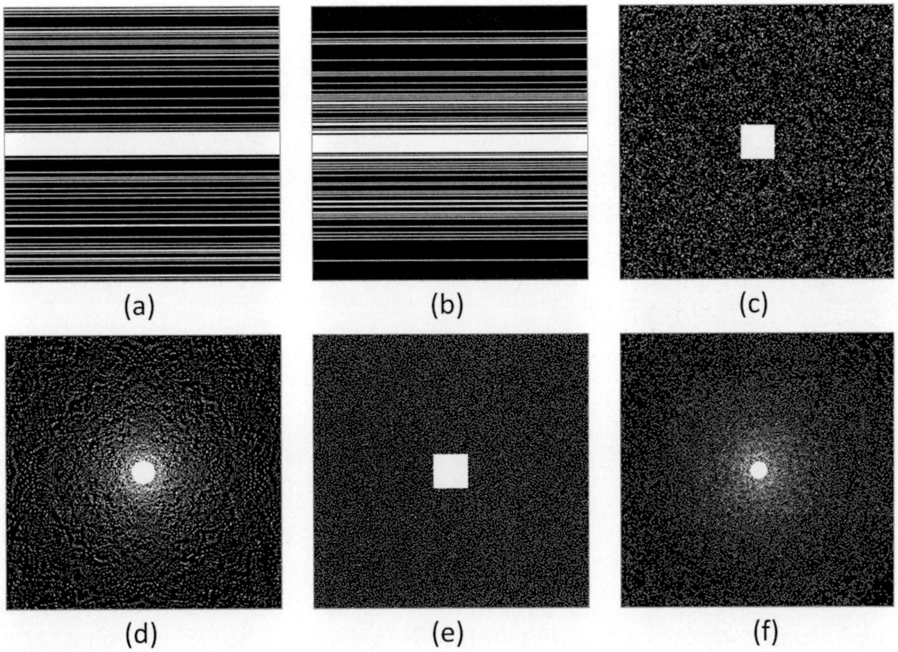

FIGURE 4.6

Different Cartesian SPs for CS MRI acquisition. In (a)1D random, (b) 1D variable density, (c) 2D random, (d) 2D variable density, (e) 2D Poisson-disc, and (f) a combined 2D Poisson-disc with variable density.

4.2.2 Accelerated MRI with non-Cartesian acquisition

There are many reasons why a particular k-space trajectory is chosen. Classical non-Cartesian trajectories, such as radial [6] and spiral [42], have many advantages in some applications, due to acquiring the center of the k-space multiple times in different instants. However, for accelerated MRI acquisition, the biggest advantage of non-Cartesian trajectories is their ability to cover more k-space area at each repetition of the pulse sequence (each readout), ultimately leading to fewer repetitions. The only limit here is how strong the magnetic gradient system can be set and how fast it can vary [43–45].

Besides having interesting advantages, non-Cartesian acquisitions have their challenges. Due to departing from the regular spacing of the Cartesian grid, some non-Cartesian trajectories do not cover the k-space entirely and uniformly, creating undersampling in some areas (e.g., high frequency) and oversampling in other areas (e.g., low frequency) of the k-space. This leads to what is called variable sampling density, which has to be handled later by the reconstruction method.

In Fig. 4.7, several non-Cartesian trajectories are shown, for example: (a) radial [6], (b) linogram radial [46], (c) *Periodically Rotated Overlapping Parallel Lines*

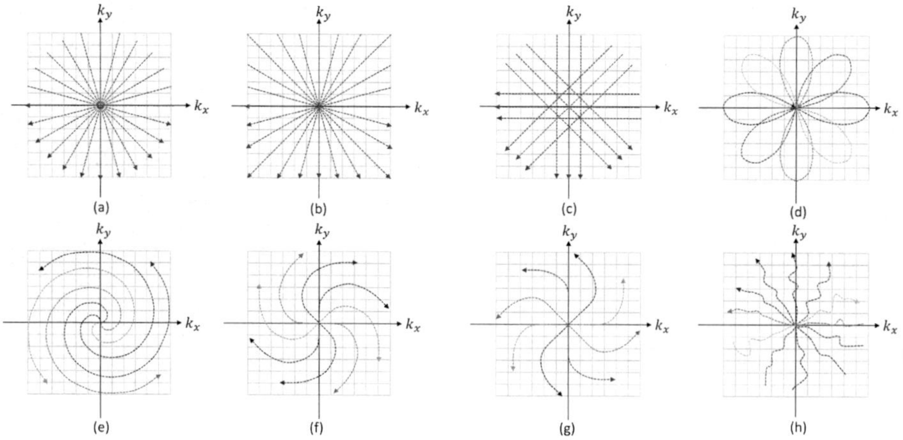

FIGURE 4.7

Different types of non-Cartesian trajectories for MRI acquisition. In (a) radial (polar) trajectories, (b) linogram trajectories, (c) PROPELLER, (d) rosette, (e) spiral, (f) WHIRL, (g) TWIRL, and (h) free trajectories.

with Enhanced Reconstruction (PROPELLER) [47], (d) rosette [48], (e) spiral [42], (f) *Winding Hybrid Interleaved Radial Lines* (WHIRL) [49], (g) *Twisting Radial Lines* (TWIRL) [50], and (h) free non-Cartesian trajectories [51–54]. There are many variations on these examples. Trajectories may start in the center of the k-space and move outwards; they may start outside (by using phase encoding gradients first) and move to the center; or cross the center from one end to the other. However, most non-Cartesian acquisitions result in oversampling central areas of the k-space and undersampling outer areas of the k-space.

In dynamic imaging, modifications of these trajectories include different k-space coverage over time to properly capture the dynamic process. This is important to effectively exploit temporal redundancies.

One effective temporal ordering scheme for dynamic MRI is the *Golden Angle* (GA) scheme, applied to radial and spiral trajectories [42,57]. As seen in Fig. 4.8a, GA uses an angle increment of 111.25 degrees between each radial readout (sometimes called spoke in the radial context), instead of small angles (10 degrees in this example) as in regular radial acquisitions [55]. After the scan is finished, the k-space data can be sorted to reconstruct images of different time points of the dynamic process. In this case, the GA acquisition provides very uniform k-space coverage, no matter how many spokes are sorted for each time point, as seen in Fig. 4.8b.

The actual non-Cartesian trajectories are sometimes slightly different from the desired trajectories. Eddy currents, mechanical effects, and other hardware limitations may cause the actual k-space trajectory to deviate from the nominal trajectory [44], as seen in Fig. 4.8c, leading to artifacts in the reconstructed images. Deviation

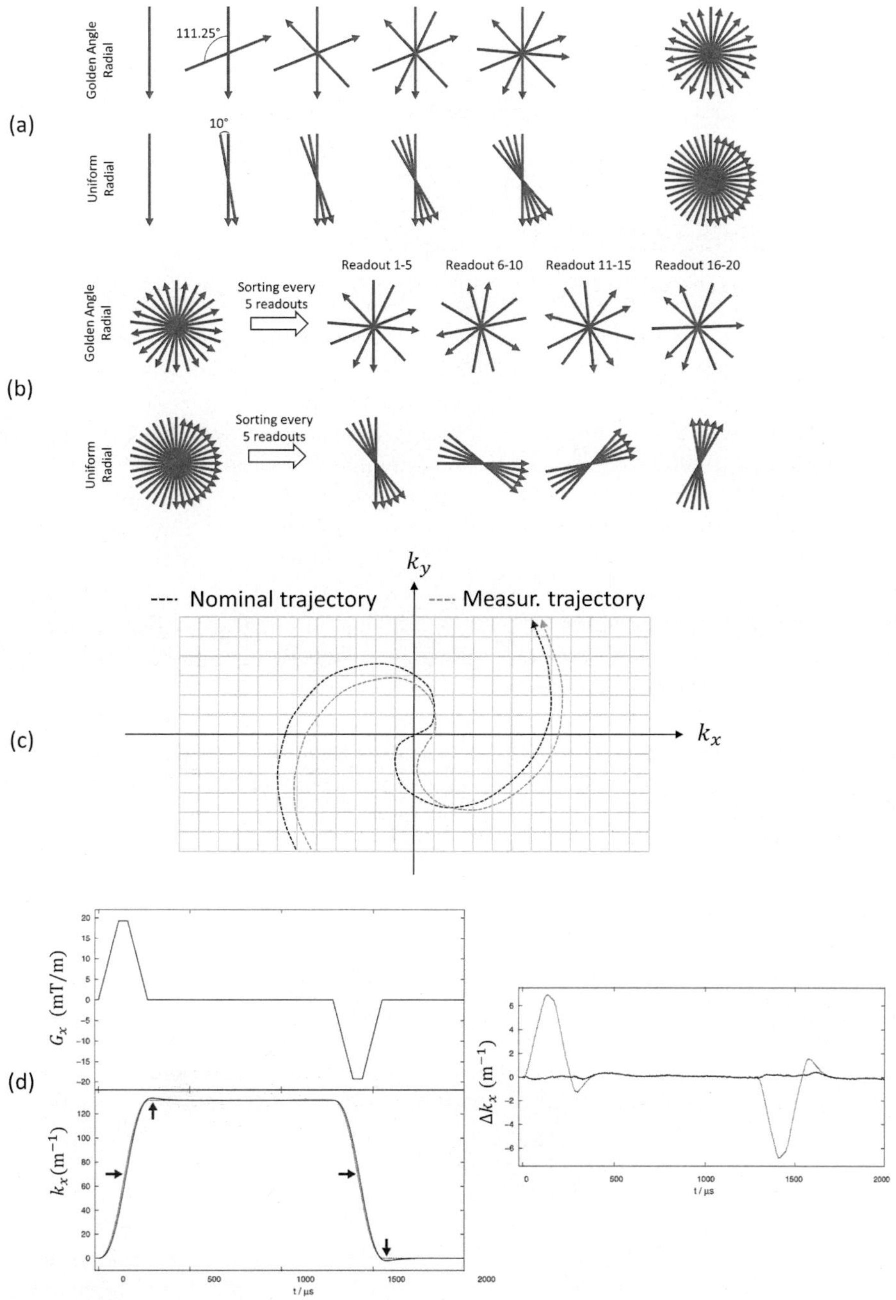
(a)
Golden Angle Radial
111.25°
Uniform Radial
10°
(b)
Golden Angle Radial
Sorting every 5 readouts
Readout 1-5
Readout 6-10
Readout 11-15
Readout 16-20
Uniform Radial
Sorting every 5 readouts
(c)
k_y
Nominal trajectory
Measur. trajectory
k_x
(d)
G_x (mT/m)
k_x(m^{-1})
t / µs
Δk_x (m^{-1})
t / µs

FIGURE 4.8

of the nominal trajectory is much easier to correct with linear trajectories than with non-linear trajectories [43,58] because it only results in small shifts of the sample positions over the linear trajectory. These trajectory errors are usually more significant when the gradient signal is changing rapidly, as seen in Fig. 4.8d.

4.3 Optimizing MRI data acquisitions

4.3.1 Automatic adjustment of SPs and k-space trajectories

Over the years, researchers have gained a better understanding of the main properties relevant to a sampling pattern (SP) or k-space trajectory for accelerated MRI problems. In Cartesian acquisitions, it was observed that variable density [35–37], with more samples close to the center of the k-space, and relatively even spacing between samples, such as in Poisson-disc [33,38], brought improvements in image quality for CS-parallel MRI reconstructions. However, many details remain unclear, such as how exact the "variable" density should be or how much undersampling space between samples is acceptable in parallel MRI. In addition, there is the question of how these SPs and k-space trajectories can be calculated. These questions opened the possibility of optimizing or, on a simpler level, automatically adjusting the SP or k-space trajectory to achieve some predefined property that human researchers have found beneficial over the years.

For Cartesian MRI, adaptive SP approaches [59–61] seek to adjust the SP to obtain a sampling density that follows the distribution of the magnitude of the k-space. This empirically assumes that more samples should be placed where the k-space magnitude is higher, that is, the center of the k-space, and fewer samples are necessary where the k-space signal is lower, i.e., outer k-space areas. However, other aspects follow typical requirements of CS, such as nearly random distribution of sampling points, and parallel MRI, with samples not too close or too far from each other. One example from Ref. [59] is shown in Fig. 4.9a.

For non-Cartesian acquisitions, one of the pioneer optimization approaches was proposed in Ref. [62] for linear MRI reconstruction. Recently, the *Spreading Projection Algorithm for Rapid K-space sampLING* (SPARKLING) [56,63] approach was proposed to optimize non-Cartesian trajectories for CS reconstructions, focusing on achieving a desired sampling density with better k-space coverage and relatively

Example of (a) golden angle radial readouts are collected to acquire data of a dynamic MRI problem, and (b) how readouts closed in time can be sorted into groups of five for image sequence reconstruction (see Ref. [55]). In (c) an example of the deviation between the nominal trajectory and measured trajectory (see more in Ref. [56]). In (d), the nominal gradient signal G_x, the nominal and measured k_x position, and their difference Δk_x (see more in Ref. [3]).

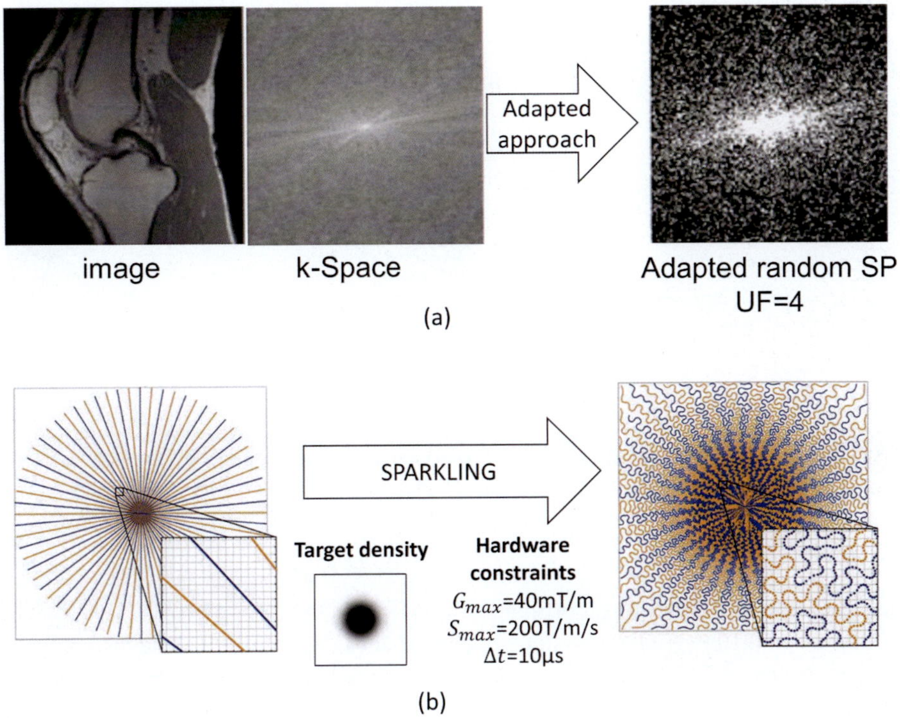

FIGURE 4.9

Comparison of different approaches to automatically adjust sampling. In (a), adapted random sampling for Cartesian acquisitions using [59]. In (b), optimized k-space trajectory using SPARKLING, using non-Cartesian radial trajectories as an initial trajectory (This figure uses part of figures in Ref. [56]).

uniform sampling space, respecting the restrictions $\mathcal{Q}$ of the gradient system of the MRI scanner. The restrictions $\mathcal{Q}$ include the maximum k-space speed:

$$\|\dot{\boldsymbol{k}}[i]\| \triangleq \left\|\frac{\boldsymbol{k}[i]-\boldsymbol{k}[i-1]}{dt}\right\| \leq \min\left(\frac{\gamma}{2\pi}G_{max}, \frac{1}{FOV.\Delta t}\right), \tag{4.6}$$

limited by the maximum gradient G_{max}, where dt is the gradient raster time (which is the time constant for the discrete representation of gradient signals in the MRI scanner) and Δt is the sampling time (also known as dwell time, in Refs. [56,64] typically $\Delta t < dt$), and by the maximum k-space acceleration:

$$\left\|\ddot{\boldsymbol{k}}[i]\right\| \triangleq \left\|\frac{\boldsymbol{k}[i+1]+2\boldsymbol{k}[i]-\boldsymbol{k}[i-1]}{dt^2}\right\| \leq \frac{\gamma S_{max}}{2\pi k_{max}}, \tag{4.7}$$

limited by the maximum slew rate S_{max}, where k_{max} is the maximum k-space frequency. Note that in Eq. (4.2) we define $\boldsymbol{k}(t)$ as a function of time, here we just express $\boldsymbol{k}[i] = big[k_x[i]\ k_y[i]\ k_z[i]]$ as a function of the discrete-time index i. However, we also extend our definition of $\boldsymbol{k}$ to the collection of all $\boldsymbol{k}[i]$ in the k-space trajectories, as a $M \times 1$ vector where each component is a 3D or 2D k-space position.

Then, the k-space trajectory is optimized by minimizing:

$$\widehat{\boldsymbol{k}} = \arg \min_{\boldsymbol{k} \in Q} \mathscr{L}_r(\boldsymbol{k}) - \mathscr{L}_a(\boldsymbol{k}). \tag{4.8}$$

That is composed of two terms:

$$\mathscr{L}_r(\boldsymbol{k}) = \frac{1}{2M^2} \sum_{1 \leq i,j \leq M} H(\boldsymbol{k}[i] - \boldsymbol{k}[j]), \tag{4.9}$$

that creates a repulsion between points, and:

$$\mathscr{L}_a(\boldsymbol{k}) = \frac{1}{M} \sum_{1 \leq i \leq M} \int H(\boldsymbol{l} - \boldsymbol{k}[i]) p(\boldsymbol{l}) d\boldsymbol{l}, \tag{4.10}$$

that attracts the samples to a pre-defined k-space sampling density given by $p(\boldsymbol{k})$, using H a well-chosen radial function [56]. The k-space sample positions are optimized according to:

$$\boldsymbol{k}_{t+1} = \Pi_Q(\boldsymbol{k}_t - \beta_t [\nabla \mathscr{L}_r(\boldsymbol{k}_t) - \nabla \mathscr{L}_a(\boldsymbol{k}_t)]), \tag{4.11}$$

that is a projected gradient algorithm.

By pushing the gradient system operation close to its limits and using long readout times, SPARKLING can obtain excellent k-space coverage per unit of time on each repetition of the pulse sequence blocks, as illustrated in Fig. 4.9b.

Note that $\boldsymbol{k}$ describes the optimized trajectory, however, it is the discrete gradient signals $\boldsymbol{G} = 2\pi \dot{\boldsymbol{k}}/\gamma$ that must be transferred to the scanner. The sampling pattern Ω is a discrete subset of the continuous trajectory produced by the scanner. However, since $\boldsymbol{k}$ is discrete and in most cases $\Delta t < dt$, one needs to interpolate $\boldsymbol{k}$ to obtain the sampling positions in Ω.

4.3.2 Machine learning approaches to automatically learn the SP or k-space trajectory

In the previous approaches, the sampling design tries to obtain the best possible SP according to a criterion inspired by good sampling practices learned by humans over the years. This means humans learned that variable density, incoherent sampling, and regular spacing between samples, among other things, are good properties for an SP. However, this does not necessarily mean that this is the best possible sampling approach because there are still things to be learned about sampling. To go beyond this, we need to design machine learning algorithms that can learn how the SP or k-space trajectory should be.

A possible approach is optimizing the *Cramér-Rao Lower Bound* (CRLB) [65]. Optimal CRLB has been proven useful in many applications, and it has also been useful in learning the SP as shown in Ref. [65]. This is optimized for CS problems where images have a known distribution of sparse coefficients, such as In *Oracle-based Experiment Design for Imaging Parsimoniously Under Sparsity Constraints* (OEDIPUS).

For parallel MRI using SENSE-based methods, the CRLB leads to:

$$\text{cov}(\widehat{\boldsymbol{\rho}}) \succcurlyeq \left(\boldsymbol{E}^H\boldsymbol{E}\right)^{-1}. \tag{4.12}$$

where $\text{cov}(\widehat{\boldsymbol{\rho}})$ is the covariance matrix, and $\succcurlyeq$ means the matrix $\text{cov}(\widehat{\boldsymbol{\rho}}) - (\boldsymbol{E}^H\boldsymbol{E})^{-1}$ is always a positive semidefinite matrix.

For some CS approaches, the model $\boldsymbol{\rho} = \boldsymbol{\Psi}\overline{\boldsymbol{c}}$ is assumed, where $\overline{\boldsymbol{c}}$ is a sparse vector, and $\boldsymbol{\Psi}$ is a dictionary used to compose the image. To make an easy connection with Eq. (4.12), assume we can represent the image using $\boldsymbol{\rho} = \boldsymbol{\Psi}\boldsymbol{U}\boldsymbol{c}$, where $\boldsymbol{c}$ is a smaller vector with only the sparse components, and $\boldsymbol{U}$ a matrix that fills the zero elements and distributes the sparse components in their exact places. In this case:

$$\text{cov}(\widehat{\boldsymbol{c}}) \succcurlyeq \left(\boldsymbol{U}^H\boldsymbol{\Psi}^H\boldsymbol{E}^H\boldsymbol{E}\boldsymbol{\Psi}\boldsymbol{U}\right)^{-1} \tag{4.13}$$

is the covariance matrix for the coefficients, and:

$$\text{cov}(\widehat{\boldsymbol{\rho}}) \succcurlyeq \boldsymbol{\Psi}\mathbf{U}\left(\boldsymbol{U}^H\boldsymbol{\Psi}^H\boldsymbol{E}^H\boldsymbol{E}\boldsymbol{\Psi}\boldsymbol{U}\right)^{-1}\boldsymbol{U}^H\boldsymbol{\Psi}^H, \tag{4.14}$$

for the image. To optimize the CRLB, the following cost function is minimized:

$$\widehat{\Omega} = \arg \min_{\substack{\Omega \subset \Gamma \\ |\Omega| = M}} \sum_{n=1}^{N_d} \text{trace}(\boldsymbol{D}_n(\Omega)), \tag{4.15}$$

where N_d is the number of exemplar images and

$$\boldsymbol{D}_n(\Omega) \triangleq \left(\sum_{m \in \Omega} \boldsymbol{U}_n^H\boldsymbol{\Psi}^H\left((\boldsymbol{e}_m)^H(\boldsymbol{e}_m)\right)\boldsymbol{\Psi}\boldsymbol{U}_n\right)^{-1}, \tag{4.16}$$

This cost function is optimized with greedy search methods in Ref. [65]. Note that, essentially, only the position of the sparse components is used to create $\boldsymbol{U}_n$. Some resulting SPs examples of OEDIPUS are shown in Fig. 4.10a.

Before [65], a Bayesian experimental design was proposed in Ref. [66] to learn the SPs for CS reconstructions. The resulting method is also a greedy search using differential entropy as the loss function. The assumptions lead to a similar cost function, that can be approximately described by $\text{cov}(\widehat{\boldsymbol{\rho}}) \gtrsim (\boldsymbol{E}^H\boldsymbol{E} + \boldsymbol{S}^H\boldsymbol{\Lambda}_\rho\boldsymbol{S})^{-1}$. In Ref. [66], the analysis sparse model $\boldsymbol{b} = \boldsymbol{S}\boldsymbol{\rho}$ is used instead of the synthesis model in $\boldsymbol{\rho} = \boldsymbol{\Psi}\overline{\boldsymbol{c}}$, [67], where $\boldsymbol{S}$ is a sparsifying transform and $\boldsymbol{b}$ is the sparse representation of $\boldsymbol{\rho}$, and $\boldsymbol{\Lambda}_\rho$ is a diagonal weighting matrix resulting from the linear approximation when $R(\boldsymbol{\rho}) = \|\boldsymbol{S}\boldsymbol{\rho}\|_1 \cong \boldsymbol{\rho}^H\boldsymbol{S}^H\boldsymbol{\Lambda}_\rho\boldsymbol{S}\boldsymbol{\rho}$ is used in Eq. (4.5).

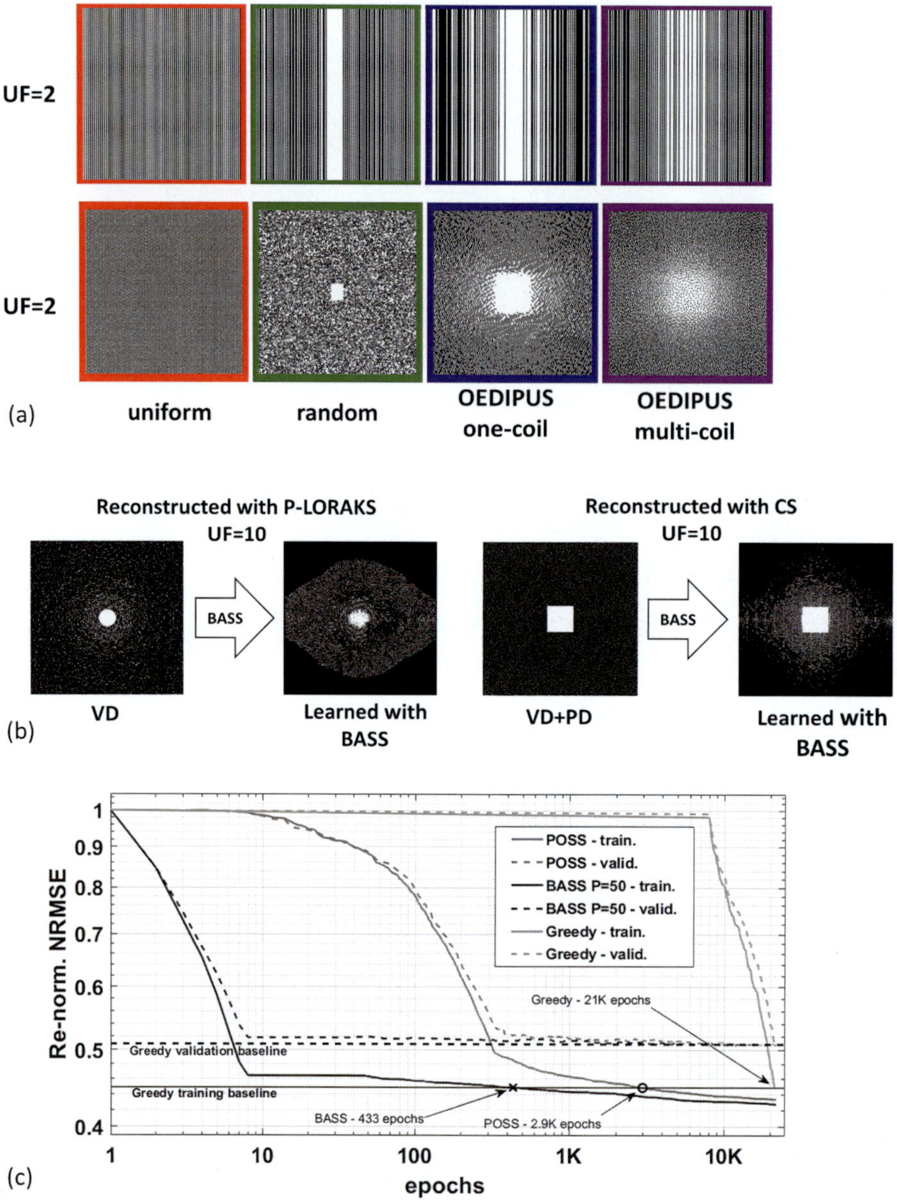

FIGURE 4.10

Comparison of different approaches to learning the SP or sampling trajectory. In (a) the results of OEDIPUS for 1D and 2D SPs. In (b), Some results of BASS, learning a different SP according to the reconstruction method. In (c), the convergence speed of the learning process with BASS compared to POSS and Greedy approaches (see more in Refs. [65,72]).

In [68–70], the cost function is generalized by replacing the CRLB for the reconstruction error:

$$\widehat{\Omega} = \arg \min_{\substack{\Omega \subset \Gamma \\ |\Omega| = M}} \mathscr{L}(\Omega), \tag{4.17}$$

being

$$\mathscr{L}(\Omega) = \frac{1}{N_d} \sum_{n=1}^{N_d} d(\boldsymbol{\rho}_n, \widehat{\boldsymbol{\rho}}_n) \tag{4.18}$$

One is free to choose an appropriate distance measurement $d(\boldsymbol{\rho}_n, \widehat{\boldsymbol{\rho}}_n)$. Note that $\widehat{\boldsymbol{\rho}} = \Theta(\Omega, \boldsymbol{s})$, where Θ represents the reconstruction method. For example in parallel MRI, let's assume $\Theta(\Omega, \boldsymbol{s}) = (\boldsymbol{E}^H\boldsymbol{E})^{-1}\boldsymbol{E}^H\boldsymbol{s} = (\boldsymbol{E}^H\boldsymbol{E})^{-1}\boldsymbol{E}^H(\boldsymbol{E}\boldsymbol{\rho} + \boldsymbol{\eta}) = \boldsymbol{\rho} + (\boldsymbol{E}^H\boldsymbol{E})^{-1}\boldsymbol{E}^H\boldsymbol{\eta}$, where $\boldsymbol{\eta}$ is a Gaussian noise, and $d(\boldsymbol{\rho}, \widehat{\boldsymbol{\rho}}) = \|\boldsymbol{\rho} - \widehat{\boldsymbol{\rho}}\|_2^2$, then $d(\boldsymbol{\rho}, \widehat{\boldsymbol{\rho}}) = \left\|(\boldsymbol{E}^H\boldsymbol{E})^{-1}\boldsymbol{E}^H\boldsymbol{\eta}\right\|_2^2 \propto (\boldsymbol{\eta}^H\boldsymbol{\eta})(\boldsymbol{E}^H\boldsymbol{E})^{-1}$, which is closely connected to the CRLB criterion used in OEDIPUS [65], shown in Eq. (4.12). The main difference is that the image reconstruction $\Theta(\Omega, \boldsymbol{s})$ is used instead of computing inverse matrix $(\boldsymbol{E}^H\boldsymbol{E})^{-1}$. The use of $\Theta(\Omega, \boldsymbol{s})$ generalizes the criterion to any reconstruction method, including when the matrix $(\boldsymbol{E}^H\boldsymbol{E})$ is not inversible, among other advantages.

In [68–70], greedy search is used, as described in Algorithm 4.1. Previous methods in Refs. [65,66] also use greedy search methods. The most challenging aspect of greedy search methods is the amount of cost function evaluations per iteration. Essentially, the cost function has to be computed for all candidates Ω' in the set $\Omega \cup \Omega_c$ at each iteration, where Ω_c is an element of $\Gamma \backslash \Omega$, significantly increasing the computational cost for large sampling patterns.

In [71], a pioneering approach that used a similar generalization of the cost function as shown in Eq. (4.17), a fast algorithm to improve the SPs was proposed. Even though much faster than greedy approaches, no convergence properties were demonstrated, which left the greedy approaches as more reliable at the time, even though they are much slower.

ALGORITHM 4.1 GREEDY

1. $\Omega \leftarrow \varnothing$
2. **While** $|\Omega| \leq M$ **do**
3. **For all** $\Omega_c \in \Gamma \backslash \Omega$
4. $\Omega' \leftarrow \Omega \cup \Omega_c$
5. $\mathscr{L}(\Omega') = \frac{1}{N_d} \sum_{n=1}^{N_d} d(\boldsymbol{\rho}_n, \Theta(\Omega', \boldsymbol{E}_{\Omega'}\boldsymbol{\rho}_n))$
6. **End For**
7. $\Omega_a = \arg \min_{\substack{\Omega_c \in \Gamma \backslash \Omega \\ |\Omega \cup \Omega_c| \leq M}} \frac{\mathscr{L}(\Omega \cup \Omega_c) - \mathscr{L}(\Omega)}{|\Omega \cup \Omega_c| - |\Omega|}$
8. $\Omega \leftarrow \Omega \cup \Omega_a$
9. return Ω

The problem of large computational cost to optimize the SP was only solved recently by Ref. [72]. Instead of greedy search, the authors investigated bitwise subset selection methods that were potentially faster and with better convergence properties than greedy approaches. Iterative methods such as *Pareto Optimization for Subset Selection* (POSS) [73] were a good fit for this problem, with robust convergence properties and better scalability. POSS was faster than greedy search, but not fast enough (since the changes in the SP elements used by POSS are unitary and random). In Ref. [72], the authors used a more effective way to select the k-space samples that should be added or removed from the SP, creating the *Bias-Accelerated Subset Selection* (BASS), described in Algorithm 4.2. This makes the optimization of the SP nearly 50 times faster than greedy approaches, as seen in Fig. 4.10c, with computational time scaling linearly with problem size, while it scales quadratically with greedy approaches. The method is also very robust, finding results at a lower cost than greedy search methods. With BASS, an SP for large 2D or 2D + time Cartesian SPs can be learned in dozens of iterations. Some examples are shown in Fig. 4.10b.

To achieve fast convergence, BASS selects the elements of Ω_a and Ω_r, using the functions **select-add**$\big(\Omega, P, p_a(P,M,N,l)\big)$ and **select-remove**$\big(\Omega, P, p_r(P,M,l)big\big)$, illustrated in Algorithm 4.2 and detailed in Fig. 4.11. These functions modify Ω in the composition of a new candidate Ω'.

Following Fig. 4.11, the first step is a random pre-selection of a number larger than P of elements. The actual number of elements is given by the probability given by $P/M < p_r(P,M,l) \leq 1$ and $P/(N-M) < p_a(P,M,N,l) \leq 1$, where l corresponds to the iteration number. Higher probability means more random pre-selected elements in the first step. The random pre-selection is filtered by keeping the elements with the best measure of importance (MI) that respect the positional constraints (PC) rules, as defined in Ref. [72].

The MI is a user-defined criterion, usually related to $\mathscr{L}(\Omega)$. Assuming the cost function $\mathscr{L}(\Omega)$ had to be computed, then the k-space error for the training dataset,

ALGORITHM 4.2 BASS I *NON-MONOTONE BASS*

$\Omega \leftarrow \Omega_{init}$
$P \leftarrow P_{init}$
$l \leftarrow 1$
While $l \leq L$ **do**
  $\Omega_r \leftarrow$ **select-remove**$(\Omega, P, p_r(P,M,l))$
  $\Omega_a \leftarrow$ **select-add**$(\Omega, P, p_a(P,M,N,l))$
$\Omega' \leftarrow (\Omega \backslash \Omega_r) \cup \Omega_a$
if $|\Omega| \neq M$, $\Omega \leftarrow \Omega'$
if $|\Omega| = M$
if $\mathscr{L}(\Omega') \leq \mathscr{L}(\Omega)$, $\Omega \leftarrow \Omega'$ | *or* 10. $\Omega \leftarrow \Omega'$
  else $P \leftarrow \lfloor (P-1)\alpha \rfloor + 1$ | *or* 11. $P \leftarrow \lfloor (P-1)\alpha \rfloor + 1$
$l \leftarrow l + 1$
return Ω

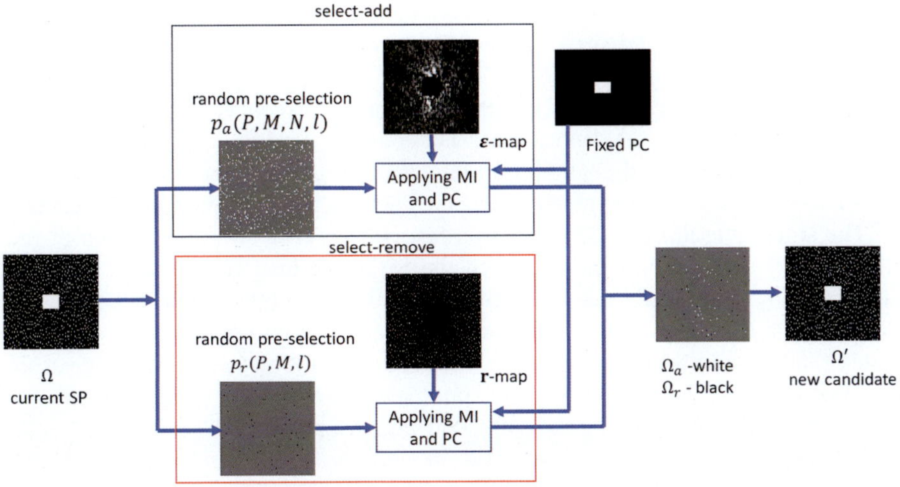

FIGURE 4.11

Illustration of how BASS chooses the k-space elements at each iteration *l*.

where $1 \leq n \leq N_d$, can be expressed as $\mathbf{e}_n = \boldsymbol{E}_\Gamma(\boldsymbol{\rho}_n - \Theta(\Omega, \boldsymbol{E}_\Omega \boldsymbol{\rho}_n))$, where $\boldsymbol{E}_\Gamma$ is the fully sampled (and multi-coil) encoding matrix. Note that in the fully-sampled case the $NN_c \times 1$ vector $\mathbf{e}_n$ can be reordered as an N-dimensional vector where each component is an N_c-dimensional vector. For $1 \leq k \leq N$, and $1 \leq c \leq N_c$, where $[\mathbf{e}_n]_{k,c}$ denotes the c-th component (corresponding to the coil) of the k-th component (corresponding to the k-space position) of $\mathbf{e}_n$.

For **select-add**, the defined MI, for $1 \leq k \leq N$, is:

$$\varepsilon_k = \frac{1}{N_d N_c} \sum\nolimits_{n=1}^{N_d} \sum\nolimits_{c=1}^{N_c} \left| [\mathbf{e}_n]_{k,c} \right|^2, \tag{4.19}$$

which is referred to as the ε-map, essentially a measure of the k-space error for each k-space position (averaged for all N_c coils and all N_d training images), as seen in Fig. 4.11. The purpose of **select-add** is to select P elements from $\Gamma\Omega$ in a randomly biased manner [72], and its MI prioritizes the inclusion of k-space elements with large error (actually large MI), which may significantly reduce the cost function. Note that in the non-monotone version $\mathscr{L}(\Omega)$ is never computed. Then, the convergence relies on the correlation between $\mathscr{L}(\Omega)$ and the chosen MI.

For **select-remove**, a sequence with P elements that are in Ω, is generated in the same way, but using $\mathbf{r}_k$ as the MI, instead of ε_k. The MI for **select-remove** is

$$\mathbf{r}_k = \frac{1}{N_d N_c} \sum\nolimits_{n=1}^{N_d} \frac{\sum_{c=1}^{N_c} \left| [\mathbf{e}_n]_{k,c} \right|^2 + \delta}{\sum_{c=1}^{N_c} \left| [\mathbf{s}_n]_{k,c} \right|^2 + \delta}, \tag{4.20}$$

for $1 \leq k \leq N$, which is referred to as the **r**-map, with δ a small constant to avoid zero/infinity in the definition of $\mathbf{r}_k$. This MI prioritizes the removal of k-space sampled elements that still have large errors compared to the magnitude of the k-space signal $\mathbf{s}_n$. Essentially, the large elements of $\mathbf{r}_k$ are elements that do not fit well in the recovery, and their removal may not significantly increase the cost function.

The positional constraints (PC) rules [72] are used to avoid changing elements in fixed positions, such as the central area, that could be used for coil sensitivity calibration, and also to avoid changing, at the same iteration, elements that reside in areas that could interfere too much with each other, such as neighboring frequencies.

The strategy used in BASS leads to a fast convergence at the beginning of the algorithm, but it may bias the choice of the SP in the long run. Due to this bias, the algorithm must reduce the probability of p_a and p_r and the number of changing elements until it reaches $P = 1$ throughout the iterations, making the selection purely random after some time. In other words, the algorithm uses bias on the subset selection to accelerate the convergence, but later in the iterative process, BASS behaves like POSS [73], sharing the same convergence properties.

4.4 Joint learning of sampling and reconstruction

All the approaches discussed before considered that the reconstruction method is fixed and does not change throughout learning the SP or k-space trajectory. This is true for many parallel and CS methods, but for more recently learned reconstructions, such as the ones based on deep learning, the reconstruction and the SP can be learned together.

$$\widehat{\Omega}, \widehat{\theta} = \underset{\substack{\Omega \subset \Gamma, \theta \in \Phi \\ s.t.\ |\Omega| = M}}{\operatorname{argmin}} \frac{1}{N_d} \sum_{n=1}^{N_d} d(\boldsymbol{\rho}_n, \Theta_\theta(\Omega, \mathbf{s}_n)), \tag{4.21}$$

where $\Theta_\theta(\Omega, \mathbf{s}_n)$ represents a learned reconstruction with learned parameters $\theta \in \Phi$. One pioneering approach for joint SP and CS reconstruction is [74]. Where they adapt the SP for a CS reconstruction and, also adjust the regularization parameter as the learned parameter.

4.4.1 Joint learning of SP and reconstruction for Cartesian acquisition

Cartesian SP learning, whether represented by discrete sampling masks or sampling positions in a discrete grid, is a non-differentiable problem. As such, it cannot simply be integrated into a neural network architecture as a component and learned using the same stochastic gradient methods used to train neural networks. The solution to this problem involves either using alternating learning, where SP learning is treated as a discrete optimization while reconstruction learning is treated as a differentiable optimization problem, or using a differentiable approximation for the SP, making it possible to integrate SP learning as a component of a neural network.

In *Learning-based Optimization of the Under-sampling PattErn* (LOUPE) [75,76], the authors decided to learn the sampling density, instead of the SP, and then generate random SPs from it. This way, the SP generator can use a differentiable sigmoid function, while the sampling density can be learned. This is only partially useful because the model used for sampling density assumes that the probability of each sampling position is independent. Consequently, LOUPE cannot learn the importance of the relative position between samples, as exploited in partial Fourier and parallel MRI. The model used in LOUPE is illustrated in Fig. 4.12a.

The sampling mechanism used in LOUPE was improved in Ref. [77] by using a new sequential generator, called *Deep Probabilistic Subsampling* (DPS). This improved the generation to have an SP with exact M samples, which is only approximately obtained in LOUPE. The DPS approach was extended to multicontrast MRI in Ref. [78]. A similar probabilistic approach was presented in Ref. [79]. Note that these approaches are also limited to the independence assumption of the k-space sampling, not learning the importance of the relative position of each sample in the k-space.

In Ref. [80], an alternating learning approach is used for learning the SP and reconstruction parameters. Here the learned SP is deterministic and valid for the entire dataset, and the algorithm is free to learn all aspects of the SP. In this joint approach, the optimization alternates between learning the SP with BASS [72] and learning the DL reconstruction with [81]. In Ref. [80], the problem in Eq. (4.21) is written as:

$$\Omega_{m+1} = \underset{\substack{\Omega \subset \Gamma \\ s.t. |\Omega| = M}}{\operatorname{argmin}} \frac{1}{N_d} \sum_{n=1}^{N_d} d\left(\rho_n, \Theta_{\theta_m}(\Omega, s_n)\right), \tag{4.22}$$

$$\theta_{m+1} = \underset{\theta \in \Phi}{\operatorname{argmin}} \frac{1}{N_d} \sum_{d=1}^{N_d} d(\rho_n, \Theta_\theta(\Omega_{m+1}, s_n)), \tag{4.23}$$

Also in Ref. [80], a non-monotone version of BASS was proposed (See Algorithm 4.2, non-monotone BASS), illustrating that monotonicity is not a requirement for successfully obtaining good SPs with BASS. In Ref. [82], a stochastic version of BASS was proposed, to keep the alternating learning fast even with very large datasets.

4.4.2 Joint learning of free k-space trajectories for non-Cartesian acquisition

The formulation of k-space trajectory learning for non-Cartesian acquisitions is differentiable for the k-space positions. This makes it feasible to solve the joint learning of deep learning reconstruction and a non-Cartesian k-space trajectory using deep learning optimization tools.

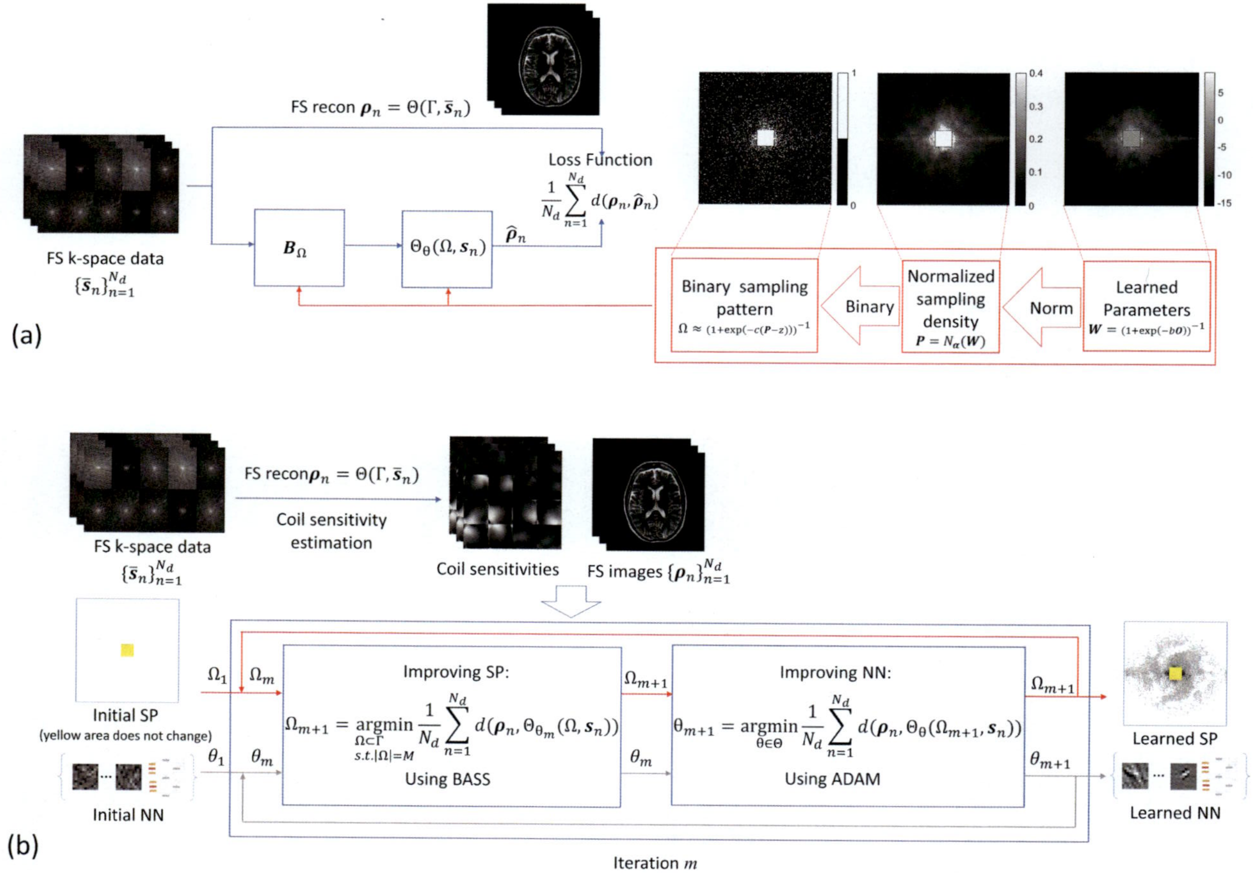

FIGURE 4.12

Comparison of different joint learning approaches for Cartesian trajectories. In (a) LOUPE, from Ref. [76], is shown. In (b), the alternating approach from Ref. [80] is shown, where BASS is used to solve SP optimization and ADAM is used to train a DL reconstruction.

Essentially, all the approaches in this section extend the idea presented in Ref. [56] with a modification of the cost function in (8) to a general form that involves image reconstruction and can be jointly solved by deep learning tools:

$$\widehat{\boldsymbol{k}}, \widehat{\theta} = \underset{\boldsymbol{k} \in \mathcal{Q}, \theta \in \Phi}{\operatorname{argmin}} \mathscr{L}(\boldsymbol{k}, \theta). \tag{4.24}$$

Also respecting the constraints of maximum gradient, in Eq. (4.6), and maximum slew rate, in Eq. (4.7), modeled as a constraint $\mathcal{Q}$.

In [83], the J-MoDL is proposed. In this study, the authors used linear trajectories, as in Cartesian acquisitions, but modeled the phase encoding positions in the continuous domain. The cost function proposed in Ref. [83] is:

$$\mathscr{L}(\boldsymbol{k}, \theta) = \frac{1}{N_d} \sum_{n=1}^{N_d} d(\boldsymbol{\rho}_n, \Theta_\theta(\Omega(\boldsymbol{k}), \boldsymbol{E}_{\Omega(\boldsymbol{k})}\boldsymbol{\rho}_n)), \tag{4.25}$$

where $\Omega(\boldsymbol{k})$ is the SP interpolated from $\boldsymbol{k}$, with $d(\boldsymbol{\rho}, \widehat{\boldsymbol{\rho}}) = \|\boldsymbol{\rho} - \widehat{\boldsymbol{\rho}}\|_2^2$. Note that the k-space data is also synthetically generated by $\boldsymbol{E}_{\Omega(\boldsymbol{k})}\boldsymbol{\rho}_n$.

Another pioneer approach is the *Physics-Informed Learned Optimized Trajectories* (PILOT) [84], where the cost function is formulated as:

$$\mathscr{L}(\boldsymbol{k}, \theta) = \frac{1}{N_d} \sum_{n=1}^{N_d} d(\boldsymbol{\rho}_n, \Theta_\theta(\Omega(\boldsymbol{k}), \boldsymbol{E}_{\Omega(\boldsymbol{k})}\boldsymbol{\rho}_n)) + \mathscr{L}_{\mathcal{Q}}(\boldsymbol{k}), \tag{4.26}$$

where the gradient constraints $\mathcal{Q}$ are included as a side penalty $\mathscr{L}_{\mathcal{Q}}(\boldsymbol{k})$.

In both studies, [83,84], instability in the joint optimization is mentioned, particularly in finding the k-space trajectory. In Ref. [83], the reduction of the number of parameters was used to stabilize the optimization. In this work, instead of estimating free k-space trajectories, they constrained the trajectories to be horizontal or vertical straight lines, as seen in Fig. 4.3b, and the phase encoding positions are continuous free parameters. In Ref. [84], several different starting configurations were used to stabilize the optimization problem, since it was noticed that the optimization always converges to a local solution (this was also observed before in Ref. [56]).

There are two main reasons for the instability in learning a free k-space trajectory. One reason is the lack of differentiability when the NUFFT is used in $\boldsymbol{E}_{\Omega(\boldsymbol{k})}$. Even though the Fourier transform is differentiable with respect to the k-space positions, the NUFFT is not always differentiable, mainly due to the use of frequency-domain interpolation inside the operator, as mentioned in Ref. [85]. The use of auto-differentiability of the NUFFT operator as a function of the trajectory parameters is not the best option.

In Ref. [84], the auto-differentiation on NUFFT, based on [86], was used with some limited success only because bilinear interpolation was used with the NUFFT. In Ref. [85], the authors addressed this limitation by showing how to compute the exact Jacobian matrices of $\boldsymbol{E}_{\Omega(\boldsymbol{k})}$. The proposed approach is more accurate, fast, and memory efficiency, improving the stability and the learning speed for non-Cartesian trajectories. Recently, in Ref. [87], the authors reported that Kaiser-Bessel functions with specific parameters for the NUFFT performed reasonably even with automatic differentiation tools, but good initialization is important, as reported in Ref. [84].

The other reason for the instability of the learning procedure is related to the number of free parameters, as clarified in Refs. [83,84]. In Ref. [88], the *B-Spline Parameterized Joint Optimization of Reconstruction and K-Space Trajectories* (BJORK) is proposed, where B-splines are used to express the k-space trajectory with a reduced number of parameters, similar to how straight lines were used in Ref. [83]. This way, only a few control points are optimized and the whole trajectory is produced by interpolation, giving more freedom and stability to the learned trajectory, particularly avoiding local minima. A multi-scale optimization approach is used in Refs. [64,84,88].

In [64], the *PROjection for Jointly learning non-Cartesian Trajectories while Optimizing Reconstructor* (PROJeCTOR) was proposed. The authors formulated the gradient constraints $\mathcal{Q}$ as hard constraints, following [56], instead of modeling them using side penalties as in Eq. (4.26), used in Refs. [84,88]. This is done with a constrained optimization algorithm, similar to the one shown in Eq. (4.11), the advantage is that PROJeCTOR pushes the gradients closer to their limits, without surpassing them, compared to the other approaches. A comparison between PILOT, BJORK, and PROJeCTOR is shown in Table 4.1 and in Fig. 4.13.

4.4.3 Extension to non-Cartesian 3D acquisitions

The latest joint learning approach for non-Cartesian acquisitions, the PROJeCTOR [64], has already been proposed for 2D and 3D problems. The other two approaches were extended to 3D problems separately. In Ref. [91], the team that developed PILOT [84], extended the approach to 3D, resulting in a method called 3D Feasible Learned Acquisition Trajectories (FLAT), but only considered one-coil problems. In Ref. [92], the team that developed BJORK [88], extended their approach to 3D problems with a method called Stochastic optimization framework for 3D NOn-Cartesian samPling trajectorY (SNOPY). In SNOPY [92], they also included a term in the loss function to reduce peripheral nerve stimulation [3,93], one of the factors that limit the use of high amplitude and high slew rates in gradient systems when scanning human subjects.

The most challenging aspect of the 3D joint learning problem for non-Cartesian 3D acquisitions, as shown in Fig. 4.4c, is the memory resources required by the computational system used for learning because the 3D reconstruction is not separable. Due to this, the number of control points in the trajectory has to be reduced, as in BJORK [88], and only reconstructions that do not use much memory are employed (many new DL methods have to be excluded), and the batch size of stochastic gradients have to be reduced to only one data element at a time.

Table 4.1 Comparing different approaches for jointly learning free non-Cartesian trajectories and deep learning reconstruction.

Method	Jacobian of E_k w. r.t trajectory	Optimization strategy	Gradient limits	Type of DL reconstruction	Validated with real acquisitions
PILOT [84]	Bilinear with auto-diff. From [86]	Multi-scale	Side penalty	UNET (no data-discrepancy layers)	No
BJORK [88]	From [85]	Multi-scale with B-splines	Site penalty	MoDL [89] (unrolled network)	Yes
PROJeCTOR [64]	From [85]	Multi-scale	Constraints	NC-PDNet [90] (unrolled network with density compensation)	Yes

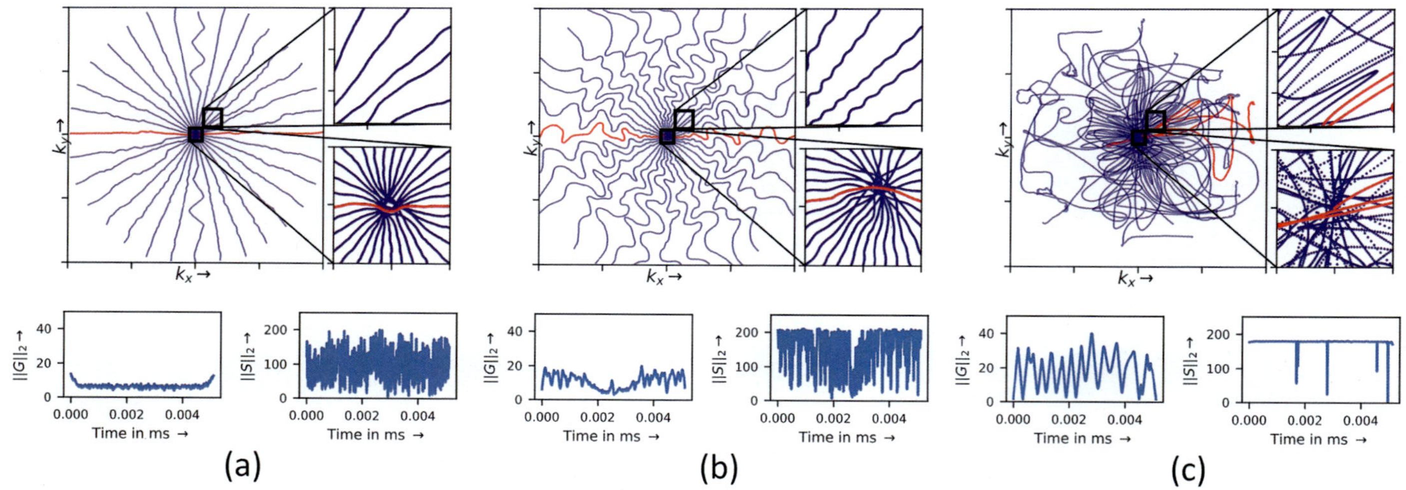

FIGURE 4.13

Comparison of different approaches for non-Cartesian trajectories. In (a) PILOT, (b) BJORK, and (c) PROJeCTOR. In all methods the initial trajectory was radial. (see more in Ref. [64]).

4.5 What the machine can learn about sampling patterns and k-space trajectories

Some aspects of the system model, $\boldsymbol{E}_{\Omega}$, and aspects of the sparsity model were learned in Refs. [65,66], where Bayesian and CRLB optimizations were used. However, no reconstruction-specific or data-specific learning was demonstrated. This is expected because these approaches are directed to learn reconstruction- and data-agnostic SPs that have a good performance on general data, but are not able to extract more specific information from the data to fully exploit the advantages of the reconstruction method employed.

In Ref. [56], more k-space coverage per readout was learned useful by optimization of non-Cartesian trajectories. This was also observed later in Refs. [64,84,88]. Non-Cartesian trajectories are more effective in covering k-space when magnetic gradients operate close to their limits, leading to stretched readouts with many twists and turns.

The dependency of the sampling density on the data was learned in Refs. [75–79]. This improved previous heuristic assumptions about the sampling density used in Refs. [56,59]. Unfortunately, these probabilistic approaches assume independence of the sampling in each location, because of this they are not able to learn the advantages of specific k-space sampling positions.

More recent approaches demonstrated that multiple data-related and reconstruction method-related aspects can be learned simultaneously by the machine. In Refs. [72,80,88], it was shown that the machine was able to learn by itself to exploit complex-conjugacy properties of the k-space, as learned by humans in Partial Fourier, as seen in Fig. 4.14a for Cartesian acquisitions and Fig. 4.14b for non-Cartesian acquisitions. In examples using 2D + time data in Refs. [72,80], it was shown that the machine was able to learn time correlation and exploit it in SP approaches such as in keyhole, only resampling the center of the k-space in following acquisitions over time, as seen in Fig. 4.14c. The increased space between k-space samples, exploited in parallel MRI and Poisson-Disc SPs, was learned by the machine when the multi-coil model was used [72,80,82]. Variable sampling density, as in many CS methods was also learned by the machine, as seen in all the examples in Fig. 4.14. Some recent research [87] has reported similar observations.

4.6 Future directions

Learned non-Cartesian acquisitions are the current most promising approaches due to the freedom to cover more k-space area at each repetition of the pulse sequence. It can achieve even better results as MRI hardware improves, particularly gradient systems. However, the deviation between the actual and the nominal trajectories should be considered in the optimization models.

Learned Cartesian acquisitions have also room to improve when non-linear trajectories are used, such as with EPI or Wave-CAIPI schemes [94]. The use of learned

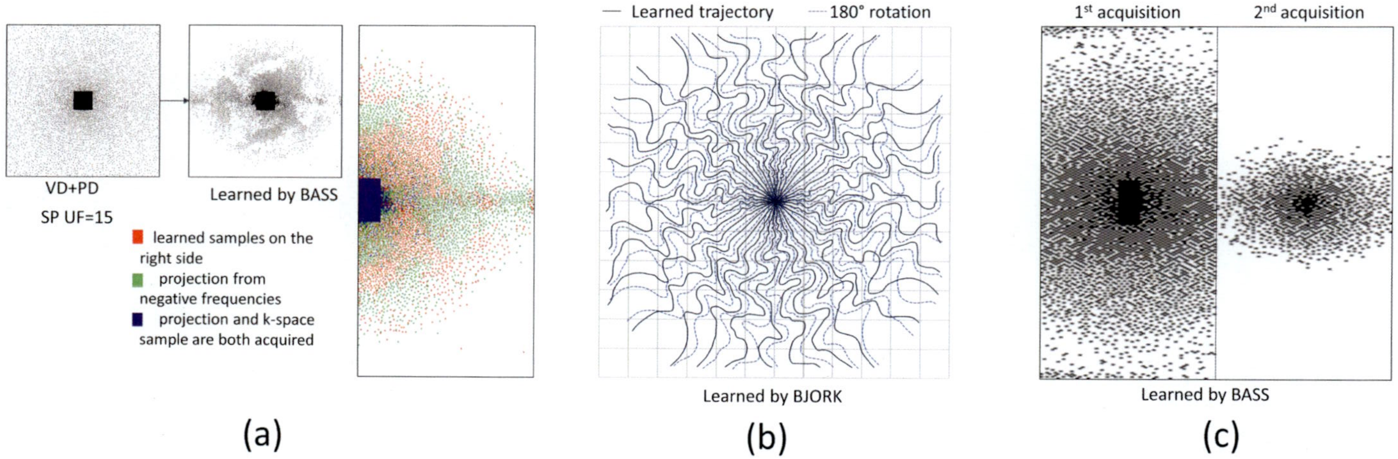

FIGURE 4.14

Sampling aspects learned by the machine with some methods discussed in this chapter.

Adapted from Refs. [80,82,88].

non-linear trajectories that pass on and samples on the Cartesian grid has not been fully exploited. Part of this challenge lies in solving the problem of reducing the differences between actual and nominal trajectories.

Object-adaptive sampling [95–100], also known as active sampling, is an interesting approach for the future of MRI. Essentially, an active predictor algorithm is used to predict the next k-space samples to be acquired by the pulse sequence while the sequence is running and the scanner is acquiring the data. The idea assumes the following hypotheses: 1) a sequentially-predicted object-specific SP is better than an anatomy-specific optimized SP (as assumed by the methods discussed in this chapter); 2) the predictor and the MRI system are fast enough to update the parameters of the pulse sequence while it is running. While these preliminary studies have shown advantages against classical fixed SPs, superiority against anatomy-specific optimized SPs has not been demonstrated yet with deep learning reconstructions. In addition, the second hypothesis is not yet realizable in current MRI systems, however, future MRI systems may be capable of exploiting these approaches.

Note that data sampling is only part of the MRI acquisition process. To effectively accelerate MRI, the overall scan time must be reduced. This requires accelerating all the steps of the pulse sequence, which may necessitate rethinking the structure and components of the sequence, including the shape of the pulses. Some examples in this direction can be found in Refs. [1,9–11,101].

4.7 Conclusion

This chapter discussed how MRI data sampling can be learned or optimized, highlighting the transition from human-inspired methods to sophisticated machine-learning techniques. We can recognize many aspects learned by the machine that were also learned by humans as beneficial to fast MRI acquisition, either for sampling patterns or non-Cartesian trajectories. An efficient MRI acquisition depends on many factors, including those related to the MRI physics, the object being scanned, and the assumptions used for reconstruction. While current methods have replicated many human-learned benefits, there remains vast potential for discovering new efficiencies. This is just the beginning, and we still need to understand what else the machine is learning as beneficial that we have not seen yet. As MRI technology and computational capabilities advance, these methodologies will likely yield even greater enhancements, pushing the boundaries of fast and efficient MRI acquisition.

Acknowledgments

The authors thank the contributions of Gabor T. Herman in the prior work [46,72] and Florian Knoll in the joint work [80]. The authors thank the contributions of Jeffrey Fessler, Guanhua Wang, Philippe Ciuciu, Chaithya Radhakrishna, Alex Bronstein, and Tomer Weiss for providing raw material for some of the figures. Results shown here were in part supported by NIH grants R01-AR076328-01A1, R01-AR076985-01A1, and R01-AR078308-01A1.

References

[1] Bernstein M, King K, Zhou X. Handbook of MRI pulse sequences. Burlington, MA: Elsevier; 2004.

[2] Liang ZP, Lauterbur PC. Principles of magnetic resonance imaging: a signal processing perspective. IEEE Press; 2000.

[3] Gudino N, Littin S. Advancements in gradient system performance for clinical and research MRI. J Magn Reson Imag 2023;57(1):57–70. https://doi.org/10.1002/jmri.28421.

[4] Schmitt F, Stehling MK, Turner R. Echo-planar imaging: theory, technique and applications. Springer-Verlag; 1998.

[5] Zhu Y, Gao S, Cheng L, Bao S. Review: K-space trajectory development. IEEE Int Conf Med Imaging Phys Eng 2013;(2011):356–60. https://doi.org/10.1109/ICMIPE.2013.6864568.

[6] Lauterbur PC. Image Formation by induced local interactions: examples employing nuclear magnetic resonance. Nature 1973;242(5394):190–1. https://doi.org/10.1038/242190a0.

[7] Fessler JA. On NUFFT-based gridding for non-Cartesian MRI. J Magn Reson 2007; 188(2):191–5. https://doi.org/10.1016/j.jmr.2007.06.012.

[8] Seiberlich N, Breuer FA, Blaimer M, Barkauskas K, Jakob PM, Griswold MA. Non-Cartesian data reconstruction using GRAPPA operator gridding (GROG). Magn Reson Med 2007;58(6):1257–65. https://doi.org/10.1002/mrm.21435.

[9] Zibetti MVW, De Moura HL, Keerthivasan MB, Regatte RR. Optimizing variable flip angles in magnetization-prepared gradient-echo sequences for efficient 3D-T1ρ mapping. Magn Reson Med Oct. 2023;90(4):1465–83. https://doi.org/10.1002/mrm.29740.

[10] Jordan SP, et al. Automated design of pulse sequences for magnetic resonance fingerprinting using physics-inspired optimization. Proc Natl Acad Sci USA 2021;118(40). https://doi.org/10.1073/pnas.2020516118.

[11] Busse RF, Hariharan H, Vu A, Brittain JH. Fast spin echo sequences with very long echo trains: design of variable refocusing flip angle schedules and generation of clinical T2 contrast. Magn Reson Med May 2006;55(5):1030–7. https://doi.org/10.1002/mrm.20863.

[12] Paschal CB, Morris HD. K-space in the clinic. J Magn Reson Imag 2004;19(2): 145–59. https://doi.org/10.1002/jmri.10451.

[13] Oppenheim AV, Willsky AS, Nawab S. Signals and systems. 2nd ed. Prentice Hall; 1997.

[14] Uecker M, Lustig M. Estimating absolute-phase maps using ESPIRiT and virtual conjugate coils. Magn Reson Med 2017;77(3):1201–7. https://doi.org/10.1002/mrm.26191.

[15] Haldar JP. Low-rank modeling of local k-space neighborhoods (LORAKS) for constrained MRI. IEEE Trans Med Imag Mar. 2014;33(3):668–81. https://doi.org/10.1109/TMI.2013.2293974.

[16] Haldar JP, Zhuo J. P-LORAKS: low-rank modeling of local k-space neighborhoods with parallel imaging data. Magn Reson Med Apr. 2016;75(4):1499–514. https://doi.org/10.1002/mrm.25717.

[17] Hu X. On the 'keyhole' technique. J Magn Reson Imag Mar. 1994;4(2):231. https://doi.org/10.1002/jmri.1880040223.

[18] Pruessmann KP. Encoding and reconstruction in parallel MRI. NMR Biomed May 2006;19(3):288–99. https://doi.org/10.1002/nbm.1042.

[19] Sodickson DK, Manning WJ. Simultaneous acquisition of spatial harmonics (SMASH): fast imaging with radiofrequency coil arrays. Magn Reson Med Oct. 1997;38(4):591–603. https://doi.org/10.1002/mrm.1910380414.

[20] Pruessmann KP, Weiger M, Scheidegger MB, Boesiger P. SENSE: sensitivity encoding for fast MRI. Magn Reson Med Nov. 1999;42(5):952–62. https://doi.org/10.1002/(SICI)1522-2594(199911)42:5<952::AID-MRM16>3.0.CO;2-S.

[21] Pruessmann KP, Weiger M, Börnert P, Boesiger P. Advances in sensitivity encoding with arbitrary k -space trajectories. Magn Reson Med Oct. 2001;46(4):638–51. https://doi.org/10.1002/mrm.1241.

[22] Griswold MA, et al. Generalized autocalibrating partially parallel acquisitions (GRAPPA). Magn Reson Med Jun. 2002;47(6):1202–10. https://doi.org/10.1002/mrm.10171.

[23] Sodickson DK, McKenzie CA, Ohliger MA, Yeh EN, Price MD. Recent advances in image reconstruction, coil sensitivity calibration, and coil array design for SMASH and generalized parallel MRI. Magn Reson Mater Phys Biol Med Jan. 2002;13(3):158–63.

[24] Barrett HH, Myers KJ. Foundations of image science. In: Wiley series in pure and applied optics. Hoboken, NJ: John Wiley and Sons, Inc.; 2004.

[25] Lustig M, Donoho DL, Pauly JM. Sparse MRI: the application of compressed sensing for rapid MR imaging. Magn Reson Med Dec. 2007;58(6):1182–95. https://doi.org/10.1002/mrm.21391.

[26] Candes EJ, Romberg J. Sparsity and incoherence in compressive sampling. Inverse Probl Jun. 2007;23(3):969–85. https://doi.org/10.1088/0266-5611/23/3/008.

[27] Candes EJ, Tao T. Near optimal signal recovery from random projections: universal encoding strategies? IEEE Trans Inf Theor 2006;52(12):5406–25.

[28] Adcock B, Hansen AC, Poon C, Roman B. Breaking the coherence barrier: a new theory for compressed sensing. Forum Math Sigma Feb. 2017;5:1–84. https://doi.org/10.1017/fms.2016.32.

[29] Unser M. Sampling-50 years after shannon. Proc IEEE 2000;88(4):569–87. http://ieeexplore.ieee.org/xpls/abs_all.jsp?arnumber=843002.

[30] Donoho D, Elad M, Temlyakov VN. Stable recovery of sparse overcomplete representations in the presence of noise. IEEE Trans Inf Theor Jan. 2006;52(1):6–18. https://doi.org/10.1109/TIT.2005.860430.

[31] Liu B, Zou YM, Ying L. Sparsesense: application of compressed sensing in parallel MRI. In: International conference on technology and applications in biomedicine. IEEE; May 2008. p. 127–30. https://doi.org/10.1109/ITAB.2008.4570588.

[32] Feng L, Benkert T, Block KT, Sodickson DK, Otazo R, Chandarana H. Compressed sensing for body MRI. J Magn Reson Imag Apr. 2017;45(4):966–87. https://doi.org/10.1002/jmri.25547.

[33] Murphy M, Alley M, Demmel J, Keutzer K, Vasanawala S, Lustig M. Fast l1-SPIRiT compressed sensing parallel imaging MRI: scalable parallel implementation and

clinically feasible runtime. IEEE Trans Med Imag Jun. 2012;31(6):1250–62. https://doi.org/10.1109/TMI.2012.2188039.
[34] Jin KH, Lee D, Ye JC. A general framework for compressed sensing and parallel MRI using annihilating filter based low-rank Hankel matrix. IEEE Trans Comput Imaging Dec. 2016;2(4):480–95. https://doi.org/10.1109/TCI.2016.2601296.
[35] Cheng JY, Zhang T, Alley MT, Lustig M, Vasanawala SS, Pauly JM. Variable-density radial view-ordering and sampling for time-optimized 3D Cartesian imaging. Proceedings of the ISMRM workshop on data sampling and image reconstruction. ISMRM; 2013.
[36] Wang Z, Arce GR. Variable density compressed image sampling. IEEE Trans Image Process Jan. 2010;19(1):264–70. https://doi.org/10.1109/TIP.2009.2032889.
[37] Puy G, Vandergheynst P, Wiaux Y. On variable density compressive sampling. IEEE Signal Process Lett Oct. 2011;18(10):595–8. https://doi.org/10.1109/LSP.2011.2163712.
[38] Kaldate A, Patre BM, Harsh R, Verma D. MR image reconstruction based on compressed sensing using Poisson sampling pattern. In: Sec inter conf cogn comp infor proc (CCIP). IEEE; Aug. 2016. https://doi.org/10.1109/CCIP.2016.7802884.
[39] Levine E, Daniel B, Vasanawala S, Hargreaves B, Saranathan M. 3D Cartesian MRI with compressed sensing and variable view sharing using complementary Poisson-disc sampling. Magn Reson Med May 2017;77(5):1774–85. https://doi.org/10.1002/mrm.26254.
[40] Dwork N, Baron CA, Johnson EMI, O'Connor D, Pauly JM, Larson PEZ. Fast variable density Poisson-disc sample generation with directional variation for compressed sensing in MRI. Magn Reson Imaging Apr. 2021;77(April 2020):186–93. https://doi.org/10.1016/j.mri.2020.11.012.
[41] Uecker M, et al. ESPIRiT-an eigenvalue approach to autocalibrating parallel MRI: where SENSE meets GRAPPA. Magn Reson Med Mar. 2014;71(3):990–1001. https://doi.org/10.1002/mrm.24751.
[42] Meyer CH, Hu BS, Nishimura DG, Macovski A. Fast spiral coronary artery imaging. Magn Reson Med Dec. 1992;28(2):202–13. https://doi.org/10.1002/mrm.1910280204.
[43] Robison RK, Li Z, Wang D, Ooi MB, Pipe JG. Correction of B0 eddy current effects in spiral MRI. Magn Reson Med 2019;81(4):2501–13. https://doi.org/10.1002/mrm.27583.
[44] Vannesjo SJ, et al. Gradient system characterization by impulse response measurements with a dynamic field camera. Magn Reson Med 2013;69(2):583–93. https://doi.org/10.1002/mrm.24263.
[45] Duyn JH, Yang Y, Frank JA, Van Der Veen JW. Simple correction method for k-space trajectory deviations in MRI. J Magn Reson 1998;132(1):150–3. https://doi.org/10.1006/jmre.1998.1396.
[46] Helou ES, Zibetti MVW, Axel L, Tobias Block K, Regatte RR, Herman GT. The discrete Fourier transform for golden angle linogram sampling. Inverse Probl Dec. 2019;35(12):125004. https://doi.org/10.1088/1361-6420/ab44ee.
[47] Pipe JG. Motion correction with PROPELLER MRI: application to head motion and free-breathing cardiac imaging. Magn Reson Med 1999;42(5):963–9. https://doi.org/10.1002/(SICI)1522-2594(199911)42:5<963::AID-MRM17>3.0.CO;2-L.
[48] Noll DC. Multishot rosette trajectories for spectrally selective mr imaging. IEEE Trans Med Imag 1997;16(4):372–7. https://doi.org/10.1109/42.611345.

[49] Pipe JG. An optimized center-out k-space trajectory for multishot MRI: comparison with spiral and projection reconstruction. Magn Reson Med 1999;42(4):714–20. https://doi.org/10.1002/(SICI)1522-2594(199910)42:4<714::AID-MRM13>3.0.CO;2-G.

[50] Jackson JI, Nishimura DG, Macovski A. Twisting radial lines with application to robust magnetic resonance imaging of irregular flow. Magn Reson Med 1992;25(1):128–39. https://doi.org/10.1002/mrm.1910250113.

[51] Chauffert N, Weiss P, Kahn J, Ciuciu P. A projection algorithm for gradient waveforms design in magnetic resonance imaging. IEEE Trans Med Imag 2016;35(9):2026–39. https://doi.org/10.1109/TMI.2016.2544251.

[52] Lustig M, Kim S-J, Pauly JM. A fast method for designing time-optimal gradient waveforms for arbitrary k-space trajectories. IEEE Trans Med Imag Jun. 2008;27(6):866–73. https://doi.org/10.1109/TMI.2008.922699.

[53] Hao S, Fessler JA, Noll DC, Nielsen J-F. Joint design of excitation k-space trajectory and RF pulse for small-tip 3D tailored excitation in MRI. IEEE Trans Med Imag Feb. 2016;35(2):468–79. https://doi.org/10.1109/TMI.2015.2478880.

[54] Boyer C, Chauffert N, Ciuciu P, Kahn J, Weiss P. On the generation of sampling schemes for magnetic resonance imaging. SIAM J Imag Sci Jan. 2016;9(4):2039–72. https://doi.org/10.1137/16M1059205.

[55] Feng L. Golden-angle radial MRI: basics, advances, and applications. J Magn Reson Imag Apr. 2022:1–18. https://doi.org/10.1002/jmri.28187.

[56] Lazarus C, et al. SPARKLING: variable-density k-space filling curves for accelerated T2*-weighted MRI. Magn Reson Med Jun. 2019;81(6):3643–61. https://doi.org/10.1002/mrm.27678.

[57] Ahn CB, Kim JH, Cho ZH. High-speed spiral-scan echo planar NMR imaging-I. IEEE Trans Med Imag Mar. 1986;5(1):2–7. https://doi.org/10.1109/TMI.1986.4307732.

[58] Block KT, Frahm J. Spiral imaging: a critical appraisal. J Magn Reson Imag Jun. 2005;21(6):657–68. https://doi.org/10.1002/jmri.20320.

[59] Knoll F, Clason C, Diwoky C, Stollberger R. Adapted random sampling patterns for accelerated MRI. Magn Reson Mater Phys Biol Med Feb. 2011;24(1):43–50. https://doi.org/10.1007/s10334-010-0234-7.

[60] Choi J, Kim H. Implementation of time-efficient adaptive sampling function design for improved undersampled MRI reconstruction. J Magn Reson Dec. 2016;273:47–55. https://doi.org/10.1016/j.jmr.2016.10.006.

[61] Vellagoundar J, Machireddy RR. A robust adaptive sampling method for faster acquisition of MR images. Magn Reson Imaging Jun. 2015;33(5):635–43. https://doi.org/10.1016/j.mri.2015.01.008.

[62] Sabat S, Mir R, Guarini M, Guesalaga A, Irarrazaval P. Three dimensional k-space trajectory design using genetic algorithms. Magn Reson Imaging 2003;21(7):755–64. https://doi.org/10.1016/S0730-725X(03)00174-7.

[63] Chaithya G, Weiss P, Daval-Frerot G, Massire A, Vignaud A, Ciuciu P. Optimizing full 3D SPARKLING trajectories for high-resolution magnetic resonance imaging. IEEE Trans Med Imag Aug. 2022. https://doi.org/10.1109/TMI.2022.3157269. 1–1.

[64] Radhakrishna CG, Ciuciu P. Jointly learning non-cartesian k-space trajectories and reconstruction networks for 2D and 3D MR imaging through projection. Bioengineering Jan. 2023;10(2):158. https://doi.org/10.3390/bioengineering10020158.

[65] Haldar JP, Kim D. OEDIPUS: an experiment design framework for sparsity-constrained MRI. IEEE Trans Med Imag Jul. 2019;38(7):1545–58. https://doi.org/10.1109/TMI.2019.2896180.
[66] Seeger M, Nickisch H, Pohmann R, Schölkopf B. Optimization of k-space trajectories for compressed sensing by Bayesian experimental design. Magn Reson Med 2010;63(1):116–26. https://doi.org/10.1002/mrm.22180.
[67] Elad M, Milanfar P, Rubinstein R. Analysis versus synthesis in signal priors. Inverse Probl Jun. 2007;23(3):947–68. https://doi.org/10.1088/0266-5611/23/3/007.
[68] Gözcü B, et al. Learning-based compressive MRI. IEEE Trans Med Imag 2018;37(6):1394–406. https://doi.org/10.1109/TMI.2018.2832540.
[69] Gözcü B, Sanchez T, Cevher V. Rethinking sampling in parallel MRI: a data-driven approach. In: Eur signal process conf IEEE; Sep. 2019. p. 1–5. https://doi.org/10.23919/EUSIPCO.2019.8903150.
[70] Sanchez T, et al. Scalable learning-based sampling optimization for compressive dynamic MRI. In: IEEE inter conf acous speech sig proc. IEEE; May 2020. p. 8584–8. https://doi.org/10.1109/ICASSP40776.2020.9053345.
[71] Ravishankar S, Bresler Y. Adaptive sampling design for compressed sensing MRI. In: Inter conf IEEE eng med bio soc. IEEE; Aug. 2011. p. 3751–5. https://doi.org/10.1109/IEMBS.2011.6090639.
[72] Zibetti MVW, Herman GT, Regatte RR. Fast data-driven learning of parallel MRI sampling patterns for large scale problems. Sci Rep Dec. 2021;11(1):19312. https://doi.org/10.1038/s41598-021-97995-w.
[73] Qian C, Yu Y, Zhou ZH. "Subset selection by Pareto optimization," in advances in neural information processing systems. In: Advances in neural information processing systems. vol 2015; 2015. p. 1774–82.
[74] Sherry F, et al. Learning the sampling pattern for MRI. IEEE Trans Med Imag Dec. 2020;39(12):4310–21. https://doi.org/10.1109/TMI.2020.3017353.
[75] Bahadir CD, Wang AQ, Dalca AV, Sabuncu MR. Deep-learning-based optimization of the under-sampling pattern in MRI. IEEE Trans Comput Imaging 2020;6(c):1139–52. https://doi.org/10.1109/TCI.2020.3006727.
[76] Zhang J, et al. Extending LOUPE for K-space under-sampling pattern optimization in multi-coil MRI. Lect Notes Comput Sci 2020;12450(LNCS):91–101. https://doi.org/10.1007/978-3-030-61598-7_9.
[77] Huijben IAM, Veeling BS, van Sloun RJG. Learning sampling and model-based signal recovery for compressed sensing MRI. In: IEEE international conference on acoustics, speech and signal processing. IEEE; May 2020. p. 8906–10. https://doi.org/10.1109/ICASSP40776.2020.9053331.
[78] Wang J, Yang Q, Yang Q, Xu L, Cai C, Cai S. Joint optimization of Cartesian sampling patterns and reconstruction for single-contrast and multi-contrast fast magnetic resonance imaging. Comput Methods Progr Biomed Nov. 2022;226:107150. https://doi.org/10.1016/j.cmpb.2022.107150.
[79] Xie J, Zhang J, Zhang Y, Ji X. PUERT: probabilistic under-sampling and explicable reconstruction network for CS-MRI. IEEE J Sel Top Signal Process Jun. 2022;16(4):737–49. https://doi.org/10.1109/JSTSP.2022.3170654.
[80] Zibetti MVW, Knoll F, Regatte RR. Alternating learning approach for variational networks and undersampling pattern in parallel MRI applications. IEEE Trans Comput Imaging Oct. 2022;8:449–61. https://doi.org/10.1109/TCI.2022.3176129.

[81] Kingma DP, Ba J. Adam: a method for stochastic optimization. arXiv Prepr Dec. 2014: 1–15.

[82] Zibetti MVW, Regatte RR. Stochastic bias-accelerated subset selection algorithm for joint learning of sampling pattern and reconstruction in accelerated MRI. In: IEEE international symposium on biomedical imaging. IEEE; Apr. 2023. p. 1–5. https://doi.org/10.1109/ISBI53787.2023.10230762.

[83] Aggarwal HK, Jacob M. J-MoDL: joint model-based deep learning for optimized sampling and reconstruction. IEEE J Sel Top Signal Process Oct. 2020;14(6):1151–62. https://doi.org/10.1109/JSTSP.2020.3004094.

[84] Weiss T, Senouf O, Vedula S, Michailovich O, Zibulevsky M, Bronstein A. PILOT: physics-informed learned optimized trajectories for accelerated MRI. J Mach Learn Biomed Imaging Sep. 2021;6:1–23.

[85] Wang G, Fessler JA. Efficient approximation of jacobian matrices involving a non-uniform fast fourier transform (NUFFT). IEEE Trans Comput Imaging 2023;9: 43–54. https://doi.org/10.1109/TCI.2023.3240081.

[86] Paszke A, et al. PyTorch: an imperative style, high-performance deep learning library. Adv Neural Inf Process Syst 2019;32(NeurIPS).

[87] Alkan C, Mardani M, Liao C, Li Z, Vasanawala SS, Pauly JM. AutoSamp: autoencoding k-space sampling via variational information maximization for 3D MRI. IEEE Trans Med Imag 2024. https://doi.org/10.1109/TMI.2024.3443292.

[88] Wang G, Luo T, Nielsen J-F, Noll DC, Fessler JA. B-spline parameterized joint optimization of reconstruction and K-space trajectories (BJORK) for accelerated 2D MRI. IEEE Trans Med Imag Sep. 2022;41(9):2318–30. https://doi.org/10.1109/TMI.2022.3161875.

[89] Aggarwal HK, Mani MP, Jacob M. MoDL: model-based deep learning architecture for inverse problems. IEEE Trans Med Imag Feb. 2019;38(2):394–405. https://doi.org/10.1109/TMI.2018.2865356.

[90] Ramzi Z, Chaithya GR, Starck JL, Ciuciu P. NC-PDNet: a density-compensated unrolled network for 2D and 3D non-cartesian MRI reconstruction. IEEE Trans Med Imag 2022;41(7):1625–38. https://doi.org/10.1109/TMI.2022.3144619.

[91] Alush-Aben J, Ackerman-Schraier L, Weiss T, Vedula S, Senouf O, Bronstein A. 3D FLAT: feasible learned acquisition trajectories for accelerated MRI. Lect Notes Comput Sci 2020;12450(LNCS):3–16. https://doi.org/10.1007/978-3-030-61598-7_1.

[92] Wang G, Nielsen J-F, Fessler JA, Noll DC. Stochastic optimization of three-dimensional non-Cartesian sampling trajectory. Magn Reson Med Aug. 2023;90(2): 417–31. https://doi.org/10.1002/mrm.29645.

[93] Tan ET, et al. Peripheral nerve stimulation limits of a high amplitude and slew rate magnetic field gradient coil for neuroimaging. Magn Reson Med 2020;83(1): 352–66. https://doi.org/10.1002/mrm.27909.

[94] Bilgic B, et al. Wave-CAIPI for highly accelerated 3D imaging. Magn Reson Med Jun. 2015;73(6):2152–62. https://doi.org/10.1002/mrm.25347.

[95] Huang Z, Ravishankar S. Single-pass object-adaptive data undersampling and reconstruction for MRI. IEEE Trans Comput Imaging 2022;8:333–45. https://doi.org/10.1109/TCI.2022.3167454.

[96] Jin KH, Unser M, Yi KM. Self-supervised deep active accelerated MRI. arXiv Prepr Jan. 2019:1–13.

[97] Bakker T, Muckley M, Romero-Soriano A, Drozdzal M, Pineda L. On learning adaptive acquisition policies for undersampled multi-coil MRI reconstruction. arXiv Prepr Mar. 2022;(1993):1–23.

[98] Yin T, Wu Z, Sun H, Dalca AV, Yue Y, Bouman KL. End-to-End sequential sampling and reconstruction for MRI. In: Proceedings of machine learning research; 2021. p. 261–81.

[99] Zhang Z, Romero A, Muckley MJ, Vincent P, Yang L, Drozdzal M. Reducing uncertainty in undersampled MRI reconstruction with active acquisition. In: The IEEE conference on computer vision and pattern recognition (CVPR); 2019.

[100] Pineda L, Basu S, Romero A, Calandra R, Drozdzal M. Active MR k-space sampling with reinforcement learning. Lect Notes Comput Sci Jul. 2020;12262(LNCS):23–33. https://doi.org/10.1007/978-3-030-59713-9_3.

[101] Zibetti MVW, Menon R, De Moura HL, Keerthivasan MB, Regatte RR. Optimized MR pulse sequence for high-resolution brain 3D-T1ρ mapping with weighted spin-lock acquisitions. Magn Reson Med Dec. 2024. https://doi.org/10.1002/mrm.30412.

CHAPTER

5 MR imaging in the low-field: Leveraging the power of machine learning

Andreas Kofler[1, *], Dongyue Si[2], David Schote[1], Rene M. Botnar[2, 3], Christoph Kolbitsch[1] and Claudia Prieto[2, 3]

[1]*Physikalisch-Technische Bundesanstalt (PTB), Braunschweig and Berlin, Germany;* [2]*King's College London, London, United Kingdom;* [3]*Pontificia Universidad Católica de Chile, Santiago de Chile, Chile*

[*]*Corresponding author E-mail: andreas.kofler@ptb.de*

5.1 Introduction

Magnetic resonance imaging (MRI) is an essential tool for the early detection, risk stratification, prognosis, treatment selection, and monitoring of many diseases, including cancer, cardiovascular disease, metabolic, musculoskeletal, and brain disorders, among many others. Its ability to produce multi-contrast and multi-parametric images of soft tissues, coupled with its non-invasive and radiation-free nature, makes it a highly valuable tool in clinical practice.

Over the past 5 decades, the technology behind MRI has undergone significant advancements, especially in terms of the magnetic field strengths used for imaging. Early MRI systems operated at low field strengths (0.15 T to 0.35 T) [1–3], and while they offered important diagnostic insights, they were limited by low signal-to-noise ratio (SNR) and image resolution. Over time, several advancements led to the development of systems operating at higher field strengths, such as 1.5 T and 3 T, which are now considered the clinical standard due to their superior SNR and image quality [4,5]. Recent developments have even pushed field strengths to ultra-high levels ($\geq$ 3 T), including 5 T, 7 T and beyond, further enhancing the spatial and temporal resolution of MRI [4–7]. However, high-field MRI has its challenges [8]. While it offers improved imaging quality, higher fields are associated with certain drawbacks, including increased costs, higher specific absorption rates (SAR), and higher radiofrequency-induced heating, which are particularly problematic for patients with implants. Although MR-safe implants exist, they typically lead to more severe image artifacts at higher fields, see e.g., Ref. [9]. However, advances in state-of-the-art MRI are driven not solely by the higher

Machine Learning in MRI. https://doi.org/10.1016/B978-0-443-14109-6.00005-5

magnetic field strength but also by a wide range of other remarkable developments in hardware and software, including improved coil arrays, digital receiver technology, and reconstruction algorithms, among many others. In recent years, there has been renewed interest in low-field MRI (< 1 T) and ultra-low-field MRI (< 0.1 T), taking full advantage of these cutting-edge developments. With lower power consumption, reduced SAR, decreased field inhomogeneity, and lower installation and maintenance costs, low-field MRI presents an opportunity to make MRI more accessible and cost-effective, particularly in resource-limited settings. Furthermore, ultra-low-field MRI offers the potential for small, mobile systems without special installation requirements, opening the door to bringing this technology to point-of-care settings, such as emergency rooms and general practitioner practices.

Despite its advantages, low-field and ultra-low-field MRI faces several technical challenges, particularly in terms of reduced SNR, spatial resolution, and scan times compared to high-field systems. These limitations could restrict its broader adoption and application, especially in cases where high-resolution and high-sensitivity imaging is required. Recently, several Machine Learning (ML) techniques have been proposed that can aid in overcoming these limitations and enhance the performance of low-field MRI. ML-based undersampled image reconstruction techniques, denoising methods, and super-resolution algorithms offer the potential to compensate for the reduced SNR inherent in low-field systems. Moreover, ML can assist in automating image analysis and optimizing acquisition protocols, as well as in image post-processing and analysis, thereby improving the diagnostic utility and efficiency of low-field MRI.

In this chapter, we explore the challenges associated with low-field and ultra-low-field MRI and highlight the growing role of ML in addressing these limitations. A brief introduction to deep neural networks is provided to give context for the discussed methods. A summary of ML-based solutions for image reconstruction acquisition and post-processing in low-field MRI is also included. Finally, we discuss the potential of integrating ML-based solutions with low-field MRI to broaden its use in clinical practice.

5.2 Challenges of low-field MRI

In recent years, contemporary low-field (< 1 T) and ultra-low-field (< 0.1 T) MRI has been shown to provide potential solutions to common historical limitations of lower field MRI, with several advantages including low power, low SAR, less field inhomogeneity and lower costs [10–16]. Furthermore, lower-field MRI has advantages in comparison to high-field strengths for some applications including MRI-guided interventions, imaging anatomy near air-tissue interfaces, and efficient image acquisition methods such as spiral out acquisition [17–19]. An early attempt at contemporary low-field MRI was conducted by modifying a 1.5 T scanner to operate at 0.55 T [20]. This study demonstrated that standard pulse sequences can provide adequate image quality in routine imaging on the high-performance ramped-down 0.55 T system in comparison with 1.5 T. Commercial low-field systems such as 0.5 T Synaptive Evry intraoperative scanner, 0.55 T Siemens Magnetom Free.Max general-purpose scanners and ultra-low-field systems such as 64 mT Hyperfine's Swoop portable MRI system have also been introduced subsequently. With lower costs, they have the potential to increase the adoption of MRI over the world. In addition to commercial scanners, there are several research groups and consortia developing ultra-low-field prototype systems for brain- or extremities-only imaging or whole-body imaging, which can be portable and more patient-friendly, enabling point-of-care applications [15,16]. Although low-field MRI has many advantages in comparison to high-field systems and can bring new applications to the clinic, it still faces many challenges [13,14].

5.2.1 Signal-to-noise ratio (SNR)

SNR is a critical factor in determining image quality in MRI, especially in clinical settings where the visualization of fine anatomical details is vital for accurate diagnosis. The main drawback of lower field imaging is the inherent reduction in SNR due to the weaker magnetic field, see Fig. 5.1, as the magnetization signal is proportional to the B0 field strength [12]. At lower magnetic fields, the net magnetization of protons in the body is reduced, leading to a weaker signal. This diminished signal must compete with the noise inherent to the systems, which remains relatively comparable regardless of field strength, assuming the same acquisition parameters and hardware [21]. As a result, the overall SNR decreases, making it more challenging to produce high-quality images. The loss of SNR at low-field can be partly compensated by using different acquisition sequences or modifying imaging parameters such as bandwidth and flip angle [19,22]. Nevertheless, these adjustments can introduce changes in contrast and potential artifacts that must be carefully considered.

High spatial resolution is critical in many clinical scenarios, such as in neuroimaging, where fine details of brain structures must be discerned. The lower SNR (per unit time) and lower scan efficiency of lower field systems often translate into lower achievable spatial resolution. Low-resolution MRI could be sufficient for broad assessments and follow-up scans. However, to enable the utility in cases requiring detailed high-resolution imaging, specific optimization may be required for low-field especially ultra-low-field MRI to match the image quality and diagnostic ability of higher-field MRI systems [23].

Fig. 5.1 shows an exemplary comparison of a brain MR image obtained from a scanner and its potential counterpart obtained by simulating the different degradation processes that are inherently linked to a low-field image reconstruction problem.

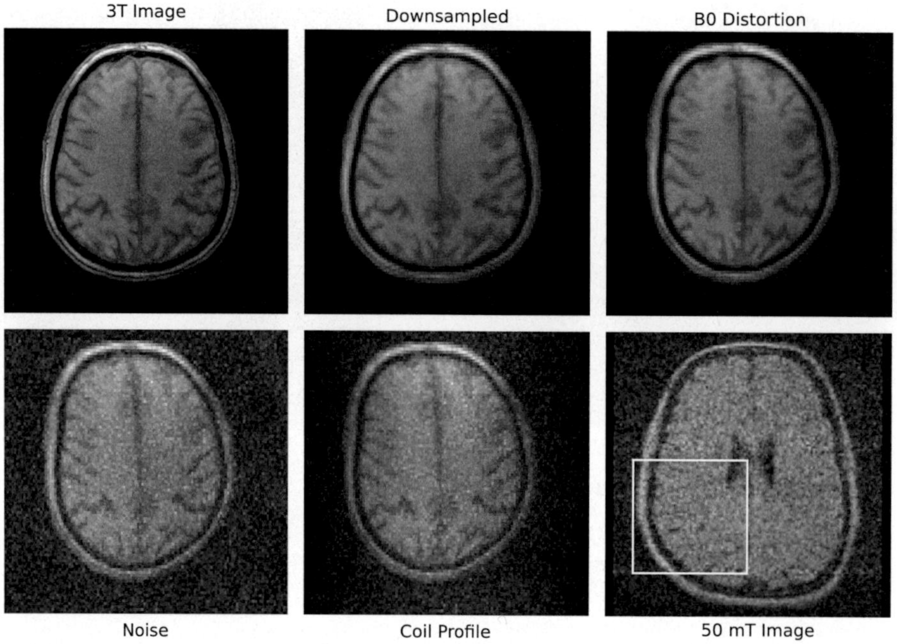

FIGURE 5.1

Comparison of a high-field (3 T) and a corresponding simulated ultra-low-field (50 mT) MR image. Lower field systems typically only achieve a decreased spatial resolution (*downsampled*), are exposed to strong B0-inhomogeneities (*B0 distortion*), poor SNR (*noise*) and use receiver coils with high Q-factors (*coil profile*). The upper left image is an example taken from the fastMRI brain dataset [26], the intermediate images were obtained by applying the just mentioned different degradation steps. The lower right image is taken from Ref. [27] and was acquired with the 50 mT ultra low-field scanner described in Ref. [28].

Image courtesy of Andrew Webb.

5.2.2 Scan time

Scan time is an important practical factor in clinical applications. However, low-field MRI usually exhibits longer scan times. On the one hand, as the SNR decreases at low field, the SNR per unit time is also reduced. Therefore, to achieve sufficient SNR, low-field imaging usually requires more time compared to higher field strengths. For example, using a lower acquisition bandwidth can improve the SNR, but at the same increases the repetition time and thus reduces the scan time efficiency. Additionally, k-space undersampling at low-field may achieve lower acceleration factors, and repeating scan and averaging signals may also be necessary to obtain high-SNR images [24].

On the other hand, the efficiency of MRI is also influenced by the performance of gradient and radiofrequency (RF) systems [12,25]. Gradients are used for spatially encoding the MRI signal. Less powerful gradients lead to longer repetition and echo times, thus lower scan efficiency. Similarly, the RF transmitter and receiver system, responsible for exciting the protons and receiving the emitted signals, also influence the duration of RF pulses and repetition time. In low-field MRI, these systems often face significant limitations. Commercial low-field MRI scanners, often come with more basic gradient and RF systems. These limitations can result in longer acquisition times, making the scanning process less efficient and more prone to motion artifacts.

5.2.3 Relaxation differences

Relaxation times (T1 and T2) are fundamental to the contrast mechanisms in MRI, influencing how different tissues appear in the resulting images. However, these relaxation times vary with field strength, leading to different image contrast at low-field compared to high-field systems [29,30]. For example, see Ref. [20] for a detailed comparison of T1, T2 and T2* values as well as the corresponding T2*-to-T2 ratio of different organs at 0.55 mT and 1.5 T. T1 relaxation times tend to decrease at lower fields, resulting in reduced contrast between tissues that would normally appear distinct in high-field imaging [29]. Chemical shift, the difference in resonance frequencies of different molecular environments, is also reduced at lower field strengths [31]. This may make the differentiation between fat and water in tissues difficult, potentially affecting the accuracy of diagnoses in areas such as liver imaging or breast MRI, where fat suppression techniques are commonly used [32]. In addition, the efficacy of gadolinium-based contrast agents is field-dependent. The relaxivity of gadolinium is reduced at lower magnetic fields especially for ultra-low-field, diminishing its contrast-enhancing effects [13]. Thus, further experiments are required to validate the utility of gadolinium-based contrast-enhanced MRI studies in low-field systems, which may require higher contrast dose or the use of new contrast agents with higher relaxivity [33].

5.2.4 Advanced imaging techniques

Advanced imaging techniques such as functional MRI (fMRI), diffusion tensor imaging (DTI), and MR spectroscopy require higher field strengths to achieve the SNR and sensitivity needed for meaningful clinical and research applications [5,34,35]. These techniques rely on the subtle differences in signal that are more readily detected at higher fields. For example, fMRI, which maps brain activity by detecting changes in blood oxygenation, requires high spatial and temporal resolution to accurately localize brain functions. DTI, used to map white matter tracts in the brain, requires high gradient strength and SNR to track the diffusion of water molecules along nerve fibers [34]. Similarly, MR spectroscopy, which analyzes the chemical composition of tissues, requires a strong signal to detect the low-concentration metabolites of interest [35]. Low-field MRI systems can struggle to achieve the necessary resolution and sensitivity for these advanced techniques, which may limit their application in both clinical practice and research [36].

5.2.5 Low-field MRI in a high-field world

As high-field MRI has become the standard in clinical practice, regulatory bodies and industry standards have been developed with these systems in mind [3]. Low-field MRI, by contrast, often falls outside these established norms, presenting challenges for its broader adoption and integration into healthcare systems. The deployment of low-field MRI systems may require different standards for safety, performance, and image quality assessment. For example, the SAR differs between high- and low-field systems, necessitating different safety guidelines [11]. In addition, regulatory approval processes for low-field MRI systems and their associated hardware and software may be more complex, potentially slowing their adoption. Moreover, the lack of standardized protocols for low-field MRI can lead to variability in image quality and diagnostic accuracy between different systems and institutions. This variability poses a challenge for radiologists, who must interpret images that may not conform to the high standards set by their high-field counterparts. As a result, there is a pressing need for the development of tailored guidelines and standards for low-field MRI to ensure consistent and reliable imaging outcomes [37].

5.3 Opportunities for AI-based solutions to advance low-field MRI

ML- and artificial intelligence (AI)-based solutions promise to address the inherent limitations of low-field MRI and unlock its potential for broader, more accessible clinical use, by exploiting algorithms developed to enhance image quality and improve diagnostic accuracy [13]. These algorithms are designed to compensate for the lower SNR, reduced spatial resolution, and other challenges specific to low-field imaging [38]. For example, AI-driven reconstruction algorithms can optimize the use of available data, reducing the need for long scan times and improving the overall efficiency of low-field MRI [39]. Deep learning models, can significantly improve the quality of low-field MR images by enhancing resolution (with super-resolution techniques), denoising, and correcting for artifacts, making them more comparable to higher-field images [40,41]. These techniques can also assist in the automatic post-processing, such as segmentation of tissue, improving the accuracy and efficiency of image analysis [42]. This is crucial in settings where radiologist expertise may be limited or unavailable, broadening access to MRI diagnostics. AI-based approaches have also been proposed to eliminate electromagnetic interference and enable RF-shielding-free ultra-low-field MRI scanners [15,16]. Furthermore, by leveraging AI to optimize pulse sequences specifically for low-field environments, it is possible to maximize image quality and diagnostic value, tailoring protocols to the constraints of these systems. By leveraging the power of AI, low-field MRI systems can produce images that are closer in quality to those obtained with high-field systems, thereby expanding their utility in clinical practice. As AI technology continues to evolve, it holds the potential to significantly mitigate the challenges of low-field MRI, paving the way for broader adoption and application of these systems.

Different applications of AI-based solutions to advance low-field MRI are discussed in the following. Before discussing these methods in detail, we provide the reader with a brief introduction to basic concepts and terms of neural networks (NNs) and deep learning (DL) that are needed to understand some aspects of the methods discussed in this Chapter. A more detailed description of these concepts can be found in Chapter 2.

5.3.1 Deep neural networks in a nutshell

Deep neural networks are functions f_θ between two different spaces V and W that are parameterized by a set of parameters θ (typically $\theta \in \mathbb{R}^{\ell}$, although networks with complex-valued parameters exist as well) that one wants to learn based on some data. Thereby, the function $f_\theta : V \rightarrow W, v \leftarrow w := f_\theta(v)$ consists of a composition of multiple functions (hence the term *deep* learning) of the form f_{θ_i} for $i = 1, \ldots, N_{\text{layers}}$ that have the form $f_{\theta_i} = \phi_i(\mathbf{W}_i \cdot + \mathbf{b}_i)$ for appropriately sized $\mathbf{W}_i$ and $\mathbf{b}_i$. The variables $\mathbf{W}_i$ and $\mathbf{b}_i$ are typically referred to as the weights and biases of the i-th layer, respectively, while the function ϕ_i is typically a non-linear activation function that is applied component-wise to the resulting vector. Then, a (feed-forward) neural network f_θ with trainable parameters θ can be constructed as the composition of the respective layers

$$f_\theta := \bigcirc_{i=1}^{N_{\text{layers}}} f_{\theta_i}, \quad \text{with} \quad \theta := \bigcup_{i=1}^{N_{\text{layers}}} \theta_i, \tag{5.1}$$

A large variety of different neural network architectures exists nowadays. Because the presentation of various neural network architectures is out of scope for this chapter, here we only mention the U-Net [43] and refer the reader to Chapter 2 for other architectures. The U-Net features a symmetric "U" shape, consisting of a contracting path (encoder) that captures context by downsampling and an expanding path (decoder) that enables feature localization through upsampling. Skip connections between the corresponding layers of the encoder and decoder paths allow for the preservation of high-resolution features. The U-Net is currently, without a doubt, one of the most widely used architectures and ubiquitous in the field of deep learning-based image processing.

5.3.2 Network training

Training a deep neural network f_θ refers to the process of tuning the parameters θ on a dataset $\mathscr{D}$. This is typically pursued to later employ f_θ as a model for prediction, i.e., after having obtained a suitable set of parameters θ, one is interested in being able to predict new meaningful outputs using the learned model f_θ.

Assuming the existence of a dataset of input-target pairs $\mathscr{D} := \{(v_i, w_i)_{i=1}^{N_{\text{samples}}}\}$, training a neural network f_θ is typically carried out employing gradient descent-like methods, that is, one constructs a sequence of parameters $\{\theta_k\}_{k\in\mathbb{N}}$ by

$$\theta_{k+1} := \theta_k - \tau_k(\nabla_\theta \mathscr{L})(\theta_k), \tag{5.2}$$

where $\mathscr{L}$ represents a so-called loss function and τ_k a (possibly iteration-dependent) step-size. The loss function $\mathscr{L}$ typically takes the form

$$\mathscr{L}(\theta) := \frac{1}{|\mathscr{D}|} \sum_{(v,w)\in\mathscr{D}} l(f_\theta(v), w) + r(\theta), \tag{5.3}$$

where $l(\cdot,\cdot)$ is an appropriately chosen similarity/discrepancy measure that is used to measure how close the predicted output is to its corresponding target, and r is some regularization term that can be further employed to avoid overfitting to the data, that is, to avoid that the network is not able to generalize to unseen examples.

As the set of parameters θ is typically rather large, training a neural network is carried out by the following procedure. One splits the available dataset in three disjoint subsets $\mathscr{D}_{\text{train}}$, $\mathscr{D}_{\text{validation}}$ and $\mathscr{D}_{\text{test}}$ and attempts to approximately minimize Eq. (5.3) over the training set $\mathscr{D}_{\text{train}}$, while at the same time monitoring the value of the loss-function $\mathscr{L}$ when evaluated over the validation set. The latter is an important step that is used to ensure that the model generalizes. The underlying idea is that the performance of the model when measured with respect to the chosen metric l over the validation set can be used as a proxy for the (in practice) unknown performance of the model at prediction time. The validation set is typically also used to choose different hyper-parameters of the neural network f_θ, e.g., the number of layers or the size of the weights and biases, or hyper-parameters of the learning procedure, e.g., the specific optimization routine, the learning rate or the number of gradient-descent steps.

Note that, for computational reasons, in Eq. (5.2), one often does not compute the exact gradient of the loss function in Eq. (5.3), as it involves the summation over the entire dataset, but rather approximates the loss function using a small number of (randomly chosen) samples at each iteration. This approach is referred to as stochastic gradient descent and further introduces some form of implicit regularization.

Further, note that, instead of employing plain gradient descent techniques as mentioned in Eq. (5.2), more sophisticated training algorithms are nowadays used to train modern architectures, for example, RMSprop [44], AdaDelta [45] or Adam [46], to name a few. See Ref. [47] for an exhaustive review.

Last, we note that the terms *validation* and *test* datasets are sometimes interchangeably used but their exact meaning typically emerges from the context. Here, we always refer to the test data as the dataset that is used to evaluate the performance of the final model for some metric.

Next, we give an overview of different AI methods that have been applied or can potentially be applied to low-field MRI to enhance image quality and diagnosis. We start with methods that can be used to adapt the acquisition tasks such as automatic planning of imaging planes that could potentially be applied for low-field MRI. We then proceed with image reconstruction methods and discuss the basic properties of deep learning methods to accelerate the acquisition and/or enhance image quality. We categorize the different deep learning-based image reconstruction approaches in (1) post-processing reconstruction techniques, which are applied to an initial (low quality) image estimate, (2) model-unaware techniques, which do not take known information about the image process into account, and (3) physics-informed approaches, which in contrast do include this information. As the vast majority of deep learning image reconstruction methods have been developed for high-field MRI, some of these methods are also mentioned here as they are needed to understand the current research landscape. Further details about image reconstruction can be found in Part III of this book, Chapters 7–11. We then finally briefly discuss deep learning approaches for post-processing analysis tasks (such as segmentation, image quality control, radiomics, and classification). Further description of AI-based image post-processing can be found in Section IV, Chapters 12–15.

5.3.3 AI solutions for image planning and acquisition in low-field MRI

One main advantage of MRI compared to most other medical imaging techniques is the ability to adapt the image orientation to the organ of interest. This can strongly improve the diagnostic quality and help with standardized diagnosis. Depending on the organ of interest, the planning of the correct image orientation can be very time-consuming. Significant training is also required to ensure accurate and reproducible scan orientations.

One particularly challenging application is fetal imaging. In Ref. [48], the authors presented an approach for automatic scan planning for fetal brain imaging at 0.55 T using a 3D U-Net architecture. The brain of the fetus is automatically detected in the womb of the mother and a set of landmarks (e.g., eyes) are identified. Based on these landmarks, scan planning is carried out. This approach reduced scans planning from several minutes down to a few seconds. Training was carried out on high-field data. A similar approach was also used to allow for automated scan planning for fetal flow imaging of the aorta and the umbilical vein [49].

An important advantage of ultra-low-field MR devices is their portability. Rather than bringing patients to a dedicated MR scanner room, the MR scanner can travel to the patient. This offers many advantages and new possibilities for healthcare. One challenge is though, that the MR scans are not carried out in a controlled environment anymore. Electromagnetic noise, in particular, can severely impair the quality of images. Approaches that measure electromagnetic noise using dedicated coils and then utilize CNN or U-Net architectures to predict the noise contribution to the acquired data have been presented [50,51]. Denoising is then a rather simple step of subtracting the estimated noise contribution prior to image reconstruction.

In AUTOSEQ [52], although not specifically designed and applied to low-field MRI, the authors explore a method for the automatic development of new non-trivial pulse sequences for the data acquisition in MRI. They achieve this by employing a reinforcement learning framework and recasting the problem of pulse sequence development as a game of perfect information that an AI agent learns to play. As initial results, the authors reported that their agent, apart from being able to learn a canonical gradient echo pulse sequence, also succeeded in learning to generate more non-intuitive pulse sequences. Due to its general nature, the approach presented in Ref. [52] holds the promise to be applicable to low-field MRI as well.

5.3.4 AI solutions for image reconstruction in low-field MRI

Here, we first briefly formalize the image reconstruction problem of interest and then present different machine learning-based approaches that can be employed to obtain a meaningful approximation of the unknown sought image.

5.3.4.1 Forward problem formulation

A discretized version of the MR image acquisition (i.e., the forward problem) can be expressed as

$$\mathbf{s} = \mathbf{E}\boldsymbol{\rho}_{\text{true}} + \boldsymbol{\eta}, \tag{5.4}$$

where $\boldsymbol{\rho}_{\text{true}} \in \mathbb{C}^N$ denotes the vector representation of the (unknown) complex-valued image with N being the total number of voxels, e.g., $N = N_x \cdot N_y$ for a 2D image or $N = N_x \cdot N_y \cdot N_z$ for a 3D volume. The forward operator $\mathbf{E}$ models the data acquisition and involves several quantities such as coil sensitivity maps, the set of k-space trajectories along which the image is sampled, field maps inhomogeneities and possibly other relaxation effects as well as other set up- and sequence-specific parameters. Thus, note that, in general, the operator $\mathbf{E}$ in Eq. (5.4) is non-linear, but the vast majority of the literature considers an idealized set-up to be able to work with a linear acquisition model. The vector $\boldsymbol{\eta} \in \mathbb{C}^K \sim \mathcal{N}(\boldsymbol{\mu}, \boldsymbol{\Sigma})$ with K being the total number of measured k-space points denotes complex-valued Gaussian noise with mean vector $\boldsymbol{\mu}$ and covariance matrix $\boldsymbol{\Sigma}$ (which is often modeled by $\sigma^2\mathbf{I}$ with variance σ^2) and $\mathbf{s} \in \mathbb{C}^M$ denotes the acquired k-space data.

For example, in a multi-coil setting, when N_c receiver coils are used, the forward operator is often modeled by

$$\mathbf{E} := (\mathbf{I}_{N_c} \otimes \mathbf{F})\mathbf{C}, \tag{5.5}$$

$$\mathbf{C} := [\mathbf{C}_1, \ldots, \mathbf{C}_{N_c}]^{\mathbf{T}}, \quad \text{with} \quad \mathbf{C}^{\mathbf{H}}\mathbf{C} = \mathbf{I}_N \tag{5.6}$$

$$\mathbf{C}_j := \text{diag}(\mathbf{c}_j), \quad \mathbf{c}_j \in \mathbb{C}^{N \times N}, \quad j = 1, \ldots, N_c, \tag{5.7}$$

where $\mathbf{I}_{N_c}$ denotes the $N_c \times N_c$ identity matrix, $\mathbf{F}$ is a Fourier-encoding operator and $\mathbf{C}_j$ represents the j-th coil sensitivity map. For single-coil acquisitions, $N_c = 1$ and $\mathbf{C}$ in Eq. (5.6) reduces to the identity matrix.

Further, in so-called super-resolution applications [27,53–57] (see also Chapter 8), the forward problem in Eq. (5.4) is extended by

$$\mathbf{s} = \mathbf{E}\mathbf{R}\boldsymbol{\rho}_{\text{true}} + \boldsymbol{\eta}, \tag{5.8}$$

with the understanding that $\boldsymbol{\rho}_{\text{true}}$ denotes the high-resolution image that is sub-sampled by a low-resolution operator $\mathbf{R} : \mathbb{C}^N \to \mathbb{C}^{N'}$ and whose k-space data is acquired. The goal is then to recover the high-resolution image from the k-space data of the low-resolution image.

Further, since in MRI, data acquisition is a time-consuming process, it is sometimes desirable to accelerate the measurement by only acquiring a portion of the k-space that would be required to be sampled. In this case, the forward problem is given as in Eq. (5.4), but we emphasize the acquisition of only a sub-portion of the required k-space data by writing $\mathbf{E}_J$ instead of $\mathbf{E}$ and $\mathbf{s}_J$ instead of $\mathbf{s}$ with the understanding that $\mathbf{E}_J$ samples k-space data only at locations indexed by an index set J, that is, the Fourier-encoding operator in Eq. (5.5) is analogously denoted by $\mathbf{F}_J$ instead of $\mathbf{F}$.

Regardless of the operator $\mathbf{E}$ being linear or non-linear, the reconstruction of the image $\boldsymbol{\rho}_{\text{true}}$ from the observed k-space data $\mathbf{s}$ often represents an ill-posed problem. The number of measured data points K and its relation to the number of sought voxels N implicitly defines the nature of the ill-posedness. For $K < N$, e.g., for accelerated undersampled acquisitions or super-resolution applications, problem Eq. (5.4) is under-determined, meaning that even in the linear case, an infinite number of solutions exists. Since $N' < N$, the super-resolution problem in Eq. (5.8) is clearly ill-posed by being under-determined. Even for $K > N$, problem Eq. (5.4) is ill-conditioned for non-Cartesian sampling trajectories.

For the considered case of low-field MRI, an additional problem is that due to the lower field strength of the static magnetic field B0, the signal generated by the transverse magnetization of the object to be imaged is inherently lower than when measured at higher field strengths. This results in a systematically lower SNR, which can be modeled by the relationship of the mean of $\boldsymbol{\rho}_{\text{true}}$ and the noise variance σ^2 in Eq. (5.4), and poses a further challenge to the image reconstruction process.

A widely followed approach to still be able to reconstruct images useful for clinical diagnosis is to formulate a variational problem of the type

$$\min_{\boldsymbol{\rho}} \frac{1}{2}\|\mathbf{E}\boldsymbol{\rho} - \mathbf{s}\|_2^2 + \mathscr{R}(\boldsymbol{\rho}), \tag{5.9}$$

where the first term in Eq. (5.9) tackles the data-consistency of the sought solution and $\mathscr{R}(\boldsymbol{\rho})$ is a regularization term that is used to impose desirable properties on the image. Note that, although in principle different data-discrepancy measures can be employed as well, the use of the squared ℓ_2-norm is motivated by the fact that Gaussian noise is the dominant noise present in the MRI acquisition [58].

The study of suitable regularization functional has been an active field of research in the last decades. Deep learning-based methods that implicitly learn features from data have also been proposed as discussed for low-field imaging in the following and more generally in Chapter 7.

5.3.4.2 Supervised versus self-supervised training for image reconstruction

In Subsection 3.2, we have implicitly assumed the existence of a dataset consisting of input-target pairs, e.g., $\mathscr{D} := \{(\mathbf{s}^{(i)}, \boldsymbol{\rho}^{(i)}_{\text{true}}) \mid \; \mathbf{s}^{(i)} = \mathbf{E}\boldsymbol{\rho}^{(i)}_{\text{true}} + \boldsymbol{\eta}^{(i)}\}$, which means the network can be trained by adjusting the parameters θ such that $f_\theta(\mathbf{s}^{(i)}) \approx \boldsymbol{\rho}^{(i)}_{\text{true}}$ for each pair. For example, by setting the discrepancy function l in Eq. (5.3) to $l(\cdot, \cdot) = \|\cdot - \cdot\|_2^2$, one can minimize the supervised loss-function

$$\mathscr{L}^{\text{sup}}(\theta) := \frac{1}{|\mathscr{D}|} \sum_{(\mathbf{s}, \boldsymbol{\rho}_{\text{true}}) \in \mathscr{D}} \|f_\theta(\mathbf{s}) - \boldsymbol{\rho}_{\text{true}}\|_2^2. \tag{5.10}$$

However, deep neural networks are known to require a rather large dataset to be properly trained. In the context of image reconstruction, this corresponds to having access to many target images of good quality, which can be costly, challenging, or, in some cases, even impossible to obtain, which is especially the case for low-field MRI.

This issue is often circumvented by relying on simulations and the retrospective data generation of input-target pairs, see e.g., Refs. [27,54,59], but can possibly yield a performance gap in the obtained model, especially when the distribution shifts from the training to the test data is rather large [60]. Further, to enrich the set of target image data used for training, it is usually advised to employ data augmentation techniques such as simple image rotations, image flipping, contrast changes, or more involved techniques, for example, elastic deformations [61].

Self-supervised training strategies as the ones discussed next can overcome these challenges by exploiting the structure of the considered problem. Recall that for image reconstruction tasks, our available input data (e.g., acquired k-space data) is in general generated according to a known forward model as in Eq. (5.4). Rather than comparing the output of the network to a known target, we can apply the forward model to the network output and compare the obtained result to the available input data. Although intuitive, this approach can suffer from overfitting problems because the employed neural networks are typically overparameterized. Thus, a mechanism that prevents the learning procedure from yielding sub-optimal parameters θ is required. Strategies to avoid overfitting are typically based either on early stopping techniques [62], on restricting the family of mappings that can be learned by imposing certain constraints on the network architectures, i,e. on the network's topology [63], invertibility of the obtained mapping [64], employing regularization on the weights, e.g., dropout [65], batch-normalization [66], spectral-normalization [67], to only name a few, or by employing appropriate loss functions [68–71] that systematically avoid overfitting.

As mentioned above, for low-field MRI, it is often challenging or sometimes even impossible to obtain high-quality training data and/or paired image data. Therefore, self-supervised training approaches can be of high interest for this application.

The *Noise2Noise* framework [68] relies on different realizations of noisy samples where the noise needs to be statistically the same between different samples. Noise2Noise is an attractive approach when different realizations of the same object can be easily acquired. For example, the approach was adopted in Ref. [72], where a cascaded model of sparsity transforms is used to reduce noise in images obtained on a 0.5 T scanner.

Noise2Noise requires different measurements of the same object, which can be challenging for MRI due to physiological changes occurring between scans. The *Noisier2Noise* approach [69] overcomes this limitation by adding additional noise to the considered sample and using the original noisy sample as target data for training. At inference time, a scalar correction needs to be applied, which compensates for the learned ability of the network to reduce noise that is higher than the target noise of interest. A later discussed example of the application of Noisier2Noise to low-field MRI was presented in Ref. [73].

Finally, in *Noise2Self* [70], network training for the task of image-denoising is carried out by self-supervision employing the same target data that is used as network input. However, to avoid overfitting, the image pixels are split into two disjoint index sets, i.e., for $\boldsymbol{\rho}^{\eta} := \boldsymbol{\rho}_{\text{true}} + \boldsymbol{\eta} \in \mathbb{R}^N$, one chooses a partition of the index set $I = \{1, \ldots, N\}$, i.e., $I := I_1 \cup I_2$ with $I_1 \cap I_2 = \varnothing$ and defines the image $\boldsymbol{\rho}_{I_1}$ by $[\boldsymbol{\rho}^{\eta}_{I_1}]_j = [\boldsymbol{\rho}^{\eta}]_j$ for $j \in I_1$ and $[\boldsymbol{\rho}^{\eta}_{I_1}]_j = 0$ for $j \notin I_1$. The input of the network is then restricted to pixels indexed by I_1 and the output of the network is compared to the input image restricted to pixels indexed by I_2.

Fig. 5.2 summarizes the just described schemes for a simple image-denoising example.

In Ref. [71], the authors propose a self-supervision method that is closely linked to the above-discussed Noise2Self and Noisier2Noise, termed *self-supervision by data undersampling* (SSDU). More precisely, the method aims at training a network to reconstruct images from accelerated MR data acquisitions where the k-space data in Eq. (5.4) is undersampled. Thereby, similar to Noise2Self, the (already undersampled) k-space data $\mathbf{s}_J$ is split into two disjoint sets $J := J_1 \cup J_2$ with $J_1 \cap J_2 = \varnothing$ and, by for example again choosing the squared ℓ_2-norm as discrepancy measure, one can train the network by minimizing a loss function of the form

$$\mathscr{L}_{\text{SSDU}}(\theta) := \frac{1}{N_{\text{train}}} \sum_{i=1}^{N_{\text{train}}} \left\| \mathbf{E}_{J_2} f_\theta\left(\mathbf{s}_{J_1}^{(i)}\right) - \mathbf{s}_{J_2}^{(i)} \right\|_2^2. \tag{5.11}$$

Further, in Ref. [74], the connection between SSDU and Noisier2Noise is theoretically investigated. We further also note that differences exist in the specific way of partitioning the data. For example, in Refs. [71,74], the employed loss function resembles the Noisier2Noise framework, i.e., during training, the network learns to reduce artifacts from a higher undersampling than the one that is considered at test time. In contrast, in Ref. [60], the distribution of the undersampling masks during training matches the one of the mask used for inference.

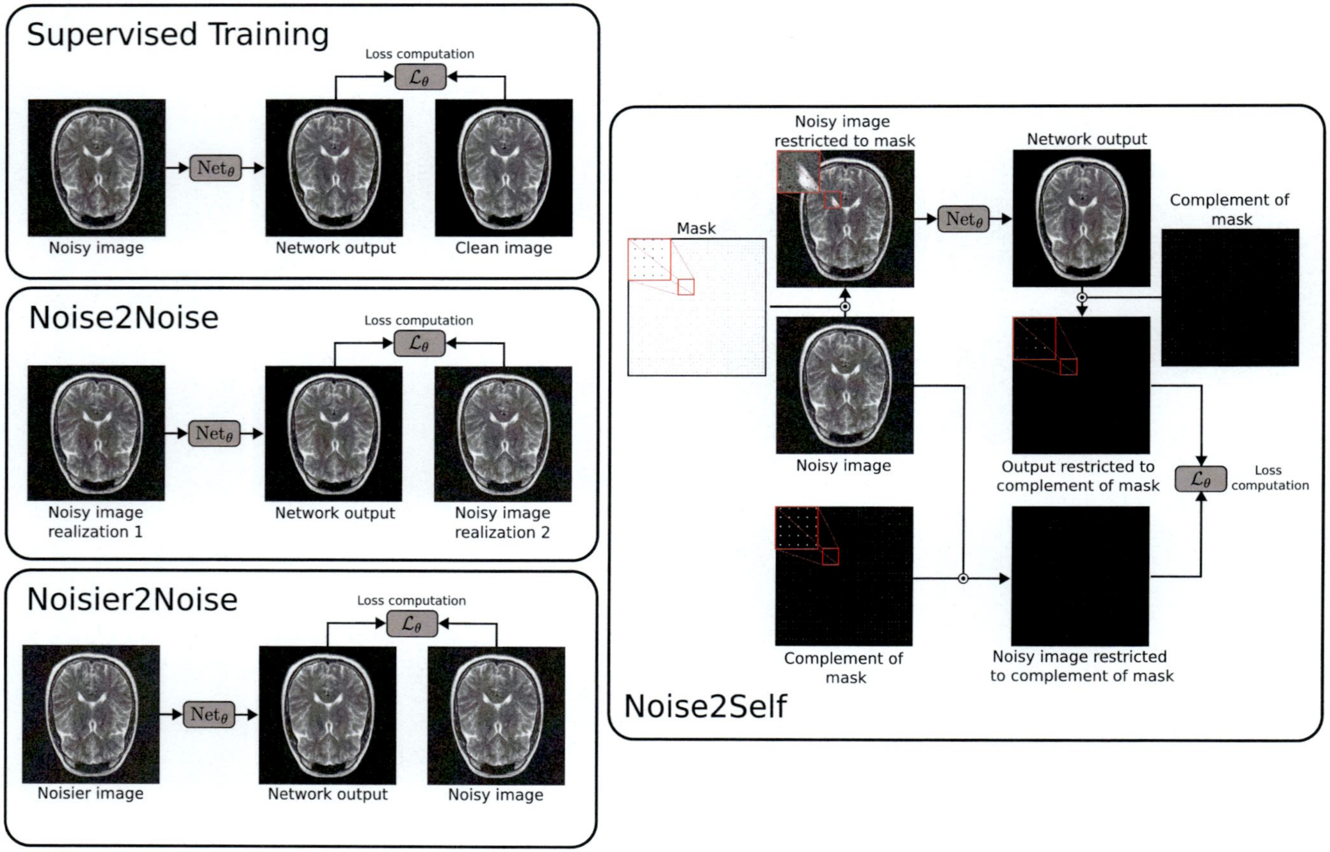

FIGURE 5.2

Different training techniques for a simple image-denoising example. Supervised training requires access to input-target image pairs and the network Net$_\theta$ is trained to estimate the clean image from the noisy input. In the self-supervised Noise2Noise framework [68], two noisy samples with identically distributed noise are required. For Noisier2Noise [69], the input sample is generated by further corrupting the given noisy sample with additional noise. For Noise2Self [70], the input image is restricted to a pre-defined mask and the estimated output is compared against the same input image projected on the complement of the employed mask. The principle of the framework self-supervision by data undersampling (SSDU) [71] (not shown in this Figure) that can be used for MR image reconstruction is similar to Noise2Self with the difference that the mask is applied in the Fourier domain of the considered image.

Fig. 5.3 shows an example of results that can be obtained by training an image-denoising network by self-supervised training on solely noisy image data using the Noise2Noise approach with the networks proposed in Ref. [72].

In the previously described methods, although no target images are assumed to be given, the workflow is still to train on a relatively large corpus of data and *then* to apply the obtained model to new unseen samples.

Here, we also mention the possibility of applying so-called *zero-shot methods*, that is, where no dataset is used for training, but instead, the parameters of the model are adapted to the considered sample one aims to reconstruct, for example, the well-known deep image prior (DIP) [62,75], the deep decoder [76–78] as well as methods that employ self-supervised loss functions, see Refs. [79–81]. Also, see Ref. [82] for an extensive review of unsupervised deep learning-based methods for image reconstruction and image enhancement.

Thereby, the idea is to reparameterize the sought image by a neural network, that is, to set $\boldsymbol{\rho} := f_\theta(\mathbf{z})$ for some (typically randomly chosen) input vector $\mathbf{z}$. Then, instead of minimizing a variational problem of the form Eq. (5.9), one directly approaches the problem

$$\min_\theta \frac{1}{2} \| \mathbf{E} f_\theta(\mathbf{z}) - \mathbf{s} \|_2^2 \tag{5.12}$$

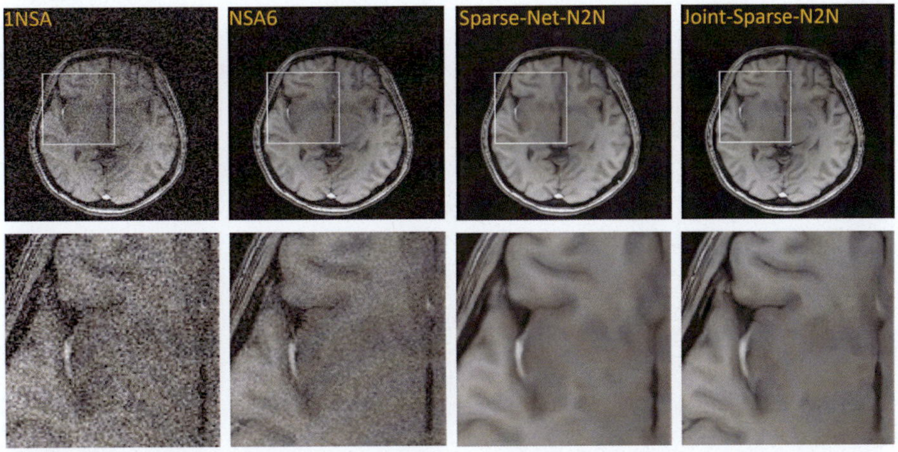

FIGURE 5.3

An example of images reconstructed with two deep neural networks for image denoising (Sparse Net and Joint Sparse-Net) that were trained using the Noise2Noise (N2N) framework on low-field MR image data acquired at 0.5 T. The obtained images show a strong noise reduction compared to the images reconstructed from one and six averaged acquisitions (1NSA and NSA6), respectively. The image was adapted from Ref. [72] and reproduced with the permission of the ISMRM.

Image courtesy of Hua Guo.

for each new considered measurement **s** with one of the previously described gradient descent-based methods, i.e., the loss-function corresponds to the data-discrepancy term and the employed dataset consists of only one sample. Typically, however, as observed by the authors in Ref. [62], overfitting occurs eventually while minimizing Eq. (5.12), meaning that at some point, the network f_θ is able to approximate the noise present in the measurement **s**, thus not providing any implicit regularization effect. Thus, it is typically necessary to employ some regularization technique, for example, early stopping as in the original paper [62], underparameterizing the network f_θ [63] or combining DIP with regularization methods, for example, with low-rank methods [83] or with total variation-minimization (TV) approaches [84].

In Ref. [83], for example, the DIP method combined with a low-rank (LR) approach was used to reconstruct cardiac cine data obtained at 0.55 T. Fig. 5.4 shows an example of different reconstructions. The cine images obtained with the proposed LR-DIP show significantly higher image quality compared to a compressed sensing-based reconstruction (ℓ_1-ESPIRIT) [85] and a low-rank + sparsity-based (L + S) [86].

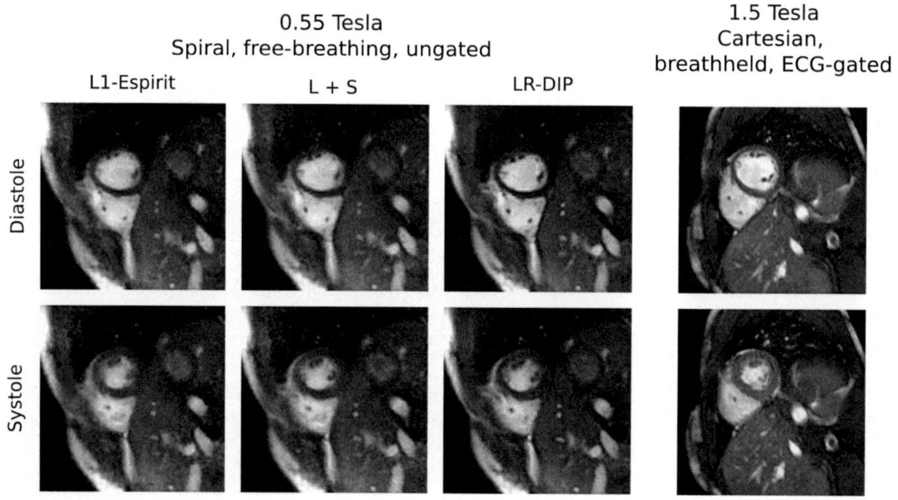

FIGURE 5.4

An example of cine images acquired at 0.55 T using spiral acquisitions during free breathing and reconstructed with ℓ_1-ESPIRIT [85], a low-rank + sparsity-based method (L + S) [86], the proposed low-rank deep image prior (LR-DIP) [83] and a reference image of the same subject obtained at 1.5 T acquired during a breath hold with an ECG-gated acquisition. The neural networks-based approach LR-DIP improves image quality over the two compressed sensing-based approaches and yields visually improved noise and artifacts suppression.

The image was adapted from Ref. [83]. Image courtesy of Jesse Hamilton.

5.3.4.3 Learned reconstruction methods

After having introduced the basic notions of neural networks and their training, we are ready to discuss different strategies that can be employed for image reconstruction with deep learning-based methods in low-field MRI.

One of the perhaps most intuitive approaches is to construct a neural network f_θ that corresponds to a *fully learned* mapping from the observed data to the unknown target image, that is, to have a learned reconstruction mapping $\mathbf{E}_\theta^\sharp$ with $\mathbf{E}_\theta^\sharp(\mathbf{s}) \approx \boldsymbol{\rho}_{\text{true}}$. Thereby, the method is fully learned in the sense that one explicitly abstains from incorporating knowledge about the considered problem (e.g., the acquisition model $\mathbf{E}$ in 4) but instead aims at learning the entire mapping from the available training data. This approach is particularly suitable when one lacks a deep understanding of the forward model, which however does usually not apply to MR applications. For MRI, a fully learned inversion method corresponds to a function that directly maps the measured k-space data to an estimate of the ground-truth image without making use of the forward/inverse Fourier transform, for example.

In Ref. [87], the authors proposed AUTOMAP, which allows for the inversion of arbitrary forward models using the same network architecture, provided that there exist input-target pairs to train the model on. Specifically for ultra-low-field MRI, AUTOMAP was applied in the work [88], where the authors demonstrated the capability of the AUTOMAP architecture to invert 3D k-space data of the human brain measured at 6.5 mT using a bSSFP sequence, see Fig. 5.5. The authors compared their method to two other well-known denoising methods that were applied after a simple inverse Fourier transform of the k-space data, namely block matching 3D (BM3D) [89] with collaborative filtering [90] and the deep learning-based denoiser DnCNN [91] and found that AUTOMAP surpassed the methods of comparison both in terms of visual image quality by preserving more image details.

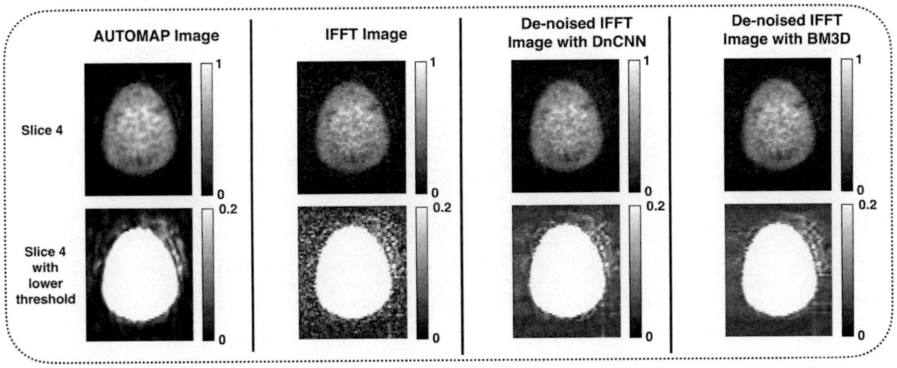

FIGURE 5.5

An example of ultra-low-field (6.5 mT) MR images of the brain reconstructed with different methods. From left to right: AUTOMAP [87,88], simple inverse FFT, inverse FFT image denoised with DnCNN [91], inverse FFT image denoised with block-matching 3D [89,90].

Figure adapted from Ref. [88]. Image courtesy of Neha Koonjoo.

Fully learned inversion methods can be applied to super-resolution problems as well. Thereby, to invert the low-resolution operator $\mathbf{R}$ in Eq. (5.8), the most common strategy is to first obtain an initial estimate of $\mathbf{R}\rho_{\text{true}}$, i.e., to invert the model $\mathbf{E}$ and then to train a network to estimate the underlying ρ_{true}.

In Ref. [27], for example, the authors trained a CNN with dense skip connections termed Super-Resolution Dense Network (SRDenseNet), see Ref. [92], to map low-resolution images to their high-resolution counterpart by directly inverting the low-resolution operator, see Fig. 5.6. Thereby, the target images were taken from the fastMRI brain dataset and the low-resolution images were retrospectively simulated by applying the four-fold low-resolution operator and further corrupting them with noise. The authors reported a noticeable improvement compared to the images obtained by zero-padding *k*-space data.

Another similar approach for portable ultra-low-field brain MRI at 64 mT [53] investigated the possibility of directly inverting the sub-sampling operator using a U-Net following the approach in Ref. [93], which also yields image segmentations of 39 regions of interest of the brain. Thereby, the obtained results on their brain MR image data were also evaluated in terms of the ability of different segmentation tools to generate accurate region-of-interest volumes from the obtained images. The

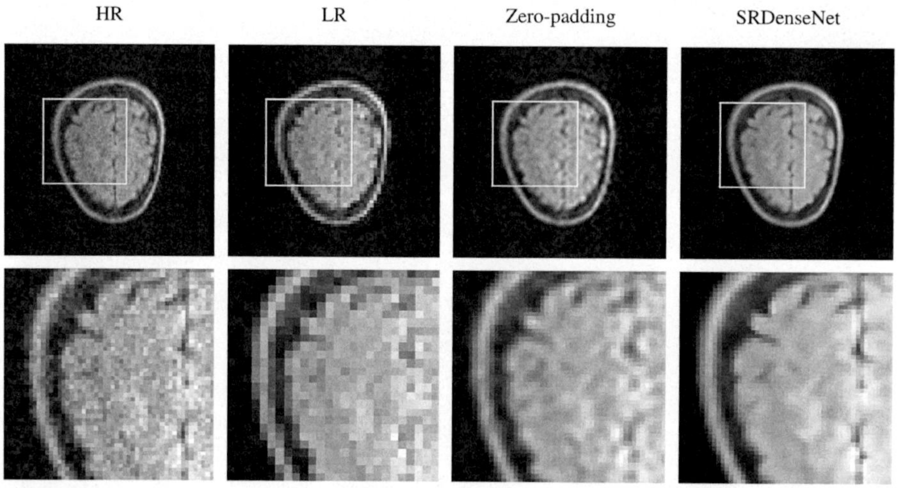

FIGURE 5.6

Example results obtained through a super-resolution network at ultra-low-field (50 mT). The image HR denotes a high-resolution slice of a brain image acquired an ultra-low-field MRI system, which serves as the reference image. The second image shows the low-resolution image simulated from the HR image using a sub-sampling forward model, see Ref. [27] for details. The third image is obtained by inverting the zero-padded *k*-space data with a simple IFFT and the fourth is the result generated by the proposed super-resolution network (SRDenseNet).

Figure adapted from Ref. [27]. Image courtesy of Andrew Webb.

authors reported good agreements between the segmentations obtained from the reference high-field strength and the ones obtained from their learned upsampling method, which demonstrates the applicability of these methods with respect to subsequent downstream tasks, for example., segmentation.

In super-resolution applications, the networks are usually trained on retrospectively simulated data, i.e., high-resolution images are corrupted by down-sampling to generate the input-target image pairs to be used for training. In Ref. [55], the authors further explored a method that employs a domain-adaptation module based on unpaired image translation from brain MR images acquired at 64 mT to high-field images obtained from the Connectome project [94]. Their method comprises a GAN [95] and a patch-based contrastive learning approach [96]. Since these two modules can be trained in an unsupervised fashion, the approach presented in Ref. [55] eliminates the need to have access to paired image data. Similar to the previously discussed approaches, the method aims at inverting the low-resolution operator $\mathbf{R}$ in Eq. (5.8).

Fully learned inversion methods, however, have, first of all, the inherent conceptual limitation of ignoring all available and valuable information that the (at least partially) known physical model provides. Second, they might lack data consistency, i.e., it is not clear how well the estimated quantities match the observed data when the forward model is applied. Third, since they typically consist of at least a few fully connected layers (i.e., some of the matrices $\mathbf{W}$ in Eq. (5.1) are dense matrices) to allow for high expressiveness to be able to learn to invert the forward model, their application is limited to input data that has the same dimension as the data the network was trained with. This, in contrast, is not the case for networks consisting of solely convolutional layers and thus represents a limitation from a practical point of view as either a new network has to be trained for each input dimension or data must be pre-processed accordingly. Additionally, although the authors in Refs. [87,88] report that AUTOMAP is inherently robust, other works have shown that AUTOMAP was the least robust compared to other methods with respect to adversarial attacks [97], especially compared to deep learning methods that employ the data acquisition model as part of their architecture.

Post-processing reconstruction or *image enhancement* methods, as opposed to fully learned methods, do not entirely ignore the structure of the given problem at hand at inference time. Instead, one typically obtains an initial estimate of the unknown ground-truth image from the noisy and/or undersampled k-space data, by applying some reconstruction operator, for example, the adjoint $\mathbf{E}_I{}^{\mathrm{H}}$ or the pseudo-inverse $\mathbf{E}^{\dagger}$, and subsequently trains a network to reduce the resulting noise and artifacts. As discussed in Chapter 7, this type of approach has been extensively investigated for MRI in a large number of works for different organs (e.g., knee [98,99], brain [100–104], cardiac [105–107]) and has shown promising results. In addition, we note that there exists the alternative, where, instead of learning the denoising/artifacts reduction mapping in the image space, one can learn a suitable transformation for filling the k-space and subsequently applying a simple inverse Fourier transform [108–110].

Specifically for low-field MRI, the authors in Ref. [111] used a 2D U-Net with residual connections for image denoising and artifacts reduction from undersampled low-field MRI scans acquired at 0.1 T, where, by the use of data-augmentation techniques, they were able to successfully train the network even on a relatively small dataset only comprising 10 human wrist images.

Further, the authors in Ref. [112] proposed a method for simulating scanner-specific images from the publicly available dataset of the Cancer Imaging Archive [113] as well as a post-processing reconstruction method for reducing different types of artifacts related to B0-inhomogeneities, nonlinear gradients, and undersampling. The post-processing reconstruction method, which is based on a stacked U-Net architecture, was then applied to images obtained from an ultra-low-field Promaxo MR scanner with a low and non-uniform B0 field strength of 60 – 67 mT.

In another work [114], the authors investigated the applicability of the Noise2-Void [115] framework to train a denoising U-Net on patches of T1-and T2-weighted images of a kiwi that was scanned as a phantom on a 0.35 T MRI system.

Another approach [116] employs a generative adversarial network (GAN) that was learned on paired brain MR image data acquired at 64 mT and 3 T, respectively. The authors evaluated the obtained synthesized images with respect to image quality, brain morphometry, and white matter lesions and found that quantitative measures such as thalamic, lateral ventricle, and total cortical volumes in the synthesized images did not significantly differ from the images obtained at 3 T.

In Ref. [73], the authors explored the previously discussed Noisier2Noise framework to learn to denoise an image in an self-supervised fashion. The work in Ref. [117] followed a two-step approach by first pre-training a DnCNN denoiser in a supervised fashion on a large corpus of data given by the Human Connectome Project (HCP) [94]. Then, in a second step, the pre-trained is applied to the noisy DWI images obtained from a 64 mT scanner to obtain pseudo-labels to be added to the initial set for a new training that is performed from scratch using the extended dataset.

One main challenge of employing deep neural networks that are trained in a supervised manner is that they must exhibit good generalization properties. Typically, since they contain a large number of trainable parameters, the dataset used for training must be rather large. However, for some applications, for example, low-field MRI, obtaining high-quality target images is simply not feasible. Thus, one possibility is to employ one of the previously described self-supervised training techniques. Another option is to rely on the availability of high-quality image data obtained from MR scanners at higher magnetic fields and to retrospectively simulate the required data pairs. However, depending on one's capability to accurately simulate the image degradation process the network's performance at test time might suffer more or less from inherent distribution shifts between the training and the target test data.

Another case where such a performance decrease attributable to a distribution shift is observable is when training and test data differ from each other in terms of anatomy and image contrast. Fig. 5.7 demonstrates this on an image-denoising

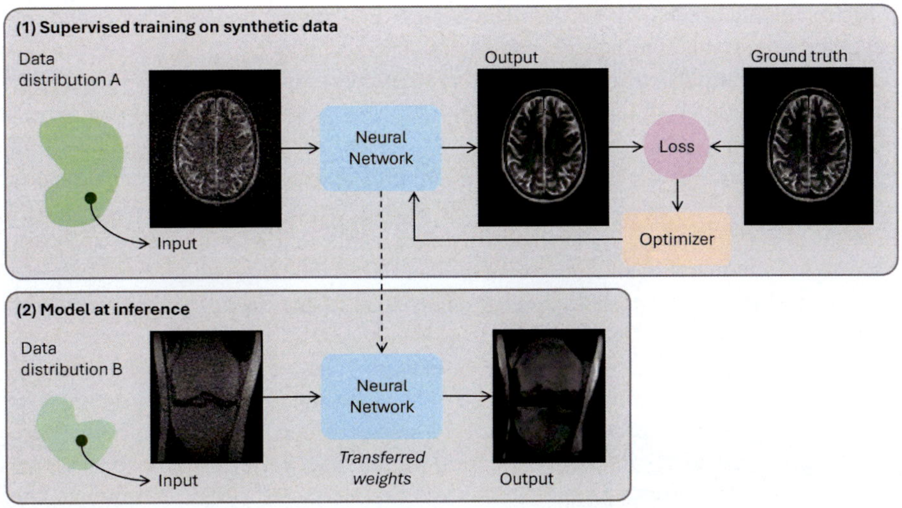

FIGURE 5.7

Training data for low-field MRI denoising problems is usually rare. Therefore, many approaches use retrospectively simulated low-field data to train models and transfer the results to real data. If the two data distributions do not match (i.e., there is a noticeable distribution shift) the output of the network can lead to severe artifacts. In this example, applying a denoising network that was trained solely on T_2-weighted brain images and then applying the obtained model to T_1-weighted knee images leads to severe signal voids within the bone marrow.

example for low-field data. A denoising network is trained on T2-weighted brain images and the network is then directly applied to T1-weighted knee images. Here, training and test data strongly differ due to their different contrast as well as the different anatomy, leading to severe signal voids in the denoised knee images. We point out that this example is extreme in the sense that on average, such drastic examples are fortunately much rarer to be found, but it highlights the importance of exposing the network to a large variety of different images. So-called test-time-training [60] can help close the performance gap that is typically observed when applying neural networks to highly different test data. A recent study shows that, in general, to achieve a small gap between the performance on different datasets in the first place, it seems to be beneficial in terms of robustness to employ rich and diverse training data, even when the target data is only a sub-class of the data used for training [118].

Post-processing reconstruction methods, however, have one major limitation. Because from a theoretical point of view, the exact functioning mechanism of neural networks is not well understood, this type of approach rightfully raises the question of how reliable they can expected to be. These concerns are further underpinned by some works that have reported hallucinations of deep neural networks for the task of image reconstruction [119]. Further details regarding the robustness of deep learning reconstruction and hallucinations can be found in Chapter 11.

The last mentioned limitation of image enhancement methods can partially be addressed by employing reconstruction methods that employ neural networks only as one of several distinct steps, often referred to as *model-unaware* or *decoupled methods*. For example, when aiming to regularize inverse problems, so-called Plug and Play (PnP) methods [120] or regularization by denoising (RED) methods [121] are image reconstruction frameworks that alternate between steps motivated from model-based reconstruction and the application of general Gaussian denoisers, which can, in particular, be given by state-of-the-art deep neural networks such as the well-known DnCNN [91].

These methods are attractive for one main reason. Training the Gaussian network denoiser is completely decoupled from the considered image reconstruction process, which makes it computationally relatively cheap to accomplish. Further, image denoising is, due to its inherent structural simplicity, without doubt, one of the most well-studied inverse problems and has thus attracted large attention of the signal processing community, yielding a large amount of available "off-the-shelf" Gaussian denoisers that can be directly integrated into the PnP or RED framework. Compared to purely post-processing-based reconstruction methods, they in addition increase the data consistency of the obtained solution since the non-learned algorithm components typically involve the use of the measured k-space data as well as the adjoint/forward operator. Additionally, this type of method was also reported to be applicable with networks that reduce artifacts instead of Gaussian denoisers, see e.g., Ref. [122].

Interestingly, to the best of our knowledge, these types of methods, that is, PnP or RED, have not been applied to low-field or ultra-low-field MRI yet, although they naturally fit well with the given low SNR regime.

Further, we note that model-unaware or decoupled methods can not only be employed to learn artifacts reduction- or denoising mappings but in general to estimate any relevant quantity to be used within the image reconstruction pipeline. For ultra-low-field MRI, field inhomogeneities leading to image distortions are often challenging. Therefore, approaches that estimate and correct for these inhomogeneities are of specific interest.

For instance, the work in Ref. [123], although not specifically designed for low-field MRI, but rather for non-Cartesian susceptibility-weighted imaging (SWI), which also suffers from field-inhomogeneities induced by the patients during long readouts, proposed a reconstruction method that compensates for the field-inhomogeneity correction utilizing a neural network. The studied neural network architectures consist of reconstruction methods based on the non-Cartesian primal-dual network architecture [124]. For the training, they employ target images that were obtained by a self-corrected method [125]. The network is trained in a so-called greedy fashion, i.e., the learnable components of the networks are trained using the current estimates of the reconstruction algorithm as input images.

Other works [126], albeit not for low-field MRI applications but for pediatric scanning, proposed a network termed Off-Resonance Correction Network (Off-ResNet) that can be used to estimate the field-inhomogeneity-corrected image map from an input image. Similarly, the authors in Ref. [127] propose to employ the same network architecture but to train it to estimate the field-inhomogeneity map itself such that a subsequent correction with a B0-informed method [128] can be applied.

A third important category of state-of-the-art reconstruction algorithms using deep learning are so-called physics-informed methods based on algorithm unrolling [129], see Chapter 7 and the review [130] for an in-depth discussion. In contrast to the methods described above, the learned networks themselves already constitute reconstruction algorithms that contain both learnable parameters as well as model-based blocks describing the physics of the considered problem. As such, the learned components are not only related to the data they were trained with but also to the reconstruction pipeline that the network defines. Thus, these networks are sometimes also referred to as physics-informed neural networks.

In Ref. [131], the authors presented an end-to-end trainable reconstruction network that addresses undersampling and hardware artifacts for an ultra-low-field Hyperfine MRI scanner at 64 mT. Their network consists of a 2D non-uniform variational network [132] and shares similarities to the cascade presented in Ref. [61]. It was trained on images taken from the Human Connectome Project [94]. Those images were retrospectively resampled to match the resolution of their Hyperfine T2-weighted fast spin echo images and field of view. Additionally, phase and coil-sensitivity weighting was introduced before retrospectively generating k-space data. They compared their method to a standard 2D U-Net post-processing reconstruction method as well as to a k-space-based U-Net [110], and reported improvements.

Similar studies have been carried out for 0.55 T. In Ref. [133] a standard k-space-based denoising and super-resolution approach was compared to a variational network approach [134] for lumbar spine imaging at 0.55 T. The variational network approach allowed for a scan time reduction of 40% while ensuring high image quality compared to the standard method. The same variational network approach was also compared for knee imaging at 0.55 T [135]. The variational network approach yielded comparable diagnostic confidence as the images obtained at 3 T. Another study comparing knee imaging at 0.55 T and 1.5 T came to similar results [136].

Note, however, that although these methods incorporate the physical model in blocks of their network architecture, their considered model for example ignores B0--field inhomogeneities, which is an important challenge for ultra-low-field MRI.

In contrast, the work in Ref. [59] corresponds to an end-to-end trainable reconstruction method that estimates both the image as well as a field map describing the underlying B0-field inhomogeneities which are responsible for image distortions and which was applied to retrospectively simulated ultra-low-field knee MR data. Thereby, for the estimation of the field map, the authors use a neural network architecture that, instead of estimating the entire field map directly as in Ref. [127], estimates the coefficients of the spherical harmonic functions [137,138] that are known to well-approximate the B0-field inhomogeneities [139].

Fig. 5.8 shows a comparison of the field map estimated from a noisy and distorted input low-field image using the previously mentioned Off-Resonance Network (ORC-Net) and the Spherical Harmonic Network (SH-Net) proposed in Ref. [59]. Additionally, the field map that is estimated from the phase difference of two different simulated low-field MR acquisitions is shown. As we can see, the latter is highly non-smooth and contains a noticeable amount of noise, while the ones estimated by the CNNs are superior in terms of smoothness. Further, the one calculated by the SH-Net is smoother than the one obtained by the ORC-Net as the output of the network is constructed to be a linear combination of spherical harmonic functions up to a certain order.

In Ref. [59], a second U-Net serves as a denoising network and its output is used to regularize the problem employing a Tikhonov regularization term $\mathscr{R}(\boldsymbol{\rho}) := \|\boldsymbol{\rho} - \boldsymbol{\rho}_{\mathrm{NN}}\|_2^2$, as used in many other works, e.g., Refs. [140,141]. Thereby, the output $\boldsymbol{\rho}_{\mathrm{NN}}$ is obtained by applying the U-Net to the noisy image for which the distortion has been corrected by using the field map estimated by the previously discussed SH-Net. The entire physics-informed network in Ref. [59] was trained on retrospectively simulated data from the knee fastMRI dataset [26] and the authors reported improvements when comparing their method to other non-learned methods. This approach falls in the category of algorithm unrolling as the entire network resembles an iterative scheme for the joint estimation of the B0 field map and the image, but, in contrast to other works, e.g., Ref. [142], uses learned regularization terms.

Fig. 5.9 shows a comparison of knee images from the fastMRI dataset [26] for which a retrospective low-field data acquisition was simulated, see Ref. [59] for details. The images were reconstructed from the noisy k-space data with the methods presented in Refs. [142,143], which both rely on non-learned image priors and the

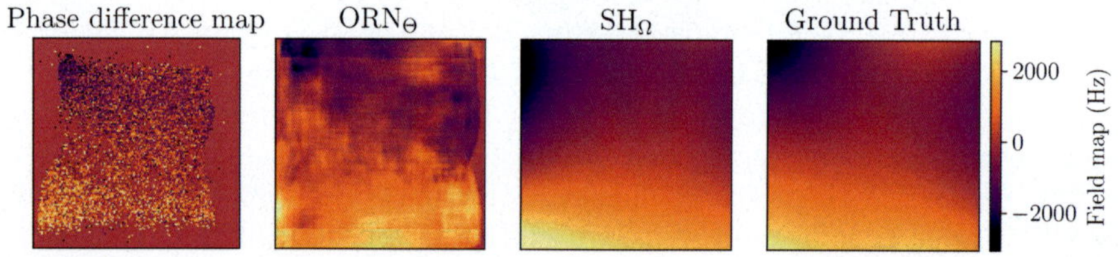

FIGURE 5.8

Results of different approaches for estimating the field map from low-field MRI data. The conventional phase difference map obtained from two acquisitions with a shift of the acquisition window results in a very noisy field map. The off-resonance net (ORN_{Θ}) [126,127] developed for higher field strengths combines different convolutional layers with residual connections but leads to patch-like artifacts in the field map. Using spherical-harmonic basis functions (SH_{Ω}) [137,138] as a decoder and learning the spherical harmonic coefficients restricts the solution to a smooth representation.

Image adapted from Ref. [59].

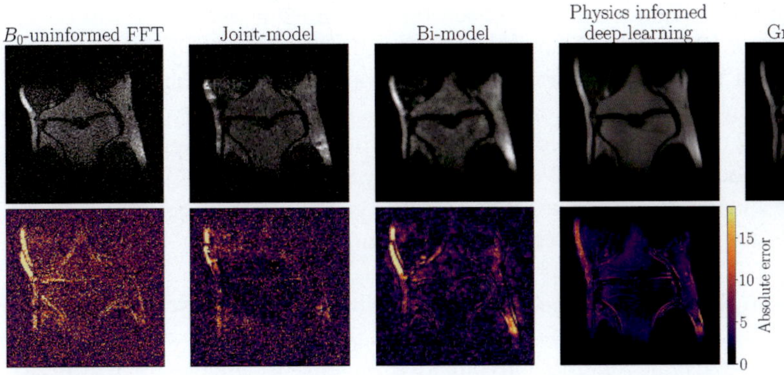

FIGURE 5.9

Comparison of different joint image and field-map reconstruction methods for simulated low-field data. The B0-uninformed IFFT reconstruction reveals the image distortion caused by B0-inhomogeneities if not corrected. The conventional approaches provide iterative reconstruction results by involving one joint model or two separate models. Due to the low SNR, the physics-informed deep learning approach results in better contrast and fewer image distortions.

end-to-end trainable network [59], which uses CNNs for the regularization of both the image and the field-map estimates. All three methods provide estimations of the field map and the image.

5.3.5 AI solutions for image processing in low-field MRI

Last, we briefly discuss some recent works that explore opportunities given by deep learning methods for different image processing tasks, e.g., image segmentation or image classification.

An important quality control step in clinical practice is the identification of artifacts in the obtained images. In addition to detecting artifacts, it is also important to correctly classify them as this allows the user to adapt the MR acquisition to reduce or avoid the artifact in a repeated acquisition. The authors in Ref. [144] presented an approach that detects and classifies Cartesian undersampling artifacts and artifacts due to Gibbs ringing for brain imaging at 0.36 T. Their method also yields heat maps that highlight the regions in the image affected by artifacts.

Automatic segmentation of different organs is of high interest to speed up medical examinations and allow for automatic assessment of diagnostic parameters. The authors in Ref. [145] utilized a U-Net trained on high-field and low-field fetal image data to segment organs of interest (e.g., lungs, liver, spleen, and kidneys). The method was evaluated on 0.55 T fetal images and used to automatically assess T2* values in these organs as a biomarker for blood oxygenation.

In Ref. [146], a segmentation approach for knee imaging at 0.18 T was developed based on a *k*-nearest neighbors algorithm trained on manual segmentations. Segmentation of tibial and femoral cartilage was achieved in healthy subjects and patients suffering from osteoarthritis.

In Ref. [147], the authors explore the applicability of deep learning-based image segmentation models trained on 3 T knee MR image data to 0.55 T MR images. The results suggested that bone segmentations were in general slightly better at 3 T, and the quality of the cartilage segmentations was comparable between the 0.55 T and 3 T.

The work in Ref. [148] investigated the applicability of a Cycle-GAN-based approach [95] for the segmentation of lung cysts at 0.55 T MRI. The employed dataset consisted of 65 lymphangioleiomyomatosis patients for images that were acquired on a 0.55 T MRI scanner as well as on a CT. Thereby, one part of the network first generates synthetic CT images from the MR images for which then a second module based on residual U-Nets segments lung volume and cysts. The network was trained on 12089 CT slices and 1258 MRI slices using an adversarial loss as well as a cycle-consistency loss. The authors reported good agreements between the segmentations obtained from the CT images and the synthetic CT images generated from the low-field MR images, which shows the potential of employing an ionizing radiation-free alternative for the quantification of lung cysts.

5.4 Conclusions

The revival of low-field and ultra-low-field MRI presents an exciting opportunity to address key limitations in the accessibility, cost, and safety of MRI technology. While low-field MRI still faces significant challenges, particularly in SNR, spatial resolution, and scan efficiency, ML/AI-driven solutions hold the potential to overcome these barriers. By leveraging deep learning techniques for image reconstruction, denoising, automatic segmentation, and post-processing, low-field MRI systems can produce higher-quality images that approach the diagnostic capability of high-field systems. Moreover, ML-based innovations in acquisition planning and scan time further enhance the practicality of low-field MRI in clinical environments.

As ML technologies continue to evolve, their integration with low-field MRI has the potential to transform the landscape of medical imaging, making MRI more accessible and cost-effective worldwide. However, future research is needed to further advance these ML-based methods and establish regulatory standards that ensure consistent image quality and safety. Through continued innovation, low-field and ultra-low-field MRI could become vital tools in global healthcare, expanding access to critical diagnostic imaging, particularly in regions where cost and accessibility are major concerns and/or enabling timely assessments in point-of-care settings.

References

[1] Hori M, Hagiwara A, Goto M, Wada A, Aoki S. Low-field magnetic resonance imaging its history and renaissance. Investig Radiol 2021;56(11):669–79 (Cited on page 1).

[2] Sarracanie M, Salameh N. Low-field MRI: how low can we go? A fresh view on an old debate. Front Phys 2020;8(June):1–14 (Cited on page 1).

[3] Hennig J. An evolution of low-field strength MRI. Magnet Reson Mater Phys Biol Med 2023;36(3):335–46. https://doi.org/10.1007/s10334-023-01104-z (Cited on pages 1 and 5).

[4] Niendorf T, Schulz-Menger J, Paul K, Huelnhagen T, Ferrari VA, Hodge R. High field cardiac magnetic resonance imaging. Circ Cardiovasc Imaging June 2017;10(6): e005460. https://doi.org/10.1161/CIRCIMAGING.116.005460 (Cited on page 1).

[5] Dumoulin SO, Fracasso A, van der Zwaag W, Siero JCW, Petridou N. Ultra-high field MRI: advancing systems neuroscience towards mesoscopic human brain function. Neuroimage 2018;168(September 2016):345–57. https://doi.org/10.1016/j.neuroimage.2017.01.028 (Cited on pages 1 and 4).

[6] Wei Z, Chen Q, Han S, Zhang S, Zhang N, Zhang L, et al. 5T magnetic resonance imaging: radio frequency hardware and initial brain imaging. Quant Imag Med Surg May 2023;13(5):3222–40 (Cited on page 1).

[7] Reiter T, Lohr D, Hock M, Ankenbrand MJ, Stefanescu MR, Kosmala A, et al. On the way to routine cardiac MRI at 7 Tesla - a pilot study on consecutive 84 examinations. PLoS One 2021;16(7):e0252797 (Cited on page 1).

[8] Winter L, Oberacker E, Özerdem C, Ji Y, von Knobelsdorff-Brenkenhoff F, Weidemann G, et al. On the RF heating of coronary stents at 7.0 Tesla MRI. Magn Reson Med October 2015;74(4):999–1010 (Cited on page 1).

[9] Van Speybroeck C, O'Reilly T, Teeuwisse W, Arnold P, Webb A. Characterization of displacement forces and image artifacts in the presence of passive medical implants in low-field (< 100 mT) permanent magnet-based MRI systems, and comparisons with clinical MRI systems. Phys Med 2021;84:116–24 (Cited on page 2).

[10] Simonetti OP, Ahmad R. Low-field cardiac magnetic resonance imaging. Circ Cardiovasc Imaging June 2017;10(6) (Cited on page 2), https://www.ahajournals.org/doi/10.1161/CIRCIMAGING.117.005446.

[11] Gilk T, Kanal E. MRI safety considerations associated with low-field MRI: mostly good news. Magnet Reson Mater Phys Biol Med 2023;36(3):427–8. https://doi.org/10.1007/s10334-023-01079-x (Cited on pages 2 and 5).

[12] Marques JP, Simonis FFJ, Webb AG. Low-field MRI: an MR physics perspective. J Magn Reson Imag June 2019;49(6):1528–42. https://doi.org/10.1002/jmri.26637 (Cited on pages 2 and 3).

[13] Arnold TC, Freeman CW, Litt B, Stein JM. Low-field MRI: clinical promise and challenges. J Magn Reson Imag January 2023;57(1):25–44 (Cited on pages 2, 4, and 6), https://onlinelibrary.wiley.com/doi/10.1002/jmri.28408.

[14] Shetty AS, Ludwig DR, Ippolito JE, Andrews TJ, Narra VR, Fraum TJ. Low-field-strength body MRI: challenges and opportunities at 0.55 T. Radiographics December 2023;43(12) (Cited on page 2), http://pubs.rsna.org/doi/10.1148/rg.230073.

[15] Liu Y, Leong ATL, Zhao Y, Xiao L, Mak HKF, Tsang ACO, et al. A low-cost and shielding-free ultra-low-field brain MRI scanner. Nat Commun 2021;12(1):1–14 (Cited on pages 2 and 6).

[16] Zhao Y, Ding Y, Lau V, Man C, Su S, Xiao L, et al. Whole-body magnetic resonance imaging at 0.05 Tesla. Science (New York, NY) 2024;384(6696):eadm7168 (Cited on pages 2 and 6).
[17] Rogers T, Campbell-Washburn AE, Ramasawmy R, Yildirim DK, Bruce CG, Grant LP, et al. Interventional cardiovascular magnetic resonance: state-of-the-art. J Cardiovasc Magn Reson 2023;25(1):48. https://doi.org/10.1186/s12968-023-00956-7 (Cited on page 2).
[18] Atalay MK, Poncelet BP, Kantor HL, Brady TJ, Weisskoff RM. Cardiac susceptibility artifacts arising from the heart-lung interface. Magn Reson Med February 2001;45(2):341–5 (Cited on page 2).
[19] Restivo MC, Ramasawmy R, Bandettini WP, Herzka DA, Campbell-Washburn AE. Efficient spiral in-out and EPI balanced steady-state free precession cine imaging using a high-performance 0.55T MRI. Magn Reson Med November 2020;84(5):2364–75 (Cited on pages 2 and 3).
[20] Campbell-Washburn AE, Ramasawmy R, Restivo MC, Bhattacharya I, Basar B, Herzka DA, et al. Opportunities in interventional and diagnostic imaging by using high-performance low-field-strength MRI. Radiology November 2019;293(2):384–93 (Cited on pages 2 and 5), http://pubs.rsna.org/doi/10.1148/radiol.2019190452.
[21] Macovski A. Noise in MRI. Magn Reson Med September 1996;36(3):494–7. https://doi.org/10.1002/mrm.1910360327 (Cited on page 3).
[22] Bieri O, Scheffler K. Fundamentals of balanced steady state free precession MRI. J Magn Reson Imag 2013;38(1):2–11 (Cited on page 3), https://onlinelibrary.wiley.com/doi/abs/10.1002/jmri.24163.
[23] Deoni SCL, Bruchhage MMK, Beauchemin J, Volpe A, D'Sa V, Huentelman M, et al. Accessible pediatric neuroimaging using a low field strength MRI scanner. Neuroimage 2021;238:118273 (Cited on page 3), https://www.sciencedirect.com/science/article/pii/S1053811921005498.
[24] Varghese J, Jin N, Giese D, Chen C, Liu Y, Pan Y, et al. Building a comprehensive cardiovascular magnetic resonance exam on a commercial 0.55 T system: a pictorial essay on potential applications. Front Cardiovasc Med March 2023;10 (Cited on page 3), https://www.frontiersin.org/articles/10.3389/fcvm.2023.1120982/full.
[25] Roemer PB, Edelstein WA, Hayes CE, Souza SP, Mueller OM. The NMR phased array. Magn Reson Med November 1990;16(2):192–225. https://doi.org/10.1002/mrm.1910160203 (Cited on pages 4, 20, and 21).
[26] Zbontar J, Knoll F, Sriram A, Murrell T, Huang Z, Muckley MJ, et al. fastMRI: an open dataset and benchmarks for accelerated MRI. arXiv preprint arXiv:181108839 2018.
[27] de Leeuw Den Bouter M, Ippolito G, O'Reilly T, Remis R, Van Gijzen M, Webb A. Deep learning-based single image super-resolution for low-field MR brain images. Sci Rep 2022;12(1):6362 (Cited on pages 4, 9, 10, 15, and 16).
[28] O'Reilly T, Teeuwisse WM, de Gans D, Koolstra K, Webb AG. In vivo 3D brain and extremity MRI at 50 mT using a permanent magnet Halbach array. Magn Reson Med 2021;85(1):495–505 (Cited on page 4).
[29] Rooney WD, Johnson G, Li X, Cohen ER, Kim SG, Ugurbil K, et al. Magnetic field and tissue dependencies of human brain longitudinal 1H2O relaxation in vivo. Magn Reson Med February 2007;57(2):308–18. https://doi.org/10.1002/mrm.21122 (Cited on page 3).

[30] Peters AM, Brookes MJ, Hoogenraad FG, Gowland PA, Francis ST, Morris PG, et al. T2* measurements in human brain at 1.5, 3 and 7 T. Magn Reson Imaging 2007;25(6): 748–53 (Cited on page 3).

[31] Delfaut EM, Beltran J, Johnson G, Rousseau J, Marchandise X, Cotten A. Fat suppression in MR imaging: techniques and pitfalls. Radiographics 1999;19(2):373–82 (Cited on page 4).

[32] Tian Y, Nayak KS. New clinical opportunities of low-field MRI: heart, lung, body, and musculoskeletal. Magnet Reson Mater Phys Biol Med 2024;37(1):1–14. https://doi.org/10.1007/s10334-023-01123-w (Cited on page 4).

[33] Waddington DEJ, Boele T, Maschmeyer R, Kuncic Z, Rosen MS. High-sensitivity in vivo contrast for ultra-low field magnetic resonance imaging using superparamagnetic iron oxide nanoparticles. Sci Adv September 2024;6(29):eabb0998. https://doi.org/10.1126/sciadv.abb0998 (Cited on page 4).

[34] Tournier JD. Diffusion MRI in the brain – theory and concepts. Prog Nucl Magn Reson Spectrosc 2019;112–113:1–16 (Cited on pages 4 and 5).

[35] Rodgers CT, Clarke WT, Snyder C, Vaughan JT, Neubauer S, Robson MD. Human cardiac 31P magnetic resonance spectroscopy at 7 Tesla. Magn Reson Med 2014;72(2): 304–15 (Cited on pages 4 and 5).

[36] Bhat SS, Fernandes TT, Poojar P, da Silva Ferreira M, Rao PC, Hanumantharaju MC, et al. Low-field MRI of stroke: challenges and opportunities. J Magn Reson Imag August 2021;54(2):372–90. https://doi.org/10.1002/jmri.27324 (Cited on page 5).

[37] Islam KT, Zhong S, Zakavi P, Chen Z, Kavnoudias H, Farquharson S, et al. Improving portable low-field MRI image quality through image-to-image translation using paired low- and high-field images. Sci Rep 2023;13(1):1–13. https://doi.org/10.1038/s41598-023-48438-1 (Cited on page 6).

[38] Lopez Schmidt I, Haag N, Shahzadi I, Frohwein LJ, Schneider C, Niehoff JH, et al. Diagnostic image quality of a low-field (0.55T) knee MRI protocol using deep learning image reconstruction compared with a standard (1.5T. Knee MRI Protocol 2023 (Cited on page 6).

[39] Zhu B, Liu JZ, Cauley SF, Rosen BR, Rosen MS. Image reconstruction by domain-transform manifold learning. Nature 2018;555(7697):487–92 (Cited on page 6).

[40] de Leeuw den Bouter ML, Ippolito G, O'Reilly TPA, Remis RF, van Gijzen MB, Webb AG. Deep learning-based single image super-resolution for low-field MR brain images. Sci Rep 2022;12(1):6362. https://doi.org/10.1038/s41598-022-10298-6 (Cited on page 6).

[41] Le DBT, Sadinski M, Nacev A, Narayanan R, Kumar D. Deep learning–based method for denoising and image enhancement in low-field MRI. In: 2021 IEEE International Conference on Imaging Systems and Techniques (IST); 2021. p. 1–6 (Cited on page 6).

[42] Bernard O, Lalande A, Zotti C, Cervenansky F, Yang X, Heng PA, et al. Deep learning techniques for automatic MRI cardiac multi-structures segmentation and diagnosis: is the problem solved? IEEE Trans Med Imag 2018;37(11):2514–25 (Cited on page 6).

[43] Ronneberger O, Fischer P, Brox T. U-net: convolutional networks for biomedical image segmentation. In: Medical Image Computing and Computer-Assisted Intervention–MICCAI 2015: 18th International Conference Proceedings, Part III 18. Springer; 2015. p. 234–41 (Cited on page 7).

[44] Hinton G, Srivastava N, Swersky K. Neural networks for machine learning lecture 6a overview of mini-batch gradient descent. Online slides 2012 (Cited on page 7).

[45] Zeiler MD. Adadelta: an adaptive learning rate method. arXiv preprint arXiv: 12125701 2012 (Cited on page 7).
[46] Kingma DP, Ba J. Adam: a method for stochastic optimization. In: Bengio Y, LeCun Y, editors. 3rd International Conference on Learning Representations, ICLR 2015 Proceedings; 2015 (Cited on page 7).
[47] Ruder S. An overview of gradient descent optimization algorithms. arXiv preprint arXiv:160904747 2016 (Cited on page 7).
[48] Silva SN, McElroy S, Verdera JA, Colford K, Clair KS, Tomi-Tricot R, et al. Fully automated planning for anatomical fetal brain MRI on 0.55T. Magn Reson Med 2024;92:1263–76 (Cited on page 8).
[49] Silva SN, Woodgate T, McElroy S, Cleri M, Clair KS, Verdera JA, et al. AutOmatic floW planning for fetaL MRI (OWL). ArXiv 2024 (Cited on page 8), http://arxiv.org/abs/2408.06326.
[50] Su J, Pellicer-Guridi R, Edwards T, Fuentes M, Rosen MS, Vegh V, et al. A CNN based software gradiometer for electromagnetic background noise reduction in low field MRI applications. IEEE Trans Med Imag 2022;41:1007–16 (Cited on page 8).
[51] Zhao Y, Xiao L, Hu J, Wu EX. Robust EMI elimination for RF shielding-free MRI through deep learning direct MR signal prediction. Magn Reson Med 2024 7;92: 112–27 (Cited on page 8), https://onlinelibrary.wiley.com/doi/10.1002/mrm.30046.
[52] Zhu B, Liu J, Koonjoo N, Rosen BR, Rosen MS. AUTOmated pulse SEQuence generation (AUTOSEQ) using Bayesian reinforcement learning in an MRI physics simulation environment. In: Proceedings of joint annual meeting ISMRM-ESMRMB; 2018 (Cited on page 8).
[53] Iglesias JE, Schleicher R, Laguna S, Billot B, Schaefer P, McKaig B, et al. Quantitative brain morphometry of portable low-field-strength MRI using super-resolution machine learning. Radiology 2022;306(3):e220522 (Cited on pages 9 and 15).
[54] Iglesias JE, Schleicher R, Laguna S, Billot B, Schaefer P, McKaig B, et al. Accurate super-resolution low-field brain MRI. arXiv preprint arXiv:220203564. 2022 (Cited on pages 9 and 10).
[55] Laguna S, Schleicher R, Billot B, Schaefer P, Mckaig B, Goldstein JN, et al. Super-resolution of portable low-field MRI in real scenarios: integration with denoising and domain adaptation. In: Medical imaging with deep learning; 2022 (Cited on pages 9 and 16).
[56] Lau V, Xiao L, Zhao Y, Su S, Ding Y, Man C, et al. Pushing the limits of low-cost ultra-low-field MRI by dual-acquisition deep learning 3D superresolution. Magn Reson Med 2023;90(2):400–16 (Cited on page 9).
[57] Donnay C, Okar SV, Tsagkas C, Gaitán MI, Poorman M, Reich DS, et al. Super resolution using sparse sampling at portable ultra-low field MR. Front Neurol 2024;15: 1330203 (Cited on page 9).
[58] Brown RW, Cheng YCN, Haacke EM, Thompson MR, Venkatesan R. Magnetic resonance imaging: physical principles and sequence design. John Wiley & Sons; 2014 (Cited on page 10).
[59] Schote D, Winter L, Kolbitsch C, Rose G, Speck O, Kofler A. Joint B_0 and image reconstruction in low-field MRI by physics-informed deep-learning. IEEE Trans Biomed Eng 2024:1–12 (Cited on pages 10, 20, and 21).
[60] Darestani MZ, Liu J, Heckel R. Test-time training can close the natural distribution shift performance gap in deep learning based compressed sensing. In: International

Conference on Machine Learning. PMLR; 2022. p. 4754–76 (Cited on pages 10, 11, and 18).

[61] Schlemper J, Caballero J, Hajnal JV, Price A, Rueckert D. A deep cascade of convolutional neural networks for MR image reconstruction. In: Information Processing in Medical Imaging: 25th International Conference, IPMI 2017, Proceedings 25. Springer; 2017. p. 647–58 (Cited on pages 10 and 19).

[62] Ulyanov D, Vedaldi A, Lempitsky V. Deep image prior. In: Proceedings of the IEEE/CVF Conference on Computer Vision and Pattern Recognition; 2018. p. 9446–54 (Cited on pages 10 and 13).

[63] Heckel R, Hand P. Deep decoder: concise image representations from untrained non-convolutional networks. In: International Conference on Learning Representations; 2019 (Cited on pages 11 and 13).

[64] Etmann C, Ke R, Schönlieb CB. iUNets: learnable invertible up-and downsampling for large-scale inverse problems. In: 2020 IEEE 30th International Workshop on Machine Learning for Signal Processing (MLSP). IEEE; 2020. p. 1–6 (Cited on page 11).

[65] Srivastava N, Hinton G, Krizhevsky A, Sutskever I, Salakhutdinov R. Dropout: a simple way to prevent neural networks from overfitting. J Mach Learn Res 2014;15(1): 1929–58 (Cited on page 11).

[66] Ioffe S. Batch normalization: accelerating deep network training by reducing internal covariate shift. arXiv preprint arXiv:150203167. 2015 (Cited on page 11).

[67] Miyato T, Kataoka T, Koyama M, Yoshida Y. Spectral normalization for generative adversarial networks. arXiv preprint arXiv:180205957 2018 (Cited on page 11).

[68] Lehtinen J, Munkberg J, Hasselgren J, Laine S, Karras T, Aittala M, et al. Noise2-Noise: learning image restoration without clean data. In: Dy JG, Krause A, editors. Proceedings of the 35th International Conference on Machine Learning, ICML 2018. vol. 80 of Proceedings of Machine Learning Research. PMLR; 2018. p. 2971–80 (Cited on pages 11 and 12).

[69] Moran N, Schmidt D, Zhong Y, Coady P. Noisier2noise: learning to denoise from unpaired noisy data. In: Proceedings of the IEEE/CVF Conference on Computer Vision and Pattern Recognition; 2020. p. 12064–72 (Cited on pages 11 and 12).

[70] Batson J, Royer L. Noise2self: blind denoising by self-supervision. In: International Conference on Machine Learning. PMLR; 2019. p. 524–33 (Cited on pages 11 and 12).

[71] Yaman B, Hosseini SAH, Moeller S, Ellermann J, Uğurbil K, Akçakaya M. Self-supervised learning of physics-guided reconstruction neural networks without fully sampled reference data. Magn Reson Med 2020;84(6):3172–91 (Cited on pages 11 and 12).

[72] Lian Y, Ye X, Luo H, Wu Z, Guo H. A deep learning network for low-field MRI denoising using group sparsity information and a Noise2Noise method. In: Joint Annual Meeting ISMRM-ESMRMB ISMRT 31st Annual Meeting. ISMRM; 2022. p. 4345 (Cited on pages 11 and 13).

[73] Schlemper J, Dey N, Salehi SSM, Lazarus C, O'Halloran R, Kundu P, et al. Unsupervised denoising for low-field diffusion MRI. Proc Int Soc Magnet Reson Med 2021: 2190 (Cited on pages 11 and 17).

[74] Millard C, Chiew M. A theoretical framework for self-supervised MR image reconstruction using sub-sampling via variable density Noisier2Noise. IEEE Transact Comput Imaging 2023 (Cited on page 11).

[75] Yoo J, Jin KH, Gupta H, Yerly J, Stuber M, Unser M. Time-dependent deep image prior for dynamic MRI. IEEE Trans Med Imag 2021;40(12):3337–48 (Cited on page 13).

[76] Arora S, Roeloffs V, Lustig M. Untrained modified deep decoder for joint denoising parallel imaging reconstruction. In: ISMRM Annual Meeting; 2020 (Cited on page 13).
[77] Heckel R, Hand P. Deep decoder: concise image representations from untrained non-convolutional networks. In: 7th International Conference on Learning Representations, ICLR 2019; 2019 (Cited on page 13).
[78] Darestani MZ, Heckel R. Accelerated MRI with un-trained neural networks. IEEE Transact Comput Imaging 2021;7:724–33 (Cited on page 13).
[79] Yaman B, Hosseini SAH, Akçakaya M. Zero-shot self-supervised learning for MRI reconstruction. ICLR 2022 2022 (Cited on page 13).
[80] Lequyer J, Philip R, Sharma A, Hsu WH, Pelletier L. A fast blind zero-shot denoiser. Nat Mach Intell 2022;4(11):953–63 (Cited on page 13).
[81] Mansour Y, Heckel R. Zero-shot noise2noise: efficient image denoising without any data. In: Proceedings of the IEEE/CVF Conference on Computer Vision and Pattern Recognition; 2023. p. 14018–27 (Cited on page 13).
[82] Akçakaya M, Yaman B, Chung H, Ye JC. Unsupervised deep learning methods for biological image reconstruction and enhancement: an overview from a signal processing perspective. IEEE Signal Process Mag 2022;39(2):28–44 (Cited on page 13).
[83] Hamilton JI, Truesdell W, Galizia M, Burris N, Agarwal P, Seiberlich N. A low-rank deep image prior reconstruction for free-breathing ungated spiral functional CMR at 0.55 T and 1.5 T. Magnetic Resonance Materials in Physics. Biol Med 2023;36: 451–64 (Cited on pages 13 and 14).
[84] Liu J, Sun Y, Xu X, Kamilov US. Image restoration using total variation regularized deep image prior. In: ICASSP 2019-2019 IEEE International Conference on Acoustics, Speech and Signal Processing (ICASSP). Ieee; 2019. p. 7715–9 (Cited on page 13).
[85] Uecker M, Lai P, Murphy MJ, Virtue P, Elad M, Pauly JM, et al. ESPIRiT–an eigenvalue approach to autocalibrating parallel MRI: where SENSE meets GRAPPA. Magn Reson Med 2014;71(3):990–1001 (Cited on page 14).
[86] Otazo R, Candes E, Sodickson DK. Low-rank plus sparse matrix decomposition for accelerated dynamic MRI with separation of background and dynamic components. Magn Reson Med 2015;73(3):1125–36 (Cited on page 14).
[87] Zhu B, Liu JZ, Cauley SF, Rosen BR, Rosen MS. Image reconstruction by domain-transform manifold learning. Nature 2018;555(7697):487–92 (Cited on pages 15 and 16).
[88] Koonjoo N, Zhu B, Bagnall GC, Bhutto D, Rosen M. Boosting the signal-to-noise of low-field MRI with deep learning image reconstruction. Sci Rep 2021;11(1):8248 (Cited on pages 15 and 16).
[89] Dabov K, Foi A, Katkovnik V, Egiazarian K. Image denoising with block-matching and 3D filtering. In: Image processing: algorithms and systems, neural networks, and machine learning, vol 6064. SPIE; 2006. p. 354–65 (Cited on page 15).
[90] Mäkinen Y, Azzari L, Foi A. Collaborative filtering of correlated noise: exact transform-domain variance for improved shrinkage and patch matching. IEEE Trans Image Process 2020;29:8339–54 (Cited on page 15).
[91] Zhang K, Zuo W, Chen Y, Meng D, Zhang L. Beyond a Gaussian denoiser: residual learning of deep CNN for image denoising. IEEE Trans Image Process 2017;26(7): 3142–55 (Cited on pages 15 and 19).

[92] Tong T, Li G, Liu X, Gao Q. Image super-resolution using dense skip connections. In: Proceedings of the IEEE International Conference on Computer Vision; 2017. p. 4799–807 (Cited on page 15).

[93] Iglesias JE, Billot B, Balbastre Y, Tabari A, Conklin J, González RG, et al. Joint super-resolution and synthesis of 1 mm isotropic MP-RAGE volumes from clinical MRI exams with scans of different orientation, resolution and contrast. Neuroimage 2021;237:118206 (Cited on page 15).

[94] Van Essen DC, Smith SM, Barch DM, Behrens TE, Yacoub E, Ugurbil K, et al. The WU-Minn human connectome project: an overview. Neuroimage 2013;80:62–79 (Cited on pages 16, 17 and 19).

[95] Zhu JY, Park T, Isola P, Efros AA. Unpaired image-to-image translation using cycle-consistent adversarial networks. In: Proceedings of the IEEE International Conference on Computer Vision; 2017. p. 2223–32 (Cited on pages 16 and 22).

[96] Park T, Efros AA, Zhang R, Zhu JY. Contrastive learning for unpaired image-to-image translation. In: Computer Vision–ECCV 2020: 16th European Conference, Glasgow, UK, August 23–28, 2020, Proceedings, Part IX 16. Springer; 2020. p. 319–45 (Cited on page 16).

[97] Antun V, Renna F, Poon C, Adcock B, Hansen AC. On instabilities of deep learning in image reconstruction and the potential costs of AI. Proc Natl Acad Sci 2020;117(48): 30088–95 (Cited on page 16).

[98] Jin KH, McCann MT, Froustey E, Unser M. Deep convolutional neural network for inverse problems in imaging. IEEE Trans Image Process 2017;26(9):4509–22 (Cited on page 17).

[99] Ding PLK, Li Z, Zhou Y, Li B. Deep residual dense U-Net for resolution enhancement in accelerated MRI acquisition. In: Medical imaging 2019: image processing, vol 10949. SPIE; 2019. p. 110–7 (Cited on page 17).

[100] Wang S, Su Z, Ying L, Peng X, Zhu S, Liang F, et al. Accelerating magnetic resonance imaging via deep learning. In: 2016 IEEE 13th International Symposium on Biomedical Imaging (ISBI). IEEE; 2016. p. 514–7 (Cited on page 17).

[101] Hyun CM, Kim HP, Lee SM, Lee S, Seo JK. Deep learning for undersampled MRI reconstruction. Phys Med Biol 2018;63(13):135007 (Cited on page 17).

[102] Lee D, Yoo J, Tak S, Ye JC. Deep residual learning for accelerated MRI using magnitude and phase networks. IEEE Trans Biomed Eng 2018;65(9):1985–95 (Cited on page 17).

[103] Kawamura M, Tamada D, Funayama S, Kromrey ML, Ichikawa S, Onishi H, et al. Accelerated acquisition of high-resolution diffusion-weighted imaging of the brain with a multi-shot echo-planar sequence: deep-learning-based denoising. Magn Reson Med Sci 2021;20(1):99–105 (Cited on page 17).

[104] Li S, Zhou J, Liang D, Liu Q. MRI denoising using progressively distribution-based neural network. Magn Reson Imaging 2020;71:55–68 (Cited on page 17).

[105] Kofler A, Dewey M, Schaeffter T, Wald C, Kolbitsch C. Spatio-temporal deep learning-based undersampling artefact reduction for 2D radial cine MRI with limited training data. IEEE Trans Med Imag 2019;39(3):703–17 (Cited on page 17).

[106] Hauptmann A, Arridge S, Lucka F, Muthurangu V, Steeden JA. Real-time cardiovascular MR with spatio-temporal artifact suppression using deep learning–proof of concept in congenital heart disease. Magn Reson Med 2019 (Cited on page 17).

[107] Sandino CM, Cheng JY. Deep convolutional neural networks for accelerated dynamic magnetic resonance imaging. Stanford University CS231N, Course project; 2017 (Cited on page 17).
[108] Cheng JY, Mardani M, Alley MT, Pauly JM, Vasanawala SS. DeepSPIRiT: generalized parallel imaging using deep convolutional neural networks. In: Annual Meeting of the International Society of Magnetic Resonance in Medicine; 2018 (Cited on page 17).
[109] Akçakaya M, Moeller S, Weingärtner S, Uğurbil K. Scan-specific robust artificial-neural-networks for k-space interpolation (RAKI) reconstruction: database-free deep learning for fast imaging. Magn Reson Med 2019;81(1):439–53 (Cited on page 17).
[110] Han Y, Sunwoo L, Ye JC. k-space deep learning for accelerated MRI. IEEE Trans Med Imag 2019;39(2):377–86 (Cited on pages 17 and 19).
[111] Ayde R, Senft T, Salameh N, Sarracanie M. Deep learning for fast low-field MRI acquisitions. Sci Rep 2022;12(1):11394 (Cited on page 17).
[112] Le DBT, Sadinski M, Nacev A, Narayanan R, Kumar D. Deep learning–based method for denoising and image enhancement in low-field MRI. In: 2021 IEEE International Conference on Imaging Systems and Techniques (IST). IEEE; 2021. p. 1–6 (Cited on page 17).
[113] Clark K, Vendt B, Smith K, Freymann J, Kirby J, Koppel P, et al. The Cancer Imaging Archive (TCIA): maintaining and operating a public information repository. J Digit Imag 2013;26:1045–57 (Cited on page 17).
[114] Kojima S, Ito T, Hayashi T. Denoising using Noise2Void for low-field magnetic resonance imaging: a phantom study. J Med Phys 2022;47(4):387–93 (Cited on page 17).
[115] Krull A, Buchholz TO, Jug F. Noise2void-learning denoising from single noisy images. In: Proceedings of the IEEE/CVF Conference on Computer Vision and Pattern Recognition; 2019. p. 2129–37 (Cited on page 17).
[116] Lucas A, Arnold TC, Okar SV, Vadali C, Kawatra KD, Ren Z, et al. Multi-contrast high-field quality image synthesis for portable low-field MRI using generative adversarial networks and paired data. medRxiv 2023 (Cited on page 17).
[117] Schlemper J, Dey N, Salehi SSM, Sheth K, Kimberly WT, Sofka M. Two-step semi-supervised denoising for low-field diffusion MRI. Proc Int Soc Magnet Reson Med 2022:4057 (Cited on page 17).
[118] Lin K, Heckel R. Robustness of deep learning for accelerated MRI: benefits of diverse training data. In: Salakhutdinov R, Kolter Z, Heller K, Weller A, Oliver N, Scarlett J, et al., editors. Proceedings of the 41st International Conference on Machine Learning. vol. 235 of Proceedings of Machine Learning Research. PMLR; 2024. p. 30018–41 (Cited on page 18), https://proceedings.mlr.press/v235/lin24h.html.
[119] Muckley MJ, Riemenschneider B, Radmanesh A, Kim S, Jeong G, Ko J, et al. Results of the 2020 fastMRI challenge for machine learning MR image reconstruction. IEEE Trans Med Imag 2021;40(9):2306–17 (Cited on page 18).
[120] Venkatakrishnan SV, Bouman CA, Wohlberg B. Plug-and-play priors for model based reconstruction. In: 2013 IEEE Global Conference on Signal and Information Processing. IEEE; 2013. p. 945–8 (Cited on page 18).
[121] Romano Y, Elad M, Milanfar P. The little engine that could: regularization by denoising (RED). SIAM J Imag Sci 2017;10(4):1804–44 (Cited on page 18).
[122] Liu J, Sun Y, Eldeniz C, Gan W, An H, Kamilov US. RARE: image reconstruction using deep priors learned without groundtruth. IEEE J Selected Topics Signal Proces 2020;14(6):1088–99 (Cited on page 19).

[123] Daval-Frérot G, Massire A, Mailhé B, Nadar M, Bapst B, Luciani A, et al. Deep learning-assisted model-based off-resonance correction for non-Cartesian SWI. Magn Reson Med 2023;90(4):1431–45 (Cited on page 19).

[124] Ramzi Z, Chaithya G, Starck JL, Ciuciu P. NC-PDNet: a density-compensated unrolled network for 2D and 3D non-Cartesian MRI reconstruction. IEEE Trans Med Imag 2022;41(7):1625–38 (Cited on page 19).

[125] Daval-Frérot G, Massire A, Mailhe B, Nadar M, Vignaud A, Ciuciu P. Iterative static field map estimation for off-resonance correction in non-Cartesian susceptibility weighted imaging. Magn Reson Med 2022;88(4):1592–607 (Cited on page 19).

[126] Zeng DY, Shaikh J, Holmes S, Brunsing RL, Pauly JM, Nishimura DG, et al. Deep residual network for off-resonance artifact correction with application to pediatric body MRA with 3D cones. Magn Reson Med 2019;82(4):1398–411 (Cited on pages 19 and 20), https://onlinelibrary.wiley.com/doi/abs/10.1002/mrm.27825.

[127] Haskell M, Cao AA, Noll DC, Fessler JA. Deep learning field map estimation with model-based image reconstruction for off-resonance correction of brain images using a spiral acquisition. In: ISMRM Workshop on Data Sampling and Image Reconstruction; 2020 (Cited on pages 19 and 20).

[128] Fessler JA, Lee S, Olafsson VT, Shi HR, Noll DC. Toeplitz-based iterative image reconstruction for MRI with correction for magnetic field inhomogeneity. IEEE Trans Signal Process 2005;53(9):3393–402 (Cited on page 19).

[129] Monga V, Li Y, Eldar YC. Algorithm unrolling: interpretable, efficient deep learning for signal and image processing. IEEE Signal Process Mag 2021;38(2):18–44 (Cited on page 19).

[130] Hammernik K, Küstner T, Yaman B, Huang Z, Rueckert D, Knoll F, et al. Physics-driven deep learning for computational magnetic resonance imaging: combining physics and machine learning for improved medical imaging. IEEE Signal Process Mag 2023;40(1):98–114 (Cited on page 19).

[131] Schlemper J, Salehi SSM, Lazarus C, Dyvorne H, O'Halloran R, de Zwart N, et al. Deep learning MRI reconstruction in application to point-of-care MRI. Proc Int Soc Mag Reson Med 2020;28:0991 (Cited on page 19.

[132] Schlemper J, Salehi SSM, Kundu P, Lazarus C, Dyvorne H, Rueckert D, et al. Nonuniform variational network: deep learning for accelerated nonuniform MR image reconstruction. In: Medical Image Computing and Computer Assisted Intervention–MICCAI 2019: 22nd International Conference, Shenzhen, China, October 13–17, 2019, Proceedings, Part III 22. Springer; 2019. p. 57–64 (Cited on page 19).

[133] Schlicht F, Vosshenrich J, Donners R, Seifert AC, Fenchel M, Nickel D, et al. Advanced deep learning-based image reconstruction in lumbar spine MRI at 0.55 T – Effects on image quality and acquisition time in comparison to conventional deep learning-based reconstruction. Eur J Radiol Open 2024;12:100567 (Cited on page 20).

[134] Hammernik K, Klatzer T, Kobler E, Recht MP, Sodickson DK, Pock T, et al. Learning a variational network for reconstruction of accelerated MRI data. Magn Reson Med 2018;79(6):3055–71 (Cited on page 20).

[135] Donners R, Vosshenrich J, Gutzeit A, Bach M, Schlicht F, Obmann MM, et al. New-generation 0.55 T MRI of the knee—initial clinical experience and comparison with 3 T MRI. Investig Radiol 2024;59:298–305 (Cited on page 20).

[136] Schmidt IL, Haag N, Shahzadi I, Frohwein LJ, Schneider C, Niehoff JH, et al. Diagnostic image quality of a low-field (0.55T) knee MRI protocol using deep learning

image reconstruction compared with a standard (1.5T) knee MRI protocol. J Clin Med 2023;12:1916 (Cited on page 20).
[137] Schote D, Winter L, Kolbitsch C, Zimmermann F, O'Reilly T, Webb A, et al. Physics-informed deep learning for image distortion correction from B 0-inhomogeneities in low-field MRI. Proc Int Soc Mag Reson Med 2021;31:1816 (Cited on page 20).
[138] Schote D, Winter L, Kolbitsch C, Kofler A. Reliable low-field B0-maps by deep learning with physical constraints. Proc Int Soc Mag Reson Med 2023;32:4344 (Cited on page 20).
[139] Wenzel K, Alhamwey H, O'Reilly T, Riemann LT, Silemek B, Winter L. B0-shimming methodology for affordable and compact low-field magnetic resonance imaging magnets. Front Phys 2021;9:704566 (Cited on page 20).
[140] Aggarwal HK, Mani MP, Jacob M. MoDL: model-based deep learning architecture for inverse problems. IEEE Trans Med Imag 2018;38(2):394–405 (Cited on page 20).
[141] Kofler A, Haltmeier M, Schaeffter T, Kolbitsch C. An end-to-end-trainable iterative network architecture for accelerated radial multi-coil 2D cine MR image reconstruction. Med Phys 2021;48(5):2412–25 (Cited on page 20).
[142] Usman M, Kakkar L, Matakos A, Kirkham A, Arridge S, Atkinson D. Joint B0 and image estimation integrated with model based reconstruction for field map update and distortion correction in prostate diffusion MRI. Magn Reson Imaging 2020;65: 90–9 (Cited on page 21).
[143] Koolstra K, O'Reilly T, Börnert P, Webb A. Image distortion correction for MRI in low field permanent magnet systems with strong B0 inhomogeneity and gradient field nonlinearities. Magnet Reson Mater Phys Biol Med 2021:1–12 (Cited on page 21).
[144] Jimeno MM, Ravi KS, Jin Z, Oyekunle D, Ogbole G, Geethanath S. ArtifactID: identifying artifacts in low-field MRI of the brain using deep learning. Magn Reson Imaging 2022;89:42–8 (Cited on page 21).
[145] Payette K, Uus A, Verdera JA, Zampieri CA, Hall M, Story L, et al. In: An automated pipeline for quantitative T2* fetal body MRI and segmentation at low field; 2023. p. 358–67 (Cited on page 21).
[146] Folkesson J, Dam EB, Olsen OF, Pettersen PC, Christiansen C. Segmenting articular cartilage automatically using a voxel classification approach. IEEE Trans Med Imag 2007;26:106–15 (Cited on page 21).
[147] Bhattacharjee R, Akkaya Z, Luitjens J, Su P, Yang Y, Pedoia V, et al. Feasibility of translating 3.0 T-trained deep-learning segmentation models out-of-the-box on low-field MRI 0.55 T knee-MRI of healthy controls. arXiv preprint arXiv:231017152 2023 (Cited on page 22).
[148] Bhattacharya I, Chen MY, Moss J, Campbell-Washburn A, Xue H. Automated quantification of lung cysts at 0.55 T MRI with image synthesis from CT using deep learning. In: Proceedings of the International Society for Magnetic Resonance in Medicine Annual Meeting, vol 3513; 2020 (Cited on page 22).

CHAPTER

The smart spin: Machine learning for magnetic resonance spectroscopy

6

Stanislav Motyka[1, 2, 3] **and Wolfgang Bogner**[1, 2, 3, *]
[1]*High-Field MR Center, Department of Biomedical Imaging and Image-Guided Therapy, Medical University of Vienna, Vienna, Austria;* [2]*Comprehensive Center for AI in Medicine (CAIM), Medical University of Vienna, Vienna, Austria;* [3]*Christian Doppler Laboratory for MR Imaging Biomarkers (BIOMAK), Department of Biomedical Imaging and Image-Guided Therapy, Medical University of Vienna, Vienna, Austria*
[*]*Corresponding author: E-mail: wolfgang.bogner@meduniwien.ac.at*

6.1 Introduction

MR spectroscopy (MRS) and MR spectroscopic imaging (MRSI) are noninvasive methods for quantifying the amount of biomolecules within living organisms, typically in a localized fashion. They have their origin in nuclear magnetic resonance (NMR), which is essentially the same technique, but is used extensively by chemists to examine and characterize the structure of metabolic compounds in liquid probes [1]. In this chapter, we will focus on human applications in vivo [2], as MRS and MRSI are highly potent techniques for evaluating human metabolism, particularly in the human brain. While others have previously summarized more chemistry-focused deep learning applications of MRS/NMR [3,4], the applications to in vivo data involve significantly more and different challenges [5]. We will focus also mostly on deep learning (DL), which has nowadays completely replaced traditional ML approaches.

Despite their potential clinical value, MRS and MRSI face several challenges including low signal-to-noise ratio (SNR), overlapping metabolite signals, experimental artifacts, and comparatively long acquisition durations. Effectively analyzing spectroscopic data necessitates considerations such as the measurement technique (i.e., pulse sequence selection), experimental factors (e.g., field inhomogeneities and instabilities), as well as preprocessing and analysis methods. Given the intricacy of these factors, implementing and managing MRS and MRSI can be daunting for individuals lacking specialized expertise, impeding their widespread clinical adoption. The increasing popularity of ML techniques in various disciplines over the past decade stems from their capacity to derive model-agnostic features from data and the potential to offer more user-friendly solutions. In the realm of magnetic resonance imaging (MRI), ML methods have found diverse applications encompassing image reconstruction, quality enhancement, image analysis, and clinical diagnostics

Machine Learning in MRI. https://doi.org/10.1016/B978-0-443-14109-6.00022-5

[6]. This trend has extended—with some delay—to MRS and MRSI, where several ML approaches have been proposed to tackle long-standing technical challenges: for example, denoising/super-resolution reconstruction of MRS data, artifacts detection and removal or spectral fitting. Beyond technical applications, researchers have also targeted clinical use cases, such as improving clinical diagnosis via direct classification of pathologic spectral/biochemical patterns to make more efficient use of the wealth of spectroscopic information (detailed below).

In the following sections, we discuss and synthesize information regarding architectures, input-output configurations, training methodologies, and intended applications for a selection of studies. However, this chapter does not and cannot provide a comprehensive review or introduction to ML methodology. For a comprehensive overview of ML, DL, and other artificial intelligence techniques we have to refer to previous chapters.

6.2 MRS data acquisition

Although the main strength of ML lies certainly in data processing, the old saying "Fix it before it's broken," certainly also applies for MRS and, thus, the first critical step where ML techniques can be employed are already during data collection.

6.2.1 Volume selection and other calibration scans

MRS generally requires users to prospectively define a volume-of-interest (VOI) in brain tissue, which warrants new and easy-to-use methods for reliable and reproducible data collection. Conducting investigations without precisely placed voxels limits the utility and interpretability of MRS data for use in evaluating pathologic changes and interventions [7,8]. In recent years, automatic repositioning using standardized localizers has therefore become widely available [9]. However, these approaches allow only reliable and consistent VOI placement for repeated scans of the same subject. It is already somewhat limited by the heterogeneity across brain structures of different healthy subjects. Finally, they often fail to identify local pathologic alterations such as brain tumors and automatically place VOIs in such regions. An attempt to overcome this limitation via DL has been proposed [10] by training a convolutional neural network (CNN) on anatomical (i.e., T2w fluid-attenuated inversion recovery) MRI and their manually segmented lesions, to automatize single-voxel MRS voxel placement, demonstrating promising early results. Such approaches could be developed to automate also other aspects of voxel placement such as optimization of suppression pulses in the future [11].

Performing shimming of the static magnetic field (B0) is imperative for acquiring MRS data of high quality and several conventional approaches have been proposed [12]. Consequently, B0 field optimization has become a target of current DL efforts. For instance, Becker et al. have proposed and compared different DL approaches to automate the shimming procedure. They initially focused on simpler architectures and linear shims in orthogonal directions before extending their methodology to CNNs and LSTM. Their results show that both individual weak learners and an ensemble model approach were able to forecast shim values that enhanced the spectral quality [13,14], the methods are depicted in Fig. 6.1.

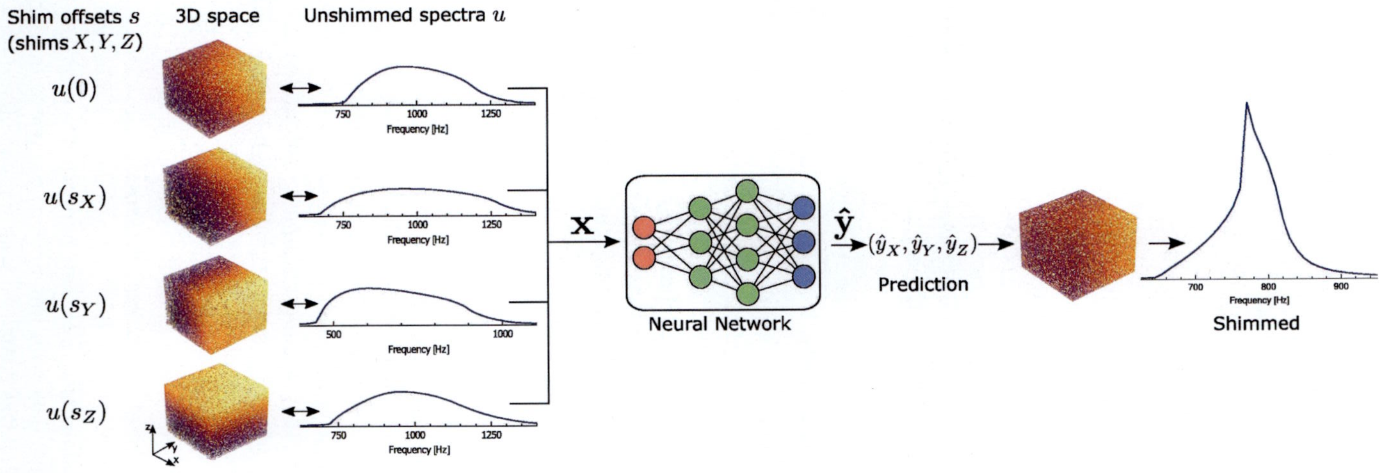

FIGURE 6.1

Deep regression with ensembles [13], three-dimensional distortions (illustrated as a 3D "inhomogeneity cube") of the sample volume collapse to a one-dimensional signal. By systematic offsets of the available shim currents, a batch of distinguishable spectra is obtained, which serves as input to a deep neural network. The prediction contains the shim values to achieve a more homogeneous field and thus a spectrum of higher quality.

6.2.2 Motion and other temporal instabilities

However, B0 fields are not necessarily constant over time. Physiological changes, especially subject motion, will translate into significant changes in the B0 field, which would warrant continuous updating of B0 shim currents or at least retrospective correction [15,16]. While the necessary information can be of course continuously measured [16], such a "navigator" approach is not for free (i.e., consumes scan time) and can impose undesirable limitations for the MR sequence design. Motyka et al. have shown that changes in the B0 field can be predicted, if the geometry of the tracked object (i.e., in their case the human head) and its motion pattern are known [17]. Another important aspect that contributes to an improved MRS data quality is of course motion and B0/B1 field instability correction [15]. Traditionally, these methods are divided into retrospective and prospective (or real-time) correction approaches. Both can potentially gain from the implementation of ML, and DL in particular. While motion correction approaches have been recently proposed for MRI [18], the application to MRSI is still subject to ongoing research.

6.3 MRS data postprocessing

The next steps after the data acquisition are the processing of the acquired raw data. Typical processing steps for MRS and MRSI include at least the following consecutive steps: coil combination, Fourier transformation, gridding, motion correction, eddy current correction, frequency and phase correction, and identification of spurious echoes [19,20]. All of these traditional steps could be in principle replaced by machine learning-based alternatives, but the actual gain in performance differs substantially depending on the task.

6.3.1 Spectral denoising

The acquisition time of SVS can be linearly reduced by decreasing the number of signal averages, which are obtained to enhance the SNR. Thus, by learning the mapping between spectra with a low number of signal averages to high number of averages can be used to denoise and improve the SNR of MR spectra.

For example, Lei et al. [21] have proposed a stack auto-encoder model to denoise MR spectra. For that purpose, they acquire multiple phantom and in vivo spectra with a high and a low number of averages, representing high and low SNR. The proposed encoder–decoder network uses low-average spectra as the input in the frequency domain and outputs high-average spectra with reduced noise. To enlarge the data variability a patch-based input was used in combination with data augmentation. This resulted in an ~40%–47% increase in SNR.

An alternative approach employed a DL model based on a Refusion Long Short-Term Memory (ReLSTM) architecture [22]. This network was designed to learn the mapping from the time domain data with low SNR to the one with high SNR. The method is depicted in Fig. 6.2. It was compared with state-of-the-art Low-Rank (LR)

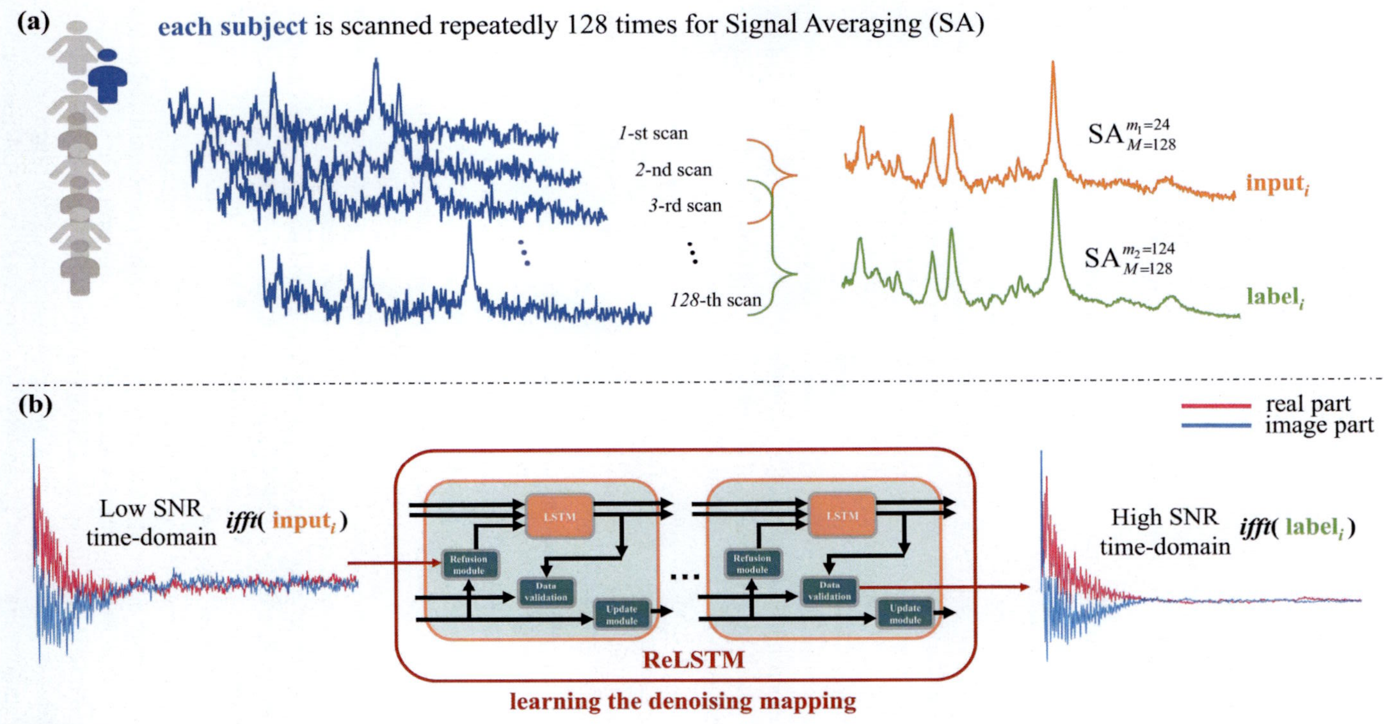

FIGURE 6.2

The entire training phase of the ReLSTM. (a) One training input was the average of 24 repeated samples (low SNR) which were randomly selected out of 128 repeated scans. The average of all 128 repeated scans (high SNR) was a target. (b) The proposed denoising ReLSTM was trained for the mapping between the low SNR and the corresponding high SNR time domain MRS signal [22].

denoising [23,24]. In contrast, Chen et al. have enhanced the capabilities of low-rank approximation and could show that denoising effectiveness was significantly improved by incorporating prelearned spectral basis and spatial priors [25].

6.3.2 Reconstruction

Single-voxel MRS methods acquire only spectral information, thus do not require special efforts to accelerate their acquisition. Traditional MRSI requires sampling also spatial information which was commonly done by phase encoding. The main disadvantage is inherently slow acquisition. Therefore, numerous more time-efficient alternatives to sample the spatial and spectral domain have been gradually developed [20,26]. Today, MRSI sequences rely heavily on spatial-spectral encoding and strong undersampling, which give rise to a number of artifacts that have to be mitigated using advanced reconstruction techniques. Many of these could be replaced by more efficient ML alternatives. Another challenge is the rapid acquisition and reconstruction of time-resolved MRSI data, which are obtained in hyperpolarized 13C-MRSI [27], or more recently deuterium metabolic imaging [28,29] and which warrant time-efficient reconstruction of large datasets.

For instance, when the free induction decay (FID) signal to fully encode the spectral domain is truncated, this results in first-order phase errors or reduced spectral resolution. These truncation artifacts can be efficiently removed using different NN architectures such as U-Nets [30]. These new approaches are faster and more accurate than established missing-point prediction using linear back prediction with the auto-regressive Burg method [31].

A more frequent reconstruction challenge is the recovery of missing k-space points (for introduction of image encoding and k-space formalisms please refer to Section 6.3.3) that would otherwise result in spatial aliasing artifacts due to violation of the Nyquist criterion. A number of traditional acceleration techniques have been proposed for MRSI [32], involving undersampling of the k-space and reconstruction using compressed-sensing [33,34] or parallel imaging techniques [35,36]. Nassirpour et al. [37] proposed a DL approach to replace conventional generalized autocalibrating partial parallel acquisition (GRAPPA) reconstruction by estimating k-space weightings via multiple NNs with superior results. However, undersampling alone is typically not providing sufficient acceleration for high-resolution 3D-MRSI. Thus, the combination of undersampling and spatial-spectral encoding have been proposed [38]. An example for this is the combination of non-Cartesian random undersampled ECCENTRIC trajectories [39]. Since reconstruction of non-Cartesian undersampled data is a computationally very expensive iterative process, DL-based and hence more time-efficient alternatives have been proposed [40]. In terms of reconstruction of high-dimensional MR Spatio-spectral imaging, DL models can be utilized to learn the low-dimensional representations for an ensemble of spectroscopic signals, which can be further used in the spatio-spectral reconstruction [41].

Other obstacles in the reconstruction of high-resolution MRSI data involve the immense computations costs that frequently involve extensive RAM requirements and long reconstruction times. To circumvent this problem coil compression approaches are traditionally employed [42] to reduce the memory and computational load immediately after data acquisition. To mitigate this, an alternative approach to compress the typically large MRSI data at an early stage within the reconstruction process was proposed using a shallow graph NN. This approach combines the signals of individual coil elements already in k-space rather than at the last reconstruction step in image space [43].

In contrast to the common 1D spectra, more advanced two-dimensional (2D) MRS techniques, which feature two spectral dimensions exist such as the localized correlated spectroscopy (L-COSY) method [44]. L-COSY is able to better separate the different spectral components that typically overlap in a normal 1D experiment. Due to the acquisition of an additional spectral dimension, L-COSY experiments are rather time consuming. Therefore, Luo et al. [45] recently introduced an encoder–decoder NN for accelerated reconstruction of nonuniformly undersampled 2D spectra using simulated training data. A similar approach was followed by Iqbal et al., who proposed a U-net architecture that performed much better than conventional compressed-sensing alternatives [46].

Finally, highly undersampled time-resolved MRSI such as for hyperpolarized 13C-MRSI benefit from the ability of DL-based reconstruction to enable very high temporal resolution mapping of 13C-labeled compounds, which is a very challenging task due to the short relaxation properties of 13C-labeled compounds [47].

6.3.3 Super-resolution

Long acquisition times and high field strengths enable the acquisition of high-resolution MRSI data. The spatial resolution can be, in addition, improved by advanced postprocessing techniques to enhance the acquired resolution of spectroscopic imaging data, thereby, making it possible to achieve the desired higher spatial resolution.

Iqbal et al. propose a supervised DL method for super-resolving MRSI data using a U-Net architecture. This method takes a T1-weighted (T1w) image and corresponding low-resolution MRSI as inputs, producing a high-resolution MRSI as output [48]. Synthetic data were generated by an MRSI simulator, which produces T1w images and pairs of low- and high-resolution MRSI data, which were used for training and testing. The whole approach is presented in Fig. 6.3. The models were ultimately tested on downsampled in vivo high-resolution MRSI data. A direct comparison with standard methods like zero-filling and bicubic interpolation indicated that the DL method outperforms standard methods at all noise levels, demonstrating its potential as a denoising or acceleration technique.

Dong et al. introduce another super-resolution model incorporating in vivo MRI and MRSI data [49]. They acquired a dataset of high-resolution MRSI data and down-sampled it to obtain low-resolution data. Metabolic maps from different patients were used as ground truth (GT). This DL method employs a U-Net with four different input modalities: low-resolution MRSI data, T1-weighted images,

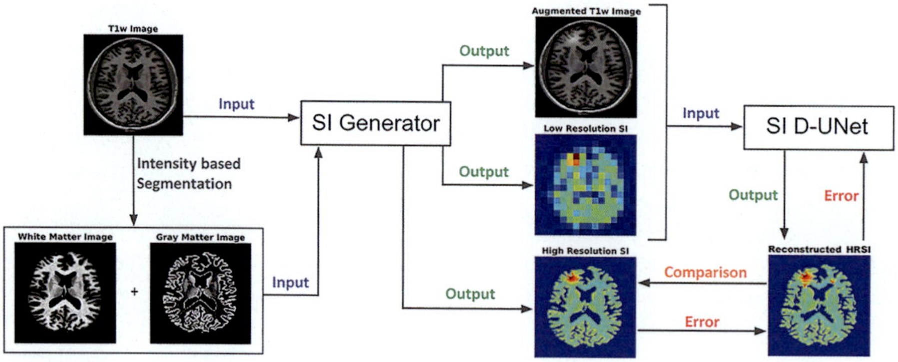

FIGURE 6.3

The workflow for training the D-UNet model is shown. The SI generator provides a dataset consisting of an augmented T1w image, a low-resolution spectroscopic image, and a ground truth high-resolution spectroscopic image. The spectroscopic images already show the distribution of a particular metabolite (or the distribution of a particular spectral point), such as choline, and therefore do not contain a spectral dimension. Then, the network transforms the aT1w (128 × 128 pixels) and LRSI (128 × 128 pixels after nearest-neighbor interpolation) into an initial HRSI reconstruction (128 × 128 pixels). In the example above, the LRSI and HRSI reconstruction have in-plane spatial resolutions of 1.4 × 1.4 cm^2 and 1.7 × 1.7 mm^2, respectively. This reconstruction is compared with the ground truth, and the mean squared error is calculated. Utilizing this error, the model changes the weighting parameters for the features, and continues training by using a different dataset. After training on 102,000 datasets, the model weights are refined and the reconstruction errors are minimized [48].

fluid-attenuated inversion recovery (FLAIR) images, and contrast-enhanced T1--weighted images. The network's decoding part includes spatial attention modules that automatically calculate spatial weight maps, focusing on significant features in each input modality. The high-resolution MRSI output is used to compute the loss function, which comprises three components. The study examines the contributions of different loss terms and input modalities by training multiple models.

Li et al. have aimed to achieve HRSI using DL to map tumor metabolism in glioma patients with IDH mutations [50]. In their study, a DL method based on a generative adversarial network (GAN) was developed and trained on simulated images from 75 glioma patients, then evaluated on data from 20 glioma patients and 10 healthy controls. The method significantly improved image quality and resolution, indicating its potential for precise tumor delineation and image-guided precision oncology therapy.

6.3.4 Frequency and phase correction

The correction of frequency and phase offsets is an important preprocessing step in MRS. While the latter can be also performed directly as part of the spectral fitting step, frequency alignment is typically managed before and both can be difficult to perform accurately for MR spectra of low SNR. This can be, for instance, the case when individual averages of the same scan need to be frequency and phase aligned prior to summing them to minimize temporal instabilities in longer measurements or in functional MRS experiments. The sources of these errors can be scanner frequency drift, subject motion, or other inconsistencies, which could otherwise impair spectral quality (e.g., lower SNR; line broadening, subtraction artifacts in spectral editing).

Tapper et al. [51] propose a method for correcting both phase and frequency via two separate NNs. For this a simulated dataset was used, the simulation parameters of which were chosen to match in vivo data from the Big GABA repository [52] as closely as possible. Both corrections were implemented with two sequentially placed fully connected NNs, where the frequency shifts were estimated prior to estimation of the phase offsets. Both models were trained individually taking the magnitude or real-only spectra, respectively, as the input to predict the frequency or phase offsets. The DL based approaches showed similar performance on in vivo data as a model-based spectral registration method, however at much faster processing speed.

Based on this work, Ma et al. [53] propose an alternative DL approach for frequency and phase correction using CNNs. They use identical parameter configurations to simulate MEGA-point resolved spectroscopy (PRESS) data for training and validation. However, they create additional spectra with lower SNRs by adding Gaussian noise. The authors observed less subtraction artifacts with the CNN and showed an overall better performance than the approach of Tapper et al. [51] Shamaei et al. [54] developed unsupervised methods for frequency/phase correction including two different convolutional encoder-model decoder (CEMD) models in which the input spectra are used as targets during training. One model focuses on reference peak fitting of creatine and uses a convolutional encoder to construct a lower-dimensional latent representation of the input. The latent parameters were then used as input for a Lorentzian lineshape model decoder. The second CEMD model used the same encoder, but a spectral registration function is used as decoder. Training and validation were done on a simulated dataset with known frequency and phase offsets and evaluation was done on phantom data and in vivo data from the Big GABA repository. The proposed unsupervised method was compared with conventional and the DL methods mentioned above [51,53]. In contrast to the non-DL FPC methods, the CEMD models performed equally well when trained and applied on a limited frequency range and their performance was less influenced by the presence of nuisance peaks. Compared with the previous DL methods, the CEMD is able to train in an unsupervised fashion and requires only one network for both frequency and phase correction.

6.3.5 Artifact removal

Spectral artifacts in MRS can come in various ways and have different origins [55]. Especially, artifacts from insufficiently suppressed water and extracranial lipids are a major problem for spectral quantification [56]. There are two ways how to deal with this issue: either removing the artifact (discussed below) or filtering poor quality spectra (in cases where option one does not work; see 6.4.1)

Ghosting artifacts, or so-called spurious echos, are caused by insufficient spoiling gradients in combination with local susceptibility variations. These artifacts negatively influence the reliability of metabolite quantification when they overlap with metabolite peaks. Kyathanahally et al. [57] evaluate two networks based on fully connected NNs and CNNs for the detection and an autoencoder network with residual blocks for the removal of ghosting artifacts. For both tasks, brain metabolite spectra were simulated and ghosting artifacts with varying line-widths and amplitudes were added randomly. The fully connected NN was trained on 1D spectra as input and class probabilities as output. An alternative classification approach was implemented with a CNN, taking the real and imaginary part of a 2D spectrogram as input. To remove the ghosting artifacts, an autoencoder network was implemented, taking the 2D corrupted spectrograms as input and outputting artifact-free spectrograms. The authors report a classification accuracy between 50% and 75% for the fully connected NN, whereas the CNN approach showed an accuracy of 94%–99%.

In contrast, Lee and Kim [58] propose a CNN architecture taking spectra contaminated with artifacts as input and predicting noise-free, metabolite-only spectra. In their case, the NN was trained using simulated data with metabolite phantom spectra as a basis set and knowledge of in vivo data for specific linewidth, SNR, and baseline ranges. For both simulated and in vivo data, the CNN clearly manages to obtain spectra with removed noise, narrow linewidth, removed frequency and phase shifts, and without spectral baseline.

6.4 MRS data analysis

6.4.1 Quality filtering

For various reasons, MRS data quality can be poor and the responsible artifacts cannot be removed. Frequently it is even hard for experts to judge whether a certain spectrum would result in poor estimates of metabolite levels and should therefore be disregarded. Manual rejection of poor quality spectra is tedious at best and for MRSI an entirely hopeless endeavor altogether. Therefore, a currently widely used approach is to filter after spectral fitting based on outliers on the metabolic maps or using measures of uncertainty for the spectral fit (i.e., Cramer Rao Lower Bound, CRLBs), even though this is generally considered questionable as it can result in analysis bias [59].

Machine learning approaches have a relatively long history in detecting low-quality spectra, dating back to initial attempts in 2009 [60] and more recently advanced versions based on convex Non-Negative Matrix Factorization models that were used as a dimensionality reduction procedure together with several classifiers to discriminate between good and bad quality data as [61]. Recent DL based alternatives offer faster and objective approaches to spectral quality filtering.

Vaziri et al. evaluated different ML models for automating quality control of MRSI data [62]. Using a mixed dataset of 53 patient/control subjects, they compared three ML models (random forest, CNN, and inception CNN) and hybrid models thereof demonstrated high accuracy. The CNN model performed best in neurological diseases and random forest excelled in brain tumors. All methods achieved rapid QC inference, significantly reducing the time compared with manual labeling.

Gurbani et al. [63] developed a deep learning model using a tiled convolutional neural network to identify and filter out poor-quality spectra, achieving high sensitivity and specificity (AUC of 0.95) compared with MRS experts. This supervised approach was attempting to match the consensus of the training data from three independent experts.

An entirely different and more general approach to this problem was suggested by Jang et al. [64], which is depicted in Fig. 6.4. In their study the use of generative adversarial networks (GANs) for unsupervised anomaly detection (AnoGAN) to ensure the quality of 1H-MRS human brain spectra was proposed. Trained only on simulated normal brain spectra, the AnoGAN filtered out abnormal spectra with various abnormalities, achieving an accuracy of ~80% in detecting low SNR, abnormal NAA levels, and spectra contaminated with artifacts like ghosting, residual water, and lipid. This method, though needing further refinement for linewidth abnormalities and in vivo validation, presents a promising unsupervised alternative for spectral quality assurance compared with supervised approaches. This method can be used for quality assurance because it effectively identifies and filters out abnormal spectra, ensuring only high-quality data is used for clinical analysis. Its ability to detect a wide range of abnormalities and artifacts, despite being trained only on normal spectra, demonstrates its robustness and potential reliability in maintaining the integrity of spectroscopic data.

Finally, another option is to obtain uncertainty estimates from DL based quantification and perform an automated quality control based on these after spectral fitting as suggested recently by Lee et al. [65]. First, they developed a CNN-based method for metabolite quantification with simultaneous uncertainty estimation. This method isolated target metabolite signals from simulated rat brain spectra and estimated intermediate spectral parameters to measure uncertainty. It outperformed traditional methods (LCModel and jMRUI). Later they introduced a Bayesian CNN with Monte Carlo dropout sampling for metabolite quantification and uncertainty estimation [65]. Trained on simulated human brain spectra, this network generated noise-free, line-narrowed metabolite spectra along with uncertainty measures. The Bayesian CNN demonstrated high accuracy and strong correlation between predicted uncertainty and ground-truth error, outperforming traditional methods like LCModel, especially under conditions of spectral degradation.

6.4.2 Quantification

The interpretation of MR spectra or in other words turning MR spectra into estimates of metabolite levels (or potentially even concentration estimates) is the most important

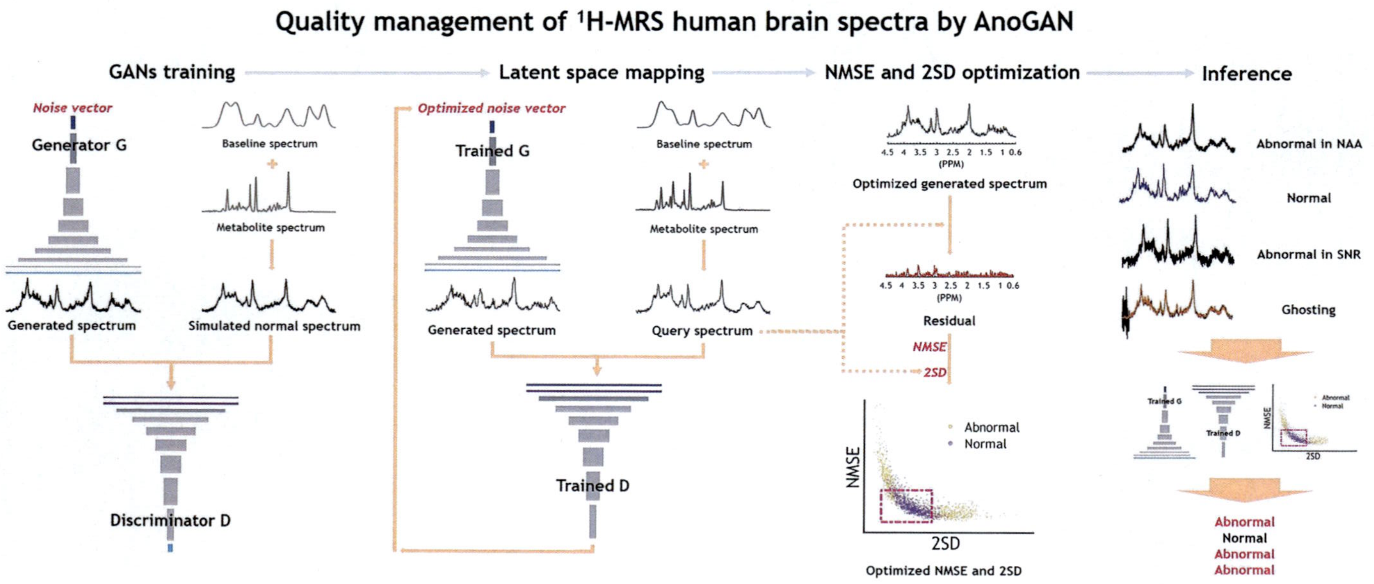

FIGURE 6.4

A schematic of the optimization of the AnoGAN. First, the generator G and the discriminator D are simultaneously trained according to a two-player minimax game. Second, using the trained G and D, the noise vector in the latent space is iteratively optimized by latent space mapping such that the spectrum generated by G from the noise vector is optimized as close to the query spectrum as possible. Third, such latent space mapping is performed for each individual spectra in the validation sets of the normal spectra group and the abnormal spectra group that contains the all types of the abnormalities considered in this study. The normalized mean squared error (NMSE) between the optimized generated spectrum and the query spectrum, and the 2× standard deviation of noise in the query spectrum (2SD) are obtained for all data in the validation sets. Then, the optimal threshold NMSE and 2SD values in the classification of spectra into normal and abnormal classes are determined from the NMSE-2SD space (red box) obtained from the validation sets. Finally, the optimal threshold NMSE and 2SD values are used on the test sets with different types of abnormalities to evaluate the performance of the proposed method. (For interpretation of the references to color in this figure legend, the reader is referred to the web version of this article.) [64].

processing step [19] in MRS, but not only are some approaches prone to poor spectral quality, it is also one of the most time-demanding part for large 3D MRSI datasets [20]. A recent multisite spectral fitting challenge has taught us several lessons about the different parameters that influence good or bad spectral quantification and how big the variety of results can be for different software packages and even when using the same fitting software, but slightly different parameters [66].

Several DL based MRS fitting approaches have been recently proposed. They do not only promise more accurate, but in particular much faster quantification, which is of major importance for state-of-the-art 3D-MRSI datasets. Hundreds of thousands of MR spectra that all require spectral fitting and established single-voxel MRS fitting approaches such as LCmodel require several seconds to quantify a single spectrum. Even when using CPU-based parallel computing this results in at least several hours of processing time. The following research groups have come up with different solutions:

An early attempt based on more traditional ML was proposed in 2017 by Das et al. [67] and a bit later in 2018 Hatami et al. [68] introduced the first CNN-based quantification approach. Das et al. showed that already using a conventional random forest regression framework and training the model on a diverse dataset of simulated and in vivo brain spectra, learning spectral features to enhance metabolite quantification, leads to significantly improved performance compared with the LCModel [67]. Hatami et al. trained a CNN model on a large-scale simulated dataset with various human brain spectra and SNRs, demonstrating superior accuracy in estimating spectral parameters compared with the standard QUEST method for 20 metabolites and macromolecules. While these first approaches offer end-to-end prediction of metabolite levels, others have proposed two-spec quantification methods. For instance, Lee et al. suggested to predict noise-free, line-narrowed, baseline-removed intact metabolite spectra from initially low-quality spectra degraded by low SNR, line broadening, and spectral baseline variation prior actual quantification [58]. In contrast, Chen et al. proposed to integrate Linear Least Squares (LLS) with deep learning to improve quantification by predicting imperfection factors and macromolecule signals, allowing analytical metabolite quantification and error feedback during training [69]. Their study showed lower quantification errors and greater stability than LCModel and other deep learning methods, potentially making it a robust and reliable approach for MRS quantification.

A very promising fitting approach was recently suggested by Shamaei et al. who suggested a physics-informed DL approach to quantification [70]. The approach is depicted in Fig. 6.5. In contrast to previous supervised approaches, their study investigated the feasibility of a self-supervised deep learning (DL) method, constrained by linear combination model (LCM) fitting, for analyzing MR spectra. Using quantum-mechanics simulated metabolite basis sets, the proposed DL-based method quantified relative metabolite concentrations, showing comparable performance to traditional methods in low-SNR simulated data and demonstrating in vivo applicability. Another, similar unsupervised approach that focused on a convolutional encoder-model decoder, but also combines the strengths of convolutional networks with physics-based models was previously proposed by Gurbani et al. [71] Both

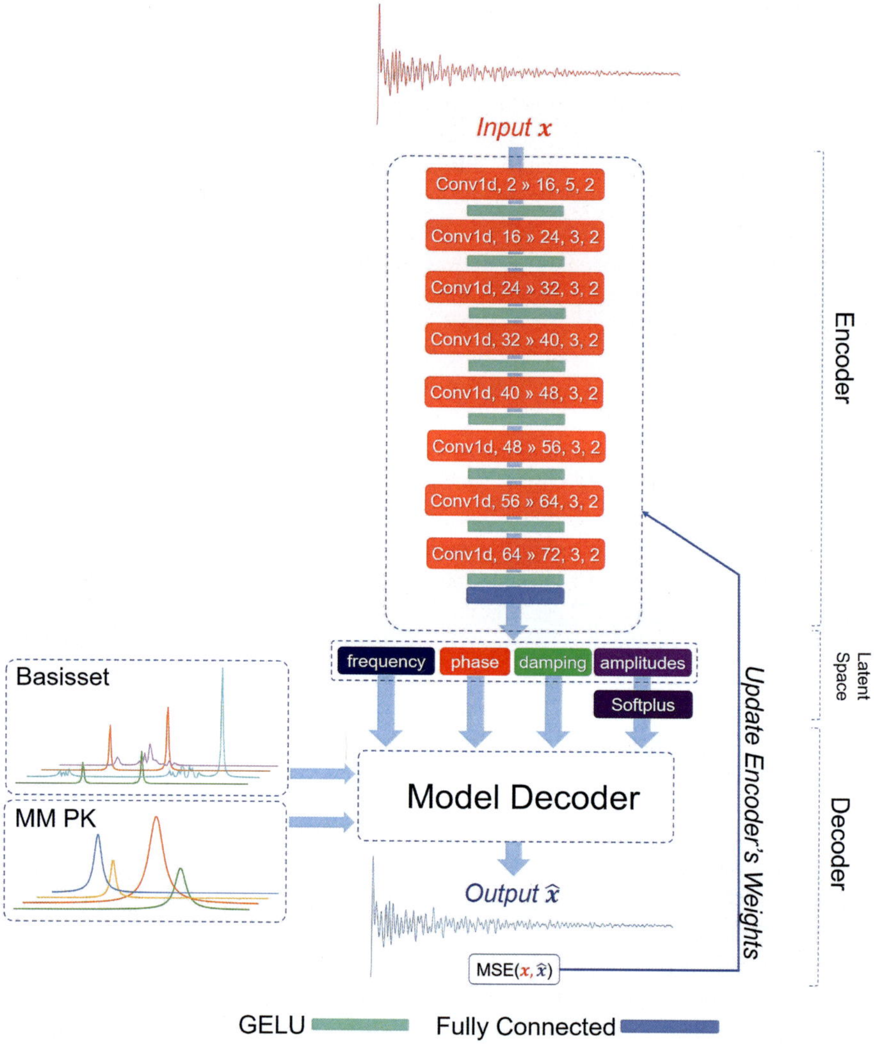

FIGURE 6.5

(a) Illustration of the proposed convolutional encoder–model decoder network. The input of the network is a complex signal in the time domain, which is fed to the encoder. The encoder consisted of eight convolutional blocks and an FC layer. A convolutional block is composed of a 1D convolution (Conv1d) layer followed by a GELU layer. The model decoder reconstructs the output signal using a parametric function, basis set, and MM prior knowledge. The DAE was trained to encode the input vector in the time domain into parameters (the amplitudes, damping factor, resonance frequency, and zero-order phase of the basis spectra) that can be used to reconstruct the output vector in the time domain. The proposed network is trained by minimizing the mean square error (MSE) between the input and output vectors [70].

these approaches significantly reduce computation time, suggesting potential to facilitate clinical practice by enabling faster processing of large MRSI datasets.

However, as shown by Ma et al. [72], spectral fitting does not necessarily have to be performed as a follow-up step to MRSI data reconstruction. Instead, they suggested a manifold learning-based method using linear tangent space alignment (LTSA) to leverage the intrinsic low-dimensional structure of MRSI signals for joint spectral quantification. The LTSA model aligns local subspace coordinates of each metabolite to global coordinates, enhancing quantification accuracy by fitting the model to noisy MRSI data with spatial smoothness and sparsity constraints. This method outperformed both QUEST and conventional subspace-based methods in reducing noise and artifacts, achieving superior spectral quantification accuracy.

While this is one of the most rapidly developing fields of DL in MRS, there has been little comparison between different approaches so far and rarely any clinical validation. Rizzo et al. have explored a range of various DL architectures, input types, and learning designs to optimize MRS quantification for simulated pathological spectra, demonstrating that ensembles of networks using both 1D and 2D spectrograms perform best [73]. While DL methods show improved precision at low SNR, they seem to suffer from bias toward the training distribution, which can be problematic in clinical settings. In their analysis traditional model fitting generally performed better in accuracy, although active learning and network ensembles enhance DL prediction performance, highlighting the importance of sufficient SNR for unbiased results.

6.4.3 Recognizing pathology

The ultimate goal after all the acquisition and processing steps described above is to translate MRS data into clinically (or scientifically) meaningful conclusions. Typically, this involves recognizing and better characterizing a pathologic metabolic deviation, which can in the end be used to make a better diagnosis or treatment decisions. Since recognizing specific patterns is a major application of machine learning algorithms, a number of research groups have proposed possible ML-based solutions.

Given the increasing data radiologists must handle, ML offers a promising alternative to traditional invasive methods like histopathological analysis. Therefore, already in 2013 large multicenter trials have used simple ML approaches to accurately classify different brain tumor types [74].

Also Ranjith et al. [75] analyzed MRI data from 28 patients with gliomas, using four classical machine learning algorithms—multilayer perceptrons, support vector machines, random forests, and locally weighted learning—to classify the tumors and concluded that random forests achieved the best overall performance with an AUC of 0.911, while locally weighted learning reached the highest sensitivity at 86.1%. This early paper suggests that ML based on MRS can effectively aid in tumor classification, with further improvements if additional MRI features are integrated.

After the advent of DL, several DL alternatives have been suggested for clinical automated characterization of tumor characteristics. For example, for MRSI data, it is particularly difficult, but important to separate voxels containing tumors from healthy tissue, which has been a study by Bjorkeli et al. [76] They developed two

CNN models capable of accurately distinguishing between healthy and pathological tissues directly from "raw" spectra (i.e., free induction decay signal) prior quantification, demonstrating its potential for improving diagnostic accuracy. The study's results suggest that this approach could significantly enhance clinical evaluations of brain tumors, with ongoing research focused on further validation and standardization.

A very similar brain tumor study, but for classification of single-voxel MRS data, was published even earlier by Dikaios et al. [77] where they trained different networks (SVM, ANN, CNN) to model the relationship between diseases and their metabolic fingerprints using synthetically generated MRS data for various scanners and protocols. The approach was tested on the problem of differentiating metastasis from glioblastoma across multiple institutions and scanner manufacturers, achieving high diagnostic accuracy with receiver operating characteristic area under the curve values of up to 0.97.

Another important clinical task is the differentiation of different brain tumor grades, which was targeted in studies by Dandil et al. [78]. Using training data from the international MRS database (INTERPRET), they trained a long short-term memory (LSTM) neural network to differentiate tumors of different grade, reaching an average area under the receiver operating characteristic (ROC) curve as high as 0.99. In another study they Dandil and collegues used a negative selection algorithm and clonal selection algorithm to differentiate benign and malignant brain tumors, classification of normal brain tissue and brain tumor, and detection of metastasis and primary brain tumors using the INTERPRET database achieving high accuracy [79]. CNNs were also utilized for the differentiation of spectra with neurodegenerative disorders (e.g., Alzheimer's), for which they performed similarly well [80].

This is just a selection of recent studies (mostly brain tumor related), but it is a rapidly evolving field with several innovative approaches. One more worth mentioning is the approach of unsupervised anomaly detection based on generative adversarial networks (GANs). Jang et al. [64] have used unsupervised anomaly detection (AnoGAN) trained solely on "normal" synthetic MRS spectra of the human brain and the AnoGAN was afterward capable of filtering out any "abnormal" spectra with a broad range of abnormalities including on the one hand spectra of unusually low spectral quality, but on the contrary can be applied also to the detection of spectra with unusual (pathologic) metabolic deviations.

6.5 Future outlook

In the long run, it is not unlikely that advanced ML methods will be a critical ingredient that will finally make the currently available MRS techniques sufficiently fast, automated, robust, and user-friendly to be useful on the broad scale needed for clinical use. While this does not automatically mean that MRS will establish its place in routine clinical practice, this step will at least ensure that a long-lasting major barrier toward clinical adaptation will be finally overcome.

In the short run, there are several prerequisites that still need to be met to facilitate and accelerate this progress. For example, for conventional MRI large data repositories are freely available to serve as an excellent foundation for DL [81]. They provide the training data that is absolutely necessary to train advanced artificial neural networks. For MRS, large multisite data repositories are rare, apart from a few yet limited exceptions [52,82]. Synthetic data generation would be another promising route to provide large amounts of training data, but the domain shift when training networks on purely simulated data and later test and apply them on real in vivo data is often large, resulting in poor performance during inference on real data. Yet, further progress in the field of synthetic data generation could in the end pay off, because realistic data cannot only be measured in infinite amounts, and a true ground truth such as known input parameters (e.g., concentrations) are available, which is not the case for measured in vivo MRS datasets. The latter aspect is a particularly important one, because it is the only way for a newly developed method to become more accurate and not just reach similar performance as an already established gold standard. Finally, different methods can be compared on the same data for which the results are perfectly known, allowing the establishment of widely accepted benchmark tests for the MRS community.

6.6 Conclusion

ML and deep learning methods in particular have diffused into the traditional MRS domain with a certain delay compared with MRI and medical imaging in general. What has been shown so far is mostly very preliminary and, thus, there is not only still significant potential for further improvements, but there is a substantial need for independent validation and generalization of these new ML methods before they can be used in a more widespread fashion, let alone clinically. In general, MRSI is likely to benefit more from this rapid development than single-voxel MRS, simply because reconstruction is so much more complicated and time-consuming, consistently high spectral quality is near impossible to maintain over very large volumes of interest, and analysis of this inherently multiparametric data source is challenging using traditional approaches. Ultimately, DL will therefore play a critical role in the translation of MRSI methods into the clinical realm.

References

[1] Bovey FA, Mirau PA, Gutowsky HS. In: Nuclear Magnetic Resonance Spectroscopy. Google Books; 2024. https://books.google.at/books?hl=de&lr=&id=htY0BX3aMNwC&oi=fnd&pg=PP1&dq=nuclear+magnetic+resonance&ots=mhbgREL941&sig=3DWUSsOthv60K5kF011SCDTcP_U&redir_esc= y#v=onepage&q=nuclear%20magnetic%20resonance&f=false. [Accessed 2 September 2024].

[2] De Graaf RA. In: In vivo NMR spectroscopy: principles and techniques. 3rd 2019.

[3] Chen D, Wang Z, Guo D, Orekhov V, Qu X. Review and prospect: deep learning in nuclear magnetic resonance spectroscopy. Chem Eur J 2020;26(46):1039110401. https://doi.org/10.1002/CHEM.202000246.

[4] Debus B, Parastar H, Harrington P, Kirsanov D. Deep learning in analytical chemistry. TrAC Trends Anal Chem 2021;145:116459. https://doi.org/10.1016/J.TRAC.2021.116459.

[5] van de Sande DMJ, Merkofer JP, Amirrajab S, et al. A review of machine learning applications for the proton MR spectroscopy workflow. Magn Reson Med 2023;90(4):1253–70. https://doi.org/10.1002/MRM.29793.
[6] Lundervold AS, Lundervold A. An overview of deep learning in medical imaging focusing on MRI. Z Med Phys 2019;29(2):102–27. https://doi.org/10.1016/j.zemedi.2018.11.002.
[7] Bishop JH, Geoly A, Khan N, et al. Real-time semi-automated and automated voxel placement using fMRI targets for repeated acquisition magnetic resonance spectroscopy. J Neurosci Methods 2023;392:109853. https://doi.org/10.1016/J.JNEUMETH.2023.109853.
[8] Öz G, Deelchand DK, Wijnen JP, et al. Advanced single voxel 1 H magnetic resonance spectroscopy techniques in humans: experts' consensus recommendations. NMR Biomed 2020;34(5). https://doi.org/10.1002/NBM.4236.
[9] Van Der Kouwe AJW, Benner T, Fischl B, et al. On-line automatic slice positioning for brain MR imaging. Neuroimage 2005;27(1):222–30. https://doi.org/10.1016/J.NEUROIMAGE.2005.03.035.
[10] Bolan PJ, Branzoli F, Di Stefano AL, et al. Automated acquisition planning for magnetic resonance spectroscopy in brain cancer. Med Image Comput Comput Assist Interv 2020;12267:730–9. https://doi.org/10.1007/978-3-030-59728-3_71.
[11] Martínez-Ramón M, Gallardo-Antolín A, Cid-Sueiro J, et al. Automatic placement of outer volume suppression slices in MR spectroscopic imaging of the human brain. Magn Reson Med 2010;63(3):592–600. https://doi.org/10.1002/MRM.22275.
[12] Juchem C, Cudalbu C, de Graaf RA, et al. B0 shimming for in vivo magnetic resonance spectroscopy: experts' consensus recommendations. NMR Biomed 2021;34(5). https://doi.org/10.1002/NBM.4350.
[13] Becker M, Lehmkuhl S, Kesselheim S, Korvink JG, Jouda M. Acquisitions with random shim values enhance AI-driven NMR shimming. J Magn Reson 2022;345:107323. https://doi.org/10.1016/J.JMR.2022.107323.
[14] Becker M, Jouda M, Kolchinskaya A, Korvink JG. Deep regression with ensembles enables fast, first-order shimming in low-field NMR. J Magn Reson 2022;336:107151. https://doi.org/10.1016/J.JMR.2022.107151.
[15] Andronesi OC, Bhattacharyya PK, Bogner W, et al. Motion correction methods for MRS: experts' consensus recommendations. NMR Biomed 2021;34(5). https://doi.org/10.1002/NBM.4364.
[16] Bogner W, Hess AT, Gagoski B, et al. Real-time motion- and B0-correction for LASER-localized spiral-accelerated 3D-MRSI of the brain at 3T. Neuroimage 2014;88:22–31. https://doi.org/10.1016/J.NEUROIMAGE.2013.09.034.
[17] Motyka S, Weiser P, Bachrata B, et al. Predicting dynamic, motion-related changes in B0 field in the brain at a 7T MRI using a subject-specific fine-trained U-net. Magn Reson Med 2024;91(5):2044–56. https://doi.org/10.1002/MRM.29980.
[18] Spieker V, Eichhorn H, Hammernik K, et al. Deep learning for retrospective motion correction in MRI: a comprehensive review, IEEE Transact Med Imaging.
[19] Near J, Harris AD, Juchem C, et al. Preprocessing, analysis and quantification in single-voxel magnetic resonance spectroscopy: experts' consensus recommendations. NMR Biomed 2021;34(5). https://doi.org/10.1002/NBM.4257.
[20] Maudsley AA, Andronesi OC, Barker PB, et al. Advanced magnetic resonance spectroscopic neuroimaging: experts' consensus recommendations. NMR Biomed 2021;34(5). https://doi.org/10.1002/NBM.4309.

[21] Lei Y, Ji B, Liu T, Curran WJ, Mao H, Yang X. Deep learning-based denoising for magnetic resonance spectroscopy signals 2021;11600:16–21. https://doi.org/10.1117/12.2580988.

[22] Chen D, Hu W, Liu H, et al. Magnetic resonance spectroscopy deep learning denoising using few in vivo data. IEEE Trans Comput Imaging 2023;9:448–58. https://doi.org/10.1109/TCI.2023.3267623.

[23] Abdoli A, Stoyanova R, Maudsley AA. Denoising of MR spectroscopic imaging data using statistical selection of principal components. Magn Reson Mater Phys Biol Med 2016;29(6):811–22. https://doi.org/10.1007/S10334-016-0566-Z/TABLES/4.

[24] Clarke WT, Chiew M. Uncertainty in denoising of MRSI using low-rank methods. Magn Reson Med 2022;87(2):574–88. https://doi.org/10.1002/MRM.29018.

[25] Chen Y, Li Y, Xu Z. Improved low-rank filtering of MR spectroscopic imaging data with PreLearnt subspace and spatial constraints. IEEE Trans Biomed Eng 2020;67(8):2381–8. https://doi.org/10.1109/TBME.2019.2961698.

[26] Bogner W, Otazo R, Henning A. Accelerated MR spectroscopic imaging—a review of current and emerging techniques. NMR Biomed 2021;34(5):e4314. https://doi.org/10.1002/NBM.4314.

[27] Gordon JW, Chen HY, Dwork N, Tang S, Larson PEZ. Fast imaging for hyperpolarized MR metabolic imaging. J Magn Reson Imag 2021;53(3):686–702. https://doi.org/10.1002/JMRI.27070.

[28] Bednarik P, Goranovic D, Svatkova A, et al. 1H magnetic resonance spectroscopic imaging of deuterated glucose and of neurotransmitter metabolism at 7 T in the human brain. Nat Biomed Eng 2023;7(8):1001–13. https://doi.org/10.1038/S41551-023-01035-Z.

[29] Niess F, Strasser B, Hingerl L, et al. Whole-brain deuterium metabolic imaging via concentric ring trajectory readout enables assessment of regional variations in neuronal glucose metabolism. Hum Brain Mapp 2024;45(6). https://doi.org/10.1002/HBM.26686.

[30] Lee H, Lee HH, Kim H. Reconstruction of spectra from truncated free induction decays by deep learning in proton magnetic resonance spectroscopy. Magn Reson Med 2020;84(2):559–68. https://doi.org/10.1002/MRM.28164.

[31] Kay SM. Modern spectral estimation: theory and application. 1988. p. 543. Published online.

[32] Bogner W, Otazo R, Henning A. Accelerated MR spectroscopic imaging-a review of current and emerging techniques. NMR Biomed 2021;34(5). https://doi.org/10.1002/NBM.4314.

[33] Nassirpour S, Chang P, Avdievitch N, Henning A. Compressed sensing for high-resolution nonlipid suppressed 1 H FID MRSI of the human brain at 9.4T. Magn Reson Med 2018;80(6):2311–25. https://doi.org/10.1002/MRM.27225.

[34] Klauser A, Klauser P, Grouiller F, Courvoisier S, Lazeyras F. Whole-brain high-resolution metabolite mapping with 3D compressed-sensing SENSE low-rank 1 H FID-MRSI. NMR Biomed 2022;35(1). https://doi.org/10.1002/NBM.4615.

[35] Strasser B, Považan M, Hangel G, et al. (2 + 1)D-CAIPIRINHA accelerated MR spectroscopic imaging of the brain at 7T. Magn Reson Med 2017;78(2):429–40. https://doi.org/10.1002/MRM.26386.

[36] Hangel G, Strasser B, Považan M, et al. Lipid suppression via double inversion recovery with symmetric frequency sweep for robust 2D-GRAPPA-accelerated MRSI of the brain at 7 T. NMR Biomed 2015;28(11):1413–25. https://doi.org/10.1002/NBM.3386.

[37] Nassirpour S, Chang P, Henning A. MultiNet PyGRAPPA: multiple neural networks for reconstructing variable density GRAPPA (a 1H FID MRSI study). Neuroimage 2018;183:336–45. https://doi.org/10.1016/J.NEUROIMAGE.2018.08.032.

[38] Moser P, Bogner W, Hingerl L, et al. Non-Cartesian GRAPPA and coil combination using interleaved calibration data - application to concentric-ring MRSI of the human brain at 7T. Magn Reson Med 2019;82(5):1587–603. https://doi.org/10.1002/MRM.27822.

[39] Klauser A, Strasser B, Bogner W, Hingerl L, Courvoisier S, Schirda C, Rosen BR, Lazeyras F, Andronesi OC. ECCENTRIC: A fast and unrestrained approach for high-resolution in vivo metabolic imaging at ultra-high field MR. Imaging Neurosci (Camb) 2024; 14(2):1–20. https://doi.org/10.1162/imag_a_00313. eCollection 2024 Oct 1.

[40] Weiser PJ, Langs G, Bogner W, Motyka S, Strasser B, Golland P, Singh N, Dietrich J, Uhlmann E, Batchelor T, Cahill D, Hoffmann M, Klauser A, Andronesi OC. Deep-ER: Deep learning ECCENTRIC reconstruction for fast high-resolution neurometabolic imaging. Neuroimage 2025;309:121045. https://doi.org/10.1016/j.neuroimage.2025.121045. Epub 2025 Feb 1.

[41] Lam F, Peng X, Liang ZP. High-dimensional MR spatiospectral imaging by integrating physics based modeling and data-driven machine learning: current progress and future directions. IEEE Signal Process Mag 2023;40(2):101–15. https://doi.org/10.1109/MSP.2022.3203867.

[42] Buehrer M, Pruessmann KP, Boesiger P, Kozerke S. Array compression for MRI with large coil arrays. Magn Reson Med 2007;57(6):1131–9. https://doi.org/10.1002/MRM.21237.

[43] Motyka S, Hingerl L, Strasser B, et al. k-Space-based coil combination via geometric deep learning for reconstruction of non-Cartesian MRSI data. Magn Reson Med 2021;86(5):23532367. https://doi.org/10.1002/MRM.28876.

[44] Albert Thomas M, Yue K, Binesh N, et al. Localized two-dimensional shift correlated MR spectroscopy of human brain. Magn Reson Med 2001;46(1):58–67. https://doi.org/10.1002/MRM.1160.

[45] Luo J, Zeng Q, Wu K, Lin Y. Fast reconstruction of non-uniform sampling multidimensional NMR spectroscopy via a deep neural network. J Magn Reson 2020;317. https://doi.org/10.1016/J.JMR.2020.106772.

[46] Iqbal Z, Nguyen D, Thomas MA, Jiang S. Deep learning can accelerate and quantify simulated localized correlated spectroscopy. Sci Rep 2021;11(1):8727. https://doi.org/10.1038/S41598-02188158-Y.

[47] Wang Z, Luo G, Li Y, Cao P. Using a deep learning prior for accelerating hyperpolarized 13C MRSI on synthetic cancer datasets. Magn Reson Med 2024;92(3):945–55. https://doi.org/10.1002/MRM.30053.

[48] Iqbal Z, Nguyen D, Hangel G, Motyka S, Bogner W, Jiang S. Super-resolution 1H magnetic resonance spectroscopic imaging utilizing deep learning. Front Oncol 2019;9. https://doi.org/10.3389/FONC.2019.01010.

[49] Dong S, Hangel G, Bogner W, et al. High-resolution magnetic resonance spectroscopic imaging using a multi-encoder attention U-net with structural and adversarial loss. In: Proceedings of the Annual International Conference of the IEEE Engineering in Medicine and Biology Society. EMBS; 2021. p. 2891–5. https://doi.org/10.1109/EMBC46164.2021.9630146. Published online.

[50] Li X, Strasser B, Neuberger U, et al. Deep learning super-resolution magnetic resonance spectroscopic imaging of brain metabolism and mutant isocitrate dehydrogenase glioma. Neurooncol Adv 2022;4(1). https://doi.org/10.1093/NOAJNL/VDAC071.

[51] Tapper S, Mikkelsen M, Dewey BE, et al. Frequency and phase correction of J-difference edited MR spectra using deep learning. Magn Reson Med 2021;85(4): 1755–65. https://doi.org/10.1002/MRM.28525.

[52] Mikkelsen M, Barker PB, Bhattacharyya PK, et al. Big GABA: edited MR spectroscopy at 24 research sites. Neuroimage 2017;159:32–45. https://doi.org/10.1016/J.NEUROIMAGE.2017.07.021.

[53] Ma DJ, Le HAM, Ye Y, et al. MR spectroscopy frequency and phase correction using convolutional neural networks. Magn Reson Med 2022;87(4):1700–10. https://doi.org/10.1002/MRM.29103.

[54] Shamaei A, Starcukova J, Pavlova I, Starcuk Z. Model-informed unsupervised deep learning approaches to frequency and phase correction of MRS signals. Magn Reson Med 2023;89(3):1221–36. https://doi.org/10.1002/MRM.29498.

[55] Kreis R. Issues of spectral quality in clinical 1H-magnetic resonance spectroscopy and a gallery of artifacts. NMR Biomed 2004;17(6):361–81. https://doi.org/10.1002/NBM.891.

[56] Tkáč I, Deelchand D, Dreher W, et al. Water and lipid suppression techniques for advanced 1 H MRS and MRSI of the human brain: experts' consensus recommendations. NMR Biomed 2021;34(5). https://doi.org/10.1002/NBM.4459.

[57] Kyathanahally SP, Döring A, Kreis R. Deep learning approaches for detection and removal of ghosting artifacts in MR spectroscopy. Magn Reson Med 2018;80(3):851–63. https://doi.org/10.1002/MRM.27096.

[58] Lee HH, Kim H. Intact metabolite spectrum mining by deep learning in proton magnetic resonance spectroscopy of the brain. Magn Reson Med 2019;82(1):33–48. https://doi.org/10.1002/MRM.27727.

[59] Kreis R. The trouble with quality filtering based on relative Cramér-Rao lower bounds. Magn Reson Med 2016;75(1):15–8. https://doi.org/10.1002/MRM.25568.

[60] Slotboom J, Nirkko A, Brekenfeld C, Van Ormondt D. Reliability testing of in vivo magnetic resonance spectroscopy (MRS) signals and signal artifact reduction by order statistic filtering. Meas Sci Technol 2009;20(10). https://doi.org/10.1088/0957-0233/20/10/104030.

[61] Mocioiu V, Kyathanahally SP, Arús C, Vellido A, Julià-Sapé M. Automated quality control for proton magnetic resonance spectroscopy data using convex non-negative matrix factorization. Lect Notes Comput Sci 2016;9656:719–27. https://doi.org/10.1007/978-3-319-31744-1_62.

[62] Vaziri S, Liu H, Xie E, et al. Evaluation of deep learning models for quality control of MR spectra. Front Neurosci 2023;17:1219343. https://doi.org/10.3389/FNINS.2023.1219343/BIBTEX.

[63] Gurbani SS, Schreibmann E, Maudsley AA, et al. A convolutional neural network to filter artifacts in spectroscopic MRI. Magn Reson Med 2018;80(5):1765–75. https://doi.org/10.1002/MRM.27166.

[64] Jang J, Lee HH, Park JA, Kim H. Unsupervised anomaly detection using generative adversarial networks in 1H-MRS of the brain. J Magn Reson 2021;325:106936. https://doi.org/10.1016/J.JMR.2021.106936.

[65] Lee HH, Kim H. Bayesian deep learning–based 1H-MRS of the brain: metabolite quantification with uncertainty estimation using Monte Carlo dropout. Magn Reson Med 2022;88(1):38–52. https://doi.org/10.1002/MRM.29214.

[66] Marjańska M, Deelchand DK, Kreis R, et al. Results and interpretation of a fitting challenge for MR spectroscopy set up by the MRS study group of ISMRM. Magn Reson Med 2022;87(1):11–32. https://doi.org/10.1002/MRM.28942.

[67] Das D, Coello E, Schulte RF, Menze BH. Quantification of metabolites in magnetic resonance spectroscopic imaging using machine learning. Lect Notes Comput Sci 2017; 10435 LNCS:462–70. https://doi.org/10.1007/978-3-319-66179-7_53/FIGURES/5.

[68] Hatami N, Sdika M, Ratiney H. Magnetic resonance spectroscopy quantification using deep learning. Lect Notes Comput Sci 2018;11070 LNCS:467–75. https://doi.org/10.1007/978-3-030-00928-1_53/TABLES/1.

[69] Chen D, Lin M, Liu H, et al. Magnetic resonance spectroscopy quantification aided by deep estimations of imperfection factors and macromolecular signal. IEEE Trans Biomed Eng 2024;71(6):1841–52. https://doi.org/10.1109/TBME.2024.3354123.

[70] Shamaei A, Starcukova J, Starcuk Z. Physics-informed deep learning approach to quantification of human brain metabolites from magnetic resonance spectroscopy data. Comput Biol Med 2023;158:106837. https://doi.org/10.1016/J.COMPBIOMED.2023.106837.

[71] Gurbani SS, Sheriff S, Maudsley AA, Shim H, Cooper LAD. Incorporation of a spectral model in a convolutional neural network for accelerated spectral fitting. Magn Reson Med 2019;81(5):3346–57. https://doi.org/10.1002/MRM.27641.

[72] Ma C, Han PK, Zhuo Y, Djebra Y, Marin T, El Fakhri G. Joint spectral quantification of MR spectroscopic imaging using linear tangent space alignment-based manifold learning. Magn Reson Med 2023;89(4):1297–313. https://doi.org/10.1002/MRM.29526.

[73] Rizzo R, Dziadosz M, Kyathanahally SP, Shamaei A, Kreis R. Quantification of MR spectra by deep learning in an idealized setting: investigation of forms of input, network architectures, optimization by ensembles of networks, and training bias. Magn Reson Med 2023;89(5):1707–27. https://doi.org/10.1002/MRM.29561.

[74] Vicente J, Fuster-Garcia E, Tortajada S, et al. Accurate classification of childhood brain tumours by in vivo ^{1}H MRS - a multi-centre study. Eur J Cancer 2013;49(3):658–67. https://doi.org/10.1016/J.EJCA.2012.09.003.

[75] Ranjith G, Parvathy R, Vikas V, Chandrasekharan K, Nair S. Machine learning methods for the classification of gliomas: initial results using features extracted from MR spectroscopy. NeuroRadiol J 2015;28(2):106–11. https://doi.org/10.1177/1971400915576637/ASSET/IMAGES/LARGE/10.1177_1971400915576637FIG6.JPEG.

[76] Bjørkeli EB, Johannessen K, Geitung JT, Karlberg AM, Esmaeili M. Deep neural network modeling for brain tumor classification using magnetic resonance spectroscopic imaging. Published online February 28, 2024. doi:10.21203/RS.3.RS-3951768/V1

[77] Dikaios N. Deep learning magnetic resonance spectroscopy fingerprints of brain tumours using quantum mechanically synthesised data. NMR Biomed 2021;34(4):e4479. https://doi.org/10.1002/NBM.4479.

[78] Dandil E, Biçer A. Automatic grading of brain tumours using LSTM neural networks on magnetic resonance spectroscopy signals. IET Image Process 2020;14(10):1967–79. https://doi.org/10.1049/IET-IPR.2019.1416.

[79] Dandil E. aiMRS: a feature extraction method from MRS signals based on artificial immune algorithms for classification of brain tumours. IET Signal Process 2020;14(6):361–73. https://doi.org/10.1049/IET-SPR.2019.0576.

[80] Acquarelli J, Van Laarhoven T, Postma GJ, et al. Convolutional neural networks to predict brain tumor grades and Alzheimer's disease with MR spectroscopic imaging data. PLoS One 2022;17(8):e0268881. https://doi.org/10.1371/JOURNAL.PONE.0268881.

[81] Van Essen DC, Ugurbil K, Auerbach E, et al. The Human Connectome Project: a data acquisition perspective. Neuroimage 2012;62(4):2222–31. https://doi.org/10.1016/J.NEUROIMAGE.2012.02.018.

[82] Považan M, Mikkelsen M, Berrington A, et al. Comparison of multivendor single-voxel MR spectroscopy data acquired in healthy brain at 26 sites. Radiology 2020;295(1):171–80. https://doi.org/10.1148/RADIOL.2020191037.

PART

3

MR image reconstruction

CHAPTER

Get the image: Machine learning for MR image reconstruction

7

Shanshan Wang[1,*], Ruoyou Wu[1,2], Reinhard Heckel[3], Mathews Jacob[4], Efrat Shimron[5,6]

[1]*Paul C. Lauterbur Research Center for Biomedical Imaging, Shenzhen Institutes of Advanced Technology, Chinese Academy of Sciences, Shenzhen, China;* [2]*University of Chinese Academy of Sciences, Beijing, China;* [3]*Department of Computer Engineering, Technical University of Munich, Munich, Germany;* [4]*Department of Electrical and Computer Engineering, University of Lowa, Lowa, IA, United States;* [5]*Department of Electrical and Computer Engineering, Technion-Israel Institute of Technology, Haifa, Israel;* [6]*Department of Biomedical Engineering, Technion-Israel Institute of Technology, Haifa, Israel*

[*]*Corresponding author: E-mail: ss.wang@siat.ac.cn*

7.1 Basics of linear and non-linear inverse problems

Magnetic resonance imaging (MRI) has long played a significant role in medical imaging, providing high-quality information on both the anatomical structure and function of the human body [1,2]. However, the main limitation of MRI is its long scan duration, which hinders clinical use and reduces patient comfort. MRI scans are commonly accelerated by sub-Nyquist sampling, also known as *undersampling*. While useful for acceleration, undersampling introduces aliasing artifacts and hence requires image reconstruction algorithms that remove artifacts and recover clean images from the measurements. A vast body of work has been dedicated to developing reconstruction methods. Classical approaches include parallel imaging techniques that utilize sensitivity maps in multi-coil MRI acquisitions [3–7] and compressed sensing methods, which leverage sparsity in transform domains, for example, the wavelet or discrete cosine transform domains [8–14]. More recently, machine learning approaches have been instrumental in solving MRI reconstruction problems, providing state-of-the-art results [1,2,15–18]. This chapter focuses on machine learning methods for image reconstruction.

We begin with a basic description of the MRI inverse problem. The term *forward model* refers to a mathematical description of the relationship between the image and measurements. Depending on the variations in the encoding operators within the imaging physics model, the forward model can be either linear or nonlinear. For

Machine Learning in MRI. https://doi.org/10.1016/B978-0-443-14109-6.00007-9

classical structural magnetic resonance imaging, the forward model can be represented as:

$$\mathbf{y} = \mathbf{A}\mathbf{x} + \boldsymbol{\eta} \tag{7.1}$$

Where $\mathbf{y} \in \mathbb{C}^m$ represents the k-space measurements, $\mathbf{x} \in \mathbb{C}^n$ represents the desired clean image to be recovered, $\boldsymbol{\eta} \in \mathbb{C}^m$ represents the system noise generated during the MRI scan. Moreover, $\mathbf{A} : \mathbb{C}^n \rightarrow \mathbb{C}^m$ represents the forward linear encoding operator. For single-coil imaging, we have $\mathbf{A} = \boldsymbol{KF}$, where $\boldsymbol{K}$ represents the sampling matrix and $\boldsymbol{F}$ represents the Fourier transform. For multi-coil imaging, $\mathbf{A} = \boldsymbol{KFC}$, where $\boldsymbol{C}$ represents the coil sensitivity, which characterizes the sensitivity maps of the coils. In quantitative magnetic resonance imaging (qMRI), the process of obtaining parameter maps can be represented as:

$$\mathbf{x} = \mathcal{A}(\boldsymbol{\rho}) \tag{7.2}$$

Where $\mathcal{A}$ is a nonlinear function that represents the relations between the contrast-weighted images to the acquisition parameters (e.g., TR, TE), and $\boldsymbol{\rho}$ represents the biophysical parameters (T_1, T_2, etc). $\boldsymbol{\rho}$ is estimated from a series of measured contrast-weighted images, and this process is typically nonlinear. Due to the typically long scan times of MRI, it is common to sample k-space at sub-Nyquist rates to accelerate the scanning process. In such cases, the forward model matrix $\mathcal{A}$ is typically rank-deficient, leading to an ill-posed inverse problem for recovering $\mathbf{x}$.

7.2 Basic formulation of the image reconstruction problem

Recovering the desired image from measurements is computationally challenging, particularly in the presence of high noise levels and high acceleration factors. Traditional methods attempt to solve this problem using iterative optimization algorithms. In the case of structural MRI, the recovery of $\mathbf{x}$ is commonly formulated as a regularized least-squares problem:

$$\widehat{\mathbf{x}} = \arg\min_{\mathbf{x}} \frac{1}{2} \|\mathbf{A}\mathbf{x} - \mathbf{y}\|_2^2 + \lambda \mathscr{R}(\mathbf{x}) \tag{7.3}$$

where the first term is commonly referred to as a *data fidelity* (i.e., data consistency) term, the second term $\mathscr{R}(\mathbf{x})$ is the regularization term, and λ is a hyperparameter that controls the tradeoff between them.

For classical compressed sensing approaches, the regularization term is typically defined explicitly, e.g., as the l_1-norm of the image coefficients in a sparse transform domain [19,20], whereas for machine learning methods the regularization term is learned from the training data.

Eq. (7.3) can be solved using various optimization algorithms [20]. A popular method is the *iterative gradient descent* algorithm:

$$\mathbf{x}_{k+1} = \mathbf{x}_k - \eta\left(\mathbf{A}^{\mathrm{H}}(\mathbf{A}\mathbf{x}_k - \mathbf{y}) + \nabla_{\mathbf{x}} \mathscr{R}(\mathbf{x}_k)\right) \tag{7.4}$$

where $\mathbf{x}_k$ is the estimated image at the k-th iteration, η is the step size, $\mathbf{A}^{\mathrm{H}}$ denotes the Hermitian transpose of the forward operator $\mathbf{A}$, and $\nabla_{\mathbf{x}}$ represents the gradient operator with respect to $\mathbf{x}$. Eq. (7.4) relies on the gradient of the regularization term. Eq.

(7.4) can be viewed as a denoising process, where the aliasing and noise components are removed from $\mathbf{x}_k$ at each iteration step. The *proximal gradient descent* algorithm [21] is often a good alternative to standard gradient descent (Eq. 7.4) and can lead to faster convergence:

$$\begin{cases} \mathbf{z}_{k+1} = \mathbf{x}_k - \eta \mathbf{A}^{\mathrm{H}}(\mathbf{A}\mathbf{x}_k - \mathbf{y}) \\ \mathbf{x}_{k+1} \triangleq prox_{\lambda\mathcal{R}}(\mathbf{z}_{k+1}) \end{cases} \tag{7.5}$$

where $\mathbf{z}_{k+1}$ represents the intermediate image estimate during the iterative process, and $prox_{\lambda\mathcal{R}}(*)$ denotes the proximal operator. When the regularization term is the l_1-norm of a sparse transform $\mathbf{S}$ (such as the wavelet transform), proximal gradient descent becomes:

$$\begin{cases} \mathbf{z}_{k+1} = \mathbf{x}_k - \eta \mathbf{A}^{\mathrm{H}}(\mathbf{A}\mathbf{x}_k - \mathbf{y}) \\ \mathbf{x}_{k+1} = \mathbf{S}^{\mathrm{T}} soft(\mathbf{S}\mathbf{z}_{k+1}, \lambda) \end{cases} \tag{7.6}$$

where $soft(*, \lambda) = sign(*)\max(|*| - \lambda, 0)$ represents the soft-thresholding operator, and is often called the *iterative shrinkage-thresholding algorithm (ISTA)* [22].

Another fast iterative algorithm commonly used to solve the optimization problem in Eq. (7.3) is the *Alternating Direction Method of Multipliers (ADMM)* [23]. ADMM can decompose a complex problem into multiple simpler subproblems, thereby simplifying the solution process. It addresses the following problem, which is equivalent to the optimization problem in Eq. (7.3):

$$\mathbf{x} = \arg\min_{\mathbf{x}} \min_{\mathbf{z}} \|\mathbf{A}\mathbf{x} - \mathbf{y}\|_2^2 + \lambda\mathcal{R}(\mathbf{z}), \ \text{s.t.} \mathbf{z} = \mathbf{x} \tag{7.7}$$

$$\begin{cases} \mathbf{x}_{k+1} = \arg\min_{\mathbf{x}} \frac{\beta}{2} \| \mathbf{x} - (\mathbf{z}_k - \mathbf{u}_k)\|_2^2 + \frac{1}{2}\|\mathbf{A}\mathbf{x} - \mathbf{y}\|_2^2 \\ \mathbf{z}_{k+1} = \arg\min_{\mathbf{z}} \frac{\beta}{2} \| \mathbf{z} - (\mathbf{x}_{k+1} - \mathbf{u}_k)\|_2^2 + \lambda\mathcal{R}(\mathbf{z}) \\ \mathbf{u}_{k+1} = \mathbf{u}_k + (\mathbf{x}_{k+1} - \mathbf{z}_{k+1}) \end{cases} \tag{7.8}$$

where $\mathbf{z}_k$ is the auxiliary variable, $\mathbf{u}_k$ is the Lagrange multiplier, and λ is the regularization parameter as before. The three steps correspond to the data consistency, proximal operator, and dual update subproblems, respectively.

Bayesian approach. From a Bayesian perspective, the above formulation can be regarded as an a-posterior estimate [9,24]. The optimization process then aims to find an image $\mathbf{x}$ that maximizes the posterior distribution $\mathrm{p}(x|y) = p(y|x) * \frac{p(x)}{p(y)}$ through minimization of the negative log posterior: $-\ log\, \mathrm{p}(x|y) = -logp(y|x) - logp(x)$. Here, the first term is the data consistency term and the second term represents the prior. We thus have $\nabla\mathcal{R}(\mathbf{x}) = \nabla log\, p(\mathbf{x})$, which typically represents the score of the distribution. This term can be regarded as an estimate of the noise, and subtracting it from the original image results in a clearer image.

7.2.1 Basic architecture of neural networks

By capitalizing the power of neural network architectures, such as convolutional neural networks (CNNs) [25–30], UNet [31–34], ResNet [35–39], generative adversarial networks (GANs) [40–45], recurrent neural networks (RNNs) [46–50], and Transformers [51–59], the field of MRI reconstruction has advanced rapidly.

One class of deep learning-based reconstruction methods learn the mapping from undersampled data $\mathbf{y}$ to fully sampled data $\mathbf{x}$ using a network f, which is parameterized by θ. The most widely used technique is *residual learning* [35], where $\mathbf{A}^{\mathrm{H}}\mathbf{y}$ is the input to the network, and $\widehat{\mathbf{x}}$ is the output.

These carefully designed models each have unique characteristics. Specifically, CNNs widely utilize architectures such as residual learning [35] and U-Net [31] in MRI reconstruction, offering distinct advantages in local feature learning.

A different architecture is a GAN [40], which consists of a generator and a discriminator. The role of the generator is to generated images from the inputs – ideally, those should be highly similar to real images - while the role of the discriminator is to distinguish between generated and real images. GANs have advantages in generating images with high perceptual quality.

$$v\left(\theta^{(g)}, \theta^{(d)}\right) = \mathbb{E}_{x \sim p_{data}} \log d(x) + \mathbb{E}_{x \sim p_{model}} \log(1 - d(x)) \tag{7.9}$$

A third class of architectures is based on RNNs, which are suitable for learning from time series data and can be used, for example, for dynamic cine imaging [47], or 3D-derived 2D sequence reconstruction [60]. Finally, to enhance the learning of global features, the recently proposed Transformer architecture has been widely applied in the field of image reconstruction [52–54]. Transformers incorporate multi-head self-attention and cross-attention mechanisms; hence, they can capture long-range dependencies in images very well. They are widely applied in MRI reconstruction and super-resolution tasks [53]. For instance, Feng et al. [52] utilized task-specific cross-attention modules for MRI reconstruction and super-resolution tasks, while Huang et al. [53] introduced the Swin Transformer for MRI reconstruction tasks, achieving excellent performance.

It is worth noting that these models can also be integrated with each other to further improve MRI reconstruction. For example, integrating the local feature extraction capabilities of CNNs with the long-range dependency-capturing ability of Transformers can enhance the learning of both local and global features. In Ref. [53], a CNN and a Transformer were combined to enhance the quality of MRI reconstruction. In addition, Qin et al. [47] achieved good results in dynamic cine imaging by leveraging the feature extraction capability of CNNs and the time-series processing ability of RNNs. The integration of these models effectively improves magnetic resonance imaging. Furthermore, these models can also be integrated with traditional iterative optimization algorithms.

These architectures can be implemented for handling data in different approaches, for example, using image-domain computations or unrolled physics-driven networks with data-consistency blocks, as reviewed next. Moreover, the complex-valued MRI data can be handled in different approaches, for example, using complex-valued data and operations [61] or through two-channel networks, where the real and imaginary parts of the data are stored in different channels [62].

7.2.2 Early fundamental approaches

Deep learning-based methods have become important tools for MRI reconstruction, each with its unique advantages. Here, we briefly introduce some of the most well-known models in each class of methods (Fig. 7.1).

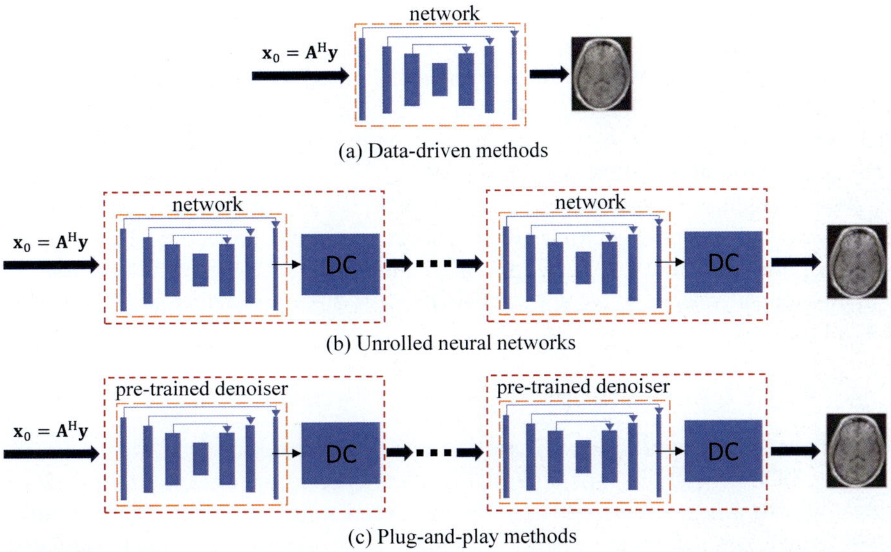

FIGURE 7.1

Three reconstruction methods. (a) Data-driven methods. (b) Unrolled neural networks. (c) Plug-and-play methods. y denotes the undersampled k-space data, A^H represents the inverse Fourier transform, x_0 denotes the undersampled image-domain data, and DC refers to the data consistency (DC) module.

Data-driven methods. We begin with end-to-end trained, data-driven neural networks; those typically map undersampled data with noise or artifacts to ground truth images [25–30]. For example, Wang et al. [25] were the first to introduce deep learning into the field of magnetic resonance imaging. Their approach utilizes big data for offline training, enabling faster online reconstruction. AUTOMAP [27] is another early work that directly learns the mapping from k-space to images. It treats image reconstruction as a data-driven supervised learning task, allowing the mapping between the sensor and image domains to emerge from appropriate training datasets.

Unrolled neural networks. Given the success of deep learning in various image-processing tasks, researchers have developed methods that combine iterative optimization algorithms with neural networks for various applications related to MRI reconstruction [63–70] and broader computational imaging tasks [71]. For example, one of the first studies to introduce this approach was that of Gregor and LeCun [63], who combined traditional model-driven algorithms with deep networks using the technique of deep unfolding.

This technique typically iterates between two blocks: one is a data fidelity block, which can be computed using various methods [67], and the other consists of neural networks that perform specific tasks such as image denoising, artifact removal, or image restoration. An early work in this regard is ADMM-Net [64], which utilizes the ADMM algorithm to solve the optimization problem presented in Eq. (7.3). In another well-known study, Hammernik et al. [65] proposed a variational network, which uses a total variation norm as a replacement for the regularization term and minimizes the optimization objective using a gradient descent algorithm. The original variational network used a relatively small number of layers, specifically designed as a domain-specific expert model.

Another approach was taken by the DCCNN [30] and MoDL [66], which introduced residual-based regularization terms into iterative networks. Those terms enable optimizing the network by minimizing the difference between the denoised, artifact-reduced images and the target images. In this approach, Eq. (7.3) can be rewritten as:

$$\hat{\mathbf{x}} = \arg\min_{\mathbf{x}} \frac{1}{2} \| \mathbf{A}\mathbf{x} - \mathbf{y} \|_2^2 + \lambda \| \mathbf{x} - f(\mathbf{x};\theta) \| \tag{7.10}$$

where the neural network $f(*;\theta)$ parameterized by θ is used for regularization. Specifically, DCCNN [30] extends proximal gradient descent by replacing the proximal operator with a neural network to solve the optimization problem. During model training, network parameters can be optimized in an end-to-end manner. Similarly, MoDL [66] decomposes the optimization objective into the following two subproblems:

$$\begin{cases} \mathbf{z}_n = f(\mathbf{x}_n;\theta) \\ \mathbf{x}_{n+1} = \left(\mathbf{A}^{\mathrm{H}}\mathbf{A} + \lambda \boldsymbol{I}\right)^{-1}\left(\mathbf{A}^{\mathrm{H}}\mathbf{y} + \lambda \mathbf{z}_n\right) \end{cases} \tag{7.11}$$

Each unit of the iterative network consists of a CNN block, which performs image denoising and artifact removal, and a data consistency blocks, which ensures consistency with the acquired measurements. The conjugate gradient algorithm is used to solve the data consistency sub-problem. The network parameters are shared across iteration steps, improving the consistency and efficiency of parameter updates.

Traditional iterative optimization algorithms used for MRI reconstruction require manual definition of regularizers and hyperparameters calibration, which may affect the model's performance. Model-based unfolding methods, on the other hand, utilize neural networks to implement the iterative steps of optimization algorithms, thereby enhancing the model's representational capacity. To some extent, these models also offer better interpretability.

Plug-and-play and generative methods. Early compressed sensing methods relied on convex prior assumptions, such as total variation regularization. Plug-and-play (PnP) methods have made it possible to solve inverse problems using pre-trained denoisers [72]. Among them, RED [73] and RARE [74] are two classic PnP methods. Generative model-based methods have been impactful for MRI reconstruction, demonstrating strong performance in generating high-quality images. Finally, untrained neural networks represent an effective approach that does not rely on training data, except for adjusting hyperparameters. Among them, Deep Image Prior (DIP) is one of the representative methods [75–77]. In addition, implicit neural representations provide an interesting approach that learns the mapping from coordinate grids to image pixel values [78–81]. The detailed information can be found in Section 7.4.

7.2.3 Advanced methods: Reconstruction in the presence of motion artifacts

Motion remains one of the main challenges in the field of MRI, and numerous methods have been proposed to address motion correction issues [82–87] (see also Chapter 9). Much of this work has focused on brain MRI scans, which involve rigid body motion of the head. Such motion is typically incorporated into the MRI forward model [82] by,

$$\widehat{\mathbf{y}} = \sum_{t=1}^{T} \mathbf{M}_t \mathbf{F} \mathbf{U}_t \mathbf{x} \tag{7.12}$$

where $\mathbf{U}_t$ represents the motion transform. For rigid-body motion, the motion transform $\mathbf{U}_t = \mathbf{T}_t \mathbf{R}_t$ consists of rotation and translation operations. Mathematically, describing such motion in the image space is equivalent to describing it in k-space [87]. Simulating such motion in the image domain can hence be achieved by rotating and translating the image while updating the corresponding k-space lines. In k-space, the rotation of the imaged object corresponds to an equivalent rotation, while translation corresponds to the multiplication by a linear phase ramp, which depends on the translation parameter a and the k-space coordinates in the readout direction k_{RO}:

$$T(\widehat{\mathbf{y}}) = \widehat{\mathbf{y}} \cdot \exp(-i2\pi a k_{RO}) \tag{7.13}$$

Recently, deep learning has also been used for correcting non-rigid motion [88–92], e.g., with integration of linearized encoding of such motion [93–97]. These methods are applied in areas such as abdominal imaging [88] and dynamic cine imaging [90]. Manifold learning has been introduced into MRI to achieve unsupervised motion-compensated reconstruction [98]. These methods have effectively improved image quality in the presence of motion (Fig. 7.2).

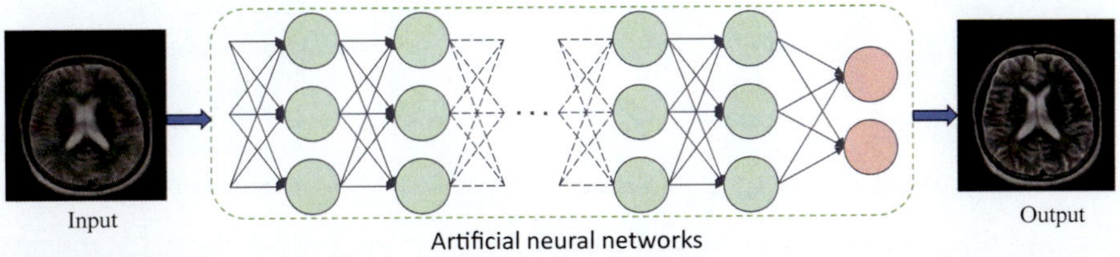

FIGURE 7.2

Schematic diagram of deep learning-based motion correction.

7.2.4 Advanced application: Diffusion MRI

Diffusion magnetic resonance imaging (dMRI) is a widely used neuroimaging technique that leverages the microscopic motion of water molecules within neuronal tissue as the signal source [99–105]. For reliable fiber reconstruction with fine anatomical details, it is often necessary to consider both k-space and q-space (diffusion wavevector domain) resolutions simultaneously. In practice, k-q acceleration can be easily implemented on MRI scanners, as there is significant redundancy in the combined measurements of q-space points, which can be leveraged for joint reconstruction. The general form of joint reconstruction is [101]:

$$\mathbf{x} = \underset{\mathbf{x}}{\operatorname{argmin}} \|\mathbf{A}\mathbf{x} - \mathbf{y}\|_2^2 + \lambda_1 \|\mathscr{R}_q\{\mathbf{x}\}\| + \lambda_2 \|\mathscr{R}_r\{\mathbf{x}\}\| \tag{7.14}$$

where $\mathscr{R}_q\{\mathbf{x}\}$ is a general q-space regularization that aids in recovery along the q-dimension, and the q-space regularization can be further enhanced through $\mathscr{R}_r\{\mathbf{x}\}$ (Fig. 7.3).

7.2.5 Advanced application: Quantitative MRI

Quantitative magnetic resonance imaging (qMRI) derives physical models from the evolution of magnetic resonance spin magnetization to obtain quantitative biophysical parameters, enabling precise measurements that reflect the microstructure and composition of biological tissues [106,107]. In rapid qMRI, it is possible to directly estimate the maps of biophysical parameters from undersampled k-space data [108] by solving an optimization problem such as

$$\boldsymbol{\rho} = \arg \min_{\boldsymbol{\rho}} \frac{1}{2} \|\mathbf{A}\mathcal{M}(\boldsymbol{\rho}, t) - \mathbf{y}_t\|_F^2 + \mathscr{R}(\boldsymbol{\rho}; \mathbf{x}_t). \tag{7.15}$$

where $\mathscr{R}(\boldsymbol{\rho}; \mathbf{x}_t)$ is a regularization term that incorporates both the parameter maps and images. Deep learning methods have been widely used in qMRI to enhance image reconstruction, estimate parameter maps directly from raw or undersampled k-space data, and model complex physical processes more efficiently than traditional approaches [107,109]. One prominent qMRI approach, magnetic resonance fingerprinting (MRF) [110], uses randomized acquisition schemes and pattern-

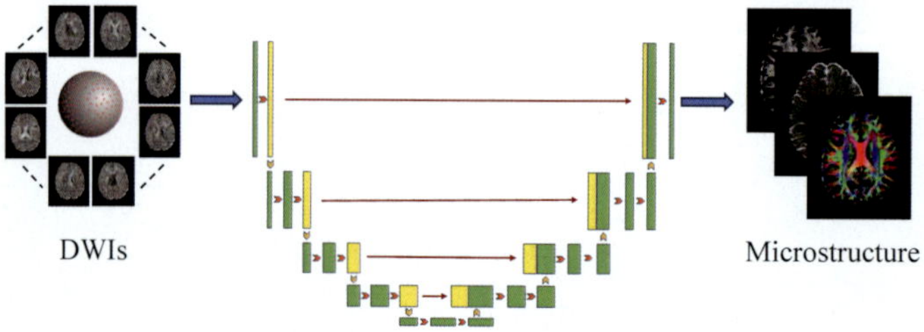

FIGURE 7.3

Flowchart of deep learning-based microstructure. DWIs stand for diffusion-weighted images.

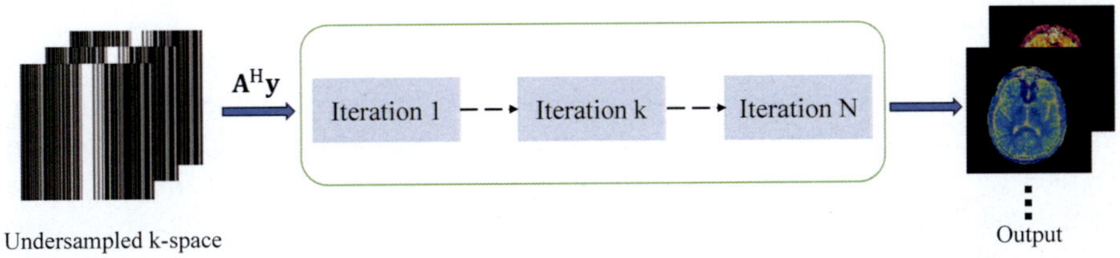

FIGURE 7.4

MR parameter mapping network with iterative unfolding architecture. y denotes the undersampled k-space data, A^H represents the inverse Fourier transform.

matching algorithms to derive multiple quantitative parameter maps from a single scan. Deep learning methods have been successfully applied to MRF to accelerate reconstruction, improve robustness, and replace traditional dictionary-matching algorithms with neural networks for efficient and accurate parameter estimation [111–115]. For more details about these topics, please refer to Chapter 10 (Fig. 7.4).

7.3 Parallel imaging

The core idea of parallel imaging is to use information about the spatial positioning of multiple coil elements to improve the reconstruction [3–5]. Different coils contain unique spatial sensitivity information, which reduces the need for traditional, time-consuming spatial encoding inference.

7.3.1 Classical approaches

We begin by describing the basic formulation of the MRI reconstruction problem in a parallel imaging setup and reviewing several classical approaches. Eq. (7.1) can be reformulated for parallel imaging by incorporating the coil-sensitivity maps. In this case, the forward operator becomes:

$$\mathbf{A} = \begin{bmatrix} \mathbf{KFC}_1 \\ \mathbf{KFC}_2 \\ \cdots \\ \mathbf{KFC}_{n_c} \end{bmatrix} \tag{7.16}$$

where n_c represents the number of coils, and $\mathbf{C}_i$ denotes the sensitivity map of the i-th coil. Broadly speaking, there are two methods of parallel imaging: image-domain methods and k-space methods. Image-domain methods, primarily represented by the well-known SENSE [3] technique, solve the inverse problem using coil sensitivity maps. In contrast, k-space methods estimate missing data using linear combinations of data from different coils, with each coil's sensitivity distribution acts as a distinct convolution kernel. Classical methods like GRAPPA [4] and SPIRiT [116] utilize fully sampled data from the center of k-space (sometimes referred to as auto-calibration data) to estimate these kernels. However, estimating or interpolating from very few k-space lines may lead to noise amplification and degrade image quality. Therefore, accurate estimation of sensitivity maps plays a crucial role in parallel imaging reconstruction. A highly popular method for estimating coil sensitivity maps is ESPIRiT [6], which calculates sensitivity maps by performing an eigenvalue decomposition on calibration data from the center of *k*-space.

7.3.2 Deep-learning approaches for parallel imaging

Recently, deep learning has also been employed to solve reconstruction problems in multi-coil MRI [117–120]. For example, Peng et al. [119] utilized deep learning to learn coil sensitivity functions for SENSE reconstruction. In addition, deep learning has been employed for the joint estimation of coil sensitivity maps and the reconstructed image, to enhance image quality [121–124]. For example, the E2E-VarNet [121] framework was proposed for such joint estimation; it integrates a sensitivity estimation module into an unfolded network and utilizes end-to-end training. Jun et al. [122] proposed a joint deep model for magnetic resonance (MR) image and coil sensitivity reconstruction, named Joint-ICNet. It integrates MR physics models to jointly reconstruct MR images and coil sensitivity maps from undersampled multi-coil k-space data. These methods have demonstrated that utilizing deep learning for sensitivity maps computation can improve the reconstruction quality.

7.4 Generative models: Deep image priors and diffusion models

Another popular approach is to learn image representations parameterized by generative neural networks, for example, generative adversarial networks [40–45], variational autoencoders [125,126], and recently acclaimed diffusion models [127–132]. The main advantages of generative-based methods include flexibility in selecting forward models and strong performance in reconstructing high-quality images from undersampled data. Moreover, their probabilistic nature enables the quantification of uncertainty, which is crucial in clinical settings [133,134].

Generative adversarial networks can preserve perceptual details well [40,135]. In the field of MRI reconstruction, Yang et al. [43] proposed a deep de-aliasing conditional generative adversarial network and introduced a refinement learning approach. They constructed a comprehensive loss function incorporating adversarial loss, perceptual loss, and frequency domain loss. Quan et al. [44] proposed a novel generative adversarial model, RefineGAN, which combines an autoencoder with a GAN and introduces a cycle data consistency loss to perform meaningful interpolation in undersampled k-space. Mardani et al. [42] used generative adversarial networks to model the low-dimensional manifold of MRI images, allowing the generator to learn how to remove aliasing artifacts by projecting onto the image manifold, and trained the model using a hybrid loss function. In addition, generative adversarial networks have also been used for unsupervised zero-shot MRI reconstruction tasks. For example, Korkmaz et al. [54] proposed an unsupervised MRI reconstruction method based on zero-shot learning with adversarial transformers, effectively combining MRI priors with imaging operators.

Diffusion models, celebrated for their generative prowess, are becoming a game-changer in computer vision, transforming tasks from image synthesis to restoration [129]. Recently, they have attracted considerable interest from researchers focusing on MRI reconstruction [130–134,139]. By gradually eliminating noise through reverse diffusion, these models can produce high-quality, detail-rich images. These methods include denoising diffusion probabilistic models [131], denoising score matching [136], and stochastic differential equation-based methods [137]. Training involves two phases: the forward process, where noise is gradually added to the image until it approximates a normal distribution, and the reverse process, where noise is progressively removed to reconstruct the original data distribution. The neural network approximates the reversible score function, enabling the transformation from a simple distribution (such as a Gaussian distribution) to a more complex one.

In MRI, Jalal et al. [138] applied a compressed sensing-based generative model for reconstruction, using publicly available data to train the model and generate priors. High-quality reconstruction was achieved through posterior sampling using Langevin dynamics. Chung et al. [132] applied score-based diffusion models for accelerated MRI reconstruction. They initially trained a continuous time-dependent score function with denoising score matching. During the inference phase, reconstruction was achieved iteratively using a numerical SDE solver along with data consistency steps, which enabled the reconstruction of complex-valued data using only magnitude data for training. Luo et al. [133] described a method for MRI reconstruction using data-driven Markov chains, which not only adapts the reconstruction process to various sampling patterns but also generates uncertainty maps, which are crucial for clinical applications. While diffusion models have many advantages, the inference phase typically involves iterating through numerous steps, resulting in longer inference times. In addition, noise in MRI does not follow a normal distribution, which means that models trained based on Gaussian noise may not be entirely suitable for MRI reconstruction.

7.5 k-space learning, dual domain networks and low-rank priors

Leaning features directly in k-space offers advantages for preserving high-frequency information, particularly anatomical details and texture information [140].

Traditional k-space learning primarily relies on GRAPPA [4], which uses fully sampled k-space lines and interpolation to estimate the unsampled k-space lines. Akçakaya et al. [141] proposed a robust k-space interpolation method called RAKI, which trains a neural network using fully sampled k-space center data. The trained neural network is then employed to estimate the missing k-space lines, enhancing noise recovery capability.

In Ref. [50], researchers proposed an sRAKI-RNN method that integrates sRAKI and calibration reconstruction stages to jointly learn self-consistent rules and employs an RNN for iterative reconstruction. Inspired by the combination of convolutional neural networks and Hankel matrix decomposition, Han et al. [32] proposed a fully data-driven deep learning method for k-space interpolation. By incorporating additional gridding layers, this method can also be applied to non-Cartesian sampling trajectories. In addition, Zhang et al. [142] improved RAKI with a residual RAKI method, blending linear and nonlinear techniques to achieve machine learning-based reconstruction tailored to specific scans, effectively reducing noise and artifacts. Zhao et al. [143] introduced implicit neural representation learning into k-space learning. Using spatial coordinates as input, they dynamically query sparse sampling points to reconstruct spectra, thereby capturing inductive biases in k-space.

Recently, dual-domain architectures, which involve image processing in both k-space and image-domain, have emerged as a powerful new class of architectures [28,144–149]. By leveraging the complementary strengths of these domains, these networks enhance data consistency and preserve structural fidelity. For example, Kiki-Net [28] includes three main components – a CNN operating in k-space, another CNN operating in the image domain, and data consistency blocks. By applying these alternatingly, this approach effectively reconstructs images from undersampled data. The recent A-LINK [144] framework extends this concept by incorporating a parallel-branch structure for simultaneous processing of k-space and image domains, coupled with low-rank modeling and attention mechanisms. This design achieves superior reconstructions, particularly in dynamic MRI settings such as single breath-hold cardiac cine imaging. Low-rank priors are particularly useful as they exploit the inherent redundancy and correlations within dynamic MRI data, enabling efficient representation of spatiotemporal patterns. By decomposing the data into low-rank and sparse components, it is possible to efficiently model background and dynamic features. This leads to enhanced image reconstruction quality, as demonstrated for both compressed sensing [150] and AI reconstruction methods [144–149].

7.6 Self-supervised learning for MRI

The previously introduced deep learning methods are primarily trained using supervised learning, which typically requires paired input-output measurement and fully sampled k-space data for supervision. However, in practice, fully sampled data are often difficult to obtain due to scan time limitations and patient compliance. As a result, self-supervised learning methods have gradually gained research attention. These methods train the network using only under-sampled k-space data, by generating paired data from those measurements [151–155].

One class of methods is based on Stein's Unbiased Risk Estimator for (SURE) [156]. These methods commonly utilize a loss function:

$$\mathrm{SURE}(f_\theta(\mathbf{v}), \mathbf{v}) = \|f_\theta(\mathbf{v}) - \mathbf{v}\|_2^2 + 2\sigma^2 \nabla_{\mathbf{v}} \cdot f_\theta(\mathbf{v}) - N\sigma^2 \tag{7.17}$$

where $\mathbf{v}$ is the noisy images, θ denotes the network parameters. This an effective method for recovering clean images from noisy measurements, The SURE framework has been extended to other frameworks, e.g., the generalized SURE (GSURE) [157] and the Ensemble SURE [158], which have shown improved performance.

Noise2Noise [159] is another renowned framework for solving computational inverse problems in a self-supervised manner. The original Noise2Noise framework utilizes independent noisy measurements of the same object (i.e., measurements from several acquisitions) to construct a self-supervised loss. In the context of MRI, Yaman et al. [151] proposed the self-supervised learning method (SSDU), which uses data from only a single acquisition. In SSDU, the measurement data are divided into two parts: one for training the network and the other for constructing the loss function. This classic work in self-supervised MRI reconstruction offers valuable insights for subsequent methods.

However, dividing the measurement data reduces the amount of information available for network training, which may result in suboptimal reconstruction performance, particularly in cases of high acceleration. To address this, an extension of the SSDU method introduced training with a multi-mask approach [153]. Recently, Millard et al. [160] introduced a general theoretical framework extending Noisier2Noise [161] and providing an analysis of SSDU's performance. A line of work extended the SSDU method to training with multiple sampling masks [153], motion-corrupted data [162], and non-Cartesian sampling [163,164].

Recently, Wang et al. [165] introduced the *k-band* framework, which enables self-supervised MRI reconstruction with decoupled train/test forward operators, i. e., using different acquisition strategies for collecting the training and test data. Its key idea is training on *k*-space bands, which are locally fully sampled and can hence be retrospectively undersampled flexibly, with various schemes, during the training phase. This enables using various test-time strategies. Furthermore, the acquisition of k-space bands is rapid; hence, it is highly suitable for real-world applications. To facilitate test-time generalization from the limited-resolution k-space bands data to higher-resolution data, *k-band* introduced an optimization technique called *stochastic gradient descent over k-space subsets*. Furthermore, the authors proposed randomizing the band angles across subjects to create a versatile training dataset and a loss-weighting mask that facilitates equal learning across all k-space areas [165]. The *k-band* framework demonstrates reconstruction quality comparable to that of fully supervised methods and SSDU, while also enabling diverse sampling strategies and test-time flexibility.

7.7 Deep learning for joint optimization of k-space sampling and image reconstruction

Recent advancements in MRI have also seen the development of deep learning frameworks that jointly optimize k-space sampling trajectories and image reconstruction. For example, Weiss et al. introduced the PILOT framework [166], which employs physics-informed learning to design optimal k-space trajectories under hardware constraints. Bahadir et al. developed LOUPE [167], an end-to-end learning approach that optimizes under-sampling patterns in MRI, focusing on 2D Cartesian sampling to improve reconstruction quality. Ciuciu and colleagues utilized deep neural networks to learn the target sampling densities for SPARKLING, an optimization method that designs non-Cartesian k-space trajectories tailored for compressed sensing MRI in compliance with MRI hardware constraints [168,169]. The same group has also later introduced PROJeCTOR, a framework that employs projected gradient descent to jointly learn non-Cartesian k-space trajectories and reconstruction networks [170]. Wang et al. introduced BJORK, a framework which utilizes B-spline parameterization for joint optimization of reconstruction and k-space trajectories, effectively reducing the number of parameters and accommodating hardware constraints [171]. They have also later introduced SNOPY, a generalized framework for optimizing 3D non-Cartesian sampling patterns, which enables using both gradient-based and data-driven methods for trajectory design [172]. Xie et al. introduced PUERT, a method that incorporates probabilistic sampling strategies and adaptive gradient estimation to achieve flexible, interpretable reconstructions [173]. Ravula et al. have leveraged pre-trained diffusion models to optimize sampling patterns for compressed sensing MRI [174]. These frameworks represent significant strides in the integration of sampling and reconstruction processes in MRI, leveraging deep learning to enhance imaging efficiency and quality.

7.8 Deep learning for low-field MRI reconstruction

Low-field magnetic resonance imaging (MRI) is an emerging technology, providing cost-effective portable imaging solutions, especially beneficial in resource-limited settings [175–179]. However, challenges such as low signal-to-noise ratio (SNR), prolonged scan times, and reduced image quality hinder its widespread clinical adoption. To address these issues, learning-based techniques have been employed for image enhancement tasks, for example, denoising, super-resolution, and contrast enhancement [180–184].

Deep learning methods have also shown significant promise for the task of image reconstruction from low-field MRI data. For example, Koonjoo et al. showed that the AUTOMAP method provides enhanced SNR in the reconstructed images [185]. Shimron et al. analyzed the robustness of compressed sensing, data-driven, and unrolled methods to various SNR levels, which correspond to different field strengths and environmental settings, and found that unrolled neural networks outperform the other methods and exhibit high robustness across the SNR spectrum [186]. Zhou et al. have presented a self-supervised dual-domain network that utilizes supervision in both k-space and image domains for reconstruction from non-Cartesian data [164]. Man et al. have introduced a deep learning method for 3D brain MRI accelerated by partial Fourier sampling and demonstrated it on 0.055T data [187]. These studies demonstrate the growing potential of deep learning methods to enhance MRI reconstruction and accelerate imaging in low-field settings. Further discussion is provided in Chapter 5.

7.9 Loss functions

Loss functions play a crucial role in model training. In general, the loss is commonly calculated between the reconstructed image and reference image. The Mean Squared Error (MSE) is the most commonly used loss function, and the Mean Absolute Error (MAE) loss is also frequently employed. They are defined by,

$$\text{MSE} = \frac{\sum_i^n \|\mathrm{x}_i - \widehat{\mathrm{x}}_i\|^2}{n} \tag{7.18}$$

$$\text{MAE} = \frac{\sum_i^n |\mathrm{x}_i - \widehat{\mathrm{x}}_i|}{n} \tag{7.19}$$

Compared to MAE loss [188], MSE loss may produce overly smooth results [33,189]. In addition, there are several loss functions based on image-domain metrics, such as the Structural Similarity Index Measure (SSIM) [190], and Normalized Root Mean Square Error (NRMSE) [191]. Alternatively, to enhance fidelity, some studies aim to minimize the data consistency loss, which measures the difference between the original undersampled data and the corresponding k-space data after reconstruction.

A different strategy, which can effectively enhance the perceptual quality of images is to calculate the loss in the feature space using pre-trained neural networks like VGG [192] and ResNet [35] as feature extractors. This approach minimizes the difference in image embeddings between reconstructed images and reference images [43,193]:

$$\mathscr{L}_{perc} = MSE\left(f_\theta{}^l(\mathrm{x}), f_\theta{}^l(\widehat{\mathrm{x}})\right) \tag{7.20}$$

where f_θ represents a pretrained neural network, and l denotes a specific layer of the neural network. Some studies also combine the aforementioned loss functions to simultaneously leverage their advantages. For example, in Ref. [43], DAGAN incorporates multiple loss functions, including mean squared error losses in both the image and frequency domains, perceptual loss based on VGG-16, and adversarial loss based on GANs. Experimental results indicate that integrating losses can indeed improve performance. However, there are challenges in selecting appropriate weighting coefficients. We also refer the reader to Chapter 13 for more discussions about image quality metrics.

7.10 Pitfalls and limitations of deep learning

In this section we discuss robustness aspects and challenges in deep learning-based imaging.

7.10.1 Hallucinations

Where deep learning offers substantial advantages for MRI reconstruction, it also encounters significant challenges and potential risks. One of the most prominent issues is its sensitivity to hallucinations, where networks generate non-existent features that appear plausible but are not present in the true data. Such hallucinations can arise from overfitting, data inconsistencies, or artifacts, and they pose a significant risk in medical imaging, as they may lead to misinterpretation of diagnostic images [194–197]. For example, subtle artifacts may be mistaken for real anatomical structures, potentially leading to false diagnoses. The problem can become critical when the model "hallucinates" features based on the patterns it has learned that may not correspond to reality. These issues are further discussed in Chapter 11.

7.10.2 Bias and "data crimes"

Another major concern is bias, which can stem from the data or the training methodology. For example, if the dataset used to train a deep learning model is not representative of the broader population—for example, if it over-represents certain anatomical structures, diseases, or demographic groups—the model will perform poorly when applied to under-represented populations. This bias can lead to inequities in healthcare, as certain groups may receive less accurate diagnoses or treatment recommendations [198–200].

Furthermore, bias can arise from the misuse of publicly available datasets, a phenomenon recently termed as "data crimes" [201]. This occurs when datasets originally intended for one task (e.g., segmentation) are used to train deep learning models for a different task (e.g., image reconstruction), without accounting for the hidden processing pipelines that were applied before the data were stored online. The work of Shimron et al. [201] demonstrated that this could have substantial effects on the results and algorithmic evaluation. For instance, MRI databases containing images that have already been reconstructed are sometimes used to generate synthetic raw data for training deep learning models. However, these reconstructed images often undergo preprocessing steps—such as zero-padding, coil compression, or conversion to lossy data-saving formats—that alter the data. Models trained on such processed data may exhibit misleadingly good performance, such as high-performance metrics, as the preprocessing of the training data makes the reconstruction problem better conditioned (loosely speaking, easier to solve). In such cases, error metrics like SSIM or NRMSE may fail to detect these biases, as both the reconstructed and reference images are affected by the same hidden processing pipeline. Therefore, in such scenarios, deep learning models could produce biased and overly optimistic results for processed data, failing to generalize well to unprocessed, real-world clinical data. In extreme cases, important details indicating pathology may be overlooked.

7.10.3 Adversarial attacks, instabilities, and generalization issues

Deep learning models can be vulnerable to adversarial attacks, where slight, often imperceptible changes to the input data can drastically alter the model's output; however not more than classical methods [202]. If triggered maliciously or accidently, these attacks could lead to poor results and perhaps incorrect diagnosis. The development of techniques to detect and mitigate such attacks is an ongoing research area.

Another notable issue is model instability [203]. For example, particularly in scenarios involving high levels of noise, deep learning models can produce unpredictable outputs, leading to poor reconstructions that may not be clinically usable. This is a major concern even in state-of-the-art models. Furthermore, studies indicate that deep learning models may struggle to generalize in cases of domain shifts, that is, when there is a substantial difference between the training and test sets. Poor generalization could lead to misleading results for out-of-distribution data [204]. For example, Johnson et al. [205] analyzed models that obtained high scores in the fastMRI challenge [197] and found that many of them were highly sensitive to small structure changes or the addition of noise to k-space data. Other studies, such as those by Darestani et al. [202] and Avidan et al. [206], found that learning-based models are sensitive to distribution shifts related to anatomical structures or k-space sampling patterns. Potential solutions include, for example, training with noise or sampling augmentations [207,208].

In summary, addressing the reliability of deep learning models in clinical MRI requires robust evaluation protocols, clinical validation, and methods to ensure outputs reflect true anatomy without hallucinations. Efforts to improve generalization and detect subtle biases are also required. Furthermore, publication of open-access datasets, development of reference-free image quality metrics, and adoption of reproducible research practices can contribute to enhancing trust and accuracy in real-world applications.

7.11 Conclusion

This chapter has explored the transformative impact of machine learning on MRI reconstruction, highlighting advancements in architectures, training paradigms, and applications. As MRI research progresses, several future directions hold great promise. First, integrating domain knowledge into deep learning models can enhance interpretability and robustness, particularly for clinical applications. The continued development of dual-domain and physics-informed networks will enable improved utilization of MRI data's inherent structure. Second, self-supervised and unsupervised learning frameworks can assist in reducing dependency on large annotated datasets, hence broadening opportunities for developing of advanced reconstruction methods. In addition, advancements in generative models, such as diffusion models, offer exciting opportunities for high-fidelity image synthesis, and enhancement. Finally, addressing challenges in robustness, generalization, and bias will be critical to translating these methods into widespread clinical practice, ensuring reliable MRI diagnostics.

References

[1] Heckel R, Jacob M, Chaudhari A, Perlman O, Shimron E. Deep learning for accelerated and robust MRI reconstruction. Magn Reson Mater Phys Biol Med 2024;37:335–68.

[2] Wang S, Wu R, Jia S, Diakite A, Li C, Liu Q, Ying L. Knowledge-driven deep learning for fast MR imaging: undersampled MR image reconstruction from supervised to unsupervised learning. Magn Reson Med 2024;92(2):496–518.

[3] Pruessmann KP, Weiger M, Scheidegger MB, Boesiger P. SENSE: sensitivity encoding for fast MRI. Magn Reson Med Off J Int Soc Magnetic Reson Med 1999;42(5):952–62.

[4] Griswold MA, Jakob PM, Heidemann RM, Nittka M, Jellus V, Wang J, Haase A. Generalized autocalibrating partially parallel acquisitions (GRAPPA). Magn Reson Med Off J Int Soc Magnetic Reson Med 2002;47(6):1202–10.

[5] McKenzie CA, Ohliger MA, Yeh EN, Price MD, Sodickson DK. Coil-by-coil image reconstruction with SMASH. Magn Reson Med Off J Int Soc Magnetic Reson Med 2001;46(3):619–23.

[6] Uecker M, Lai P, Murphy MJ, Virtue P, Elad M, Pauly JM, Lustig M. ESPIRiT—an eigenvalue approach to autocalibrating parallel MRI: where SENSE meets GRAPPA. Magn Reson Med 2014;71(3):990–1001.

[7] Wang S, Tan S, Gao Y, Liu Q, Ying L, Xiao T, Liang D. Learning joint-sparse codes for calibration-free parallel MR imaging. IEEE Trans Med Imag 2017;37(1):251–61.

[8] Lustig M, Donoho D, Pauly JM. Sparse MRI: the application of compressed sensing for rapid MR imaging. Magn Reson Med Off J Int Soc Magnetic Reson Med 2007;58(6):1182–95.

[9] Ravishankar S, Ye JC, Fessler JA. Image reconstruction: from sparsity to data-adaptive methods and machine learning. Proc IEEE 2019;108(1):86–109.

[10] Geethanath S, Reddy R, Konar AS, Imam S, Sundaresan R, DR RB, Venkatesan R. Compressed sensing MRI: a review. Crit Rev Biomed Eng 2013;41(3).

[11] Liang D, Liu B, Wang J, Ying L. Accelerating SENSE using compressed sensing. Magn Reson Med Off J Int Soc Magnetic Reson Med 2009;62(6):1574–84.

[12] Feng L, Benkert T, Block KT, Sodickson DK, Otazo R, Chandarana H. Compressed sensing for body MRI. J Magn Reson Imag 2017;45(4):966–87.

[13] Feng L, Grimm R, Block KT, Chandarana H, Kim S, Xu J, Otazo R. Golden-angle radial sparse parallel MRI: combination of compressed sensing, parallel imaging, and golden-angle radial sampling for fast and flexible dynamic volumetric MRI. Magn Reson Med 2014;72(3):707–17.

[14] Vasanawala SS, Alley MT, Hargreaves BA, Barth RA, Pauly JM, Lustig M. Improved pediatric MR imaging with compressed sensing. Radiology 2010;256(2):607–16.

[15] Hammernik K, Küstner T, Yaman B, Huang Z, Rueckert D, Knoll F, Akçakaya M. Physics-driven deep learning for computational magnetic resonance imaging: combining physics and machine learning for improved medical imaging. IEEE Signal Process Mag 2023;40(1):98–114.

[16] Wang S, Ke Z, Cheng H, Jia S, Ying L, Zheng H, Liang D. DIMENSION: dynamic MR imaging with both k-space and spatial prior knowledge obtained via multi-supervised network training. NMR Biomed 2022;35(4):e4131.

[17] Wang S, Cao G, Wang Y, Liao S, Wang Q, Shi J, Shen D. Review and prospect: artificial intelligence in advanced medical imaging. Front Radiol 2021;1:781868.

[18] Li C, Li W, Liu C, Zheng H, Cai J, Wang S. Artificial intelligence in multiparametric magnetic resonance imaging: a review. Med Phys 2022;49(10):e1024–54.

[19] Lustig M, Donoho DL, Santos JM, Pauly JM. Compressed sensing MRI. IEEE Signal Process Mag 2008;25(2):72–82.

[20] Fessler JA. Optimization methods for magnetic resonance image reconstruction: key models and optimization algorithms. IEEE Signal Process Mag 2020;37(1):33–40.

[21] Combettes PL, Pesquet JC. Proximal splitting methods in signal processing. Fixed-point algorithms for inverse problems in science and engineering. 2011. p. 185–212.

[22] Zhang J, Ghanem B. ISTA-Net: interpretable optimization-inspired deep network for image compressive sensing. In: Proceedings of the IEEE Conference on Computer Vision and Pattern Recognition; 2018. p. 1828–37.

[23] Boyd S, Parikh N, Chu E, Peleato B, Eckstein J. Distributed optimization and statistical learning via the alternating direction method of multipliers. Foundations Trends Machine learning 2011;3(1):1–122.

[24] Bilgic B, Goyal VK, Adalsteinsson E. Multi-contrast reconstruction with Bayesian compressed sensing. Magn Reson Med 2011;66(6):1601–15.

[25] Wang S, Su Z, Ying L, Peng X, Zhu S, Liang F, Liang D. Accelerating magnetic resonance imaging via deep learning. IEEE 13th International Symposium on Biomedical Imaging (ISBI). IEEE; 2016. p. 514–7.

[26] Wang S, Chen Y, Xiao T, Zhang L, Liu X, Zheng H. LANTERN: learn analysis transform network for dynamic magnetic resonance imaging. Inverse Probl Imag 2020;15 (6):1363–79.

[27] Zhu B, Liu JZ, Cauley SF, Rosen BR, Rosen MS. Image reconstruction by domain-transform manifold learning. Nature 2018;555(7697):487–92.

[28] Eo T, Jun Y, Kim T, Jang J, Lee HJ, Hwang D. KIKI-net: cross-domain convolutional neural networks for reconstructing undersampled magnetic resonance images. Magn Reson Med 2018;80(5):2188–201.

[29] Wu R, Li C, Zou J, Liu X, Zheng H, Wang S. Generalizable reconstruction for accelerating MR imaging via federated learning with neural architecture search. IEEE Trans Med Imag 2024.

[30] Schlemper J, Caballero J, Hajnal JV, Price AN, Rueckert D. A deep cascade of convolutional neural networks for dynamic MR image reconstruction. IEEE Trans Med Imag 2017;37(2):491–503.
[31] Ronneberger, O., Fischer, P., & Brox, T. (2015). U-net: Convolutional networks for biomedical image segmentation. In Medical image computing and computer-assisted intervention-MICCAI 2015: 18th international conference, Munich, Germany, October 5-9, 2015, proceedings, part III 18 (pp. 234-241). Springer International Publishing.
[32] Han Y, Sunwoo L, Ye JC. *k*-space deep learning for accelerated MRI. IEEE Trans Med Imag 2019;39(2):377–86.
[33] Lee D, Yoo J, Tak S, Ye JC. Deep residual learning for accelerated MRI using magnitude and phase networks. IEEE Trans Biomed Eng 2018;65(9):1985–95.
[34] Han Y, Yoo J, Kim HH, Shin HJ, Sung K, Ye JC. Deep learning with domain adaptation for accelerated projection-reconstruction MR. Magn Reson Med 2018;80(3):1189–205.
[35] He K, Zhang X, Ren S, Sun J. Deep residual learning for image recognition. In: Proceedings of the IEEE Conference on Computer Vision and Pattern Recognition; 2016. p. 770–8.
[36] Shi J, Liu Q, Wang C, Zhang Q, Ying S, Xu H. Super-resolution reconstruction of MR image with a novel residual learning network algorithm. Phys Med Biol 2018;63(8):085011.
[37] Bao L, Ye F, Cai C, Wu J, Zeng K, van Zijl PC, Chen Z. Undersampled MR image reconstruction using an enhanced recursive residual network. J Magn Reson 2019;305:232–46.
[38] Du J, He Z, Wang L, Gholipour A, Zhou Z, Chen D, Jia Y. Super-resolution reconstruction of single anisotropic 3D MR images using residual convolutional neural network. Neurocomputing 2020;392:209–20.
[39] Wu Y, Ma Y, Capaldi DP, Liu J, Zhao W, Du J, Xing L. Incorporating prior knowledge via volumetric deep residual network to optimize the reconstruction of sparsely sampled MRI. Magn Reson Imaging 2020;66:93–103.
[40] Goodfellow I, Pouget-Abadie J, Mirza M, Xu B, Warde-Farley D, Ozair S, Bengio Y. Generative adversarial nets. Adv Neural Inf Process Syst 2014;27.
[41] Shende P, Pawar M, Kakde S. A brief review on: MRI images reconstruction using GAN. In: 2019 International Conference on Communication and Signal Processing (ICCSP). IEEE; April 2019. p. 0139–42.
[42] Mardani M, Gong E, Cheng JY, Vasanawala SS, Zaharchuk G, Xing L, Pauly JM. Deep generative adversarial neural networks for compressive sensing MRI. IEEE Trans Med Imag 2018;38(1):167–79.
[43] Yang G, Yu S, Dong H, Slabaugh G, Dragotti PL, Ye X, Firmin D. DAGAN: deep de-aliasing generative adversarial networks for fast compressed sensing MRI reconstruction. IEEE Trans Med Imag 2017;37(6):1310–21.
[44] Quan TM, Nguyen-Duc T, Jeong WK. Compressed sensing MRI reconstruction using a generative adversarial network with a cyclic loss. IEEE Trans Med Imag 2018;37(6):1488–97.
[45] Murugesan, B., Vijaya Raghavan, S., Sarveswaran, K., Ram, K., & Sivaprakasam, M. (2019). Recon-GLGAN: a global-local context based generative adversarial network for MRI reconstruction. In Machine Learning for Medical Image Reconstruction: Second International Workshop, MLMIR 2019, Held in Conjunction with MICCAI 2019,

Shenzhen, China, October 17, 2019, Proceedings 2 (pp. 3-15). Springer International Publishing.

[46] Liang M, Hu X. Recurrent convolutional neural network for object recognition. In: Proceedings of the IEEE Conference on Computer Vision and Pattern Recognition; 2015. p. 3367–75.

[47] Qin C, Schlemper J, Caballero J, Price AN, Hajnal JV, Rueckert D. Convolutional recurrent neural networks for dynamic MR image reconstruction. IEEE Trans Med Imag 2018;38(1):280–90.

[48] Oh, C., Kim, D., Chung, J. Y., Han, Y., & Park, H. (2018). ETER-net: end to end MR image reconstruction using recurrent neural network. In Machine Learning for Medical Image Reconstruction: First International Workshop, MLMIR 2018, Held in Conjunction with MICCAI 2018, Granada, Spain, September 16, 2018, Proceedings 1 (pp. 12-20). Springer International Publishing.

[49] Zhang, T., Jackson, L. H., Uus, A., Clough, J. R., Story, L., Rutherford, M. A., ... & Deprez, M. (2019). Self-supervised recurrent neural network for 4D abdominal and in-utero MR imaging. In Machine Learning for Medical Image Reconstruction: Second International Workshop, MLMIR 2019, Held in Conjunction with MICCAI 2019, Shenzhen, China, October 17, 2019, Proceedings 2 (pp. 16-24). Springer International Publishing.

[50] Hosseini, S. A. H., Zhang, C., Uğurbil, K., Moeller, S., & Akcakaya, M. (2019, September). sRAKI-RNN: accelerated MRI with scan-specific recurrent neural networks using densely connected blocks. In Wavelets and Sparsity XVIII (Vol. 11138, pp. 393-399). SPIE.

[51] Liu Z, Lin Y, Cao Y, Hu H, Wei Y, Zhang Z, Guo B. Swin transformer: hierarchical vision transformer using shifted windows. In: Proceedings of the IEEE/CVF International Conference on Computer Vision; 2021. p. 10012–22.

[52] Feng CM, Yan Y, Fu H, Chen L, Xu Y. Task transformer network for joint MRI reconstruction and super-resolution. In: Medical Image Computing and Computer Assisted Intervention–MICCAI 2021: 24th International Conference, Strasbourg, France, September 27–October 1, 2021, Proceedings, Part VI 24. Springer International Publishing; 2021. p. 307–17.

[53] Huang J, Fang Y, Wu Y, Wu H, Gao Z, Li Y, Yang G. Swin transformer for fast MRI. Neurocomputing 2022;493:281–304.

[54] Korkmaz Y, Dar SU, Yurt M, Özbey M, Cukur T. Unsupervised MRI reconstruction via zero-shot learned adversarial transformers. IEEE Trans Med Imag 2022;41(7): 1747–63.

[55] Ekanayake M, Pawar K, Harandi M, Egan G, Chen Z. McSTRA: a multi-branch cascaded swin transformer for point spread function-guided robust MRI reconstruction. Comput Biol Med 2024;168:107775.

[56] Darestani MZ, Nath V, Li W, He Y, Roth HR, Xu Z, Zhao C. IR-FRestormer: iterative refinement with fourier-based restormer for accelerated MRI reconstruction. In: Proceedings of the IEEE/CVF Winter Conference on Applications of Computer Vision; 2024. p. 7655–64.

[57] Sun K, Wang Q, Shen D. Joint cross-attention network with deep modality prior for fast MRI reconstruction. IEEE Trans Med Imag 2023.

[58] Guo P, Mei Y, Zhou J, Jiang S, Patel VM. ReconFormer: accelerated MRI reconstruction using recurrent transformer. IEEE Trans Med Imag 2023.

[59] Zhou B, Dey N, Schlemper J, Salehi SSM, Liu C, Duncan JS, Sofka M. DSFormer: a dual-domain self-supervised transformer for accelerated multi-contrast MRI reconstruction. In: Proceedings of the IEEE/CVF Winter Conference on Applications of Computer Vision; 2023. p. 4966–75.

[60] Guo Y, Wang C, Zhang H, Yang G. Deep attentive wasserstein generative adversarial networks for MRI reconstruction with recurrent context-awareness. In: Medical Image Computing and Computer Assisted Intervention–MICCAI 2020: 23rd International Conference, Lima, Peru, October 4–8, 2020, Proceedings, Part II 23. Springer International Publishing; 2020. p. 167–77.

[61] Küstner T, Fuin N, Hammernik K, Bustin A, Qi H, Hajhosseiny R, Prieto C. CINENet: deep learning-based 3D cardiac CINE MRI reconstruction with multi-coil complex-valued 4D spatio-temporal convolutions. Sci Rep 2020;10(1):13710.

[62] Sandino CM, Lai P, Vasanawala SS, Cheng JY. Accelerating cardiac cine MRI using a deep learning-based ESPIRiT reconstruction. Magn Reson Med 2021;85(1):152–67.

[63] Gregor K, LeCun Y. Learning fast approximations of sparse coding. In: Proceedings of the 27th International Conference on International Conference on Machine Learning; June 2010. p. 399–406.

[64] Sun J, Li H, Xu Z. Deep ADMM-Net for compressive sensing MRI. Adv Neural Inf Process Syst 2016;29.

[65] Hammernik K, Klatzer T, Kobler E, Recht MP, Sodickson DK, Pock T, Knoll F. Learning a variational network for reconstruction of accelerated MRI data. Magn Reson Med 2018;79(6):3055–71.

[66] Aggarwal HK, Mani MP, Jacob M. MoDL: model-based deep learning architecture for inverse problems. IEEE Trans Med Imag 2018;38(2):394–405.

[67] Hammernik K, Schlemper J, Qin C, Duan J, Summers RM, Rueckert D. Systematic evaluation of iterative deep neural networks for fast parallel MRI reconstruction with sensitivity-weighted coil combination. Magn Reson Med 2021;86(4):1859–72.

[68] Wang S, Wu R, Li C, Zou J, Zhang Z, Liu Q, Zheng H. PARCEL: physics-based unsupervised contrastive representation learning for multi-coil mr imaging. IEEE ACM Trans Comput Biol Bioinf 2022;20(5):2659–70.

[69] Zou J, Pei T, Li C, Wu R, Wang S. Self-supervised federated learning for fast MR imaging. IEEE Trans Instrum Meas 2023.

[70] Wu R, Li C, Zou J, Liang Y, Wang S. Model-based federated learning for accurate MR image reconstruction from undersampled k-space data. Comput Biol Med 2024;180:108905.

[71] Dong J, Valzania L, Maillard A, Pham TA, Gigan S, Unser M. Phase retrieval: from computational imaging to machine learning: a tutorial. IEEE Signal Process Mag 2023;40(1):45–57.

[72] Ahmad R, Bouman CA, Buzzard GT, Chan S, Liu S, Reehorst ET, Schniter P. Plug-and-play methods for magnetic resonance imaging: using denoisers for image recovery. IEEE Signal Process Mag 2020;37(1):105–16.

[73] Romano Y, Elad M, Milanfar P. The little engine that could: regularization by denoising (RED). SIAM J Imag Sci 2017;10(4):1804–44.

[74] Liu J, Sun Y, Eldeniz C, Gan W, An H, Kamilov US. RARE: image reconstruction using deep priors learned without groundtruth. IEEE J Select Topics Signal Process 2020;14(6):1088–99.

[75] Ulyanov D, Vedaldi A, Lempitsky V. Deep image prior. In: Proceedings of the IEEE Conference on Computer Vision and Pattern Recognition; 2018. p. 9446–54.

[76] Darestani MZ, Heckel R. Accelerated MRI with un-trained neural networks. IEEE Transact Comput Imaging 2021;7:724–33.
[77] Heckel R, Soltanolkotabi M. Compressive sensing with un-trained neural networks: gradient descent finds a smooth approximation. In: International Conference on Machine Learning. PMLR; November 2020. p. 4149–58.
[78] Huang W, Li HB, Pan J, Cruz G, Rueckert D, Hammernik K. Neural implicit k-space for binning-free non-cartesian cardiac MR imaging. In: International Conference on Information Processing in Medical Imaging. Cham: Springer Nature Switzerland; June 2023. p. 548–60.
[79] Kunz JF, Ruschke S, Heckel R. Implicit neural networks with fourier-feature inputs for free-breathing cardiac MRI reconstruction. IEEE Transact Comput Imaging 2024.
[80] Shen L, Pauly J, Xing L. NeRP: implicit neural representation learning with prior embedding for sparsely sampled image reconstruction. IEEE Transact Neural Networks Learn Syst 2022;35(1):770–82.
[81] Wu R, Cheng J, Li C, Zou J, Yang J, Fan W, Wang S. CSR-dMRI: continuous super-resolution of diffusion MRI with anatomical structure-assisted implicit neural representation learning. In: International Workshop on Machine Learning in Medical Imaging. Cham: Springer Nature Switzerland; October 2024. p. 114–23.
[82] Spieker V, Eichhorn H, Hammernik K, Rueckert D, Preibisch C, Karampinos DC, Schnabel JA. Deep learning for retrospective motion correction in MRI: a comprehensive review. IEEE Trans Med Imag 2023.
[83] Chang Y, Li Z, Saju G, Mao H, Liu T. Deep learning-based rigid motion correction for magnetic resonance imaging: a survey. Meta-Radiology 2023:100001.
[84] Haskell MW, Cauley SF, Bilgic B, Hossbach J, Splitthoff DN, Pfeuffer J, Wald LL. Network accelerated motion estimation and reduction (NAMER): convolutional neural network guided retrospective motion correction using a separable motion model. Magn Reson Med 2019;82(4):1452–61.
[85] Levac B, Jalal A, Tamir JI. Accelerated motion correction for MRI using score-based generative models. In: 2023 IEEE 20th International Symposium on Biomedical Imaging (ISBI). IEEE; April 2023. p. 1–5.
[86] Atkinson D. Retrospective correction of motion in MR images. In: Advances in magnetic resonance technology and applications. 6. Academic Press; 2022. p. 259–67.
[87] Loktyushin A, Nickisch H, Pohmann R, Schölkopf B. Blind retrospective motion correction of MR images. Magn Reson Med 2013;70(6):1608–18.
[88] Singh A, Salehi SSM, Gholipour A. Deep predictive motion tracking in magnetic resonance imaging: application to fetal imaging. IEEE Trans Med Imag 2020;39(11):3523–34.
[89] Qi H, Fuin N, Cruz G, Pan J, Kuestner T, Bustin A, Prieto C. Non-rigid respiratory motion estimation of whole-heart coronary MR images using unsupervised deep learning. IEEE Trans Med Imag 2020;40(1):444–54.
[90] Wang X, Uecker M, Feng L. Fast real-time cardiac MRI: a review of current techniques and future directions. Invest Magnetic Reson Imaging 2021;25(4):252–65.
[91] Murray V, Siddiq S, Crane C, El Homsi M, Kim TH, Wu C, Otazo R. Movienet: deep space–time-coil reconstruction network without k-space data consistency for fast motion-resolved 4D MRI. Magn Reson Med 2024;91(2):600–14.
[92] Klug T, Wang K, Ruschke S, Heckel R. MotionTTT: 2D test-time-training motion estimation for 3D motion corrected MRI 2024. arXiv preprint arXiv:2409.09370.

[93] Hammernik K, Pan J, Rueckert D, Küstner T. Motion-guided physics-based learning for cardiac MRI reconstruction. In: 2021 55th Asilomar Conference on Signals, Systems, and Computers. IEEE; October 2021. p. 900–7.
[94] Pan J, Hamdi M, Huang W, Hammernik K, Kuestner T, Rueckert D. Unrolled and rapid motion-compensated reconstruction for cardiac CINE MRI. Med Image Anal 2024;91: 103017.
[95] Phair A, Fotaki A, Felsner L, Fletcher TJ, Qi H, Botnar RM, Prieto C. A motion-corrected deep-learning reconstruction framework for accelerating whole-heart magnetic resonance imaging in patients with congenital heart disease. J Cardiovasc Magn Reson 2024;26(1):101039.
[96] Yang J, Küstner T, Hu P, Liò P, Qi H. End-to-end deep learning of non-rigid groupwise registration and reconstruction of dynamic MRI. Front Cardiovasc Med 2022;9: 880186.
[97] Ghoul A, Pan J, Lingg A, Kübler J, Krumm P, Hammernik K, Küstner T. Attention-aware non-rigid image registration for accelerated MR imaging. IEEE Trans Med Imag 2024.
[98] Zou Q, Torres LA, Fain SB, Higano NS, Bates AJ, Jacob M. Dynamic imaging using motion-compensated smoothness regularization on manifolds (MoCo-SToRM). Phys Med Biol 2022;67(14):144001.
[99] Tournier JD, Mori S, Leemans A. Diffusion tensor imaging and beyond. Magn Reson Med 2011;65(6):1532.
[100] Le Bihan D, Mangin JF, Poupon C, Clark CA, Pappata S, Molko N, Chabriat H. Diffusion tensor imaging: concepts and applications. J Magn Reson Imag Off J Int Soc Magnetic Reson Med 2001;13(4):534–46.
[101] Mani M, Magnotta VA, Jacob M. qModeL: a plug-and-play model-based reconstruction for highly accelerated multi-shot diffusion MRI using learned priors. Magn Reson Med 2021;86(2):835–51.
[102] Wu R, Cheng J, Li C, Zou J, Fan W, Ma X, Wang S. Spherical harmonics representation learning for high-fidelity and generalizable super-resolution in diffusion MRI. IEEE Trans on Biomed Eng 2025.
[103] Yang J, Cheng J, Li C, Fan W, Zou J, Wu R, Wang S. Simultaneous Q-space sampling optimization and reconstruction for fast and high-fidelity diffusion magnetic resonance imaging. In: 2024 IEEE International Symposium on Biomedical Imaging (ISBI). IEEE; May 2024. p. 1–4.
[104] Xiao T, Cheng J, Fan W, Yang J, Li C, Dong E, Wang S. RobNODDI: robust NODDI parameter estimation with adaptive sampling under continuous representation. In: Chamberland, M., et al. Computational Diffusion MRI. CDMRI 2024. Lecture Notes in Computer Science, vol 15171. Springer, Cham. 2025.
[105] Fan W., Cheng J., Li C., Yang J., Wu R., Zou J., Wang S. (2025). AID-DTI: accelerating high-fidelity diffusion tensor imaging with detail-preserving model-based deep learning. In: Chamberland, M., et al. Computational Diffusion MRI. CDMRI 2024. Lecture Notes in Computer Science, vol 15171. Springer, Cham.
[106] Liu F, Feng L, Kijowski R. MANTIS: model-augmented neural network with incoherent k-space sampling for efficient MR parameter mapping. Magn Reson Med 2019;82(1):174–88.
[107] Zhao B, Setsompop K, Ye H, Cauley SF, Wald LL. Maximum likelihood reconstruction for magnetic resonance fingerprinting. IEEE Trans Med Imag 2016;35(8): 1812–23.

[108] Zhu Y, Cheng J, Cui ZX, Zhu Q, Ying L, Liang D. Physics-driven deep learning methods for fast quantitative magnetic resonance imaging: performance improvements through integration with deep neural networks. IEEE Signal Process Mag 2023;40(2): 116–28.

[109] Seiberlich N, Gulani V, Campbell-Washburn A, Sourbron S, Doneva MI, Calamante F, Hu HH, editors. Quantitative magnetic resonance imaging. Academic Press; 2020.

[110] Ma D, Gulani V, Seiberlich N, Liu K, Sunshine JL, Duerk JL, Griswold MA. Magnetic resonance fingerprinting. Nature 2013;495(7440):187–92.

[111] Cohen O, Zhu B, Rosen MS. MR fingerprinting deep reconstruction network (DRONE). Magn Reson Med 2018;80(3):885–94.

[112] Balsiger, F., Shridhar Konar, A., Chikop, S., Chandran, V., Scheidegger, O., Geethanath, S., & Reyes, M. (2018). Magnetic resonance fingerprinting reconstruction via spatiotemporal convolutional neural networks. In Machine Learning for Medical Image Reconstruction: First International Workshop, MLMIR 2018, Held in Conjunction with MICCAI 2018, Granada, Spain, September 16, 2018, Proceedings 1 (pp. 39-46). Springer International Publishing.

[113] Hamilton JI, Currey D, Rajagopalan S, Seiberlich N. Deep learning reconstruction for cardiac magnetic resonance fingerprinting T1 and T2 mapping. Magn Reson Med 2021;85(4):2127–35.

[114] Golbabaee M, Chen D, Gómez PA, Menzel MI, Davies ME. Geometry of deep learning for magnetic resonance fingerprinting. In: ICASSP 2019-2019 IEEE International Conference on Acoustics, Speech and Signal Processing (ICASSP). IEEE; May 2019. p. 7825–9.

[115] Perlman O, Zhu B, Zaiss M, Rosen MS, Farrar CT. An end-to-end AI-based framework for automated discovery of rapid CEST/MT MRI acquisition protocols and molecular parameter quantification (AutoCEST). Magn Reson Med 2022;87(6):2792–810.

[116] Lustig M, Pauly JM. SPIRiT: iterative self-consistent parallel imaging reconstruction from arbitrary k-space. Magn Reson Med 2010;64(2):457–71.

[117] Knoll F, Hammernik K, Zhang C, Moeller S, Pock T, Sodickson DK, Akcakaya M. Deep-learning methods for parallel magnetic resonance imaging reconstruction: a survey of the current approaches, trends, and issues. IEEE Signal Process Mag 2020;37 (1):128–40.

[118] Sriram A, Zbontar J, Murrell T, Zitnick CL, Defazio A, Sodickson DK. GrappaNet: combining parallel imaging with deep learning for multi-coil MRI reconstruction. In: Proceedings of the IEEE/CVF Conference on Computer Vision and Pattern Recognition; 2020. p. 14315–22.

[119] Peng X, Sutton BP, Lam F, Liang ZP. DeepSENSE: learning coil sensitivity functions for SENSE reconstruction using deep learning. Magn Reson Med 2022;87(4): 1894–902.

[120] Chen, Y., Xiao, T., Li, C., Liu, Q., & Wang, S. (2019). Model-based convolutional dealiasing network learning for parallel MR imaging. In Medical Image Computing and Computer Assisted Intervention-MICCAI 2019: 22nd International Conference, Shenzhen, China, October 13-17, 2019, Proceedings, Part III 22 (pp. 30-38). Springer International Publishing.

[121] Sriram A, Zbontar J, Murrell T, Defazio A, Zitnick CL, Yakubova N, Johnson P. End-to-end variational networks for accelerated MRI reconstruction. In: Medical Image Computing and Computer Assisted Intervention–MICCAI 2020: 23rd International

Conference, Lima, Peru, October 4–8, 2020, Proceedings, Part II 23. Springer International Publishing; 2020. p. 64–73.
[122] Jun Y, Shin H, Eo T, Hwang D. Joint deep model-based MR image and coil sensitivity reconstruction network (joint-ICNet) for fast MRI. In: Proceedings of the IEEE/CVF Conference on Computer Vision and Pattern Recognition; 2021. p. 5270–9.
[123] Luo G, Wang X, Roeloffs V, Tan Z, Uecker M. Joint estimation of coil sensitivities and image content using a deep image prior. In: Proceedings of the 29th Annual Meeting of ISMRM; 2021. p. 280.
[124] Zhang M, Xu J, Arefeen Y, Adalsteinsson E. Zero-shot self-supervised joint temporal image and sensitivity map reconstruction via linear latent space. In: Medical imaging with deep learning. PMLR; January 2024. p. 1713–25.
[125] Kingma DP. Auto-encoding variational bayes. arXiv preprint arXiv:1312.6114 2013.
[126] Tezcan KC, Baumgartner CF, Luechinger R, Pruessmann KP, Konukoglu E. MR image reconstruction using deep density priors. IEEE Trans Med Imag 2018;38(7):1633–42.
[127] Yang L, Zhang Z, Song Y, Hong S, Xu R, Zhao Y, Yang MH. Diffusion models: a comprehensive survey of methods and applications. ACM Comput Surv 2023;56(4): 1–39.
[128] Kawar B, Elad M, Ermon S, Song J. Denoising diffusion restoration models. Adv Neural Inf Process Syst 2022;35:23593–606.
[129] Croitoru FA, Hondru V, Ionescu RT, Shah M. Diffusion models in vision: a survey. IEEE Trans Pattern Anal Mach Intell 2023;45(9):10850–69.
[130] Güngör A, Dar SU, Öztürk S, Korkmaz Y, Bedel HA, Elmas G, Çukur T. Adaptive diffusion priors for accelerated MRI reconstruction. Med Image Anal 2023;88: 102872.
[131] Ho J, Jain A, Abbeel P. Denoising diffusion probabilistic models. Adv Neural Inf Process Syst 2020;33:6840–51.
[132] Chung H, Ye JC. Score-based diffusion models for accelerated MRI. Med Image Anal 2022;80:102479.
[133] Luo G, Blumenthal M, Heide M, Uecker M. Bayesian MRI reconstruction with joint uncertainty estimation using diffusion models. Magn Reson Med 2023;90(1):295–311.
[134] Zach M, Knoll F, Pock T. Stable deep MRI reconstruction using generative priors. IEEE Trans Med Imag 2023.
[135] Bengio Y, Goodfellow I, Courville A. Deep learning, 1. Cambridge, MA, USA: MIT Press; 2017.
[136] Song Y, Ermon S. Improved techniques for training score-based generative models. Adv Neural Inf Process Syst 2020;33:12438–48.
[137] Song Y, Sohl-Dickstein J, Kingma DP, Kumar A, Ermon S, Poole B. Score-based generative modeling through stochastic differential equations. arXiv preprint arXiv: 2011.13456 2020.
[138] Jalal A, Arvinte M, Daras G, Price E, Dimakis AG, Tamir J. Robust compressed sensing MRI with deep generative priors. Adv Neural Inf Process Syst 2021;34: 14938–54.
[139] Kazerouni A, Aghdam EK, Heidari M, Azad R, Fayyaz M, Hacihaliloglu I, Merhof D. Diffusion models in medical imaging: a comprehensive survey. Med Image Anal 2023; 88:102846.
[140] Wang S, Xiao T, Liu Q, Zheng H. Deep learning for fast MR imaging: a review for learning reconstruction from incomplete k-space data. Biomed Signal Process Control 2021;68:102579.

[141] Akçakaya M, Moeller S, Weingärtner S, Uğurbil K. Scan-specific robust artificial-neural-networks for k-space interpolation (RAKI) reconstruction: database-free deep learning for fast imaging. Magn Reson Med 2019;81(1):439–53.

[142] Zhang C, Moeller S, Demirel OB, Uğurbil K, Akçakaya M. Residual RAKI: a hybrid linear and non-linear approach for scan-specific k-space deep learning. Neuroimage 2022;256:119248.

[143] Zhao Z, Zhang T, Xie W, Wang Y, Zhang Y. K-space transformer for undersampled MRI reconstruction. arXiv preprint arXiv:2206.06947 2022.

[144] Xu S, Hammernik K, Lingg A, Kübler J, Krumm P, Rueckert D, Küstner T. Attention incorporated network for sharing low-rank, image and k-space information during MR image reconstruction to achieve single breath-hold cardiac Cine imaging. Comput Med Imag Graph 2024:102475.

[145] Singh NM, Iglesias JE, Adalsteinsson E, Dalca AV, Golland P. Joint frequency and image space learning for MRI reconstruction and analysis. J Machine Learning Biomed Imaging 2022;2022.

[146] El-Rewaidy H, Fahmy AS, Pashakhanloo F, Cai X, Kucukseymen S, Csecs I, Nezafat R. Multi-domain convolutional neural network (MD-CNN) for radial reconstruction of dynamic cardiac MRI. Magn Reson Med 2021;85(3):1195–208.

[147] Ran M, Xia W, Huang Y, Lu Z, Bao P, Liu Y, Zhang Y. MD-Recon-Net: a parallel dual-domain convolutional neural network for compressed sensing MRI. IEEE Transact Radiation Plasma Med Sci 2020;5(1):120–35.

[148] Liu X, Pang Y, Jin R, Liu Y, Wang Z. Dual-domain reconstruction network with V-Net and K-Net for fast MRI. Magn Reson Med 2022;88(6):2694–708.

[149] Souza R, Bento M, Nogovitsyn N, Chung KJ, Loos W, Lebel RM, Frayne R. Dual-domain cascade of U-nets for multi-channel magnetic resonance image reconstruction. Magn Reson Imaging 2020;71:140–53.

[150] Otazo R, Candes E, Sodickson DK. Low-rank plus sparse matrix decomposition for accelerated dynamic MRI with separation of background and dynamic components. Magn Reson Med 2015;73(3):1125–36.

[151] Yaman B, Hosseini SAH, Moeller S, Ellermann J, Uğurbil K, Akçakaya M. Self-supervised learning of physics-guided reconstruction neural networks without fully sampled reference data. Magn Reson Med 2020;84(6):3172–91.

[152] Hu C, Li C, Wang H, Liu Q, Zheng H, Wang S. Self-supervised learning for MRI reconstruction with a parallel network training framework. In: Medical Image Computing and Computer Assisted Intervention–MICCAI 2021: 24th International Conference, Strasbourg, France, September 27–October 1, 2021, Proceedings, Part VI 24. Springer International Publishing; 2021. p. 382–91.

[153] Yaman B, Gu H, Hosseini SAH, Demirel OB, Moeller S, Ellermann J, Akçakaya M. Multi-mask self-supervised learning for physics-guided neural networks in highly accelerated magnetic resonance imaging. NMR Biomed 2022;35(12):e4798.

[154] Akçakaya M, Yaman B, Chung H, Ye JC. Unsupervised deep learning methods for biological image reconstruction and enhancement: an overview from a signal processing perspective. IEEE Signal Process Mag 2022;39(2):28–44.

[155] Zou J, Li C, Jia S, Wu R, Pei T, Zheng H, Wang S. SelfCoLearn: self-supervised collaborative learning for accelerating dynamic MR imaging. Bioengineering 2022; 9(11):650.

[156] Stein CM. Estimation of the mean of a multivariate normal distribution. Ann Stat 1981:1135–51.

[157] Eldar YC. Generalized SURE for exponential families: applications to regularization. IEEE Trans Signal Process 2008;57(2):471–81.
[158] Aggarwal HK, Pramanik A, John M, Jacob M. ENSURE: a general approach for unsupervised training of deep image reconstruction algorithms. IEEE Trans Med Imag 2022;42(4):1133–44.
[159] Lehtinen J. Noise2Noise: learning image restoration without clean data. arXiv preprint arXiv:1803.04189 2018.
[160] Millard C, Chiew M. A theoretical framework for self-supervised MR image reconstruction using sub-sampling via variable density Noisier2Noise. IEEE Transact Comput Imaging 2023.
[161] Moran N, Schmidt D, Zhong Y, Coady P. Noisier2noise: learning to denoise from unpaired noisy data. In: Proceedings of the IEEE/CVF Conference on Computer Vision and Pattern Recognition; 2020. p. 12064–72.
[162] Gan W, Sun Y, Eldeniz C, Liu J, An H, Kamilov US. Deformation-compensated learning for image reconstruction without ground truth. IEEE Trans Med Imag 2022;41(9):2371–84.
[163] Blumenthal M, Fantinato C, Unterberg-Buchwald C, Haltmeier M, Wang X, Uecker M. Self-supervised learning for improved calibrationless radial MRI with NLINV-Net. Magn Reson Med 2024;92(6):2447–63.
[164] Zhou B, Schlemper J, Dey N, Salehi SSM, Sheth K, Liu C, Sofka M. Dual-domain self-supervised learning for accelerated non-Cartesian MRI reconstruction. Med Image Anal 2022;81:102538.
[165] Wang F, Qi H, De Goyeneche A, Heckel R, Lustig M, Shimron E. K-Band: self-supervised MRI reconstruction via stochastic gradient descent over K-space subsets. arXiv preprint arXiv:2308.02958 2023.
[166] Weiss T, Senouf O, Vedula S, Michailovich O, Zibulevsky M, Bronstein A. PILOT: physics-informed learned optimized trajectories for accelerated MRI. arXiv preprint arXiv:1909.05773 2019.
[167] Bahadir CD, Wang AQ, Dalca AV, Sabuncu MR. Deep-learning-based optimization of the under-sampling pattern in MRI. IEEE Transact Comput Imaging 2020;6:1139–52.
[168] Lazarus C, Weiss P, Chauffert N, Mauconduit F, El Gueddari L, Destrieux C, Ciuciu P. SPARKLING: variable-density k-space filling curves for accelerated T2*-weighted MRI. Magn Reson Med 2019;81(6):3643–61.
[169] Chaithya GR, Ramzi Z, Ciuciu P. Learning the sampling density in 2D SPARKLING MRI acquisition for optimized image reconstruction. In: 2021 29th European Signal Processing Conference (EUSIPCO). IEEE; August 2021. p. 960–4.
[170] Radhakrishna CG, Ciuciu P. Jointly learning non-cartesian k-space trajectories and reconstruction networks for 2D and 3D MR imaging through projection. Bioengineering 2023;10(2):158.
[171] Wang G, Luo T, Nielsen JF, Noll DC, Fessler JA. B-spline parameterized joint optimization of reconstruction and k-space trajectories (BJORK) for accelerated 2D MRI. IEEE Trans Med Imag 2022;41(9):2318–30.
[172] Wang G, Nielsen JF, Fessler JA, Noll DC. Stochastic optimization of 3D non-cartesian sampling trajectory (SNOPY). arXiv preprint arXiv:2209.11030 2022.
[173] Xie J, Zhang J, Zhang Y, Ji X. PUERT: probabilistic under-sampling and explicable reconstruction network for CS-MRI. IEEE J Selected Top Signal Process 2022;16 (4):737–49.

[174] Ravula S, Levac B, Jalal A, Tamir JI, Dimakis AG. Optimizing sampling patterns for compressed sensing MRI with diffusion generative models. arXiv preprint arXiv: 2306.03284 2023.

[175] Sarracanie M, LaPierre CD, Salameh N, Waddington DE, Witzel T, Rosen MS. Low-cost high-performance MRI. Sci Rep 2015;5(1):15177.

[176] Marques JP, Simonis FF, Webb AG. Low-field MRI: an MR physics perspective. J Magn Reson Imag 2019;49(6):1528–42.

[177] Wald LL, McDaniel PC, Witzel T, Stockmann JP, Cooley CZ. Low-cost and portable MRI. J Magn Reson Imag 2020;52(3):686–96.

[178] Zhao Y, Ding Y, Lau V, Man C, Su S, Xiao L, Wu EX. Whole-body magnetic resonance imaging at 0.05 Tesla. Science 2024;384(6696):eadm7168.

[179] Salehi A, Mach M, Najac C, Lena B, O'Reilly T, Dong Y, Webb A. Denoising low-field MR images with a deep learning algorithm based on simulated data from easily accessible open-source software. J Magn Reson 2024:107812.

[180] Lau V, Xiao L, Zhao Y, Su S, Ding Y, Man C, Wu EX. Pushing the limits of low-cost ultra-low-field MRI by dual-acquisition deep learning 3D superresolution. Magn Reson Med 2023;90(2):400–16.

[181] de Leeuw Den Bouter ML, Ippolito G, O'Reilly TPA, Remis RF, Van Gijzen MB, Webb AG. Deep learning-based single image super-resolution for low-field MR brain images. Sci Rep 2022;12(1):6362.

[182] Kim S, Tregidgo HF, Eldaly AK, Figini M, Alexander DC. A 3D conditional diffusion model for image quality transfer–an application to low-field MRI. arXiv preprint arXiv:2311.06631 2023.

[183] Islam KT, Zhong S, Zakavi P, Chen Z, Kavnoudias H, Farquharson S, Chen Z. Improving portable low-field MRI image quality through image-to-image translation using paired low-and high-field images. Sci Rep 2023;13(1):21183.

[184] Jimeno MM, Ravi KS, Jin Z, Oyekunle D, Ogbole G, Geethanath S. ArtifactID: Identifying artifacts in low-field MRI of the brain using deep learning. Magn Reson Imaging 2022;89:42–8.

[185] Koonjoo N, Zhu B, Bagnall GC, Bhutto D, Rosen MS. Boosting the signal-to-noise of low-field MRI with deep learning image reconstruction. Sci Rep 2021;11(1):8248.

[186] Shimron E, Shan S, Grover J, Koonjoo N, Shen S, Boele T, Waddington DE. Accelerating low-field MRI: compressed sensing and AI for fast noise-robust imaging. arXiv preprint arXiv:2411.06704 2024.

[187] Man C, Lau V, Su S, Zhao Y, Xiao L, Ding Y, Wu EX. Deep learning enabled fast 3D brain MRI at 0.055 tesla. Sci Adv 2023;9(38):eadi9327.

[188] Wang S, Cheng H, Ying L, Xiao T, Ke Z, Zheng H, Liang D. DeepcomplexMRI: exploiting deep residual network for fast parallel MR imaging with complex convolution. Magn Reson Imaging 2020;68:136–47.

[189] Li Y, Cheng X, Gui G. Co-robust-ADMM-net: joint ADMM framework and DNN for robust sparse composite regularization. IEEE Access 2018;6:47943–52.

[190] Recht MP, Zbontar J, Sodickson DK, Knoll F, Yakubova N, Sriram A, Zitnick CL. Using deep learning to accelerate knee MRI at 3 T: results of an interchangeability study. Am J Roentgenol 2020;215(6):1421–9.

[191] Liu Y, Liu Q, Zhang M, Yang Q, Wang S, Liang D. IFR-Net: iterative feature refinement network for compressed sensing MRI. IEEE Transact Comput Imaging 2019;6: 434–46.

[192] Simonyan K, Zisserman A. Very deep convolutional networks for large-scale image recognition. arXiv preprint arXiv:1409.1556 2014.

[193] Wang K, Tamir JI, De Goyeneche A, Wollner U, Brada R, Yu SX, Lustig M. High fidelity deep learning-based MRI reconstruction with instance-wise discriminative feature matching loss. Magn Reson Med 2022;88(1):476–91.
[194] Bhadra S, Kelkar VA, Brooks FJ, Anastasio MA. On hallucinations in tomographic image reconstruction. IEEE Trans Med Imag 2021;40(11):3249–60.
[195] Cohen JP, Luck M, Honari S. Distribution matching losses can hallucinate features in medical image translation. In: Medical Image Computing and Computer Assisted Intervention–MICCAI 2018: 21st International Conference, Granada, Spain, September 16-20, 2018, Proceedings, Part I. Springer International Publishing; 2018. p. 529–36.
[196] Gottschling NM, Antun V, Hansen AC, Adcock B. The troublesome kernel–On hallucinations, no free lunches and the accuracy-stability trade-off in inverse problems. arXiv preprint arXiv:2001.01258 2020.
[197] Muckley MJ, Riemenschneider B, Radmanesh A, Kim S, Jeong G, Ko J, Knoll F. Results of the 2020 fastMRI challenge for machine learning MR image reconstruction. IEEE Trans Med Imag 2021;40(9):2306–17.
[198] Varoquaux G, Cheplygina V. Machine learning for medical imaging: methodological failures and recommendations for the future. NPJ Digit Med 2022;5(1):48.
[199] Seyyed-Kalantari L, Zhang H, McDermott MB, Chen IY, Ghassemi M. Underdiagnosis bias of artificial intelligence algorithms applied to chest radiographs in under-served patient populations. Nat Med 2021;27(12):2176–82.
[200] Chen RJ, Wang JJ, Williamson DF, Chen TY, Lipkova J, Lu MY, Mahmood F. Algorithmic fairness in artificial intelligence for medicine and healthcare. Nat Biomed Eng 2023;7(6):719–42.
[201] Shimron E, Tamir JI, Wang K, Lustig M. Implicit data crimes: machine learning bias arising from misuse of public data. Proc Natl Acad Sci 2022;119(13):e2117203119.
[202] Darestani MZ, Chaudhari AS, Heckel R. Measuring robustness in deep learning based compressive sensing. In: International Conference on Machine Learning. PMLR; July 2021. p. 2433–44.
[203] Antun V, Renna F, Poon C, Adcock B, Hansen AC. On instabilities of deep learning in image reconstruction and the potential costs of AI. Proc Natl Acad Sci 2020;117(48): 30088–95.
[204] Wang J, Lan C, Liu C, Ouyang Y, Qin T, Lu W, Philip SY. Generalizing to unseen domains: a survey on domain generalization. IEEE Trans Knowl Data Eng 2022;35(8): 8052–72.
[205] Johnson PM, Jeong G, Hammernik K, Schlemper J, Qin C, Duan J, Knoll F. Evaluation of the robustness of learned MR image reconstruction to systematic deviations between training and test data for the models from the fastMRI challenge. In: Machine Learning for Medical Image Reconstruction: 4th International Workshop, MLMIR 2021, Held in Conjunction with MICCAI 2021, Strasbourg, France, October 1, 2021, Proceedings 4. Springer International Publishing; 2021. p. 25–34.
[206] Avidan N, Freiman M. Physically-primed deep-neural-networks for generalized undersampled MRI reconstruction. arXiv preprint arXiv:2209.00462 2022.
[207] Desai AD, Ozturkler BM, Sandino CM, Boutin R, Willis M, Vasanawala S, Chaudhari AS. Noise2Recon: enabling SNR-robust MRI reconstruction with semi-supervised and self-supervised learning. Magn Reson Med 2023;90(5):2052–70.
[208] Liu F, Samsonov A, Chen L, Kijowski R, Feng L. SANTIS: sampling-augmented neural network with incoherent structure for MR image reconstruction. Magn Reson Med 2019;82(5):1890–904.

CHAPTER

8 Enhance the image: Super resolution in MRI

Ziyu Li[1,2], Zihan Li[1], Haoxiang Li[1], Qiuyun Fan[3], Karla L. Miller[2], Wenchuan Wu[2], Akshay S. Chaudhari[4,5] and Qiyuan Tian[1,*]

[1]School of Biomedical Engineering, Tsinghua University, Beijing, China; [2]Wellcome Centre for Integrative Neuroimaging, FMRIB, Nuffield Department of Clinical Neurosciences, University of Oxford, Oxford, United Kingdom; [3]Academy of Medical Engineering and Translational Medicine, Medical College, Tianjin University, Tianjin, China; [4]Department of Radiology, Stanford University, Stanford, CA, United States; [5]Department of Biomedical Data Science, Stanford University, Stanford, CA, United States

[*]Corresponding author: E-mail: qiyuantian@tsinghua.edu.cn

8.1 Introduction

MRI with higher spatial resolution provides more detailed insights into the structure and function of living human bodies noninvasively, which is highly desirable for accurate clinical diagnosis and image analysis. The spatial resolution of MRI is characterized by in-plane and through-plane resolutions (Fig. 8.1). The in-plane spatial resolution along a dimension (Δx) depends on the maximal extent of the k-space ($k_x^{\max}$):

$$\Delta x = \frac{1}{2k_x^{\max}}, \tag{8.1}$$

where a broader coverage of k-space $k_x^{\max}$ corresponds to a higher $\frac{1}{\Delta x}$ and therefore a higher in-plane resolution (Fig. 8.1a). On the contrary, the through-plane resolution, also referred to as slice thickness, is determined differently for 2D and 3D imaging. In 2D imaging, the slice thickness is defined by the full width at half maximum (FWHM) of the slice-selection radiofrequency (RF) pulse profile. In 3D imaging, the slice-selection direction is encoded by another phase encoding gradient. Consequently, the through-plane resolution is determined similarly to the in-plane resolution by the maximal extent of the k-space along slice-selection direction as in Eq. (8.1).

Both RF pulse with large FWHM and limited k-space coverage can degrade resolution. In 2D imaging, RF pulse directly affects the through-plane resolution in the image domain by modulating the signal with the RF profile along the slice selection

Machine Learning in MRI. https://doi.org/10.1016/B978-0-443-14109-6.00008-0

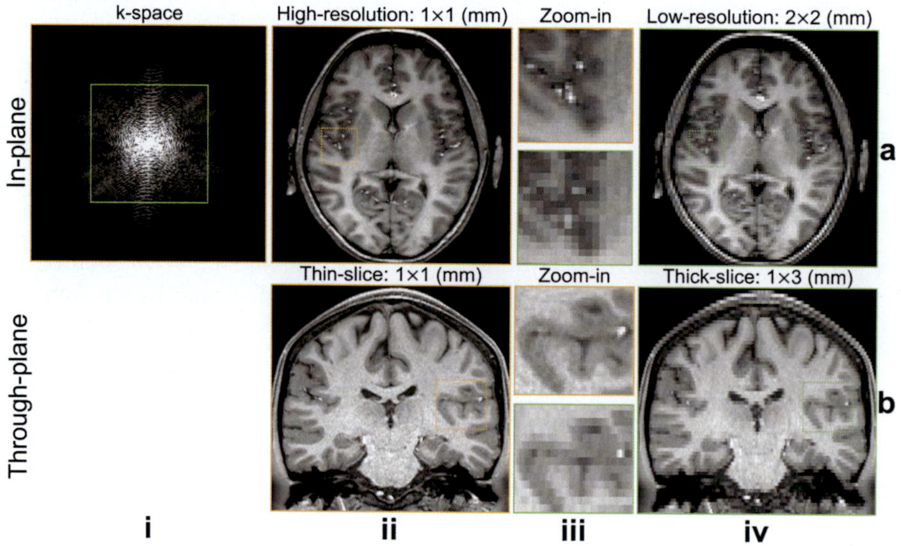

FIGURE 8.1 Spatial resolution of MRI.

The in-plane resolution is dictated by the k-space coverage, and a larger k-space coverage brings higher spatial resolution (a). The slice thickness is determined by the slice-selective RF pulse for 2D imaging, and by k-space extent along slice-selection direction for 3D imaging (b).

direction. Because an ideal rectangular RF profile is unattainable in practice, nonrectangular profiles can further introduce blurring or crosstalk (aliasing) between adjacent slices. Limited k-space coverage reduces the high-frequency components acquired, leading to degraded spatial resolution and the potential for Gibbs ringing artifacts due to abrupt truncation of k-space data. Consequently, low-resolution images may suffer from artifacts such as blurring, ringing, and aliasing, in addition to reduced spatial detail, which all undermine diagnostic accuracy and analysis reliability.

However, higher spatial resolution in MRI is achieved at the cost of prolonged scan times. Depending on the scheme used for image formation, the pursuit of higher in-plane resolution through a broader k-space coverage generally necessitates an extended readout time, more phase-encoding steps, and/or more excitations. Regarding the through-plane resolution, for 2D imaging, a reduction in slice thickness escalates the number of slices to acquire, leading to increased scan time. More importantly, the ability of 2D imaging to resolve very thin slice (e.g., 1 mm or thinner) is intrinsically limited by the sharpness of RF profiles. Consequently, the demand for achieving slender slice thickness and superior, isotropic resolution often mandates the adoption of 3D imaging that is often more time-consuming due to an additional phase-encoding dimension and substantially increased phase-encoding steps. Moreover, to compensate the decreased SNR for high-resolution scans with smaller voxel size, averaging multiple repetitions of data is often required. The prolonged acquisition imposes challenges such as increased scan costs, subject discomfort, susceptibility to motion artifacts, and impracticality for subjects intolerant to extended scans, such as certain populations of patients and children.

Image super-resolution technology offers a promising venue to harness the advantages of higher resolution for a short scan. It is a well-established problem which aims to restore a high-resolution image from a given low-resolution input [1], often referred to as single-image super-resolution and formulated as an ill-posed inverse problem that incorporates appropriate priors. Early efforts in computer vision indicated that priors derived from high-resolution example patches effectively enhance low-resolution patches sharing similar textures (example-based super-resolution) [2,3]. In the context of MRI, various priors, including high-resolution example-based priors [4], nonlocal patches [5], self-similarity [5,6], sparse encoding [7], total variation, and low-rankness [8] have been proposed. However, these methods only demonstrated moderate resolution improvement due to the intrinsically limited high-frequency information in peripheral k-space and the difficulty in solving the nonlinear, ill-posed problem. An alternative category of MRI super-resolution methods uses multiple acquired images with different slice shifts [9], FOV orientations [10,11], or RF encodings [12–14]. Van Reeth et al. comprehensively reviewed this type of approaches [15]. Despite their appealing performance especially for improving the through-plane resolutions, they require extended scan times to acquire multiple images. The primary focus of this chapter is on single-image super-resolution techniques that do not necessitate additional scan time.

Recent strides in machine learning and deep learning have exhibited significant potential for achieving substantial resolution improvement from a single input image. It has been demonstrated that even a shallow 3-layer convolutional neural network (CNN) can outperform state-of-the-art traditional methods, such as sparse encoding, for super-resolving 2D natural images [16]. Increasing the depth of CNNs [17] and employing advanced training strategies, such as adversarial training [18], can further enhance their performance. Impressively, these deep learning methods are also faster and easier to deploy once the networks are trained. Their success in natural image super-resolution, coupled with these advantages, positions them as a promising method for achieving high-fidelity MRI super-resolution without the need for additional scan time.

In this chapter, we offer a comprehensive overview of current deep learning-based MRI super-resolution techniques. Our discussion encompasses not only the methodologies but also the impact of super-resolved images on clinical and neuroscientific assessments. Delving into practical considerations, we cover various aspects such as network architectures, image evaluation metrics, network loss functions, and pertinent training considerations that range from methods for simulating low-resolution images to dataset selection. Finally, we evaluate existing challenges and propose potential future directions for deep learning-based MRI super-resolution, consolidating insights for researchers and practitioners in this evolving field.

8.2 Deep learning-based MRI super-resolution

8.2.1 Standard CNN

After achieving notable success in natural image super-resolution within the field of computer vision, CNNs have found widespread applications in MRI super-resolution. The fundamental framework of CNN-based MRI super-resolution is depicted in Fig. 8.2.

Similar to its application in natural image super-resolution, CNN in MRI is employed to transform low-resolution input with thick slice thickness and/or low in-plane resolution into super-resolved images. Low-resolution inputs are often interpolated to match the dimensions of the high-resolution target using methods like trilinear or cubic spline interpolation. The optimization process involves minimizing the discrepancy, quantified through metrics such as mean squared error (MSE, L2-loss) or mean absolute error (MAE, L1-loss), between the network's output and the high-resolution training target. Taking MSE as an example loss function, the training objective is formulated as:

$$\min_{\theta} \mathscr{L}_{L2}(\theta) = ||\widehat{x}_{SR} - x_{HR}||_2^2 = ||f_\theta(x_{LR}) - x_{HR}||_2^2, \tag{8.2}$$

where x_{LR}, $\widehat{x}_{SR}$, and x_{HR} are the interpolated low-resolution input, the CNN-synthesized super-resolved image, and target high-resolution reference, respectively. f_θ is the CNN parametrized by θ.

Given the 3D nature of MRI data, CNNs with 3D convolutional kernels are commonly used to exploit redundancy along all three dimensions. Notably, a shallow 3-layer 3D CNN [19] has demonstrated enhanced performance for brain MRI super-resolution when compared to traditional methods such as nonlocal

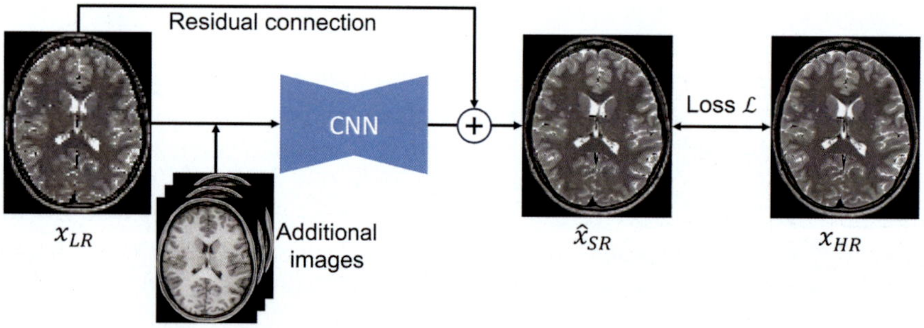

FIGURE 8.2 CNN-based MRI super-resolution.

A basic framework for super-resolution MRI using CNN is demonstrated. The CNN is optimized by minimizing the loss between the high-resolution target x_{HR} and the super-resolved image $\widehat{x}_{SR}$ generated from the low-resolution input x_{LR} (for supervised learning). Residual learning strategy can be adopted for faster convergence and boosted performance. Additional images that are available can be also included to leverage redundant information.

upsampling, total variation and low-rank-based super-resolution, and their 2D CNN counterparts.

The efficacy of deeper networks has been established, with the addition of more layers further enhancing performance [20]. Drawing inspiration from ResNet [21] that introduced residual connections to facilitate the training of deep CNNs and unleash their full potential for nonlinear mapping in natural image classification, studies have confirmed the beneficial impact of residual connections on both natural image [17] and MRI super-resolution [22–24]. Intuitively, CNNs with residual connections can focus on learning to estimate high-frequency components, requiring less information and leading to enhanced performance, particularly considering the number of CNN parameters and available training resources. Specifically, instead of directly generating x_{HR}, residual learning estimates the difference between x_{LR} and x_{HR}:

$$\min_{\theta} \left\| f_{\theta}^{R}(x_{LR}) - (x_{HR} - x_{LR}) \right\|_2^2, \tag{8.3}$$

where f_{θ}^{R} is the CNN parametrized by θ optimized using residual learning strategy. The final super-resolved output $\widehat{x}_{SR}$ can be obtained by adding the interpolated low-resolution input to the network output:

$$\widehat{x}_{SR} = x_{LR} + f_{\theta}^{R}(x_{LR}). \tag{8.4}$$

Moreover, the inclusion of complementary images has been identified as a means to potentially enhance performance. For brain MRI, Zeng et al. showed that including high-resolution T1-weighted images contributes to the better performance of T2-weighted images super-resolution [25]. Lyu et al. proposed a progressive network that also incorporates high-resolution images of proton density and T1--weighted contrast for T2-weighted image super-resolution [26]. In cardiac imaging, combining multiple time frames [27,28] or images acquired from different slice directions [29] as input has been shown to improve super-resolution accuracy. In addition, cardiac phase information has proven useful in further refining cardiac MRI super-resolution [30]. Since many of these complementary images are routinely acquired in practice, their integration does not necessarily increase the scan time. This approach of leveraging additional images to enhance super-resolution performance has also been explored in other medical imaging modalities. For example, the inclusion of high-resolution MRI has been found beneficial for deep learning-based positron emission tomography (PET) super-resolution [31].

8.2.2 Generative adversarial network (GAN)

While standard CNNs trained using metrics like MSE or MAE have demonstrated promising results in MRI super-resolution, with high peak signal-to-noise ratios (PSNR) compared with native high-resolution images, these super-resolved images do not necessarily exhibit satisfying visual quality. The limitation lies in the use of pixel-wise loss functions such as MSE, which face challenges in effectively addressing the inherent uncertainty associated with recovering lost high-frequency texture details. The minimization of MSE tends to favor pixel-wise averages of plausible solutions, leading to outcomes that are often oversmoothed and, consequently, exhibit limited perceptual quality [18].

GAN has proven to be an effective solution to this problem. GAN consists of a generator and a discriminator [32] (Fig. 8.3). For MRI super-resolution, the generator super-resolves the input low-resolution images. CNNs discussed in Section 8.2.1 can be considered as a generator. Meanwhile, the discriminator's role is to classify an image as either real or generated. The generator and the discriminator are trained alternatively to compete against each other. As training progresses, the discriminator becomes highly proficient in distinguishing between real and generated images. Simultaneously, the generator refines its ability to produce realistic-looking images based on the feedback from the discriminator. Eventually, the discriminator becomes an adept classifier, while the generator excels at generating images that closely resemble real ones, as guided by the learned criteria of the discriminator.

Mathematically, when the generator is trained, the discriminator is fixed, and the optimizer tries to solve:

$$\min_{\theta_G} \mathscr{L}^G_{con} + \lambda \mathscr{L}^G_{adv}, \tag{8.5}$$

where θ_G is the generator network parameters, $\mathscr{L}^G_{con}$ is the content loss between the generated super-resolved image and the high-resolution target (e.g., MSE as in Eqs. 8.1 and 8.2), and $\mathscr{L}^G_{adv}$ is the adversarial loss derived from the discriminator. This loss reflects the probability, as determined by the discriminator, that a super-resolved image could be classified as an actual high-resolution image. λ is the hyper-parameter which determines the contribution of the adversarial loss.

The adversarial loss can be defined as:

$$\mathscr{L}^G_{adv} = -\log D_{\theta_D}(\hat{x}_{SR}) = -\log D_{\theta_D}\left(G_{\theta_G}(x_{LR})\right), \tag{8.6}$$

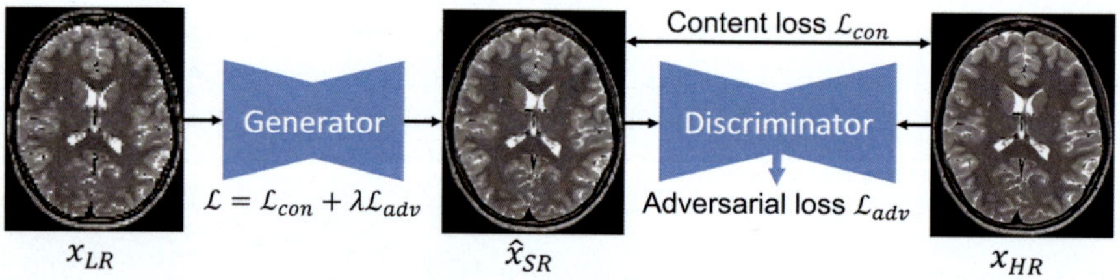

FIGURE 8.3 GAN for MRI super-resolution.

The GAN comprises a generator which super-resolves the input low-resolution image x_{LR}, and a discriminator which discerns between the generated image $\hat{x}_{SR}$ and the original high-resolution image x_{HR}. The generator and discriminator are trained to compete against each other. This adversarial training dynamic aims to converge toward the generation of realistic high-resolution images indistinguishable from the native high-resolution counterparts.

where G_{θ_G} and D_{θ_D} denote the generator and discriminator parametrized by θ_G and θ_D respectively. θ_D is fixed during the training of the generator. $D_{\theta_D}(\widehat{x}_{SR})$ represents the probability assigned by the discriminator to the classification of a super-resolved image as a genuine high-resolution image.

During the training of the discriminator, θ_G is fixed. The training process tries to optimize the parameters of the discriminator (θ_D) such that it could accurately discriminate the generated and actual images. Using binary cross-entropy loss which can classify generated and target images, as originally proposed for GAN [32], the training objective of the discriminator can be formulated as:

$$\min_{\theta_D} \mathscr{L}^D = -\left(\log D_{\theta_D}(x_{HR}) - \log D_{\theta_D}(\widehat{x}_{SR})\right) = -\left(\log D_{\theta_D}(x_{HR}) - \log D_{\theta_D}\left(G_{\theta_G}(x_{LR})\right)\right). \tag{8.7}$$

In practice, GANs with binary cross-entropy discriminator loss are notoriously challenging to train, often suffering from problems such as gradient vanishing, model collapsing, nonconvergence, etc. Regularization techniques (e.g., spectral normalization [33]) as well as other variants of GANs with different loss function (e.g., least-square GAN [34], Wasserstein GAN [35,36]) have been proposed to stabilize the training.

Studies leverage GAN to improve the visual quality of the super-resolved MRI. For example, in brain MRI, a 3D Wasserstein GAN has been proposed for T1--weighted structural image super-resolution, which produces sharper and more realistic textures compared with its counterpart trained without adversarial loss (i.e., $\lambda = 0$ in Eq. 8.5) [37,38]. In addition, GAN has proven beneficial for simultaneous super-resolution and geometric distortion correction in diffusion-weighted images [39]. In cardiac imaging, GANs have been employed to improve the resolution of whole-heart MRI [40] and facilitate high-fidelity coronary MR angiography (CMRA) super-resolution, achieving 3D isotropic CMRA in under a minute [41]. For knee MRI, an ensembled GAN has been proposed to transform images obtained from multiple conventional super-resolution methods into a more accurate super-resolved image with improved image fidelity [42].

8.2.3 Transformer

An inherent limitation of CNNs is that their receptive field is constrained by kernel size and network architecture, which hinders their ability to capture long-range information for MRI super-resolution. Transformers have emerged as a promising architecture to address this challenge by modeling long-range dependencies within the image using the self-attention mechanism [43,44] (see also Chapter 2). The self-attention mechanism operates by transforming an input vector (e.g., a flattened patch of an image) into three vectors: the query vector $\boldsymbol{q}$, the key vector $\boldsymbol{k}$, and the value vector $\boldsymbol{v}$. The vectors from all inputs are then stacked to form matrices Q, K, and V. This allows for the calculation of the attention function between all combinations of input vectors using an inner product and a Softmax function (Fig. 8.4) to indicate

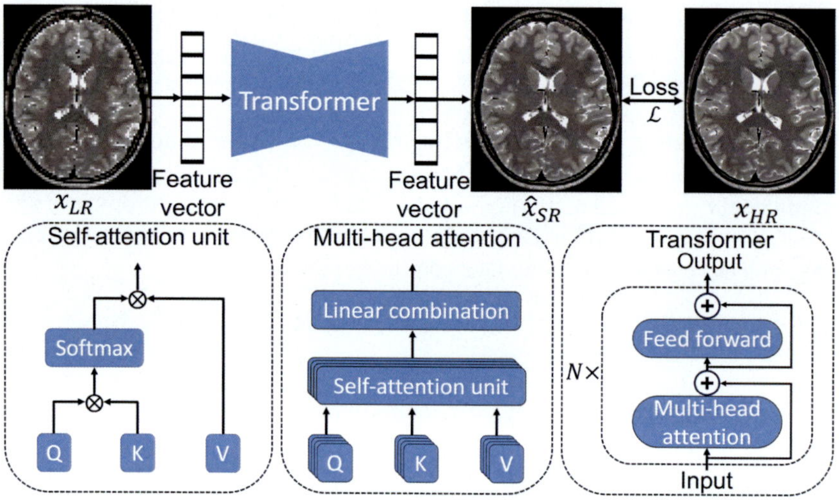

FIGURE 8.4 Transformer framework for MRI super-resolution.

The transformer operates on the feature vectors extracted from the low-resolution input x_{LR} and produces output feature vectors from which the super-resolved image $\hat{x}_{SR}$ can be generated. A self-attention unit is illustrated, where Q, K, V denote the query, key, and value matrices, respectively, and ⊗ represents the inner product operation. Multiple self-attention units can be stacked in parallel to form a multihead attention unit. The transformer comprises N multihead attention and feed-forward blocks with residual connections.

the importance of each input vector to each output. Essentially, the key $\boldsymbol{k}$ identifies features, the query $\boldsymbol{q}$ specifies the feature of interest, and the value $\boldsymbol{v}$ is the response corresponding to a matching query-key pair.

Mathematically, given an input sequence of vectors X, the self-attention mechanism calculates the output feature vector Y following:

$$Y = \text{softmax}\left(\frac{QK^T}{\sqrt{d_k}}\right)V = \text{softmax}\left(\frac{(W_QX)(W_KX)^T}{\sqrt{d_k}}\right)(W_VX), \tag{8.8}$$

where d_k is the dimension of the key vector $\boldsymbol{k}$, and W_Q, W_K, and W_V are learnable weight matrices for query, key, and value projections, respectively. Scaling the inner product of QK^T with $1/\sqrt{d_k}$ is beneficial when the dimension of key vector is large [43]. Instead of performing a single attention function with multidimensional queries, keys, and values, it has been found beneficial to linearly project queries, keys, and values and parallelly perform self-attention function (i.e., having multiple W_Q, W_K, and W_V), known as multihead attention (Fig. 8.4). The transformer comprises multiple multihead attention and feed-forward blocks with residual connections to boost performance (Fig. 8.4). In addition, the information about the position of each input/output element within the sequence can be leveraged using positional encoding, which is typically achieved by adding vectors to the sequence, where each vector encodes the relative position of the element within the sequence [43].

After initially being proposed for natural language processing tasks [43], transformers have found applications in vision tasks as well. For instance, the Vision Transformer (ViT) [45] flattens 2D patches to obtain vectors, which are then input to a transformer for image classification. However, pure transformers like ViT have not shown significant performance improvement compared with CNNs for vision tasks, possibly due to a limited ability to leverage local information [44]. For image super-resolution, the integration of convolutional layers with transformers emerges as a promising solution [46], where convolutional layers can be used to extract local features from input images as well as to reconstruct output images from features.

Transformers have demonstrated promise in MRI super-resolution, often incorporating auxiliary contrasts for enhanced performance. For example, Feng et al. employed a multimodality transformer for super-resolution and reconstruction of brain and knee MRI, which introduces a cross-attention module to fully exploit information in each modality, outperforming CNN-based methods [47]. Fang et al. proposed a transformer for multimodality MRI super-resolution, leveraging inter-modality correlation as well as high-frequency structure priors [48]. Li et al. introduced a transformer-based multiscale network for multicontrast MRI super-resolution [49]. Despite improved performance compared with CNNs, these studies also report increased model complexity and potential larger training data requirements for transformer-based models.

8.2.4 Diffusion model

Diffusion models, inspired by principles from nonequilibrium thermodynamics [50], establish a sequential diffusion process to gradually introduce random noise to the data. They then learn to reverse this process, ultimately reconstructing the original data from the noise [50,51]. Diffusion models address the instability often associated with GANs in synthesizing high-fidelity images and have proven effective in generating realistic, high-quality natural images [51]. In the forward Markovian diffusion process q gradually adds noise to the high-resolution image x_{HR}^{0} over T iterations (Fig. 8.5), ultimately generating a noise map which serves as a latent representation of the original data:

$$q\left(x_{HR}^{1:T} \middle| x_{HR}^{0}\right) = \prod_{t=1}^{T} q\left(x_{HR}^{t} \middle| x_{HR}^{t-1}\right), \tag{8.9}$$

$$q\left(x_{HR}^{t} \middle| x_{HR}^{t-1}\right) = \mathcal{N}\left(x_{HR}^{t} \middle| \sqrt{\alpha_t}\, x_{HR}^{t-1}, (1-\alpha_t)I\right), \tag{8.10}$$

where $\alpha_t \in (0,1)$ is the hyperparameter determining the variance of the added noise and $\mathcal{N}(0,I)$ is the normal distribution. x_{HR}^{t-1} is attenuated by $\sqrt{\alpha_t}$ to ensure the random variable remains bounded when $t \rightarrow \infty$. To shorten the calculation time of

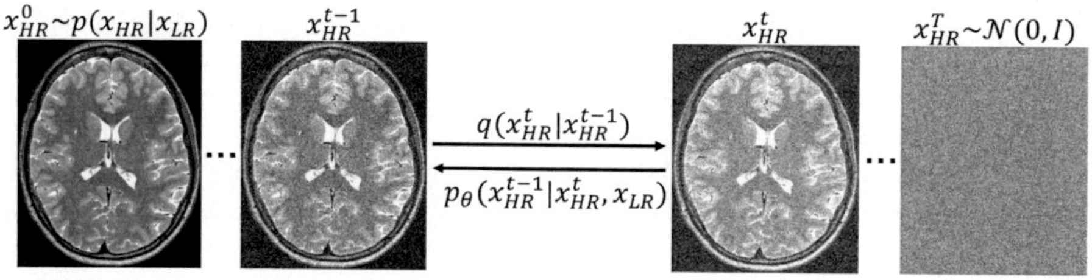

FIGURE 8.5 Forward and reverse diffusion process.

The forward diffusion process q (*left* to *right*) gradually adds Gaussian noise to the target high-resolution image x_{HR}. The reverse process p (*right* to *left*) iteratively denoises x_{HR}, conditioned on a source low-resolution image x_{LR}, which is not shown.

x_{HR}^t, the reparameterization trick which sets $\overline{\alpha}_t := \prod_{s=1}^{t} \alpha_s$ will allow the sampling of x_{HR}^t at any arbitrary time step from the following distribution [51]:

$$q\left(x_{HR}^t \middle| x_{HR}^0\right) = \mathcal{N}\left(x_{HR}^t \middle| \sqrt{\overline{\alpha}_t}\, x_{HR}^0, (1 - \overline{\alpha}_t)I\right). \tag{8.11}$$

The reverse process aims to recover the high-resolution image by gradually removing the noise (p in Fig. 8.5). Without conditioning this will lead to the generation of random high-resolution images with similar distribution as the training data. For image super-resolution, we leverage the low-resolution image as conditioning [52,53] to recover the super-resolved image:

$$p_\theta\left(x_{HR}^{0:T} \middle| x_{LR}\right) = p\left(x_{HR}^T\right) \prod_{t=1}^{T} p_\theta\left(x_{HR}^{t-1} \middle| x_{HR}^t, x_{LR}\right), \tag{8.12}$$

$$p_\theta\left(x_{HR}^{t-1} \middle| x_{HR}^t, x_{LR}\right) = \mathcal{N}\left(x_{HR}^{t-1} \middle| \mu_\theta\left(x_{HR}^t, x_{LR}, t\right), \Sigma_\theta\left(x_{HR}^t, x_{LR}, t\right)\right), \tag{8.13}$$

where $x_{HR}^T \sim \mathcal{N}(0, I)$. Essentially, the training of the diffusion model estimates the noise distribution $\mathcal{N}(\mu_\theta, \Sigma_\theta)$ for given timestep t, noisy image x_{HR}^t, and the low-resolution conditioning x_{LR} using a neural network parametrized by θ. Each training step takes a gradient descent step on:

$$\nabla_\theta \left|\left| \epsilon - \epsilon_\theta\left(x_{HR}^t, x_{LR}, t\right) \right|\right|_2^2 = \nabla_\theta \left|\left| \epsilon - \epsilon_\theta\left(\sqrt{\overline{\alpha}_t}\, x_{HR}^0 + (1 - \overline{\alpha}_t)\epsilon, x_{LR}, t\right) \right|\right|_2^2, \tag{8.14}$$

where $\epsilon \sim \mathcal{N}(0,I)$. In practice, U-Net can be employed for learning the distribution, and the conditioning on x_{LR} can be achieved by concatenating x_{LR} to x_{HR}^t along the channel dimension [52].

T networks are trained to reverse x_{HR}^T back to x_{HR}^0. During inference, x_{HR}^{t-1} is calculated from:

$$x_{HR}^{t-1} = \frac{1}{\sqrt{\alpha_t}}\left(x_{HR}^t - \frac{1 - \alpha_t}{\sqrt{1 - \overline{\alpha}_t}} \epsilon_\theta\left(x_{HR}^t, x_{LR}, t\right)\right) + \sigma_t z, \tag{8.15}$$

where $z \sim \mathcal{N}(0, I)$ when $t > 1$ (otherwise $z = 0$) and $\sigma_t^2 I = \Sigma_\theta(x_{HR}^t, x_{LR}, t)$. In practice σ_t can be set to $\sqrt{1 - \alpha_t}$ [51,52].

Diffusion models have proven effective in synthesizing high-fidelity, high-resolution MRI from low-resolution inputs. Wu et al. demonstrated the model's capability in synthesizing realistic-looking brain T1-weighted and diffusion-weighted MRI from 64× down-sampled low-resolution input [54]. Wang et al. [55] utilized a denoising diffusion implicit model [56] for optimal latent representations of low-resolution thick-slice brain MRI and employed a brain image synthesis diffusion model [57] pretrained on UK Biobank data to recover high-resolution brain images from the latent representations. Chung et al. [58] proposed a score-based diffusion model [59] which denoises knee and liver MRI and super-resolves the denoised images. The super-resolution is achieved using the same diffusion model by applying a data consistency term which takes blurring into account.

8.2.5 Implicit neural representation (INR)

As opposed to conventional pixel-based representation, INR aims to obtain a continuous representations of images, which theoretically allows the accurate generation of images in arbitrary resolutions by mapping any continuous image coordinates to the corresponding signal [60,61].

For image super-resolution with INR (Fig. 8.6), the representation can be learned using an encoder, typically implemented as a CNN to capture local information [61,62]:

$$v = E_{\varphi}(x), \tag{8.16}$$

where x is the input image, v is the obtained feature map, and E is the encoding network parametrized by φ.

Subsequently, high-resolution images can be reconstructed from the encoded implicit features, which spatially correspond to the object. The decoding step is often

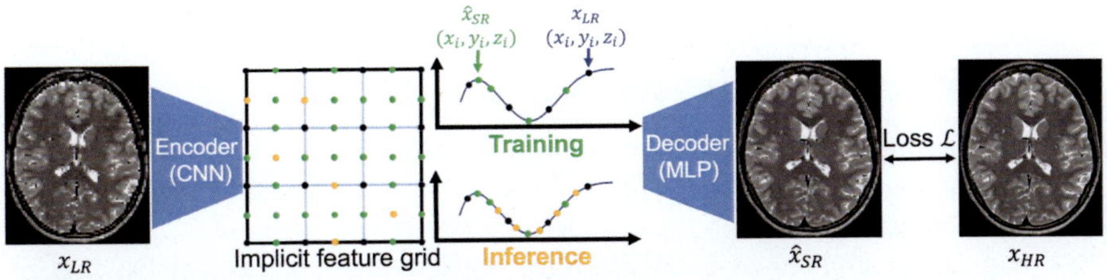

FIGURE 8.6 Implicit neural representation for MRI super-resolution.

The input low-resolution image x_{LR} is encoded as an implicit feature map using a convolutional neural network (CNN). The super-resolved image $\hat{x}_{SR}$ is reconstructed by finding the intensity of each voxel according to the coordinate and the implicit features using the decoder (often implemented as a multilayer perceptron (MLP)). The encoder and decoder are jointly optimized by minimizing the difference between $\hat{x}_{SR}$ and the target high-resolution image x_{HR}.

executed voxel by voxel through a multilayer perceptron (i.e., fully-connected neural network), taking the desired coordinate and corresponding feature as inputs:

$$\widehat{x}_q = \mathscr{F}_\theta(v_q, q), \tag{8.17}$$

where q is the image coordinate which can take arbitrary values and does not need to be integers, $\widehat{x}_q$ and v_q are the network estimated image intensity and feature at q, respectively, and $\mathscr{F}_\theta$ is the decoder network parametrized by θ. For the query of v_q with noninteger q, interpolation (e.g., nearest neighbor [61], trilinear [63]) can be applied to the feature map v. The spatial coordinate q requires normalization for low- and high-resolution images.

The encoder and decoder can be optimized simultaneously by solving:

$$\min_{\varphi,\theta} ||\widehat{x}_{SR,q} - x_{HR,q}||_2^2 = ||\mathscr{F}_\theta(E_\varphi(x_{LR})_q, q) - x_{HR,q}||_2^2. \tag{8.18}$$

The training process aims to fit a continuous implicit function that links spatial location to image intensity, enabling the acquisition of intensity at any desired location during inference (Fig. 8.6).

INR's superior performance for MRI super-resolution has been shown in various studies. Wu et al. leveraged INR for the super-resolution of thick-slice T1-weighted brain MRI, yielding high-fidelity, high-resolution brain images in both simulations and prospective in-vivo experiments [64]. They further extended their method to achieve robust arbitrary scale super-resolution, producing images with improved quality and more accurate brain region segmentations compared with CNN-based methods [63]. Li et al. showed that INR super-resolved brain quantitative susceptibility mapping (QSM) and T1-weighted images benefit the localization and delineation of human pedunculopontine nucleus [65]. Xu et al. proposed an INR-based framework which facilitates 3D volume reconstruction from thick-slice acquisition of fetal MRI, demonstrating improved image quality and shortened processing time compared with conventional methods [66].

8.3 Clinical and neuroscientific applications

These deep learning-based MRI super-resolution techniques hold significant promise in enhancing image quality while reducing scan time, making them valuable tools across various clinical and neuroscientific applications (Fig. 8.7).

A number of clinical assessments have underscored the clinical benefits of learning-based MRI super-resolution. Even early example-based super-resolution methods demonstrated significant improvements, such as enhancing the detection of coronary artery stenoses in 1.5T whole-heart CMRA, where the area under the receiver operating characteristic curve (AUC) increased from 0.79 to 0.84 across three observers in a cohort of 46 patients [68]. Later, Rudie et al. [67] conducted a systematic clinical evaluation of deep learning super-resolved brain MRI. Four board-certified neuroradiologists participated in the study, assessing 3D

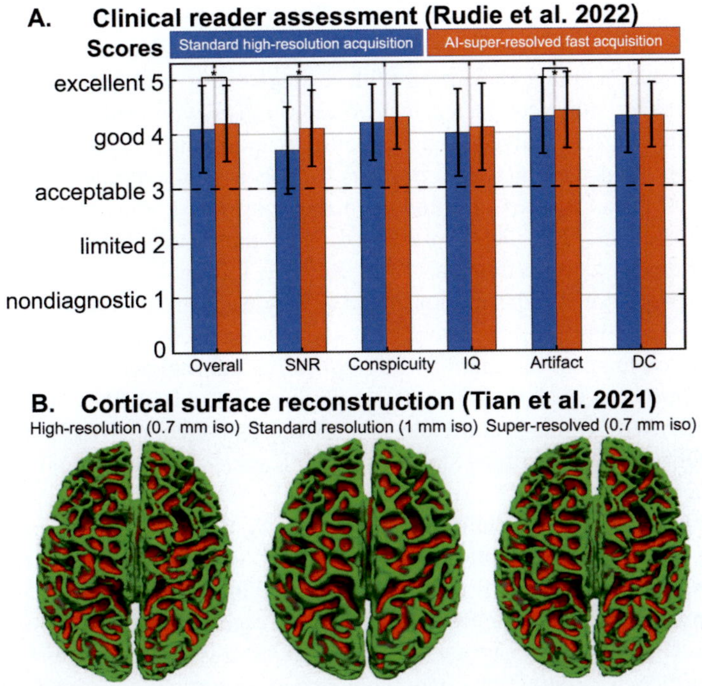

FIGURE 8.7 Example clinical and neuroscientific applications of deep learning-based MRI super-resolution.

Rudie et al. conducted a comprehensive clinical assessment of brain MRI from fast acquisition with deep learning super-resolution and standard, longer high-resolution acquisition, with regard to overall evaluation, SNR, anatomical conspicuity, image quality (IQ), artifact [67]. The scores for T2-weighted FLAIR images are shown and comparisons with statistical significance are marked with asterisks (A). Tian et al. demonstrated deep learning-based super-resolution effectively improved the accuracy of cortical surface reconstruction and thickness estimation [24] (B).

T1-weighted precontrast, T1-weighted postcontrast, and T2-weighted FLAIR clinical brain MRI scans from 32 patients. The images were acquired through both standard high-resolution and a faster acquisition with lower resolution (~45% shorter scan time), which was subsequently super-resolved using a pretrained U-Net. Evaluation criteria included SNR, anatomical conspicuity, image quality, artifact, and diagnostic confidence (Fig. 8.7A). Despite the ~45% reduction in scan time, the deep learning-enhanced images exhibited no significant compromise in image quality. On the contrary, they surpassed standard acquisition, particularly in terms of SNR and image artifact reduction. The authors attributed this improvement to the denoising effect of the network trained on large samples and the reduction in motion artifacts from the faster acquisition. Chaudhari et al. proposed and clinically evaluated "DeepResolve," a deep learning super-resolution method for knee MRI [22,69].

This approach utilized a CNN with a residual learning strategy to map tricubic-interpolated thick-slice low-resolution images, obtained from 2D acquisition, to 3D high-resolution images acquired with a longer scan time. Initial assessments involved a reader study conducted by two radiologists who evaluated interpolated low-resolution, super-resolved, and native high-resolution images. The evaluation revealed a preference among radiographers for super-resolved images over interpolated low-resolution images in terms of image contrast, sharpness, SNR, artifacts, and overall quality [22]. They further extended their evaluation to qualitatively and quantitatively demonstrate that DeepResolve minimally biases cartilage and osteophyte biomarkers with superior image quality. Super-resolved images achieved reader scores, cartilage segmentation, and osteophyte detection comparable to native high-resolution images, substantially outperforming interpolated low-resolution images [69]. Masutani et al. demonstrated that for cardiac MRI, CNN super-resolved images exhibited similar ventricular volumetry and ejection fraction to native high-resolution images, with no statistical significance, supporting their use in clinical cardiac imaging [27].

Deep learning super-resolution techniques have also shown great potential for advancing neuroscientific applications. Along this line of research, Tian et al. utilized a CNN (i.e., VDSR [17]) to generate T1-weighted anatomical images at sub-millimeter isotropic resolution from standard T1-weighted images at 1 mm isotropic resolution. The resulting images exhibited superior gray–white contrast, improved cortical surface reconstruction, and more accurate cortical thickness estimation compared with interpolated standard 1 mm images (Fig. 8.7B). This development holds great promise in depicting neurological development, degeneration, and pathology, opening up new possibilities for cortical morphometry applications by significantly reducing scan times (from ~1 hour to less than 5 minutes) [24]. Iglesias et al. leveraged a 3D U-Net trained on synthetic data to achieve joint super-resolution and synthesis of 1 mm isotropic MP-RAGE volumes, allowing FreeSurfer-type [70–72] analyses such as cortical surface reconstruction and brain region segmentation. The method reliably estimated hippocampal volumes and right hemisphere thickness, maintained significance between Alzheimer's disease and control groups, and proved valuable in registration for CNN-synthesized volumes super-resolved from thick-slice acquisition [73]. Li et al. employed a 3D U-Net to convert diffusion data to higher-resolution anatomical data, enabling various neuroscientific diffusion analyses, such as region-specific analysis, cortical surface-based analysis, and tractography [74]. Liu et al. also validated the reliability of CNN-super-resolved T1-weighted images in various neuroimaging tasks, including region-specific analysis, brain morphometry analysis, and structural covariance analysis [75]. It has also been demonstrated that CNN-based super-resolution for diffusion MRI reveals more detailed and accurate depiction of fiber distribution and orientation and enables more accurate tractography [39,76–78]. Moreover, deep learning-based super-resolution has demonstrated potential in improving the image quality of fetal brain MRI, a critical area of study where high-resolution imaging is often hindered by motion and limited acquisition time [79–81].

The benefits of CNN-based MRI super-resolution for clinical and neuroscientific applications have been comprehensively validated. The super-resolved images are generally preferred by radiologists and can provide more accurate neuroscientific analyses compared with the low-resolution images. However, for more recent advances such as transformers, diffusion models, and INR-based methods, comprehensive evaluations and assessments are still limited, highlighting a potential direction for future studies.

8.4 Practical considerations

8.4.1 Network architecture

The application of CNN in image super-resolution started with a simple 3-layer architecture (i.e., SRCNN, Fig. 8.8A) which already demonstrated improved performance compared with state-of-the-art traditional methods [16]. Each layer of SRCNN consists of a convolution module followed by a nonlinear activation module. For natural images, 2D convolution kernels are routinely adopted, which are extended to 3D convolution kernels for MRI to harness the redundancy along an additional dimension for performance improvement [19,23]. The kernel size dictates the network's receptive field, with larger kernels increasing the receptive field at the expense of more parameters. In practice, smaller kernels (e.g., $3 \times 3 \times 3$) are typically chosen. A batch normalization module can be added after the convolution operation for easier training and faster convergence [82]. ReLU [83] and Parametric ReLU (PReLU) [84] are widely employed as activation modules due to their robust performance.

It has been demonstrated that augmenting the number of layers and introducing residual connections significantly enhances the performance of super-resolution networks [17] (Fig. 8.8B). The increased number of layers improves the network's capacity for nonlinear mapping, while residual connections alleviate training difficulties by simplifying the problem to learning the difference between high-resolution target and low-resolution input. The success of VDSR has spurred exploration of network architectures for further performance enhancement. DenseNet, which extensively connects all layers with numerous shortcuts [23,37,85] (Fig. 8.8C) was proposed to alleviate the vanishing-gradient problem, strengthen feature propagation, encourage feature reuse, and reduce the number of parameters. Another widely used network for image super-resolution is U-Net [86] (Fig. 8.8D), originally proposed for biomedical image segmentations. U-Net's multilevel architecture, utilizing max pooling layers in the downward path and upsampling in the upward path, along with shortcuts connecting within each level, allows it to capture both global information in the lower levels and local details in the upper levels. It demonstrates robust performance in various super-resolution applications [27,73,74,87].

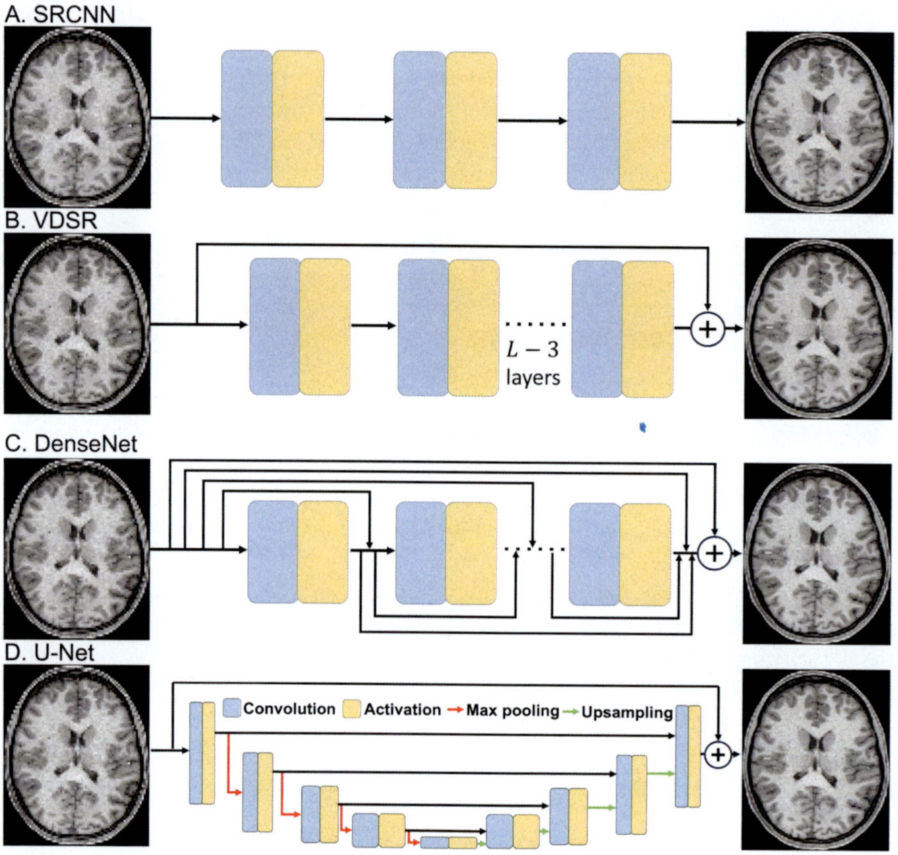

FIGURE 8.8 Network architecture.

The architectures of several widely used networks for image super-resolution including SRCNN (A), VDSR (B), DenseNet (C), and U-Net (D) are illustrated. L denotes the total number of layers for VDSR (L = 20 in Ref. [17]).

While task-specific modifications of network architectures can bring minor performance improvement [20,25,27,37,75,88], widely adopted representative networks such as VDSR and U-Net generally achieve robust and decent performance, serving as excellent starting points for most applications. When considering network architectures beyond basic CNNs, representative designs from previous studies also offer valuable references. For instance, for GANs, the discriminator in SRGAN [18] has proven useful in various applications [37,39,74]. Transformer architectures are more complex, integrating self-attention mechanisms, and often employ convolutional layers alongside transformer modules for efficient feature extraction. The original transformer [43] remains a solid reference for designing transformer modules. For diffusion models, U-Net is a popular choice for estimating the noise mean and variance at each time step [52–55]. Regarding INR, the encoder (CNN) and decoder (MLP) setup in Ref. [61] may provide insights for future designs.

8.4.2 Image evaluation metrics and loss functions

Another important aspect is the selection of the metrics for evaluating the similarity between super-resolved MRI and the target high-resolution image. Table 8.1 provides an overview of commonly employed evaluation metrics for image super-resolution. MSE, MAE, and peak SNR (PSNR) stand out as widely used metrics owing to their simplicity, clear physical interpretation, and simplicity of integration into mathematical optimization frameworks [89]. Despite their prevalence, these metrics, based purely on data consistency, are recognized as suboptimal indicators of image visual quality [18,89,90]. To address this limitation, structural similarity (SSIM) was introduced, aiming to better capture perceptual quality by considering

Table 8.1 Image evaluation metrics.

Evaluation metric	Equation
Mean squared error (MSE)	$MSE = \|\|x_{HR} - \widehat{x}_{SR}\|\|_2^2 = \frac{1}{N}\sum_{i=1}^{N}(x_{HRi} - \widehat{x}_{SRi})^2$
Mean absolute error (MAE)	$MAE = \|\|x_{HR} - \widehat{x}_{SR}\|\|_1 = \frac{1}{N}\sum_{i=1}^{N}\|x_{HRi} - \widehat{x}_{SRi}\|$
Peak signal-to-noise ratio (PSNR)	$PSNR = 20\log_{10}\left(\frac{MAX_I}{\sqrt{MSE}}\right)$
Structural similarity (SSIM)	$SSIM = \frac{(2\mu(x_{HR})\mu(\widehat{x}_{SR})+c_1)(2\sigma(x_{HR},\widehat{x}_{SR})+c_2)}{(\mu(x_{HR})^2+\mu(\widehat{x}_{SR})^2+c_1)(\sigma(x_{HR})^2+\sigma(\widehat{x}_{SR})^2+c_2)}$
VGG perceptual loss (VGG loss)	$VGG\ loss = \|\|VGG(x_{HR}) - VGG(\widehat{x}_{SR})\|\|^2$
Fréchet inception distance (FID)	$FID = \|\|\mu - \mu_w\|\|^2 + tr\left(\Sigma + \Sigma_w - 2(\Sigma\Sigma_w)^{\frac{1}{2}}\right)$

Commonly used image evaluation metrics include mean squared error (MSE), mean absolute error (MAE), peak signal-to-noise ratio (PSNR), structural similarity (SSIM), VGG perceptual loss (VGG loss), and Fréchet inception distance (FID). $||x_{HR} - \hat{x}_{SR}||_1$ *and* $||x_{HR} - \hat{x}_{SR}||_2^2$ *denote the L1 and L2 distance between* x_{HR} *and* $\hat{x}_{SR}$*, respectively. N is the total number of pixels inside the image.* MAX_I *is the maximum possible value of the image.* (μx_{HR}) *and* (σx_{HR}) *are the mean and variance of* x_{HR}*, respectively.* $(\sigma x_{HR}, \hat{x}_{SR})$ *denotes the covariance of* x_{HR} *and* $\hat{x}_{SR}$*.* $\left(c_1 = k_1 MAX_I^2\right)$ *and* $\left(c_2 = k_2 MAX_I^2\right)$ *are the two variables to stabilize the division with weak denominator, where* $k_1 = 0.01, k_2 = 0.03$ *by default.* $(VGGx_{HR})$ *is the feature vector calculated from pretrained VGG net with* x_{HR} *as input.* $(N\mu, \Sigma)$ *is the multivariate normal distribution estimated from Inception v3 features calculated on real life images and* $(N\mu_w, \Sigma_w)$ *is the multivariate normal distribution estimated from Inception v3 features calculated on generated images.*

luminance, contrast, and structural information within the images [89]. However, subsequent research has revealed a direct empirical and analytical link between SSIM and MSE-based metrics [91].

More recently, evaluation metrics derived from neural network-extracted features have been proposed, which attempt to better emulate human perceptions. The VGG perceptual loss, based on feature vectors from a VGG network pretrained on the ImageNet natural image dataset for object classification [92], is a notable example. Leveraging the extensive training samples in ImageNet, VGG perceptual loss is believed to capture visual features akin to those perceived by humans. Typically, feature vectors are extracted from deeper layers of the VGG network to encompass higher-level features [18]. Another metric, the Fréchet Inception Distance (FID), has been proposed for reference-free image quality assessment [93]. It quantifies the distance between feature vectors of real and network-generated images using the pretrained Inception v3 [94], another image recognition model. A lower FID score is indicative of higher visual quality.

Loss functions for network training are crafted based on image evaluation metrics, and by minimizing specific loss functions, the network can be optimized in terms of corresponding metrics. MSE-related loss functions are most popular choices thanks to the ease of their optimization and their ability to achieve optimized PSNR and data consistency, which however might introduce potential blurring to the generated images as discussed in Section 8.2.2. Recently, it is reported that minimizing MAE would lead to better image super-resolution performance compared with minimizing MSE [95]. SSIM-based loss functions have also been proposed, intending to produce images with improved SSIM [27,95,96]. Nevertheless, these images still exhibit visual similarities to those generated from MSE-related loss functions, given the direct relation between SSIM and MSE [91].

In the pursuit of better image visual quality, more advanced loss functions have been designed. GAN is a representative example which leverages adversarial loss to push synthesized images toward the manifold of target images, generating images with perceptually indistinguishable features such as contrast and textural details [18], as discussed in Section 8.2.1. In practice, a weighted summation of MSE-related loss and adversarial loss is often adopted to strike a balance between visual quality and data consistency (Eq. 8.5). Fig. 8.9 visually illustrates the impact of adversarial loss, showcasing improved visual quality at the expense of reduced data consistency, as quantified by PSNR and SSIM (Fig. 8.9, iii) [38]. Similarly, VGG-based loss has proven effective in enhancing image visual quality while compromising data consistency [18].

In the context of MRI super-resolution, the selection of evaluation metrics and loss functions requires careful consideration. Beyond the output images themselves, downstream analyses and diagnoses are arguably more critical. As discussed in Section 3, comprehensive evaluation metrics tailored for specific clinical and neuroscientific applications have been proposed. Reader assessments by radiologists have been identified as supportive for the clinical value of deep learning-based super-resolved MRI [22,27,67,69]. Furthermore, quantitative analyses of super-resolved images, such as

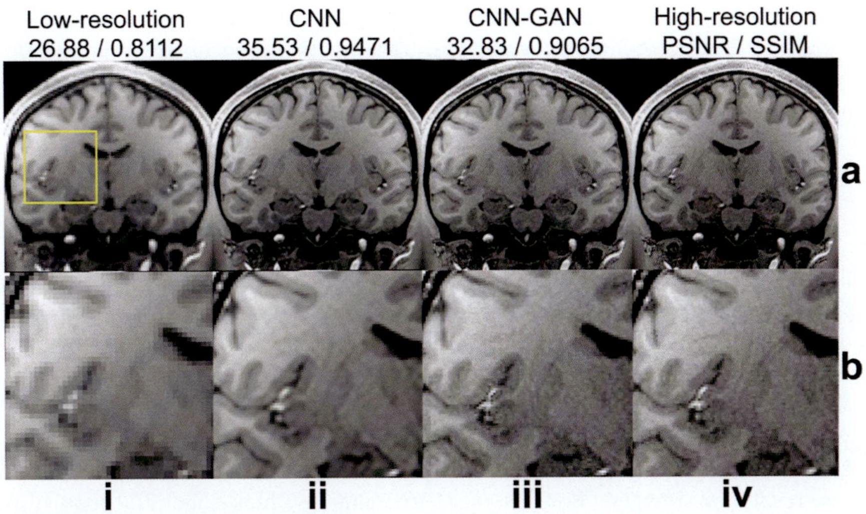

FIGURE 8.9 Effect of adversarial loss.

The nearest neighbor-interpolated low-resolution (i), CNN trained with MAE loss super-resolved (ii), CNN trained with MAE and adversarial loss super-resolved (iii), and target high-resolution images (iv) (a) and an enlarged region (b) are demonstrated. The peak signal-to-noise ratio (PSNR) and structural similarity (SSIM) between the low-resolution/super-resolved images and the target high-resolution images are listed.

The figure is reproduced from Fig. 1 in Ref. [38] with permission.

brain region segmentation [73,74] and cortical surface reconstruction [24,70], provide valuable evidence of their benefits for neuroscientific research. Task-specific evaluations, derived from clinical and neuroscientific needs, are deemed more compelling than relying solely on traditional metrics like MSE and PSNR.

These task-specific evaluations can also offer insights into the selection of appropriate loss functions. Notably, it has been demonstrated that despite GAN-synthesized images exhibiting more visually appealing textures (as shown in Fig. 8.9), their quantitative analyses (e.g., brain region segmentation and cortical surface reconstruction) may not show improvement compared with those from CNN trained without adversarial loss. This discrepancy may be attributed to compromised data consistency introduced by the inclusion of adversarial loss [74]. The inclusion of adversarial loss in GANs also demands more training data and complicates network fine-tuning and generalization. Nevertheless, radiologists have expressed a preference for GAN-synthesized images with sharper and more realistic textures for brain MRI denoising [97]. Hence, the benefits of GANs for MRI super-resolution need to be carefully evaluated with respect to specific tasks.

In addition, it is feasible to design tailored loss functions to generate images suitable for specific tasks. For instance, Iglesias et al. proposed the inclusion of a segmentation loss in the training of a super-resolution network for low-field brain MRI. This segmentation loss was calculated using a pretrained 3D U-Net for brain region segmentation of high-resolution brain images. The incorporation of this segmentation loss enabled the network to generate high-resolution brain images with excellent brain segmentation accuracy amenable to automated quantitative morphometry [98].

8.4.3 Training data

The supervised deep learning methods discussed earlier necessitate paired low-resolution and high-resolution images for effective training. While such data can be prospectively acquired using low- and high-resolution protocols, this approach faces challenges such as long scan time and motion-induced inconsistency between the two scans. In light of this, we discuss an alternative approach that obtains training data through the retrospective downsampling of native high-resolution images, along with an examination of potential datasets suitable for method development and validation.

Downsampling strategies should ideally emulate the forward acquisition model of low-resolution images. For instance, to simulate thick-slice 2D acquisition (Fig. 8.1b), image-space downsampling is implemented as a 1D finite impulse response low-pass filter, mimicking a realistic RF profile [22,69]. Alternatively, for simulating low in-plane resolution in 2D acquisition (Fig. 8.1a) or low isotropic resolution in 3D acquisition, k-space truncation can be utilized. This involves Fourier transforming the complex image data to k-space, cropping the central region, zero-padding, and then inversely Fourier transforming to obtain the interpolated low-resolution images [23,27,37,38]. In addition, low isotropic resolution images can also be obtained through image-space downsampling with anti-aliasing filtering and interpolation [24]. A recent advance in achieving more realistic downsampling leveraged GAN to synthesize low-resolution images with textures indistinguishable from other low-resolution images from native high-resolution inputs [99].

The selection of public datasets for MRI super-resolution depends on the chosen downsampling strategy. For image-space downsampling, public datasets with high-resolution magnitude images are useful, including Human Connectome Project (HCP) [100], UK Biobank [101,102], Alzheimer's Disease Neuroimaging Initiative (ADNI) [103], Massachusetts General Hospital Connectome Diffusion Microstructure Dataset (MGH CDMD) [104], etc. In contrast, k-space downsampling requires complex image data with both magnitude and phase information, which are typically found in datasets designed for MRI reconstruction, such as fastMRI [105].

8.5 Challenges and future directions

Deep learning-based MRI super-resolution has demonstrated promising performance potentially valuable for a variety of clinical and neuroscientific applications, but faces challenges in terms of the practicality, feasibility, and reliability. These challenges limit its broader adoption while offering potential directions for future research.

First of all, the practicality and feasibility of deep learning-based super-resolution are limited by the requirement of training data. The training of supervised learning methods necessitates paired low- and high-resolution images. Pham et al. used data from 21 subjects to train a 3D SRCNN with three layers and small number of parameters [19]. For a deeper network with more parameters such as VDSR, data from more subjects are usually involved in training (e.g., 64 training subjects in Ref. [24]). The inclusion of adversarial loss for GAN training brings larger training data requirement. Chen et al. incorporated over 1000 subjects for training a 3D GAN [37] presumably due to the difficulty for training a 3D discriminator which performs classification in a volume-by-volume manner. The training data requirement for GAN can be potentially reduced using a hybrid architecture with 3D generator and 2D discriminator, as demonstrated by Li et al. for 3D brain MRI denoising [97]. Data requirements for transformers and diffusion models are also significant (300 subjects for training a 2D transformer in Ref. [49] and 900 subjects for training a 2D diffusion model in Ref. [54]). While leveraging public datasets and retrospective downsampling, as discussed in Section 8.4.3, might address some of these issues, leveraging deep learning super-resolution models for tailored applications presents a significant challenge. Even for a relatively modest set of training data (e.g., 20 subjects), the associated time and financial costs for data acquisition are substantial.

Transfer learning is a promising strategy to reduce the training data requirement. It is demonstrated that CNNs such as VDSR and U-Net is highly generalizable for MRI super-resolution [24,74]. Tian et al. presented compelling evidence by applying a pretrained VDSR, originally trained on 7T data, directly to 3T data, yielding satisfying results (Fig. 8.10, iii). Fine-tuning this pretrained model with just one additional subject further enhanced the image sharpness (Fig. 8.10, iv) [24]. Similarly, Steeden et al. successfully employed a 3D U-Net pretrained on retrospectively under-sampled whole-heart MRI for prospectively acquired low-resolution data, resulting in high-quality images comparable to those acquired with prospective high-resolution acquisition [106]. The decent generalizability suggests a promising approach—pretraining a super-resolution model on large-scale public datasets from HCP and UK Biobank and fine-tuning the pretrained model using a relatively small dataset tailored to specific applications. This approach aligns with the concept of image quality transfer [107], postulating that rich information from high-quality images (e.g., high-resolution images from public datasets) can be effectively transferred to low-quality data (e.g., low-resolution images acquired for specific applications) using machine learning techniques. To date, the investigation of the transfer learning for more advanced deep learning-based super-resolution techniques beyond standard CNNs (e.g., GANs, transformers, diffusion models) is still limited, presenting a compelling direction for future exploration.

Self-supervised learning methods also offer a data-efficient solution, eliminating the need for external high-resolution training data. For instance, Zhao et al. introduced SMORE, a self-supervised approach for anisotropic resolution MRI with high in-plane but low through-plane resolution [108]. SMORE simulates low-

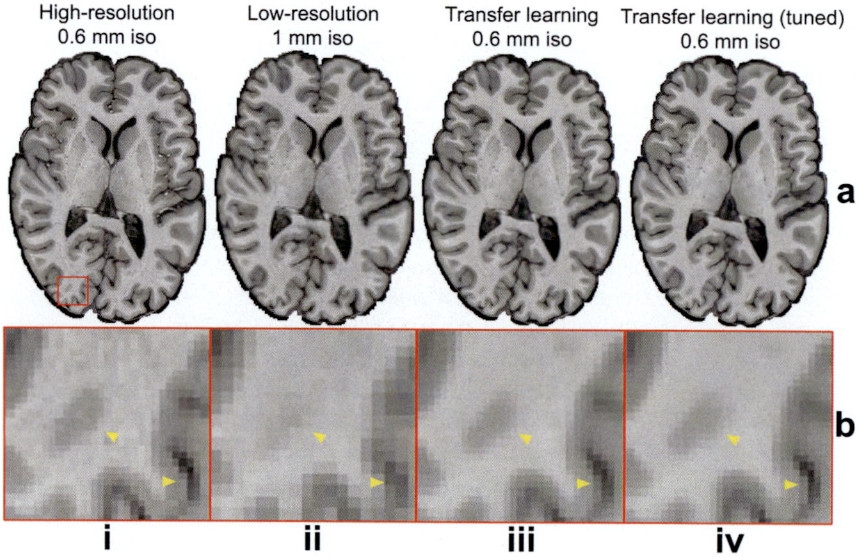

FIGURE 8.10 Network generalization.

A representative axial slice (a) and an enlarged region (b) from native high-resolution (i), standard-resolution (b), and super-resolved (iii, iv) T1-weighted images at 3T are demonstrated by Tian et al. [24]. The super-resolved images are obtained by directly applying a VDSR pretrained using 7T data (iii) and fine-tuning the pretrained network using one subject of the 3T data (iv).

resolution images from high-resolution in-plane slices and trains a model to super-resolve these simulated low-resolution images, which is applied to the low-resolution through-plane slices to improve their resolutions. This approach was extended to fetal MRI [80] and holds promise especially for modalities where 2D acquisitions are routinely adopted such as diffusion MRI and functional MRI.

INR introduced in Section 8.2.5 stands out as another low-data-demanding MRI super-resolution technique. By obtaining continuous image representations, INR enables mapping from continuous image coordinates to corresponding signals [60,61]. Its coordinate-based design treats each pixel/coordinate as a training sample, coupled with straightforward encoder and decoder network architectures, making it trainable with limited data. Wu et al. successfully trained an INR for thick-slice brain MRI super-resolution with fewer than 10 training subjects, yielding decent image quality [64]. The potential of self-supervised INR for anisotropic MRI super-resolution has also been demonstrated, which constructs low- and high-resolution pairs from anisotropic MRI data, efficiently training the INR model [109]. The promising capabilities of INR to achieve high-quality super-resolution with a manageable amount of training data merit exploration in more neuroscientific and clinical applications.

Second, the reliability of deep learning-based methods is another hurdle for their wider adoptions. The inherent black-box nature of these algorithms raises concerns about the interpretability and trustworthiness of their outputs. Studies indicate that models trained with distribution matching losses (e.g., GANs) carry the risk of hallucinating features or even removing structures from the output images [110]. Given the super-resolution problem is intrinsically ill-posed, it is crucial to exercise caution to prevent such hallucinations, which could introduce ambiguity in scientific research conclusions and complicate clinical diagnostic, prognostic, and interventional decisions.

To enhance reliability, large-scale reader assessments of deep learning super-resolved MRI from clinicians are instrumental. To date, such assessments have been conducted for basic CNN-based super-resolutions [22,27,67,69], supporting their use in clinical settings. The clinical adoption of more advanced super-resolution models, especially for those trained with distribution matching losses such as GANs and diffusion models, require more comprehensive evaluations from radiologists.

Efforts to introduce uncertainty quantifications have been made to bolster the robustness, safety, and interpretability of deep learning-based super-resolution. For example, Tanno et al. modeled the uncertainty in terms of intrinsic and parameter uncertainty [76,77]. Intrinsic uncertainty reflects the inherent ambiguity linked to the ill-posed nature of the super-resolution problem, quantified through the variance of the target conditional distribution estimated from a separate network. Parameter uncertainty accounts for ambiguity in selecting model parameters, reduced using variational dropouts [111]. This approach captures various settings of network parameters for the given training data, providing a more robust result and an uncertainty estimation represented by the standard deviation. The efficacy of dropout layers for estimating uncertainty and improving robustness has been demonstrated in various applications, such as MRI reconstruction [112] and quantitative MRI [113]. A more thorough evaluation of the uncertainty estimation framework on advanced super-resolution models, along with a comprehensive assessment of generated uncertainty maps, especially in the context of patient data, may offer a promising avenue to further enhance the reliability and robustness of deep learning-based MRI super-resolution.

Finally, the ill-posed nature of super-resolution introduces the possibility of artifacts, hallucinations, and unrecoverable details during image translation [110]. This issue becomes particularly pronounced when discrepancies arise between the distributions of training and testing data, such as when applying models trained on healthy subjects to patient data. Currently, the demonstration and evaluation of such artifacts and their impact on downstream analysis remain limited. A thorough analysis of these artifacts could yield valuable insights into the potential risks associated with super-resolution and potential strategies for their mitigation. Furthermore, the partial volume effect inherent in low-resolution images presents an additional challenge in recovering fine details, even with advanced super-resolution methods. For instance, attempting to recover lesions smaller than 2 mm from a 2 mm resolution image would be nearly impossible. In such cases, an alternative approach may involve denoising high-resolution, low-SNR images [97,114] rather than attempting to super-resolve low-resolution, high-SNR images. Both image denoising and super-resolution represent promising strategies for enabling accelerated MRI without significantly compromising image quality. However, further evaluations and discussions are warranted to determine their optimal use in various scenarios.

8.6 Conclusions

In conclusion, MRI super-resolution driven by deep learning techniques from CNNs to advanced models such as GANs, transformers, diffusion models, and INR models holds great promise in revolutionizing clinical and neuroscientific applications. These techniques respond to challenges posed by limited MRI resolution, enhancing image quality, analysis, and diagnostic accuracy in research and clinical settings. Practically, classical architectures such as VDSR and U-Net prove robust. The use of nondata-consistency loss such as adversarial loss requires careful considerations especially for quantitative imaging tasks. Task-specific customization of loss functions and metrics is vital, extending assessment beyond traditional image metrics to downstream analyses. Despite these advances, the feasibility and reliability of deep learning-based super-resolution MRI are confronted with several hurdles before it can be truly adopted broadly. Enhancing feasibility calls for innovative solutions like transfer learning and self-supervised learning to overcome limited training data. Ensuring reliability of AI-synthesized super-resolved MR images demands extensive assessments, robust uncertainty quantification, and comprehensive artifact analysis. While deep learning-based MRI super-resolution is rapidly evolving, this chapter provides a concise framework for navigating the current state of the field and identifies key avenues for future endeavors. Ongoing collaboration between researchers, clinicians, and neuroscientists is essential to unlock the full potential of AI-driven advancements in medical imaging.

Acknowledgment

The authors would like to thank Dr. Fredrick Lange, Minhao Hu, and Haoruo Cheng for their thoughts on specific techniques.

References

[1] Milanfar P. Super-resolution imaging. CRC press; 2017.
[2] Freeman WT, Jones TR, Pasztor EC. Example-based super-resolution. IEEE Comput Graphics Appl 2002;22(2):56–65.
[3] Glasner D, Bagon S, Irani M. Super-resolution from a single image. In: 2009 IEEE 12th International Conference on Computer Vision. IEEE; 2009.

[4] Rousseau F. Brain hallucination. In: Computer Vision–ECCV 2008: 10th European Conference on Computer Vision, Marseille, France, October 12-18, 2008, Proceedings, Part I 10. Springer; 2008.
[5] Manjón JV, et al. Non-local MRI upsampling. Med Image Anal 2010;14(6):784–92.
[6] Manjón JV, et al. MRI superresolution using self-similarity and image priors. J Biomed Imaging 2010;2010:1–11.
[7] Wang Y-H, et al. Sparse representation-based MRI super-resolution reconstruction. Measurement 2014;47:946–53.
[8] Shi F, et al. LRTV: MR image super-resolution with low-rank and total variation regularizations. IEEE Trans Med Imag 2015;34(12):2459–66.
[9] Greenspan H, et al. MRI inter-slice reconstruction using super-resolution. Magn Reson Imaging 2002;20(5):437–46.
[10] Vis G, et al. Accuracy and precision in super-resolution MRI: enabling spherical tensor diffusion encoding at ultra-high b-values and high resolution. Neuroimage 2021;245: 118673.
[11] Gholipour A, et al. Maximum a posteriori estimation of isotropic high-resolution volumetric MRI from orthogonal thick-slice scans. In: Medical Image Computing and Computer-Assisted Intervention–MICCAI 2010: 13th International Conference, Beijing, China, September 20-24, 2010, Proceedings, Part II 13. Springer; 2010.
[12] Setsompop K, et al. High-resolution in vivo diffusion imaging of the human brain with generalized slice dithered enhanced resolution: Simultaneous multislice (g S lider-SMS). Magn Reson Med 2018;79(1):141–51.
[13] Liao C, et al. Distortion-free, high-isotropic-resolution diffusion MRI with gSlider BUDA-EPI and multicoil dynamic B0 shimming. Magn Reson Med 2021;86(2): 791–803.
[14] Liao C, et al. High-fidelity mesoscale in-vivo diffusion MRI through gSlider-BUDA and circular EPI with S-LORAKS reconstruction. Neuroimage 2023;275:120168.
[15] Van Reeth E, et al. Super-resolution in magnetic resonance imaging: a review. Concepts Magn Reson 2012;40(6):306–25.
[16] Dong C, et al. Image super-resolution using deep convolutional networks. IEEE Trans Pattern Anal Mach Intell 2015;38(2):295–307.
[17] Kim J, Lee JK, Lee KM. Accurate image super-resolution using very deep convolutional networks. In: Proceedings of the IEEE Conference on Computer Vision and Pattern Recognition; 2016.
[18] Ledig C, et al. Photo-realistic single image super-resolution using a generative adversarial network. In: Proceedings of the IEEE Conference on Computer Vision and Pattern Recognition; 2017.
[19] Pham C-H, et al. Brain MRI super-resolution using deep 3D convolutional networks. In: 2017 IEEE 14th International Symposium on Biomedical Imaging (ISBI 2017). IEEE; 2017.
[20] Qiu D, et al. Super-resolution reconstruction of knee magnetic resonance imaging based on deep learning. Comput Methods Progr Biomed 2020;187:105059.
[21] He K, et al. Deep residual learning for image recognition. In: Proceedings of the IEEE Conference on Computer Vision and Pattern Recognition; 2016.
[22] Chaudhari AS, et al. Super-resolution musculoskeletal MRI using deep learning. Magn Reson Med 2018;80(5):2139–54.

[23] Chen Y, et al. Brain MRI super resolution using 3D deep densely connected neural networks. In: 2018 IEEE 15th International Symposium on Biomedical Imaging (ISBI 2018). IEEE; 2018.
[24] Tian Q, et al. Improving in vivo human cerebral cortical surface reconstruction using data-driven super-resolution. Cerebr Cortex 2021;31(1):463–82.
[25] Zeng K, et al. Simultaneous single-and multi-contrast super-resolution for brain MRI images based on a convolutional neural network. Comput Biol Med 2018;99:133–41.
[26] Lyu Q, et al. Multi-contrast super-resolution MRI through a progressive network. IEEE Trans Med Imag 2020;39(9):2738–49.
[27] Masutani EM, Bahrami N, Hsiao A. Deep learning single-frame and multiframe super-resolution for cardiac MRI. Radiology 2020;295(3):552–61.
[28] Basty N, Grau V. Super resolution of cardiac cine MRI sequences using deep learning. In: Image Analysis for Moving Organ, Breast, and Thoracic Images: Third International Workshop, RAMBO 2018, Fourth International Workshop, BIA 2018, and First International Workshop, TIA 2018, Held in Conjunction with MICCAI 2018, Granada, Spain, September 16 and 20, 2018, Proceedings 3. Springer; 2018.
[29] Oktay O, et al. Multi-input cardiac image super-resolution using convolutional neural networks. In: Medical Image Computing and Computer-Assisted Intervention-MICCAI 2016: 19th International Conference, Athens, Greece, October 17-21, 2016, Proceedings, Part III 19. Springer; 2016.
[30] Lin J-Y, Chang Y-C, Hsu WH. Efficient and phase-aware video super-resolution for cardiac MRI. Cham: Springer International Publishing; 2020.
[31] Song T-A, et al. Super-resolution PET imaging using convolutional neural networks. IEEE Transact Comput Imaging 2020;6:518–28.
[32] Goodfellow I, et al. Generative adversarial networks. Commun ACM 2020;63(11): 139–44.
[33] Miyato T, et al. Spectral normalization for generative adversarial networks. arXiv preprint arXiv:1802.05957 2018.
[34] Mao X, et al. Least squares generative adversarial networks. In: Proceedings of the IEEE International Conference on Computer Vision; 2017.
[35] Arjovsky M, Chintala S, Bottou L. Wasserstein generative adversarial networks. In: International Conference on Machine Learning. PMLR; 2017.
[36] Gulrajani I, et al. Improved training of wasserstein gans. Adv Neural Inf Process Syst 2017;30.
[37] Chen Y, et al. Efficient and accurate MRI super-resolution using a generative adversarial network and 3D multi-level densely connected network. In: International Conference on Medical Image Computing and Computer-Assisted Intervention. Springer; 2018.
[38] Chen Y, et al. MRI super-resolution with GAN and 3D multi-level DenseNet: smaller, faster, and better. arXiv preprint arXiv:2003.01217 2020.
[39] Ye X, et al. Simultaneous superresolution reconstruction and distortion correction for single-shot EPI DWI using deep learning. Magn Reson Med 2023;89(6):2456–70.
[40] Zhao M, et al. Super-resolution of cardiac magnetic resonance images using Laplacian Pyramid based on Generative Adversarial Networks. Comput Med Imag Graph 2020; 80:101698.
[41] Küstner T, et al. Deep-learning based super-resolution for 3D isotropic coronary MR angiography in less than a minute. Magn Reson Med 2021;86(5):2837–52.

[42] Lyu Q, Shan H, Wang G. MRI super-resolution with ensemble learning and complementary priors. IEEE Transact Comput Imaging 2020;6:615–24.

[43] Vaswani A, et al. Attention is all you need. Adv Neural Inf Process Syst 2017;30.

[44] Han K, et al. A survey on vision transformer. IEEE Trans Pattern Anal Mach Intell 2022;45(1):87–110.

[45] Dosovitskiy A, et al. An image is worth 16x16 words: transformers for image recognition at scale. arXiv preprint arXiv:2010.11929 2020.

[46] Yang F, et al. Learning texture transformer network for image super-resolution. In: Proceedings of the IEEE/CVF Conference on Computer Vision and Pattern Recognition; 2020.

[47] Feng C-M, et al. Multi-modal transformer for accelerated mr imaging. IEEE Trans Med Imag 2022.

[48] Fang C, et al. Cross-modality high-frequency transformer for mr image super-resolution. In: Proceedings of the 30th ACM International Conference on Multimedia; 2022.

[49] Li G, et al. Transformer-empowered multi-scale contextual matching and aggregation for multi-contrast MRI super-resolution. In: Proceedings of the IEEE/CVF Conference on Computer Vision and Pattern Recognition; 2022.

[50] Sohl-Dickstein J, et al. Deep unsupervised learning using nonequilibrium thermodynamics. In: International Conference on Machine Learning. PMLR; 2015.

[51] Ho J, Jain A, Abbeel P. Denoising diffusion probabilistic models. Adv Neural Inf Process Syst 2020;33:6840–51.

[52] Saharia C, et al. Image super-resolution via iterative refinement. IEEE Trans Pattern Anal Mach Intell 2022;45(4):4713–26.

[53] Li H, et al. Srdiff: single image super-resolution with diffusion probabilistic models. Neurocomputing 2022;479:47–59.

[54] Wu Z, et al. Super-resolution of brain MRI images based on denoising diffusion probabilistic model. Biomed Signal Process Control 2023;85:104901.

[55] Wang J, et al. InverseSR: 3D brain MRI super-resolution using a latent diffusion model. In: International Conference on Medical Image Computing and Computer-Assisted Intervention. Springer; 2023.

[56] Song J, Meng C, Ermon S. Denoising diffusion implicit models. arXiv preprint arXiv: 2010.02502 2020.

[57] Pinaya WH, et al. Brain imaging generation with latent diffusion models. In: MICCAI Workshop on Deep Generative Models. Springer; 2022.

[58] Chung H, Lee ES, Ye JC. MR image denoising and super-resolution using regularized reverse diffusion. IEEE Trans Med Imag 2022;42(4):922–34.

[59] Song Y, et al. Score-based generative modeling through stochastic differential equations. arXiv preprint arXiv:2011.13456 2020.

[60] Sitzmann V, et al. Implicit neural representations with periodic activation functions. Adv Neural Inf Process Syst 2020;33:7462–73.

[61] Chen Y, Liu S, Wang X. Learning continuous image representation with local implicit image function. In: Proceedings of the IEEE/CVF Conference on Computer Vision and Pattern Recognition; 2021.

[62] Peng S, et al. Convolutional occupancy networks. In: Computer Vision–ECCV 2020: 16th European Conference, Glasgow, UK, August 23–28, 2020, Proceedings, Part III 16. Springer; 2020.

[63] Wu Q, et al. An arbitrary scale super-resolution approach for 3d mr images via implicit neural representation. IEEE J Biomed Health Informatics 2022;27(2):1004–15.

[64] Wu Q, et al. IREM: high-resolution magnetic resonance image reconstruction via implicit neural representation. In: Medical Image Computing and Computer Assisted Intervention–MICCAI 2021: 24th International Conference, Strasbourg, France, September 27–October 1, 2021, Proceedings, Part VI 24. Springer; 2021.
[65] Li J, et al. Direct localization and delineation of human pedunculopontine nucleus based on a self-supervised magnetic resonance image super-resolution method. Hum Brain Mapp 2023.
[66] Xu J, et al. NeSVoR: implicit neural representation for slice-to-volume reconstruction in MRI. IEEE Trans Med Imag 2023.
[67] Rudie JD, et al. Clinical assessment of deep learning–based super-resolution for 3D volumetric brain MRI. Radiol Artif Intell 2022;4(2):e210059.
[68] Ishida M, et al. Learning-based super-resolution technique significantly improves detection of coronary artery stenoses on 1.5 T whole-heart coronary MRA. J Cardiovasc Magn Reson 2014;16:1–2.
[69] Chaudhari AS, et al. Utility of deep learning super-resolution in the context of osteoarthritis MRI biomarkers. J Magn Reson Imag 2020;51(3):768–79.
[70] Dale AM, Fischl B, Sereno MI. Cortical surface-based analysis: I. Segmentation and surface reconstruction. Neuroimage 1999;9(2):179–94.
[71] Fischl B, Sereno MI, Dale AM. Cortical surface-based analysis: II: inflation, flattening, and a surface-based coordinate system. Neuroimage 1999;9(2):195–207.
[72] Fischl B. FreeSurfer. Neuroimage 2012;62(2):774–81.
[73] Iglesias JE, et al. Joint super-resolution and synthesis of 1 mm isotropic MP-RAGE volumes from clinical MRI exams with scans of different orientation, resolution and contrast. Neuroimage 2021;237:118206.
[74] Li Z, et al. Diffusion MRI data analysis assisted by deep learning synthesized anatomical images (DeepAnat). Med Image Anal 2023;86:102744.
[75] Liu G, et al. Recycling diagnostic MRI for empowering brain morphometric research--critical & practical assessment on learning-based image super-resolution. Neuroimage 2021;245:118687.
[76] Tanno R, et al. Bayesian image quality transfer with CNNs: exploring uncertainty in dMRI super-resolution. In: Medical Image Computing and Computer Assisted Intervention– MICCAI 2017: 20th International Conference, Quebec City, QC, Canada, September 11-13, 2017, Proceedings, Part I 20. Springer; 2017.
[77] Tanno R, et al. Uncertainty modelling in deep learning for safer neuroimage enhancement: demonstration in diffusion MRI. Neuroimage 2021;225:117366.
[78] Tian Q, et al. SRDTI: deep learning-based super-resolution for diffusion tensor MRI. arXiv preprint arXiv:2102.09069 2021.
[79] McDonagh S, et al. Context-sensitive super-resolution for fast fetal magnetic resonance imaging. In: Molecular Imaging, Reconstruction and Analysis of Moving Body Organs, and Stroke Imaging and Treatment: Fifth International Workshop, CMMI 2017, Second International Workshop, RAMBO 2017, and First International Workshop, SWITCH 2017, Held in Conjunction with MICCAI 2017, Québec City, QC, Canada, September 14, 2017, Proceedings 5. Springer; 2017.
[80] Xu J, et al. STRESS: super-resolution for dynamic fetal MRI using self-supervised learning. In: Medical Image Computing and Computer Assisted Intervention–MICCAI 2021: 24th International Conference, Strasbourg, France, September 27–October 1, 2021, Proceedings, Part VII 24. Springer; 2021.

[81] Song L, et al. Deep robust residual network for super-resolution of 2D fetal brain MRI. Sci Rep 2022;12(1):406.
[82] Ioffe S, Szegedy C. Batch normalization: accelerating deep network training by reducing internal covariate shift. In: International Conference on Machine Learning. pmlr; 2015.
[83] Nair V, Hinton GE. Rectified linear units improve restricted boltzmann machines. In: Proceedings of the 27th International Conference on Machine Learning (ICML-10); 2010.
[84] He K, et al. Delving deep into rectifiers: surpassing human-level performance on imagenet classification. In: Proceedings of the IEEE International Conference on Computer Vision; 2015.
[85] Huang G, et al. Densely connected convolutional networks. In: Proceedings of the IEEE Conference on Computer Vision and Pattern Recognition; 2017.
[86] Ronneberger O, Fischer P, Brox T. U-net: convolutional networks for biomedical image segmentation. In: International Conference on Medical Image Computing and Computer-assisted Intervention. Springer; 2015.
[87] Li Z, et al. Deepvolume: brain structure and spatial connection-aware network for brain MRI super-resolution. IEEE Trans Cybern 2019;51(7):3441–54.
[88] Wang J, et al. Enhanced generative adversarial network for 3D brain MRI super-resolution. In: Proceedings of the IEEE/CVF Winter Conference on Applications of Computer Vision; 2020.
[89] Wang Z, et al. Image quality assessment: from error visibility to structural similarity. IEEE Trans Image Process 2004;13(4):600–12.
[90] Wang Z, Bovik AC, Lu L. Why is image quality assessment so difficult?. In: 2002 IEEE International Conference on Acoustics, Speech, and Signal Processing. IEEE; 2002.
[91] Dosselmann R, Yang XD. A comprehensive assessment of the structural similarity index. Signal Image Video Process 2011;5:81–91.
[92] Simonyan K, Zisserman A. Very deep convolutional networks for large-scale image recognition. arXiv preprint arXiv:1409.1556 2014.
[93] Heusel M, et al. Gans trained by a two time-scale update rule converge to a local nash equilibrium. Adv Neural Inf Process Syst 2017;30.
[94] Szegedy C, et al. Rethinking the inception architecture for computer vision. In: Proceedings of the IEEE Conference on Computer Vision and Pattern Recognition; 2016.
[95] Zhao H, et al. Loss functions for image restoration with neural networks. IEEE Transact Comput Imaging 2016;3(1):47–57.
[96] Snell J, et al. Learning to generate images with perceptual similarity metrics. In: 2017 IEEE International Conference on Image Processing (ICIP). IEEE; 2017.
[97] Li Z, et al. High-fidelity fast volumetric brain MRI using synergistic wave-controlled aliasing in parallel imaging and a hybrid denoising generative adversarial network (HDnGAN). Med Phys 2022;49(2):1000–14.
[98] Iglesias JE, et al. Quantitative brain morphometry of portable low-field-strength MRI using super-resolution machine learning. Radiology 2022;306(3):e220522.
[99] Huang B, et al. MRI super-resolution via realistic downsampling with adversarial learning. Phys Med Biol 2021;66(20):205004.
[100] Van Essen DC, et al. The WU-Minn human connectome project: an overview. Neuroimage 2013;80:62–79.

[101] Miller KL, et al. Multimodal population brain imaging in the UK Biobank prospective epidemiological study. Nat Neurosci 2016;19(11):1523–36.
[102] Sudlow C, et al. UK biobank: an open access resource for identifying the causes of a wide range of complex diseases of middle and old age. PLoS Med 2015;12(3): e1001779.
[103] Mueller SG, et al. The Alzheimer's disease neuroimaging initiative. Neuroimaging Clinics 2005;15(4):869–77.
[104] Tian Q, et al. Comprehensive diffusion MRI dataset for in vivo human brain microstructure mapping using 300 mT/m gradients. Sci Data 2022;9(1):7.
[105] Zbontar J, et al. fastMRI: an open dataset and benchmarks for accelerated MRI. arXiv preprint arXiv:1811.08839 2018.
[106] Steeden JA, et al. Rapid whole-heart CMR with single volume super-resolution. J Cardiovasc Magn Reson 2020;22(1):56.
[107] Alexander DC, et al. Image quality transfer and applications in diffusion MRI. Neuroimage 2017;152:283–98.
[108] Zhao C, et al. SMORE: a self-supervised anti-aliasing and super-resolution algorithm for MRI using deep learning. IEEE Trans Med Imag 2020;40(3):805–17.
[109] Zhang H, et al. Self-supervised arbitrary scale super-resolution framework for anisotropic MRI. arXiv preprint arXiv:2305.01360 2023.
[110] Cohen JP, Luck M, Honari S. Distribution matching losses can hallucinate features in medical image translation. In: Medical Image Computing and Computer Assisted Intervention–MICCAI 2018: 21st International Conference, Granada, Spain, September 16-20, 2018, Proceedings, Part I. Springer; 2018.
[111] Kingma DP, Salimans T, Welling M. Variational dropout and the local reparameterization trick. Adv Neural Inf Process Syst 2015;28.
[112] Schlemper J, et al. Bayesian deep learning for accelerated MR image reconstruction. In: Machine Learning for Medical Image Reconstruction: First International Workshop, MLMIR 2018, Held in Conjunction with MICCAI 2018, Granada, Spain, September 16, 2018, Proceedings 1. Springer; 2018.
[113] Avci MY, et al. Quantifying the uncertainty of neural networks using Monte Carlo dropout for deep learning based quantitative MRI. arXiv preprint arXiv:2112.01587 2021.
[114] Tian Q, et al. Improved cortical surface reconstruction using sub-millimeter resolution MPRAGE by image denoising. Neuroimage 2021;233:117946.

CHAPTER

Freeze the motion: Machine learning for motion correction

9

Haikun Qi[1,*] and Chen Qin[2]

[1]*School of Biomedical Engineering, ShanghaiTech University, Shanghai, China;* [2]*I-X and Department of Electrical and Electronic Engineering, Imperial College London, London, United Kingdom*

[*]*Corresponding author: E-mail: qihk@shanghaitech.edu.cn*

9.1 Short introduction to motion and motion correction

As introduced in Chapter 4, the spatial encoding in MRI to fill up k-space data can be a time-consuming process, during which the movement of the imaging subject can cause severe motion artifacts and result in nondiagnostic images. Therefore, besides acceleration, motion correction is another hot research topic in MRI. The motion occurring in in vivo MR generally falls into two categories: rigid motion and deformable nonrigid motion. Rigid motion can be fully characterized by three translations and three rotations. It could occur in all parts of the body, especially in the brain. Rigid brain motion can either be voluntary in uncooperative subjects such as children, or be involuntary in subjects with neurodegenerative disorders, such as Parkinson's. Nonrigid motion is typically observed in the thoracic and abdominal parts caused by physiological factors, such as respiratory motion, cardiac motion, and bowel movement. Nonrigid motion is more complex than rigid motion, as different locations experience different displacements. The physiological types of motion are unavoidable and quasi-periodic, with typical period of 5s and 1s for respiratory motion and cardiac motion, respectively. Example MR images of brain with rigid motion and heart with respiratory motion artifacts are shown in Fig. 9.1. A detailed introduction of motion artifacts and the state-of-the-art in motion detection and correction in MRI can be found in the book by Van Der Kouwe et al. [1] (Fig. 9.1).

Breath-holding is an effective way of suppressing respiratory motion, which however has limited duration of 10–20s and is commonly used for fast 2D imaging. In addition, triggering or gating of the motion cycle to time the MR data acquisition to certain respiratory and cardiac motion state is another way of mitigating the influence of physiological motions [2,3]. Triggering or gating requires the measurement of the physiological motion signal. Electrocardiogram electrodes and pulse oximeters can be used to monitor cardiac motion. Respiratory bellows affixed to

Machine Learning in MRI. https://doi.org/10.1016/B978-0-443-14109-6.00009-2

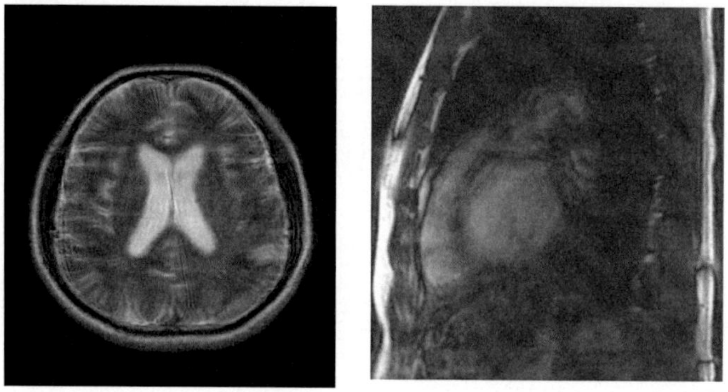

(a) (b)

FIGURE 9.1

Example motion artifacts: (a) rigid brain motion; (b) respiratory motion in cardiac cine MRI.

the subject's belly can reflect respiratory motion, which is not commonly used in practice. Instead, the diaphragm navigator that estimates the respiration-induced diaphragm motion [4] without additional devices is a more reliable and easier-to-use method for respiratory triggering or gating. Recently, the Pilot Tone [5,6] has been proposed as a contactless technique to monitor respiratory and cardiac motion.

Different from the above-mentioned motion prevention which attempts to minimize the cooccurrence of motion during MR imaging, motion correction aims to estimate motion and correct it in a prospective or a retrospective way. Prospective motion correction eliminates the influence of motion by tracking the moving subject in real time during the MR scan and then adjust the radio-frequency pulse central frequency and imaging gradients according to the estimated motion [7] with external sensors or navigators [8,9]. Navigators are predominantly used for tracking respiratory motion in cardiac imaging, including 1D diaphragm navigator, 2D and 3D image navigators [10–12] as well as self-navigator [13–16] which extracts motion from some repeatedly measured imaging data. However, prospective methods can only compensate global motion such as translations and rotations as current MRI systems adopt linear gradients for spatial encoding. More complex motion with local deformations can be corrected with retrospective methods, which invert the effect of motion during the reconstruction stage after data acquisition. The motion-compensated (MC) image reconstruction can either be performed with preestimated motion parameters or motion fields, or be optimized jointly for motion estimation and reconstruction. MC reconstruction typically requires time-consuming iterative optimizations, especially the MC reconstruction methods optimizing the large number of unknown motion parameters and the underlying clean image simultaneously which are computationally burdensome. Furthermore, conventional methods may not accurately estimate the complex and unpredictable motion, leading to poor MC reconstruction performance.

Deep learning (DL) has achieved great success in various medical imaging tasks including MR motion correction, where DL methods have been extensively investigated to address the challenges facing conventional motion correction methods. For example, a DL neural network can be trained to learn the complex relationship between motion corrupted and motion clean images and then used to remove motion artifacts. Besides motion artifacts reduction, DL methods can also be used to estimate motion to assist MC reconstruction. Comprehensive reviews about DL MR motion correction can be found in Refs. [17–19].

The motion artifacts depend on the motion types and the imaging sequences. The motion artifact manifestations such as blurring, ghosting, and distortion in MR images will be introduced in Section 9.2. A common way of estimating motion parameters or motion fields is image registration. Pair-wise registration estimating motion between two images and group-wise registration involving measuring motion for a group of images are elucidated in Section 9.3. Deep learning-based MR motion correction methods, including the motion artifacts reduction and registration networks, are introduced in Section 9.4. For motion-compensated image reconstruction, the general problem formulation, conventional methods, and DL methods for rigid motion and nonrigid motion are described in Section 9.5. Finally, a brief summary of this chapter and existing challenges and future directions is provided in Section 9.6.

9.2 Motion artifact manifestation in the image

Motion represents a formidable challenge in MRI due to the relatively long acquisition times. It can occur at any time point during the acquisition process, such as between acquisition of volumes (inter-image), within a volume between slices (inter-scan) or between excitation and signal acquisition (intra-scan) [22]. Therefore, the motion not only translates or rotates the MRI slices, but it also corrupts the signals. The exact manifestation of motion-induced artifacts in MRI is in fact a complex problem and is related to how inconsistent raw data acquired in different motion states are combined into k-space before being transformed to image space [21]. They typically result from complicated interactions between image structures, types of motion, specifics of MR pulse sequences, and k-space acquisition strategies [23], leading to motion artifacts manifestation such as blurring, ghosting, geometric distortion, or signal loss (see Fig. 9.1).

In MRI acquisition, spatial information is encoded using a Fourier basis and subject motion part way through the acquisition of the Fourier lines will disrupt the spatial encoding and affect the entire image. When motion is present, k-space data from different positions are combined, which can lead to signal amplitude variation due to spin history effect, as well as signal phase variation and incorrect phase accumulation [22]. For instance, in Fig. 9.2, it can be seen that after a rotation, the spin history effect leads to striping throughout the image and there are shifts between odd and even slices of opposite direction anteriorly and posteriorly. When subject rotates during the k-space acquisition process, it results in inhomogeneously sampled k-space and the inconsistencies between the effective read-out directions for different k-space lines (Fig. 9.3) [23], and will lead to artifacts such as ghosting and streaking that are hard to correct even if the motion is known. Another corruption results that manifest on the intensity variation of the signal is the region's distance from the receiver coils. If the region moves toward or away from the coils, intensity changes will be introduced leading to blurring effects. Magnetic field inhomogeneity alterations are another source of effects related with motion. The static magnetic field is assumed to be constant across body parts, but variations in the magnetic properties of air, bone, and tissues lead to inhomogeneities that distort images. If the body part moves, these inhomogeneities will change correspondingly, producing geometric distortions and noises [21]. Motion artifact manifestation on MRI also depends on the level and type of motion and they are generally combinations of the various types of artifacts described above. Fig. 9.4 shows examples of brain motion with different levels of nodding movement [24] as well as respiratory motion with various levels of breath-holding behaviors [25], where different levels of ghosting, intensity variation, and geometric distortion artifacts can be observed. For more detail on motion artifact manifestation in MRI and their underlying physical mechanisms, please refer to the reviews on the topic [22,23].

Most current machine learning methods for motion correction rely on supervised learning, which typically requires pairs of motion-corrupted and motion-free images for training (see Section 9.4). However, in practice, it is costly and not always feasible to acquire such pairs of data. Motion simulation is therefore necessary for building a large dataset of those for learning-based approaches. Typically, a forward model of MRI acquisition in the presence of motion can be formulated as [18]:

$$y = \sum_{t=1}^{T} \mathbf{E}_t \mathbf{M}_t x, \tag{9.1}$$

where the forward MR encoding operator $\mathbf{E}_t$ and motion transform $\mathbf{M}_t$ are applied to the motion-free image x for each time point t to generate the motion-corrupted

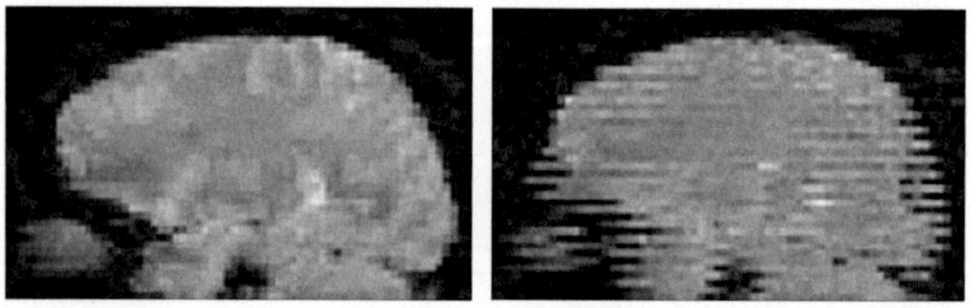

FIGURE 9.2

Motion and spin history effect. Top-left: volume free of distortions and spin-history effects; top-right: volume acquired in the presence of left-right rotation [20,21].

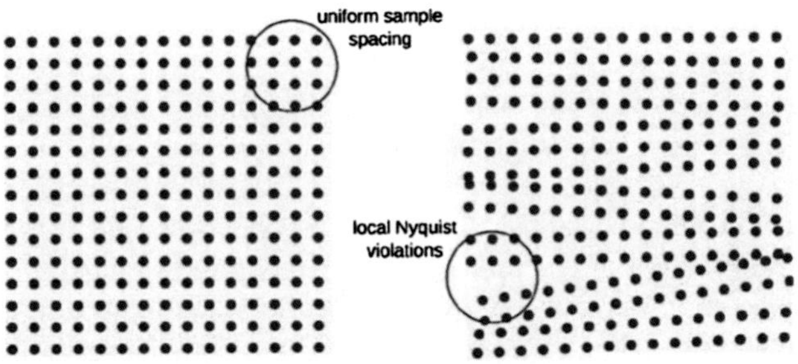

FIGURE 9.3

The effect of rotations on the k-space sample locations in object coordinates. Left: in absence of rotations; Right: in presence of rotations [23].

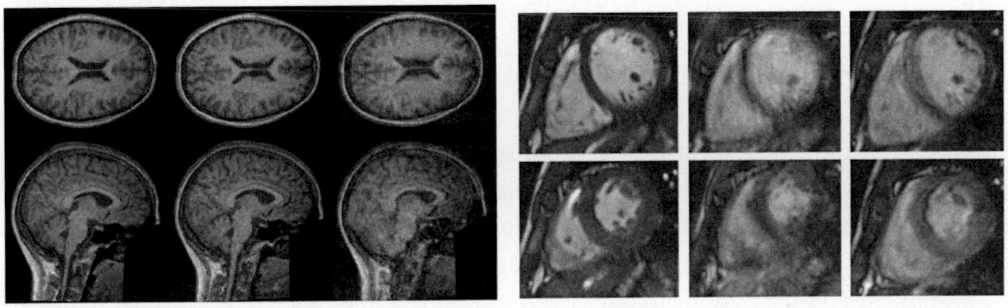

(a) Clean (left) and motion corrupted images with low (middle) and high (right) levels of motion [44].

(b) Same participant performing full breath-hold (left), half breath-hold (middle) and free breath (right) [85].

FIGURE 9.4

Examples of different motion artifacts under various levels of motion.

k-space signals y. In rigid motion such as brain movement, $\mathbf{M}_t$ mainly consists of object translation and rotation. According to the Fourier shift theorem, object rotation in image space corresponds to equivalent rotation in k-space and object translation results in a linear phase ramp in k-space in the direction of motion [18,22]. In contrast, simulation of nonrigid deformable motion such as respiratory motion is more complex, and one possible way of simulating that is to approximate the deformable movement with varying translations in k-space [18]. To account for motion artifacts induced by magnetic field inhomogeneities, B_0 inhomogeneities can be incorporated in Eq. (9.1) as a second-order motion effect for more realistic simulations [26]. Based on Bloch equation, there are also some available numerical simulation tools for simulating motion effects in MRI. Some of the most used open-source MRI simulators include POSSUM [27], JEMRIS [28], and Koma [29], which have considered many properties of interest in MRI for simulations, and they can be useful resources for creating synthetic data to train machine learning models.

9.3 Image registration

Image registration serves as an important tool for motion analysis and has been widely used in motion-related applications such as retrospective motion correction, motion tracking, and motion compensation. Image registration is a process of aligning images or volumes to establish spatial correspondences, and is fundamental to relate spatial information of anatomical structures within patients [30]. We will review some of the basic principles of image registration in this section together with some conventional image registration techniques for motion correction. More recent learning-based registration methods will be discussed in Section 9.4.

Image registration is an inverse problem, where transformations are estimated between given images. Depending on whether the registration is performed on a pair of images or on a group of images, it can be further categorized into pair-wise and group-wise registration, whereas the principles of them are shared. In pair-wise registration, the problem can be formulated as a minimization problem between a pair of a moving image $\boldsymbol{x}_m$ and a fixed image $\boldsymbol{x}_f$:

$$\underset{\boldsymbol{u} \in \mathscr{D}(\Omega)}{\operatorname{argmin}} \mathcal{L}_{sim}\left(\boldsymbol{x}_f, \boldsymbol{x}_m \circ \boldsymbol{u}\right) + \lambda \mathcal{R}(\boldsymbol{u}), \tag{9.2}$$

where $\mathcal{L}_{sim}$ stands for the image dissimilarity measure, $\boldsymbol{u}$ denotes the transformation from $\boldsymbol{x}_m$ to $\boldsymbol{x}_f$, Ω is the image domain, $\mathcal{D}(\Omega)$ is the group of feasible transformations and λ is a hyper-parameter that trades-off between the image similarity and deformation regularity. Similarly, in group-wise registration, the problem can be formulated as

$$\underset{\boldsymbol{u} \in \mathscr{D}(\Omega)}{\operatorname{argmin}} \sum_i \mathcal{L}_{sim}\left(\boldsymbol{x}_T, \boldsymbol{x}_m^i \circ \boldsymbol{u}^i\right) + \lambda \mathcal{R}(\boldsymbol{u}), \tag{9.3}$$

where $\boldsymbol{x}_m^i$ represents each individual moving image in the group and $\boldsymbol{u}$ denotes a vector of each separate transformation $[\boldsymbol{u}^0, \boldsymbol{u}^0, \ldots, \boldsymbol{u}^n]$ between the corresponding moving images and the template image $\boldsymbol{x}_T$. The template image can be explicitly designated as one of the images in the group or optimized inherently, and the set of transformations are commonly jointly optimized.

In general, registration techniques mainly consist of three components: a similarity metric $\mathcal{L}_{sim}$, a transformation $\boldsymbol{u} \in \mathcal{D}(\Omega)$, and a regularization term $\mathcal{R}$. The similarity metric measures the similarity between moving and fixed images. One of the simplest and widely used similarity measures in mono-modal registration is the sum of squared differences (SSD) between image intensities. These are typically computed between the fixed image $\boldsymbol{x}_f$ and the transformed moving image $\boldsymbol{x}_m^T = \boldsymbol{x}_m \circ \boldsymbol{u}$. Other commonly used similarity metrics also include normalized cross-correlation (NCC) [31], which assumes a linear intensity relationship between images when well aligned, as well as mutual information (MI) [32] and normalized mutual information (NMI) [33] where distributions of image intensities are taken into account. The latter metrics are often used in multimodal registration as they do not assume linear intensity relationship between images.

A transformation model defines a geometric transformation between images. There are several transformation models, ranging from simple (e.g., rigid or affine transformations) to more complex transforms (e.g., deformable transforms), see Fig. 9.5, where the spectrum of the complexity is usually characterized by the number of parameters that describe the Degrees of Freedom (DoF) of the transformation [35]. The rigid transformation is the simplest one, and it can be defined by 6 parameters in 3D space, including three translational and three rotational parameters. Affine transformation, in addition, includes scaling and shear compared with rigid transformation, giving 9 and 12 DoF correspondingly. However, tissues usually deform in a complicated way, and thus deformable transformation is necessary to describe the movement, where the moving image is elastically deformed to fit the fixed image. State-of-the-art deformable transformation models include B-spline based free form deformations [36] that are parameterized by a regular mesh of uniformly spacing control points, Demons algorithm [37] that considers nonrigid registration problem as a diffusion process, and optical flow-based methods [38] which assume that image intensities of a particular point do not change between images. Recent learning-based image registration approaches have built on these different transformation models and developed them into learning-based ones (see Section 9.4.2). More details of the transformation models can be referred to the review on medical image registration [34,35].

Since image registration is an ill-posed inverse problem where many viable solutions could exist, regularization on the deformation fields is commonly imposed to make sure that the solution is unique and with desired properties. To incorporate the prior assumption on the deformation fields, the majority of methods either have built-in regularization constraints offered by the transformation model, or directly regularize the deformation field or its updates [39,40]. For instance, spline-based models compactly represent deformation fields in a low-dimensional representation space which inherently ensures the smoothness of the deformation [36]. For ensuring transformation invertibility, some diffeomorphic registration methods have been proposed to parameterize the displacement field as the integral of a time-varying

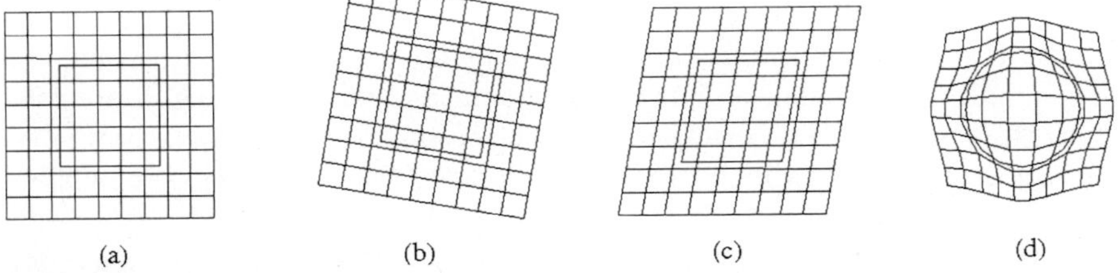

FIGURE 9.5

Example of different types of transformations of a square: (a) identity transformation, (b) rigid transformation, (c) affine transformation, and (d) nonrigid transformation.

or stationary velocity field [41,42]. Other variants also considered enforcing volumetric preservation for soft tissue tracking [43]. In explicit regularization with $\mathcal{R}$, the form of L2 norm or bending energy that penalizes the first- or second-order derivatives of the deformation can be used to ensure smoothness [44], and some other constraints, for example, tissue incompressibility, can also be imposed [45].

Image registration techniques have been widely investigated in image-based motion correction. These methods assume that significant motion only occurs between acquisitions of volumes or slices. In a sequence of MRI, the movement can be calculated using image registration methods for every volume or frame with respect to a reference volume or frame, such as the first or a time-averaged one [22,46]. For correcting motion that occurs between slice acquisitions, some more refined approaches have also been proposed to perform slice-to-volume registration. To improve on the quality of the reference image, group-wise registration methods have also been investigated which simultaneously take the intensity distributions of all images into account [47]. Other group-wise registration methods also include algorithms using multivariant similarity measure [48], voxel-wise variance [49], and stack entropy cost function [50], which can also be used for the tasks of motion correction. Currently, most image-based motion correction methods employ learning-based image registration approaches [18], which will be discussed in the following section.

9.4 Motion correction networks

The learning-based motion correction methods including motion artifacts reduction and image registration networks as well as their training strategies are detailed in the following.

9.4.1 Motion artifacts reduction

Leveraging the powerful learning capability of DL neural networks, motion artifacts reduction (MAR) networks have been designed to learn the nonlinear mapping between motion-corrupted and motion-free images. As shown in Fig. 9.6, the network has input of image with motion artifacts and output of image with reduced motion, with the training being performed by minimizing the error between the output and the ground-truth motion-clean image. With sufficiently large training dataset, the DL network is able to learn to invert the motion corruption process and remove motion artifacts for images with the same motion type as to the training images without any supplementary data acquisition.

Paired images with and without motion artifacts are essential for supervised learning of MAR networks, which, however, are difficult and costly to obtain, so the motion artifacts are mostly simulated. The motion artifacts depend on the motion types and imaging sequences as explained in Sections 9.1 and 9.2, which should be considered during motion simulation. The motion simulation follows the forward

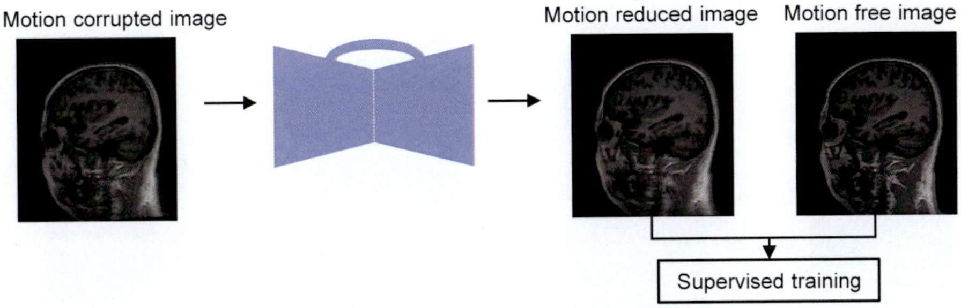

FIGURE 9.6

The flowchart of motion artifacts reduction network. In a supervised learning setting, the network learns the nonlinear mapping between motion-corrupted and motion-free images by minimizing the error between predicted and motion-free images.

acquisition model of Eq. (9.1). For rigid motion, the motion operator includes translations and rotations, and the motion simulation can be performed either in image space or in k-space, as shown in Fig. 9.7. When simulating in image space, the images after rotations and translations are Fourier transformed to obtain a set of k-space which are concatenated according to the sampling pattern to form the motion-corrupted k-space and image. When simulating in k-space, according to the Fourier transform properties, translational vector $\boldsymbol{t} = [t_x, t_y, t_z]$ affects the k-space data acquired $\boldsymbol{k} = [k_x, k_y, k_z]$ at by a phase shift: $s_t = s{\cdot}e^{-i2\pi \boldsymbol{kt}}$. s and s_t indicates the k-space data before and after translational shifts, respectively. Rotations in the image domain are equivalent to the same rotations in k-space.

For nonrigid motion such as the respiratory and cardiac motion, the motion simulation is typically performed in image space. During simulation, the motion operator M warps the reference image according to the provided motion field, $\boldsymbol{u}$. Specifically, the warping operation interpolates the image to the new coordinates, $\boldsymbol{r} + \boldsymbol{u}(\boldsymbol{r})$, where $\boldsymbol{r}$ indicates the original spatial coordinates. Bilinear and trilinear interpolation is commonly used for warping 2D and 3D images, respectively. However, the realistic nonrigid motion field can be challenging to simulate with a possible way of registering images of different motion states. Nonrigid motion simulation can also be performed in k-space by applying multiple translations and rotations to randomly selected k-space lines [51] or by swapping k-space lines acquired at different motion states [52].

Convolutional neural networks Regarding the MAR models, convolutional neural networks (CNNs) have been widely used. Sommer et al. [53] trained a CNN to remove rigid brain motion artifacts in the magnitude MR images. The training data were simulated with translational and rotational motions. Moreover, residual learning and multiresolution learning were adopted to improve motion correction performance. Quantitative neuroradiological ratings indicated that the network filtered images had reduced motion level. However, the 2D MAR model may not work well for 3D bulk motion where there will be through-plane motion. Duffy

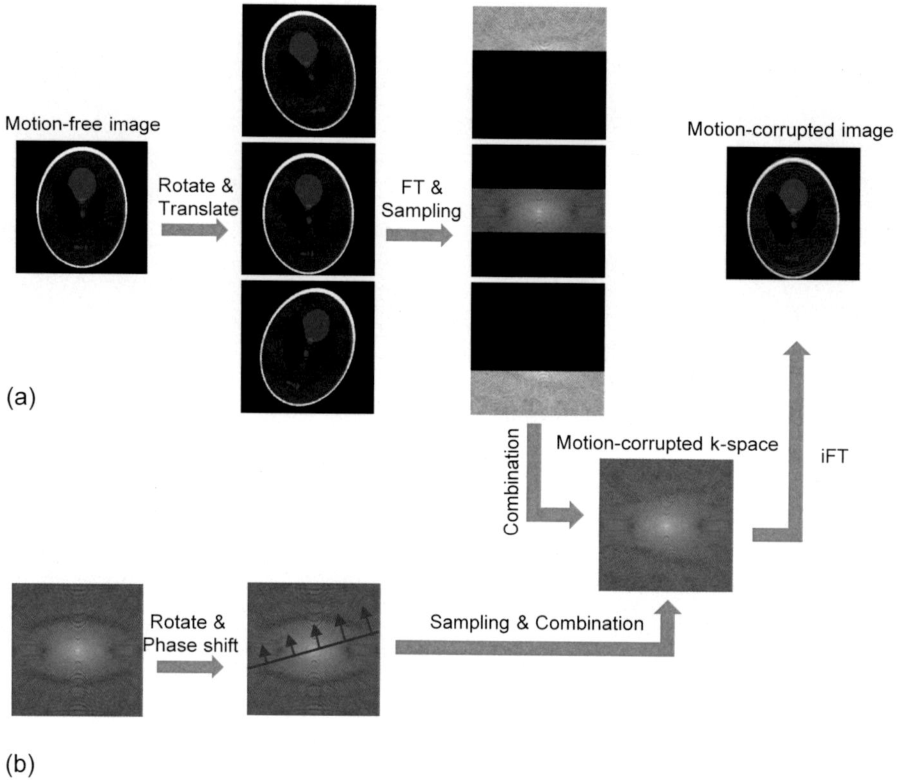

FIGURE 9.7

Rigid motion simulation. (a) In image space, the translations and rotations are applied to a motion-free image, followed by Fourier transform (FT) and concatenation according to the sampling pattern to generated motion-corrupted k-space, which is inverse Fourier transformed to get motion-corrupted image. (b) In k-space, phase shifts and rotations are applied for translational and rotational motion, respectively. The motion-transformed k-space is aggregated according to the sampling pattern to obtain motion-corrupted k-space and image.

et al. [54] developed a 3D CNN model with a modified U-Net [55] which has an encoder-decoder architecture to remove brain motion artifacts retrospectively. The motion simulation was performed in the Autism Brain Imaging Data Exchange dataset by adding 3D translations and rotations in the k-space domain. Instead of inputting the whole 3D volume into the network, patch-wise training was adopted by randomly selecting 3D image patches as inputs to reduce the GPU memory consumption and augment the training data to prevent overfitting. The trained model based on simulated motion data was shown to be able to remove motion in real acquired data from challenging subjects such as Parkinson's patients and children.

Pawar et al. [56] proposed the MoCoNet which adopted the encoder-decoder U-Net to correct motion artifacts in T1-weighted brain images. The MAR model was also trained in simulated data. The authors further improved the performance of MoCoNet by using an Inception-ResNet network as the encoder with different convolutional kernel sizes at each resolution scale to capture both local and global motion artifacts and simulating more realistic brain motion encompassing both abrupt motion and gradual motion [57]. Motion artifacts in real acquisitions can be removed by MoCoNet, while there tends to be residual motion when motion corrupts low-frequency k-space.

Multicontrast and quantitative MRI require the acquisition of a series of images, the long scan time of which makes them more prone to motion artifacts. Lee et al. [58] proposed a multicontrast brain MRI motion correction network, aiming to exploit the inter-contrast correlation to improve the motion correction performance. The MAR model consists of two parts: the contrast alignment part aims to align the multicontrast images by a self-supervised registration CNN which will be introduced in the next section; the motion correction part corrects motion artifacts in the aligned multicontrast images which has a multiinput multioutput encoder-decoder architecture with residual-learning blocks and was trained in simulated data to minimize the structural similarity loss and the VGG loss. Pirkl et al. [59] developed a 3D multiscale CNN with residual learning for motion correction of 3D multiparametric brain MRI. The realistic motion artifacts were simulated with a physics-informed method. Then, the MAR model was trained in a supervised manner to minimize the error between the filtered parameter maps and the motion-free ground truth. Validated in real-motion in vivo data, the proposed method was able to effectively remove motion artifacts and generate high-quality parameter maps.

Compared with rigid brain motion, nonrigid motion correction networks are less investigated in self-supervised learning methods as it is challenging to simulate realistic motion data. Tamada et al. [60] proposed a CNN model with residual learning to reduce respiration induced motion in liver dynamic contrast enhanced MRI. The network has multichannel inputs to handle the dynamic images, and was trained with 2D image patches to increase the training data. For supervised training, the motion data were simulated by adding phase errors to the k-space obtained from motion-free magnitude images. The trained model was tested in real acquired patients data and found to be able to substantially reduce respiration induced blurring and artifacts.

Generative models Besides CNN, generative models have also been used for MAR. Küstner et al. [61] designed and compared two generative models, variational autoencoder (VAE) and generative adversarial network (GAN), for correcting rigid brain motion and nonrigid abdominal motion. Different from aforementioned studies that adopted simulated motion data, this study performed each acquisition twice to obtain the motion-free and motion-corrupted image pairs. The GAN method outperformed VAE and generated images close to motion-free images. However, it was noted that the trained motion correction model may not generalize well to some anatomical features which were altered or hidden after motion correction. Moreover, a conditional GAN was utilized for 3D rigid brain motion correction with motion data simulated on the Human Connectome Project dataset. The proposed method was found to outperform the CNN MAR methods of 3D generic U-Net and MoCo-Net. In addition, there is recent research interest in applying diffusion probabilistic models (DPMs) for MR motion correction. DPMs as a new type of generative model have demonstrated remarkable performance in various MRI tasks [62] due to their adversarial-free training strategy and sampling with good quality and diversity. Levac et al. [63] proposed a method based on a score-based diffusion model to estimate rigid brain motion parameters for motion corrected reconstruction. The score-based DPM trained with simulated motion data was able to generate high-fidelity images on par with motion-free images. However, this method was not validated in real motion data.

Unpaired learning The above-mentioned approaches are supervised-learning methods. However, the MAR model trained with simulated motion may not generalize well to real motion data, leading to suboptimal motion correction. Therefore, the motion correction methods that do not require paired motion free and motion-affected images have been investigated. Without paired training data, the artifacts reduction problem has been considered as an unpaired image-to image translation problem, translating from the motion-corrupted domain (X_c) to the motion-free (X_f) domain. Armanious et al. [64] adopted the CycleGAN to achieve this aim, which consists of two sets of generators and discriminators. As shown in Fig. 9.8, one set of the generator and discriminator learns the mapping from X_c to X_f, and the other set of generator and discriminator learns the inverse mapping from X_f to X_c. In addition to the adversarial training with GAN loss for each set of generator and discriminator, the two generators are trained to invert each other which can be achieved with pixel-wise cycle-consistency loss. The trained CyeleGAN with unpaired data was shown to be able to remove rigid brain motion. With a similar idea, Liu et al. [65] developed a disentangled unsupervised cycle-consistent adversarial network. With the observation that the manually selected clean images may also contain certain artifacts, two kinds of encoders were designed to encode the input image belonging to X_c or X_f to a domain-invariant content space and a domain-specific artifact space. Then, two types of decoders translate the images to X_c or X_f based on the content and artifact features by adding or removing the artifacts. Several consistency losses were introduced for effective disentanglement learning. Experimental results indicate that the proposed method can effectively remove different levels of brain motion artifacts, outperforming the supervised U-Net which tends to generate blurring images and unpaired CycleGAN.

Besides GAN, the DPMs have also been adopted in the unpaired training setting, where a score-based diffusion model was trained in only motion-free images to capture the prior distribution [66]. Then, the trained model was used to remove respiratory motion artifacts in liver MRI by repeatedly applying the forward and reverse diffusion process with a low-frequency data consistency. It was found that the DPM-based motion correction method outperformed CycleGAN and the authors' bootstrap subsampling and aggregation method [67].

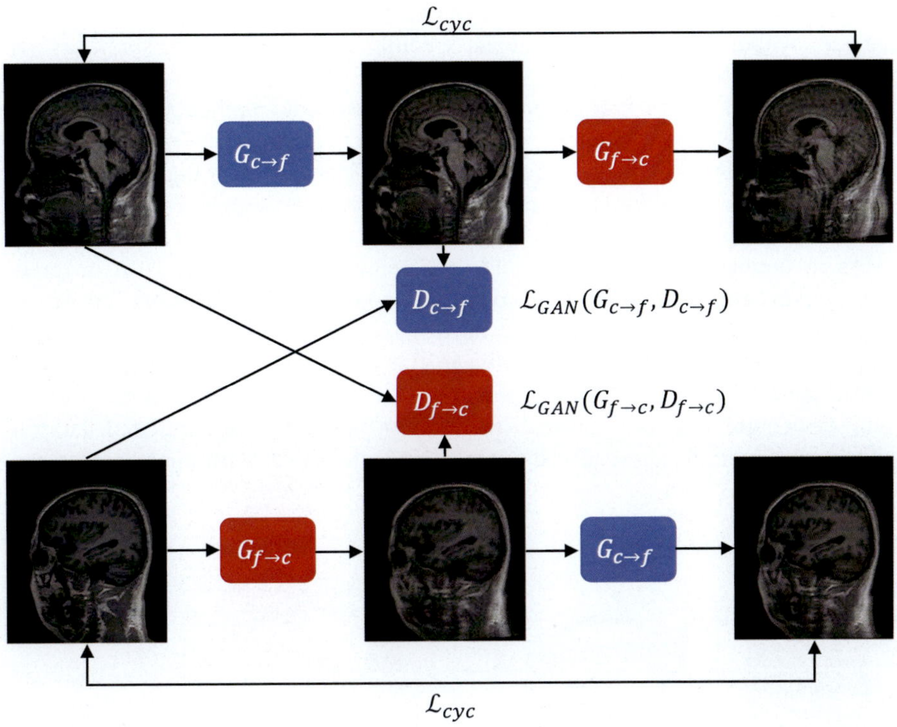

FIGURE 9.8

Unpaired training with CycleGAN for motion correction. The generator and discriminator in blue learn to translate the image from the motion-corrupted domain to the motion-free domain ($c \to f$), and the generator and discriminator in red learns the inverse translation ($f \to c$). The adversarial and cycle-consistency loss are used to train the network.

9.4.2 Registration networks

Image registration aims to estimate the displacements between different motion states, which can be used for motion-compensated reconstruction. However, conventional registration methods as introduced in Section 9.3, requiring iterative optimization are not computationally efficient. Deep learning methods have been developed to accelerate the registration process, which can be performed in the image domain or the k-space domain as detailed below.

Image domain registration DL registration is commonly performed in the image domain by optimizing a network in a set of training images to learn a global function that generates the deformation field aligning the input images. Properly trained registration model can align new images from the same distribution of training data. The training phase of the registration model can take hours to days.

However, once trained, the registration can be performed in a simple feed-forward manner without the costly subject-specific iterative optimization as in nonlearning-based methods.

DL registration was first proposed as an optical flow method (FlowNet) [68] for registering natural images and was later adapted for medical image registration. A Typical image registration network is shown in Fig. 9.9, which is a CNN with encoder–decoder architecture. The encoder extracts features of the inputted moving image and fixed image by using convolution and downsampling layers to increase the reception field. The learned representations at different spatial levels are inputted to the decoder with convolution and upsampling layers to predict the displacement field. The network can be trained with in a supervised manner by minimizing the error between the predicted motion and the ground truth motion (Fig. 9.9a). However, the ground truth motion field may be cumbersome to acquire with conventional registration methods and simulated datasets with known motion may cause domain mismatch between training and testing.

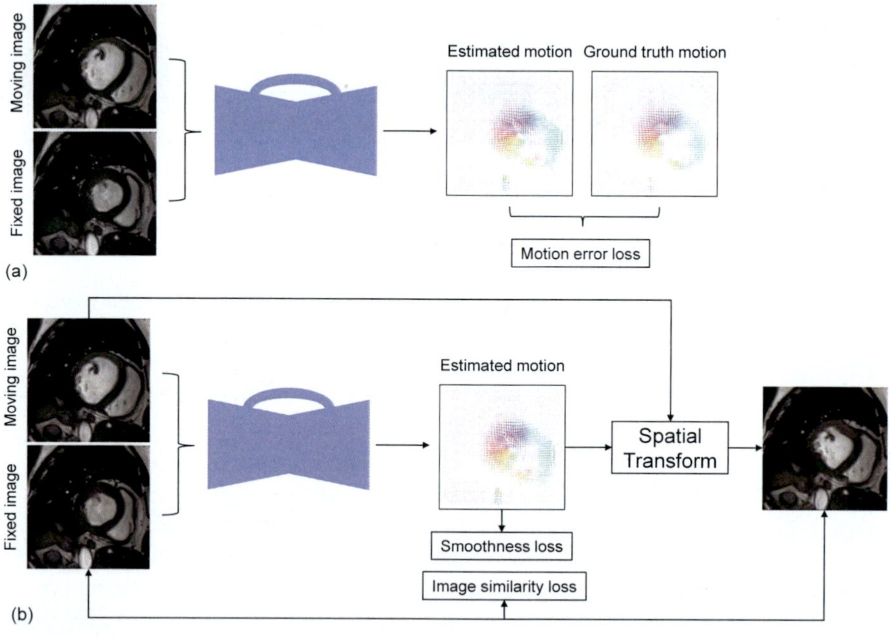

FIGURE 9.9

Learning-based image registration. (a) For supervised training, the registration network is optimized by minimizing the error between the predicted motion and the ground truth motion. (b) For self-supervised learning, the predicted motion field is inputted to a spatial transform layer to warp the moving image. The similarity between the warped moving image and the fixed image and the smoothness of the motion field are enforced to optimize the network.

Therefore, self-supervised training strategy is more popular for DL registration. Similar to the conventional registration (Eq. 9.2), the self-supervised training loss is constructed by measuring the similarity between the warped moving image ($\boldsymbol{x}_m$) with the predicted motion field ($\boldsymbol{u}$) and the fixed image ($\boldsymbol{x}_f$) as shown in Fig. 9.9b:

$$\mathcal{L}_{reg} = \mathcal{L}_{sim}\left(\boldsymbol{x}_f, \boldsymbol{x}_m \circ u\right) + \lambda \mathcal{L}_{smooth}(u) \tag{9.4}$$

where $\circ$ represents the warping operator, $\mathcal{L}_{sim}$ measures the similarity of two images, $\mathcal{L}_{smooth}$ enforces the smoothness of the motion field and λ is the regularization parameter that needs to be manually tuned for different applications. The similarity measurements in conventional registration methods such as the difference image, cross-correlation, and mutual information can be adopted in the self-supervised registration loss.

VoxelMorph is a representative MR image registration network [69]. A U-Net was adopted to take input of fixed and moving 3D volumes and output the deformation field between them. Qi et al. [70] proposed a registration network with an encoder–decoder architecture for measuring the 3D nonrigid motion between respiratory resolved cardiac images. They adopted the patch-wise training strategy to augment the training samples and reduce computation burden on GPU. Furthermore, the decoder was designed to output multiresolution motion fields, enabling coarse-to-fine learning for more accurate motion estimation. The predicted respiratory motion was subsequently used for motion-compensated reconstruction of coronary MR angiography acquired under free-breathing, achieving similar performance to conventional registration methods, while being much faster. For diffeomorphic registration that preserves topology, Dalca et al. [71] developed a probabilistic diffeomorphic registration network with a 3D U-Net architecture. It estimated the stationary velocity field, which was integrated through a scaling and squaring layer to obtain the motion field that is differentiable and invertible. For uncertainty estimates of the predicted velocity field, a probabilistic generative framework was embedded in the registration network.

Qin et al. [72] proposed a multimodal registration network, which decomposed the multimodal images into a domain-invariant latent shape representation and a domain-specific appearance code using a multimodal image-to-image translation network. Then, a similarity metric can be defined in the common latent space, reducing the challenging multimodal registration problem to a simpler mono-modal one. In addition, a discriminator network was designed to learn the implicit similarity in image space. Instead of directly measuring the dense motion field, Qiu et al. [73] proposed to parameterize the motion field with B-spline functions similar to the B-spline free-form deformation registration method [36]. Then, the network was trained to output the motion parameters of the B-spline control points, reducing the number of unknowns. A differentiable variant of the mutual information metric was designed in this work to handle both mono-modal and multimodal images.

Besides registering a pair of images, group-wise registration networks have been designed for registering a group of images. Different from pair-wise registration which has an explicit fixed image to be registered to, group-wise registration network could learn an implicit template image which lies in the mean-space of the input images, as proposed by Li et al. [74]. It takes input of a group of images and outputs the motion fields between these images and the implicit template which can be obtained by averaging the warped input images with the estimated motions. The group-wise registration network can be trained in a similar self-supervised way to pair-wise registration by enforcing the similarity between the warped input images and the template image.

k-Space domain registration Motion estimation in k-space domain typically aims to estimate motion within undersampled data. When undersampling, the aliasing artifacts in the image domain need to be suppressed with some reconstruction methods before registration, which limits the achievable acceleration. Direct motion estimation in k-space domain can avoid this reconstruction step, facilitating motion estimation with highly undersampled k-space data and enabling real-time motion estimation. Huttinga et al. [75] proposed the MR-MOTUS model that relates the motion parameters to an artifact-free reference image in a forward acquisition model and then estimates the motion parameters with highly undersampled k-space data by solving the inverse problem.

With the observation that the motion field is smooth and that its frequency domain should be band-limited, Wang proposed a network [25], termed as DeepFLASH, to estimate the frequency domain of the motion field. The network takes inputs of the real and imaginary parts of the Fourier transformed moving and fixed images and outputs the central part of the frequency domain motion. Estimation focusing on the central frequency region reduces the learning complexity and enforces the motion to be smooth. Subsequently, the frequency domain motion was zero-padded and Fourier transformed to obtain the motion in spatial space. DeepFLASH was trained with supervision from the fast geodesic shooting registration algorithm based on Fourier-approximated Lie algebras [76].

The registration accuracy of DeepFLASH was demonstrated for fully sampled data, while its ability of handling undersampled k-space has not been investigated. Küstner et al. [51] developed a network, LAPNet, built based on the extension of the local all-pass (LAP) filter based optical flow measurement method in image space [77] to the frequency domain, that could estimate 3D nonrigid respiratory motion from undersampled k-space. The inputs to LAPNet are the tapered k-space corresponding to the moving and fixed image patches, which go through a series of convolutional operations to generate the motion parameters of the central voxel location in the image patch. The 3D motion field can then be obtained by sliding through the whole 3D volumes and merging the estimated motion parameters. The motion estimated with the conventional LAP method was used to supervise the training of LAPNet.

9.5 Motion-compensated image reconstruction

9.5.1 Problem formulation

MC image reconstruction aims to compensate motion in the acquired data during the retrospective reconstruction stage. It typically solves the following inverse problem for estimating the motion field $\boldsymbol{u}$ and the motion-free image $\boldsymbol{x}$:

$$\arg\min_{\boldsymbol{x},\boldsymbol{u}} ||EM_{\boldsymbol{u}}\boldsymbol{x} - y||_2^2 + \lambda_1 \mathcal{R}_1(\boldsymbol{x}) + \lambda_2 \mathcal{R}_2(\boldsymbol{u}) \tag{9.5}$$

where M_u is the motion transformation operator based on $\boldsymbol{u}$, E is the forward MR encoding operator, y is the k-space data containing motion. $\mathcal{R}_1$ and $\mathcal{R}_2$ indicates the regularization on the prior information of $\boldsymbol{x}$ and $\boldsymbol{u}$, respectively, with λ_1 and λ_2 being the corresponding weighting parameters. If the k-space data are undersampled, R_1 can be constructed by regularizing on the sparsity of $\boldsymbol{x}$ to remove the undersampling artifacts as in the compressed sensing reconstruction, and R_2 is commonly implemented by measuring the smoothness of $\boldsymbol{u}$.

9.5.2 Conventional MC reconstruction

The MC reconstruction problem can be solved using alternating optimization algorithms such as ADMM (Alternating Direction Method of Multipliers) as described in solving the nonlinear MR reconstruction problem in Chapter 7. However, it is noted that without any prior measurement or modeling about motion, the optimization of the large nonlinear problem can be computationally prohibitive. For rigid motion, Haskell et al. [78] developed a method for TArgeted Motion Estimation and Reduction (TAMER), which jointly estimates the motion trajectory and the motion-free image by minimizing the data-consistency error. Furthermore, only a subset of target voxels were assessed at each step of the motion search, significantly reducing the computation complexity. Similarly, by minimizing the data consistency error, Polak et al. [79] proposed to use a low-resolution motion-free scout image to estimate the rigid motion trajectory from under-sampled k-space data.

Odille et al. [80] proposed to optimize the motion estimation and image reconstruction as two coupled inverse problems and adopt the motion signals estimated from navigators or external devices to reduce motion searching space and thus reduce optimization complexity. Following this work, there are a bunch of studies adopting the similar idea of leveraging the reference motion signal to make the nonrigid MC reconstruction problem tractable. Especially, a type of method has been proposed for free-breathing cardiovascular MRI, which performs separate motion estimation and reconstruction. As shown in Fig. 9.10, it adopts an image navigator [11] to estimate beat-to-beat translational motion and bins the acquired data into different respiratory motion states to reconstruct the respiratory-resolved images, from which the inter-bin nonrigid respiratory motion can be estimated. The

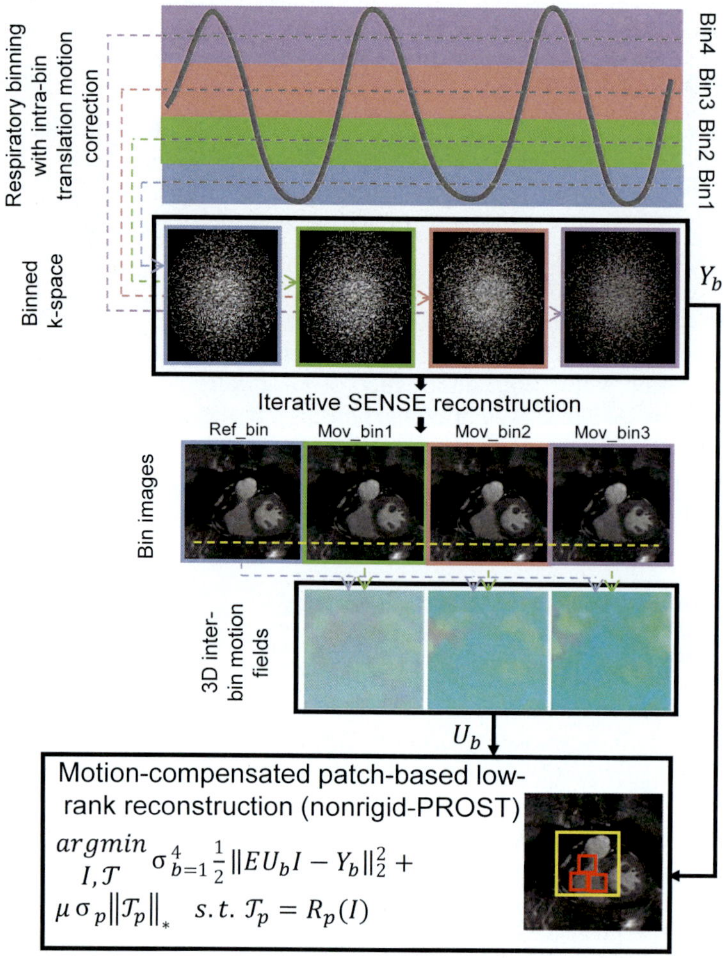

FIGURE 9.10

Overview of nonrigid respiratory motion-compensated coronary MRI reconstruction [70]. The k-space data of a 3D free-breathing acquisition is sorted into four respiratory states according to the superior-inferior respiratory motion curve estimated from image navigators. After translational motion correction for each bin and the reconstruction of the respiratory resolved images, inter-bin image registrations are performed to obtain the nonrigid motion fields which are incorporated into an iterative motion-compensated reconstruction.

motion in each respiratory bin is small and can be assumed to be translational. Then, intra-bin translational motion correction can be performed for each bin data in k-space and inter-bin nonrigid motion can be embedded in the encoding operator E and corrected in the iterative optimization process. Cruz et al. [81] applied this framework for free-breathing high-resolution coronary vessel wall imaging. Bustin et al. [82] incorporated the nonrigid motion correction framework with a patch-based low-rank reconstruction method, achieving accelerated submillimeter coronary MR imaging.

Besides using MC reconstruction to generate a single static image, motion correction can also be applied in motion-resolved reconstruction to improve the reconstruction performance. Cardiac or respiratory motion-resolved images can be used to quantify the dynamic information, and the motion dimension can be reconstructed as an additional dimension as proposed in XD-GRASP [83], which regularizes on the temporal sparsity to exploit the temporal redundancy. However, the motion caused misalignments between the dynamic images can reduce the temporal sparsity and affect the motion-resolved reconstruction quality. Rank et al. [84] proposed a motion-compensated total variation algorithm for 4D (3D + respiratory motion dimension) highly undersampled MRI reconstruction, which alternates between motion-compensated image reconstruction and artifact-robust motion estimation at multiple resolution levels. The motion-compensated total variation operator achieved better reconstruction than conventional total variation regularization without motion correction.

9.5.3 Learning-based MC reconstruction

Deep learning methods have also been investigated to address the challenging motion estimation and reconstruction problem. For rigid motion, Haskell et al. [85] improved the TAMER method by introducing a CNN to guide the motion estimation, which was pretrained on a set of images to have the ability of reducing motion artifacts. The motion artifacts-mitigated image by the CNN was used to estimate the motion parameters by minimizing the data consistency error, facilitating MC reconstruction. The reconstructed image was inputted to the CNN for another round of iteration. Hossbach et al. [86] designed a Densenet to estimate in-plane rigid brain motion from a center patch of the motion-corrupted k-space and a motion-free low-resolution reference, which was trained with simulated motion data. Feinler et al. [87] proposed to use a CNN to represent more complex motion than rigid motion explicitly. Once trained, the motion estimation network can be used for MC reconstruction, achieving better image quality than the GAN-based methods. In addition, to generate more realistic motion, Eichhorn et al. [88] developed a physics-aware motion simulation procedure by utilizing real recorded patient motion and accounting for B0 inhomogeneities in T2*-weighted brain MRI. They also trained a 3D CNN to detect individual motion-affected k-space lines.

For nonrigid motion such as respiratory and cardiac motion, the motion estimation and MC reconstruction are typically optimized in an end-to-end framework. In the introduced separate motion estimation and reconstruction framework, the motion estimation with conventional registration methods depends on the image quality of the motion-resolved images. Since these images are typically reconstructed from the subsets of k-space of a 3D free-breathing acquisition, the undersampling of the 3D acquisition cannot be high. Otherwise, the highly undersampled motion-resolved data will compromise the motion estimation. To deal with this challenge, Qi et al. [89] developed MoCo-MoDL, consisting of a motion estimation network and a motion-informed undersampled reconstruction network which were trained in an end-to-end fashion. Especially, the motion estimation network was trained to be robust to undersampling artifacts by using the corresponding good-quality motion-resolved images as registration supervision. MoCo-MoDL enabled robust nonrigid MC reconstruction for highly accelerated coronary MRI.

Nonrigid motion estimation has also been used to augment the dynamic MRI reconstruction performance, such as cardiac cine MRI. Seegoolam et al. [90] demonstrated that by fusing the data of all dynamic frames with motion estimation, the dynamic reconstruction network can work for aggressive accelerations of more than $50\times$. Specifically, a pair-wise registration network was designed to register all other dynamic frames to a target frame to generate dynamic images with each frame containing data of all frames to augment the reconstruction network performance. The motion estimation and reconstruction networks were optimized simultaneously in a cascaded CNN. However, this method is computationally intensive as a large number of pair-wise registrations are needed to generate the motion-augmented dynamic images and the motion between adjacent frames is not constrained to be invertible. Yang et al. [91] proposed to replace the pair-wise registration with a more computationally efficient group-wise registration network, which also enforces the motion to be invertible by adopting a diffeomorphic registration network. The implicit registration template learned by the network facilitated more accurate motion estimation than pair-wise registration, translating to better reconstruction.

9.6 Conclusion and outlook

This chapter introduced commonly encountered motion types and manifestation of motion artifacts in MRI, highlighting the importance and complexity of MR motion correction. Various deep learning methods have been developed for MR motion correction, including motion artifacts reduction, motion estimation in image space and k-space and motion-compensated reconstruction. It should be noted that, instead of exhaustive search of all relevant papers, this chapter described representative methods for different motion types (rigid and nonrigid) and in various imaging scenarios. In the following, the challenges and aspects that require further investigation are discussed.

Firstly, training complex deep learning models for MR motion correction often requires large amounts of paired motion-corrupted and motion-free data, which, however, is challenging and costly to acquire. Developing more sophisticated data simulation techniques, especially for continuous nonrigid motion, to generate realistic motion artifacts in MR images can help overcome the limitations of real-world data acquisition. Data augmentation strategies can also be employed to increase the diversity of training data, improving the generalization capabilities of deep learning models. Secondly, with the rapid development of artificial intelligence, new deep learning networks can be expected to provide more efficient and effective motion correction in MRI. For example, investigating unsupervised or self-supervised learning models can help reduce the dependency on paired training data. Thirdly, the role of MR in guided treatment is being increasingly recognized, and developing deep learning models for 3D real time motion estimation is useful for MR-guided therapies in thorax and abdomen. Finally, it is noted that there is still no universal method that is applicable to all situations due to the complex and unpredictable nature of motion and the versatile MR imaging sequences. The appropriate motion correction method should be determined according to the specific application.

References

[1] Van Der Kouwe A, Andre JB. Motion correction in MR: correction of position, motion, and dynamic field changes. Academic Press; 2022.
[2] Ehman RL, McNamara MT, Pallack M, Hricak H, Higgins CB. Magnetic resonance imaging with respiratory gating: techniques and advantages. Am J Roentgenol 1984;143 (6):1175–82.
[3] Lanzer P, Barta C, Botvinick EH, Wiesendanger HU, Modin G, Higgins CB. ECG-synchronized cardiac MR imaging: method and evaluation. Radiology 1985; 155(3):681–6.
[4] Wang Y, Riederer SJ, Ehman RL. Respiratory motion of the heart: kinematics and the implications for the spatial resolution in coronary imaging. Magn Reson Med 1995;33 (5):713–9.
[5] Chen C, Liu Y, Simonetti OP, Tong M, Jin N, Bacher M, Speier P, Ahmad R. Cardiac and respiratory motion extraction for MRI using pilot tone–a patient study. Int. J. Cardiovasc. Imaging 2023:1–13.
[6] Speier P, Fenchel M, Rehner R. PT-Nav: a novel respiratory navigation method for continuous acquisitions based on modulation of a pilot tone in the MR-receiver. Magn Reson Mater Phys Biol Med 2015;28:S97–8.
[7] Kay N, Börnert P. Prospective correction of affine motion for arbitrary MR sequences on a clinical scanner. Magn Reson Med 2005;54(5):1130–8.
[8] Aranovitch A, Haeberlin M, Gross S, Dietrich BE, Reber J, Schmid T, Pruessmann KP. Motion detection with NMR markers using real-time field tracking in the laboratory frame. Magn Reson Med 2020;84(1):89–102.
[9] Barnwell JD, Smith JK, Castillo M. Utility of navigator-prospective acquisition correction technique (pace) for reducing motion in brain MR imaging studies. Am J Neuroradiol 2007;28(4):790–1.
[10] Okai Addy N, Ingle RR, Luo J, Baron CA, Yang PC, Hu BS, Nishimura DG. 3d image-based navigators for coronary mr angiography. Magn Reson Med 2017;77(5):1874–83.
[11] Henningsson M, Botnar RM. Advanced respiratory motion compensation for coronary mr angiography. Sensors 2013;13(6):6882–99.

[12] Wu HH, Gurney PT, Hu BS, Nishimura DG, McConnell MV. Free-breathing multiphase whole-heart coronary mr angiography using image-based navigators and three-dimensional cones imaging. Magn Reson Med 2013;69(4):1083–93.
[13] Larson AC, White RD, Laub G, McVeigh ER, Li D, Simonetti OP. Self-gated cardiac cine MRI. Magn Reson Med 2004;51(1):93–102.
[14] Piccini D, Littmann A, Nielles-Vallespin S, Zenge MO. Respiratory self-navigation for whole-heart bright-blood coronary MRI: methods for robust isolation and automatic segmentation of the blood pool. Magn Reson Med 2012;68(2):571–9.
[15] Stehning C, Börnert P, Nehrke K, Eggers H, Stuber M. Free-breathing whole-heart coronary MRA with 3d radial SSFP and self-navigated image reconstruction. Magn Reson Med 2005;54(2):476–80.
[16] Uribe S, Muthurangu V, Boubertakh R, Schaeffter T, Razavi R, Hill DLG, Hansen MS. Whole-heart cine MRI using real-time respiratory self-gating. Magn Reson Med 2007;57(3):606–13.
[17] Chang Y, Li Z, Saju G, Mao H, Liu T. Deep learning-based rigid motion correction for magnetic resonance imaging: a survey. Meta-Radiology 2023:100001.
[18] Spieker V, Eichhorn H, Hammernik K, Rueckert D, Preibisch C, Karampinos DC, Schnabel JA. Deep learning for retrospective motion correction in MRI: a comprehensive review. arXiv preprint arXiv:2305.06739 2023.
[19] Zhou Z, Hu P, Qi H. Stop moving: MR motion correction as an opportunity for artificial intelligence. In: Magnetic resonance materials in physics, biology and medicine; 2024. p. 1–13.
[20] Speck O, Hennig J, Zaitsev M. Prospective real-time slice-by-slice motion correction for fmri in freely moving subjects. Magnet Reson Mater Phys Biol Med 2006;19:55–61.
[21] Zaitsev M, Akin B, LeVan P, Knowles BR. Prospective motion correction in functional MRI. Neuroimage 2017;154:33–42.
[22] Godenschweger F, Kägebein U, Stucht D, Yarach U, Sciarra A, Yakupov R, Lüsebrink F, Schulze P, Speck O. Motion correction in MRI of the brain. Phys Med Biol 2016;61(5):R32.
[23] Zaitsev M, Maclaren J, Herbst M. Motion artifacts in MRI: a complex problem with many partial solutions. J Magn Reson Imag 2015;42(4):887–901.
[24] Nárai Á, Hermann P, Auer T, Kemenczky P, Szalma J, Homolya I, Somogyi E, Vakli P, Weiss B, Vidnyánszky Z. Movement-related artefacts (MR-ART) dataset of matched motion-corrupted and clean structural MRI brain scans. Sci Data 2022;9(1):630.
[25] Wang J, Zhang M. Deepflash: an efficient network for learning-based medical image registration. In: Proceedings of the IEEE/CVF conference on computer vision and pattern recognition; 2020. p. 4444–52.
[26] Eichhorn H, Hammernik K, Spieker V, Epp SM, Rueckert D, Preibisch C, Schnabel JA. Physics-aware motion simulation for t2*-weighted brain MRI. In: International Workshop on Simulation and Synthesis in Medical Imaging. Springer; 2023. p. 42–52.
[27] Drobnjak I, Gavaghan D, Süli E, Pitt-Francis J, Jenkinson M. Development of a functional magnetic resonance imaging simulator for modeling realistic rigid-body motion artifacts. Magn Reson Med 2006;56(2):364–80.
[28] Stöcker T, Vahedipour K, Pflugfelder D, Shah NJ. High-performance computing MRI simulations. Magn Reson Med 2010;64(1):186–93.
[29] Castillo-Passi C, Coronado R, Varela-Mattatall G, Alberola-López C, Botnar R, Irarrazaval P. Komamri. JL: an open-source framework for general MRI simulations with GPU acceleration. Magn Reson Med 2023;90(1):329–42.

[30] Chen Q. Machine learning for magnetic resonance image reconstruction and analysis. 2019.

[31] Sarvaiya JN, Patnaik S, Bombaywala S. Image registration by template matching using normalized cross-correlation. In: 2009 international conference on advances in computing, control, and telecommunication technologies. IEEE; 2009. p. 819–22.

[32] Maes F, Collignon A, Vandermeulen D, Marchal G, Paul S. Multimodality image registration by maximization of mutual information. IEEE Trans Med Imag 1997;16(2): 187–98.

[33] Studholme C, Hill DLG, Hawkes DJ. An overlap invariant entropy measure of 3d medical image alignment. Pattern Recogn 1999;32(1):71–86.

[34] Rueckert D. Nonrigid registration: Concepts, algorithms, and applications. Medical Image Registration; 2001. p. 281–301.

[35] Rueckert D, Schnabel JA. Medical image registration. In: Biomedical image processing. Springer; 2010. p. 131–54.

[36] Rueckert D, Sonoda LI, Hayes C, Hill DLG, Leach MO, Hawkes DJ. Nonrigid registration using free-form deformations: application to breast MR images. IEEE Trans Med Imag 1999;18(8):712–21.

[37] Thirion J-P. Image matching as a diffusion process: an analogy with maxwell's demons. Med Image Anal 1998;2(3):243–60.

[38] Horn BKP, Schunck BG. Determining optical flow. Artif Intell 1981;17(1–3):185–203.

[39] Qin C, Wang S, Chen C, Bai W, Rueckert D. Generative myocardial motion tracking via latent space exploration with biomechanics-informed prior. Med Image Anal 2023;83: 102682.

[40] Rueckert D, Schnabel JA. Model-based and data-driven strategies in medical image computing. Proc IEEE 2019;108(1):110–24.

[41] Ashburner J. A fast diffeomorphic image registration algorithm. Neuroimage 2007;38 (1):95–113.

[42] Faisal Beg M, Miller MI, Trouvé A, Younes L. Computing large deformation metric mappings via geodesic flows of diffeomorphisms. Int J Comput Vis 2005;61:139–57.

[43] Fidon L, Ebner M, Garcia-Peraza-Herrera LC, Modat M, Ourselin S, Vercauteren T. Incompressible image registration using divergence-conforming b-splines. In: Medical Image Computing and Computer Assisted Intervention–MICCAI 2019: 22nd International Conference, Shenzhen, China, October 13–17, 2019, Proceedings, Part II 22. Springer; 2019. p. 438–46.

[44] Sotiras A, Davatzikos C, Paragios N. Deformable medical image registration: a survey. IEEE Trans Med Imag 2013;32(7):1153–90.

[45] Shi W, Zhuang X, Wang H, Duckett S, Luong DVN, Tobon-Gomez C, Tung KP, Edwards PJ, Rhode KS, Razavi RS, et al. A comprehensive cardiac motion estimation framework using both untagged and 3-d tagged MR images based on nonrigid registration. IEEE Trans Med Imag 2012;31(6):1263–75.

[46] Chen Q, Bai W, Schlemper J, Petersen SE, Piechnik SK, Neubauer S, Rueckert D. Joint learning of motion estimation and segmentation for cardiac MR image sequences. In: Medical Image Computing and Computer Assisted Intervention–MICCAI 2018: 21st International Conference, Granada, Spain, September 16-20, 2018, Proceedings, Part II 11. Springer; 2018. p. 472–80.

[47] Huizinga W, Metz CT, Poot DHJ, de Groot M, Niessen WJ, Leemans A, Klein S. Groupwise registration for correcting subject motion and eddy current distortions in diffusion MRI using a PCA based dissimilarity metric. In: Computational Diffusion

MRI and Brain Connectivity: MICCAI Workshops, Nagoya, Japan, September 22nd, 2013. Springer; 2014. p. 163–74.
[48] Wachinger C, Navab N. Simultaneous registration of multiple images: similarity metrics and efficient optimization. IEEE Trans Pattern Anal Mach Intell 2012;35(5): 1221–33.
[49] Metz CT, Klein S, Schaap M, Walsum Tvan, Niessen WJ. Nonrigid registration of dynamic medical imaging data using nd+ t b-splines and a groupwise optimization approach. Med Image Anal 2011;15(2):238–49.
[50] Miller EG, Matsakis NE, Viola PA. Learning from one example through shared densities on transforms. Proceedings IEEE Conference on Computer Vision and Pattern Recognition. CVPR 2000 (Cat. No. PR00662) 2000;1:464–71.
[51] Küstner T, Pan J, Qi H, Cruz G, Gilliam C, Blu T, Yang B, Gatidis S, Botnar R, Prieto C. LAPNet: non-rigid registration derived in k-space for magnetic resonance imaging. IEEE Trans Med Imag 2021;40(12):3686–97.
[52] Oksuz I, Ruijsink B, Puyol-Antón E, Clough JR, Cruz G, Bustin A, Prieto C, Botnar R, Rueckert D, Schnabel JA, King AP. Automatic CNN-based detection of cardiac MR motion artefacts using k-space data augmentation and curriculum learning. Med Image Anal 2019;55:136–47.
[53] Sommer K, Brosch T, Wiemker R, Harder T, Saalbach A, Hall CS, Andre JB. Correction of motion artifacts using a multi-resolution fully convolutional neural network. In: Proceedings of the ISMRM Scientific Meeting & Exhibition, Paris, 1175; 2018.
[54] Duffy BA, Zhang W, Tang H, Zhao L, Law M, Toga AW, Kim H. Retrospective correction of motion artifact affected structural MRI images using deep learning of simulated motion. In: Medical imaging with deep learning; 2022.
[55] Ronneberger O, Fischer P, Brox T. U-net: convolutional networks for biomedical image segmentation. In: Medical Image Computing and Computer-Assisted Intervention–MICCAI 2015: 18th International Conference, Munich, Germany, October 5-9, 2015, Proceedings, Part III 18. Springer; 2015. p. 234–41.
[56] Pawar K, Chen Z, Jon Shah N, Egan GF. MoCoNet: motion correction in 3d MPRAGE images using a convolutional neural network approach. arXiv preprint arXiv: 1807.10831 2018.
[57] Pawar K, Chen Z, Jon Shah N, Egan GF. Suppressing motion artefacts in MRI using an inception-ResNet network with motion simulation augmentation. NMR Biomed 2022; 35(4):e4225.
[58] Lee J, Kim B, Park HW. Mc2-net: motion correction network for multi-contrast brain MRI. Magn Reson Med 2021;86(2):1077–92.
[59] Pirkl CM, Cencini M, Kurzawski JW, Waldmannstetter D, Li H, Sekuboyina A, Endt S, Peretti L, Donatelli G, Rosa P, et al. Learning residual motion correction for fast and robust 3d multiparametric MRI. Med Image Anal 2022;77:102387.
[60] Tamada D, Kromrey M-L, Ichikawa S, Onishi H, Motosugi U. Motion artifact reduction using a convolutional neural network for dynamic contrast enhanced MR imaging of the liver. Magn Reson Med Sci 2020;19(1):64–76.
[61] Küstner T, Karim A, Yang J, Yang B, Schick F, Gatidis S. Retrospective correction of motion-affected MR images using deep learning frameworks. Magn Reson Med 2019; 82(4):1527–40.
[62] Fan Y, Liao H, Huang S, Luo Y, Fu H, Qi H. A survey of emerging applications of diffusion probabilistic models in MRI. arXiv preprint arXiv:2311.11383 2023.

[63] Levac B, Jalal A, Tamir JI. Accelerated motion correction for MRI using score-based generative models. In: 2023 IEEE 20th International Symposium on Biomedical Imaging (ISBI). IEEE; 2023. p. 1–5.
[64] Armanious K, Tanwar A, Abdulatif S, Küstner T, Gatidis S, Yang B. Unsupervised adversarial correction of rigid MR motion artifacts. In: 2020 IEEE 17th International Symposium on Biomedical Imaging (ISBI). IEEE; 2020. p. 1494–8.
[65] Liu S, Thung K-H, Qu L, Lin W, Shen D, Yap P-T. Learning MRI artefact removal with unpaired data. Nat Mach Intell 2021;3(1):60–7.
[66] Oh G, Lee JE, Ye JC. Annealed score-based diffusion model for MR motion artifact reduction. arXiv preprint arXiv:2301.03027 2023.
[67] Oh G, Lee JE, Ye JC. Unpaired MR motion artifact deep learning using outlier-rejecting bootstrap aggregation. IEEE Trans Med Imag 2021;40(11):3125–39.
[68] Dosovitskiy A, Fischer P, Ilg E, Hausser P, Hazirbas C, Golkov V, Van Der Smagt P, Cremers D, Brox T. Flownet: learning optical flow with convolutional networks. In: Proceedings of the IEEE International Conference on Computer Vision; 2015. p. 2758–66.
[69] Balakrishnan G, Zhao A, Sabuncu MR, Guttag J, Dalca AV. Voxelmorph: a learning framework for deformable medical image registration. IEEE Trans Med Imag 2019; 38(8):1788–800.
[70] Qi H, Fuin N, Cruz G, Pan J, Kuestner T, Bustin A, Botnar RM, Prieto C. Non-rigid respiratory motion estimation of whole-heart coronary MR images using unsupervised deep learning. IEEE Trans Med Imag 2020;40(1):444–54.
[71] Dalca AV, Balakrishnan G, Guttag J, Sabuncu MR. Unsupervised learning for fast probabilistic diffeomorphic registration. In: Medical Image Computing and Computer Assisted Intervention–MICCAI 2018: 21st International Conference, Granada, Spain, September 16-20, 2018, Proceedings, Part I. Springer; 2018. p. 729–38.
[72] Chen Q, Shi B, Liao R, Mansi T, Rueckert D, Ali K. Unsupervised deformable registration for multi-modal images via disentangled representations. In: International Conference on Information Processing in Medical Imaging. Springer; 2019. p. 249–61.
[73] Qiu H, Qin C, Schuh A, Hammernik K, Rueckert D. Learning diffeomorphic and modality-invariant registration using b-splines. In: Medical imaging with deep learning; 2021.
[74] Li B, Niessen WJ, Klein S, Ikram MA, Vernooij MW, Bron EE. Learning unbiased group-wise registration (LUGR) and joint segmentation: evaluation on longitudinal diffusion MRI. Medical Imaging 2021: Image Process 2021;11596:136–44.
[75] Huttinga NRF, Van den Berg CAT, Luijten PR, Sbrizzi A. MR-MOTUS: model-based non-rigid motion estimation for MR-guided radiotherapy using a reference image and minimal k-space data. Phys Med Biol 2020;65(1):015004.
[76] Zhang M, Fletcher PT. Fast diffeomorphic image registration via fourier-approximated lie algebras. Int J Comput Vis 2019;127:61–73.
[77] Gilliam C, Blu T. Local all-pass filters for optical flow estimation. In: 2015 IEEE International Conference on Acoustics, Speech and Signal Processing (ICASSP). IEEE; 2015. p. 1533–7.
[78] Haskell MW, Cauley SF, Wald LL. Targeted motion estimation and reduction (TAMER): data consistency based motion mitigation for MRI using a reduced model joint optimization. IEEE Trans Med Imag 2018;37(5):1253–65.
[79] Polak D, Nicolas Splitthoff D, Clifford B, Lo W-C, Huang SY, Conklin J, Wald LL, Setsompop K, Cauley S. Scout accelerated motion estimation and reduction (SAMER). Magn Reson Med 2022;87(1):163–78.

[80] Odille F, Vuissoz P-A, Marie P-Y, Felblinger J. Generalized reconstruction by inversion of coupled systems (GRICS) applied to free-breathing MRI. Magn Reson Med 2008;60 (1):146–57.
[81] Cruz G, Atkinson D, Henningsson M, Botnar RM, Prieto C. Highly efficient nonrigid motion-corrected 3d whole-heart coronary vessel wall imaging. Magn Reson Med 2017;77(5):1894–908.
[82] Bustin A, Rashid I, Cruz G, Hajhosseiny R, Correia T, Neji R, Rajani R, Ismail TF, Botnar RM, Prieto C. 3d whole-heart isotropic sub-millimeter resolution coronary magnetic resonance angiography with non-rigid motion-compensated prost. J Cardiovasc Magn Reson 2020;22(1):1–16.
[83] Feng L, Axel L, Chandarana H, Block KT, Sodickson DK, Otazo R. XD-GRASP: golden-angle radial MRI with reconstruction of extra motion-state dimensions using compressed sensing. Magn Reson Med 2016;75(2):775–88.
[84] Rank CM, Heußer T, Buzan MTA, Wetscherek A, Freitag MT, Dinkel J, Kachelrieß M. 4d respiratory motion-compensated image reconstruction of free-breathing radial MR data with very high undersampling. Magn Reson Med 2017;77(3):1170–83.
[85] Haskell MW, Cauley SF, Bilgic B, Hossbach J, Splitthoff DN, Pfeuffer J, Setsompop K, Wald LL. Network accelerated motion estimation and reduction (NAMER): convolutional neural network guided retrospective motion correction using a separable motion model. Magn Reson Med 2019;82(4):1452–61.
[86] Hossbach J, Nicolas Splitthoff D, Cauley S, Clifford B, Polak D, Lo W-C, Meyer H, Maier A. Deep learning-based motion quantification from k-space for fast model-based magnetic resonance imaging motion correction. Med Phys 2023;50(4):2148–61.
[87] Feinler MS, Hahn BN. Retrospective motion correction in gradient echo MRI by explicit motion estimation using deep CNNS. arXiv preprint arXiv:2303.17239 2023.
[88] Eichhorn H, Hammernik K, Spieker V, Epp SM, Rueckert D, Preibisch C, Schnabel JA. Deep learning-based detection of motion-affected k-space lines for t2*-weighted MRI. arXiv preprint arXiv:2303.10987 2023.
[89] Qi H, Hajhosseiny R, Cruz G, Kuestner T, Kunze K, Neji R, Botnar R, Prieto C. End-to-end deep learning nonrigid motion-corrected reconstruction for highly accelerated free-breathing coronary MRA. Magn Reson Med 2021;86(4):1983–96.
[90] Seegoolam G, Schlemper J, Qin C, Price A, Hajnal J, Rueckert D. Exploiting motion for deep learning reconstruction of extremely-undersampled dynamic MRI. In: International Conference on Medical Image Computing and Computer-Assisted Intervention. Springer; 2019. p. 704–12.
[91] Yang J, Küstner T, Hu P, Liò P, Qi H. End-to-end deep learning of non-rigid groupwise registration and reconstruction of dynamic MRI. Front Cardiovasc Med 2022;9:880186.

CHAPTER

Map the image: Machine learning for quantitative MR mapping

10

Li Feng[1,*], Haoyang Pei[1,2] and Fang Liu[3,4]

[1]*Center for Advanced Imaging Innovation and Research (CAI2R), New York University Grossman School of Medicine, New York, NY, United States;* [2]*Department of Electrical and Computer Engineering, NYU Tandon School of Engineering, New York, NY, United States;* [3]*Athinoula A. Martinos Center for Biomedical Imaging, Massachusetts General Hospital, Charlestown, MA, United States;* [4]*Harvard Medical School, Boston, MA, United States*

[*]*Corresponding author Email: Li.Feng@nyulangone.org*

10.1 Introduction

Magnetic resonance imaging (MRI) is a cornerstone of modern clinical diagnosis, playing a vital role in a wide range of clinical applications. MRI not only captures detailed structural information about the underlying anatomy, like many other imaging modalities, but also offers unique capabilities to quantitatively assess different physical and biochemical properties of tissues [1]. The latter approach, known as quantitative MRI or quantitative MR mapping, provides a detailed view of tissue composition across various clinical settings without radiation exposure.

Over the past decades, quantitative MRI has undergone significant advancements, driven by technological innovations in both software and hardware. Various quantitative MRI methods have been developed, refined, and optimized for a growing range of clinical applications. For example, more precise and reliable techniques for T1, T2, T2*, and T1rho mapping have been developed, which enable more accurate analysis of tissue composition and function [2–12]. MR fingerprinting (MRF) has emerged as a revolutionary approach to enable rapid and comprehensive evaluation of multiple tissue properties simultaneously [13–16]. Perfusion MRI now achieves unprecedented spatial and temporal resolution for pharmacokinetic quantification, providing detailed insights into blood flow and tissue vascular permeability [17–21]. Novel chemical exchange saturation transfer (CEST) MRI markers have been introduced to provide more details about tissue metabolism and cellular changes associated with various diseases [22–26]. Other examples include advanced diffusion modeling methods to probe tissue microstructure [27–32] and quantitative susceptibility mapping (QSM) techniques for assessing iron content and blood oxygenation levels in the brain and other organs [33–38]. These advancements

Machine Learning in MRI. https://doi.org/10.1016/B978-0-443-14109-6.00010-9

have all proven valuable in various clinical applications, including early disease detection, more thorough and objective evaluation of treatment response, and understanding of different pathological conditions in cancer, neurological disorders, and musculoskeletal diseases [39–46].

Despite its advantages and clinical benefits, the widespread adoption of quantitative MRI in clinical practice is hindered by several challenges [47]. One of the most significant issues is the time-intensive nature of both data acquisition and analysis. Traditional quantitative MRI techniques often require lengthy scan times, which may be impractical for daily clinical use. In addition, the complexity of data analysis and the need for specialized expertise to interpret quantitative measures present major barriers in the clinical workflow. Another challenge is the standardization of quantitative MRI protocols [48]. Variations of imaging parameters, acquisition schemes, and hardware configurations from MRI scanners of different vendors or even across different software versions can lead to inconsistent results. This lack of standardization impedes the generation of reproducible data for comparative studies across institutions or over time, thereby limiting its broader applications in routine clinical practice, multicenter studies, and longitudinal disease assessment.

The integration of deep learning technologies into MRI has demonstrated a profound impact across various stages of the MRI workflow, from data collection to image interpretation [49–51]. Deep learning is also well-suited for quantitative MRI, presenting great potential to address longstanding challenges in its clinical implementation [52]. For example, deep learning can be used to enable accelerated data acquisition, rapid image reconstruction, and efficient generation of quantitative parameter maps, all of which are essential for the successful integration of quantitative MRI into clinical practice.

This chapter presents an overview of the application of deep learning in quantitative MRI. Following the introduction, Section 10.2 provides a concise review of the general quantitative MRI workflow and the challenges currently limiting its clinical implementation. Section 10.3 discusses how deep learning can be applied to improve different stages of this workflow. Section 10.4 addresses the challenges associated with implementing deep learning for quantitative MRI, including issues such as data diversity, model generalizability, and the need for large, annotated datasets to train robust models. Finally, Section 10.5 explores the future prospects of deep learning in quantitative MRI, including potential solutions to overcome the challenges discussed in Section 10.4. By the end of this chapter, readers are expected to have a clear understanding of how deep learning techniques can be applied to quantitative MRI and their potential to transform quantitative MRI into a powerful, efficient, and valuable imaging tool in clinical practice and medical imaging research.

10.2 The general workflow of quantitative MRI

The general workflow of quantitative MRI typically involves three main steps: data acquisition, image reconstruction, and parameter fitting, as illustrated in Fig. 10.1. A brief overview of these steps is provided below.

10.2.1 Data acquisition

The first stage of quantitative MRI involves acquiring data using a specialized pulse sequence designed to encode specific tissue characteristics of interest, such as T1, T2, fat/water composition, or water diffusion. Selecting an appropriate sequence and acquisition parameters is essential to ensure accurate quantification. Advanced techniques, such as MRF, are capable of detecting multiple tissue properties simultaneously, thus enabling the quantification of different imaging parameters from a

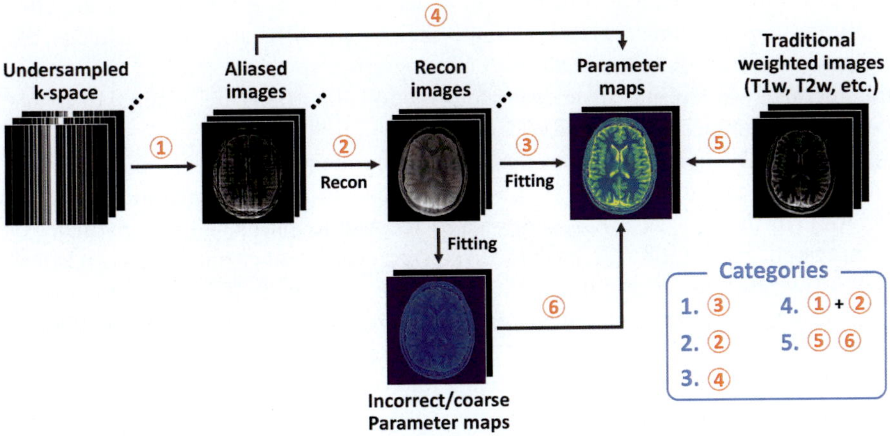

FIGURE 10.1

The general imaging workflow of quantitative MRI, which typically consists of three main steps, including data acquisition, image reconstruction, and parameter fitting. Deep learning techniques can be applied to improve different stages of the quantitative MRI workflow indicated by the arrows, and they can be classified into five distinct categories according to the role of deep learning in the quantitative MRI workflow.

single acquisition [13]. Data acquisition in quantitative MRI can be implemented using either a standard Cartesian sampling trajectory or non-Cartesian sampling schemes, such as radial or spiral trajectories [53–55]. Over the past decade, non-Cartesian sampling has gained substantial attention due to its unique incoherent imaging behavior, which can be combined with sparsity-based or deep learning-based reconstruction techniques to achieve higher acceleration rates. Regardless of the sampling scheme, all quantitative MRI acquisitions share a common process of collecting multiple images with varying contrast weightings. This is achieved by adjusting relevant imaging parameters, such as inversion recovery times for T1 mapping, echo times for T2/T2* mapping, or diffusion preparations for diffusion-weighted imaging.

10.2.2 Image reconstruction

The need to acquire multiple images with varying contrast weightings in quantitative MRI results in longer scan times compared with traditional qualitative MRI, which typically requires only a single image per contrast type. To address this, accelerated data acquisition, often achieved through undersampling, is essential to ensure clinically feasible scan times. However, undersampling in MRI acquisition violates the Nyquist sampling theorem, leading to aliasing artifacts using standard reconstruction methods. As a result, advanced reconstruction algorithms are required to enable fast imaging in quantitative MRI.

Early fast imaging techniques aimed at overcoming this challenge included parallel imaging and its variants [56–63], as well as various k-t acceleration methods [64–67]. In the last decade, sparsity-based reconstruction techniques, particularly compressed sensing [68–77], have received significant interest and have been extensively investigated for rapid quantitative MRI acquisition [78–82]. Compressed sensing is particularly well-suited for quantitative MRI reconstruction, as the dynamic dimension of quantitative images, characterized by variation in contrast weighting, provides abundant spatiotemporal correlations that can be exploited to improve image reconstruction performance at high acceleration rates.

The general optimization problem of sparsity-based iterative reconstruction can be formulated as Eq. (10.1):

$$\rho = \text{argmin}_{\rho} \left\| \text{E}\rho - \text{s} \right\|_2^2 + \lambda R(\rho) \tag{10.1}$$

where E = BFC represents a multicoil encoding operator for image reconstruction combining the undersampling pattern B, the fast Fourier transform (FFT) operator F, and the coil sensitivity maps C. s denotes the undersampled multicoil k-space, and ρ represents the series of contrast-weighted images to be reconstructed for quantification. $R(\cdot)$ denotes a regularization function to enforce a constraint (e.g., total variations) on the reconstructed images ρ by incorporating prior knowledge to improve reconstruction performance.

Nonetheless, these conventional methods have their drawbacks. Parallel imaging, while widely used in clinical practice, offers only modest acceleration rates.

Sparsity-based iterative reconstruction faces challenges such as computationally intensive processing, the need for careful tuning of sparsifying transforms and regularization parameters, and inconsistent performance across different patient anatomies, pathologies, and MRI sequences. These limitations of sparsity-based approaches have hindered their widespread clinical adoption.

10.2.3 Parameter fitting

In the final stage of quantitative MRI, MR parameters of interest are derived from a series of reconstructed images with varying contrast weightings. This process involves modeling the signal intensity across contrast-varied images, either within a specific region of interest (ROI) or on a pixel-by-pixel basis, using a signal model that incorporates relevant quantitative parameters, such as relaxation rates (e.g., T1, T2, T2*), diffusion coefficients, and others. The fitting process is commonly performed using a least-squares algorithm, as shown in Eq. (10.2), which minimizes the difference between the reconstructed images ρ and synthesized contrast-weighted images from the estimated parameter p according to the underlying signal model ϕ.

$$p = \text{argmin}_p \, \|\phi(p) - \rho\|_2^2 \tag{10.2}$$

When modeling is performed on a pixel-wise basis, quantitative parameter maps (or multiple parameter maps) of the same dimensions as the acquired images can be generated. However, since the fitting process is carried out at each pixel location, it is more computationally intensive and may be sensitive to image noise, thus posing challenges for achieving robust quantification.

In addition to parameter fitting using an analytic signal model, novel quantitative MRI approaches, such as MRF, propose to generate parameter maps through dictionary-based matching, where a preestimated dictionary, generated based on the Bloch equations with corresponding imaging parameters, is used for efficient parameter estimation using pattern matching algorithms [13,83].

In contrast to the sequential approach of image reconstruction followed by parameter fitting, MR parameter maps can also be directly reconstructed from undersampled multicoil k-space data or images using a model-based reconstruction approach [81,82]. This technique combines image reconstruction and parameter fitting into a single optimization step, as formulated in Eq. (10.3).

$$p = \text{argmin}_p \left\| \text{E}\phi(p) - \text{s} \right\|_2^2 + \lambda R(p) \tag{10.3}$$

In this formulation, iterative reconstruction directly estimates MR parameter maps of interest, with an additional regularization function applied to ensure reconstruction stability and robustness. This integrated approach holds promise for improving the efficiency and accuracy of quantitative MRI, particularly in scenarios where traditional methods face limitations.

10.3 Application of deep learning in quantitative MRI

Over the past years, deep learning has emerged as a powerful tool for overcoming different challenges presented in the quantitative MRI workflow [52]. As shown in Fig. 10.1, deep learning can be applied to improve different stages of the quantitative MRI workflow, including more intelligent data acquisition (arrow 1), faster and more reliable image reconstruction (arrow 2), and more efficient and robust parameter fitting (arrow 3). Moreover, deep learning also facilitates direct reconstruction of quantitative parameter maps from undersampled images (arrow 4), a model-based reconstruction strategy that combines image reconstruction and parameter fitting into a single step as described in Section 10.2. In addition, deep learning can also be utilized to generate parameter maps from conventional contrast-weighted images (arrow 5) or from inaccurate parameter maps (arrow 6). In this chapter, deep learning-based quantitative MRI methods are classified into five distinct categories according to their roles in the quantitative MRI workflow, as outlined in Fig. 10.1. The subsequent subsections will describe each category in detail.

10.3.1 Deep learning-based quantitative parameter fitting

In this category, deep learning replaces the traditional parameter fitting step shown in Eq. (10.2) to enable more efficient and robust generation of parameter maps. This approach overcomes the limitation of conventional parameter fitting or dictionary-based parameter matching that requires a time-consuming process. Fig. 10.2 shows an example of using deep learning to replace conventional least-squares fitting for T1 map estimation from an inversion recovery Look-Locker acquisition, where deep learning achieves a substantial reduction in parameter fitting time. In addition, deep learning also enables accurate generation of parameter maps from a smaller set of contrast-weighted images, such as a reduced range of echo times for T2 mapping or fewer spin-lock times for T1rho mapping. This in turn helps reduce scan times, increase volumetric coverage, and/or lower the Specific Absorption Rate (SAR).

Deep learning models in this category can be trained using two different approaches. The first one performs network training on the entire images to exploit their spatiotemporal correlations within the series of contrast-weighted images. While this approach ensures optimal model performance, it demands a considerable amount of quantitative images for network training, which might not be readily available. The second approach focuses on pixel-level training without relying on spatial correlations between pixels. In this approach, individual pixels within the same image are considered separate datasets, which allows for network training with a limited set of image series. Moreover, since spatial image correlations are not utilized, this approach also enables network training on simulated signal curves derived from corresponding fitting models, with extra noise added to the simulated

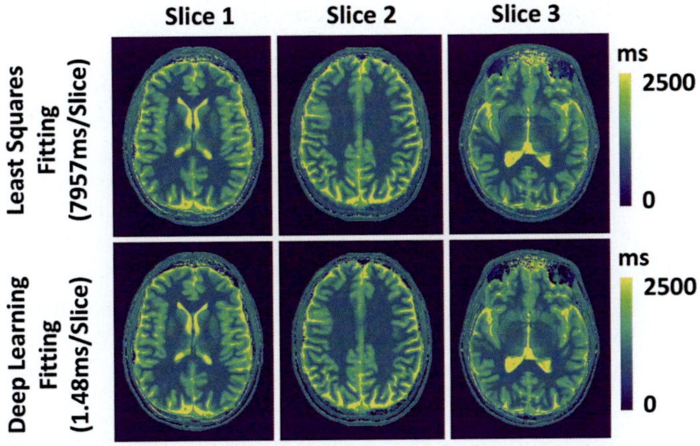

FIGURE 10.2

An example showing the use of using deep learning to replace conventional least-squares fitting for T1 map estimation from an inversion recovery Look-Locker acquisition, where deep learning achieves a substantial reduction in parameter fitting time.

curves to better replicate realistic signal evolutions. This strategy eliminates the need for training datasets to develop deep learning models for parameter fitting.

In deep learning-based parameter fitting, supervised training is the most commonly employed training scheme to date. It involves creating a loss function to ensure consistency between the estimated parameter maps and the reference parameter maps, as shown in Eq. (10.4). Here, the reference parameter maps (p) can be obtained with traditional fitting algorithms, and f_θ is a trained neural network parameterized by θ. In practical implementation, the difference can be quantified using a specific metric (denoted as m), such as the L1-norm, the L2-norm, or the structural similarity index (SSIM).

$$\theta = \text{argmin}_\theta \left\| f_\theta(\rho) - p \right\|_m \tag{10.4}$$

In addition, a loss function to ensure model consistency can be further incorporated by minimizing the difference between the input images and the synthesized contrast-weighted images generated from the estimated parameter maps, as shown in Eq. (10.5).

$$\theta = \text{argmin}_\theta \left\| f_\theta(\rho) - p \right\|_m + \left\| \phi(f_\theta(\rho)) - \rho \right\|_n \tag{10.5}$$

Here, the synthesized images are generated from the estimated parameter maps based on the underlying physical signal model (ϕ). The difference in the model consistency loss can be quantified using a metric (denoted as n) that is the same as or different from the first loss function. This model consistency loss enhances the robustness and fidelity of the trained model. Moreover, the use of a model consistency loss also enables network training in a self-supervised manner, as shown in

Eq. (10.6), which eliminates the need for reference parameter maps. In this scenario, an additional regularization function (denoted by $R(\cdot)$), such as a spatial total variation constraint, can be integrated to improve the training and reconstruction performance.

$$\theta = \operatorname{argmin}_{\theta} \|\phi(f_{\theta}(\rho)) - \rho\|_{n} + R(f_{\theta}(\rho)) \tag{10.6}$$

Instead of simply using deep learning as a substitute for parameter fitting, a more compelling and practical approach is leveraging deep learning to generate accurate parameter maps from a limited set of contrast-weighted images, such as those acquired with fewer echoes. In this context, reference parameter maps can still be derived from a complete set of contrast-weighted images. This process is formulated as Eq. (10.7) below, where ρ_r represents the reduced set of contrast-weighted images.

$$\theta = \operatorname{argmin}_{\theta} \|f_{\theta}(\rho_r) - p\|_{m} + \|\phi(f_{\theta}(\rho_r)) - \rho\|_{n} \tag{10.7}$$

This approach can also be potentially extended to self-supervised training without the need for reference parameter maps as shown in Eq. (10.6).

10.3.1.1 Representative studies

Various studies have proposed to generate quantitative parameter maps using deep learning, either from a full set or a reduced set of contrast-varying images [84–90]. For example, Menezes et al. proposed a deep learning-based approach named Multi-Decoder Water-Fat Separation Network (MDWF-Net) to estimate the proton density fat fraction (PDFF) of the liver from chemical shift-encoded MRI using a reduced number of echoes (e.g., from 6 echoes to 3 echoes) [84]. MDWF-Net employs a convolutional neural network (CNN) to generate water/fat-separated images, an R2* map, and an off-resonance map (Δf) for PDFF calculation. This network was trained in a supervised manner using ground truth parameter maps derived from chemical shift-encoded MRI with all echoes. MDWF-Net demonstrated accurate estimation of liver PDFF using only 3 echoes to allow for a shorter repetition time (TR) and thus reduction in scan times.

A number of studies have also demonstrated deep learning-based quantification using simulated dictionaries for network training [91–95]. For example, Cohen et al. proposed a deep learning method to streamline the time-consuming dictionary-based matching process in MRF for efficient generation of parameter maps [91]. This approach, named the MR Fingerprinting Deep Reconstruction Network (DRONE), employs a fully connected network trained on simulated MRF data computed with the extended phase graph formalism at the pixel level. Moreover, DRONE has also shown improved robustness to noise and dictionary precision compared with standard dictionary-based matching in MRF. However, a significant drawback of DRONE, and other techniques that perform network training at the pixel level, is its dependence on artifact-free images for parameter generation, since the training occurs at the pixel level without leveraging spatial correlations.

Several studies have demonstrated the feasibility of deep learning-based quantification using self-supervised learning based on the model-consistency loss [96–99]. For example, Jun et al. proposed a self-supervised learning approach for efficient estimation of T1, T2, and inversion efficiency maps without using reference parameter maps [96]. This technique, called SSL-QALAS, is based on the 3D interleaved Look-Locker acquisition sequence with a T2 preparation pulse (3D-QALAS) for data acquisition. SSL-QALAS employs a convolutional neural network (CNN) to derive multiple parameter maps from five QALAS contrast images and a B1+ map. The deep learning model is trained by minimizing the difference between the acquired images and synthetic images generated from the estimated parameter maps. This approach enables rapid generation of multiparametric maps from 3D-QALAS measurements, eliminating the requirement for an external dictionary or labeled ground-truth training data.

10.3.2 Deep learning-based image reconstruction

In this category, deep learning replaces the traditional reconstruction step to generate clean contrast-weighted images reconstructed from undersampled k-space, while the rest of the workflow remains the same. The main advantages of utilizing deep learning over traditional reconstruction algorithms are its capabilities to significantly expedite reconstruction speed and achieve higher acceleration rates. Given that quantitative MRI comprises a series of contrast-weighted images, deep learning-based reconstruction usually exploits their spatiotemporal correlations, like other dynamic MRI reconstruction tasks as described in Chapter 7. As shown in Eq. (10.8) below, this approach typically involves supervised training, where a loss function is designed to ensure consistency between the reference and reconstructed image series. A data consistency loss can also be incorporated to improve the reconstruction performance.

$$\theta = \text{argmin}_{\theta} \left\| f_{\theta}(\rho_u) - \rho \right\|_m + \left\| \mathrm{E^H E} f_{\theta}(\rho_u) - \rho_u \right\|_n \qquad (10.8)$$

In Eq. (10.8), ρ_u represents the undersampled image series to be reconstructed for parameter fitting, and f_θ is a reconstruction neural network parameterized by θ. $\mathrm{E^H}$ denotes the Hermitian transpose of E, and $\mathrm{E^H E} = \mathrm{C^H F^H BFC}$.

In addition to supervised training, quantitative image reconstruction can also be implemented using self-supervised models. For example, the self-supervised learning via data undersampling (SSDU) method [100], which does not require artifact-free reference images for network training, holds great potential for application in quantitative MRI. Moreover, quantitative images can also be reconstructed from a reduced number of contrast-weighted undersampled images to further reduce scan times.

Deep learning-based image reconstruction and parameter fitting can be combined together using two distinct deep learning models. These models are typically trained simultaneously, with different loss functions applied to contrast-weighted images and parameter maps separately. During inference, the trained reconstruction and fitting models can be concatenated to directly estimate parameter maps in a single step. This method combines the benefits of the first two categories to enable efficient generation of parameter maps from undersampled images. Fig. 10.3 shows an example comparing sparsity-based iterative reconstruction with deep learning reconstruction for highly accelerated T1 mapping using the Look-Locker imaging method, where the deep learning method achieves superior performance in reconstructing both quantitative images and T1 maps.

10.3.2.1 Representative studies

Different studies have demonstrated the application of deep learning for reconstructing quantitative MR images [101–103] or for joint image reconstruction and parameter fitting [104–108]. For example, Liu et al. proposed a reconstruction and generation network, referred to as RG-Net, for highly accelerated T1rho mapping by undersampling k-space and reducing the acquired contrast-weighted images simultaneously [101]. The proposed RG-Net contains a reconstruction module and a generative module. The reconstruction module reconstructs MR images from undersampled k-space data, while the generative module synthesizes the missing contrast-weighted images from the reconstructed images. The T1rho maps can then be estimated from the reconstructed images using standard parameter fitting. This model, as described in Ref. [101], can achieve a high acceleration rate while maintaining the reconstruction quality through simultaneous undersampling of k-space and reduction in the number of contrast-weighted images.

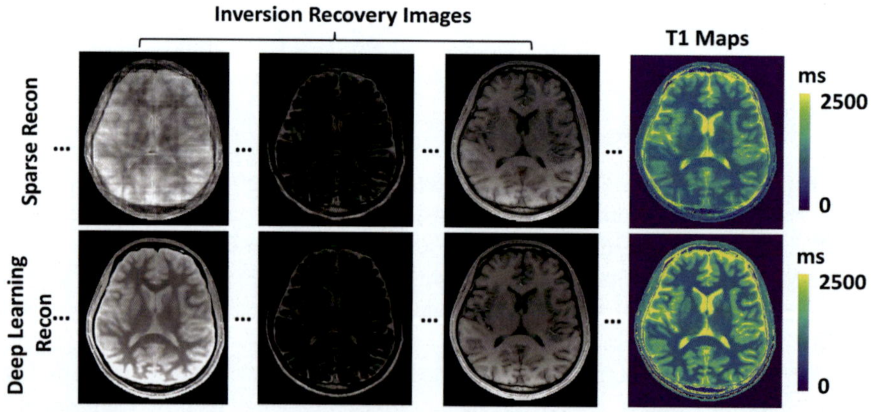

FIGURE 10.3

An example comparing sparsity-based iterative reconstruction with deep learning reconstruction for highly accelerated T1 mapping using the Look-Locker imaging method, where deep learning method achieves superior performance in reconstructing both quantitative images and T1 maps.

Wu et al. proposed a multistep framework for accelerating the acquisition of T1 mapping with reduced variable flip angle images [104]. This framework is structured around two main modules: a reconstruction module and a parameter mapping module. It first reconstructs contrast-weighted images with fewer flip angles from undersampled k-space data and then proceeds to generate 3D T1 maps from these reconstructed images. The training for both image reconstruction and parameter generation is performed simultaneously through supervised learning, in which a loss function is enforced between the reconstructed and ground truth images/maps to jointly optimize both networks.

10.3.3 Model-based quantitative MRI reconstruction using deep learning

Model-based quantitative MRI reconstruction has been well demonstrated using traditional iterative reconstruction algorithms to directly reconstruct parameter maps from undersampled k-space or images, as described above in Section 10.2. However, conventional model-based methods are often computationally intensive, and their performance varies depending on optimization factors, which limits their widespread clinical application. Deep learning can be an efficient alternative for implementing model-based quantitative MRI reconstruction. In this approach, a deep learning model can be trained to directly estimate parameter maps from undersampled images or k-space data, and the underlying MR signal model can be further incorporated to enforce model consistency, as shown in Fig. 10.4.

There are two main approaches for model-based network training. The first one is based on supervised training, which uses reference data to provide reference parameter maps. As shown in Eq. (10.9), the training can be performed using two loss functions, including one to ensure consistency between the estimated parameter maps and the reference maps and another one to ensure model consistency for improving reconstruction performance and robustness.

$$\theta = \text{argmin}_\theta \left\| f_\theta(\rho_u) - p \right\|_m + \left\| \mathrm{E}^{\mathrm{H}}\mathrm{E}\phi(f_\theta(\rho_u)) - \rho_u \right\|_n \tag{10.9}$$

The second approach is based on self-supervised training, which employs only a single loss function to ensure model and data consistency between input undersampled images (or k-space data) and the undersampled images (or k-space data) synthesized using corresponding MR physics model from the estimated parameter maps, as shown in Eq. (10.10). In addition, an optional regularization function may be applied to the reconstructed parameter maps,

$$\theta = \text{argmin}_\theta \left\| \mathrm{E}^{\mathrm{H}}\mathrm{E}\phi(f_\theta(\rho_u)) - \rho_u \right\|_n + \lambda R(f_\theta(\rho_u)) \tag{10.10}$$

There are different ways of implementation for model-based quantitative MRI reconstruction using deep learning. One approach is to use deep learning to directly convert quantitative images to parameter maps. This can be achieved through end-to-end network models such as the U-Net, which enables efficient generation of

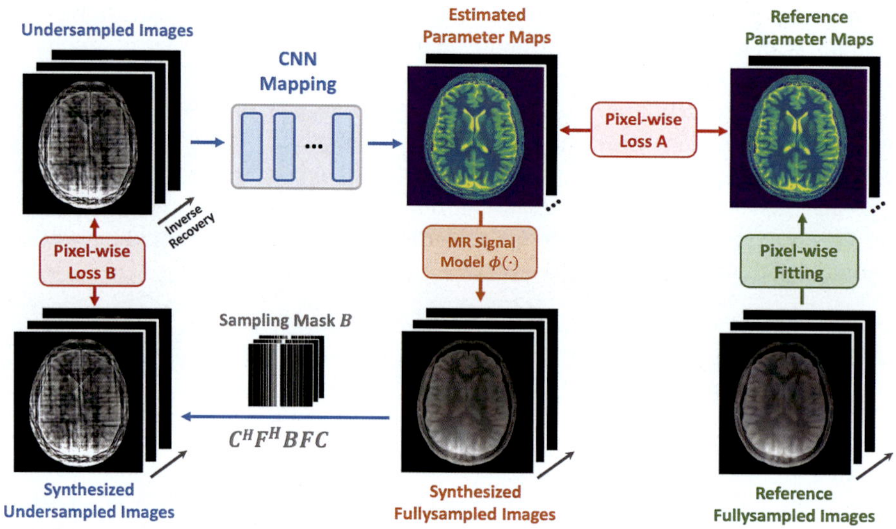

FIGURE 10.4

A model-based quantitative MRI reconstruction pipeline for accelerated quantitative MRI. A neural network is employed to transform undersampled contrast-weighted image series to parameter maps. A loss function is enforced between the estimated and reference parameter maps. In addition, a model consistency loss is further incorporated between the input images and synthetic images generated from the estimated parameter maps.

input images in parameter maps through cross-domain learning. Advanced network structures such as deep generative adversarial networks (GANs) can further improve the efficiency of end-to-end mapping and reconstruction accuracy. However, end-to-end mapping requires a large amount of training datasets, which may be challenging to obtain in practice. Recent studies have explored integrating traditional optimization algorithms into deep learning networks. Specifically, iterative reconstruction algorithms leveraging optimization methods can be unrolled and represented in the mapping network to help network training, thereby resulting in faster and more stable model convergence with lower training data demand.

10.3.3.1 Representative studies

Over the past few years, many model-based reconstruction techniques within this category have been demonstrated using deep learning [109–117]. One of the early studies to showcase this feasibility was conducted by Liu et al. who have developed a supervised deep learning approach called Model-Augmented Neural neTwork with Incoherence Sampling (MANTIS) for accelerated T2 mapping [109]. MANTIS employs a U-Net architecture to transform undersampled multiecho spin-echo (MESE) images into T2 maps. A T2 signal model is further integrated into the model training to enhance training robustness. As shown in Fig. 10.5, MANTIS demonstrated the

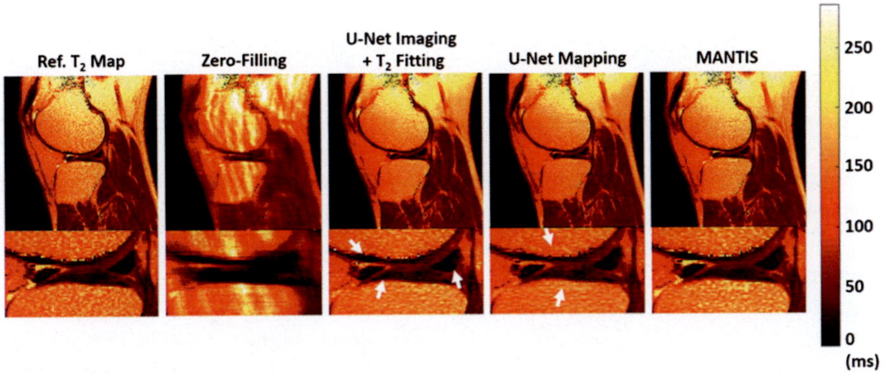

FIGURE 10.5

Demonstration of accelerated T2 mapping using MANTIS. Model-based reconstruction enables improved accuracy compared with standard U-Net without the model consistency loss and also the separate steps for image reconstruction and parameter fitting.

Image was reproduced from Fig. 4 in Magn Reson Med. 2019 Jul; 82(1):174–188 with permission from John Wiley & Sons, Inc.

capability of producing accurate T2 maps of the knee to enable efficient and accurate T2 quantification across the entire knee. It has also been demonstrated to outperform traditional iterative reconstruction algorithms that rely on low-rank and sparsity constraints. In addition, MANTIS has also been augmented with GANs to further improve mapping accuracy and detail preservation in critical anatomical structures [110].

In a subsequent study, Liu et al. extended the MANTIS framework to a self-supervised version, named Reference-free Latent Map Extraction (RELAX), for accelerated quantitative parameter mapping [111]. RELAX directly converts undersampled k-space data into MR parameter maps and relies on a model consistency loss (as shown in Eq. 10.10) to ensure accurate generation of parameter maps without using reference datasets. Its performance has been demonstrated for rapid T1 and T2 mapping in the knee and brain. Recently, Bian et al. further extended this method into a new iteration known as RELAX with MOdel REinforcement (RELAX-MORE), aiming to address the challenges posed by limited training data and the domain shift problem in deep learning [118]. The domain shift problem, which can lead to suboptimal reconstruction results due to discrepancies between training and testing data, is mitigated by RELAX-MORE through subject-specific self-supervised learning. RELAX-MORE employs a bi-level optimization design within the deep network. Similar to RELAX, its upper level employs self-supervised learning to transform undersampled k-space data into MR parameter maps, guided by the underlying MR signal model. The lower level of RELAX-MORE is different from RELAX and replaces the end-to-end mapping with a forward encoding process to solve for the optimal solution of model-based quantitative MRI reconstruction using a proximal gradient descent algorithm.

10.3.4 Joint learning of data acquisition and image reconstruction

Techniques in this category employ deep learning to simultaneously optimize the undersampling pattern, acquisition schemes, and image reconstruction process. The objective is to improve the accuracy of the reconstructed images at a given undersampling rate while minimizing the number of contrast-weighted images required to generate accurate parameter maps. Note that techniques within this category bear similarities to those discussed in Chapter 4, highlighting the potential of deep learning as an effective tool to optimize data acquisition in quantitative MRI.

10.3.4.1 Representative studies

While several studies have demonstrated joint learning of data acquisition and image reconstruction for static imaging [119–121], its extension to quantitative MRI is relatively limited to date. Zhang et al. recently presented a new framework, called Learned Acquisition and Reconstruction Optimization (LARO), to accelerate multiecho gradient echo (mGRE) imaging for quantitative susceptibility mapping (QSM) [122]. The LARO network architecture features a deep Alternating Direction Method of Multipliers (ADMM) backbone for undersampled k-space reconstruction, a Sampling Pattern Optimization (SPO) module designed to determine the optimal k-space undersampling pattern, and a Temporal Feature Fusion (TFF) module integrated into the deep ADMM network to capture the signal evolution across echoes. The optimization of these modules occurred in two phases. In the first phase, the network parameters of the deep ADMM network and the SPO module were updated jointly. In the second phase, the deep ADMM network, pretrained in the first phase, was fine-tuned using the learned undersampling pattern. The results from the LARO study indicate that the optimized sampling pattern, alongside the proposed reconstruction strategy, can significantly improve the quality of multiecho image reconstruction for QSM.

10.3.5 Generation of parameter maps from conventional weighted images or inaccurate maps using deep learning

In this category, deep learning is used to directly generate parameter maps from conventional weighted images (e.g., T1-weighted and/or T2-weighted images) without adhering to the standard quantitative MRI workflow. It can be formulated as Eq. (10.11) below, where ρ_w represents conventional weighted images.

$$\theta = \text{argmin}_\theta \left\| f_\theta(\rho_w) - p \right\|_m \qquad (10.11)$$

This approach is based on the premise that since conventional weighted images are already acquired in current clinical practice, it is feasible to synthesize quantitative parameter maps from these images, thereby obviating the need for acquiring additional quantitative images for parameter fitting. Two distinct methods can be employed in this category. In the first approach, deep learning can be applied to identify a potential mapping from traditional weighted images to parameter maps by applying a loss function between the estimated maps and reference maps derived from a series of reference quantitative images. In the second approach, deep learning can first transform traditional weighted images into a series of contrast-weighted images that are needed for parameter quantification, from which a standard fitting process is then performed to produce the corresponding parameter maps. To improve performance, network training can be performed on multicontrast images, such as combined T1-weighted and T2-weighted images, to synthesize desired parameter maps of interest. However, it is important to recognize that deep learning acts as a "black box" in this scenario without relying on underlying physical principles.

Another application of deep learning in this category involves refining MR parameter maps derived from previously inaccurate ones. In quantitative MRI, inaccuracies in parameters frequently can arise either from oversimplified models or from less precise imaging parameters attributable to system imperfections. For instance, it is a recognized issue that an incorrect T2 map might result from applying a mono-exponential fitting model to MESE images. Similarly, B1 inhomogeneity, a common occurrence in MRI scans, can compromise the accuracy of T1 mapping. In such scenarios, deep learning offers a potential solution for efficiently adjusting inaccurate parameter maps to produce accurate ones. Nonetheless, a significant limitation of this approach is the necessity for accurate parameter maps, derived from precise signal models or imaging parameters, to serve as references for network training.

10.3.5.1 Representative studies

Qiu et al. developed a deep learning-based method for quantifying multiple parameters in the brain directly from conventional contrast-weighted images [123]. This technique utilized a 2D U-Net to simultaneously estimate T1 and T2 maps from T1 MPRAGE (magnetization-prepared rapid gradient echo), T1 GRE (gradient echo), and T2 FLAIR (fluid attenuated inversion recovery) images. The model employs supervised training, using reference quantitative maps generated using a multiparametric MRI technique called MR multitasking [124]. This approach enables both qualitative and quantitative MRI analysis concurrently, eliminating the need for acquiring additional quantitative images specifically.

Pei et al. recently developed a technique to demonstrate the application of deep learning to produce accurate T1 maps from previously inaccurate T1 maps in 3D Look-Locker inversion recovery (LLIR) [125]. Typically, 3D LLIR T1 mapping requires data acquisition across multiple repetitions to reconstruct images at various inversion times for T1 fitting. To ensure signal recovery and B1 robustness, a sufficient delay time (TD) is usually required between repetitions, which extends the total scan duration. Eliminating the TD can shorten scan times, but this leads to underestimation in T1 maps. In this study, the authors developed a method to create

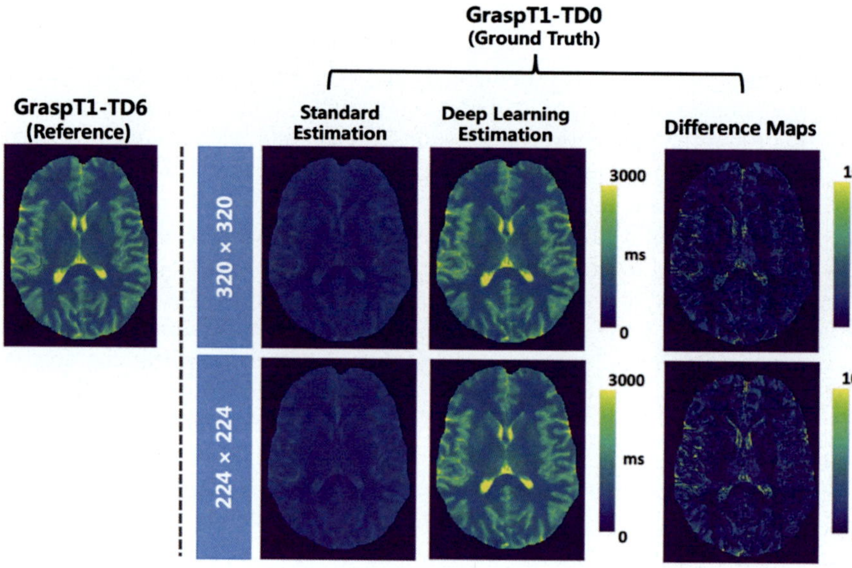

FIGURE 10.6

Generation of accurate T1 maps from inaccurate T1 maps in 3D LLIR imaging using deep learning. The reference T1 map is generated with sufficient delay time between consecutive inversion recovery preparations to ensure accurate T1 quantification. Eliminating the delay results in biased T1 estimation using the standard method, and deep learning is able to generate accurate T1 maps from the inaccurate ones.

Image was reproduced from Fig. 7 in Magn Reson Med. 2023 Aug; 90(2):569–582 with permission from John Wiley & Sons, Inc.

accurate T1 maps from those inaccurately generated without a TD to reduce overall scan times, as shown in Fig. 10.6 for an example.

10.4 Challenges in deep learning-based quantitative MRI

There has long been interest in integrating quantitative MRI into routine clinical practice to improve disease assessment. However, most quantitative MRI techniques remain confined to research settings to date. Deep learning holds great promise to advance quantitative MRI, but its practical implementation faces several significant challenges. A primary hurdle is the limited availability of MRI databases suitable for training deep learning-based reconstruction models. Although efforts in recent years have led to the growing availability of public datasets for static MRI reconstruction, this challenge remains particularly pronounced in quantitative MRI. The primary reason is that quantitative MRI is not routinely performed at most hospitals, making it difficult to accumulate large datasets from clinical scans. As a result, there is an unmet need to develop dedicated MRI databases specifically for quantitative MRI to facilitate further advancements in the field.

Moreover, the computational demands of quantitative MRI pose another challenge. Since quantitative MRI requires multiple images with varying contrast weightings for parameter fitting, it demands greater computational resources and more extensive network training to process these high-dimensional datasets compared with standard static image reconstruction. This increased demand for computing power and storage can be particularly problematic for institutions with limited research resources. Therefore, developing more efficient computational strategies and algorithms that reduce resource burdens while preserving the accuracy of quantitative MRI reconstruction is crucial for improving its accessibility and broader clinical adoption.

Another challenge in deep learning-based quantitative MRI is how to ensure the robustness and generalizability of trained network models. At the time of writing this chapter, quantitative MRI protocols remain unstandardized across institutions, vendors, and software versions, particularly in terms of imaging parameters, acquisition settings, and undersampling schemes. This lack of standardization introduces variability that complicates the development of universally applicable models, posing a significant obstacle to model generalizability and deployment. Therefore, developing novel strategies that can enable network training to account for these variations without losing reconstruction accuracy is important for advancing the application of deep learning in quantitative MRI.

The clinical translation of quantitative MRI techniques remains another significant challenge, even in the era of deep learning. While deep learning is expected to address many obstacles within the quantitative MRI workflow and facilitate its clinical adoption, the full potential and clinical value of many quantitative MRI techniques have yet to be thoroughly explored and validated. A crucial step toward this goal is integrating deep learning-based quantitative MRI methods into existing clinical workflows for large-scale patient studies to assess their effectiveness and reliability.

In recent years, various software packages have been developed to facilitate the clinical translation of deep learning models [126,127]. These tools are expected to play a key role in supporting this transition and promoting the adoption of quantitative MRI in clinical practice.

Finally, data privacy and security remain pressing concerns in the era of deep learning, particularly when handling medical data that contain sensitive patient information. This challenge extends beyond quantitative MRI and is a broader issue in medical imaging [128–130]. Ensuring the confidentiality and integrity of patient data while enabling the training and deployment of deep learning models requires careful adherence to strict data management protocols to balance innovation with ethical and regulatory considerations.

10.5 Future directions

Deep learning has shown great promise in advancing quantitative MRI in recent years. This section explores additional potential directions for deep learning-based quantitative MRI. One particularly promising avenue is self-supervised training [131–133], as demonstrated by several pioneering studies highlighted in this chapter. Unlike traditional supervised learning, which requires large annotated datasets for model training, self-supervised learning leverages the inherent physical model of quantitative MRI to eliminate the need for labeled reference data. This approach is especially valuable in quantitative MRI, where acquiring large volumes of annotated data is challenging due to the complexity and cost of data collection.

Another promising direction involves leveraging deep learning to enhance standardization in quantitative MRI. Deep learning algorithms can identify inconsistencies in images acquired from different MRI scanners, vendors, or software versions, as demonstrated in early studies [134–136]. Achieving uniform quantitative measurements across various scanners is essential to ensure consistency and comparability of results. Such standardization is particularly important for multicenter studiesto enable reliable comparisons of quantitative MRI data across diverse research and clinical environments.

In addition, developing universal deep learning models is a broader objective in deep learning-based MRI reconstruction [137], including quantitative MRI. Training these models on diverse datasets from multiple sources, scanners, organs, and undersampling schemes aims to establish a more generalized approach to quantitative MRI analysis. Such models could provide consistent quantitative assessments regardless of the MRI data source, which is crucial for multicenter and longitudinal studies focused on disease monitoring and treatment evaluation. However, training universal model architectures presents several challenges, including optimizing hyperparameters across a wide range of applications and ensuring access to diverse, high-quality training datasets. The feasibility of these models remains under investigation and requires careful validation to determine their clinical applicability.

10.6 Conclusion

Deep learning has demonstrated a profound impact on quantitative MRI, providing solutions to longstanding challenges and unlocking new opportunities in medical imaging. This chapter explored the application of deep learning in quantitative MRI from multiple perspectives. By accelerating image acquisition, reducing processing times, and improving image quality, deep learning is not only reshaping how quantitative MRI is performed in clinical practice but also expanding its role in disease diagnosis, treatment planning, and general medical imaging research.

References

[1] Seiberlich N, Gulani V, Campbell A, Sourbron S, Doneva MI, Calamante F, et al., editors. Quantitative magnetic resonance imaging. 1st ed., Vol 1; 2020. p. 1–1066 Available from: https://www.elsevier.com/books/quantitative-magnetic-resonance-imaging/seiberlich/978-0-12-817057-1.

[2] Messroghli DR, Radjenovic A, Kozerke S, Higgins DM, Sivananthan MU, Ridgway JP. Modified Look-Locker inversion recovery (MOLLI) for high-resolution T1 mapping of the heart. Magn Reson Med 2004;52(1):141–6. Available from: https://pubmed.ncbi.nlm.nih.gov/15236377/.

[3] Chow K, Flewitt JA, Green JD, Pagano JJ, Friedrich MG, Thompson RB. Saturation recovery single-shot acquisition (SASHA) for myocardial T(1) mapping. Magn Reson Med 2014;71(6):2082–95. Available from: https://pubmed.ncbi.nlm.nih.gov/23881866/.

[4] Lota AS, Gatehouse PD, Mohiaddin RH. T2 mapping and T2* imaging in heart failure. Heart Fail Rev July 1, 2017;22(4):431.

[5] Wolf M, De Boer A, Sharma K, Boor P, Leiner T, Sunder-Plassmann G, et al. Magnetic resonance imaging T1- and T2-mapping to assess renal structure and function: a systematic review and statement paper. Nephrol Dial Transplant September 1, 2018;33 (Suppl. 1_2):ii41–50. https://doi.org/10.1093/ndt/gfy198.

[6] Puntmann VO, Peker E, Chandrashekhar Y, Nagel E. T1 mapping in characterizing myocardial disease. Circ Res July 8, 2016;119(2):277–99. Available from: https://www.ahajournals.org/doi/abs/10.1161/CIRCRESAHA.116.307974.

[7] Jellis CL, Kwon DH. Myocardial T1 mapping: modalities and clinical applications. Cardiovasc Diagn Ther April 2014;4(2):12637. Available from: https://cdt.amegroups.org/article/view/2775/html.

[8] Taylor AJ, Salerno M, Dharmakumar R, Jerosch-Herold M. T1 mapping: basic techniques and clinical applications. JACC Cardiovasc Imaging January 1, 2016;9(1): 67–81.

[9] Serai SD. Basics of magnetic resonance imaging and quantitative parameters T1, T2, T2*, T1rho and diffusion-weighted imaging. Pediatr Radiol February 1, 2022;52(2): 217–27. Available from: https://link.springer.com/article/10.1007/s00247-021-05042-7.

[10] Wáng YXJ, Zhang Q, Li X, Chen W, Ahuja A, Yuan J. T1ρ magnetic resonance: basic physics principles and applications in knee and intervertebral disc imaging. Quant Imaging Med Surg 2015;5(6):858.

[11] Feng L, Liu F, Soultanidis G, Liu C, Benkert T, Block KT, et al. Magnetization-prepared GRASP MRI for rapid 3D T1 mapping and fat/water-separated T1 mapping. Magn Reson Med July 1, 2021;86(1):97–114. Available from: https://onlinelibrary.wiley.com/doi/full/10.1002/mrm.28679.

[12] Assländer J, Novikov DS, Lattanzi R, Sodickson DK, Cloos MA. Hybrid-state free precession in nuclear magnetic resonance. Commun Phys 2019;2(1):1–12. Available from: https://www.nature.com/articles/s42005-019-0174-0.

[13] Ma D, Gulani V, Seiberlich N, Liu K, Sunshine JL, Duerk JL, et al. Magnetic resonance fingerprinting. Nature 2013;495(7440):187–92.

[14] Poorman ME, Martin MN, Ma D, McGivney DF, Gulani V, Griswold MA, et al. Magnetic resonance fingerprinting Part 1: potential uses, current challenges, and

recommendations. J Magn Reson Imag March 1, 2020;51(3):675–92. Available from: https://pubmed.ncbi.nlm.nih.gov/31264748/.
[15] McGivney DF, Boyacıoğlu R, Jiang Y, Poorman ME, Seiberlich N, Gulani V, et al. Magnetic resonance fingerprinting review part 2: technique and directions. J Magn Reson Imag April 1, 2020;51(4):993–1007. Available from: https://pubmed.ncbi.nlm.nih.gov/31347226/.
[16] Cloos MA, Knoll F, Zhao T, Block KT, Bruno M, Wiggins GC, et al. Multiparametric imaging with heterogeneous radiofrequency fields. Nat Commun 2016;7.
[17] Jahng GH, Li KL, Ostergaard L, Calamante F. Perfusion magnetic resonance imaging: a comprehensive update on principles and techniques. Korean J Radiol September 12, 2014;15(5):554–77. Available from: https://synapse.koreamed.org/articles/1026982.
[18] Cha S. Perfusion MR imaging: basic principles and clinical applications. Magn Reson Imaging Clin N Am August 1, 2003;11(3):403–13. Available from: http://www.mri.theclinics.com/article/S1064968903000667/fulltext.
[19] Sourbron S. Technical aspects of MR perfusion. Eur J Radiol December 1, 2010;76(3):304–13.
[20] Luypaert R, Boujraf S, Sourbron S, Osteaux M. Diffusion and perfusion MRI: basic physics. Eur J Radiol April 1, 2001;38(1):19–27.
[21] Feng L, Benkert T, Block KT, Sodickson DK, Otazo R, Chandarana H. Compressed sensing for body MRI. J Magn Reson Imag April 1, 2017;45(4):966–87.
[22] Kogan F, Hariharan H, Reddy R. Chemical exchange saturation transfer (CEST) imaging: description of technique and potential clinical applications. Curr Radiol Rep June 14, 2013;1(2):102–14. Available from: https://link.springer.com/article/10.1007/s40134-013-0010-3.
[23] Kim J, Wu Y, Guo Y, Zheng H, Sun PZ. A review of optimization and quantification techniques for chemical exchange saturation transfer MRI toward sensitive in vivo imaging. Contrast Media Mol Imaging May 1, 2015;10(3):163–78. Available from: https://onlinelibrary.wiley.com/doi/full/10.1002/cmmi.1628.
[24] Vinogradov E, Sherry AD, Lenkinski RE. CEST: from basic principles to applications, challenges and opportunities. J Magn Reson April 1, 2013;229:155–72.
[25] Jones KM, Pollard AC, Pagel MD, Pagel M, Jones K, Pollard A. Clinical applications of chemical exchange saturation transfer (CEST) MRI. J Magn Reson Imag January 1, 2018;47(1):11–27. Available from: https://onlinelibrary.wiley.com/doi/full/10.1002/jmri.25838.
[26] Wu B, Warnock G, Zaiss M, Lin C, Chen M, Zhou Z, et al. An overview of CEST MRI for non-MR physicists. EJNMMI Phys December 1, 2016;3(1):1–21. Available from: https://ejnmmiphys.springeropen.com/articles/10.1186/s40658-016-0155-2.
[27] Alexander DC, Dyrby TB, Nilsson M, Zhang H. Imaging brain microstructure with diffusion MRI: practicality and applications. NMR Biomed April 1, 2019;32(4):e3841. Available from: https://onlinelibrary.wiley.com/doi/full/10.1002/nbm.3841.
[28] Novikov DS, Fieremans E, Jespersen SN, Kiselev VG. Quantifying brain microstructure with diffusion MRI: theory and parameter estimation. NMR Biomed April 1, 2019;32(4):e3998. Available from: https://onlinelibrary.wiley.com/doi/full/10.1002/nbm.3998.
[29] Kiselev VG. Microstructure with diffusion MRI: what scale we are sensitive to? J Neurosci Methods January 1, 2021;347:108910.

[30] Afzali M, Pieciak T, Newman S, Garyfallidis E, Özarslan E, Cheng H, et al. The sensitivity of diffusion MRI to microstructural properties and experimental factors. J Neurosci Methods January 1, 2021;347:108951.

[31] Fadzillah DM, Rosli MI, Talib MZM, Kamarudin SK, Daud WRW. Review on microstructure modelling of a gas diffusion layer for proton exchange membrane fuel cells. Renew Sustain Energy Rev September 1, 2017;77:1001–9.

[32] Nilsson M, Van Westen D, Ståhlberg F, Sundgren PC, Lätt J, Nilsson M, et al. The role of tissue microstructure and water exchange in biophysical modelling of diffusion in white matter. Magnetic Reson Mat Phys Biol Med 2013;26(4):345–4370. Available from: https://link.springer.com/article/10.1007/s10334-013-0371-x.

[33] Deistung A, Schweser F, Reichenbach JR. Overview of quantitative susceptibility mapping. NMR Biomed April 1, 2017;30(4):e3569. Available from: https://onlinelibrary.wiley.com/doi/full/10.1002/nbm.3569.

[34] Haacke EM, Liu S, Buch S, Zheng W, Wu D, Ye Y. Quantitative susceptibility mapping: current status and future directions. Magn Reson Imaging January 1, 2015;33(1):1–25.

[35] Reichenbach JR, Schweser F, Serres B, Deistung A. Quantitative susceptibility mapping: concepts and applications. Clin Neuroradiol October 30, 2015;25(2):225–30. Available from: https://link.springer.com/article/10.1007/s00062-015-0432-9.

[36] Liu C, Wei H, Gong NJ, Cronin M, Dibb R, Decker K, et al. Quantitative susceptibility mapping: contrast mechanisms and clinical applications. Tomography 2015;1(1):3–17. Available from: https://www.mdpi.com/2379-139X/1/1/3.

[37] Jung W, Bollmann S, Lee J. Overview of quantitative susceptibility mapping using deep learning: current status, challenges and opportunities. NMR Biomed April 1, 2022;35(4):e4292. Available from: https://onlinelibrary.wiley.com/doi/full/10.1002/nbm.4292.

[38] Harada T, Kudo K, Fujima N, Yoshikawa M, Ikebe Y, Sato R, et al. Quantitative susceptibility mapping: basic methods and clinical applications. Radiographics July 1, 2022;42(4):1161–76. Available from: https://pubs.rsna.org/doi/10.1148/rg.210054.

[39] Granziera C, Wuerfel J, Barkhof F, Calabrese M, De Stefano N, Enzinger C, et al. Quantitative magnetic resonance imaging towards clinical application in multiple sclerosis. Brain June 22, 2021;144(5):1296–311. https://doi.org/10.1093/brain/awab029.

[40] Li X, Majumdar S. Quantitative MRI of articular cartilage and its clinical applications. J Magn Reson Imag November 1, 2013;38(5):991–1008. Available from: https://onlinelibrary.wiley.com/doi/full/10.1002/jmri.24313.

[41] Margaret Cheng HL, Stikov N, Ghugre NR, Wright GA. Practical medical applications of quantitative MR relaxometry. J Magn Reson Imag October 1, 2012;36(4):805–24. Available from: https://onlinelibrary.wiley.com/doi/full/10.1002/jmri.23718.

[42] Curtis WA, Fraum TJ, An H, Chen Y, Shetty AS, Fowler KJ. Quantitative MRI of diffuse liver disease: current applications and future directions. Radiology January 1, 2019;290(1):23–30. Available from: https://pubs.rsna.org/doi/10.1148/radiol.2018172765.

[43] Wehrli FW, Song HK, Saha PK, Wright AC. Quantitative MRI for the assessment of bone structure and function. NMR Biomed November 1, 2006;19(7):731–64. Available from: https://onlinelibrary.wiley.com/doi/full/10.1002/nbm.1066.

[44] Seiler A, Nöth U, Hok P, Reiländer A, Maiworm M, Baudrexel S, et al. Multiparametric quantitative MRI in neurological diseases. Front Neurol March 8, 2021;12:640239. Available from: www.frontiersin.org.

[45] Teipel SJ, Grothe M, Lista S, Toschi N, Garaci FG, Hampel H. Relevance of magnetic resonance imaging for early detection and diagnosis of alzheimer disease. Med Clinics May 1, 2013;97(3):399–424. Available from: http://www.medical.theclinics.com/article/S0025712512002398/fulltext.

[46] McEvoy LK, Brewer JB. Quantitative structural MRI for early detection of Alzheimer's disease. Expert Rev Neurother November 2010;10(11):1675–88. Available from: https://www.tandfonline.com/doi/abs/10.1586/ern.10.162.

[47] Gulani V, Seiberlich N. Quantitative MRI: rationale and challenges. Adv Magnetic Reson Technol Appl 2020;1.

[48] Keenan KE, Biller JR, Delfino JG, Boss MA, Does MD, Evelhoch JL, et al. Recommendations towards standards for quantitative MRI (qMRI) and outstanding needs. J Magn Reson Imaging June 1, 2019;49(7):e26.

[49] Liu J, Pan Y, Li M, Chen Z, Tang L, Lu C, et al. Applications of deep learning to MRI Images: a survey. Big Data Mining Analy March 1, 2018;1(1):1–18.

[50] Mazurowski MA, Buda M, Saha A, Bashir MR. Deep learning in radiology: an overview of the concepts and a survey of the state of the art with focus on MRI. J Magn Reson Imag April 1, 2019;49(4):939–54. Available from: https://onlinelibrary.wiley.com/doi/full/10.1002/jmri.26534.

[51] Lundervold AS, Lundervold A. An overview of deep learning in medical imaging focusing on MRI. Z Med Phys May 1, 2019;29(2):102–27.

[52] Feng L, Ma D, Liu F. Rapid MR relaxometry using deep learning: an overview of current techniques and emerging trends. NMR Biomed April 1, 2022;35(4):e4416. Available from: https://onlinelibrary.wiley.com/doi/full/10.1002/nbm.4416.

[53] Feng L. Golden-angle radial MRI: basics, advances, and applications. J Magn Reson Imag April 9, 2022. Available from: https://onlinelibrary.wiley.com/doi/full/10.1002/jmri.28187.

[54] Zi R, Zhu D, Qin Q. Quantitative T2 mapping using accelerated 3D stack-of-spiral gradient echo readout. Magn Reson Imaging November 1, 2020;73:138–47.

[55] Chen Y, Lo WC, Hamilton JI, Barkauskas K, Saybasili H, Wright KL, et al. Single breath-hold 3D cardiac T1 mapping using through-time spiral GRAPPA. NMR Biomed June 1, 2018;31(6):e3923. Available from: https://onlinelibrary.wiley.com/doi/full/10.1002/nbm.3923.

[56] Sodickson DK, Manning WJ. Simultaneous acquisition of spatial harmonics (SMASH): fast imaging with radiofrequency coil arrays. Magn Reson Med October 1997;38(4):591–603.

[57] Jakob PM, Grisowld MA, Edelman RR, Sodickson DK. AUTO-SMASH: a self-calibrating technique for SMASH imaging. SiMultan Acquis Spatial Harm MAGMA November 1998;7(1):42–54. Available from: https://pubmed.ncbi.nlm.nih.gov/9877459/.

[58] Heidemann RM, Griswold MA, Haase A, Jakob PM. VD-AUTO-SMASH imaging. Magn Reson Med June 1, 2001;45(6):1066–74. Available from: https://onlinelibrary.wiley.com/doi/full/10.1002/mrm.1141.

[59] Sodickson DK, McKenzie CA. A generalized approach to parallel magnetic resonance imaging. Med Phys August 2001;28(8):1629–43.

[60] Pruessmann KP, Weiger M, Scheidegger MB, Boesiger P. SENSE: sensitivity encoding for fast MRI. Magn Reson Med November 1999;42(5):952–62.

[61] Kellman P, Epstein FH, McVeigh ER. Adaptive sensitivity encoding incorporating temporal filtering (TSENSE). Magn Reson Med 2001;45(5):846–52.

[62] Pruessmann KP, Weiger M, Börnert P, Boesiger P. Advances in sensitivity encoding with arbitrary k-space trajectories. Magn Reson Med October 1, 2001;46(4):638–51. Available from: https://onlinelibrary.wiley.com/doi/full/10.1002/mrm.1241.

[63] Breuer FA, Kellman P, Griswold MA, Jakob PM. Dynamic autocalibrated parallel imaging using temporal GRAPPA (TGRAPPA). Magn Reson Med April 2005;53(4): 981–5.

[64] Tsao J, Boesiger P, Pruessmann KP. K-t BLAST and k-t SENSE: dynamic MRI with high frame rate exploiting spatiotemporal correlations. Magn Reson Med November 2003;50(5):1031–42.

[65] Huang F, Akao J, Vijayakumar S, Duensing GR, Limkeman M. K-t GRAPPA: a k-space implementation for dynamic MRI with high reduction factor. Magn Reson Med November 2005;54(5):1172–84.

[66] Lai P, Huang F, Larson AC, Li D. Fast four-dimensional coronary MR angiography with k-t GRAPPA. J Magn Reson Imaging March 2008;27(3):659.

[67] Tsao J, Kozerke S. MRI temporal acceleration techniques. J Magn Reson Imaging September 2012;36(3):543–60. Available from: https://pubmed.ncbi.nlm.nih.gov/22903655/.

[68] Lustig M, Donoho D, Pauly JM. Sparse MRI: the application of compressed sensing for rapid MR imaging. Magn Reson Med December 2007;58(6):1182–95.

[69] Lustig M, Pauly JM. SPIRiT: iterative self-consistent parallel imaging reconstruction from arbitrary k-space. Magn Reson Med August 2010;64(2):457–71.

[70] Lustig M, Santos JM, Donoho DL, Pauly JM. k-t SPARSE: high frame rate dynamic MRI exploiting spatio-temporal sparsity. In: ISMRM 2006 Annual Meeting Proceedings p2420.

[71] Lustig M, Donoho DL, Santos JM, Pauly JM. Compressed sensing MRI: a look at how CS can improve on current imaging techniques. In: Vol. 25, IEEE Signal Processing Magazine. Institute of Electrical and Electronics Engineers Inc.; 2008. p. 72–82.

[72] Uecker M, Lai P, Murphy MJ, Virtue P, Elad M, Pauly JM, et al. ESPIRiT - an eigenvalue approach to autocalibrating parallel MRI: where SENSE meets GRAPPA. Magn Reson Med March 2014;71(3):990–1001.

[73] Otazo R, Kim D, Axel L, Sodickson DK. Combination of compressed sensing and parallel imaging for highly accelerated first-pass cardiac perfusion MRI. Magn Reson Med September 2010;64(3):767–76.

[74] Feng L, Axel L, Chandarana H, Block KT, Sodickson DK, Otazo R. XD-GRASP: golden-angle radial MRI with reconstruction of extra motion-state dimensions using compressed sensing. Magn Reson Med February 1, 2016;75(2):775–88.

[75] Feng L, Otazo R, Jung H, Jensen JH, Ye JC, Sodickson DK, et al. Accelerated cardiac T2 mapping using breath-hold multiecho fast spin-echo pulse sequence with k-t FOCUSS. Magn Reson Med June 1, 2011;65(6):1661–9. Available from: https://onlinelibrary.wiley.com/doi/full/10.1002/mrm.22756.

[76] Feng L, Grimm R, Block KT obias, Chandarana H, Kim S, Xu J, et al. Golden-angle radial sparse parallel MRI: combination of compressed sensing, parallel imaging, and golden-angle radial sampling for fast and flexible dynamic volumetric MRI. Magn Reson Med September 1, 2014;72(3):707–17.

[77] Feng L, Srichai MB, Lim RP, Harrison A, King W, Adluru G, et al. Highly accelerated real-time cardiac cine MRI using k–t SPARSE-SENSE. Magn Reson Med July 1, 2013;70(1):64–74. Available from: https://onlinelibrary.wiley.com/doi/full/10.1002/mrm.24440.

[78] Smith DS, Li X, Abramson RG, Quarles CC, Yankeelov TE, Welch EB. Potential of compressed sensing in quantitative MR imaging of cancer. Cancer Imaging December 30, 2013;13(4):633.

[79] Smith DS, Welch EB, Li X, Arlinghaus LR, Loveless ME, Koyama T, et al. Quantitative effects of using compressed sensing in dynamic contrast enhanced MRI. Phys Med Biol July 19, 2011;56(15):4933. Available from: https://iopscience.iop.org/article/10.1088/0031-9155/56/15/018.

[80] Doneva M, Börnert P, Eggers H, Stehning C, Sénégas J, Mertins A. Compressed sensing reconstruction for magnetic resonance parameter mapping. Magn Reson Med October 1, 2010;64(4):1114–20. Available from: https://onlinelibrary.wiley.com/doi/full/10.1002/mrm.22483.

[81] Block KT, Uecker M, Frahm J. Model-based iterative reconstruction for radial fast spin-echo MRI. IEEE Trans Med Imag November 2009;28(11):1759–69.

[82] Wang X, Roeloffs V, Klosowski J, Tan Z, Voit D, Uecker M, et al. Model-based T1 mapping with sparsity constraints using single-shot inversion-recovery radial FLASH. Magn Reson Med February 1, 2018;79(2):730–40. Available from: https://onlinelibrary.wiley.com/doi/full/10.1002/mrm.26726.

[83] Ben-Eliezer N, Sodickson DK, Block KT. Rapid and accurate T2 mapping from multi-spin-echo data using bloch-simulation-based reconstruction. Magn Reson Med February 1, 2015;73(2):809–17. Available from: https://onlinelibrary.wiley.com/doi/full/10.1002/mrm.25156.

[84] Meneses JP, Arrieta C, della Maggiora G, Besa C, Urbina J, Arrese M, et al. Liver PDFF estimation using a multi-decoder water-fat separation neural network with a reduced number of echoes. Eur Radiol September 1, 2023;33(9):6557–68. Available from: https://pubmed.ncbi.nlm.nih.gov/37014405/.

[85] Shao J, Ghodrati V, Nguyen KL, Hu P. Fast and accurate calculation of myocardial T1 and T2 values using deep learning Bloch equation simulations (DeepBLESS). Magn Reson Med November 1, 2020;84(5):2831.

[86] Gu Y, Pan Y, Fang Z, Ma L, Zhu Y, Androjna C, et al. Deep learning-assisted preclinical MR fingerprinting for sub-millimeter T1 and T2 mapping of entire macaque brain. Magn Reson Med March 1, 2024;91(3):1149–64. Available from: https://onlinelibrary.wiley.com/doi/full/10.1002/mrm.29905.

[87] Wu Y, Alley M, Li Z, Datta K, Wen Z, Sandino C, et al. Deep learning-based water-fat separation from dual-echo chemical shift-encoded imaging. Bioengineering October 1, 2022;9(10). Available from: https://pubmed.ncbi.nlm.nih.gov/36290546/.

[88] Goldfarb JW, Craft J, Cao JJ. Water-fat separation and parameter mapping in cardiac MRI via deep learning with a convolutional neural network. J Magn Reson Imaging August 1, 2019;50(2):655–65. Available from: https://pubmed.ncbi.nlm.nih.gov/30701621/.

[89] Bollmann S, Rasmussen KGB, Kristensen M, Blendal RG, Østergaard LR, Plocharski M, et al. DeepQSM - using deep learning to solve the dipole inversion for quantitative susceptibility mapping. Neuroimage July 15, 2019;195:373–83. Available from: https://pubmed.ncbi.nlm.nih.gov/30935908/.

[90] Polak D, Chatnuntawech I, Yoon J, Iyer SS, Milovic C, Lee J, et al. Nonlinear dipole inversion (NDI) enables robust quantitative susceptibility mapping (QSM). NMR Biomed December 1, 2020;33(12). Available from: https://pubmed.ncbi.nlm.nih.gov/32078756/.

[91] Cohen O, Zhu B, Rosen MS. MR fingerprinting deep RecOnstruction NEtwork (DRONE). Magn Reson Med September 1, 2018;80(3):885–94. Available from: https://pubmed.ncbi.nlm.nih.gov/29624736/.

[92] Le JV, Mendes JK, McKibben N, Wilson BD, Ibrahim M, DiBella EVR, et al. Accelerated cardiac T1 mapping with recurrent networks and cyclic, model-based loss. Med Phys November 1, 2022;49(11):6986–7000. Available from: https://pubmed.ncbi.nlm.nih.gov/35703369/.

[93] Kim B, Schär M, Park HW, Heo HY. A deep learning approach for magnetization transfer contrast MR fingerprinting and chemical exchange saturation transfer imaging. Neuroimage November 1, 2020;221:117165.

[94] Zhang X, Duchemin Q, Liu K, Gultekin C, Flassbeck S, Fernandez-Granda C, et al. Cramér-Rao bound-informed training of neural networks for quantitative MRI. Magn Reson Med July 1, 2022;88(1):436–48. Available from: https://pubmed.ncbi.nlm.nih.gov/35344614/.

[95] Cabini RF, Barzaghi L, Cicolari D, Arosio P, Carrazza S, Figini S, et al. Fast deep learning reconstruction techniques for preclinical magnetic resonance fingerprinting. NMR Biomed January 1, 2024;37(1):e5028. Available from: https://onlinelibrary.wiley.com/doi/full/10.1002/nbm.5028.

[96] Jun Y, Cho J, Wang X, Gee M, Grant PE, Bilgic B, et al. SSL-QALAS: self-Supervised Learning for rapid multiparameter estimation in quantitative MRI using 3D-QALAS. Magn Reson Med November 1, 2023;90(5):2019–32. Available from: https://onlinelibrary.wiley.com/doi/full/10.1002/mrm.29786.

[97] Oh G, Bae H, Ahn HS, Park SH, Moon WJ, Ye JC. Unsupervised resolution-agnostic quantitative susceptibility mapping using adaptive instance normalization. Med Image Anal July 1, 2022;79. Available from: https://pubmed.ncbi.nlm.nih.gov/35605505/.

[98] Jafari R, Spincemaille P, Zhang J, Nguyen TD, Luo X, Cho J, et al. Deep neural network for water/fat separation: supervised training, unsupervised training, and no training. Magn Reson Med April 1, 2021;85(4):2263–77. Available from: https://pubmed.ncbi.nlm.nih.gov/33107127/.

[99] Kang B, Kim B, Schär M, Park HW, Heo HY. Unsupervised learning for magnetization transfer contrast MR fingerprinting: application to CEST and nuclear Overhauser enhancement imaging. Magn Reson Med April 1, 2021;85(4):2040–54. Available from: https://pubmed.ncbi.nlm.nih.gov/33128483/.

[100] Yaman B, Hosseini SAH, Moeller S, Ellermann J, Uğurbil K, Akçakaya M. Self-supervised learning of physics-guided reconstruction neural networks without fully sampled reference data. Magn Reson Med December 1, 2020;84(6):3172–91. Available from: https://pubmed.ncbi.nlm.nih.gov/32614100/.

[101] Liu S, Li H, Liu Y, Cheng G, Yang G, Wang H, et al. Highly accelerated MR parametric mapping by undersampling the k-space and reducing the contrast number simultaneously with deep learning. Phys Med Biol September 21, 2022;67(18). Available from: https://pubmed.ncbi.nlm.nih.gov/36001990/.

[102] Fu Z, Mandava S, Keerthivasan MB, Li Z, Johnson K, Martin DR, et al. A multi-scale residual network for accelerated radial MR parameter mapping. Magn Reson Imaging

November 1, 2020;73:152–62. Available from: https://pubmed.ncbi.nlm.nih.gov/32882339/.

[103] Gao Y, Cloos M, Liu F, Crozier S, Pike GB, Sun H. Accelerating quantitative susceptibility and R2* mapping using incoherent undersampling and deep neural network reconstruction. Neuroimage October 15, 2021;240:118404.

[104] Wu Y, Ma Y, Du J, Xing L. Accelerating quantitative MR imaging with the incorporation of B1 compensation using deep learning. Magn Reson Imaging October 1, 2020; 72:78–86. Available from: https://pubmed.ncbi.nlm.nih.gov/32610065/.

[105] Jeelani H, Yang Y, Zhou R, Kramer CM, Salerno M, Weller DS. A myocardial T1--mapping framework with recurrent and U-net convolutional neural networks. In: Proceedings - International Symposium on Biomedical imaging; 2020. p. 1941–4.

[106] Shih SF, Kafali SG, Calkins KL, Wu HH. Uncertainty-aware physics-driven deep learning network for free-breathing liver fat and R2* quantification using self-gated stack-of-radial MRI. Magn Reson Med April 1, 2023;89(4):1567–85. Available from: https://onlinelibrary.wiley.com/doi/full/10.1002/mrm.29525.

[107] Hong JS, Hermann I, Zöllner FG, Schad LR, Wang SJ, Lee WK, et al. Acceleration of magnetic resonance fingerprinting reconstruction using denoising and self-attention pyramidal convolutional neural network. Sensors February 1, 2022;22(3). Available from: https://pubmed.ncbi.nlm.nih.gov/35162007/.

[108] Yang Q, Lin Y, Wang J, Bao J, Wang X, Ma L, et al. MOdel-based SyntheTic data-driven learning (MOST-DL): application in single-shot T2 mapping with severe head motion using overlapping-echo acquisition. IEEE Trans Med Imag November 1, 2022;41(11):3167–81.

[109] Liu F, Feng L, Kijowski R. MANTIS: model-Augmented Neural neTwork with Incoherent k-space Sampling for efficient MR parameter mapping. Magn Reson Med July 1, 2019;82(1):174–88.

[110] Liu F, Kijowski R, Feng L, El Fakhri G. High-performance rapid MR parameter mapping using model-based deep adversarial learning. Magn Reson Imaging December 1, 2020;74:152–60.

[111] Liu F, Kijowski R, El Fakhri G, Feng L. Magnetic resonance parameter mapping using model-guided self-supervised deep learning. Magn Reson Med 2021;85(6):3211–26. Available from: https://onlinelibrary.wiley.com/doi/full/10.1002/mrm.28659.

[112] Li H, Yang M, Kim JH, Zhang C, Liu R, Huang P, et al. SuperMAP: deep ultrafast MR relaxometry with joint spatiotemporal undersampling. Magn Reson Med January 1, 2023;89(1):64–76. Available from: https://pubmed.ncbi.nlm.nih.gov/36128884/.

[113] Jun Y, Shin H, Eo T, Kim T, Hwang D. Deep model-based magnetic resonance parameter mapping network (DOPAMINE) for fast T1 mapping using variable flip angle method. Med Image Anal May 1, 2021;70. Available from: https://pubmed.ncbi.nlm.nih.gov/33721693/.

[114] Li Y, Wang Y, Qi H, Hu Z, Chen Z, Yang R, et al. Deep learning-enhanced T1 mapping with spatial-temporal and physical constraint. Magn Reson Med September 1, 2021;86(3):1647–61. Available from: https://pubmed.ncbi.nlm.nih.gov/33821529/.

[115] Tolpadi AA, Han M, Calivà F, Pedoia V, Majumdar S. Region of interest-specific loss functions improve T2 quantification with ultrafast T2 mapping MRI sequences in knee, hip and lumbar spine. Sci Rep 2022;12(1):1–17. Available from: https://www.nature.com/articles/s41598-022-26266-z.

[116] Hamilton JI, Currey D, Rajagopalan S, Seiberlich N. Deep learning reconstruction for cardiac magnetic resonance fingerprinting T1 and T2 mapping. Magn Reson Med

April 1, 2021;85(4):2127–35. Available from: https://pubmed.ncbi.nlm.nih.gov/33107162/.

[117] Cao P, Cui D, Vardhanabhuti V, Hui ES. Development of fast deep learning quantification for magnetic resonance fingerprinting in vivo. Magn Reson Imaging July 1, 2020;70:81–90. Available from: https://pubmed.ncbi.nlm.nih.gov/32276007/.

[118] Bian W, Jang A, Liu F. Improving quantitative MRI using self-supervised deep learning with model reinforcement: Demonstration for rapid T1 mapping. Magn Reson Med February 11, 2024. Available from: https://onlinelibrary.wiley.com/doi/full/10.1002/mrm.30045.

[119] Peng W, Feng L, Zhao G, Liu F. Learning optimal K-space acquisition and reconstruction using physics-informed neural networks. In: 2022 IEEE/CVF Conference on Computer Vision and Pattern Recognition (CVPR); 2022-June. p. 20762–71.

[120] Wang G, Nielsen JF, Fessler JA, Noll DC. Stochastic optimization of three-dimensional non-Cartesian sampling trajectory. Magn Reson Med August 1, 2023; 90(2):417–31. Available from: https://pubmed.ncbi.nlm.nih.gov/37066854/.

[121] Zibetti MVW, Herman GT, Regatte RR. Fast data-driven learning of parallel MRI sampling patterns for large scale problems. Sci Rep December 1, 2021;11(1). Available from: https://pubmed.ncbi.nlm.nih.gov/34588478/.

[122] Zhang J, Spincemaille P, Zhang H, Nguyen TD, Li C, Li J, et al. LARO: learned acquisition and reconstruction optimization to accelerate quantitative susceptibility mapping. Neuroimage March 1, 2023;268. Available from: https://pubmed.ncbi.nlm.nih.gov/36669747/.

[123] Qiu S, Chen Y, Ma S, Fan Z, Moser FG, Maya MM, et al. Multiparametric mapping in the brain from conventional contrast-weighted images using deep learning. Magn Reson Med January 1, 2022;87(1):488–95. Available from: https://pubmed.ncbi.nlm.nih.gov/34374468/.

[124] Christodoulou AG, Shaw JL, Nguyen C, Yang Q, Xie Y, Wang N, et al. Magnetic resonance multitasking for motion-resolved quantitative cardiovascular imaging. Nat Biomed Eng April 2018;2(4):215–26.

[125] Pei H, Xia D, Xu X, Yang Y, Wang Y, Liu F, et al. Rapid 3D T1 mapping using deep learning-assisted Look-Locker inversion recovery MRI. Magn Reson Med August 1, 2023;90(2). Available from: https://pubmed.ncbi.nlm.nih.gov/37125662/.

[126] Block KT, Grimm R, Feng L, Otazo R, Chandarana H, Bruno M, et al. Bringing compressed sensing to clinical reality: prototypic setup for evaluation in routine applications. Proc Intl Soc Mag Reson Med 2013:3809. Available from: https://cds.ismrm.org/protected/13MPresentations/abstracts/3809.pdf.

[127] Veldmann M, Ehses P, Chow K, Nielsen JF, Zaitsev M, Stöcker T. Open-source MR imaging and reconstruction workflow. Magn Reson Med December 1, 2022;88(6): 2395–407. Available from: https://onlinelibrary.wiley.com/doi/full/10.1002/mrm.29384.

[128] Ziller A, Mueller TT, Stieger S, Feiner LF, Brandt J, Braren R, et al. Reconciling privacy and accuracy in AI for medical imaging. Nat Mach Intell 2024;6(7):764–7774. Available from: https://www.nature.com/articles/s42256-024-00858-y.

[129] Ziller A, Usynin D, Braren R, Makowski M, Rueckert D, Kaissis G. Medical imaging deep learning with differential privacy. Sci Rep 2021;11(1):1–18. Available from: https://www.nature.com/articles/s41598-021-93030-0.

[130] de Aguiar EJ, Traina C, Traina AJM. Security and privacy in machine learning for health systems: strategies and challenges. Yearb Med Inform December 26, 2023;32 (1):269. Available from: https://pmc.ncbi.nlm.nih.gov/articles/PMC10751106/.

[131] Chandra SS, Lorenzana MB, Liu X, Liu S, Bollmann S, Crozier S, et al. Deep learning in magnetic resonance image reconstruction. J Med Imaging Radiat Oncol August 1, 2021;65(5):564–77. Available from: https://onlinelibrary.wiley.com/doi/full/10.1111/1754-9485.13276.

[132] Wang S, Xiao T, Liu Q, Zheng H. Deep learning for fast MR imaging: a review for learning reconstruction from incomplete k-space data. Biomed Signal Process Control July 1, 2021;68:102579.

[133] Singh D, Monga A, de Moura HL, Zhang X, Zibetti MVW, Regatte RR. Emerging trends in fast MRI using deep-learning reconstruction on undersampled k-space data: a systematic review. Bioengineering 2023;10(9):1012. Available from: https://www.mdpi.com/2306-5354/10/9/1012/htm.

[134] Marzi C, Giannelli M, Barucci A, Tessa C, Mascalchi M, Diciotti S. Efficacy of MRI data harmonization in the age of machine learning: a multicenter study across 36 datasets. Sci Data 2024;11(1):1–27. Available from: https://www.nature.com/articles/s41597-023-02421-7.

[135] Dewey BE, Zhao C, Reinhold JC, Carass A, Fitzgerald KC, Sotirchos ES, et al. DeepHarmony: a deep learning approach to contrast harmonization across scanner changes. Magn Reson Imaging December 1, 2019;64:160–70.

[136] Li XT, Huang RY. Standardization of imaging methods for machine learning in neuro-oncology. Neurooncol Adv December 31, 2020;2(Suppl. 4):iv49–55. https://doi.org/10.1093/noajnl/vdaa054.

[137] Liu X, Wang J, Liu F, Zhou SK. Universal undersampled MRI reconstruction. Lect Notes Comput Sci 2021;12906(LNCS):211–21. Available from: https://link.springer.com/chapter/10.1007/978-3-030-87231-1_21.

CHAPTER

11 Am (A)I hallucinating? Nonrobustness, hallucinations, and unpredictable performance of AI for MR image reconstruction

Vegard Antun[1], Nina M. Gottschling[2], Anders C. Hansen[3], Ben Adcock[4]

[1]*University of Oslo, Oslo, Norway;* [2]*Deutsches Zentrum für Luft- und Raumfahrt (DLR), Köln, Germany;* [3]*University of Cambridge, Cambridge, United Kingdom;* [4]*Simon Fraser University, Burnaby, BC, Canada*

11.1 Introduction

Recovering high-quality images from measurements is a task of fundamental importance in science, medicine, engineering, and industry. In the last several years, artificial intelligence (AI)—driven by deep learning (DL)—has shown great potential for significantly enhancing image reconstruction performance across a range of imaging modalities. The amount of activity in this area is perhaps unprecedented, with many forecasting that AI-inspired algorithms may transform the field.

11.1.1 Nonrobustness of AI for MR image reconstruction

However, such optimism has recently started to temper. Recent studies have raised serious concerns regarding the trustworthiness of AI-based image reconstruction algorithms, as existing approaches generally fail to produce *robust* methods [1–7]. In particular, the resulting methods may suffer from any of the following effects:

1. *Hallucinations*: AI-based algorithms may generate realistic-looking, but, ultimately, false details in the reconstructed image.
2. *Instabilities*: AI-based algorithms may be highly sensitive to perturbations in the measurements.
3. *Unpredictable/inconsistent generalization*: AI-based algorithms may perform exceedingly well on some images, but may perform significantly worse on other nearby images.

Machine Learning in MRI. https://doi.org/10.1016/B978-0-443-14109-6.00011-0

Evidence for and awareness of these issues have recently come to the forefront of scientific discussion. For example, to quote the 2020 Facebook fastMRI challenge:

Such hallucinatory features are not acceptable and, especially, problematic if they mimic normal structures that are either not present or actually abnormal. Neural network models can be unstable as demonstrated via adversarial perturbation studies [8].

Or to quote [9]:

We stress that while DL reconstruction output may provide high-quality images upon initial inspection, the reconstructions may also include hallucinations or omit small structural elements, limiting its diagnostic value. Moreover, it is important to ensure the robustness of MR image reconstruction models, given the sensitive nature of radiological data [9].

We refer also to Refs. [5,9–16] and the references therein for further discussion.

11.1.2 Trustworthiness of AI

MR image reconstruction is by no means alone as an application area of AI in which there are growing concerns regarding nonrobustness and trustworthiness. Even within the field of computational imaging, similar issues have been identified in tomographic imaging [17], fluorescence [18] and electron microscopy [19], Positron Emission Tomography (PET) [20], and various other imaging modalities.

Outside of imaging, there is now an overwhelming amount of evidence that DL produces universally unstable methods [21] for problems ranging from image classification [22,23] and audio and speech recognition [24,25] to natural language processing [26] and automated diagnosis in medicine [27]. These issues are frequently raised in news articles. Indeed, the word "hallucinate" has now become so common for it to be declared "Word of the Year 2023" by the Cambridge Dictionary. To quote:

The Cambridge Dictionary team chose hallucinate as its Word of the Year 2023 as it recognized that the new meaning gets to the heart of why people are talking about AI. Generative AI is a powerful tool but one we're all still learning how to interact with safely and effectively—this means being aware of both its potential strengths and its current weaknesses [28].

Such is the importance of this issue; many authorities are now developing legal and regularity frameworks to promote its safe use. For example, to quote the European Commission:

In the light of the recent advances in AI, the serious negative consequences of its use for EU citizens and organisations have led to multiple initiatives [...] to set up the principles of a trustworthy and secure AI [...] the concepts of robustness and explainability of AI systems have emerged as key elements for a future regulation of this technology. [29].

Due to its key role in medical diagnosis and treatment monitoring, MRI is yet another key area where the trustworthiness of AI is of key concern.

11.1.3 This chapter

In this chapter, we document the three phenomena introduced in Section 11.1.1 and discuss the theoretical explanations for why they arise in AI-inspired algorithms. After a brief section describing the setup of this chapter (Section 11.2), we commence in Section 11.3 with a series of examples demonstrating how hallucinations, instabilities, and unpredictable generalization performance can arise in different AI-inspired algorithms for MRI reconstruction. Then, in Section 11.4, we delve into the mathematics of nonrobustness for image reconstruction. We present a series of theorems that describe the different mechanisms that can result in hallucinations, instabilities, or unpredictable generalization. While our theoretical results are not specific to DL-based methods, we explain how such methods are particularly susceptible to the conditions which necessarily lead to these effects. Finally, in Section 11.5, we offer some conclusions, further topics, and future prospects. Code related to Figs. 11.3, 11.4, 11.6, and 11.7 can be found at https://github.com/vegarant/am_AI_hallucinating. For the other figures, we refer to their respective publications.

11.2 Setup

In this chapter, we consider a discrete imaging scenario where the object to be recovered is a vectorized, complex image $x \in \mathbb{C}^N$. Here, $N = n^d$, where $d = 2, 3$, is the number of image pixels or voxels. We consider noisy, linear measurements of the form

$$y = Ex + \eta \tag{11.2.1}$$

Here, $E \in \mathbb{C}^{m \times N}$ is the *sampling operator* (or *encoding operator/measurement matrix*), $y \in \mathbb{C}^m$ is the vector of measurements, and $\eta \in \mathbb{C}^m$ is an (unknown) noise term, accounting for inaccuracies in the measurement model, measurement noise, and other artifacts.

11.2.1 Subsampled MR acquisition

A key concern in MRI is to reduce the scan time by acquiring as few measurements as possible. This leads to *subsampled acquisition*. Here, we consider a subsampling mask $\Omega \subsetneq \{1, \ldots, N\}$ of cardinality $|\Omega| = m < N$ and let $P_\Omega \in \mathbb{C}^{m \times N}$ denote the corresponding projection operator which extracts the elements of a vector indexed by Ω. In other words, $(P_\Omega x)_i = x_{\Omega(i)}$ for $i = 1, \ldots, m$, where $\Omega(i)$ denotes the ith element of Ω and $x = (x_i)_{i=1}^N \in \mathbb{C}^N$.

Let $F \in \mathbb{C}^{N \times N}$ denote the matrix of the d-dimensional discrete Fourier transform (DFT). In single-coil MRI, the sampling operator

$$E = P_\Omega F. \tag{11.2.2}$$

Since $m < N$, this is an undersampled acquisition. In multicoil (parallel) MRI, multiple receiver coils simultaneously acquire k-space data. To acquire as much information as possible, each of the receiver coils is adjusted to be sensitive to a limited spatial region (see, e.g., Ref. [32]). For a scanner with c coils, one models this by introducing a diagonal matrix $S_i \in \mathbb{C}^{N \times N}$, where $i = 1, \ldots, c$ for each coil. This matrix is usually called the *sensitivity matrix* or *sensitivity profile* of the coil. Given these sensitivity profiles, the full model for the encoding operator $E \in \mathbb{C}^{m \times N}$ is

$$E = \begin{bmatrix} P_\Omega F S_1 \\ \vdots \\ P_\Omega F S_c \end{bmatrix}, \tag{11.2.3}$$

where $|\Omega| = m'$, $P_\Omega \in \mathbb{C}^{m' \times N}$, and $m = m'c$. Depending on the choice of m' and c, the total number of measurements m can exceed the number of pixels N in the image, i.e., $m > N$. However, this is still a form of undersampled acquisition, since $|\Omega| = m' < N$ and, therefore, the acquisition process involves less than full sampling in each coil.

11.2.2 The inverse problem in MRI

In MR image recovery, we consider the inverse problem of reconstructing the unknown image $x \in \mathbb{C}^N$ as accurately and robustly as possible from the measurements (11.2.1). In subsampled acquisition, this inverse problem is either ill-posed or ill-conditioned. In single-coil MRI, the matrix (11.2.2) is not of full column rank, i. e., $\operatorname{rank}(E) < N$. This means that the null space $\mathcal{N}(E) \neq \{0\}$ and, therefore, the inverse problem is ill-posed. In parallel MRI, the matrix (11.2.3) may or may not have full column rank—i.e., $\operatorname{rank}(E) = N$ or, equivalently, $\mathcal{N}(E) = \{0\}$—depending on choice of m' and the design of the sensitivity profiles. However, even when $\operatorname{rank}(E) = N$, it is often ill-conditioned. Such ill-conditioning typically gets worse as one increases the undersampling factor [7]. Note that reducing the number of measurements m (which leads to worsen ill-conditioning and, eventually, ill-posedness) is, generally speaking, a key objective. Reducing acquisition times while retaining the quality of the recovered images is one of the most important goals of MRI innovation.

11.2.3 Reconstruction maps

We consider a *reconstruction map* as a map of the form $\Psi : \mathbb{C}^m \to \mathbb{C}^N$ from the measurement domain $\mathbb{C}^m$ to the image domain $\mathbb{C}^N$ that approximately solves this inverse problem. To combat the ill-posedness or ill-conditioning of the inverse problem, this map is generally designed to promote certain approximate solutions that match the underlying ground truth image. In standard *model-based* methods, this is typically done by a carefully designed (sometimes termed *handcrafted*) model class in which one expects the ground truth to belong. A standard example is the set of images that are (approximately) sparse in some dictionary, e.g., wavelets. Then, one carefully designs the reconstruction map to promote fidelity to the model class, e.g., by solving an ℓ^1-norm minimization problem, see, e.g., Ref. [33] for an introduction to model-based methods.

Conversely, in AI-inspired algorithms or *learning-based methods*, the aim is to tackle such ill-posedness or ill-conditioning by using training data—i.e., a representative dataset for the problem at hand—to learn a suitable reconstruction map Ψ. The map Ψ is often chosen as a deep neural network (DNN) and is trained via DL.

11.3 Examples

In this section, we present a series of numerical examples illustrating the main effects described in Section 11.1.1.

11.3.1 Hallucinations

We commence in Fig. 11.1 with two examples of hallucinations stemming from the fastMRI challenge [8]. These examples are two of the best-performing DNNs from this challenge. However, in tandem with this high performance comes hallucinations. As shown in the figure, both reconstruction maps produce artifacts that are not present in the original images. In Section 11.4, we will explain how there is a fundamental trade-off between performance and hallucinations. As one strives for the better performance of a reconstruction map (e.g., via training deeper DNNs on larger training sets), the potential for hallucinations inevitably increases. Note that the features introduced in Fig. 11.1 look realistic and, therefore, are not easily dismissed as artifacts stemming from the recovery procedure. The potential for realistic-looking artifacts is a cause of particular concern in healthcare applications.

11.3.2 Instability

In the previous example, we saw that high performance can lead to hallucinations. In Figs. 11.2 and 11.3, we demonstrate that, for some trained DNNs, high performance and instability go hand-in-hand. Both the DNNs considered perform very well on certain images. However, they are both highly unstable. Specifically, a small perturbation v in the underlying image x—or, equivalently, a small perturbation $\eta = Ev$ in the measurements—leads to a much larger perturbation in the recovered image $\Psi(Ex + \eta)$. As seen in these figures, such artifacts can vary significantly in character. In Fig. 11.3, the instability results in noise-like artifacts, with increasing severity as the size of the perturbation increases. These artifacts are arguably quite easy to

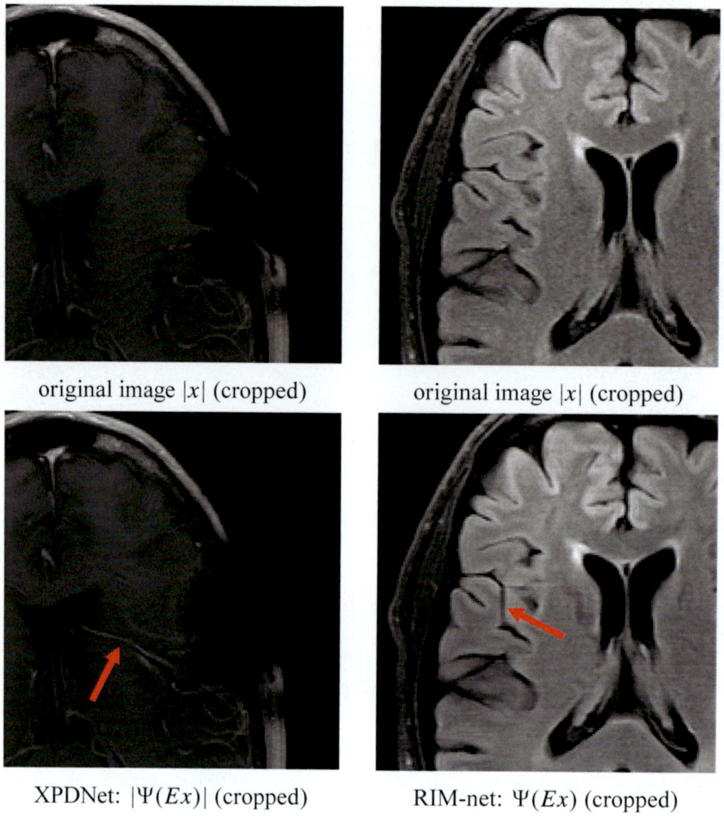

FIGURE 11.1

High-performing AI-inspired method for MR image reconstruction may hallucinate. This figure shows that two of the best-performing DNNs from the 2020 fastMRI challenge [8]—namely, XPDNet [34] and RIM-net [35]—hallucinate by adding realistic-looking features that are not present in the original image x into the reconstructed image $\Psi(Ex)$.

detect and protect against. Conversely, in Fig. 11.2, the artifacts are realistic-looking hallucinations, and, therefore, very difficult to detect and, thereby, militate against. It is particularly notable that standard, model-based reconstruction maps do not produce such dramatic artifacts. Given suitable conditions on E, the theory of *compressed sensing* implies that such maps are stable (see, e.g., [1, Chpt. 5]). For further discussion on the comparison between unstable AI-inspired methods and stable model-based methods, see [1, Chpt. 20].

As we discuss in Section 11.4, there is also a fundamental trade-off between performance and stability. Reconstruction maps that *overperform* in a certain sense are necessarily unstable. In particular, we provide a precise theoretical explanation for why the instability seen in Fig. 11.3 arises.

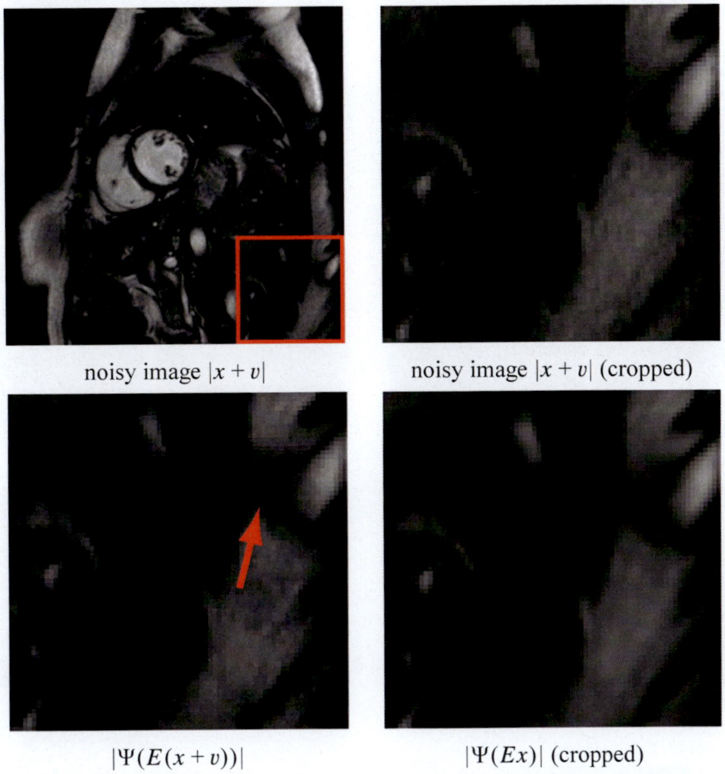

FIGURE 11.2

Unstable AI-inspired algorithms may hallucinate realistic-looking artifacts. This figure considers the reconstruction map Ψ corresponding to the DeepMRI network [36]. Small, mean-zero Gaussian noise v is added to the image, which yields small, mean-zero Gaussian noise Ev in the measurements. However, when the perturbed image $x + v$ is reconstructed from its measurements $E(x+v)$ (*bottom left*), it results in a realistic-looking artifact, i.e., a hallucination, which is not present in the reconstruction from the unperturbed measurements (*bottom right*). For full details on this experiment see [22, Sec. A.3].

Instabilities in AI-inspired algorithms for image reconstruction were first identified in Refs. [1,4]. See also [2, 3, 37] and [1, Chpt. 20] for subsequent studies. The perturbations v constructed in these studies were generally of the worst-case type, i. e., given a fixed image x and reconstruction map Ψ, they are constructed specifically to cause the most significant effect on the recovered image. They are closely related to the topic of *adversarial attacks* in DL [21,26,27,38–41].

In MRI, such worst-case perturbations may not be encountered in practice. One may, therefore, question the relevance of such studies. However, it is notable that the perturbations in Figs. 11.2 and 11.3 are due to Gaussian noise. In Fig. 11.2, the perturbation is simply mean-zero Gaussian noise—a standard noise model in

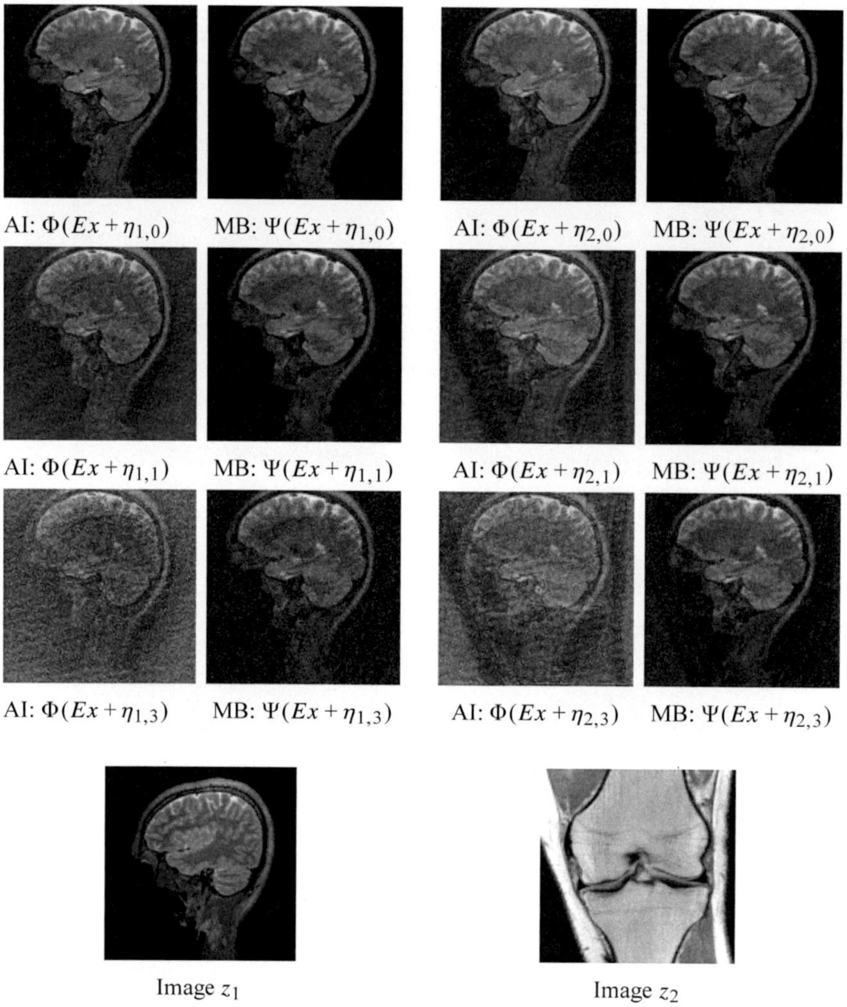

FIGURE 11.3

Unstable AI-inspired algorithms may suffer from severe noise-like artifacts. This figure considers the reconstruction map Ψ, labeled "AI", corresponding to the AUTOMAP network [30]. In this experiment, small, image-independent noise $\eta_{i,j}$, $j = 0, 1, 2$, is constructed as follows for each case $i = 1, 2$. Let $w_{i,0} = 0 \in \mathbb{C}^N$. Then, for $j = 1, 2$, we compute worst-case perturbations $w_{i,j} \in \mathbb{C}^N$, $j = 1, 2$, using the technique of [1] based on an image z_i that is distinct from the image x to be reconstructed. Finally, the perturbation is computed as $\eta_{i,j} = Ew_{i,j} + n_{i,j}$, where $n_{i,j}$ is mean-zero Gaussian noise. We also report the results for a standard model-based reconstruction map, labeled "MB", which involves solving a LASSO-type optimization problem with a wavelet penalty.

MRI. In Fig. 11.3, the perturbation is due to Gaussian noise with a nonzero, but *image-independent*, mean. It is akin to a *universal* perturbation in DL for classification. It is particularly notable that the perturbation constructed using a knee MR image, which is very far from the brain image being recovered, also leads to dramatic artifacts. Furthermore, there is more than one such image-independent mean that leads to such significant effects, as highlighted by the two very different images z used to construct the perturbations in this figure. This, combined with the fact that the noise is random, means that such bad perturbations are not rare *rare events*. In transpires that this can be shown mathematically—see Section 11.4 for further discussion.

11.3.3 Unpredictable generalization

In Fig. 11.4, we show that AI-inspired methods may generalize in inconsistent ways. This figure demonstrates that the error of an AI-inspired reconstruction map can vary substantially and unpredictably when exposed to different inputs that are outside, but very close to the training set. In this experiment, the DNN has been trained on a database of brain MR images. However, the error varies by as much as 50% from the average value. In this figure, we also report the error for a standard, model-based reconstruction map which is based on a LASSO-type optimization problem with a wavelet penalty. This more classical approach delivers much more consistent performance, with far less variation in the error from example to example. Such consistent performance is in line with the underlying compressed sensing theory for such model-based approaches.

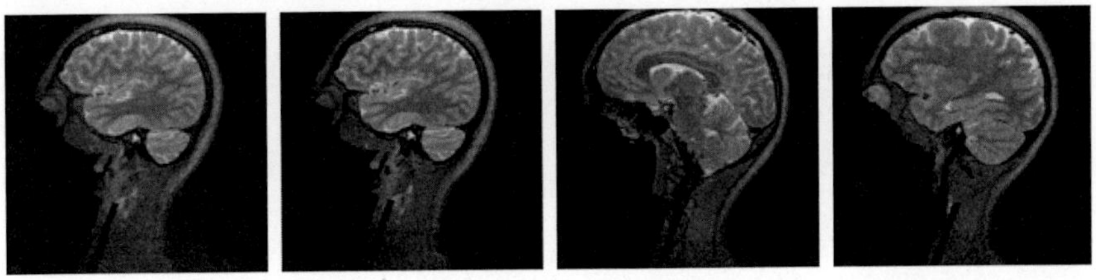

$\lvert x^{(1)} \rvert$	$\lvert x^{(2)} \rvert$	$\lvert x^{(3)} \rvert$	$\lvert x^{(4)} \rvert$
AI: 0.0238	AI: 0.0639	AI: 0.0373	AI: 0.0443
MB: 0.0097	MB: 0.0114	MB: 0.0141	MB: 0.0129

FIGURE 11.4

AI-inspired methods for image reconstruction may generalize in unpredictable ways. This figure examines the performance of the AUTOMAP network [30], labeled "AI", on four different MR brain images. The images $x^{(2)}$, $x^{(3)}$, and $x^{(4)}$ are from the MGH–USC dataset [31] used for training and/or testing of AUTOMAP. Image $x^{(1)}$ is identical to $x^{(2)}$, except for a small, visually imperceivable perturbation, which has been added to it in order to increase the performance of AUTOMAP. The first row shows the images. The second and third rows report the root mean-squared error (RMSE). We also report the error for a standard model-based reconstruction map, labeled "MB", which involves solving a LASSO-type optimization problem with a wavelet penalty.

11.4 The mathematics of nonrobustness

In this section, we present a mathematical analysis of nonrobustness for image reconstruction—in particular, we describe theoretical mechanisms that explain some of the key issues raised in Section 11.1.1, namely, hallucinations, instabilities, and unpredictable/inconsistent generalization. This section is based primarily on [3] and [1, Chpt. 20].

The theoretical results in this section consider an arbitrary encoding operator E, with a focus on the case where E has a nontrivial null space, i.e., $\mathrm{rank}(E) < N$, or where $\mathcal{N}(E) = \{0\}$ but E is ill-conditioned. Our results do not assume that E stems from either of the subsampled MR acquisition models described in Section 11.2.1. This means that the theoretical results apply not just to MR image recovery, but also to any discrete linear, inverse problem with measurements of the form (11.2.1). In particular, they apply to many other computational imaging problems.

11.4.1 Additional notation

We first require some additional notation. Throughout this section, $\|\cdot\|$ and $|||\cdot|||$ denote arbitrary norms on $\mathbb{C}^N$ and $\mathbb{C}^m$, respectively. We write $\Psi : \mathbb{C}^m \to \mathbb{C}^N$ for an arbitrary reconstruction map. We define its (global) Lipschitz constant as

$$L(\Psi) = \sup_{\substack{y,z\in\mathbb{C}^m \\ z\neq y}} \frac{\|\Psi(z) - \Psi(y)\|}{|||z - y|||}.$$

Finally, given $\varepsilon > 0$ and a set $\mathcal{S} \subseteq \mathbb{C}^m$, we write

$$\mathcal{S}^\varepsilon = \{z \in \mathbb{C}^m : \exists y \in \mathcal{S} \text{ with } |||z - y||| < \varepsilon\}.$$

11.4.2 Hallucinations due to detail transfer

The following result, which is based on [22, Thm. 4.1], shows that hallucinations can arise due to *detail transfer*.

Theorem 11.4.1. (Hallucinations due to detail transfer). *Let $E \in \mathbb{C}^{m\times N}$, $\delta, L > 0$, and $x, x_{\mathrm{Det}} \in \mathbb{C}^N$ with $|||Ex_{\mathrm{Det}}||| \leq \delta$.*

1. (Ψ hallucinates by transferring details). *Let $\Psi : \mathbb{C}^m \to \mathbb{C}^N$ be Lipschitz continuous with constant at most L and suppose that*

$$\|\Psi(E(x + x_{\mathrm{Det}})) - (x + x_{\mathrm{Det}})\| \leq \delta. \tag{11.4.1}$$

Then, for every $\eta \in \mathbb{C}^m$ with $|||\eta||| \leq \delta$, there is a $z \in \mathbb{C}^N$ with $\|z\| \leq (1 + 2L)\delta$ such that

$$\Psi(Ex + \eta) = x + x_{\mathrm{Det}} + z. \tag{11.4.2}$$

2. (There always exists a DNN that hallucinates). For any family of deep neural networks $\mathcal{NN}$, there is a $\Psi \in \mathcal{NN}$ with the Lipschitz constant at most L that satisfies (11.4.1).

The first part of this result says the following: Suppose that there is a detail x_{Det} that lies close to the null space of E, i.e., $|||Ex_{\text{Det}}||| \leq \delta$. Suppose also that a stable reconstruction map (i.e., one for which L is not too large) achieves a small error (11.4.1) over the "detail" image $x + x_{\text{Det}}$. Then, Ψ incorrectly transfers this detail when recovering the detail-free image x. Specifically, due to (11.4.2), we have $\Psi(Ex+\eta) \approx x + x_{\text{Det}}$. In essence, this result is one of *overperformance*. The map Ψ performs too well on a single image, namely, the detail image $x + x_{\text{Det}}$, and, as a result, it performs poorly on the detail-free image x.

This result is very general and places minimal assumptions on either E or Ψ. Note that it does not require Ψ to be unstable: in fact, its conclusions are most pertinent when the Lipschitz constant L is small. In general, it is meaningful when x_{Det} is some detail that is not small, i.e., $\|x_{\text{Det}}\| \gg 0$, and that is physically relevant for the image reconstruction problem at hand (i.e., not noise or texture). Note that the condition $\|x_{\text{Det}}\| \gg \delta \geq |||Ex_{\text{Det}}|||$ may readily hold when $\text{rank}(E) < N$. In this case, E has a nontrivial null space, and consequently, there are many "candidate" details x_{Det}. But it may also hold when $\text{rank}(E) = N$ and E is ill-conditioned, as ill-conditioning also implies the existence of many candidate details for which $\|x_{\text{Det}}\| \gg \delta \geq |||Ex_{\text{Det}}|||$. Thus, this theorem is relevant for both subsampled MR acquisition models discussed in Section 11.2.1.

The key point is that it is extremely easy for a reconstruction map that has been obtained via training to hallucinate. Successful training (i.e., small training error) will readily lead to many candidates in which (11.4.1) holds for elements $x + x_{\text{Det}}$ in the training set. However, there may be other candidates outside of the training set for which Ψ satisfies (11.4.1). The main conclusion is that it is exceedingly difficult to prevent (11.4.1) from occurring when training a reconstruction map from the data. The second part of the theorem also emphasizes that *any* approach that trains a DNN Ψ can potentially suffer from this effect. We omit the definition of $\mathcal{NN}$ here for brevity, but note that it is extremely general and encompasses essentially any possible DNN architecture. See [22, §3].

In Fig. 11.5, we show an illustration of this effect for a subsampled single-coil MR acquisition. In this case, the "Mickey Mouse" detail lies in the null space of E, whereas the "Thumb" detail lies far from the null space, i.e., $|||Ex_{\text{th}}||| \gg 0$. A DNN is trained to accurately recover the detail image $x_{\text{br}} + x_{\text{th}} + x_{\text{mi}}$, along with 1200 other images from the fastMRI brain dataset. Due to this training, Ψ is expected to reconstruct typical brain images well. However, in accordance with the theorem, it incorrectly transfers the "Mickey Mouse" detail when recovering both the images $x_{\text{br}} + x_{\text{th}}$ and x_{br}. The problem is that there is a detail lying close to (in this case, in) $\mathcal{N}(E)$. The map Ψ has been trained to recover brain images without this detail, plus the brain image with this detail. The inevitable result is hallucinations by detail transfer.

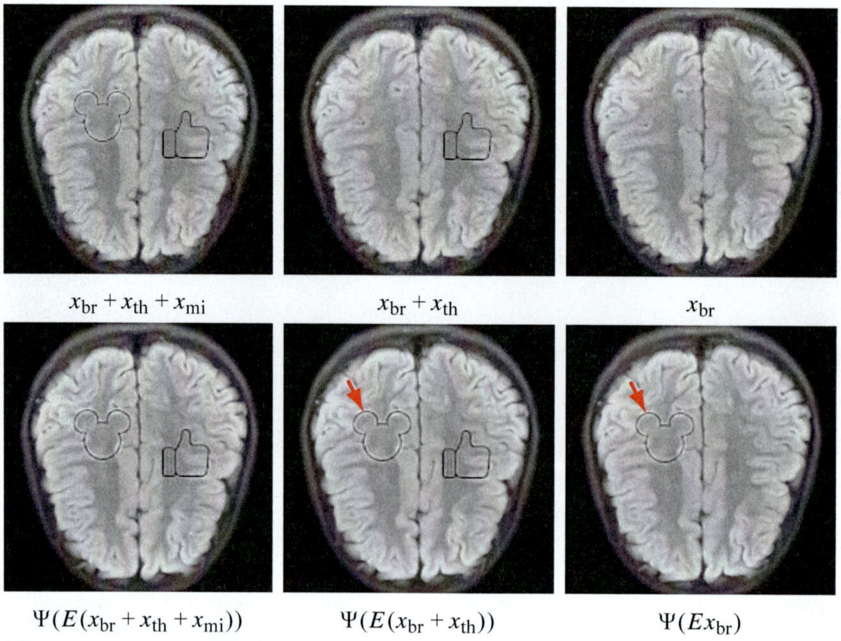

FIGURE 11.5

Overperformance on a single image can cause hallucinations. In this figure, a DNN Ψ is trained on a database consisting of the detail image $x_{br} + x_{th} + x_{mi}$, plus 1200 other images from the fastMRI brain dataset. The sampling operator E is as in (11.2.2), for a suitable mask Ω, and models a two-dimensional, single-coil MR acquisition. The subsampling percentage $m/N = 20\%$. The "Mickey mouse" detail x_{mi} was designed so that $Ex_{mi} = 0$, while the "Thumb" detail x_{th} was not. Due to Theorem 11.4.1, Ψ incorrectly transfers the former detail when reconstructing the images $x_{br} + x_{th}$ and x_{br}. For full details on this experiment, see [22, Sec. A.2].

11.4.3 Hallucinations due to overperformance

The previous result considered hallucinations arising from overperformance on a single image. In the following result, which is [22, Thm. 4.2(i)], we consider the setting where a reconstruction map Ψ overperforms on a class of images.

Theorem 11.4.2. (Overperformance implies hallucinations). *Let $E \in \mathbb{C}^{m \times N}$, $\mathcal{T} \subset \mathbb{C}^N$ be a nonempty and finite set, $\delta > 0$, $\Psi : \mathbb{C}^m \to \mathbb{C}^N$ with the Lipschitz constant $L > 0$ and $x_{\text{Det}} \in \mathbb{C}^N$ with $|||Ex_{\text{Det}}||| \leq \delta/(4L)$. Suppose that Ψ satisfies*

$$\max_{x \in \mathcal{T}} \|\Psi(Ex) - x\| \leq \delta.$$

Then, for any $\varepsilon \geq \delta/(2L)$, there is an uncountable family $\mathscr{C}$ of finite or countably infinite sets $\mathscr{M} \subset \mathbb{C}^N$ with $\mathcal{T} \subset \mathscr{M}$ and $E\mathscr{M} \subset (E\mathcal{T})^{\varepsilon}$, such that for each $\mathscr{M} \in \mathscr{C}$ the *following holds. For any probability distribution* $\mathscr{D}$ *on* $\mathscr{M}$ *with the property that* $\mathbb{P}_{X\sim\mathscr{D}}(X \in \mathscr{T}) \leq q$, *it holds that*

$$\mathbb{P}_{X\sim\mathscr{D}}(\exists\lambda \in \mathbb{C}, |\lambda| = 1 \text{ such that} \|\Psi(EX) - (X + \lambda x_{\text{Det}})\| \leq 2\delta) \geq 1 - q. \tag{11.4.3}$$

This result says the following: For any reconstruction Ψ that performs too well on a set $\mathscr{T}$ (e.g., a training and/or test set), there are infinitely many *model classes* $\mathscr{M}$ on which, with high probability, Ψ hallucinates over $\mathscr{M}$, regardless of the distribution on $\mathscr{M}$.

Another way of viewing this result is that Ψ suffers from *in-distribution* hallucinations—a potentially far more serious issue than *out-of-distribution* hallucinations—since the images X in (11.4.3) are drawn from a distribution $\mathscr{D}$ over $\mathscr{M}$. Typically, $\mathscr{M}$ represents the types of images we wish to recover in a given problem. For instance, $\mathscr{M}$ could be a set of brain MR images and $\mathscr{T}$ could be a training set. This theorem states that there are uncountably many possibilities for the model class $\mathscr{M}$ for which Ψ hallucinates.

This result implies that hallucinations may be unavoidable in a standard DL scenario where a DNN is trained to achieve a small error over a training and/or test set (notice that Theorem 11.4.2 is general, therefore $\mathscr{T}$ could consist of both the training and test sets). Put another way, to avoid hallucinations, it is not enough to ensure a small error over the training and test sets, one may also need to incorporate additional information about the relevant model class $\mathscr{M}$.

11.4.4 Hallucinations and instabilities due to over- or inconsistent performance

In the next result [22, Thm. 4.4], we focus not only on the issue of *overperformance* but also on *inconsistent performance*. Specifically, we show that over- or inconsistent performance on two images whose measurements are sufficiently close necessarily leads to instabilities and hallucinations.

Theorem 11.4.3 (Over- or inconsistent performance implies both hallucinations and instabilities). *Let* $E \in \mathbb{C}^{m\times N}$, $x, x' \in \mathbb{C}^N$, *and* $\delta > 0$. *Let* $\Psi : \mathbb{C}^m \to \mathbb{C}^N$ *be continuous and suppose that*

$$\|\Psi(Ex) - x\| < \delta, \left\|\Psi(Ex') - x''\right\| < \delta, \tag{11.4.4}$$

for some $x'' \in \mathbb{C}^N$ and that

$$\left|\left|\left|Ex - Ex'\right|\right|\right| \leq \delta. \tag{11.4.5}$$

Then, the following holds:

1. (Ψ is unstable). There is an $\eta \in \mathbb{C}^m$ with $|||\eta||| \leq \delta$ and a closed nonempty ball $\mathscr{B}_\eta \subset \mathbb{C}^m$ centered at η such that

$$\left\|\Psi\left(Ex + \tilde{\eta}\right) - \Psi(Ex)\right\| \geq \left\|x - x''\right\| - 2\delta, \forall\tilde{\eta} \in \mathscr{B}_\eta. \tag{11.4.6}$$

2. (Ψ hallucinates). *There exists* $z \in \mathbb{C}^N$ *with* $\|z\| \geq \|x - x''\|$ *(for example,* $z = x'' - x$*),* $\eta \in \mathbb{C}^m$ *with* $|||\eta||| \leq \delta$*, and closed nonempty balls* $\mathcal{B}_x$*,* $\mathcal{B}_\eta$*, and* $\mathcal{B}_z$ *centered at x,* η*, and z, respectively, such that*

$$\left\|\Psi\left(A\tilde{x}+\tilde{e}\right)-\left(\tilde{x}+\tilde{z}\right)\right\| \leq \delta, \forall\ \tilde{x} \in \mathcal{B}_x,\ \tilde{\eta} \in \mathcal{B}_\eta, \tilde{z} \in \mathcal{B}_z. \tag{11.4.7}$$

This result has several consequences. First, it implies that overperformance causes instability. Indeed, suppose that $x'' = x'$, $\|x - x'\| \gg \delta$, and that δ is small. Then, any reconstruction map Ψ that overperforms, in the sense that it recovers two distinct images x and x' sufficiently well, i.e., (11.4.4), from measurements that are close, i.e., (11.4.5), must be unstable. Specifically, its local ε-Lipschitz constant scales like $1/\delta$, i.e., (11.4.6), for any $\varepsilon \geq \delta$. Moreover, the instability scales with the difference between x and x'. A stable reconstruction map can, therefore, recover either x accurately or x' accurately, but not both. Thus, there is a trade-off between accuracy of the reconstruction map and its stability. Note that the situation described can easily occur with learning, e.g., when x and x' are elements of the training set.

Second, this result implies that inconsistent performance also causes instability. Let $x'' = \Psi(Ex')$. Then, Theorem 11.4.3 states that any map Ψ that recovers x well, i.e., (11.4.4), but reconstructs some x' having similar measurements, i.e., (11.4.5), poorly, i.e., $\|x - x''\| = \|x - \Psi(Ex')\| \gg \delta$, must be unstable. Hence, inconsistent performance—recovering x well but some nearby x' poorly—necessarily results in instability. This situation can easily occur with learning, e.g., when x is in the training set and x' is close to this set, but not in it.

Theorem 11.4.3 also shows that over- or inconsistent performance causes hallucinations. Indeed, let $x' = x + x_{\mathrm{Det}}$ and suppose that $x'' = x'$. Then, this theorem states that any reconstruction map Ψ that recovers both x and x' well must hallucinate whenever x_{Det} is close to the null space of E. Specifically, a consequence of (11.4.7) is that $\Psi(Ex + \eta) \approx x' = x + x_{\mathrm{Det}}$ for some small perturbation η, i.e., Ψ hallucinates the detail x_{Det} when recovering the detail-free image x.

It is notable that hallucinations and instabilities predicted by Theorem 11.4.3 are *stable* in the sense that there is always a ball within which each conclusion (instability and hallucinations) holds. Instabilities and hallucinations are not, therefore, rare events—an effect already witnessed in Figs. 11.2 and 11.3, where bad perturbations arise from random noise. On the other hand, this theorem says nothing about how large such balls are, and, therefore, how prevalent such perturbations may be. For a more quantitative analysis, see [22, Thm. 4.6].

It is also notable that Theorem 11.4.3 does not require E to be ill-conditioned or have $\mathrm{rank}(E) < N$. Condition (11.4.4) is purely a property of the reconstruction map Ψ and, as noted, one that can be easily achieved via training. The other condition (11.4.5) can occur even when E is a full column rank and well-conditioned, although it is more likely to arise when it is not. Indeed, suppose that $\mathrm{rank}(E) < N$ and let x be an image satisfying (11.4.4). Then, there are many (in fact, infinitely many) vectors x' of the from $x' = x + v$ for which $\|Ex - Ex'\| = \|Ev\| \leq \delta$. In other words, there are many ways in which the condition $\|Ex - Ex'\| \leq \delta$ can be achieved. If just one of those vectors x' yields $\|\Psi(Ex') - x''\| < \delta$ for some x'' not too close to x (i.e., $\|x - x''\| \geq 2\delta$), then the effects described in this theorem must arise. The key point is that these conditions are extremely weak ones. If there is E which is ill-conditioned or not full column rank, there are many ways in which (11.4.4 and 11.4.5) can occur. In particular, the existence of a single triple $\{x, x', x''\}$ satisfying (11.4.4 and 11.4.5) implies the existence of infinitely many other triples for which these conditions also hold. In other words, there are infinitely many images for which instabilities and hallucinations can occur.

In Table 11.1 and Fig. 11.6, we show that this theorem precisely predicts the instability of the AI-inspired method AUTOMAP that was previously considered in Fig. 11.3. First, in Fig. 11.6, we simulate worst-case perturbations $\eta = \eta_j$ and $j = 1, 2, 3$, of increasing sizes for this method. As is evident, the method is highly unstable: the effect of the perturbation on the output is many orders of magnitude larger than the perturbation itself. Next, in Table 11.1, we use these perturbations to empirically verify Theorem 11.4.3. After choosing appropriate values for δ and x'', we not only see that (11.4.6) holds, but also that the ratio of the left- and right-hand sides (labeled "sharpness ratio" in the table) is very close to one. This demonstrates that the theorem accurately predicts the unstable behavior seen in Fig. 11.6.

11.5 Discussion and future directions

The purpose of this chapter has been to highlight, both empirically and theoretically, a number of key issues to which AI-based methods may be particularly susceptible. Specifically, we have explored the issues of hallucinations, instabilities, and unpredictable/inconsistent generalization and the underlying mechanisms which can cause them. For applications such as MRI, where reliability is key, such issues may limit the uptake of these methods, despite their great promise. With this in mind, we close this chapter with some additional discussion.

11.5.1 Nonrobustness

First, it is worth reemphasizing that bad perturbations, which may cause reconstruction maps to exhibit severe instabilities or artifacts, are not necessarily rare events; therefore, their presence cannot be ignored without further studies. See Section 11.4.4. Moreover, the conditions of the four theorems in this chapter are very weak, and, therefore, difficult to avoid, especially with learning-based approaches.

Besides the effects studied in this paper, AI-based methods may also fail to be robust in several other ways. For example, such methods may be unstable to

Table 11.1 Theorem 11.4.3 predicts the instability of AUTOMAP. This table demonstrates that the instabilities of AUTOMAP, shown in Fig. 11.6, are accurately predicted by Theorem 11.4.3. In this table, we consider the image x, measurement matrix E, and perturbations $\eta = \eta_j = Ew_j$ and $j = 1,2,3$, from Fig. 11.6, and let $\Psi : \mathbb{C}^m \to \mathbb{C}^N$ be the AUTOMAP reconstruction map once more. To apply Theorem 11.4.3, we choose the values for $x' = x'_j = x + w_j$, $x'' = x''_j = \Psi(Ex'_j)$, and $\delta = \delta_j = \|\Psi(Ex) - x\|_{\ell^2}$ to verify conditions (11.4.4–11.4.6). Note that a sharpness ratio $P_j/Q_j \geq 1$ means that condition (1.4.6) holds. This table confirms conditions (11.4.4–11.4.6) for these choices of x', x'', and δ (notice that $\|\eta_j\|_{\ell^2} = \|Ew_j\|_{\ell^2} = \left\|Ex - Ex'_j\right\|$ in this setup). Furthermore, the sharpness ratio is close to one in all cases, meaning that the instability estimate of Theorem 11.4.3, which is given by the right-hand side of (11.4.6), very precisely describes the true instability seen in Fig. 11.6.

Condition	Quantity	$J = 1$	$J = 2$	$J = 3$
	δ_j	4.9	4.9	4.9
(11.4.4)	$\|\Psi(Ex) - x\|_{\ell^2} + \|\Psi(Ex'_j) - x''_j\|_{\ell^2}$	4.8	4.8	4.8
(11.4.5)	$\left\|Ex - Ex'_j\right\|_{\ell^2}$	2.0	3.0	4.2
	$\left\|\Psi(Ex + \eta_j) - \Psi(Ex)\right\|_{\ell^2} =: P_j$	13.7	23.9	35.7
(11.4.6)	$\left\|x - x''_j\right\|_{\ell^2} - \delta_j =: Q_j$	12.5	22.6	34.3
	Sharpness ratio P_j/Q_j	**1.11**	**1.06**	**1.04**
Other	$\left\|x'_j\right\|_{\ell^2}$	25.5	25.1	24.9
Quantities	$\left\|Ex'_j\right\|_{\ell^2}$	12.1	11.9	11.9

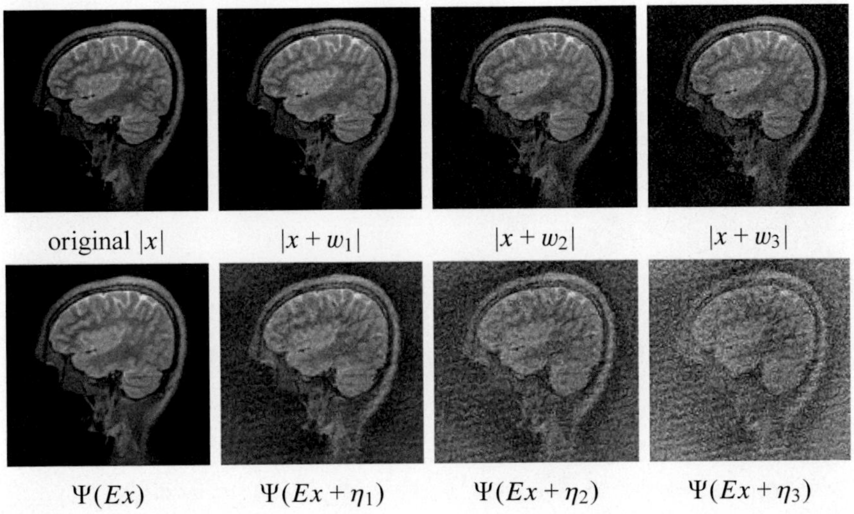

FIGURE 11.6

Worst-case perturbations for AUTOMAP. This figure shows that the reconstruction map Ψ of the AUTOMAP network [30] is unstable to worst-case perturbations. Top row: similar to Fig. 1.3, three image perturbations $w_1, w_2, w_3 \in \mathbb{C}^N$ of increasing magnitude are computed using the technique of [1]. Second row: AUTOMAP's reconstruction from the perturbed measurements $Ex + \eta_j$, where $\eta_j = Ew_j$ and $j = 1,2,3$. These reconstructions show significant artifacts, whose magnitudes greatly exceed those of the perturbations. This demonstrates the instability of Ψ.

changing the acquisition strategy, with the addition of more samples potentially leading to worse reconstructions [1]. Some partial remedies for this problem can be found in Refs. [42–44]. This phenomenon is also related to the issue of distribution shifts in machine learning. We refer to [22, Sec. 2.3.2] for further discussion.

This aside, another potential nonrobustness of AI-based methods in comparison to more standard methods is their general inability to fail gracefully. In Fig. 11.7, we demonstrate this effect. The AI-based method continues to produce seemingly realistic brain images even when the sampling rate is drastically reduced. By contrast, a model-based method based on based a LASSO-type optimization problem with a wavelet penalty produces noticeable and easily recognizable wavelet artifacts.

11.5.2 Model classes return?

When AI-based methods for image reconstruction were first proposed, there was promise that they may allow one to avoid some of the downsides of standard model-based methods, in particular, the need to carefully design a model and (convex) penalty and then tune hyperparameters. Methods such as AUTOMAP [30] aimed at learning an end-to-end DNN reconstruction map from the data,

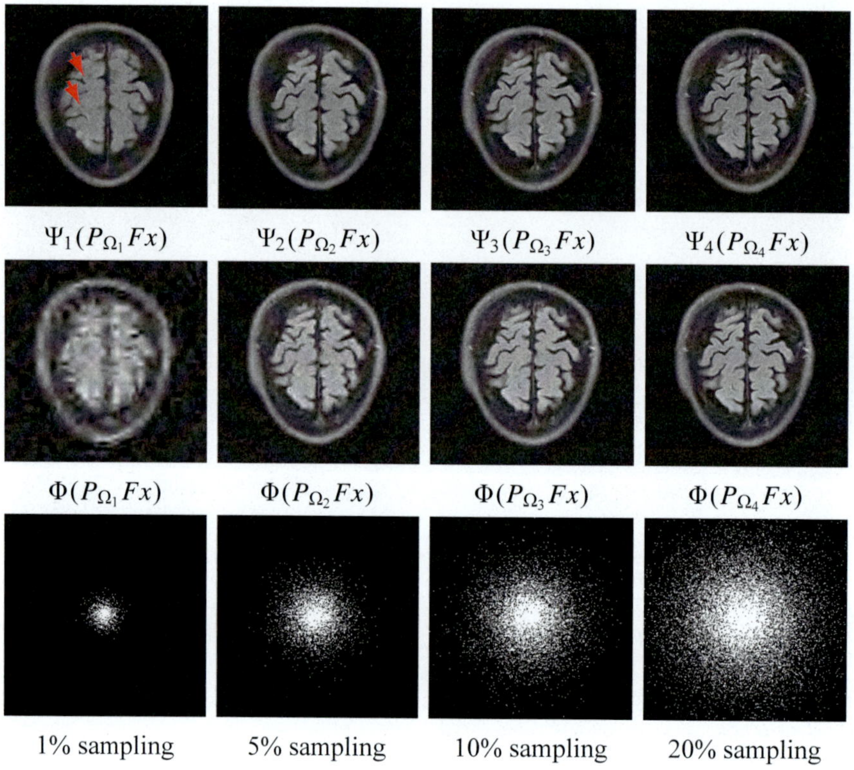

FIGURE 11.7

AI-based methods may produce realistic-looking images irrespective of the sampling rate. This figure shows the performance of four DNNs Ψ_j, $j = 1,\ldots,4$, trained on 1200 images of size $N = 256 \times 256$ from the fastMRI challenge using Fourier sampling with the sampling patterns $\Omega_j \subset \{1, \ldots, N\}$ shown in the last row. Here, each Ψ_j is trained on samples from the matrix $P_{\Omega_j}F$, as in Section 11.2.1. We compare these maps against a standard model-based reconstruction map Φ that involves solving a LASSO-type optimization problem with a wavelet penalty. As the sampling rate decreases, the model-based reconstruction produces increasingly noticeable artifacts and, for 1% subsampling, the image quality of the reconstructed image is heavily corrupted. The trained DNNs, on the other hand, reconstruct realistic-looking images at all sampling rates. However, as the upper left image shows, these images are not necessarily accurate reconstructions of the underlying brain.

without even having the knowledge of the forward operator E. The problem, as elucidated in this chapter, is that such approaches are generally nonrobust.

Take, for example, Theorem 11.4.3. In order to avoid its conclusions, one may seek to restrict the set of images one wants to recover well to some model class $\mathcal{M} \subset \mathbb{C}^N$ for which

$$\|x - x'\| \leq C\|E(x - x')\| \quad \forall x, x' \in \mathcal{M} \subset \mathbb{C}^N, \tag{11.5.1}$$

for some constant $C > 0$ that is not too small. When this holds, one avoids the mechanism that leads to instabilities and hallucinations in this theorem. In particular, it is also clear that for a fixed set $\mathcal{M}$, (11.5.1) cannot hold for arbitrary E. Therefore, conditions on E are needed to ensure a stable recovery. But ensuring (11.5.1) may be difficult for AI-based methods, where the class $\mathcal{M}$ is implicitly learned during training.

As discussed in Ref. [3], conditions such as (11.5.1) are typically built into model-based methods via the compressed sensing theory [33]. For these methods, one strives to construct E and the reconstruction method in tandem in relation to the set $\mathcal{M}$ one strives to recover. Of course, AI-based methods have shown great potential to outperform such methods, albeit at the expense of theoretical guarantees and, as things currently stand, robustness. It is nonetheless interesting to note that recent studies have shown that model-based methods, when tuned by machine learning techniques, can achieve similar performance to certain AI-based methods [45].

11.5.3 Toward trustworthy AI for MR image reconstruction

AI has great potential to improve modern image reconstruction. Yet, the trustworthiness of AI-based methods is an issue of utmost concern. This chapter aims to provide guidance on how one might feasibly avoid nonrobustness in image reconstruction [51]. In particular, necessary conditions for a method to be robust are that they must not satisfy the conditions of any of our main theorems. Deriving realistic, sufficient conditions would ensure trustworthiness. The challenge, however, is to do this while maintaining excellent recovery performance. It is, of course, easy to find a provably stable, nonhallucinating DNN: the zero network is perfectly stable and nonhallucinating, yet wholly ineffective in practice.

Naturally, there are various ways one might strive to enhance robustness when training an AI-based method. One may, for example, perform adversarial training with perturbations added to the measurements in the training process. As discussed in Ref. [3], however, how to best balance the need for performance with the need for robustness is delicate. It is straightforward increase robustness, but at the cost of losing details in the recovered images, see, e.g., [19, Fig. 14], [22, Fig. 5], or [10, Fig. 4]. Other approaches involve using interval DNNs [46] or Bayesian techniques [47–49] to quantify the uncertainty in the recovered images. However, these approaches may not necessarily succeed [50].

Our theorems also establish how difficult it is to avoid hallucinations and instabilities. In particular, even if a reconstruction does not satisfy the conditions of, say, Theorem 11.4.3 on the training data, it is extremely difficult to guarantee that these conditions are not satisfied on other nearby images. This adds theoretical justification for why the question of how to prevent hallucinations is very much open.

11.6 Conclusion

By highlighting these issues, this chapter aims to stimulate and inform future progress on the integration of AI-based methods into imaging. Empirical studies such as the fastMRI challenges provide important practical insight into these issues, especially with regard to modality-specific considerations, such as the structure of typical perturbations (which may be Gaussian or nonGaussian, mean zero or non-mean zero) and desirable quality metrics (which may or may not be in terms of the RMSE or SSIM). At the same time, mathematical examinations provide key theoretical insights into the mechanisms that cause instabilities and hallucinations in general. With hope, the combination of these two approaches will provide a path toward the development of trustworthy AI for imaging.

References

[1] Antun V, Renna F, Poon C, Adcock B, Hansen AC. On instabilities of deep learning in image reconstruction and the potential costs of AI. Proc Natl Acad Sci USA 2020;117 (48):30088–95.

[2] Genzel M, Macdonald J, Marz M. Solving inverse problems with deep neural networks – robustness included? IEEE Trans Pattern Anal Mach Intell 2023;45(1):1119–34.

[3] Gottschling NM, Antun V, Hansen AC, Adcock B. The troublesome kernel – on hallucinations, no free lunches and the accuracy-stability trade-off in inverse problems. SIAM Rev 2024. in press, https://epubs.siam.org/doi/10.1137/23M1568739. in press.

[4] Huang Y, Würfl T, Breininger K, Liu L, Lauritsch G, Maier A. Some investigations on robustness of deep learning in limited angle tomography. In: International conference on medical image computing and computer-assisted intervention; 2018. p. 145–53.

[5] Morshuis JN, Gatidis S, Hein M, Baumgartner CF. Adversarial robustness of MR image reconstruction under realistic perturbations. arXiv:2208.03161 2022. https://link.springer.com/chapter/10.1007/978-3-031-17247-2_3.

[6] Raj A, Bresler Y, Li B. Improving robustness of deep-learning-based image reconstruction. In: Daumé III Hal, Singh Aarti, editors. Proceedings of the 37th International Conference on machine learning, volume 119 of Proceedings of machine learning research. PMLR; 2020. p. 7932–42.

[7] Zhang C, Jia J, Yaman B, Moeller S, Liu S, Hong M, Akçakaya M. Instabilities in conventional multi-coil MRI reconstruction with small adversarial perturbations. In: 2021 55th Asilomar Conference on signals, systems, and computers; 2021. p. 895–9.

[8] Muckley MJ, Riemenschneider B, Radmanesh A, Kim S, Jeong G, Ko J, Jun Y, Shin H, Hwang D, Mostapha M, et al. Results of the 2020 fastMRI challenge for machine learning MR image reconstruction. IEEE Trans Med 2021. Imaging, https://ieeexplore.ieee.org/abstract/document/9420272.

[9] Noordman CR, Yakar D, Bosma J, Simonis FFJ, Huisman H. Complexities of deep learning-based undersampled MR image reconstruction. Eur Radiol Exp 2023;7(58).

[10] Laine RF, Arganda-Carreras I, Henriques R, Jacquemet G. Avoiding a replication crisis in deep-learning-based bioimage analysis. Nat Methods 2021;18(10):1136–44.

[11] Liu X, Glocker B, McCradden MM, Ghassemi M, Denniston AK, Oakden-Rayner L. The medical algorithmic audit. Lancet Digit Health 2022;4(5):e384–97.

[12] Shimron E, Tamir JI, Wang K, Lustig M. Implicit data crimes: machine learning bias arising from misuse of public data. Proc Natl Acad Sci USA 2022;119(13): e2117203119.

[13] Tölle M, Laves M-H, Schlaefer A. A mean-field variational inference approach to deep image prior for inverse problems in medical imaging. In: Heinrich M, Dou Q, de Bruijne M, Lellmann J, Schläfer A, Ernst F, editors. Proceedings of the fourth Conference on medical Imaging with deep learning, volume 143 of Proceedings of machine learning research. PMLR; 2021. p. 745–60. Lübeck.

[14] Varoquaux G, Cheplygina V. Machine learning for medical imaging: methodological failures and recommendations for the future. npj Digit Med 2022;5(1):1–8.

[15] Wu E, Wu K, Daneshjou R, Ouyang D, Ho DE, Zou J. How medical AI devices are evaluated: limitations and recommendations from an analysis of FDA approvals. Nat Med 2021;27(4):582–4.

[16] Yu T, Hilbert T, Piredda GG, Joseph A, Bonanno G, Zenkhri S, Omoumi P, Cuadra MB, Canales-Rodríguez EJ, Kober T, et al. Validation and generalizability of self-supervised image reconstruction methods for undersampled MRI. arXiv:2201 2022:12535.

[17] Bhadra S, Kelkar VA, Brooks FJ, Anastasio MA. On hallucinations in tomographic image reconstruction. IEEE Trans Med Imag 2021;40(11):3249–60.

[18] Belthangady C, Royer LA. Applications, promises, and pitfalls of deep learning for fluorescence image reconstruction. Nat Methods 2019;16(12):1215–25.

[19] Hoffman DP, Slavitt I, Fitzpatrick CA. The promise and peril of deep learning in microscopy. Nat Methods 2021;18(2):131–2.

[20] Reader AJ, Pan B. AI for PET image reconstruction. Br J Radiol 2023;96(1150): 20230292.

[21] Heaven D. Why deep-learning AIs are so easy to fool. Nature 2019;574(7777):163–6.

[22] Eykholt K, Evtimov I, Fernandes E, Li B, Rahmati A, Xiao C, Prakash A, Kohno T, Song D. Robust physical-world attacks on deep learning visual classification. In: IEEE Conference on Computer Vision and Pattern Recognition; 2018. p. 1625–34.

[23] Kurakin A, Goodfellow I, Bengio S. Adversarial examples in the physical world. In: International Conference on learning representations; 2017.

[24] Carlini N, Mishra P, Vaidya T, Zhang Y, Sherr M, Shields C, Wagner D, Zhou W. Hidden voice commands. In: 25th USENIX security symposium; 2016. p. 513–30.

[25] Carlini N, Wagner D. Audio adversarial examples: targeted attacks on speech-to-text. In: 2018 IEEE security and privacy workshop; 2018. p. 1–7.

[26] Goyal S, Doddapaneni S, Khapra MM, Ravindran B. A survey of adversarial defenses and robustness in NLP. ACM Comput Surv 2023;55(14s).

[27] Finlayson SG, Bowers JD, Ito J, Zittrain JL, Beam AL, Kohane IS. Adversarial attacks on medical machine learning. Science 2019;363(6433):1287–9.

[28] The Cambridge Dictionary. The Cambridge dictionary word of the year 2023 is. Webpage: https://dictionary.cambridge.org/editorial/woty; 2023.

[29] European Commission. Europe fit for the digital age: Commission proposes new rules and actions for excellence and trust in artificial intelligence. Press release, https://ec.europa.eu/commission/presscorner/detail/en/IP_21_1682. [Accessed 21 April 2021].

[30] Zhu B, Liu JZ, Cauley SF, Rosen BR, Rosen MS. Image reconstruction by domain-transform manifold learning. Nature 2018;555(7697):487–92.

[31] Fan Q, Witzel T, Nummenmaa A, Van Dijk K, Van Horn JD, Drews MK, Somerville LH, Sheridan MA, Santillana RM, Snyder J, Hedden T, Shaw EE, Hollinshead MO, Renvall V, Zanzonico R, Keil B, Cauley S, Polimeni JR, Tisdall D, Buckner RL, Wedeen VJ, Wald LL, Toga AW, Rosen BR. MGH–USC human connectome project datasets with ultra-high b-value diffusion MRI. Neuroimage 2016;124(Pt B):1108–14.

[32] Deshmane A, Gulani V, Griswold MA, Seiberlich N. Parallel MR imaging. J Magn Reson Imag 2012;36(1):55–72.

[33] Adcock B, Hansen AC. Compressive imaging: structure, sampling, learning. Cambridge, UK: Cambridge University Press; 2021.

[34] Ramzi Z, Ciuciu P, Starck J-L. XPDNet for MRI reconstruction: an application to the 2020 fastMRI challenge. arXiv:2010 2020:07290.

[35] Putzky P, Welling M. Recurrent inference machines for solving inverse problems. arXiv:1706.04008 2017.

[36] Schlemper J, Caballero J, Hajnal JV, Price A, Rueckert D. A deep cascade of convolutional neural networks for MR image reconstruction. In: Niethammer M, Styner M, Aylward S, Zhu H, Oguz I, Yap P-T, Shen D, editors. Information Processing in medical imaging, volume 10265 of lecture notes in computer science. Cham: Springer; 2017. p. 647–58.

[37] Darestani MZ, Chaudhari AS, Heckel R. Measuring robustness in deep learning based compressive sensing. In: International Conference on machine learning; 2021. p. 2433–44.

[38] Akhtar N, Mian A. Threat of adversarial attacks on deep learning in computer vision: a survey. IEEE Access 2018;6:14410–30.

[39] Chakraborty A, Alam M, Dey V, Chattopadhyay A, Mukhopadhyay D. Adversarial attacks and defences: a survey. arXiv:1810.00069 2018.

[40] Szegedy C, Zaremba W, Sutskever I, Bruna J, Erhan D, Ian Goodfellow J, Fergus R. Intriguing properties of neural networks. arXiv:1312.6199 2013.

[41] Wiyatno RR, Xu A, Dia O, de Berker A. Adversarial examples in modern machine learning: a review. arXiv:1911.05268 2019.

[42] Gilton D, Ongie G, Willett R. Model adaptation for inverse problems in imaging. IEEE Trans Comput Imaging 2021;7:661–74.

[43] Gossard A, Weiss P. Training adaptive reconstruction networks for inverse problems. arXiv:2202.11342 2022. https://epubs.siam.org/doi/full/10.1137/23M1545628.

[44] Johnson PM, Jeong G, Hammernik K, Schlemper J, Qin C, Duan J, Rueckert D, Lee J, Pezzotti N, De Weerdt E, et al. Evaluation of the robustness of learned MR image reconstruction to systematic deviations between training and test data for the models from the fastMRI challenge. In: International workshop on machine learning for medical image reconstruction. Springer; 2021. p. 25–34.

[45] Gu H, Yaman B, Moeller S, Ellermann J, Ugurbil K, Akçakaya M. Revisiting ℓ_1-wavelet compressed-sensing MRI in the era of deep learning. Proc Natl Acad Sci 2022;119(33):e2201062119.

[46] Oala L, Heiß C, Macdonald J, März M, Kutyniok G, Samek W. Detecting failure modes in image reconstructions with interval neural network uncertainty. Int J Comput Assist Radiol Surg 2021;16(12):2089–97.

[47] Barbano R, Zhang C, Arridge S, Jin B. Quantifying model uncertainty in inverse problems via Bayesian deep gradient descent. In: 2020 25th International conference on pattern recognition (ICPR). IEEE; 2021. p. 1392–9.

[48] He W, Jiang Z, Xiao T, Xu Z, Li Y. A survey on uncertainty quantification methods for deep learning. arXiv:2302.13425 2023.

[49] Laves M-H, Tölle M, Ortmaier T. Uncertainty estimation in medical image denoising with bayesian deep image prior. In: Uncertainty for safe utilization of machine learning in medical imaging, and graphs in biomedical image analysis. Springer; 2020. p. 81–96.

[50] Gazdag, L. E., Antun, V., Hansen, A. C. Instability in deep learning – when algorithms cannot compute uncertainty quantifications for neural networks. Preprint 2024.

[51] Colbrook MJ, Antun V, Hansen AC. The difficulty of computing stable and accurate neural networks: on the barriers of deep learning and Smale's 18th problem. Proc Natl Acad Sci USA 2022;119(12):e2107151119.

PART

4

MR image postprocessing

CHAPTER

Cut it here: Image segmentation for MRI

12

Chen (Cherise) Chen[1,] * and Zeju Li[2,] *

[1]*School of Computer Science, Insigneo Institute, University of Sheffield, Sheffield, United Kingdom;*
[2]*FMRIB Centre, Wellcome Centre for Integrative Neuroimaging, University of Oxford, Oxford, United Kingdom*

**Corresponding author: E-mail: work.cherise@gmail.com, lizeju8@gmail.com*

12.1 Introduction

Semantic segmentation refers to a process of partitioning an image into several semantically (e.g., anatomically or pathologically) meaningful regions, such as organs, tumors, and lesions. Segmenting these areas allows one to visualize, characterize, and model specific regions of interest from medical images, leading to clinically valuable measurements like location, volume, and shape information. The accurate segmentation of magnetic resonance imaging (MRI) data is essential to help the understanding of anatomical structures and/or pathological abnormalities. For example, brain MRI segmentation typically involves the segmentation of gray matter (GM), white matter (WM), and cerebrospinal fluid (CSF) as well as the identification of tumor regions if there are any. The segmentation step not only plays a pivotal role in clinical settings—such as diagnosis, treatment planning, and monitoring—but also facilitates cutting-edge scientific research. For example, characterizing changes in ventricle volumes across different stages can enhance our understanding of brain development. Yet, manually tracing the boundaries of these regions can be labor-intensive, costly, and prone to significant intra- and inter-observer inconsistencies. Machine learning-based techniques can greatly accelerate the segmentation process through automation. This approach not only reduces time and costs but also makes large-scale clinical studies feasible [1].

This chapter delves deep into the methodologies, applications, challenges, and prospects of semantic segmentation in MRI with machine learning. We will begin with an overview of traditional segmentation methods that have laid the foundation in this domain, then transition to cutting-edge deep learning-based techniques, where deep neural networks have played an pivotal role. Next, we will present an evaluation of these developed tools, analyze their challenges, investigate specific applications, and assess AI's potential to advance healthcare.

Machine Learning in MRI. https://doi.org/10.1016/B978-0-443-14109-6.00012-2

12.2 Traditional segmentation methods

Medical image segmentation has been a major focus in medical imaging research for decades. Its significance lies in its potential to offer vital clinical insights and act as a fundamental requirement for the ensuing image-based diagnostic procedures [2]. Over the years, numerous techniques have been proposed to automate this process—either partially or fully—with the overarching goal of improving precision and efficiency, as summarized in Fig. 12.1.

12.2.1 Early solutions

The initial endeavors in addressing biomedical image segmentation employed thresholding-based techniques [3]. In the simplest case of a binary segmentation task, a pixel at (x, y) is assigned to foreground or background classes based on its intensity value $I(x, y)$ relative to a threshold t:

$$f(x,y)=\begin{cases}1 & \text{if } I(x,y)\geq t;\\ 0 & \text{otherwise.}\end{cases} \tag{12.1}$$

Selecting an appropriate t is crucial for successful segmentation. The most representative threshold selection algorithm is called Otsu's method [4], which automatically searches for the threshold that minimizes the intra-class variance, defined as a weighted sum of variances of the two classes:

$$t^* = \arg\min_t \left(\omega_0(t)\sigma_0^2(t) + \omega_1(t)\sigma_1^2(t)\right), \tag{12.2}$$

where ω_0 and ω_1 represent the class probabilities separated by threshold t, and σ_0^2 and σ_1^2 denote their corresponding variances. Yet, the thresholding-based method is often sensitive to intensity variations and does not consider spatial relationships. By contrast, region-based segmentation algorithms sample a seed pixel and then grow regions by recursively including the neighboring pixels that are similar to it

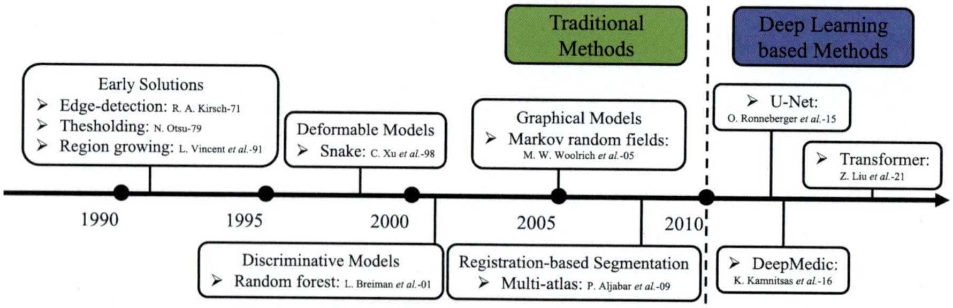

FIGURE 12.1

A historical roadmap: 50 years of medical image segmentation techniques. In this section, we briefly review traditional approaches before introducing deep learning-based methods.

(e.g., have similar intensity values). Regions of the same object can be merged using techniques such as Watershed algorithm [5], a hierarchical region merging strategy where adjacent regions are iteratively combined based on similarity metrics (e.g., intensity variance, edge strength). Another early and intuitive segmentation strategy is edge detection-based segmentation, which identifies regions with abrupt gray-level changes. These methods often utilize diverse edge detection operators, such as the Laplacian operator [6] and Canny edge detector [7]. Subsequently, adjacent edges are linked to form complete object contours. Despite their simplicity, these algorithms face limitations in low-contrast images due to their heavy reliance on regional intensity variations.

12.2.2 Deformable models

More concerted efforts in addressing medical image segmentation have centered around deformable models, exemplified by active contour techniques [8], which strive to align with predetermined deformable models through energy minimization [8]. These algorithms typically start with an initial contour placed outside the objects. The contour is then optimized using an energy function composed of internal and external energy terms. The internal energy enforces smoothness and coherence of the contour, preventing it from oscillating or deforming excessively. The external energy attracts the contour toward features of interest in the image, such as edges or intensity gradients.

Due to their inherent deformable nature, active contour models demonstrate inherent resilience, enabling automatic adaptation to varying image conditions. One major constraint of these models is their sensitivity to initial conditions and local minima. To address this limitation, the level-set method was proposed [9], reformulating curve evolution as the zero-level contour of a higher-dimensional implicit function $\phi(\mathbf{x}, t)$ governed by a partial differential equation:

$$\frac{\partial \phi}{\partial t} + F|\nabla \phi| = 0, \tag{12.3}$$

where $\mathbf{x} \in \mathbb{R}^n$ represents the spatial coordinates, t denotes time, and F combines curvature-dependent (internal) and image-derived (external) forces. This Eulerian framework naturally handles topological changes (e.g., splitting/merging contours) through continuous ϕ-field evolution, eliminating the need for explicit parametric representations while offering numerical stability. However, it is important to note that the level-set method incurs significant computational costs, as it involves solving higher-dimensional functions for curve evolution.

12.2.3 Graphical models

A graphical model is a mathematical representation that depicts the probabilistic relationships and dependencies among a set of random variables [10]. Representative works include Markov random field (MRF) [11] and conditional random field

(CRF)-based methods, which perform segmentation by capturing spatial interactions among neighboring pixels with pair-wise edge potentials. For example, CRF-based methods perform segmentation by modeling the labeling of pixels in an image as a probabilistic graphical model, where each pixel is considered a random variable whose label depends not only on its own features (like color, intensity, texture) but also on the labels of its neighboring pixels. This approach allows for the incorporation of both local pixel attributes and the contextual information from neighboring pixels, enabling more accurate and coherent segmentation [12]. Another approach, graph cut [13], involves segmenting an image into subregions by maximizing the flow within a graph, essentially minimizing the energy in the form of an MRF. Yet, these methodologies are limited in accurately tracing complex contours and can fall into local minima yielding suboptimal segmentation results.

12.2.4 Registration-based segmentation

Anatomical structures in medical images exhibit high consistency across subjects. Hence, registration techniques [14] have been widely used in segmentation to align target images with multiple atlases [15]. This method begins by optimizing a transformation field to align a target image with an atlas image. The segmentation is then obtained by applying the inverse transformation field to the corresponding anatomical prior from the atlas.

While registration-driven segmentation methods have been proved to be resilient and efficient in delineating entities with consistent and relatively large structures, like the brain and the heart, they fall short in segmenting small objects with varying appearances across cases, such as tumors and lesions.

12.2.5 Discriminative models

Segmentation can be viewed as a dense pixel-level prediction task. This perspective enables the use of discriminative models such as random forests [16] and sparse representation [17], which generate predictions by applying sliding windows to extracted features. The effectiveness of these feature extraction-then-classification approaches depends critically on the quality of the engineered features. However, optimal features may vary across specific cases, necessitating ad-hoc adjustments for different tasks. This makes it challenging to develop a universal pipeline. Notably, some traditional discriminative models also fall under the broad definition of machine learning approaches, as they involve optimizing parameterized models that can generalize to unseen data.

12.2.6 Deep learning-based segmentation methods

Deep neural networks (DNNs) can also be viewed as discriminative models. Unlike traditional approaches that rely heavily on manual feature extraction, DNNs automatically learn hierarchical visual features from input images in an end-to-end manner for specific tasks. However, initial attempts found neural networks to be resource-intensive and challenging to train [18]. Recent advances in software and hardware infrastructure, coupled with the increased availability of large labeled datasets, have enabled the widespread use of DNNs in medical image segmentation tasks [19]. DNNs have already demonstrated superior segmentation accuracy with faster processing speeds (typically under one second for a single 2D image) compared with traditional approaches discussed above. We will provide detailed discussions in the following sections, covering recent advancements, limitations, and future directions.

12.2.7 Further reading

For readers interested in implementing traditional segmentation algorithms, it is worth exploring the scikit-image library[1] and online tutorials dedicated to medical imaging data[2] provided by UCL researchers. They offer a comprehensive collection of various segmentation algorithms, providing a practical and accessible means to understand and apply these techniques.

12.3 Deep learning-based segmentation methods

In this section, we will dive into the world of deep learning-based approaches for medical image segmentation. A typical development pipeline of medical image segmentation with deep learning involves the following procedure: (1) design a deep neural network; (2) train a deep neural network on a training dataset so that the underlying network parameters are optimized for a specific task; (3) evaluate the deep neural network's performance on test sets, as shown in Fig. 12.2. We will briefly discuss standard practices for these three steps in the following sections.

12.3.1 Generic encoder-decoder structure for segmentation

For step 1, there are massive different types of neural networks for magnetic resonance (MR) segmentation tasks that can be selected and adapted from. The majority of them can be categorized into two groups: a) fully convolutional neural network (FCN)-based; b) transformer-based. Both of them can be abstracted into a generic encoder-decoder architecture, as shown in Fig. 12.3.

It generally consists of two parts: (1) an image encoder $\mathbf{E}$ parameterized by θ_e to compress the information into a lower-dimensional representation $\mathbf{z}$. The latent representation typically has reduced spatial dimensions ($h \ll H$, $w \ll W$, $d \ll D$), yet it features an increased number of feature channels compared with the input ($c > C_{\text{in}}$).[3] This expansion in the number of feature maps facilitates the capture of

[1]https://scikit-image.org/docs/stable/auto_examples/segmentation/index.html.
[2]https://mphy0026.readthedocs.io/en/latest/segmentation/segmentation_methods.html.
[3]For gray-scale images, $C_{\text{in}} = 1$, as shown in Fig. 12.3 whereas for RGB images, $C_{\text{in}} = 3$.

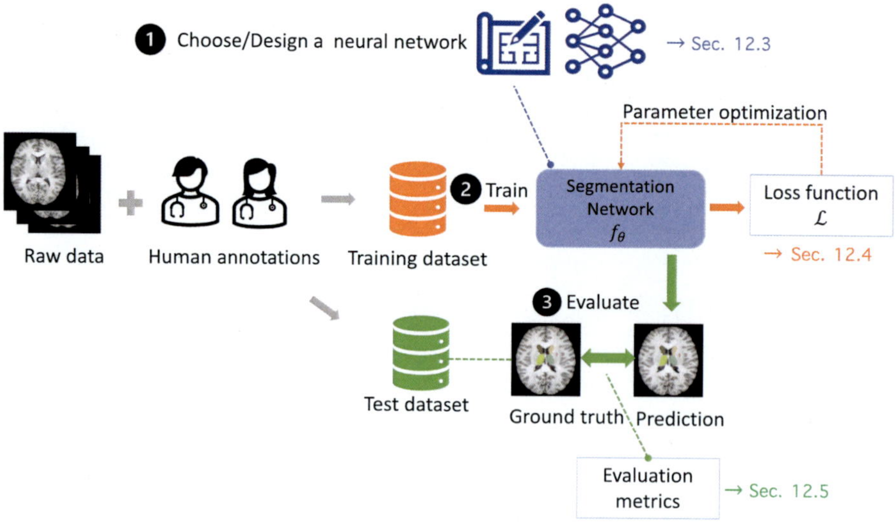

FIGURE 12.2 Development pipeline of deep learning-based medical segmentation algorithms.

(1) Selection or design of a neural network architecture for segmentation tasks, as detailed in Sec. 12.3. (2) Training of the neural network using a specified training dataset and a loss function to optimize underlying parameters, see Sec. 12.4 for a collection of different losses for segmentation tasks. (3) Evaluation of the network's segmentation performance using various metrics, see Sec. 12.5.

multiple complex, nonlinear representations inherent in the data. (2) a decoder **D** parameterized by θ_d that maps the latent representation to a probabilistic segmentation map of different regions with different semantic meanings. This structure applies to both a) 2D-based image segmentation and b) 3D-based segmentation. For 3D MR scans, the 2D-based approach will first convert the 3D volume into 2D slices and then infer each structure or abnormal tissue prediction in a slice-by-slice fashion. As shown in Fig. 12.3c, 2D networks employ 2D kernels (e.g., 2D convolutions with a kernel size of $k \times k$) to compute features across the height H and width W of the slice to make predictions [20,21] whereas 3D networks take 3D patches of the images with 3D kernels (e.g., 3D convolutions with a kernel size of $k \times k \times m$) to process and extract features along height H, width W and depth D [22].

2D versus 3D versus 2.5 Networks: 2D networks often lack interslice context, resulting in segmented volumes that contain discontinuities, which hampers accuracy improvements. Conversely, 3D approaches excel at capturing volumetric spatial details but come with higher computational costs, leading to longer inference latency. And 3D networks tend to be more computationally demanding and often necessitate a larger number of training examples due to their significantly increased parameters. A 2.5D approach aiming to balance computational efficiency and

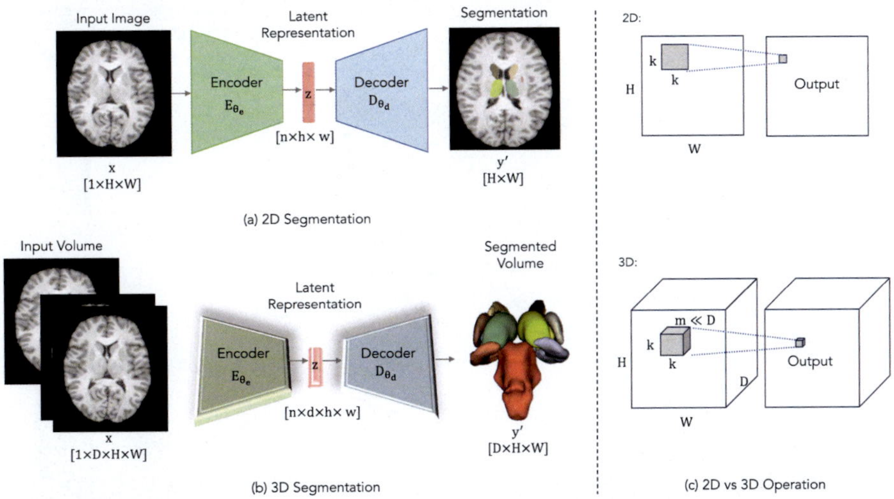

FIGURE 12.3 Schematic overview of generic encoder-decoder structure for image segmentation.

a) 2D segmentation; b) 3D segmentation; c) difference between 2D and 3D operations (e.g., convolution kernels) in the 2D and 3D networks, respectively. H: image height, W: image width, D: image depth, k: kernel height/width (e.g., convolutional kerner size), d: kernel depth.

effectiveness has therefore been proposed [23]. Unlike traditional 2D networks which use a single slice, the 2.5D network based on a 2D network incorporates multiple complementary slices as input to enrich spatial context. This enhancement might involve using slices from different planes like axial, sagittal, and coronal, or adjacent slices, to improve predictions on the target slice. Essentially, it is a hybrid approach, leveraging the simplicity of 2D processing while introducing elements of 3D context for a more accurate segmentation.

12.3.2 Fully convolutional neural networks-based segmentation networks

The most prevalent type of neural networks for image segmentation is FCNs. In general, an FCN consists of a series of convolutional layers with a number of downsampling and upsampling layers in the encoder and decoder part, respectively, as shown in Fig. 12.4a. Those convolutional layers are applied to an input image or features to produce semantic feature maps. As the network deepens, these feature maps grow increasingly abstract, capturing higher-level representations. The downsampling layers, often implemented using max-pooling, reduce the spatial dimensions of these feature maps, making them more compact and focusing on the most salient features at different scales. For segmentation tasks, since it is important to recover the spatial

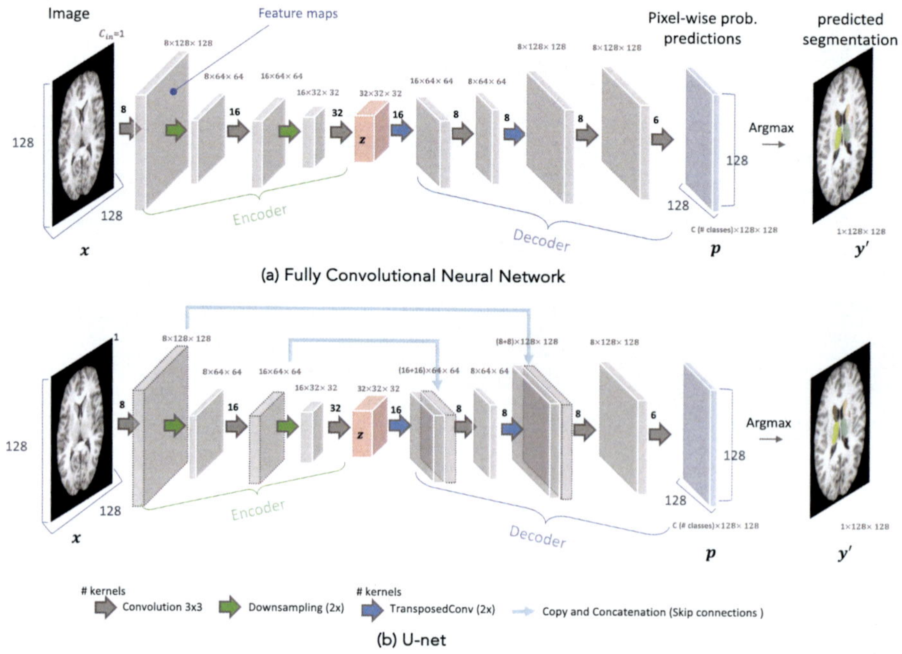

FIGURE 12.4

a) Architecture of a FCN. It takes an entire image as input, extracting image features through the encoder with a set of convolutional kernels (gray arrows) and downsampling operations (green arrows), and then progressively restoring spatial dimensions and refine features upsampling layers (e.g., transposed convolutional kernels, see blue arrows) and convolutional kernels in the decoder. The process outputs pixel-wise probabilistic maps for C classes, where C is equal to the total number of distinct structures plus one for the background. For each pixel, the class with the highest probability is assigned, resulting in the final segmentation map. b) A simplified version of a U-net. U-net enhances the FCN structure by introducing "skip connections" (illustrated as light blue arrows). These connections merge feature maps at various levels, from coarse to fine, via concatenation and convolution operations. This diagram presents a simplified version with fewer downsampling and upsampling blocks for clarity. For an in-depth understanding, it is recommended to read the original paper [21], which provides detailed information on this architecture.

resolution to match the input, upsampling layers in the decoder part are therefore used in FCNs to gradually increase the spatial dimensions of the high-level feature maps back to the original input space. The output is a C-class probability segmentation map (after soft-max) $\mathbf{p} \in \mathbb{R}^{H \times W \times C}$. At inference time, the predicted segmentation map for each image is obtained by assigning each pixel with the class of the highest probability: $\mathbf{y}'_i = \text{argmax}_c \mathbf{p}_i^c$.

The most well-known member of the FCN family is U-net [21], which was originally proposed by Olaf Ronneberger, Philipp Fischer, and Thomas Brox in 2015.

Since that, U-net has dominated medical image analysis and has been adapted for various segmentation tasks due to its effectiveness and efficiency. One main unique characteristic is skip connections, which empowers the network to generate segmentation maps that are both granular in detail and precise in segmentation, see Fig. 12.4b. The U-net architecture has become the de facto standard and serves as the backbone for many advanced neural networks [24].

12.3.3 Transformer-based segmentation networks

A common limitation of purely convolutional-based neural networks (e.g., FCNs) is that they cannot capture global, long-range semantic relationships as the convolution operation is applied to local regions one by one. In recent years, transformers, originally designed for natural language processing tasks [25], have been adapted for image classification [26] and segmentation tasks [27] due to their ability to handle long-range dependencies and capture global context.

Fig. 12.5 shows the transformer-based segmentation workflow, demonstrated in 2D for simplicity but extensible to 3D. The typical procedure follows:

- **Preprocessing Images into Tokens:** An input MR image slice $\boldsymbol{x} \in \mathbb{R}^{C_{in} \times H \times W}$ is first split into a sequence of n small, uniform patches with equal size (p,p): $\boldsymbol{x} = [\boldsymbol{x}_1, \ldots, \boldsymbol{x}_n] \in \mathbb{R}^{C_{in} \times n \times p^2}$ where $n = \frac{H \times W}{p^2}$ is the number of patches, and C_{in} is the number of input channels ($C_{in} = 1$ for grayscale images). These patches are then flattened and linearly transformed into tokens with a fixed feature dimension D_k: $\boldsymbol{x}_0 = [\mathbf{E}_{\boldsymbol{x}_1}, \ldots, \mathbf{E}_{\boldsymbol{x}_n}] \in \mathbb{R}^{n \times D_k}$.
- **Adding Spatial Context with Positional Encodings:** Positional encodings are crucial as they carry the necessary spatial information of patches (e.g.,

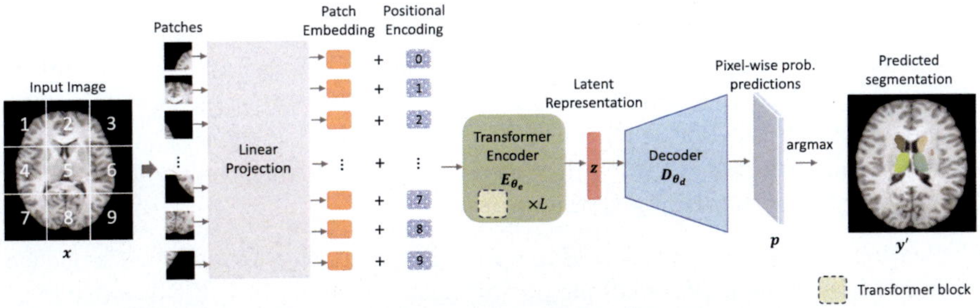

FIGURE 12.5 Architecture of a transformer-based segmentation network.

The transformer first divides an image into a sequence of fixed-size patches and then linearly embeds them, incorporating position embeddings and then sends these embedded patched into a transformer encoder E_{θ_e} to capture and refine interpatch interactions and then use a decoder D_{θ_d} to recover and refine the prediction of segmentation maps. In this example, images are firstly divided into 9 patches.

location), which is informative for the follow-up attention mechanism [25] in the transformer encoder. Positional encodings can be fixed encodings or learnable encodings: $\mathrm{PE} = [\mathrm{PE}_1, \ldots, \mathrm{PE}_n] \in \mathbb{R}^{n \times D_k}$, which can be directly added to the sequence of tokenized patches: $\boldsymbol{x}_0 = \boldsymbol{x}_0 + \mathrm{PE}$. The most commonly used fixed positional encoding is the sinusoidal position encoding, which uses sine and cosine functions of different frequencies to encode each position:

$$\mathrm{PE}\,(i, 2j) = \sin\left(i/10000^{2j/D_k}\right), \tag{12.4}$$

$$\mathrm{PE}\,(i, 2j+1) = \cos\left(i/10000^{(2j+1)/D_k}\right) \tag{12.5}$$

Here, i is an integer position index for an input patch, and j ranges from 0 to $D_k/2-1$ that spans the feature dimension.

- **Learning Global Interactions with a Transformer Encoder:** The crux of the transformer's power lies in the L stacked base transformer blocks consisting of multihead self-attention blocks [25] followed by a two-layer feed-forward network to process the sequence of patch tokens (c.f. Fig. 12.6). The multihead self-attention module [25] models interpatch relationships to adjust the contribution of each token, capturing the global context and dependencies between distant regions in the image. Mathematically, the self-attention mechanism first transforms each token in input $\mathbf{x}_0$ into three different vectors: the Query ($\mathbf{Q}$), Key ($\mathbf{K}$), and Value ($\mathbf{V}$) vectors and then computes the self-attention matrix $\mathbf{A}$ across the three sequences of input vectors:

$$\mathrm{SA}(Q, K, V) = \mathrm{softmax}\left(\frac{\mathbf{Q}\mathbf{K}^\top}{\sqrt{D_k}}\right)\mathbf{V} \tag{12.6}$$

$$= \mathbf{A}\mathbf{V}, \tag{12.7}$$

where $\mathbf{A} \in \mathbb{R}^{n \times n}$ is the attention weight matrix and D_k is the input feature dimension.[4] After that a two-layer feedforward subnetwork is applied to postprocessing and refining the calibrated features. Layer norm [28] is applied before every block, and after every block with residual connections to normalize the distribution of features for stabilized training.

- **Feature Extraction and Decoding with a Segmentation Decoder:** A sequence of patch encodings after multiple transformer blocks $\mathbf{z} \in \mathbb{R}^{n \times D_k}$ is decoded to a probability segmentation map (after softmax) $\boldsymbol{p} \in \mathbb{R}^{H \times W \times C}$ where C is the

[4] Here for simplicity, we omit the details of the multi-head mechanism, a trick to reduce computational costs by parallel computation. It first linearly projects the queries, keys, and values h times with different learnable linear projection matrices to lower dimensions (D_k/h) respectively to compute the attention matrix and re-weight the projected value matrix. After that, all the output from h heads are concatenated to recover the feature dimension D_k [25].

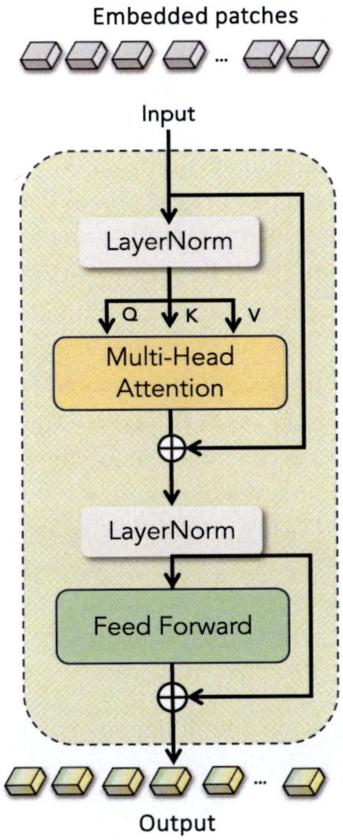

FIGURE 12.6

Architecture of a standard transformer block.

number of classes. To achieve this, similar to the function of decoder in the aforementioned FCN-based approach, the decoder is a lightweight convolutional neural network or another transformer-based structure with linear layers and/or upsampling layers to recover the spatial dimensions and generate a pixel-wise probabilistic map, similar to the procedure of FCN.

12.3.4 Advanced building blocks and hybrid architectures

Over the past years, many researchers have developed advanced building blocks (also see Chapter 2), which can be inserted into existing FCNs, such as U-net [21] and transformers. These advanced blocks can be used to enhance the feature exaction ability of network learned at different levels, for precise and robust segmentation. Several representative blocks include:

1. Inception modules [29–31], which concatenate multiple convolutional filter banks with different kernel sizes to extract multiscale features in parallel, see Fig. 12.7a; One can also apply dilated kernels [32], which are modified

convolution kernels with the same kernel size but with different kernel strides to process input feature maps at different scales or employ spatial pyramid pooling [33] with various kernel strides for multiscale feature fusion [34];

2. Residual blocks with residual connections, which add the input feature of a certain layer to its output, usually before applying an activation function [35], see Fig. 12.7b. These approaches help mitigate the vanishing gradient problem and enable efficient learning. Representative examples include Multi-ResUNet, which enhances U-net through multiscale residual connections [36];
3. Dense blocks with dense connections [37]: In dense blocks, the output of each layer is fed directly to all subsequent layers at different depths, see Fig. 12.7c. This connectivity pattern allows for direct connections between layers at different depths within the network. Representative works include Unet++ (with nested dense connections) [38], and HDenseUnet [39];
4. Adaptive feature/re-weighting blocks such as aforementioned attention module in the transformers [25,40]. Those adaptive feature blocks are integrated into FCN-based networks to reweigh features across different channels or spatial locations to better capture task-relevant features and ignore irrelevant features, see Fig. 12.7d. Representative works include Attention U-net [40,41] and Squeeze-and-Excitation (SE) fully convolutional network [42].

On the other side, there is a growing tendency of integrating transformer models with U-net architecture. This hybrid approach leverages the strengths of both

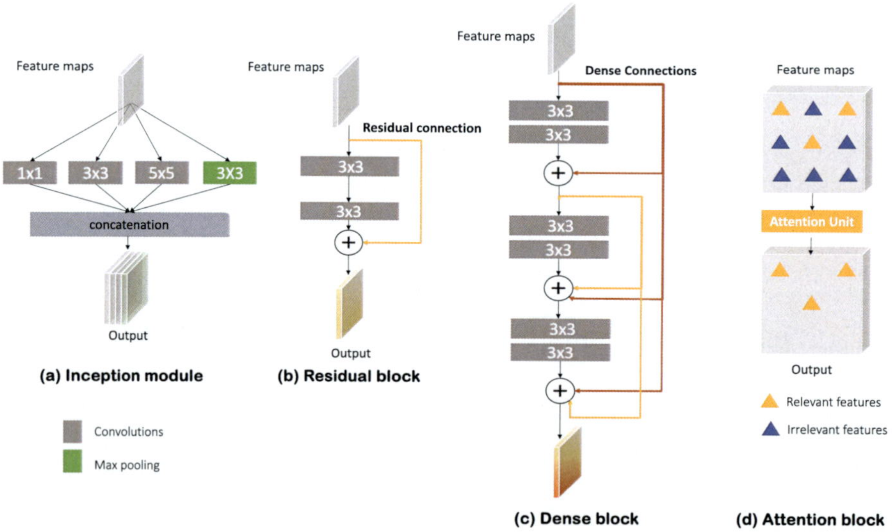

FIGURE 12.7

Schematic diagrams of modern network modules. a) Naive version of the inception module. b) Residual block. c) Dense block with dense connections. d) An attention block.

architectures: the U-net's proficiency in capturing local spatial hierarchies and the transformer's ability to model long-range dependencies. Notable examples include UnetR [43], Swin-Unet [44], and nnFormer [27], P-TransUnet [45], which replace the convolutional layers with transformer blocks in the encoder and/or decoder part or use a transformer as the secondary encoder and fuse the features from two branches smartly for better prediction such as TransFuse [46], DS-TransUNet [47]. These hybrid architectures have demonstrated superior results on some medical segmentation tasks including cardiac MR segmentation tasks and brain tumor segmentation from MR images [27], outperforming several conventional U-net-based baselines (e.g., self-configuring U-net (nn-Unet) [24]).

Yet, transformer-based networks, in comparison with traditional purely FCN-based, comprise a larger number of parameters. As a result, training transformers is more computationally demanding. In addition, they typically require a substantial volume of labeled images or advanced techniques such as pretraining (e.g., exploiting large-scale pretraining dataset with self-supervised learning and then performing transfer learning. See Sec. 12.6.1.1 for more advanced techniques to mitigate the risk of overfitting [48].

12.3.4.1 Further reading

Readers interested in a comprehensive overview of hybrid networks incorporating advanced blocks may refer to relevant reviews [48,49].

12.3.5 Other networks

Other network architectures beyond FCNs and transformers can also be used. For example, recurrent neural networks (RNNs) (e.g., long-short term memory (LSTM) [50], gated recurrent unit (GRU) [51]) and state space models (SSMs) [52], which memorize past inputs have been employed to extract knowledge learned from past inputs to inform current decisions. For example, inserting an RNN module after the encoder in a 2D FCN allows the network to capture information from past input, for example, adjacent slices or frames from previous time points, thereby enhancing the interslice/temporal coherence of segmentations [53,54].

Graph neural networks (GNNs) can also be used for segmentation tasks. Different from previous approaches where images are represented as a rectangular matrix or grid of square pixels, GNNs treat images as graphs where pixels or regions are nodes, and their spatial relationships are the edges. GNNs apply specialized layers. These include graph convolution and element-wise summation functions for information aggregation as well as fully connected layers and nonlinear layers for information abstraction to transform information from a node's neighbors to analyze data. These operations can help to extract the intricate topological and relational information from the graphs, which is useful to enhance semantic structure and spatial connectivity [55,56].

The use of GNNs for medical image segmentation is not as extensively discussed as other methods. One reason is that GNNs typically struggle with objects that have irregular edges or varied or small structures. As a result, GNNs often work together with FCNs and/or transformers to compensate for its limitation. This relative scarcity of publications in this area suggests both innovation potential and the need to address existing challenges.

12.4 Neutral network training and losses

Before performing segmentation on test images, neural networks must first be trained. At a high level, the training objective for a neural network f parameterized by $\boldsymbol{\theta}$ (e.g., encoder parameters $\boldsymbol{\theta_e}$, and decoder parameters $\boldsymbol{\theta_d}$ including weights and biases in the convolutional kernels) is an optimization problem:

$$\boldsymbol{\theta}^* = \arg\min_{\boldsymbol{\theta}} \mathbb{E}_{(\boldsymbol{x},\boldsymbol{y})\in D_{\text{train}}} \mathcal{L}(\boldsymbol{x},\boldsymbol{y};f_{\boldsymbol{\theta}}). \tag{12.8}$$

The objective of this process is to discover the optimal parameter set $\boldsymbol{\theta}^*$, which minimizes the averaged error scores quantified by the loss function $\mathcal{L}$ over a training dataset D_{train}. This training set consists of images $\boldsymbol{x}$ and their corresponding detailed annotations $\boldsymbol{y}$, such as manual segmentations. Training involves a number of iterations to update the parameters of the model. Each iteration requires to compute the loss gradient of each parameter ($\nabla_\theta \mathcal{L}$) via backpropagation [57,58], which indicates the direction for updating the parameter:

$$\theta_{\text{new}} = \theta_{\text{old}} - \eta \nabla_\theta \mathcal{L}. \tag{12.9}$$

Here, η represents the learning rate. An optimizer, such as stochastic gradient descent (SGD) or its more sophisticated variants like Adam [59] and AdamW [60], is needed to adjust both the pace (learning rate) and direction of these updates during each iteration. The loss function $\mathcal{L}$ is very crucial as it quantifies the errors in the network's predictions, guiding the optimization process at each step. Different learning schemes and dataset configurations may utilize different types of loss functions. The choice and design of the loss function are tailored to the nature of the training data and the goals of the learning task. We will discuss different training schemes with specific loss functions in the following sections.

12.4.1 Supervised learning

Supervised learning serves as the standard approach for model training. Consider a dataset consisting of images, represented as $\mathbf{x}$, each paired with detailed *pixel-wise* annotations, denoted as $\mathbf{y}$. Consider a dataset containing 2D grayscale MR image slices. For a given slice $\mathbf{x}$, which can be expressed as a matrix in $\mathbb{R}^{H\times W}$ (where H and W represent the image's height and width, respectively), its corresponding pixel-wise annotation is represented as a one-hot encoded map $\mathbf{y}$ in $\mathbb{R}^{C\times H\times W}$ where C is the number of unique classes (number of different segmented structures or tissue types plus the background class) in the annotation: $\mathbf{x} \in \mathbb{R}^{H\times W}$, $\mathbf{y} \in \mathbb{R}^{C\times H\times W}$. Each pixel in $\mathbf{x}$ corresponds to a C-dimensional vector in $\mathbf{y}$, indicating its class with a single nonzero entry (typically represented as 1):

$$\mathbf{y}^{(c,h,w)} = \begin{cases} 1, & \text{if pixel at (h, w) belongs to class } c; \\ 0, & \text{otherwise.} \end{cases} \tag{12.10}$$

This format allows for precise classification of each pixel in the MR image slices into one of the C classes.

12.4.1.1 Cross-entropy loss

As segmentation can be formulated as pixel-wise classification, one can simply apply pixel-by-pixel cross-entropy loss to quantify the errors between a predicted probabilistic output: $\boldsymbol{p}_i = softmax(f_\theta(\boldsymbol{x}_i))$ and its corresponding target segmentation map $\boldsymbol{y}_i$ in the training dataset:

$$\mathcal{L}_{\mathrm{CE}} = -\frac{1}{n}\sum_{i=1}^{n}\sum_{c=1}^{C}\boldsymbol{y}_i^c \log\left(\boldsymbol{p}_i^c\right), \tag{12.11}$$

where i is the pixel index, C is the number of all classes, and c is the class index, n is the total number of pixels in training images. $\boldsymbol{p}_i^c$ is the predicted probabilistic output for class c at i-th pixel and $\boldsymbol{y}_i^c$ is the corresponding target prediction in $\boldsymbol{y}_i$ for a particular class c (0 or 1).

12.4.1.2 Soft-dice loss

Another loss function that is specifically designed for object segmentation is called soft-Dice loss function [22], which penalizes the mismatch between a predicted segmentation map and its target map at pixel-level:

$$\mathcal{L}_{Dice} = 1 - \frac{2\sum_{i=1}^{n}\sum_{c=1}^{C}\boldsymbol{y}_i^c\boldsymbol{p}_i^c}{\sum_{i=1}^{n}\sum_{c=1}^{C}\left(\boldsymbol{y}_i^c + \boldsymbol{p}_i^c\right)}. \tag{12.12}$$

Furthermore, various modifications of the cross-entropy and soft-Dice loss functions exist, including the weighted cross-entropy loss [61,62], the weighted soft-Dice loss [63–65], and Focal loss [66]. These adaptations address class imbalance in medical image segmentation tasks, where the loss terms are adjusted in a weighted manner to give due consideration to rare classes or small objects. In the below, we introduce some most representative ones.

12.4.1.3 Generalized dice loss

Generalized Dice loss [65] introduces class weights w_c in the Dice loss function:

$$\mathcal{L}_{GDice} = 1 - \frac{2\sum_{i=1}^{n}\sum_{c=1}^{C}w_c\boldsymbol{y}_i^c\boldsymbol{p}_i^c}{\sum_{i=1}^{n}\sum_{c=1}^{C}w_c\left(\boldsymbol{y}_i^c + \boldsymbol{p}_i^c\right)}, \tag{12.13}$$

where $w_c = \frac{1}{(\sum_{i=1}^{n} y_i^c)^2}$ adjusts the contribution of each label by the factor of the inverse of the class volume so that rare classes weight more than major classes.

12.4.1.4 Focal loss

Focal loss [66] adapts the standard cross entropy so that the network is forced to focus on hard examples rather than easy examples:

$$\mathcal{L}_{\text{focal}} = -\frac{1}{n}\sum_{i=1}^{n}\sum_{c=1}^{C} w_c (1 - \boldsymbol{p}_i^c)^\gamma \boldsymbol{y}_i^c \log(\boldsymbol{p}_i^c), \tag{12.14}$$

where the introduced w_c is a reweighting factor to alleviate class imbalance, and $(1 - \boldsymbol{p}_i^c)^\gamma$ is a downweighting factor, so that the weight of easy-to-classify examples $(\boldsymbol{p}_i^c \gg 0.5)$ is reduced. If $\boldsymbol{p}_i^c$ is small and it is mis-predicted, this factor is near 1 and does not affect the training. When $\boldsymbol{p}_i^c$ goes to 1 (meaning the network is very confident at prediction, suggesting the segmented case is a easy one), the factor will go to 0 so that the loss for these well-classified examples is down-weighted. The focusing parameter γ is used to adjust the down-weighting rate. When $\gamma = 0$, $\mathcal{L}_{\text{focal}} = \mathcal{L}_{\text{CE}}$. As $\gamma > 0$ increases, the relative loss for well-classified examples is gradually reduced. In practice, γ is typically set to 2.

A limitation of the above pixel-wise losses is that they do not take high-level shape information into account. In this regard, researchers have also developed a set of topology-aware losses including boundary-aware loss [67–70], topology-preserving regularizers [71–80], or higher-order shape-prior-aware losses [81] to complement the pixel-wise loss to better guide the network optimization process. The most commonly used one is boundary loss.

12.4.1.5 Boundary loss

Boundary loss focuses errors across the boundaries of segmented regions between the prediction maps and ground truth maps. In the 2D case, the boundary loss function [67,70,82] can be implemented as:

$$\mathcal{L}_{\text{Boundary}} = \sum_{c \neq BG}^{C} \sum_{\text{Sobel} \in Sobel_x, \text{Sobel}_y} \|\text{Sobel}(\mathbf{p}^c) - \text{Sobel}(\boldsymbol{y}^c)\|_2^2. \tag{12.15}$$

Here, $\text{Sobel}_x, \text{Sobel}_y$ represent Sobel filters [83] in the $x-$ and $y-$ directions, which are used to extract object boundaries from the model's probabilistic map (the network output) and corresponding one-hot maps (ground truth) for every class except the background (BG) class.

12.4.1.6 Deep supervision loss

Deep supervision loss is a specialized loss that introduces supervision signals at multiple layers within the network, rather than solely at the final output layer [84]. This method is independent of loss function but focus on enhancing training effectiveness through multilayer supervision:

$$\mathcal{L}_{\text{total}} = \lambda_1 \cdot \mathcal{L}_1(\mathbf{p}_1, \mathbf{y}) + \lambda_2 \cdot \mathcal{L}_2(\mathbf{p}_2, \mathbf{y}) + \ldots + \lambda_n \cdot \mathcal{L}_n(\mathbf{p}_n, \mathbf{y}). \tag{12.16}$$

Here, $\mathcal{L}_1, \ldots, \mathcal{L}_n$ represent individual losses calculated at different intermediate layers of the network. $\mathbf{p}_i$ represents the network prediction using the output from the i-th layer. An upsampling operation may be applied to the intermediate branch output if the resolution of the output does not match the ground truth $\mathbf{y}$. $\lambda_1, \ldots, \lambda_n$ are coefficients reweighing of each intermediate loss in the total loss function. $\mathcal{L}$ can be any suitable loss function (e.g., cross-entropy).

In practice, combining various loss functions can enhance the training process. For example, integrating soft-Dice loss with cross-entropy loss in a deep supervision framework often yields better results. Interested readers are also suggested to read the recent survey on medical segmentation losses [85] with a more comprehensive summary of existing losses. For ongoing developments in advanced loss functions, a dedicated GitHub repository[5] is available to follow.

12.4.2 Advanced shape-aware losses for anatomical segmentation

For anatomical segmentation tasks, generating accurate masks is essential for subsequent shape analysis and mesh generation. This can be done by postrefining the predicted segmentation maps (e.g., using a denoising autoencoder (DAE)-based shape correction network to refine segmentation results [86]) or directly integrating more advanced shape-aware loss functions alongside the previously mentioned ones during the training process. This enhancement often utilizes prior knowledge extracted from label maps or atlases. Implementing these advanced loss functions typically involves introducing auxiliary networks. These networks are designed to learn and infuse higher-order shape knowledge into the training process, which in turn enhances the overall effectiveness of the model. Below, we discuss some of the most significant methods in this area.

12.4.2.1 Latent shape loss

This approach utilizes an auxiliary auto-encoder where an encoder E_{shape} is trained to encode shape knowledge derived from the ground truth or prediction maps to a low-dimension shape vector, see Fig. 12.8a. A representative work of this can be found in Ref. [81] and defines:

$$\mathcal{L}_{\text{latent\_shape}} = \text{Dist}\left[E_{\text{shape}}(\mathbf{p}), E_{\text{shape}}(\mathbf{y})\right], \tag{12.17}$$

where $\text{Dist}(\cdot)$ is a distance metric measuring the distance between the ground truth and predicted segmentation map in the latent space, for example, mean squared error (MSE). Adding a latent shape loss to a standard supervised loss $\mathcal{L}_{seg}$ (e.g., cross-

[5]https://github.com/JunMa11/SegLossOdyssey.

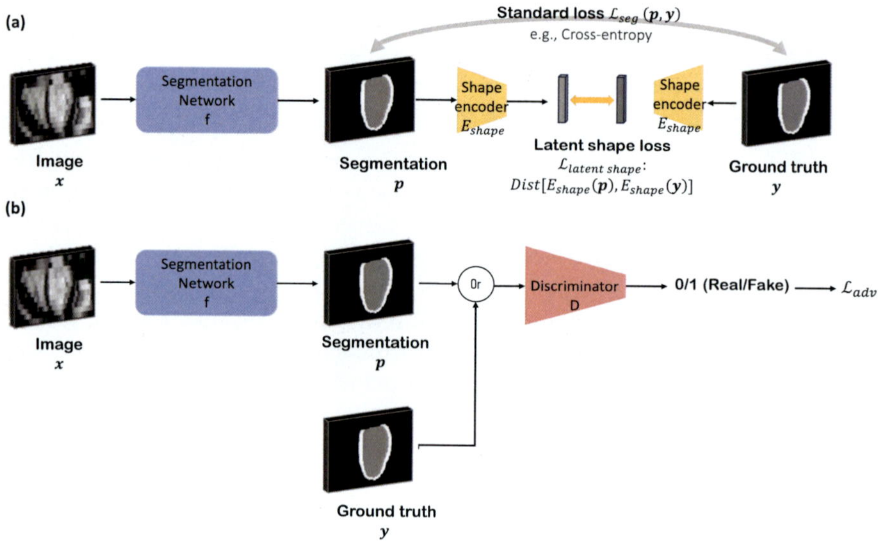

FIGURE 12.8

Schematic diagrams of a segmentation network trained with advanced shape-aware losses. a) Latent shape loss. b) Adversarial loss.

entropy) can regularize the segmentation network to follow the global anatomical properties of the underlying anatomy (e.g., shape, label structure) via learned nonlinear representations of the shape.

12.4.2.2 Adversarial loss $\mathcal{L}_{adv}$

To regularize the segmentation network produce anatomically correct segmentation maps, adversarial training with adversarial loss can be applied [87–90]. Specifically, as shown in Fig. 12.8b, an adversarial shape discriminator D is introduced, which evaluates whether the predicted shape is realistic ("truth") or not ("fake"). The output of D is a single scalar, which represents the probability of being real. Training involves a min-max optimization over the segmentation network f and the shape discriminator D, following the idea from Generative adversarial network (GAN) [87]:

$$\max_D \mathbb{E}_{\boldsymbol{x}}[\log(1 - D(f(\boldsymbol{x}))) + \log(D(\boldsymbol{y})], \tag{12.18}$$

$$\min_f \mathbb{E}_{\boldsymbol{x}}\left[\mathcal{L}_{seg}(f(\boldsymbol{x}), \boldsymbol{y}) + \lambda_{adv}\mathcal{L}_{adv}\right], \tag{12.19}$$

$$\mathcal{L}_{adv} = \log(1 - D(f(\boldsymbol{x}))) \tag{12.20}$$

The discriminator D is trained to maximize the probability that the segmentor f is correctly classifying the ground-truth segmentation as real, and the segmentation network f is trained to minimize the standard supervised segmentation loss $\mathcal{L}_{seg}$

(e.g., cross-entropy) and an adversarial loss $\mathcal{L}_{adv}$, which aims to fool the discriminator into classifying the predicted segmentation map as real. λ_{adv} is a coefficient to control the contribution of $\mathcal{L}_{adv}$. The segmentation network is thus driven to produce more anatomically accurate maps [88–90].

12.4.2.3 Atlas-based loss

One can also integrated atlases[6] as a strong shape constraint into the deep learning framework, which either formulates the segmentation task as an atlas-based segmentation task where a neural network $\mathcal{M}$ for image registration is used to estimate deformation fields from the atlas to the subject in a deep learning fashion, see Fig. 12.9a or a joint image segmentation and registration framework where the segmentation network f is jointly trained with a atlas registration network $\mathcal{M}$ for mutual benefit [91,92], see Fig. 12.9b.

Below are more details about the two methods:

- **(a) Atlas-based segmentation powered by image registration** [93]. Here the image registration module is implemented as a deep neural network (e.g., an FCN [94]), which takes the input image $\boldsymbol{x}_i$ and an atlas image $\boldsymbol{x}_A$ as input and produces an estimated deformation field $\Phi_{A\rightarrow i}$ [7] so that the warped atlas $\boldsymbol{x}_A \circ \Phi_{A\rightarrow i}$ looks similar to the subject $\boldsymbol{x}_i$. The deformation field could then be used to propagate labels from the atlas to the input subject to get a segmentation mask: $\mathbf{y}_i = \mathbf{y}_A \circ \Phi_{A\rightarrow i}$. The registration network $\mathcal{M}$ is trained via:

$$\mathcal{L}_{\text{atlas}} = \mathcal{L}_{reg}(\boldsymbol{x}_A \circ \Phi_{A\rightarrow i}, \boldsymbol{x}_i) + \mathcal{L}_{seg}(\mathbf{y}_A \circ \Phi_{A\rightarrow i}, \mathbf{y}_i) \tag{12.21}$$

  Here the registration network is trained to minimize the image registration loss $\mathcal{L}_{reg}$ and a segmentation loss $\mathcal{L}_{seg}$ as long as the ground truth segmentation $\mathbf{y}_i$ for the input image $\boldsymbol{x}_i$ is provided. $\mathcal{L}_{reg}$ can be implemented using either MSE or cosine similarity to measure pixel-wise dissimilarity between images. The segmentation loss can be cross-entropy. This method also enjoys the benefit of being data-efficient as the network can still be trained with only $\mathcal{L}_{reg}$ and then predict the segmentation map based on the registration result.

- **(b) Joint training with an image segmentation network and a registration network.** Given an input image $\boldsymbol{x}_i$ and an atlas image $\boldsymbol{x}_A$, along with its corresponding atlas segmentation mask $\mathbf{y}_A$, the process begins by feeding the input image $\boldsymbol{x}_i$ into the segmentation network to obtain a predicted segmentation $\mathbf{p}_i$. The registration network then uses both the predicted segmentation and the atlas segmentation, potentially along with their corresponding image data, to estimate deformation fields for mapping one image to the other, and vice versa.

[6]Here, the atlas(es) can be one or multiple reference image from the training set or template of the average anatomy, derived from a population, offering both intensity and anatomical label probability at each spatial location.
[7]Deformation fields describe pixel-wise displacements in x-, y- (2D) and z- (if it is 3D) directions.

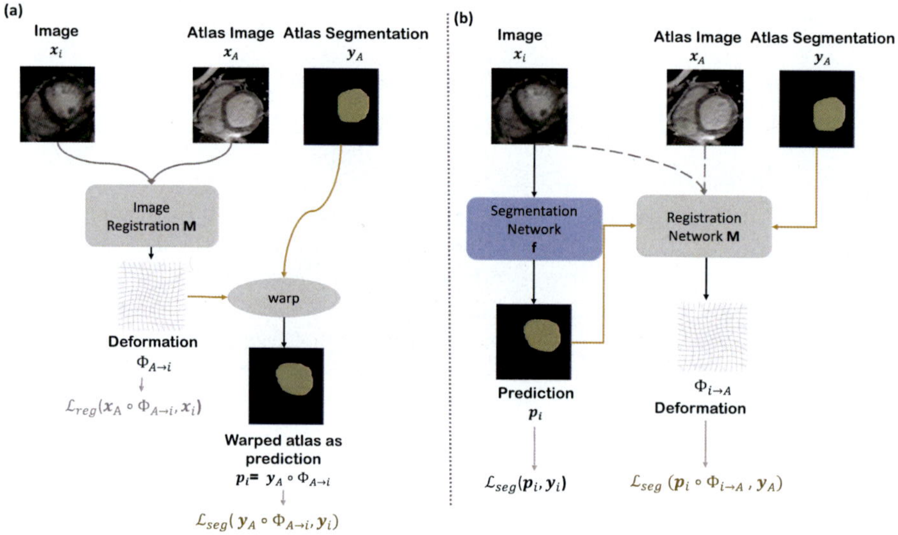

FIGURE 12.9

Training with shape-aware losses. a) Atlas-based segmentation. b) Joint image segmentation and registration pipeline.

The segmentation network f aims to minimize not only a standard supervised training loss $\mathcal{L}_{seg}$ between the prediction $\mathbf{p}_i$ and the ground truth $\mathbf{y}_i$ but also an atlas-based loss $\mathcal{L}_{\text{atlas}}$, defined as:

$$\mathcal{L}_{\text{atlas}} = \mathcal{L}_{seg}(\mathbf{p}_i \circ \Phi_{i\to A}, \mathbf{y}_A). \tag{12.22}$$

Here, $\mathcal{L}_{\text{atlas}}$ encourages the warped predicted segmentation $\mathbf{p}_i \circ \Phi_{i\to A}$, transformed by the deformation fields estimated from the registration network $\mathcal{M}$, to closely match the atlas segmentation $\mathbf{y}_A$. The loss function for training the registration network $\mathcal{M}$ mirrors that used in scenario (a), encouraging the similarity between corresponding image-atlas pairs and prediction-atlas segmentation pairs after registration. We omit the intricate details for brevity. Readers seeking deeper insights are referred to sources such as [91,92].

12.4.3 Weakly supervised learning

The supervised losses discussed in Sec. 12.4.1 (e.g., cross-entropy loss) as well as those advanced shape-aware loss except the atlas-based loss presented in Sec. 12.4.2 rely on pixel-wise annotations of training images for model training. Pixel-wise image labeling is often time-consuming and costly. To alleviate the need for pixel-level annotations, researchers look into developing novel training methods with weak annotations [95], such as class labels, points, scribbles, as well as bounding boxes (see Fig. 12.10), which are far cheaper and easier to get.

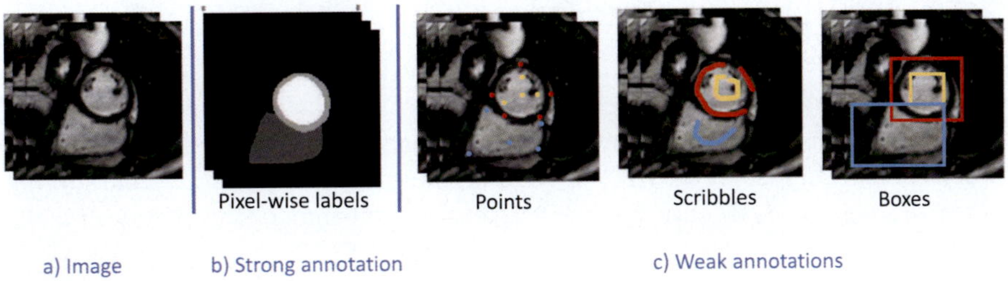

FIGURE 12.10

Illustration of strong and weak annotations for cardiac MR images.

The majority of weakly supervised approaches are based on pseudo labeling, where pseudo pixel-wise annotations are generated from weak labels for network training. Mathematically, the training objective in Eq. (12.8) becomes as follows:

$$\theta^* = \arg\min_{\theta} \mathcal{L}(\boldsymbol{x}, Gen(\boldsymbol{y}_{weak}); f_\theta), \tag{12.23}$$

where f is a segmentation network configurated with a set of parameters θ. The key to the weakly supervised learning lies in the pseudo mask generation function: $Gen(\cdot)$. For most weak labels with spatial clues, traditional interactive segmentation techniques such as region growing [96] and graph cuts [97] can be used to propagate labels from labeled pixels to neighboring pixels so that one can use these propagated labels to train the networks [98]. Similarly, for bounding boxes, one can use the GrabCut method [99] or unsupervised region proposals generation methods [100] to obtain pseudo segmentation masks from bounding boxes [101]. For image-level labels (e.g., classification labels) without any location clues, class activation maps (CAMs) can be used to infer discriminative regions and use that to generate plausible annotations [102]. To get CAMs, the predicted class score is mapped back through the network, layer by layer, until reaching the input layer, where the regions most relevant to the predicted class will be highlighted [103]. Yet, the generated labels can be extremely noisy, potentially limiting their effectiveness for training improvement.

In additional to the pseudo label-based supervision, one can also directly pose certain constraints with prior knowledge derived from weak annotations [104,105], for example, predicted segmentation masks must cover the regions of weak labels; no predicted masks across the background regions; predicted shape size must be within a reasonable range [104]. The most intuitive constraint loss for general purposed segmentation can be partial cross-entropy loss, which exclusively calculates losses within the regions labeled weakly [106]:

$$\mathcal{L}_{\text{partial\_cross\_entropy}} = -\sum_{m \in \Omega_{\text{Weakly\_labelled\_regions}}} \boldsymbol{y}_m \cdot \log(\boldsymbol{p}_m), \tag{12.24}$$

where $\boldsymbol{p}$ is the predicted output from a segmentation network and $\boldsymbol{y}$ is a provided weak annotation labeling region Ω.

12.4.4 Semisupervised learning

Semisupervised learning is strategies that leverage extensive unlabeled data alongside a limited set of labeled data. These approaches capitalize on the abundance of unlabeled information to enhance learning and improve the model's performance. At the time of writing, there are two prominent paradigms within semisupervised learning for general medical image segmentation tasks [107].

- **Self-training with pseudo labeling:** This paradigm involves iterative training a model on the labeled data and then using this model to predict labels for the unlabeled data with high confidence scores. The newly pseudo-labeled data are then added to the labeled training set for subsequent training iterations [108,109].
- **Consistency regularization:** This paradigm introduces a consistency loss on unlabeled data points to further regularize the network to produce similar predictions for similar perturbed instances of unlabeled data (e.g., by adding noise or other data perturbation or data augmentation functions to the original input images) [70,110–112]. The consistency regularization loss $\mathcal{L}_{ConReg}$ can be added to the fully supervised loss $\mathcal{L}_{\text{supervised}}$ described in Sec. 12.4.1 to form a semisupervised loss:

$$\mathcal{L}_{\text{semi}} = \mathcal{L}_{\text{supervised}} + \lambda \mathcal{L}_{ConReg}, \tag{12.25}$$

$$\mathcal{L}_{ConReg} = \frac{1}{|D_{\text{unlabelled}}|} \sum_{\mathbf{x} \in D_{\text{unlabelled}}} D\left[f_\theta(\boldsymbol{x}), f_{\theta'}\left(\boldsymbol{x}'\right)\right], \tag{12.26}$$

where D represents a distance metric measuring dissimilarity between two predictions. Commonly used metrics include Kullback-Leibler (KL) divergence, Jensen–Shannon divergence, and MSE. Here, $\boldsymbol{x}'$ refers to a perturbed version of the original input image $\boldsymbol{x}$. Predictions are derived either by utilizing a network with shared parameters ($\theta = \theta'$) [70] or employing a student-teacher model. In the latter, there are usually two models: a student model and a teacher model with a shared architecture but different parameters and updating schemes. In general, the student model is updated through directly backpropagation using both supervised and the consistency loss. Differently, the teacher model is updated as an exponential moving average (EMA) of the student model's parameters: $\theta' = m\theta' + (1 - m)\theta$, where m controls the update speed (momentum) and is typically set close to 1, for example, 0.999 [111,112]. During training, the teacher model receives the original input or weakly augmented input (e.g., with mild perturbation or transformation), while the student model might be exposed to a strong data-augmented version to make the training process more robust. The consistency regularization term aims to encourage the student model to produce similar outputs to the teacher model.

12.5 Evaluation metrics for semantic segmentation

To evaluate segmentation algorithm results, it is essential for practitioners to employ appropriate metrics for evaluation [113]. These metrics typically gauge the similarity between predictions and ground truth (GT) labels, where the latter are typically derived from manual annotations.

12.5.1 Region-based evaluation metrics

In the context of segmentation, the standard accuracy metric (measuring the proportion of correctly classified pixels in the entire image) is less used, as it can be excessively biased due to the abundance of background pixels in an image. This dominance of background samples can overshadow the performance of the model on less represented yet more critical foreground classes. As a result, metrics focus on the foreground regions are more commonly used.

The most commonly used metric in medical segmentation tasks is Dice similarity coefficient (DSC) [114], which measures the overlap between two volumes. Assuming we have a predicted segmentation region (R_A) and a GT segmentation region (R_B), see Fig. 12.11 (left), DSC (R_A, R_B) is calculated as twice the intersection divided by the sum of the sizes of the two regions:

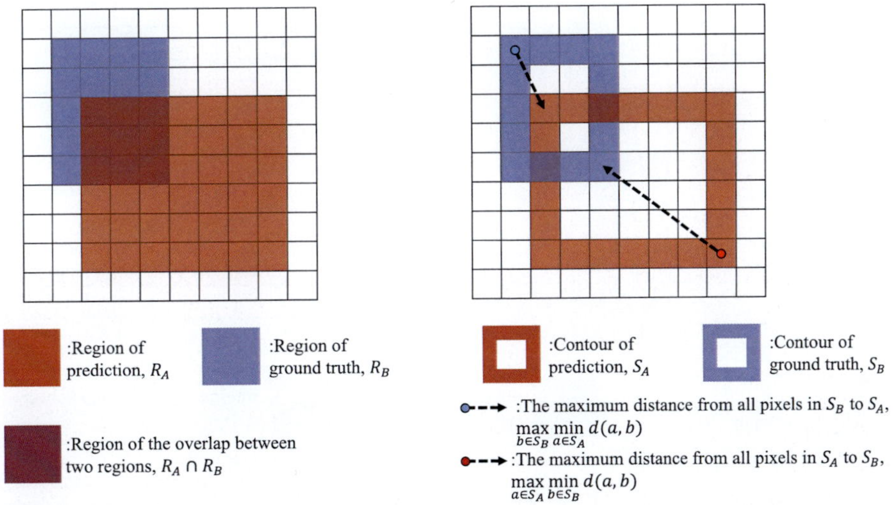

FIGURE 12.11 Illustration of region-based anad contour-based evaluation metrics.

Left: Region-based evaluation metric, which is established by measuring the overlap between two regions. Right: Contour-based evaluation metric, which is defined considering the distance between two boundaries. Specifically, here we illustrate Hausdorff distance.

$$\mathrm{DSC}(R_A, R_B) = \frac{2|R_A \cap R_B|}{|R_A| + |R_B|}. \tag{12.27}$$

The range of DSC score ranges from 0 (mismatch) to 1 (perfect match). A higher DSC indicates increased overlap between the predicted and GT regions.

The DSC provides a single scalar value (range: 0–1) measuring volumetric overlap but offers no information about boundary-specific segmentation errors. For instance, over-segmentation occurs when algorithms detect more region-of-interests (ROIs) than necessary (c.f. Fig. 12.12). Conversely, under-segmentation describes a situation where not all the ROI regions are identified. In these entirely distinct scenarios, DSC are comparable. Therefore, to comprehensively evaluate performance, careful researchers should also provide additional region-based metrics, such as sensitivity or precision, which are computed as follows:

$$\mathrm{Sensitivity}(R_A, R_B) = \frac{|R_A \cap R_B|}{|R_B|}, \mathrm{Precision}(R_A, R_B) = \frac{|R_A \cap R_B|}{|R_A|}. \tag{12.28}$$

It's important to emphasize that achieving high sensitivity alone or high precision alone does not necessarily indicate successful segmentation results. This is because

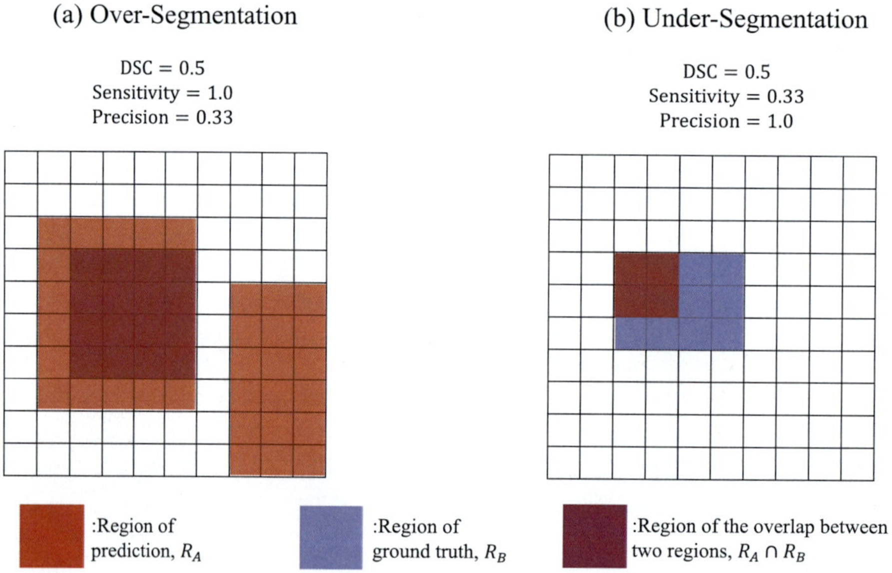

FIGURE 12.12 Illustration of over-segmentation and under-segmentation.

While the DSC quantifies spatial overlap between regions, it offers limited clinical utility by failing to differentiate among segmentation error types. For example, while cases of over-segmentation a) and under-segmentation b) may yield identical DSC values, they exhibit different biases. To provide a comprehensive assessment, additional metrics such as sensitivity and precision should be included.

practitioners can potentially enhance sensitivity by reducing the probability threshold, albeit at the expense of precision. However, the trade-off between sensitivity and precision will help the reader discern whether the model is inclined toward over-segmentation or under-segmentation. A key limitation of region-based metrics is their inability to capture shape characteristics, spatial distances, and center relationships. Furthermore, these metrics disproportionately penalize missed pixels in small object segmentation tasks.

12.5.2 Contour-based evaluation metrics

Another group consists of contour-based metrics. Among these, the Hausdorff distance (HD) [115] is one of the most widely used, measuring the similarity between point sets representing shapes in a metric space. As demonstrated on the right of Fig. 12.11, HD measures errors on two subsets of contour points derived from segmentation masks. Given predicted contour S_A and GT contour S_B, it measures the maximum distance of a set to the nearest point in the other set:

$$\mathrm{HD}(S_A, S_B) = \max\{\max_{b\in S_B} \min_{a\in S_A} d(a,b), \max_{a\in S_A} \min_{b\in S_B} d(a,b)\}, \tag{12.29}$$

where $d(a,b)$ represents the Euclidean distance between the point a and b. To enhance the robustness of HD against outliers when false positives are distant from ROIs, it is advisable to use the 95 % HD, which is calculated as:

$$\mathrm{HD95}(S_A, S_B) = \max\left\{\max^{95}_{b\in S_B} \min_{a\in S_A} d(a,b), \max^{95}_{a\in S_A} \min_{b\in S_B} d(a,b)\right\}, \tag{12.30}$$

where $\max^{95} \mathscr{X}$ calculates the value that is larger than 95 % of the elements in $\mathscr{X}$. HD may lack informativeness when two contours closely coincide, due to the maximum operation. In such instances, an alternative choice could be average symmetric surface distance (ASSD) [116], which calculates the average distance:

$$\mathrm{ASSD}(S_A, S_B) = \frac{\sum_{a\in S_A} \min_{b\in S_B} d(a,b) + \sum_{b\in S_B} \min_{a\in S_A} d(a,b)}{|S_A| + |S_B|}. \tag{12.31}$$

In general, contour-based metrics do not penalize regional overlap but are sensitive to outliers. Furthermore, these metrics are undefined for missed detections. Consequently, reporting both region-based and contour-based metrics is generally recommended. Notably, segmentation challenges frequently rank results according to weighted performance across both evaluation types [117].

12.5.3 Clinical performance metrics

After obtaining the segmentation masks, additional evaluation can be performed by assessing the performance in downstream tasks. A common approach involves comparing segmentation volumes across different groups to demonstrate that the segmentation masks are informative. For instance, in Ref. [118], ventricular volume and mass derived from segmentation maps are compared between the normal weight and obese groups. Similarly [119], compares hippocampal volumes from segmentation maps between the control and Alzheimer's disease (AD) groups.

12.5.4 Metric reliability

Practitioners should be mindful that, even if the metric calculations are accurate, the aforementioned evaluation metrics may still be unreliable, and the conclusions drawn might have limitations. A well-known drawback of the evaluation system is its sensitivity to different methods of aggregating results across all test cases [117]. For instance, it is customary to present the mean results of all cases when computing metrics. However, choosing the median value across cases may yield an alternative interpretation. Hence, it is advisable to augment the metrics with additional information, such as standard deviation, boxplots, or a Student's t-test, to provide further substantiation for the conclusions.

Another potential source of distortion arises in cases where segmentation involves multiple classes, and the aggregation results may be biased due to the rarity of minority classes [120]. This occurs because, when computing performance metrics averaged across classes, even small variations in results for a few minority classes can substantially affect the overall outcomes.

In addition, interoperator variability may impact the evaluation of segmentation results, particularly in manual segmentations of MRI, where subjective interpretations may lead to variations in the best segmentation results under different rules. The metrics' interpretation may vary based on the GTs provided by different experts [117]. Therefore, these metrics should be used with caution.

12.5.5 Further reading

To ensure a standardized and numerically stable assessment of algorithm performance (especially addressing issues like division by zero), we have curated a list of implemented evaluation metrics. These metrics, available in various programming languages, are provided for the convenience of interested readers.

- *torchmetrics*[8] [121]: a collection of 100+ metrics implementations in PyTorch including DSC;
- *EvaluateSegmentation*[9] [122]: an open-sourced tool consisting of 22 different metrics in *C* for medical volume segmentations;
- *seg_metrics*[10] [123]: a package consists of volume-based metrics and distance-based metrics written in Python;
- *miseval* [11] [124]: a metric library for medical image segmentation evaluation;
- *MedPy* [12]: a Python package that contains the calculation of commonly used medical image segmentation metrics.

[8]https://lightning.ai/docs/torchmetrics/stable/.
[9]https://github.com/Visceral-Project/EvaluateSegmentation.
[10]https://github.com/Jingnan-Jia/segmentation_metrics.
[11]https://github.com/frankkramer-lab/miseval.
[12]https://github.com/loli/medpy.

12.6 Challenges of existing deep learning-based methods and strategies to alleviate

12.6.1 Data scarcity

Medical images can vary greatly due to different imaging modalities, protocols, and patient demographics, necessitating a broad dataset for robust models. Yet, in general, multisource, large-scale labeled images with pixel-wise annotations are very difficult to obtain due to several reasons:

- **Strict Privacy and Regulatory Constraints:** Regulations and patient confidentiality often limit the availability of diverse medical images;
- **High Expertise Requirement for Annotation:** Precise annotation of medical images is time-consuming and expensive, requiring specialized expertise.
- **Interobserver Variability:** Different medical experts might interpret and annotate the same image differently, leading to variability in the data, which can affect the consistency and reliability of the training dataset;
- **Large Cost of Data Collection and Storage:** Collecting, storing, and managing large volumes of medical imaging data can be costly, especially considering the need for secure storage solutions to protect patient privacy.

These above factors collectively contribute to the difficulties in acquiring comprehensive and reliable datasets necessary for advanced medical image segmentation research and development. As a result, neural networks can be easily overfitted, resulting in poor performance at test time.

12.6.1.1 Advanced techniques to alleviate overfitting and data scarcity

Common approaches to alleviate overfitting typically include:

- **Weight Regularization**. This regularization technique, incorporated into the loss function, encourages the model to favor smaller or zero weights—particularly for less significant or irrelevant inputs. Common approaches include L1 and L2 regularization. L1 regularization penalizes the sum of the absolute values of the weights, while L2 regularization penalizes the sum of their squares.
- **Dropout**. Dropout [125] is a regularization strategy that randomly omits certain neurons from the neural network during the training phase. This process helps the network develop a more robust and sparse representation, reducing overfitting by preventing complex co-adaptations on training data.
- **Transfer Learning**. Transfer learning is a powerful approach that allows for leveraging preexisting models to achieve high performance on a different, but related task, often with less needed training data and lower computational costs. A model developed for one task is reused as the starting point for a model on a second task. It is especially popular in deep learning where large neural networks with many layers have been trained on large datasets where the weights

of these networks or the datasets are public. For example, these pretrained models can be large models trained from public ImageNet [126] or chest radiograph dataset [127]. It can also be a large foundation model trained for general segmentation tasks such as Segment Anything Model (SAM) [128].

- **Multitask Learning**. This strategy trains a model on multiple related tasks simultaneously, allowing it to share features and representations across tasks, thus enhancing performance on individual tasks. For instance, incorporating image pre/postablation tasks can improve atrial segmentation from Late gadolinium enhancement (LGE) images with varying image contrast [129]. An example of multitask learning is shown in Sec. 12.7.2.2 where a network is trained with segmentation and motion estimation tasks for mutual benefits.
- **Ensemble Learning**. This approach involves combining several independently trained models (e.g., using different training subsets, neural networks, random seeds, optimizers, loss functions, or data augmentation strategies) to enhance predictive performance [130]. It leverages the strengths of multiple models to achieve better accuracy and reliability than individual models.

Beyond these methods, advanced strategies now directly combat data/label scarcity. Data-focused solutions generally take two approaches:

- **Data Augmentation**. Data augmentation focuses on making slightly (i.e., label-preserving) modifications to existing data to create new, varied data points [131–133]. This can be done by applying data transformations such as rotation, scaling, or noise addition, contrast enhancement to existing images [134–136]. For MRI, some specific types of MR artifacts such as bias fields [137,138] can also be added. Recently, learning-based approaches such as adversarial data augmentation [138,139] and generative models (e.g., variational auto-encoder (VAE) [140], GAN [87], diffusion models [141]) have been proposed to capture complex, challenging imaging variations to augment the training data, for example, transferring the image style of a bSSFP image to be an LGE-like image [82]. Another application is to synthesize missing imaging modalities from other available imaging modalities, see an example in the application of brain segmentation with incomplete modalities (Sec. 12.7.1.1).
- **Synthetic Data Generation**: Data synthesis involves generating entirely new data points from scratch [142], often utilizing generative models such as VAE [140], GAN [87,143], diffusion models [141,144]). Images can also be generated directly from random noises, or weak conditions such as language descriptions, class labels, skeletons, and segmentation maps [119]. Synthetic data generation is particularly useful when directly sharing images is impossible or a certain population is rare. Applications include synthesizing abnormalities such as tumors, edemas, or lesions onto existing images [145]. The synthesized dataset can be used as a surrogate dataset for network training [119].

For labeling, several methods reduce the need for extensive annotations:

- **Self-Supervised Learning**: Unlike traditional supervised learning that relies on human annotations, self-supervised learning creates auto-generated labels to define a pretext task for model training. This approach enables models to learn rich representations of the data by exploiting the underlying structure within the data itself. For example, in image processing, a model might predict one part of an image based on another [146]. Contrastive learning, often used alongside, focuses on differentiating between similar and dissimilar data pairs [147,148]. Models pretrained through self-supervised learning can be fine-tuned for specific applications, maximizing both efficacy and adaptability.
- **Weakly- and Semisupervised Learning**: These approaches utilize a small set of labeled data with a larger batch of unlabeled or partially labeled data. It is efficient for situations where obtaining fully labeled datasets is challenging. In medical imaging, labels might be limited, but vast amounts of unlabeled or partially labeled images are available (c.f. Sec. 12.4.3 and Sec. 12.4.4).
- **Active Learning**: Active learning focuses on reducing labeled data points by selecting the most informative data points for expert labeling. Current studies investigate methods for detecting informative samples using uncertainty quantification, mutual information, or gradient-based metrics [149,150].
- **Few-shot learning**. Few-shot segmentation aims to train a model to perform image segmentation tasks using only a few labeled examples per category. This is challenging because segmentation requires detailed pixel-level understanding, and learning this from very few examples is inherently difficult. A common approach in few-shot segmentation is the use of prototypical networks [151–153]. These networks learn to represent each category (e.g., object type) as a prototype in a feature space. For a new segmentation task with few examples, the model quickly adapts by comparing the features of new images to these learned prototypes to make pixel-level predictions.
- **Federated Learning**. Federated learning is a distributed approach to train models across multiple decentralized devices or datasets while keeping data localized. This approach is highly pertinent in medical imaging due to the critical importance of data privacy. For instance, hospitals can collaboratively train a segmentation model, with each facility's data securely retained on-site. This enables the model to benefit from a variety of datasets, enhancing its robustness and relevance while safeguarding patient privacy. An early innovation in this field is FedBN [154], designed to tackle data heterogeneity challenges that arise during joint model updates from different sites.

From the model learning perspective, advanced model training and optimization paradigms have been developed to maximize model generalization ability considering limited access to multidomain data:

- **Domain Adaptation**. This technique involves adapting a model trained on one domain (source domain) to perform effectively on a different, but related domain (target domain). In medical image segmentation, this is crucial when applying models across different medical imaging modalities or patient populations or clinical sites. For example, a model trained on MRI scans from one hospital might be adapted to work with MRI scans from another hospital.

Existing approaches can be categorized into a) *supervised domain adaptation*: assuming both domains have labeled data for model adaptation; b) *unsupervised domain adaptation*: the model is adapted from a source domain with labeled data to a target domain without any labels; c) *source-free domain adaptation (test time adaptation)*: This represents a more recent and challenging scenario in which a model, pretrained on the source domain, must adapt to the target domain during testing without access to source domain data [155].

- **Domain Generalization**. Unlike domain adaptation, domain generalization trains models to perform well across multiple domains without requiring dedicated adaptation processes. This approach is trained on many datasets from varied domains, enabling the model to learn more universal features that are robust to domain-specific variations. Meta-learning [156] and virtual domain generation via image-space and feature-space data augmentation [157,158] facilitate the development of robust segmentation models. These models are capable of generalizing well to new, unseen medical imaging domains, thereby reducing the need for extensive domain-specific tuning.

12.6.2 Data bias

The efficacy of machine learning models is closely tied to the availability of extensive, representative datasets. However, the presence of bias in MRI datasets could distort the model, hindering its ability to generalize as anticipated.

12.6.2.1 Class imbalance

The main source of bias in MRI segmentation stems from class imbalance, reflecting an uneven distribution of samples among different categories. Fig. 12.13 presents examples of MRI segmentation tasks. Due to the rarity of particular ROIs like tumors and small abdominal organs, only a small portion of the 3D volume is represented by the foreground samples.

Even with an ample number of images, the issue of class imbalance persists, where ROIs (foreground classes) are significantly underrepresented in the training data [159,160]. The challenge of class imbalance in image segmentation differs from that in image recognition [161]. In segmentation, the dominant class is the background class with diverse characteristics, and its segmentation accuracy is highly robust. Conversely, in long-tailed image recognition, accuracy for the majority classes can degrade with common techniques such as reweighting [162]. Since segmentation performance is primarily evaluated based on foreground classes, enhancing accuracy in these classes is essential for segmentation.

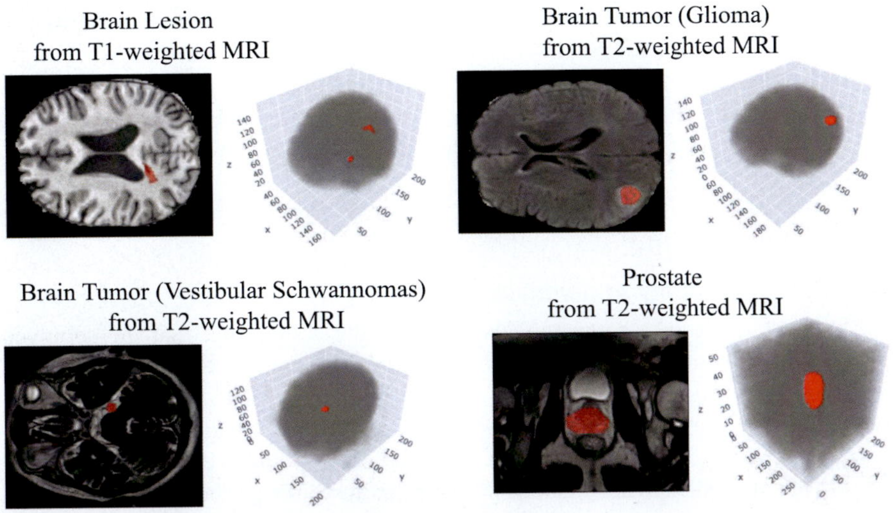

FIGURE 12.13 Class imbalance in MRI segmentation.

We generate visual representations, including both 2D slices and 3D scatter plots, for a few binary segmentations based on MRI. The ROIs are highlighted in red in the binary segmentations. Notably, these ROIs are relatively small compared with the background samples.

When convolutional neural networks (CNNs) are trained with imbalanced segmentation datasets, they are prone to overfitting the underrepresented foreground classes, memorizing the rare ROIs in the training data [163]. Practitioners should be cognizant of bias when dealing with highly imbalanced datasets and strive to mitigate biases by increasing the variance of foreground samples to counter overfitting under class imbalance. For example, it is a practical approach to apply transformations with a greater magnitude to foreground samples, as suggested in Ref. [136]. In Sec. 12.4, we also discussed some class reweighted losses that account for the class imbalance issue, for example, generalized Dice loss (Eq. 12.13).

12.6.2.2 Population bias

Fairness is an ethical principle in machine learning that algorithms and models should treat individuals or groups fairly and impartially, without exhibiting biases or discrimination based on sensitive attributes such as race, age, gender, and other demographic traits [164]. Consideration of fairness issues is essential in medical imaging applications due to the fact that existing datasets are usually gathered within specific centers and for particular demographic groups. Consequently, these datasets may not exhibit equal performance when applied to other populations. The results of fairness analysis can be influenced by various factors, including pretraining strategies and network architectures, as indicated in Ref. [165].

Fairness challenges are present in medical image segmentation tasks as well. To illustrate, investigations reveal that segmentation models for cardiac MRI, trained on the UK Biobank dataset, exhibit statistically significant variations in performance across diverse racial groups [166]. To enhance fairness in segmentation models, incorporating meta-information during the training stage is essential. For instance, improving model fairness can be achieved by resampling the training dataset to ensure a balance among different racial groups.

12.6.3 Unreliability

12.6.3.1 Uncertainty at deployment

The efficacy of deep learning models relies on the assumption that the deployed model will encounter data drawn from the same distribution as the training data. However, in practical scenarios, $\mathscr{D}_{train}$ may differ from $\mathscr{D}_{test}$. The disparity between training and test data is commonly referred to as domain shift. As a data-driven model, CNN performance can be compromised under domain shifts. This is a prevalent challenge in medical imaging, particularly as medical images may be acquired with varying parameters from diverse clinical settings [167]. An illustration of this challenge is presented in Fig. 12.14, showcasing prostate MRI examples collected from six different scanners [168].

In sec. 12.6.1.1, we have discussed that domain generalization and domain adaptation can be developed to improve the model generalization performance given limited training data. Yet, it is still difficult to ensure or evaluate the performance at a new test domain without sufficient labeled data. As neural networks produce pixel-wise probability predictions, one approach is to use the predicted probability as model confidence for performance estimation. However, it is widely acknowledged that neural networks are prone to producing overly confident predictions

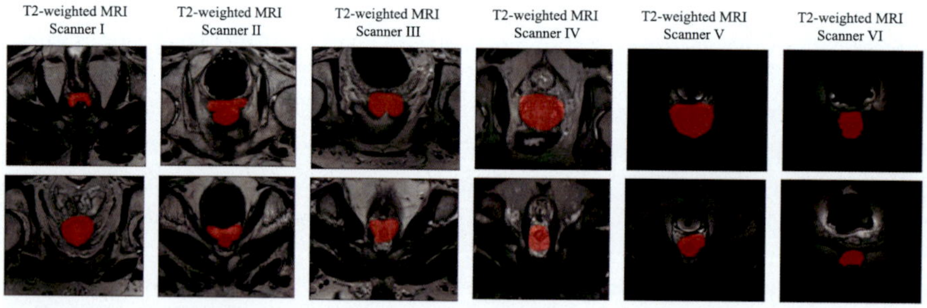

FIGURE 12.14 Effects of Scanner-Induced Domain Shifts on MRI Segmentation.

We present visualizations of T2-weighted MR images of the prostate obtained from six distinct scanners. The ROIs are identified by the use of red color. The variations in appearance and contrast across diverse domains pose challenges for CNNs when trained on data from one domain but tested on others.

[169], particularly under domain shift [170]. Consequently, deriving reliable uncertainty directly from model outputs is difficult.

To assess model performance without manual labels, various performance estimation methods, such as automated quality control (QC) have been extensively discussed [164,171,172]. These methods estimate model performance based on metrics like DSC, utilizing shape priors of segmentation masks or calibrated probabilities. For example, another neural network is trained to predict the DSC on a per-case basis [173]. In this way, clinicians can make informed decisions on whether to deploy a learned CNN for prostate MRI from local machines based on the prediction accuracy of a few unlabeled images.

12.6.3.2 Explainability

When compared with traditional medical image segmentation as described in Section 12.2, DNNs are often criticized for their lack of explainability because they are data-driven and operate in high-dimensional spaces, making it challenging to visualize and comprehend how the model makes predictions.

One common way to understand the model decision process is through the attention mechanism, as discussed in Section 12.3.4. The attention mechanism assigns weights to different spatial locations or channels in the feature maps based on their relevance to the task [174,175]. To enhance data fitting, the model tends to allocate greater weights around salient regions, directing its focus toward them. After inference, the attention map generated can offer more profound insights into the process behind model predictions.

12.7 Clinical applications of semantic segmentation in MR imaging

Deep learning-based approaches have significantly revolutionized and accelerated the segmentation process in various medical fields. Research has investigated the application of segmentation algorithms in diverse regions, including brain [176], lung [177], cardiac [178], knee [179], and abdominal organ segmentation [180]. Deep neural network can also be applied to segmenting abnormalities such as myocardial scars from contrast-enhanced MRI [181], brain tumors [182], and malignant lesions from whole-body MRI [183]. Readers seeking more research sources on the MRI segmentation of the human body can consult [184]. Besides, segmentation algorithms have been explored in the context of other animals, such as mice [185] and macaques [186]. These applications can provide valuable insights into disease modeling and offer potential translational applications for comprehending human function and disorders. For those interested, a comprehensive list of state-of-the-art (SOTA) deep learning algorithms for diverse segmentation tasks is available in a GitHub repository.[13] In the subsequent sections, we will focus on two of the

[13]https://github.com/JunMa11/SOTA-MedSeg.

most prominent applications: brain segmentation and cardiac segmentation. Here, we will explore the unique challenges each area presents and how deep neural networks are effectively used to address these challenges in practical, real-world scenarios.

12.7.1 Brain MR segmentation

The segmentation of brain MR images into distinct tissue classes, notably GM, WM, and CSF, is a crucial step with significant implications for both clinical and research purposes. This process facilitates accurate diagnosis, aids in treatment planning, and contributes to an enhanced understanding of the structure and function of the brain. Analyzing subcortical regions based on volume or shape is particularly relevant to the study of neurodegenerative disorders and psychiatric conditions [187]. For instance, techniques like voxel-based morphometry (VBM), which elucidates differences in brain tissues through voxel-wise comparisons, necessitate prior brain segmentation [188]. Nonetheless, segmentation remains consistently challenging and time-consuming, primarily owing to the diverse appearance of the ROI and the considerable size of the 3D imaging volume.

In more recent developments, CNNs have emerged as effective tools for brain segmentation [176]. Conventional segmentation methods for brain MRI are modality-agnostic as segmentation results are conditioned on anatomical priors (i.e., a statistical atlas) using generative models, as outlined in Ref. [189]. However, CNNs face challenges in generalizing across modalities. For example, a model trained on T1 MRI data may not exhibit strong generalization performance on T2 MRI. To address this limitation, SynSeg has been designed to ensure DNN compatibility with MRI data across all sequences [119]. This is achieved by training CNNs with synthesized samples generated through domain randomization.

Apart from brain segmentation, segmentation algorithms are also needed for the analysis of intracranial pathology. We illustrate an instance of brain tumor segmentation in MRI derived from diverse registered sequences using data sourced from the brain tumor segmentation (BraTS) challenge [182], as shown in Fig. 12.15. The segmentation of brain tumors plays a crucial role in assessing disease progression and the efficacy of selected treatments [190].

Segmenting brain tumors is acknowledged to be a more challenging task compared with general brain segmentation due to the diverse characteristics of pathology regions in terms of size, shape, and location [182]. Moreover, limitations in automatic segmentation algorithms for brain tumors are heightened by challenges inherent in MRI, including intensity heterogeneity and unforeseen intensity ranges of tissues [191]. Currently, researchers reach a consensus that CNN-based segmentation methods [192] perform better than previous methods such as random forests [193]. Despite the introduction of various modifications to enhance CNN architectures for brain tumor segmentation, naive U-Net continues to demonstrate competitive results in brain tumor segmentation challenges [194]. This implies that a systematic design of the processing pipeline, involving preprocessing, data

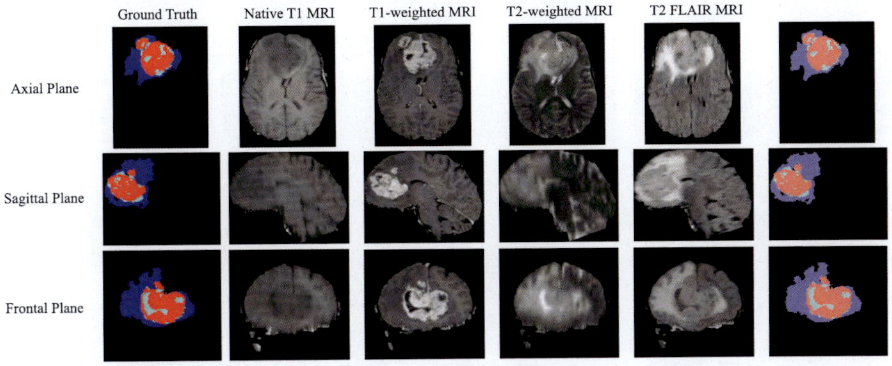

FIGURE 12.15 Multimodel MRI for brain tumor segmentation.

We demonstrate the brain tumor segmentation process using MRI data from the BraTS challenge. This involves delineating three distinct classes of brain tumors by analyzing four 3D MRI sequences. Our presentation includes MR images and their corresponding segmentations in three planes, where various colors distinguish between the different classes of brain tumors.

augmentation, normalization, and postprocessing, is essential for attaining segmentation results of high quality.

12.7.1.1 Robust inference with incomplete modalities

Utilizing deep learning-based algorithms for brain tumor segmentation may pose challenges if the model has been trained on specific MRI sequences that are unavailable in the deployment scenario. For instance, the application of a CNN becomes questionable when only one of the four modalities depicted in Fig. 12.15 is accessible during deployment. As shown in Fig. 12.16, approaches for handling missing modalities in model inference fall into two main types:

- **Imputation-Based**: A straightforward idea is to synthesize missing modalities before feeding into the segmentation networks. A baseline method is to fill the missing channels with static values which could be calculated from the training data [195]. Imputation-based methods can take advantage of advanced generative models such as GAN to generate realistic images [196]. However, the segmentation performance will rely on the quality of synthesized data. Furthermore, the inference process lacks dynamism and is unable to accommodate situations where arbitrary modalities are present.
- **Feature Fusion-Based**: Alternatively, feature fusion-based methods utilize modality-specific encoders and aggregate the multimodality information with feature fusion [197,198]. For example, HeMIS calculates the mean and variance of intermediate features. As statistical feature calculations can be applied to

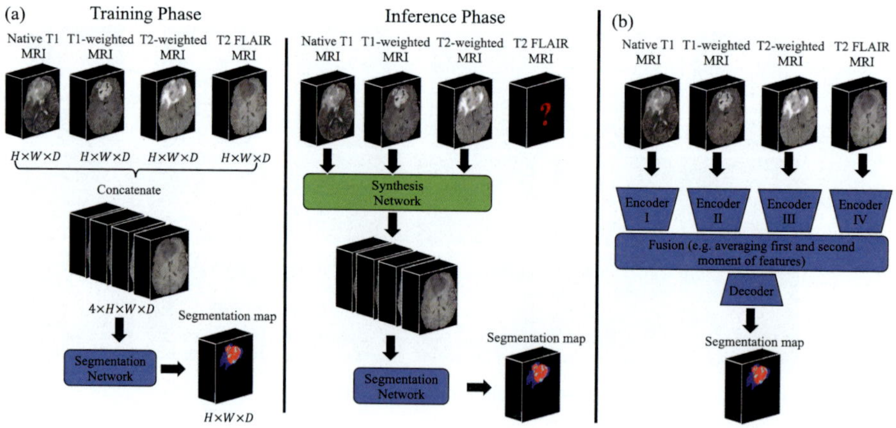

FIGURE 12.16 Brain tumor segmentation with missing modalities.

Multisequence data offers valuable complementary information for brain MRI, nevertheless, practical challenges may arise if not all MRI sequences are accessible during the inference stage. There are mainly two approaches to tackle this challenge, as we illustrate in the context of brain tumor segmentation from multisequence MRI: a) Imputation-based methods which replace missing modalities with synthesized data. b) Feature fusion-based methods which employ modality-specific encoders to enable the network to work with arbitrary amounts of modalities.

various input combinations, they do not make any assumptions about the shape of the input data, making them highly versatile during testing.

12.7.2 Cardiac MR segmentation

Cardiac MR segmentation is pivotal in diagnosing, planning treatment, and managing various cardiac diseases in the realm of cardiovascular healthcare. Diverse MR sequences are employed to visualize and analyze the heart, each highlighting different aspects. Cine MRI is used to create a sequence of cardiac images over cardiac cycles, showing the structure of cardiac chambers with the temporal movement. In contrast, LGE MRI segmentation primarily identifies areas of myocardial scar tissue whereas T2 imaging highlights edema regions, see Fig. 12.17. Integrating segmentation results from these varied modalities yields a comprehensive cardiac health assessment, enhancing diagnosis, risk stratification, and therapy planning. We outline several key cardiac MR imaging types below.

- **Cine MRI**: Cine MRI is recognized as the gold standard for visualizing cardiac chamber structures in both short-axis (SAX) and long-axis (LAX). For example, cardiac ventricle segmentation in cine imaging facilitates assessment of cardiac function, including ventricular ejection fraction and myocardial

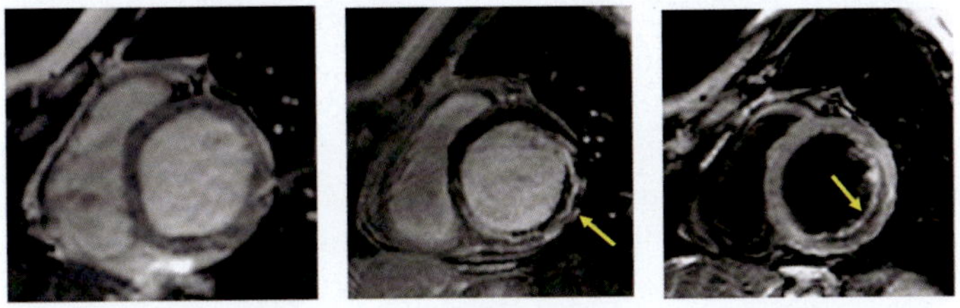

FIGURE 12.17 Comparison of heart visualization utilizing cine imaging with the bSSFP pulse sequence, LGE imaging, and T2 imaging, respectively.

The bSSFP technique excels in depicting clear cardiac anatomy and function. Conversely, LGE imaging is effective in highlighting scarred and fibrotic regions, while T2 imaging accentuates areas exhibiting edema, see yellow arrows in the images.

motion analysis. Besides, bSSFP pulse sequence is commonly used due to its exceptional balance in signal-to-noise ratio (SNR) efficiency and its ability to generate strong contrast between blood and myocardium.

- **LGE MRI**: This contrast-enhanced MRI technique accentuates myocardial scar tissue and fibrosis, often resulting from ischemic heart diseases or cardiomyopathies. Segmenting fibrosis or scarring areas offers quantitative measures for evaluating disease severity and informs treatment strategies.
- **T1 and T2 Mapping**: These imaging techniques are particularly responsive to myocardial fibrosis and edema, respectively. Segmenting T1 and T2 maps facilitates the quantitative evaluation of widespread myocardial changes, thus aiding in the early detection and characterization of cardiac conditions.
- **Tagging MRI**: Utilized to assess myocardial strain and deformation, tagging MRI creates a spatial pattern of markers (or "tags") on the myocardium. Tag segmentation enables detailed myocardial motion and functional analysis, essential for assessing cardiac mechanics and detecting subtle diseases.
- **Other Sequences**: Additional MR imaging types, including magnetic resonance angiography (MRA) and perfusion/stress MR, are instrumental in assessing conditions like coronary artery disease, aneurysms, and other vascular anomalies. For instance, perfusion MR is utilized to evaluate myocardial perfusion and ischemia, typically under pharmacological stress conditions. Segmentation in these modalities helps identify areas with reduced perfusion, offering crucial insights for diagnosis and treatment planning.

A more detailed overview of machine learning applications for cardiac MR segmentation can be found in Refs. [178,199].

Among various imaging modalities, cine imaging stands out for its capability to capture and display the heart's dynamic function. The heart's constant motion, as it beats and pumps blood, presents unique challenges in acquiring clear and high-quality images. In clinical practice, to mitigate respiratory motion artifacts, patients are instructed to perform breath-holding. During each breath-hold, typically lasting 10–20 s, the MRI scanner rapidly acquires 2D slices of the heart. Multiple breath-hold cycles may be necessary to comprehensively cover the entire heart and to capture images across different phases of the cardiac cycle. Compared with 3D imaging, 2D cine imaging is often preferred due to its lower costs and shorter scan durations. However, as indicated in Fig. 12.18, significant misalignment between slices can occur, potentially impairing the visualization of cardiac structures and subsequent motion analysis. This highlights the need for precise imaging techniques and alignment correction in cardiac cine MRI.

12.7.1.1 Joint networks for motion-correction, super-resolution, and image segmentation

Recent efforts have been made on jointly performing segmentation, motion correction, and super-resolution [200,201]. This refinement is crucial for obtaining accurate and detailed shape and/or follow-up cardiac mesh generation and motion analysis, facilitating more reliable and detailed image analysis.

As illustrated in Fig. 12.19, common approaches to enhancing cardiac image segmentation through machine learning can be categorized into three main types:

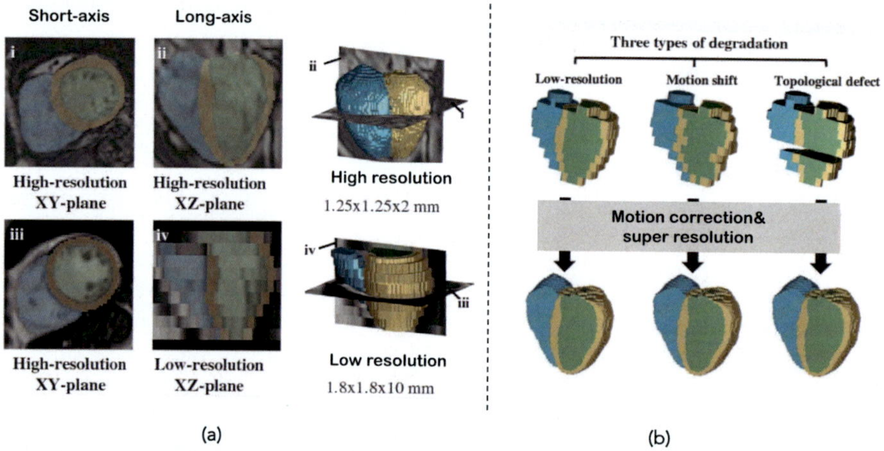

FIGURE 12.18 Challenges in cardiac MRI segmentation.

a) Difference between a high-resolution volume using 3D imaging and a low-resolution stack of slices acquired by a standard protocol with 2D cine imaging. b) The misalignment between slices in the low-resolution stacks can be caused by multiple factors, for example, low-resolution imaging techniques with large slice thickness, motion shift due to cardiac motion and respiratory motion, topological defect due to missing slices. Advanced machine learning techniques are needed to alleviate these artifacts to get high-quality, high-resolution segmentation for precise diagnosis. Credit: Image adapted from Ref. [200]. License acquired from Springer Nature under CC-BY.

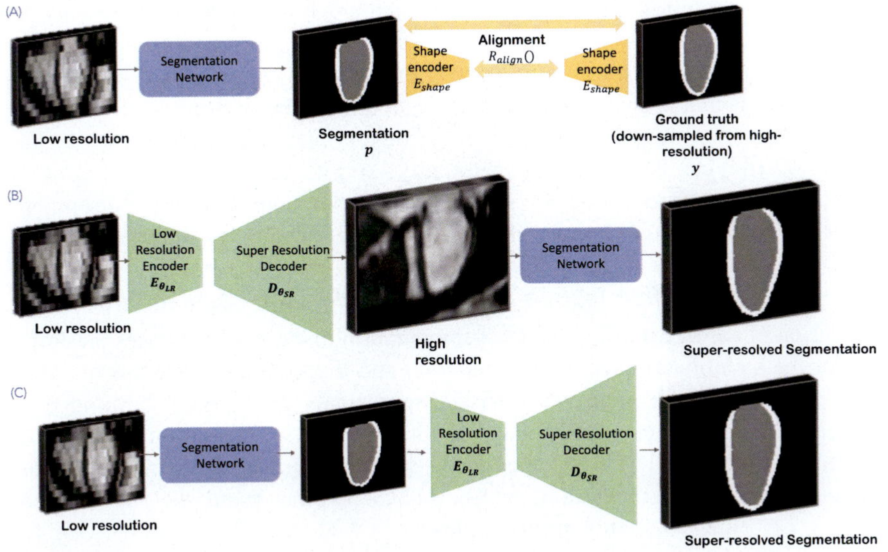

FIGURE 12.19

Three common ways for motion correction and/or super-resolution of cardiac segmentation. A) shape alignment regularization-based for motion correction. B) Image super-resolution-based. C) Shape super-resolution-based.

- **Shape Alignment-Based**: This method employs shape regularization during the training phase to enhance the continuity and anatomical plausibility of the segmented shapes during testing. Regularization is applied in both the segmentation space and the latent space: $R_{align}(\boldsymbol{p}, \boldsymbol{y})$ or $R_{align}(E_{shape}(\boldsymbol{p}), E_{shape}(\boldsymbol{y}))$ (as shown in Fig. 12.18a), which is similar to the latent shape loss $\mathscr{L}_{latent\_shape}$ we introduced earlier in Sec. 12.4.2, see Eq. (12.17). Differently, the ground truth (without motion artifacts) **y** here is actually derived by down-sampling the annotations on the corresponding high-resolution images [81].
- **Image Super-Resolution-Based**: This approach improves the resolution of cardiac images with a low-resolution encoder followed by a super-resolution decoder (as shown in Fig. 12.18b). The latter typically comprises multiple upsampling/transposed convolutional layers to increase image resolution [81,202] (also see Chapter 8). By applying super-resolution techniques, the finer details and textures of cardiac structures are more visibly enhanced, thereby improving segmentation accuracy. This method requires joint training with both low-resolution and high-resolution images for the image super-resolution network. When low-resolution images are unavailable, they are simulated using various degradation models and motion corruption simulators [203].
- **Shape Super-Resolution-Based**: This technique focuses on increasing the resolution or performing motion correction directly on the segmented shapes

from low-resolution images. It involves training a low-resolution shape encoder and a high-resolution shape decoder separately, using segmentation from low-resolution and high-resolution images, respectively. A key advantage of this approach is that it does not necessitate paired low- and high-resolution images, as the encoder, decoder, and motion correction modules (as shown in Fig. 12.18c) can be trained independently. And it can be used as a plug-in module for postprocessing the segmentation from general pretrained segmentation networks. An example of this approach is detailed in Ref. [200].

12.7.1.2 Joint cardiac motion estimation and image segmentation

Estimation of cardiac motion and the segmentation of cardiac structures are important for the quantification of cardiac function and the accurate diagnosis of cardiovascular diseases. Deep learning-based approaches have been developed for simultaneous motion estimation (see Chapter 9) and segmentation from cardiac MR image sequences [54,204]. Specifically, as shown in Fig. 12.20, a dual-branch network predicts segmentation maps for each time point in the sequence $\{\boldsymbol{p}_t\}_{t=0,\ldots,T-1}$ and computes motion fields $\Phi_t = \{\Delta_t^x, \Delta_t^y\}$ that represent displacements in both the x- and y-directions from a subsequent time point (time $t = 1, \ldots, T-1$) to a target time frame (e.g., time frame 0).

The training loss includes a standard supervised loss $\mathscr{L}_{supervised}$ such as cross-entropy loss (Sec. 12.4.1) and an unsupervised temporal consistency loss $\mathscr{R}_{\text{temporal}}$ that operates across various temporal frames and segmentation maps:

- $\frac{1}{T-1}\sum_t \mathscr{L}_{\text{cons}}(\boldsymbol{x}_t \circ \Phi_t, \boldsymbol{x}_0)$: Measuring the dissimilarity between transformed image from time t to time 0 (using the estimated motion field Φ_t, and the target image at time 0: $\mathbf{x}_0$. One can use MSE, cosine similarity, or mutual information-based metrics to measure the dissimilarity.

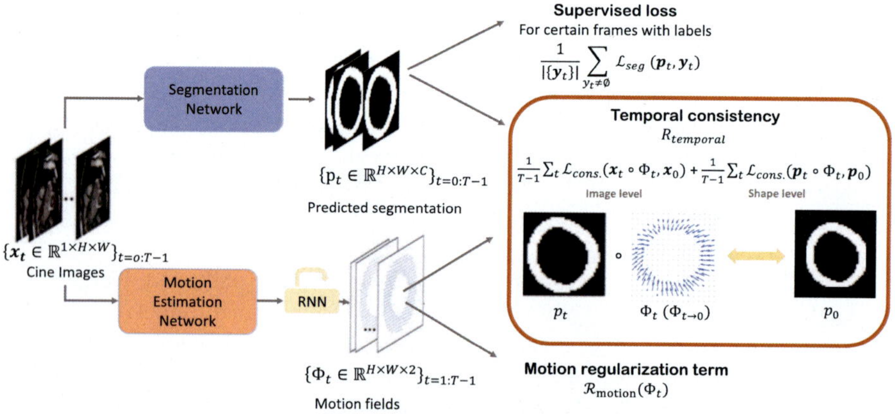

FIGURE 12.20

A general framework for joint cardiac motion estimation and segmentation.

- $\frac{1}{T-1}\sum_t \mathscr{L}_{\text{cons}}(\boldsymbol{p}_t \circ \Phi_t, \boldsymbol{p}_0)$: Measuring the dissimilarity between segmentation map from time t to time 0 and the target segmentation map at time 0: $\boldsymbol{p}_0$.

The consistency losses ensure that the model's predictions are coherent over time, both in terms of the segmented structures and the motion between frames. In the motion estimation branch, an RNN is used to propagate information along temporal dimension for better temporal coherence. In addition, a regularization term is applied to the estimated motion fields to ensure spatial smoothness. A standard regularization technique involves imposing an L2-norm on the gradients of the motion fields, which can be mathematically expressed as follows: $\mathscr{R}_{motion} = \frac{1}{T-1}\sum_t \left\| \nabla \Phi_t \right\|_2^2$. Additional regularization terms further constrain deformation fields. The literature on image registration offers a variety of other regularization strategies, which are tailored to the specific requirements of the motion estimation problem at hand and can be adapted to address challenges such as nonrigid body motion, topological changes, or large deformations [205]. Since the focus of this chapter is image segmentation, we refer interested readers to the section of image registration and the review article [205].

12.8 Future directions

Toward trustworthy deep segmentation models with robustness, uncertainty, and fairness into consideration.

- Robustness: Segmentation models should maintain consistent performance under varying imaging conditions, such as changes in imaging protocols. To enhance robustness, models can be trained on diverse datasets. In addition, the incorporation of advanced technologies like data augmentation and domain adaptation techniques, previously discussed in Sec. 12.6.1.1, can be a remedy.
- Uncertainty: Segmentation models should inform clinicians about uncertainties arising from image quality, inherent noise, and the subjective nature of image interpretation and annotation processes. This can be achieved through: a) Probabilistic modeling approaches, such as Bayesian neural networks [206], to directly model uncertainty. b) Random sampling-based approaches, like dropout [125], test-time data augmentation [207], and diffusion models [208,209], which estimate uncertainty from the variance of randomly sampled predictions. c) Model calibration, where a model's probabilistic predictions are adjusted to more accurately reflect uncertainty [210]. This information about uncertainty provides clinicians with a clearer understanding of the confidence level in the model's predictions, for better-informed decisions.
- Fairness: Segmentation models should perform equally well across different demographic groups, including patients of various ages, genders, ethnicities, and socio-economic backgrounds. Yet, recent studies have found that segmentation models can be biased to age, sex, or race due to the imbalanced

training set [211,212]. Achieving fairness typically requires optimization in dataset collection, model training, and evaluation. Researchers should use datasets representing diverse patient population with balanced contributions, and thoroughly evaluate models for biases [213].

12.8.1 Beyond imaging: semantic segmentation with multimodal pretraining

In Sec. 12.6.1, we have talked about the challenge of data scarcity, which hinders the development of deep learning-based segmentation models. We have also highlighted some advanced strategies including transfer learning and self-supervised learning, suggesting the benefits of using pretrained models. Recent advancements have evolved to incorporate multimodal pretraining, a technique that goes beyond relying solely on imaging data. This innovative approach integrates information from various sources or modalities, including medical reports, speech, audio, and signals [214]. By combining the knowledge extracted from these diverse data types, semantics related to imaging can be better understood. This holistic approach has shown segmentation improvements especially when task-specific annotated data are limited [215,216].

12.8.2 Toward adaptive segmentation models

Given the increased availability of powerful graphical processing units (GPUs) and a broad spectrum of public datasets across different tasks, it is increasingly feasible to train a universal segmentation model capable of performing well on a wide range of existing segmentation tasks. The research question now shifts toward the adaptation of these trained universal or foundational models, for example, the segment anything model (SAM) [128], for specific/new tasks. Recently, the focus has turned to using prompts (texts, points, bounding boxes, or learnable vectors) to tune existing segmentation networks for a specific novel task, combined with parameter-efficient fine-tuning techniques [217] for fast and efficient network fine-tuning.

12.9 Open-source tools for medical image segmentation

Open-source tools serve as pivotal accelerators for advancing the development of medical image segmentation algorithms, benefiting both research endeavors and clinical applications. Below, we have compiled a list of popular open-sourced software and segmentation tools, along with a summary of their key features:

- FMRIB software library (FSL)[14]: A comprehensive library for analyzing functional magnetic resonance imaging (fMRI), MRI, and diffusion brain imaging

[14] https://fsl.fmrib.ox.ac.uk/fsl.

data [218]. It includes tools for brain extraction, segmentation, registration, and statistical analysis;

- FreeSurfer[15] [187]: FreeSurfer is widely used for processing human brain MRIs, including a full processing stream for MRI data that involves skull-stripping, bias field correction, registration, and anatomical segmentation as well as cortical surface reconstruction, registration, and parcellation;
- Statistical parametric mapping (SPM)[16]: SPM [219] focuses on the analysis of fMRI and positron emission tomography (PET), but the principles of data processing generalize to other imaging modalities such as structure MRI, electroencephalogram (EEG), and magnetoencephalography (MEG). A fundamental feature within the realm of SPM involves the comparison of one set of subjects to another, examining distinctions such as structural differences (e.g., VBM) or neurophysiological indicators of brain functions;
- Advanced normalization tools (ANTs) [220] [17]: ANTs specializes in tasks like image registration, normalization, and segmentation, making it invaluable for aligning and comparing medical datasets. One of the distinctive features of ANTs is its adaptability to a wide range of imaging modalities, making it a versatile tool in the medical research community;

Remark: Although the brain MRI segmentation tools mentioned above can yield plausible segmentation results, it is important to note that no single approach excels across all measurements [221]. In addition, these methods may require hyperparameter tuning to align with specific research goals [222];

- ITK-Snap[18]: ITK-Snap [223] is a software application used for segmenting structures in 3D and recently supporting 3D + time medical images. The software utilizes the Insight Segmentation and Registration Toolkit (ITK) and is particularly known for its semiautomatic segmentation using active contour methods, allowing for precise delineation of anatomical structures;
- Medical Imaging Interaction Toolkit (MITK)[19]: MITK [224] is a free and versatile open-source software project for the development of interactive medical image processing applications. It can be used as a C++ toolkit or application framework for software development;
- 3D Slicer[20]: 3D Slicer is an advanced, free, and open-source software platform designed for the visualization, processing, segmentation, registration, and analysis of medical, biomedical, and other 3D images and meshes. It also facilitates planning and navigation for image-guided procedures;

[15]https://surfer.nmr.mgh.harvard.edu/.
[16]https://www.fil.ion.ucl.ac.uk/spm/.
[17]https://picsl.upenn.edu/software/ants/.
[18]http://www.itksnap.org/pmwiki/pmwiki.php.
[19]https://www.mitk.org/wiki/The_Medical_Imaging_Interaction_Toolkit_(MITK).
[20]https://www.slicer.org/.

- nn-Unet[21]: nn-Unet is an open-source neural network framework specifically designed for medical image segmentation [24]. It is an extension of the U-net architecture, which automatically configures itself to optimally handle 2D and 3D image segmentation tasks in medical imaging. nn-Unet is known for its robust performance across various datasets and challenges;
- MONAI (Medical Open Network for AI)[22]: MONAI [225] is a PyTorch-based, open-source framework with ready-to-use functionalities for annotating, building, training, deploying, and optimizing AI workflows in healthcare. It provides tutorials to help beginners learn neural network training and evaluation. For example, how to train a 3D Unet on a synthetic 3D dataset;
- Web-based segmentation tools: Online platforms such as MedSeg,[23] BioImage Suite Web,[24] and Biomedisa[25] are available for practitioners to perform segmentation tasks. MedSeg and Biomedisa fit computed tomography (CT) or MRI datasets and BioImage mainly focuses on neuroimaging analysis.

12.10 Summary

This chapter presents an in-depth analysis of the progression from traditional machine learning methods to advanced deep learning techniques for medical image segmentation. It focuses on prevalent neural network models used in segmentation, including FCNs and transformer-based networks as well as advanced building blocks within them and hybrid architectures built upon them. It covers training these neural networks under different set-ups including supervised, weakly supervised, and semisupervised learning. The chapter confronts key challenges in the field, notably the issue of data scarcity and uncertainty. It explores specific use cases and distinct challenges in deep learning applications for brain and cardiac MRI segmentation. These examples serve to illuminate the ongoing advancements and future potential of deep learning in enhancing medical imaging technologies. Furthermore, it includes a curated selection of educational resources for advanced research topics in the end of certain sections for further reading and open-source toolkits, aiding practitioners in further exploration.

[21]https://github.com/mic-dkfz/nnunet.
[22]https://monai.io/.
[23]https://www.medseg.ai/.
[24]https://bioimagesuiteweb.github.io/webapp/.
[25]https://biomedisa.info/gallery/.

References

[1] Greenspan H, Van Ginneken B, Summers RM. Guest editorial deep learning in medical imaging: overview and future promise of an exciting new technique. IEEE Trans Med Imag 2016;35(5):1153–9. https://doi.org/10.1109/TMI.2016.2553401.

[2] Rueckert D, Schnabel JA. Model-based and data-driven strategies in medical image computing. Proc IEEE 2019;108(1):110–24.

[3] Fu K-S, Mui J. A survey on image segmentation. Pattern Recogn 1981;13(1):3–16.

[4] Otsu N. A threshold selection method from gray-level histograms. IEEE Trans Sys Man Cybernet 1979;9(1):62–6. https://doi.org/10.1109/TSMC.1979.4310076.

[5] Vincent L, Soille P. Watersheds in digital spaces: an efficient algorithm based on immersion simulations. IEEE Trans Pattern Anal Mach Intell 1991;13(6):583–98.

[6] Kirsch RA. Computer determination of the constituent structure of biological images. Comput Biomed Res 1971;4(3):315–28.

[7] Canny J. A computational approach to edge detection. IEEE Trans Pattern Anal Mach Intell 1986;8(6):679–98.

[8] Xu C, Prince JL. Snakes, shapes, and gradient vector flow. IEEE Trans Image Process 1998;7(3):359–69.

[9] Li C, Xu C, Gui C, Fox MD. Level set evolution without re-initialization: a new variational formulation. In: 2005 IEEE Computer Society Conference on Computer Vision and Pattern Recognition (CVPR'05), Vol 1. IEEE; 2005. p. 430–6.

[10] Pham DL, Xu C, Prince JL. A survey of current methods in medical image segmentation. Annu Rev Biomed Eng 2000;2(3):315–37.

[11] Woolrich MW, Behrens TEJ, Beckmann CF, Smith SM. Mixture models with adaptive spatial regularization for segmentation with an application to fmri data. IEEE Trans Med Imag 2005;24(1):1–11.

[12] P. Krähenbühl, V. Koltun, Efficient inference in fully connected crfs with Gaussian edge potentials, Adv Neural Inf Process Syst 24.

[13] Boykov Y, Veksler O, Zabih R. Fast approximate energy minimization via graph cuts. IEEE Trans Pattern Anal Mach Intell 2001;23(11):1222–39.

[14] Iglesias JE, Sabuncu MR. Multi-atlas segmentation of biomedical images: a survey. Med Image Anal 2015;24(1):205–19.

[15] Aljabar P, Heckemann RA, Hammers A, Hajnal JV, Rueckert D. Multi-atlas based segmentation of brain images: atlas selection and its effect on accuracy. Neuroimage 2009;46(3):726–38.

[16] Breiman L. Random forests. Mach Learn 2001;45(1):5–32.

[17] Wright J, Yang AY, Ganesh A, Sastry SS, Ma Y. Robust face recognition via sparse representation. IEEE Trans Pattern Anal Mach Intell 2008;31(2):210–27.

[18] Sharma N, Aggarwal LM. Automated medical image segmentation techniques. J Med Phys/Assoc Med Phys India 2010;35(1):3.

[19] Moeskops P, Viergever MA, Mendrik AM, De Vries LS, Benders MJ, Išgum I. Automatic segmentation of mr brain images with a convolutional neural network. IEEE Trans Med Imag 2016;35(5):1252–61.

[20] Long J, Shelhamer E, Darrell T. Fully convolutional networks for semantic segmentation. In: Conference on Computer Vision and Pattern Recognition; 2015. p. 3431–40.

[21] Ronneberger O, Fischer P, Brox T. U-net: convolutional networks for biomedical image segmentation. In: Medical Image Computing and Computer-Assisted

Intervention–MICCAI 2015: 18th International conference, Munich, Germany, October 5-9, 2015, Proceedings, Part III 18. Springer; 2015. p. 234–41.

[22] Milletari F, Navab N, Ahmadi S. V-Net: fully convolutional neural networks for volumetric medical image segmentation. In: Fourth International Conference on 3D Vision, 3DV 2016. Stanford, CA, USA: IEEE Computer Society; 2016. p. 565–71. https://doi.org/10.1109/3DV.2016.79.

[23] A. Avesta, S. Hossain, M. Lin, M. Aboian, H. M. Krumholz, S. Aneja, Comparing 3d, 2.5d, and 2D approaches to brain image Auto-Segmentation, Bioengineering 10 (2).

[24] Isensee F, Jaeger PF, Kohl SAA, Petersen J, Maier-Hein KH. nnU-Net: a self-configuring method for deep learning-based biomedical image segmentation. Nat Methods 2021;18(2):203–11.

[25] Vaswani A, Shazeer N, Parmar N, Uszkoreit J, Jones L, Gomez AN, Kaiser ŁU, Polosukhin I. Attention is all you need. In: Guyon I, Luxburg UV, Bengio S, Wallach H, Fergus R, Vishwanathan S, Garnett R, editors. Conference on Neural Information Processing Systems. Curran Associates, Inc.; 2017. p. 5998–6008.

[26] A. Dosovitskiy, L. Beyer, A. Kolesnikov, D. Weissenborn, X. Zhai, T. Unterthiner, M. Dehghani, M. Minderer, G. Heigold, S. Gelly, J. Uszkoreit, N. Houlsby, An image is worth 16x16 words: transformers for image recognition at scale, ICLR.

[27] Zhou H-Y, Guo J, Zhang Y, Han X, Yu L, Wang L, Yu Y. nnformer: volumetric medical image segmentation via a 3d transformer. IEEE Trans Image Process 2023;32: 4036–45. https://doi.org/10.1109/TIP.2023.3293771.

[28] J. L. Ba, J. R. Kiros, G. E. Hinton, Layer normalization, Arxiv Preprint abs/1607.06450, available at: http://arxiv.org/abs/1607.06450 (Accessed September 1, 2019). arXiv:1607.06450.

[29] Szegedy C, Liu W, Jia Y, Sermanet P, Reed SE, Anguelov D, Erhan D, Vanhoucke V, Rabinovich A. Going deeper with convolutions. In: IEEE Conference on Computer Vision and Pattern Recognition, CVPR 2015, Boston, MA, USA, June 7-12, 2015; 2015. p. 1–9. https://doi.org/10.1109/CVPR.2015.7298594.

[30] Szegedy C, Vanhoucke V, Ioffe S, Shlens J, Wojna Z. Rethinking the inception architecture for computer vision. In: 2016 IEEE Conference on Computer Vision and Pattern Recognition, CVPR 2016, Las Vegas, NV, USA, June 27-30, 2016. IEEE Computer Society; 2016. p. 2818–26. https://doi.org/10.1109/CVPR.2016.308.

[31] Szegedy C, Ioffe S, Vanhoucke V, Alemi AA. Inception-v4, inception-resnet and the impact of residual connections on learning. In: Proceedings of the Thirty-First AAAI Conference on Artificial Intelligence, February 4-9, 2017, San Francisco, California, USA; 2017. p. 4278–84. URL, http://aaai.org/ocs/index.php/AAAI/AAAI17/paper/view/14806.

[32] Yu F, Koltun V. Multi-Scale context aggregation by dilated convolutions. In: International Conference on Learning Representations; 2016. p. 1–13.

[33] He K, Zhang X, Ren S, Sun J. Spatial pyramid pooling in deep convolutional networks for visual recognition. IEEE Trans Pattern Anal Mach Intell 2014. https://doi.org/10.1109/TPAMI.2015.2389824.

[34] L.-C. Chen, G. Papandreou, F. Schroff, H. Adam, Rethinking atrous convolution for semantic image segmentation, Arxiv Preprint abs/1706.05587, available at: http://arxiv.org/abs/1706.05587(Accessed September 1, 2019). arXiv: 1706.05587.

[35] He K, Zhang X, Ren S, Sun J. Deep residual learning for image recognition. In: 2016 IEEE Conference on Computer Vision and Pattern Recognition, CVPR

2016, Las Vegas, NV, USA, June 27-30, 2016. IEEE Computer Society; 2016. p. 770–8.
[36] Ibtehaz N, Rahman MS. MultiResUNet : rethinking the U-Net architecture for multimodal biomedical image segmentation. Neural Netw 2020;121:74–87.
[37] Huang X, Liu M-Y, Belongie S, Kautz J. Multimodal unsupervised image-to-image translation. In: Ferrari V, Hebert M, Sminchisescu C, Weiss Y, editors. Computer Vision - ECCV 2018 - 15th European Conference. Munich, Germany: Springer; 2018. p. 172–89.
[38] Zhou Z, Siddiquee MMR, Tajbakhsh N, Liang J. Unet++: redesigning skip connections to exploit multiscale features in image segmentation. IEEE Trans Med Imag 2019;39(6):1856–67.
[39] X. Li, H. Chen, X. Qi, Q. Dou, C.-W. Fu, P. A. Heng, H-DenseUNet: Hybrid densely connected UNet for liver and tumor segmentation from CT volumes, TMI.
[40] Schlemper J, Oktay O, Schaap M, Heinrich M, Kainz B, Glocker B, Rueckert D. Attention gated networks: learning to leverage salient regions in medical images. Med Image Anal 2019;53:197–207.
[41] Oktay O, Schlemper J, Folgoc LL, Lee M, Heinrich M, Misawa K, Mori K, McDonagh S, Hammerla NY, Kainz B, Glocker B, Rueckert D. Attention U-Net: learning where to look for the pancreas. In: Medical imaging with deep learning; 2018. p. 1804. 03999.
[42] Roy AG, Navab N, Wachinger C. Recalibrating fully convolutional networks with spatial and channel 'squeeze and excitation' blocks. IEEE Trans Med Imag 2019;38 (2):540–9.
[43] Hatamizadeh A, Tang Y, Nath V, Yang D, Myronenko A, Landman B, Roth HR, Xu D. Unetr: transformers for 3d medical image segmentation. In: Proceedings of the IEEE/CVF Winter Conference on Applications of Computer Vision; 2022. p. 574–84.
[44] Cao H, Wang Y, Chen J, Jiang D, Zhang X, Tian Q, Wang M. Swin-unet: Unet-like pure transformer for medical image segmentation. In: European Conference on Computer Vision. Springer; 2022. p. 205–18.
[45] Chong Y, Xie N, Liu X, Pan S. P-TransUNet: an improved parallel network for medical image segmentation. BMC Bioinf 2023;24(1):285.
[46] Zhang Y, Liu H, Hu Q. TransFuse: fusing transformers and CNNs for medical image segmentation. In: MICCAI; 2021.
[47] Lin A, Chen B, Xu J, Zhang Z, Lu G, Zhang D. DS-TransUNet: dual swin transformer U-Net for medical image segmentation. IEEE Trans Instrum Meas 2022;71: 1–15.
[48] Li J, Chen J, Tang Y, Wang C, Landman BA, Zhou SK. Transforming medical imaging with transformers? a comparative review of key properties, current progresses, and future perspectives. Med Image Anal 2023;85:102762.
[49] Siddique N, Paheding S, Elkin CP, Devabhaktuni V. U-net and its variants for medical image segmentation: a review of theory and applications. IEEE Access 2021;9: 82031–57. https://doi.org/10.1109/ACCESS.2021.3086020.
[50] Hochreiter S, Schmidhuber J. Long short-term memory. Neural Comput 1997;9(8): 1735–80. https://doi.org/10.1162/neco.1997.9.8.1735.
[51] Cho K, van Merrienboer B, Gulcehre C, Bahdanau D, Bougares F, Schwenk H, Bengio Y. Learning phrase representations using RNN Encoder-Decoder for statistical machine translation. In: Conference on Empirical Methods in Natural Language Processing. ACL; 2014. p. 1724–34. https://www.aclweb.org/anthology/D14-1179/.

[52] Gu A, Goel K, Ré C. Efficiently modeling long sequences with structured state spaces. In: International Conference on Learning Representations; 2022.

[53] Bai W, Suzuki H, Qin C, Tarroni G, Oktay O, Matthews PM, Rueckert D. Recurrent neural networks for aortic image sequence segmentation with sparse annotations. In: Frangi AF, Schnabel JA, Davatzikos C, Alberola-López C, Fichtinger G, editors. Medical Image Computing and Computer Assisted Intervention - MICCAI 2018 - 21st International Conference. Granada, Spain: Springer International Publishing; 2018. p. 586–94. https://doi.org/10.1007/978-3-030-00937-3_67.

[54] Qin C, Bai W, Schlemper J, Petersen SE, Piechnik SK, Neubauer S, Rueckert D. Joint learning of motion estimation and segmentation for cardiac mr image sequences. In: Frangi AF, Schnabel JA, Davatzikos C, Alberola-López C, Fichtinger G, editors. Medical Image Computing and Computer Assisted Intervention - MICCAI 2018 - 21st International Conference, Granada, Spain, September 16-20, 2018, Proceedings, Part I. Springer International Publishing; 2018. p. 472–80.

[55] Qi X, Liao R, Jia J, Fidler S, Urtasun R. 3D graph neural networks for RGBD semantic segmentation. In: 2017 IEEE International Conference on Computer Vision (ICCV). IEEE; 2017. p. 10.

[56] Saueressig C, Berkley A, Munbodh R, Singh R. A joint graph and image convolution network for automatic brain tumor segmentation. In: International MICCAI Brainlesion Workshop. Springer; 2022. p. 356–65.

[57] Rumelhart DE, Hinton GE, Williams RJ. Learning representations by back-propagating errors. Nature 1986;323(6088):533–6.

[58] Goodfellow I. Deep learning, Adaptive computation and machine learning. Cambridge, Massachusetts ; London, England: The MIT Press; 2016.

[59] Kingma DP, Ba J. Adam: A method for stochastic optimization. In: International Conference on Learning Representations; 2015. 1–1.

[60] Loshchilov I, Hutter F. Decoupled weight decay regularization. In: International Conference on Learning Representations; 2019. 1–1, https://openreview.net/forum?id=Bkg6RiCqY7.

[61] Jang Y, Hong Y, Ha S, Kim S, Chang H-J. Automatic segmentation of LV and RV in cardiac MRI. In: Pop M, Sermesant M, Jodoin P, Lalande A, Zhuang X, Yang G, Young AA, Bernard O, editors. International Workshop on Statistical Atlases and Computational Models of the Heart. Springer; 2017. p. 161–9.

[62] Baumgartner CF, Koch LM, Pollefeys M, Konukoglu E. An exploration of 2D and 3D deep learning techniques for cardiac MR image segmentation. In: Pop M, Sermesant M, Jodoin P, Lalande A, Zhuang X, Yang G, Young AA, Bernard O, editors. International Workshop on Statistical Atlases and Computational Models of the Heart, vol 10663. Springer; 2017. p. 1–8.

[63] Yang X, Bian C, Yu L, Ni D, Heng P-A. Class-Balanced deep neural network for automatic ventricular structure segmentation. In: Pop M, Sermesant M, Jodoin P, Lalande A, Zhuang X, Yang G, Young AA, Bernard O, editors. Statistical Atlases and Computational Models of the Heart. ACDC and MMWHS Challenges - 8th International Workshop, STACOM 2017, Held in Conjunction with MICCAI 2017, Quebec City, Canada, September 10-14, 2017, Revised Selected Papers. Springer International Publishing; 2017. p. 152–60. https://doi.org/10.1007/978-3-319-75541-0_16.

[64] Khened M, Kollerathu VA, Krishnamurthi G. Fully convolutional multi-scale residual DenseNets for cardiac segmentation and automated cardiac diagnosis using ensemble

of classifiers. Med Image Anal 2019;51:21–45. https://doi.org/10.1016/j.media.2018.10.004.

[65] Sudre CH, Li W, Vercauteren T, Ourselin S, Jorge Cardoso M. Generalised dice overlap as a deep learning loss function for highly unbalanced segmentations, Deep learning in medical image analysis and multimodal learning for clinical decision support. In: Third International Workshop, DLMIA 2017, and 7th International Workshop, ML-CDS 2017, held in conjunction with MICCAI 2017 Quebec City, QC,...2017; 2017. p. 240–8.

[66] Lin T-Y, Goyal P, Girshick R, He K, Dollar P. Focal loss for dense object detection. IEEE Trans Pattern Anal Mach Intell 2018;42(2):318–27.

[67] Kervadec H, Bouchtiba J, Desrosiers C, Granger E, Dolz J, Ayed IB. Boundary loss for highly unbalanced segmentation. In: International Conference on Medical Imaging With Deep Learning. PMLR; 2019. p. 285–96.

[68] Karimi D, Salcudean SE. Reducing the hausdorff distance in medical image segmentation with convolutional neural networks. IEEE Trans Med Imag 2020;39(2):499–513.

[69] Ma J, He J, Yang X. Learning geodesic active contours for embedding object global information in segmentation cnns. IEEE Trans Med Imag 2020;40(1):93–104.

[70] Chen C, Qin C, Ouyang C, Li Z, Wang S, Qiu H, Chen L, Tarroni G, Bai W, Rueckert D. Enhancing MR image segmentation with realistic adversarial data augmentation. Med Image Anal 2022;82:102597.

[71] X. Hu, F. Li, D. Samaras, C. Chen, Topology-preserving deep image segmentation, Adv Neural Inf Process Syst 32.

[72] Clough JR, Byrne N, Oksuz I, Zimmer VA, Schnabel JA, King AP. A topological loss function for deep-learning based image segmentation using persistent homology. IEEE Trans Pattern Anal Mach Intell 2020;44(12):8766–78.

[73] Clough JR, Oksuz I, Byrne N, Schnabel JA, King AP. Explicit topological priors for deep-learning based image segmentation using persistent homology. In: Information Processing in Medical Imaging: 26th International Conference, IPMI 2019, Hong Kong, China, June 2–7, 2019, Proceedings 26. Springer; 2019. p. 16–28.

[74] Li L, Ma Q, Li Z, Ouyang C, Zhang W, Price A, Kyriakopoulou V, Grande LC, Makropoulos A, Hajnal J, et al. Fetal cortex segmentation with topology and thickness loss constraints. In: Ethical and Philosophical Issues in Medical Imaging, Multimodal Learning and Fusion Across Scales for Clinical Decision Support, and Topological Data Analysis for Biomedical Imaging: 1st International Workshop, EPIMI 2022, 12th International Workshop, ML-CDS 2022, 2nd International Workshop, TDA4BiomedicalImaging, Held in Conjunction with MICCAI 2022, Singapore, September 18–22, 2022, Proceedings. Springer; 2022. p. 123–33.

[75] de Dumast P, Kebiri H, Atat C, Dunet V, Koob M, Cuadra MB. Segmentation of the cortical plate in fetal brain mri with a topological loss. In: Uncertainty for Safe Utilization of Machine Learning in Medical Imaging, and Perinatal Imaging, Placental and Preterm Image Analysis: 3rd International Workshop, UNSURE 2021, and 6th International Workshop, PIPPI 2021, Held in Conjunction with MICCAI 2021, Strasbourg, France, October 1, 2021, Proceedings 3. Springer; 2021. p. 200–9.

[76] Byrne N, Clough JR, Valverde I, Montana G, King AP. A persistent homology-based topological loss for cnn-based multiclass segmentation of cmr. IEEE Trans Med Imag 2022;42(1):3–14.

[77] N. Stucki, J. C. Paetzold, S. Shit, B. Menze, U. Bauer, Topologically faithful image segmentation via induced matching of persistence barcodes, arXiv preprint arXiv: 2211.15272.

[78] Shit S, Paetzold JC, Sekuboyina A, Ezhov I, Unger A, Zhylka A, Pluim JP, Bauer U, Menze BH. cldice-a novel topology-preserving loss function for tubular structure segmentation. In: Proceedings of the IEEE/CVF Conference on Computer Vision and Pattern Recognition; 2021. p. 16560–9.

[79] Li L, Ma Q, Ouyang C, Li Z, Meng Q, Zhang W, Qiao M, Kyriakopoulou V, Hajnal JV, Rueckert D, et al. Robust segmentation via topology violation detection and feature synthesis. In: International Conference on Medical Image Computing and Computer-Assisted Intervention. Springer; 2023. p. 67–77.

[80] Stucki N, Paetzold JC, Shit S, Menze B, Bauer U. Topologically faithful image segmentation via induced matching of persistence barcodes. In: International Conference on Machine Learning. PMLR; 2023. p. 32698–727.

[81] Oktay O, Ferrante E, Kamnitsas K, Heinrich M, Bai W, Caballero J, Cook SA, De Marvao A, Dawes T, O'Regan DP, Kainz B, Glocker B, Rueckert D. Anatomically constrained neural networks (acnns): application to cardiac image enhancement and segmentation. IEEE Trans Med Imag 2018;37(2):384–95. https://doi.org/10.1109/TMI.2017.2743464.

[82] Chen C, Ouyang C, Tarroni G, Schlemper J, Qiu H, Bai W, Rueckert D. Unsupervised multi-modal style transfer for cardiac MR segmentation. In: Statistical Atlases and Computational Models of the Heart. Multi-Sequence CMR Segmentation, CRT-EPiggy and LV Full Quantification Challenges - 10th International Workshop, STACOM 2019, Held in Conjunction with MICCAI 2019, Shenzhen, China, October 13, 2019, Revised Selected Papers; 2019.

[83] Kanopoulos N, Vasanthavada N, Baker RL. Design of an image edge detection filter using the sobel operator. IEEE J Solid State Circ 1988;23(2):358–67.

[84] Lee C-Y, Xie S, Gallagher P, Zhang Z, Tu Z. Deeply-Supervised nets. In: Lebanon G, Vishwanathan SVN, editors. Proceedings of the Eighteenth International Conference on Artificial Intelligence and Statistics, Vol. 38 of Proceedings of Machine Learning Research. San Diego, California, USA: PMLR; 2015. p. 562–70. http://proceedings.mlr.press/v38/lee15a.html.

[85] Ma J, Chen J, Ng M, Huang R, Li Y, Li C, Yang X, Martel AL. Loss odyssey in medical image segmentation. Med Image Anal 2021;71:102035.

[86] A. J. Larrazabal, C. Martínez, B. Glocker, E. Ferrante, Post-DAE: Anatomically Plausible Segmentation via Post-Processing with Denoising Autoencoders, TMI.

[87] Goodfellow IJ, Pouget-Abadie J, Mirza M, Xu B, Warde-Farley D, Ozair S, Courville A, Bengio Y. Generative adversarial networks. In: Conference on Neural Information Processing Systems. Curran Associates, Inc.; 2014. p. 2672–80. http://papers.nips.cc/paper/5423-generative-adversarial-nets.pdf.

[88] Luc P, Couprie C, Chintala S, Verbeek J. Semantic segmentation using adversarial networks. In: NIPS Workshop on Adversarial Training; 2016. p. 1–12.

[89] N. Savioli, M. S. Vieira, P. Lamata, G. Montana, A generative adversarial model for right ventricle segmentation, Arxiv Preprint abs/1810.03969, available at: http://arxiv.org/abs/1810.03969 (Accessed September 1, 2019). arXiv:1810.03969.

[90] Dai W, Dong N, Wang Z, Liang X, Zhang H, Xing EP. SCAN: structure correcting adversarial network for organ segmentation in chest X-Rays. In: Deep learning in

medical image analysis and multimodal learning for clinical decision Support. Springer International Publishing; 2018. p. 263–73.

[91] Sinclair M, Schuh A, Hahn K, Petersen K, Bai Y, Batten J, Schaap M, Glocker B. Atlas-ISTN: Joint segmentation, registration and atlas construction with image-and-spatial transformer networks. Med Image Anal 2022;78:102383.

[92] Dong S, Luo G, Tam C, Wang W, Wang K, Cao S, Chen B, Zhang H, Li S. Deep atlas network for efficient 3D left ventricle segmentation on echocardiography. Med Image Anal 2020;61:101638.

[93] Bach Cuadra M, Duay V, Thiran J-P. Atlas-based segmentation. In: Paragios N, Duncan J, Ayache N, editors. Handbook of biomedical imaging: methodologies and clinical research. Boston, MA: Springer US; 2015. p. 221–44.

[94] G. Balakrishnan, A. Zhao, M. R. Sabuncu, J. Guttag, A. V. Dalca, VoxelMorph: a learning framework for deformable medical image registration, IEEE Trans Med Imag.

[95] Syed S, Anderssen KE, Stormo SK, Kranz M. Weakly supervised semantic segmentation for MRI: exploring the advantages and disadvantages of class activation maps for biological image segmentation with soft boundaries. Sci Rep 2023;13(1):2574.

[96] Mehnert A, Jackway P. An improved seeded region growing algorithm. Pattern Recognit Lett 1997;18(10):1065–71.

[97] Boykov YY, Jolly M-P. Interactive graph cuts for optimal boundary and region segmentation of objects in N-D images. In: Proceedings Eighth IEEE International Conference on Computer Vision. ICCV 2001, 1. IEEE; 2001. p. 105–12.

[98] Lin D, Jia H, Sun SS. Scribble-Supervised convolutional networks for semantic segmentation. In: 2016 IEEE Conference on Computer Vision and Pattern Recognition (CVPR); 2016. p. 3159–67.

[99] Rother C, Kolmogorov V, Blake A. "GrabCut": interactive foreground extraction using iterated graph cuts. ACM Trans Graph 2004;23(3):309–14.

[100] Uijlings JRR, van de Sande KEA, Gevers T, Smeulders AWM. Selective search for object recognition. Int J Comput Vis 2013;104(2):154–71.

[101] Dai J, He K, Sun J. BoxSup: exploiting bounding boxes to supervise convolutional networks for semantic segmentation. In: International Conference on Computer Vision; 2015.

[102] Syed S, Anderssen KE, Stormo SK, Kranz M. Weakly supervised semantic segmentation for MRI: exploring the advantages and disadvantages of class activation maps for biological image segmentation with soft boundaries. Sci Rep 2023;13(1): 2574.

[103] Selvaraju RR, Cogswell M, Das A, Vedantam R, Parikh D, Batra D. Grad-CAM: visual explanations from deep networks via gradient-based localization. In: IJCV; 2016.

[104] H. Kervadec, J. Dolz, S. Wang, others, Bounding boxes for weakly supervised segmentation: global constraints get close to full supervision, Med Imaging.

[105] Kervadec H, Dolz J, Tang M, Granger E, Boykov Y, Ben Ayed I. Constrained-CNN losses for weakly supervised segmentation. Med Image Anal 2019;54:88–99.

[106] G. Shi, L. Xiao, Y. Chen, S. Kevin Zhou, Marginal loss and exclusion loss for partially supervised multi-organ segmentation, Med Image Anal.

[107] Jiao R, Zhang Y, Ding L, Xue B, Zhang J, Cai R, Jin C. Learning with limited annotations: a survey on deep semi-supervised learning for medical image segmentation. Comput Biol Med 2023;169:107840.

[108] Bai W, Oktay O, Sinclair M, Suzuki H, Rajchl M, Tarroni G, Glocker B, King A, Matthews PM, Rueckert D. Semi-supervised learning for Network-Based cardiac

MR image segmentation. In: Descoteaux M, Maier-Hein L, Franz A, Jannin P, Collins DL, Duchesne S, editors. Medical Image Computing and Computer Assisted Intervention - MICCAI 2017 - 20th International Conference, Quebec City, QC, Canada, September 11-13, 2017, Proceedings, Part II, vol 10434. Springer International Publishing; 2017. p. 253–60.

[109] Thompson BH, Di Caterina G, Voisey JP. Pseudo-label refinement using superpixels for semi-supervised brain tumour segmentation. In: 2022 IEEE 19th International Symposium on Biomedical Imaging (ISBI). IEEE; 2022.

[110] Z. Liu, C. Zhao, Semi-supervised medical image segmentation via geometry-aware consistency trainingarXiv:2202.06104.

[111] W. Cui, Y. Liu, Y. Li, M. Guo, Y. Li, X. Li, T. Wang, X. Zeng, C. Ye, Semi-Supervised brain lesion segmentation with an adapted mean teacher modelarXiv:1903.01248.

[112] Li X, Yu L, Chen H, Fu C-W, Xing L, Heng P-A. Transformation-Consistent Self-Ensembling model for semisupervised medical image segmentation. IEEE Transact Neural Networks Learn Syst 2021;32(2):523–34.

[113] Maier-Hein L, Reinke A, Godau P, Tizabi MD, Buettner F, Christodoulou E, Glocker B, Isensee F, Kleesiek J, Kozubek M, Reyes M, Riegler MA, Wiesenfarth M, Kavur AE, Sudre CH, Baumgartner M, Eisenmann M, Heckmann-Nötzel D, Rädsch T, Acion L, Antonelli M, Arbel T, Bakas S, Benis A, Blaschko MB, Cardoso MJ, Cheplygina V, Cimini BA, Collins GS, Farahani K, Ferrer L, Galdran A, van Ginneken B, Haase R, Hashimoto DA, Hoffman MM, Huisman M, Jannin P, Kahn CE, Kainmueller D, Kainz B, Karargyris A, Karthikesalingam A, Kofler F, Kopp-Schneider A, Kreshuk A, Kurc T, Landman BA, Litjens G, Madani A, Maier-Hein K, Martel AL, Mattson P, Meijering E, Menze B, Moons KGM, Müller H, Nichyporuk B, Nickel F, Petersen J, Rajpoot N, Rieke N, Saez-Rodriguez J, Sánchez CI, Shetty S, van Smeden M, Summers RM, Taha AA, Tiulpin A, Tsaftaris SA, Van Calster B, Varoquaux G, Jäger PF. Metrics reloaded: recommendations for image analysis validation. Nat Methods 2024;21(2):195–212.

[114] Zijdenbos AP, Dawant BM, Margolin RA, Palmer AC. Morphometric analysis of white matter lesions in mr images: method and validation. IEEE Trans Med Imag 1994;13(4):716–24.

[115] Huttenlocher DP, Klanderman GA, Rucklidge WJ. Comparing images using the hausdorff distance. IEEE Trans Pattern Anal Mach Intell 1993;15(9):850–63.

[116] Heimann T, Van Ginneken B, Styner MA, Arzhaeva Y, Aurich V, Bauer C, Beck A, Becker C, Beichel R, Bekes G, et al. Comparison and evaluation of methods for liver segmentation from ct datasets. IEEE Trans Med Imag 2009;28(8):1251–65.

[117] Maier-Hein L, Eisenmann M, Reinke A, Onogur S, Stankovic M, Scholz P, Arbel T, Bogunovic H, Bradley AP, Carass A, et al. Why rankings of biomedical image analysis competitions should be interpreted with care. Nat Commun 2018;9(1):5217.

[118] Bai W, Sinclair M, Tarroni G, Oktay O, Rajchl M, Vaillant G, Lee AM, Aung N, Lukaschuk E, Sanghvi MM, et al. Automated cardiovascular magnetic resonance image analysis with fully convolutional networks. J Cardiovasc Magn Reson 2018;20(1):1–12.

[119] B. Billot, D. N. Greve, O. Puonti, A. Thielscher, K. Van Leemput, B. Fischl, A. V. Dalca, J. E. Iglesias, Synthseg: domain randomisation for segmentation of brain mri scans of any contrast and resolution, arXiv preprint arXiv:2107.09559.

[120] A. Reinke, M. D. Tizabi, M. Baumgartner, M. Eisenmann, D. Heckmann-Nötzel, A. E. Kavur, T. Rädsch, C. H. Sudre, L. Acion, M. Antonelli, et al., Understanding metric-related pitfalls in image analysis validation, ArXiv.

[121] Detlefsen NS, Borovec J, Schock J, Jha AH, Koker T, Liello LD, Stancl D, Quan C, Grechkin M, Falcon W. Torchmetrics - measuring reproducibility in pytorch. J Open Source Softw 2022;7(70):4101. https://doi.org/10.21105/joss.04101.

[122] Taha AA, Hanbury A. Metrics for evaluating 3D medical image segmentation: analysis, selection, and tool. BMC Med Imag 2015;15:29.

[123] Jia J. A package to compute segmentation metrics: seg-metrics. https://github.com/Ordgod/segmentation_metrics; 2020.

[124] Muller D, Hartmann D, Meyer P, Auer F, Soto-Rey I, Kramer F. Miseval: a metric library for medical image segmentation evaluation. Stud Health Technol Inf 2022;294: 33–7. https://doi.org/10.3233/SHTI220391.

[125] Srivastava N, Hinton G, Krizhevsky A, Sutskever I, Salakhutdinov R. Dropout: a simple way to prevent neural networks from overfitting. J Mach Learn Res 2014;15: 1929–58.

[126] Deng J, Dong W, Socher R, Li L-J, Li K, Fei-Fei L. Imagenet: a large-scale hierarchical image database. In: 2009 IEEE Conference On Computer Vision and Pattern Recognition. Ieee; 2009. p. 248–55.

[127] J. Irvin, P. Rajpurkar, M. Ko, Y. Yu, S. Ciurea-Ilcus, C. Chute, H. Marklund, B. Haghgoo, R. Ball, K. Shpanskaya, J. Seekins, D. A. Mong, S. S. Halabi, J. K. Sandberg, R. Jones, D. B. Larson, C. P. Langlotz, B. N. Patel, M. P. Lungren, A. Y. Ng, CheXpert: a large chest radiograph dataset with uncertainty labels and expert comparisonarXiv: 1901.07031.

[128] A. Kirillov, E. Mintun, N. Ravi, H. Mao, C. Rolland, L. Gustafson, T. Xiao, S. Whitehead, A. C. Berg, W.-Y. Lo, P. Dollár, R. Girshick, Segment anything, arXiv: 2304.02643.

[129] Chen C, Bai W, Rueckert D. Multi-task learning for left atrial segmentation on GE-MRI. In: Pop M, Sermesant M, Zhao J, Li S, McLeod K, Young AA, Rhode KS, Mansi T, editors. Statistical Atlases and Computational Models of the Heart. Atrial Segmentation and LV Quantification Challenges - 9th International Workshop, STACOM 2018, Held in Conjunction with MICCAI 2018, Granada, Spain, September 16, 2018, Revised Selected Papers. Springer International Publishing; 2018. p. 292–301. https://doi.org/10.1007/978-3-030-12029-0_32.

[130] Kamnitsas K, Bai W, Ferrante E, Mcdonagh S, Sinclair M. Ensembles of multiple models and architectures for robust brain tumour segmentation. In: Brainlesion: Glioma, Multiple Sclerosis, Stroke and Traumatic Brain Injuries - Third International Workshop, BrainLes 2017, Held in Conjunction with MICCAI 2017, Quebec City, QC, Canada, September 14, 2017, Revised Selected Papers; 2017. p. 450–62.

[131] Mumuni A, Mumuni F. Data augmentation: a comprehensive survey of modern approaches. Array 2022;16:100258.

[132] Shorten C, Khoshgoftaar TM. A survey on image data augmentation for deep learning. J Big Data 2019;6(1):60.

[133] Goceri E. Medical image data augmentation: techniques, comparisons and interpretations. Artif Intell Rev 2023:1–45.

[134] Cubuk ED, Zoph B, Shlens J, Le QV. RandAugment: practical automated data augmentation with a reduced search space. In: Conference on Neural Information Processing Systems; 2019.

[135] Chen C, Bai W, Davies RH, Bhuva AN, Manisty CH, Augusto JB, Moon JC, Aung N, Lee AM, Sanghvi MM, Fung K, Paiva JM, Petersen SE, Lukaschuk E, Piechnik SK, Neubauer S, Rueckert D. Improving the generalizability of convolutional neural Network-Based segmentation on CMR images. Front Cardiovasc Med 2020.

[136] Z. Li, K. Kamnitsas, Q. Dou, C. Qin, B. Glocker, Joint optimization of class-specific training-and test-time data augmentation in segmentation, IEEE Trans Med Imag.

[137] Pérez-García F, Sparks R, Ourselin S. Torchio: a python library for efficient loading, preprocessing, augmentation and patch-based sampling of medical images in deep learning. Comput Methods Progr Biomed 2021. https://doi.org/10.1016/j.cmpb.2021.106236. https://www.sciencedirect.com/science/article/pii/S0169260721003102.

[138] Chen C, Qin C, Qiu H, Ouyang C, Wang S, Chen L, Tarroni G, Bai W, Rueckert D. Realistic adversarial data augmentation for MR image segmentation. In: Medical image computing and computer assisted intervention – MICCAI 2020. Springer International Publishing; 2020. p. 667–77.

[139] Chen C, Li Z, Ouyang C, Sinclair M, Bai W, Rueckert D. Maxstyle: adversarial style composition for robust medical image segmentation. In: International Conference on Medical Image Computing and Computer-Assisted Intervention. Springer; 2022. p. 151–61.

[140] Kingma DP, Welling M. Auto-Encoding variational bayes. In: Bengio Y, LeCun Y, editors. 2nd International Conference on Learning Representations, ICLR 2014, Banff, AB, Canada, April 14-16, 2014, Conference Track Proceedings; 2013. p. 1–14.

[141] Ho J, Jain A, Abbeel P. Denoising diffusion probabilistic models. In: NeuraIPS; 2020.

[142] Yu B, Wang Y, Wang L, Shen D, Zhou L. Medical image synthesis via deep learning. Adv Exp Med Biol 2020;1213:23–44.

[143] Osuala R, Kushibar K, Garrucho L, Linardos A, Szafranowska Z, Klein S, Glocker B, Diaz O, Lekadir K. Data synthesis and adversarial networks: a review and meta-analysis in cancer imaging. Med Image Anal 2023;84:102704.

[144] Khader F, Müller-Franzes G, Tayebi Arasteh S, Han T, Haarburger C, Schulze-Hagen M, Schad P, Engelhardt S, Baeßler B, Foersch S, Stegmaier J, Kuhl C, Nebelung S, Kather JN, Truhn D. Denoising diffusion probabilistic models for 3D medical image generation. Sci Rep 2023;13(1):7303.

[145] Wu J, Guo D, Wang L, Yang S, Zheng Y, Shapey J, Vercauteren T, Bisdas S, Bradford R, Saeed S, Kitchen N, Ourselin S, Zhang S, Wang G. TISS-net: brain tumor image synthesis and segmentation using cascaded dual-task networks and error-prediction consistency. Neurocomputing 2023;544.

[146] He K, Chen X, Xie S, Li Y, Dollár P, Girshick R. Masked autoencoders are scalable vision learners. In: Proceedings of the IEEE/CVF Conference on Computer Vision and Pattern Recognition; 2022. p. 16000–9.

[147] Chen T, Kornblith S, Norouzi M, Hinton G. A simple framework for contrastive learning of visual representations. In: International Conference on Machine Learning. PMLR; 2020. p. 1597–607.

[148] Chaitanya K, Erdil E, Karani N, Konukoglu E. Contrastive learning of global and local features for medical image segmentation with limited annotations. In: Conference on Neural Information Processing Systems; 2020.

[149] Dai C, Wang S, Mo Y, Angelini E, Guo Y, Bai W. Suggestive annotation of brain MR images with gradient-guided sampling. Med Image Anal 2022;77:102373.

[150] V. Nath, D. Yang, B. A. Landman, D. Xu, H. R. Roth, Diminishing uncertainty within the training pool: active learning for medical image segmentation, IEEE Trans Med Imag.

[151] Wang K, Liew JH, Zou Y, Zhou D, Feng J. PANet: few-Shot image semantic segmentation with prototype alignment. In: ICCV; 2019.

[152] Ouyang C, Biffi C, Chen C, Kart T, Qiu H, Rueckert D. Self-supervision with superpixels: training Few-Shot medical image segmentation without annotation. In: European Conference on Computer Vision – ECCV 2020. Springer International Publishing; 2020. p. 762–80.

[153] A. G. Roy, S. Siddiqui, S. Pölsterl, N. Navab, C. Wachinger, 'squeeze & excite'guided few-shot segmentation of volumetric images, Med Image Anal.

[154] Li X, Jiang M, Zhang X, Kamp M, Dou Q. FedBN: Federated learning on Non-IID features via local batch normalization. In: ICLR; 2021.

[155] Guan H, Liu M. Domain adaptation for medical image analysis: a survey. IEEE Trans Biomed Eng 2022;69(3):1173–85.

[156] Dou Q, Castro DC, Kamnitsas K, Glocker B. Domain generalization via Model-Agnostic learning of semantic features. In: Conference on Neural Information Processing Systems; 2019.

[157] C. Chen, Z. Li, C. Ouyang, M. Sinclair, W. Bai, D. Rueckert, MaxStyle: adversarial style composition for robust medical image segmentation, in: MICCAI, 2022. arXiv:2206.01737.

[158] Zhang L, Wang X, Yang D, Sanford T, Harmon S, Turkbey B, Wood BJ, Roth H, Myronenko A, Xu D, Xu Z. Generalizing deep learning for medical image segmentation to unseen domains via deep stacked transformation. IEEE Trans Med Imag 2020; 39(7):2531–40.

[159] Brosch T, Tang LY, Yoo Y, Li DK, Traboulsee A, Tam R. Deep 3d convolutional encoder networks with shortcuts for multiscale feature integration applied to multiple sclerosis lesion segmentation. IEEE Trans Med Imag 2016;35(5): 1229–39.

[160] Litjens G, Kooi T, Bejnordi BE, Setio AAA, Ciompi F, Ghafoorian M, Van Der Laak JA, Van Ginneken B, Sánchez CI. A survey on deep learning in medical image analysis. Med Image Anal 2017;42:60–88.

[161] Liu Z, Miao Z, Zhan X, Wang J, Gong B, Yu SX. Large-scale long-tailed recognition in an open world. In: Proceedings of the IEEE Conference on Computer Vision and Pattern Recognition; 2019. p. 2537–46.

[162] Kang B, Xie S, Rohrbach M, Yan Z, Gordo A, Feng J, Kalantidis Y. Decoupling representation and classifier for long-tailed recognition. In: International Conference on Learning Representations (ICLR); 2020. 1–1.

[163] Li Z, Kamnitsas K, Glocker B. Analyzing overfitting under class imbalance in neural networks for image segmentation. IEEE Trans Med Imag 2020;40(3):1065–77.

[164] Mehrabi N, Morstatter F, Saxena N, Lerman K, Galstyan A. A survey on bias and fairness in machine learning. ACM Comput Surv 2021;54(6):1–35.

[165] Islam M, Li Z, Glocker B. Robustness stress testing in medical image classification. In: International Workshop on Uncertainty for Safe Utilization of Machine Learning in Medical Imaging. Springer; 2023. p. 167–76.

[166] Puyol-Antón E, Ruijsink B, Piechnik SK, Neubauer S, Petersen SE, Razavi R, King AP. Fairness in cardiac mr image analysis: an investigation of bias due to data imbalance in deep learning based segmentation. In: Medical Image Computing and Computer Assisted Intervention–MICCAI 2021: 24th International Conference, Strasbourg, France, September 27–October 1, 2021, Proceedings, Part III 24. Springer; 2021. p. 413–23.

[167] Castro DC, Walker I, Glocker B. Causality matters in medical imaging. Nat Commun 2020;11(1):1–10.

[168] Liu Q, Dou Q, Heng P-A. Shape-aware meta-learning for generalizing prostate mri segmentation to unseen domains. In: International Conference on Medical Image Computing and Computer-Assisted Intervention. Springer; 2020. p. 475–85.

[169] Guo C, Pleiss G, Sun Y, Weinberger KQ. On calibration of modern neural networks. In: ICML. PMLR; 2017. p. 1321–30.

[170] Y. Ovadia, E. Fertig, J. Ren, Z. Nado, D. Sculley, S. Nowozin, J. Dillon, B. Lakshminarayanan, J. Snoek, Can you trust your model's uncertainty? evaluating predictive uncertainty under dataset shift, Adv Neural Inf Process Syst 32.

[171] Valindria VV, Lavdas I, Bai W, Kamnitsas K, Aboagye EO, Rockall AG, Rueckert D, Glocker B. Reverse classification accuracy: predicting segmentation performance in the absence of ground truth. IEEE Trans Med Imag 2017;36(8):1597–606.

[172] Li Z, Kamnitsas K, Islam M, Chen C, Glocker B. Estimating model performance under domain shifts with class-specific confidence scores. In: International Conference on Medical Image Computing and Computer-Assisted Intervention. Springer; 2022. p. 693–703.

[173] Robinson R, Oktay O, Bai W, Valindria VV, Sanghvi MM, Aung N, Paiva JM, Zemrak F, Fung K, Lukaschuk E, et al. Real-time prediction of segmentation quality. In: Medical Image Computing and Computer Assisted Intervention–MICCAI 2018: 21st International Conference, Granada, Spain, September 16-20, 2018, Proceedings, Part IV 11. Springer; 2018. p. 578–85.

[174] Schlemper J, Oktay O, Schaap M, Heinrich M, Kainz B, Glocker B, Rueckert D. Attention gated networks: learning to leverage salient regions in medical images. Med Image Anal 2019;53:197–207.

[175] Fu J, Liu J, Tian H, Li Y, Bao Y, Fang Z, Lu H. Dual attention network for scene segmentation. In: Proceedings of the IEEE/CVF Conference on Computer Vision and Pattern Recognition; 2019. p. 3146–54.

[176] Chen H, Dou Q, Yu L, Qin J, Heng P-A. Voxresnet: deep voxelwise residual networks for brain segmentation from 3d mr images. Neuroimage 2018;170:446–55.

[177] Astley JR, Biancardi AM, Hughes PJC, Marshall H, Collier GJ, Chan H-F, Saunders LC, Smith LJ, Brook ML, Thompson R, Rowland-Jones S, Skeoch S, Bianchi SM, Hatton MQ, Rahman NM, Ho L-P, Brightling CE, Wain LV, Singapuri A, Evans RA, Moss AJ, McCann GP, Neubauer S, Raman B, C-MORE/PHOSP-COVID Collaborative Group, Wild JM, Tahir BA. Implementable deep learning for multi-sequence proton MRI lung segmentation: a multi-center, multi-vendor, and multi-disease study. J Magn Reson Imag 2023;58(4):1030–44.

[178] Chen C, Qin C, Qiu H, Tarroni G, Duan J, Bai W, Rueckert D. Deep learning for cardiac image segmentation: a review. Front Cardiovas Med 2020;7:25.

[179] Aprovitola A, Gallo L. Knee bone segmentation from mri: a classification and literature review. Biocybern Biomed Eng 2016;36(2):437–49.
[180] Kavur AE, Gezer NS, Barıs M, Aslan S, Conze P-H, Groza V, Pham DD, Chatterjee S, Ernst P, Özkan S, et al. Chaos challenge-combined (ct-mr) healthy abdominal organ segmentation. Med Image Anal 2021;69:101950.
[181] Moccia S, Banali R, Martini C, Muscogiuri G, Pontone G, Pepi M, Caiani EG. Development and testing of a deep learning-based strategy for scar segmentation on CMR-LGE images, Magnetic Resonance Materials in Physics. Biol Med 2019;32 (2):187–95. https://doi.org/10.1007/s10334-018-0718-4.
[182] Menze BH, Jakab A, Bauer S, Kalpathy-Cramer J, Farahani K, Kirby J, Burren Y, Porz N, Slotboom J, Wiest R, et al. The multimodal brain tumor image segmentation benchmark (brats). IEEE Trans Med Imag 2014;34(10):1993–2024.
[183] Rockall AG, Li X, Johnson N, Lavdas I, Santhakumaran S, Prevost AT, Punwani S, Goh V, Barwick TD, Bharwani N, et al. Development and evaluation of machine learning in whole-body magnetic resonance imaging for detecting metastases in patients with lung or colon cancer: a diagnostic test accuracy study. Investig Radiol 2023:10–1097.
[184] J. Li, G. Zhu, C. Hua, M. Feng, B. Bennamoun, P. Li, X. Lu, J. Song, P. Shen, X. Xu, et al., A systematic collection of medical image datasets for deep learning, ACM Comput Surv.
[185] De Feo R, Shatillo A, Sierra A, Valverde JM, Gröhn O, Giove F, Tohka J. Automated joint skull-stripping and segmentation with multi-task u-net in large mouse brain mri databases. Neuroimage 2021;229:117734.
[186] Zhao J, Chen W, Liu C, Gao Y, Chen X, Chen G, Xia L, Dai Y, Zhang X. Automatic macaque brain segmentation based on 7t mri. Magn Reson Imaging 2022;92:232–42.
[187] Fischl B, Salat DH, Busa E, Albert M, Dieterich M, Haselgrove C, Van Der Kouwe A, Killiany R, Kennedy D, Klaveness S, et al. Whole brain segmentation: automated labeling of neuroanatomical structures in the human brain. Neuron 2002;33(3):341–55.
[188] Good CD, Johnsrude IS, Ashburner J, Henson RN, Friston KJ, Frackowiak RS. A voxel-based morphometric study of ageing in 465 normal adult human brains. Neuroimage 2001;14(1):21–36.
[189] Wells WM, Viola P, Atsumi H, Nakajima S, Kikinis R. Multi-modal volume registration by maximization of mutual information. Med Image Anal 1996;1(1):35–51.
[190] Holland EC. Progenitor cells and glioma formation. Curr Opin Neurol 2001;14(6): 683–8.
[191] Tustison NJ, Avants BB, Cook PA, Zheng Y, Egan A, Yushkevich PA, Gee JC. N4itk: improved n3 bias correction. IEEE Trans Med Imag 2010;29(6): 1310–20.
[192] Kamnitsas K, Ledig C, Newcombe VF, Simpson JP, Kane AD, Menon DK, Rueckert D, Glocker B. Efficient multi-scale 3d cnn with fully connected crf for accurate brain lesion segmentation. Med Image Anal 2017;36:61–78.
[193] S. Bakas, M. Reyes, A. Jakab, S. Bauer, M. Rempfler, A. Crimi, R. T. Shinohara, C. Berger, S. M. Ha, M. Rozycki, et al., Identifying the best machine learning algorithms for brain tumor segmentation, progression assessment, and overall survival prediction in the brats challenge, arXiv preprint arXiv:1811.02629.
[194] Isensee F, Jaeger PF, Kohl SA, Petersen J, Maier-Hein KH. nnu-net: a self-configuring method for deep learning-based biomedical image segmentation. Nat Methods 2021; 18(2):203–11.

[195] Van Tulder G, de Bruijne M. Why does synthesized data improve multi-sequence classification?. In: Medical Image Computing and Computer-Assisted Intervention–MICCAI 2015: 18th International Conference, Munich, Germany, October 5-9, 2015, Proceedings, Part I 18. Springer; 2015. p. 531–8.
[196] Sharma A, Hamarneh G. Missing mri pulse sequence synthesis using multi-modal generative adversarial network. IEEE Trans Med Imag 2019;39(4):1170–83.
[197] Havaei M, Guizard N, Chapados N, Bengio Y. Hemis: hetero-modal image segmentation. In: Medical Image Computing and Computer-Assisted Intervention–MICCAI 2016: 19th International Conference, Athens, Greece, October 17-21, 2016, Proceedings, Part II 19. Springer; 2016. p. 469–77.
[198] Chen C, Dou Q, Jin Y, Chen H, Qin J, Heng P-A. Robust multimodal brain tumor segmentation via feature disentanglement and gated fusion. In: Medical Image Computing and Computer Assisted Intervention–MICCAI 2019: 22nd International Conference, Shenzhen, China, October 13–17, 2019, Proceedings, Part III 22. Springer; 2019. p. 447–56.
[199] Peng P, Lekadir K, Gooya A, Shao L, Petersen SE, Frangi AF. A review of heart chamber segmentation for structural and functional analysis using cardiac magnetic resonance imaging, Magnetic Resonance Materials in Physics. Biol Med 2016;29(2): 155–95. https://doi.org/10.1007/s10334-015-0521-4.
[200] Wang S, Qin C, Savioli N, Chen C, O'Regan DP, Cook S, Guo Y, Rueckert D, Bai W. Joint motion correction and super resolution for cardiac segmentation via latent optimisation. In: Medical Image Computing and Computer Assisted Intervention – MICCAI 2021. Springer International Publishing; 2021. p. 14–24.
[201] Sinclair M, Bai W, Puyol-Antón E, Oktay O, Rueckert D, King AP. Fully automated segmentation-based respiratory motion correction of multiplanar cardiac magnetic resonance images for large-scale datasets. In: Medical Image Computing and Computer Assisted Intervention. Berlin, Heidelberg: Springer-Verlag; 2017. p. 332–40.
[202] Dong C, Loy CC, He K, Tang X. Image Super-Resolution using deep convolutional networks. IEEE Trans Pattern Anal Mach Intell 2016;38(2):295–307.
[203] Zhang W, Basaran B, Meng Q, Baugh M, Stelter J, Lung P, Patel U, Bai W, Karampinos D, Kainz B. MoCoSR: respiratory motion correction and Super-Resolution for 3D abdominal MRI. In: Medical Image Computing and Computer Assisted Intervention – MICCAI 2023. Springer Nature Switzerland; 2023. p. 121–31.
[204] Qin C, Bai W, Schlemper J, Petersen SE, Piechnik SK, Neubauer S, Rueckert D. Joint motion estimation and segmentation from undersampled cardiac MR image. In: Knoll F, Maier AK, Rueckert D, editors. Machine Learning for Medical Image Reconstruction - First International Workshop, MLMIR 2018, Held in Conjunction with MICCAI 2018, Granada, Spain, September 16, 2018, Proceedings. Springer International Publishing; 2018. p. 55–63. https://doi.org/10.1007/978-3-030-00129-2_7.
[205] Fu Y, Lei Y, Wang T, Curran WJ, Liu T, Yang X. Deep learning in medical image registration: a review. Phys Med Biol 2020;65(20).
[206] L. V. Jospin, W. Buntine, F. Boussaid, H. Laga, M. Bennamoun, Hands-on bayesian neural networks – a tutorial for deep learning usersarXiv:2007.06823.
[207] Wang G, Li W, Aertsen M, Deprest J, Ourselin S, Vercauteren T. Aleatoric uncertainty estimation with test-time augmentation for medical image segmentation with convolutional neural networks. Neurocomputing 2019;338:34–45.

[208] Wolleb J, Sandkühler R, Bieder F, Valmaggia P, Cattin PC. Diffusion models for implicit image segmentation ensembles. In: Medical imaging with deep learning; 2021.
[209] Wu J, Fu R, Fang H, Zhang Y, Yang Y, Xiong H, Liu H, Xu Y. MedSegDiff: medical image segmentation with diffusion probabilistic model. In: Medical imaging with deep learning; 2023.
[210] Ouyang C, Wang S, Chen C, Li Z, Bai W, Kainz B, Rueckert D. Improved post-hoc probability calibration for Out-of-Domain MRI segmentation. In: Uncertainty for safe utilization of machine learning in medical imaging. Switzerland: Springer Nature; 2022. p. 59–69.
[211] S. Ioannou, H. Chockler, A. Hammers, A. P. King, others, A study of demographic bias in cnn-based brain mr segmentation, Proceedings/sponsored by IEEE Computer Society Technical Committee on Pattern Analysis and Machine Intelligence. Workshop on Mathematical Methods in Biomedical Image Analysis.
[212] T. Lee, E. Puyol-Antón, B. Ruijsink, M. Shi, others, A systematic study of race and sex bias in CNN-based cardiac MR segmentation, International Workshop on Databases and Expert Systems Applications: Proceedings.
[213] K. Lekadir, A. Feragen, A. J. Fofanah, A. F. Frangi, others, FUTURE-AI: international consensus guideline for trustworthy and deployable artificial intelligence in healthcare, [arXiv preprint arXiv].
[214] Acosta JN, Falcone GJ, Rajpurkar P, Topol EJ. Multimodal biomedical AI. Nat Med 2022;28(9):1773–84.
[215] C. Liu, S. Cheng, C. Chen, M. Qiao, W. Zhang, others, M-flag: medical vision-language pre-training with frozen language models and latent space geometry optimization, Conference on Medical ....
[216] C. Wu, X. Zhang, Y. Zhang, Y. Wang, W. Xie, Medklip: medical knowledge enhanced language-image pre-training, Proceedings of the IEEE/CVF International Conference on Computer Vision.
[217] Mangrulkar S, Gugger S, Debut L, Belkada Y, Paul S, Bossan B. Peft: state-of-the-art parameter-efficient fine-tuning methods. https://github.com/huggingface/peft; 2022.
[218] Zhang Y, Brady M, Smith S. Segmentation of brain mr images through a hidden markov random field model and the expectation-maximization algorithm. IEEE Trans Med Imag 2001;20(1):45–57.
[219] Ashburner J, Friston KJ. Unified segmentation. Neuroimage 2005;26(3):839–51.
[220] Avants BB, Tustison NJ, Wu J, Cook PA, Gee JC. An open source multivariate framework for n-tissue segmentation with evaluation on public data. Neuroinformatics 2011;9:381–400.
[221] de Bresser J, Portegies MP, Leemans A, Biessels GJ, Kappelle LJ, Viergever MA. A comparison of mr based segmentation methods for measuring brain atrophy progression. Neuroimage 2011;54(2):760–8.
[222] Schoemaker D, Buss C, Head K, Sandman CA, Davis EP, Chakravarty MM, Gauthier S, Pruessner JC. Hippocampus and amygdala volumes from magnetic resonance images in children: assessing accuracy of freesurfer and fsl against manual segmentation. Neuroimage 2016;129:1–14.
[223] Yushkevich PA, Piven J, Cody Hazlett H, Gimpel Smith R, Ho S, Gee JC, Gerig G. User-guided 3D active contour segmentation of anatomical structures: significantly improved efficiency and reliability. Neuroimage 2006;31(3): 1116–28.

[224] Maleike D, Nolden M, Meinzer H-P, Wolf I. Interactive segmentation framework of the medical imaging interaction toolkit. Comput Methods Progr Biomed 2009;96 (1):72–83.

[225] M. Jorge Cardoso, W. Li, R. Brown, N. Ma, E. Kerfoot, Y. Wang, B. Murrey, A. Myronenko, C. Zhao, D. Yang, V. Nath, Y. He, Z. Xu, A. Hatamizadeh, A. Myronenko, W. Zhu, Y. Liu, M. Zheng, Y. Tang, I. Yang, M. Zephyr, B. Hashemian, S. Alle, M. Z. Darestani, C. Budd, M. Modat, T. Vercauteren, G. Wang, Y. Li, Y. Hu, Y. Fu, B. Gorman, H. Johnson, B. Genereaux, B. S. Erdal, V. Gupta, A. Diaz-Pinto, A. Dourson, L. Maier-Hein, P. F. Jaeger, M. Baumgartner, J. Kalpathy-Crr, M. Flores, J. Kirby, L. A. D. Cooper, H. R. Roth, D. Xu, D. Bericat, R. Floca, S. Kevin Zhou, H. Shuaib, K. Farahani, K. H. Maier-Hein, S. Aylward, P. Dogra, S. Ourselin, A. Feng, MONAI: an open-source framework for deep learning in healthcarearXiv:2211.02701.

CHAPTER

13 Quality matters: Automated MR image quality control

Nicola Pezzotti[1, 2, *], Sergey Kastryulin[3] and Mariya Doneva[4]

[1]*Eindhoven University of Technology, Eindhoven, The Netherlands;* [2]*Philips Healthcare, Eindhoven, The Netherlands;* [3]*Yandex Research, Moscow, Russia;* [4]*Philips Innovative Technologies, Hamburg, Germany*

[*]*Corresponding author: E-mail: nicola.pezzotti@proton.me*

13.1 Introduction

In the preceding chapters, we've delved into the many advantages that Magnetic Resonance Imaging (MRI) gives to the realm of medical imaging. Yet, it's crucial to recognize that the foundation of any accurate diagnosis rests upon the fidelity of the images themselves. The notion of "quality" assumes a paramount significance in the context of MRI, a large number of image degradations may occur, from motion artifacts to excessive noise. However, its definition often remains elusive. What precisely constitutes image quality? Is it merely the attainment of clear, noise-free images, or does it extend beyond—encompassing the avoidance of motion artifacts, irregularities, and other distortions? The pursuit of quality in MRI, in the end, encompasses the synthesis of images that provide clear diagnostic value, enabling radiologists to provide the correct and appropriate diagnosis.

We face several challenges when discussing image quality. Primarily, the absence of a standardized approach presents a challenge, leaving us without a definitive approach to define what truly constitutes a superior image. This brings the question of **objectivity**. Ideally, we want to measure the intrinsic value of diagnostic insight. This value isn't quantifiable by conventional means, and its measurement is based upon the interplay of contextual factors. Consider, for instance, the nuanced impact of certain artifacts in specific scenarios. Seemingly inconspicuous distortions can amplify their impact—small ghosting artifacts induced by motion may be misidentified as additional lesions, and henceforth leading to a misdiagnosis and potentially harmful follow-ups. We find ourselves in a situation where metrics for quantifying objective measures of quality, such as signal-to-noise ratio or structural similarity, are at our disposal. However, as we will delve into within this chapter, these metrics do not perfectly align with what radiologists perceive as diagnostic value.

Machine Learning in MRI. https://doi.org/10.1016/B978-0-443-14109-6.00013-4

This brings us to a second point: if radiologists were to assess the quality of images, we could assign scores based on their perceived value. However, this approach is not scalable for large-scale image quality assessment. The **scalability** of quality assessment is a pivotal consideration in enabling innovative clinical applications. By developing an automated scoring system for MRI images, a range of intriguing applications becomes feasible. For instance, we could enhance automation during the image acquisition process. For example, one may ask: "Given a specific sequence, which parameters should be configured on the machine to achieve optimal quality?" In this scenario, various parameters would be tested, and our automatic scoring system would determine the best match. At present, this is a manual task, where optimization experts iteratively test different parameters until satisfactory image quality is attained—a labor-intensive process. Establishing a reliable method to identify what constitutes image quality becomes a pivotal enabler for a plethora of AI-powered techniques. As an example, the entire process described can be fully automated using the concept of image quality as a reward signal for reinforcement learning algorithms. While examples of such approaches have been presented in the literature, they often rely on simplistic metrics such as mean squared error. As we delve deeper into this chapter, we'll come to understand that the adoption of such approaches is hindered by the absence of an appropriate metric to comprehensively evaluate image quality.

The widespread evolution of artificial intelligence, especially in the realm of machine learning, holds significant promise for this field. Envision a future where automated assessment of image quality goes beyond solely relying on image properties, incorporating the patient's clinical history as well. This amalgamation of imaging data with electronic medical records is pivotal. The development of systems that centralize the acquisition of such data could potentially pave the way for the creation of vastly improved scoring methods for MRI. In an ideal scenario, these methods might eventually serve as training losses for enhancing various aspects, such as reconstruction methods. While much research has explored this avenue, the practical application of advanced metrics for training MRI image reconstruction methods remains limited. The majority of current efforts still rely on rudimentary loss functions. In our perspective, despite the formidable challenges faced over the past decades, advancements in this field hold the potential for substantial improvements in both the quality and accessibility of MRI.

The rest of the chapter is organized as follows. In the next section we will investigate the meaning of "image quality" in an MRI context. In Section 13.1 we will present traditional metrics, including those driven by a human interpretation, for image quality assessment (IQA). In Section 13.4 the state of the art in terms of metrics is presented, while Section 13.5 presents a study to compare the correlation of such metrics to a radiologist evaluator on MRI images affected by several types of corruptions. In Section 13.6 we will discuss reproducibility considerations, while in Section 13.7 we address the robustness and trustworthiness of AI-based methods, before concluding in Section 13.8.

13.2 What do we intend with image quality in MRI?

Image quality is often used as a criterion for evaluating new MRI techniques, including new data acquisition approaches, image reconstruction techniques, image enhancement, and post-processing. Many of these techniques aim at reducing the scan time and the question is if the images acquired at a shorter scan time still provide sufficient image quality. Image enhancement approaches aim at improving image quality, therefore a clear definition of image quality is needed for assessing these approaches.

In the context of medical imaging, image quality relates most often to the diagnostic quality of an image. Since images are almost exclusively analyzed by a visual inspection of a human observer (radiologist), it is widely accepted that a radiologist assessment is the ultimate test for diagnostic image quality. Radiological image quality is frequently assessed with the use of a Likert scale, where observers are assessing the image quality typically using a scale with five levels of perceived image quality. The levels can be qualitatively described for example as "poor," "fair," "good," "very good," and "excellent" or could also be represented numerically on a scale of 1–5. This type of analysis is highly subjective and does not give insights of what constitutes good diagnostic image quality.

The assessment of diagnostic image quality should determine whether the image meets the criteria for acceptable quality, enabling a clear visualization of the anatomy of interest, without any technical flaws that could hinder its diagnostic usage. In a sense diagnostic image quality is the threshold for acceptable image quality that determines if an image can be used diagnostically. It reflects properties of the MR images but also depends on the experience of the evaluator. Both years of experience in radiology and familiarity with the specific technique being evaluated may influence the rating. There are several image properties that contribute to the (perceived) image quality. These include mainly the following aspects:

- Spatial resolution,
- Image contrast,
- Signal-to-Noise Ratio (SNR),
- Presence of Artifacts.

Dedicated test objects called **phantoms** are typically used to characterize these image properties. Standard tests using phantoms are used for quality assurance, providing objective metrics that can detect potential issues leading to image quality degradation. In recent years, there has been a growing interest in developing numerical phantoms or digital reference objects to further refine the assessment of image quality [1].

The signal-to-noise ratio (SNR) is defined as the ratio between the signal level and the noise level. The signal level is the average signal of the pixels and the noise level is the standard deviation of the pixel values. In MRI, the SNR is routinely used as a quality indicator. The SNR is typically estimated from the image using two regions of interest. The first region of interest is used to estimate the signal level by computing the mean of the pixel values, the second region of interest is in the background and the standard deviation of this ROI is used to estimate the noise level. In Kaufman et al. [2] this approach is described, including a correction factor of 0.655 to account for a reduction in the standard deviation when taking the magnitude of complex Gaussian noise. This approach is still used today, although its validity is questionable. Nowadays, most MR scans are performed using a coil array, and in most scans parallel imaging is used for scan acceleration. This leads to spatial variations in the observed noise described by the so-called g-factor. In other words, the SNR is different at different image locations. In addition, the reconstruction may suppress the background, making the estimation of background noise impossible. As a result, the measurement of SNR from images reconstructed with parallel imaging is not straightforward. Reeder et al. [3] propose a method for measuring the SNR in parallel imaging using repeated acquisition of multiple images with identical scan parameters. This approach enables accurate SNR measurements, but cannot be used for characterizing the image quality of individual images. Non-uniform k-space sampling and/or non-linear reconstruction techniques such as compressed sensing and AI-based reconstruction may change the noise statistics, further complicating the objective SNR measurements. An alternative approach for characterizing the SNR is performing noise measurements at the receive coils and analyzing the noise propagation through the reconstruction. Similarly to the multiple measurements approach, this method for SNR measurements is appropriate for reconstruction algorithm development, but not appropriate for the image quality characterization in clinical practice.

The contrast-to-noise ratio (CNR) is the relationship between signal intensity differences between two regions and the noise level. A good contrast-to-noise ratio means that one can clearly differentiate between two tissues of interest. The challenges of accurate CNR measurements are related to the difficulties related to accurate noise estimation from the images described above.

The spatial resolution is related to the size of the smallest structures that can be visualized in the image as well as the accurate representation of tissue boundaries that is often described as sharpness. Often the final images are interpolated to higher resolution or smaller voxel size as a post-processing step; therefore, we can distinguish between acquisition resolution/voxel size and reconstruction resolution/voxel size. The interpolation to a higher resolution is usually done for visualization purposes, but this process usually does not add any detail or sharpness to the images. Higher (acquisition) spatial resolution requires larger k-space coverage, which typically translates into longer scan times and/or reduced SNR. The spatial resolution can be assessed by phantom sections with closely spaced holes, lines, or edges containing material that has a strong contrast to the background.

Another type of assessment that combines the effects of SNR and resolution is the so-called low contrast detectability. Here objects with low contrast to the background and varying sizes are used for the assessment, mimicking the challenging situation of detecting a small lesion with very weak contrast. This assessment is important for characterizing the performance of advanced reconstruction approaches and denoising methods, since improved SNR at the final image may come at the price of losing small low contrast features.

Due to the complexity of the image contrast in MRI as well as the long history of development of fast MRI techniques, practically all MRI images contain some form of an artifact or information that does not accurately reflect the physical reality. Examples of such artifacts include the chemical shift artifact, parallel imaging artifacts, geometric distortions in EPI, motion artifacts, and many more. The presence of artifacts does not necessarily affect the diagnostic quality of the images. This largely depends on the severity of the artifacts, their location, and the ability of the radiologist to see through the artifacts. This renders the automatic image quality assessment based on artifact detection a very difficult task. A particular challenge in introducing new reconstruction approaches is that they may introduce new image artifacts that may not be easily detected by a human observer. Specifically, there is a concern that generative AI image reconstruction and enhancement approaches may produce very realistically looking images that can lead to misdiagnosis. To minimize the risk of such artifacts, typically an evaluation including a state of the art reference is performed. Full reference image quality metrics can be advantageous here to detect structural changes in the images. However, if the reference image is obtained in a separate scan, interscan motion may complicate the evaluation.

To better understand the goal of image quality metrics, it is helpful to consider the clinical value of MR imaging. In their paper [4], Gazelle et al. present a framework for assessing the clinical value of diagnostic imaging containing six levels of efficacy:

- Technical efficacy
- Diagnostic accuracy efficacy
- Diagnostic thinking efficacy
- Therapeutic efficacy
- Patient outcome efficacy
- Societal efficacy

Usually, automatic image quality assessment is related to the technical efficacy of a method. A radiologist evaluation also includes diagnostic accuracy. Proving the impact of image quality improvements on all six levels can only be done in large clinical studies including aspects of healthcare economics analysis.

So far, we discussed image quality in the context of diagnostics based on visual inspection. Quantitative MRI provides quantitative measurements of tissue properties, leading to a different type of assessment of the imaging information. This requires considering additional aspects of image quality, including the accuracy and precision of the quantitative parameter maps. Analyzing the impact of the imaging sequence or image reconstruction on the quantitative values is easier to automate since it is less subjective. If the maps are used to derive further (semi-) quantitative information, this may also be taken into account. This further analysis may include multiple images, parameter maps, or the type of information and a test for diagnostic image quality would need to predict how modifications in the image quality would affect the subsequent (semi-) quantitative analysis. For example, if the image or parameter map is used for radiological tumor scoring (e.g., PI-RADS), the diagnostic image quality assessment should predict if the image would affect the scoring.

Beyond its role for diagnostic purposes, MRI can also serve as a valuable tool for therapy planning and image-guided therapy. In these application cases, different aspects of image quality are relevant. For example, in image-guided radiation therapy or image-guided biopsies, geometric accuracy is of very high importance and any form of geometric distortion is problematic. If the images are used for radiation therapy planning, accurate tissue segmentation and classification are essential.

13.3 Traditional methods to assess image quality

In this section, we will explore traditional methods to assess the quality of images. Historically, the first works on perceptual image quality assessment appeared almost half a century ago, with the pioneering work of Sakrison and Mannos [5] focusing on a class of visual fidelity criteria in the context of image encoding. Over the past few decades, a number of alternative models mimicking certain functionalities of the Human Visual System (HVS) were proposed. Such approaches could not model the real human visual system, which is a complex and highly nonlinear system. Most models rely on simplifications and strong assumptions (e.g., linearity or quasi-linearity for visual stimuli) and exhibit shortcomings regarding the definition of visual quality. Despite widespread criticism and the emergence of a large number of more advanced approaches, these methods are firmly entrenched in the scientific community and are widely used to this day.

In Section 13.3.1, the most widely used numerical metrics are introduced with their formulation and their best use and shortcomings are discussed. In Section 13.3.2, we will present several human-centered approaches for image quality assessment. We will focus on radiologist evaluation of images, showing different established approaches for assessing quality without using numerical metrics. More advanced metrics, with a focus on the interplay between the assessment of natural images and MRI images, will be discussed in Section 13.4.

13.3.1 General image quality metrics

Mean Average Error (MAE or ℓ_1-norm), Mean Squared Error (MSE or ℓ_2-norm), and closely related Peak Signal-to-Noise Ratio (PSNR) are the most frequently used quality metrics, which are a de-facto standard way of measuring image quality. MAE and MSE are easy to use, have clear physical meaning (energy of image distortions), satisfies the Parseval's theorem and can be used for algorithm optimization leading to a closed-form solutions.

$$\mathrm{MAE} = \frac{1}{n}\sum_{i=1}^{n} |x_i - y_i|_1 \tag{13.1}$$

$$\mathrm{MSE} = \frac{1}{n}\sum_{i=1}^{n} (x_i - y_i)^2 \tag{13.2}$$

where n is a total number of samples, x and y are images to be compared.

PSNR is defined as a ratio between the maximum possible power of a signal and the power of corrupting noise that affects the fidelity of signal representation. PSNR is most easily defined via the MSE as follows:

$$\text{PSNR} = 20 \cdot \log_{10}(\text{MAX}_I) - 10 \cdot \log_{10}(\text{MSE}) \tag{13.3}$$

where MAX_I is the maximum possible pixel value of the image.

MSE and PSNR have been repeatedly shown to poorly correlate with human judgments in controlled experiments. The main reason for this is four strong underlying assumptions about visual quality:

1. Independence of spatial relationships between samples;
2. Independence of relationship between signal and error;
3. Independence of sign of error samples;
4. Equal importance of all signal samples and errors.

None of this assumptions hold on in real life and accurately describe HVS.

Structural similarity index measure (SSIM) was proposed to release some of the listed assumptions. SSIM is one of the most popular top-down approach which has become a de-facto standard in the field of perceptual image processing (along with PSNR) and has inspired subsequent image quality assessment methods based on feature similarity.

The main idea behind SSIM is to split image distortions into structural and not-structural and focus on the first ones, as the latter are less noticeable by HVS.

$$\text{SSIM} = [l(x,y)]^{\alpha} \cdot [c(x,y)]^{\beta} [s(x,y)]^{\gamma} \tag{13.4}$$

where $l(x,y)$ is luminance (average pixel intensity), $c(x,y)$ is contrast (standard deviation of the local image regions) and $s(x,y)$ is structure (cross correlation values between two local image regions) components defined as follows:

$$l(x,y) = \frac{2\mu_x\mu_y + c_1}{\mu_x^2 + \mu_y^2 + c_1} \tag{13.5}$$

$$c(x,y) = \frac{2\sigma_x\sigma_y + c_2}{\sigma_x^2 + \sigma_y^2 + c_2} \tag{13.6}$$

$$s(x,y) = \frac{\sigma_x y + c_3}{\sigma_x\sigma_y + c_3} \tag{13.7}$$

which are later combined into a single quality map by averaging of local quality scores assuming equal importance of the measures $\alpha = \beta = \gamma = 1$ as follows:

$$\text{SSIM} = \frac{\left(2\mu_x\mu_y + c_1\right)\left(2\sigma_{xy} + c_2\right)}{\left(\mu_x^2 + \mu_y^2 + c_1\right)\left(\sigma_x^2 + \sigma_y^2 + c_2\right)} \tag{13.8}$$

where μ_x and μ_y are mean values for images to be compared, σ_x and σ_y are standard deviation values and σ_{xy} is their covariation matrix. c_1 and c_2 are correction coefficients that are required for numerical stability in case of division with a weak denominator. Exact values of c_1 and c_2 depend on dynamic range of the pixel-values and by default are equal to 0.01 and 0.03, respectively.

The main disadvantage of SSIM is that it takes into account only single image scale, and thus can't adapt to different sets of viewing conditions.

Multi-Scale Structural Similarity Index Measure (MS-SSIM) [6] measures SSIM on five different scales, computing contrast and similarity on all levels, while measuring luminance only at the final scale. Resulting scores than combined through a weighted product using weights adjusted on human dataset of mean opinion scores.

$$\text{MS-SSIM} = [l_M(x,y)]^{\alpha_M} \cdot \prod_{j=1}^{M} \left[c_j(x,y)\right]^{\beta_j} \left[s_j(x,y)\right]^{\gamma_j} \tag{13.9}$$

where α_M, β_M and γ_M are used to adjust relative importance of visual components defined above for each of selected scales M.

13.3.2 Radiologist evaluation

At present, the only accepted approach for validation of a new imaging method is to perform a clinical study, in which the image quality of the method is assessed in a clinical setting. This includes scanning a large enough patient population to cover the scope of application of the method and estimate its reliability, and subjective image quality assessment by one or more radiologists. Radiologist evaluation can be based on the comparison between two images or the assessment of a single image. In both cases, the assessment is typically done considering multiple criteria. These may include overall image quality, diagnostic confidence, perceived SNR, perceived image sharpness, artifacts (usually multiple specific artifact types are considered), and visualization of specific anatomical structures or pathology. Blind comparison (without annotation of the method being used) is done to avoid bias in the assessment. This approach is expensive and time consuming and may need to be repeated multiple times during the development of a new imaging method.

The necessity to design mathematical models that are capable of predicting the quality evaluation of an average human observer have resulted in numerous objective IQA methods. These mathematical models are designed to be able to predict the quality of an image accurately and automatically. An ideal objective image quality assessment method or metric should be able to mimic the quality predictions of an average human observer.

The field of image quality assessment has mainly developed through the various applications in the domain of natural images. Earlier works aimed to characterize color displays and half-toning optimization methods and assess the quality of image compression algorithms [7]. Later works extended the field toward smartphone photography, scanned documents, and remote sensing [8,9].

Given the majority of metrics were not designed for the healthcare domain, it is not always clear whether they can be re-used for the evaluation of medical images in general and MRI images in particular. Recent paper [10] performed an extensive evaluation of 30+ image quality metrics, showing that some general methods can assess particular characteristics of MRI images while most of them cannot.

The development of MRI-specific quality metrics has yielded several specialized methodologies, each achieving notable recognition within their respective domains. Notable contributions include automated degradation detection frameworks [11,12], and granular quality assessment protocols for MRI images [13]. Nevertheless, advancement in this field remains constrained by both the substantial resources required for human expert annotations and the inherent heterogeneity of imaging data and use cases.

13.4 Advanced image quality metrics

In this section, we present a list of advanced metrics, categorized into three main classes: full-reference, no-reference, and distribution-based.

Full-reference metrics necessitate the presence of a reference image or ground truth for comparison. These metrics analyze the differences between the image at hand and the reference image, providing a numerical measure of their similarity. In general, full-reference metrics are the most reliable in terms of numerical stability and representation power. However, they require a reference image, which can be challenging to obtain in many cases. For example, when assessing the performance of a denoising algorithm on an MRI image, a noise-free reference image is essential. Acquiring such noise-free images can be a non-trivial task in this context unless phantoms are employed. Multiple signal averaging can yield the reference image, but this approach increases the risk of motion in the scanner, potentially introducing artifacts and misalignments, thus affecting the reference image's quality.

To address the limitations of full-reference metrics, **no-reference** metrics can be employed. These metrics attempt to assess image quality without the need for a reference image. For example, they may compute localized noise levels to derive a global quality assessment. **Distribution-based** metrics go even further by evaluating image quality based on how well an image fits into the distribution of all high-quality images that can be obtained from an MRI scanner.

In general, full-reference metrics are more established compared to no-reference and distribution-based metrics. This is evident in the greater number of full-reference metrics available in the literature and their wider adoption as the gold standard for algorithm testing.

13.4.1 Full-reference

MSE [14], the Mean Squared Error (ℓ_2-norm), and its closely related counterpart Peak Signal-to-Noise Ratio (**PSNR**) [15], are the most frequently used quality metrics, which are a de-facto standard way of measuring image quality. MSE is easy to use, has clear physical meaning, capturing the energy of image distortions, satisfies the Parseval's theorem and can be used for algorithm optimization leading to a closed-form solutions [15,16]. PSNR is defined as a ratio between the maximum possible power of a signal and the power of corrupting noise that affects the fidelity of signal representation [15].

As discussed in Section 3.1, MSE and PSNR have limitations that a number of subsequent papers addressed the weaknesses of PSNR and modified it to better suit for image quality assessment. **WSNR** [17], the Weighted Signal-to-Noise Ratio, uses a contrast sensitivity function to approximate the human visual system and assign different weights to signal and noise components, leading to a linear quality measure. **NQM** [17], the noise quality measure, also uses non-linear quasi-local processing model of the human visual system to deliver an assessment that is closer to human perception. **IW-PSNR** [18], the Information Weighted PSNR, builds on the idea of applying additional "weights" to address the importance of different areas in the image. More specifically, it uses information theoretic principles to add weights for regions of visual content, that are perceptually more important than others, either due to the visual attention property of the human visual system or due to the influence of distortions.

Going beyond extensions and adaptations of PSNR derived metrics, **MAD** [19] or the most apparent distortion metric, explicitly models adaptive strategies that can be observed in the human visual system. For high-quality images with only near-threshold distortions a detection-based strategy is employed, and an appearance-based strategy is activated if the distortions are clearly visible. The results are then combined into a single score by weighting scheme, where importance of each strategy is dependent on the distortions strength.

Following the success in image classification and recent developments in deep learning, the usage of convolutional neural networks (CNN) became a popular way for constructing image quality assessment methods. A double-path CNN was introduced by the pioneering work of Liang et al. [20]. In this work, the authors proposed to crop images into small patches and propagate them through dedicated network branches to obtain a patch score that should capture a notion of quality. The scores determined the final overall image quality score by averaging the predicted patch-wise values. Of course, the quality of the patch should derive from some notion of human derived image quality. To this end, the CNN is trained on a dataset, more specifically the TID2013 dataset [21], to regress quality scores that were annotated by human readers.

Another popular approach, proven to be useful for many image processing tasks, is the usage of generic features obtained from pre-trained deep-neural networks, such as AlexNet [22], VGG [23], SqueezeNet [24], and others. **Perceptual Loss** and **Style Score** were introduced by Johnson et al. [25]: it uses a single-path CNN trained on ImageNet ILSVRC 2012 dataset [22] to extract deep feature representations of reference and distorted images. Those features were then compared by taking **MSE** and averaging the error between spatial and feature dimensions. Such an approach is based on the fact that, deeper in layers of image classification networks, feature representations captures high-level semantic concepts [26] and a numerical comparison on such features should be able to capture distortions at this level.

In **LPIPS** [27], the Learned Perceptual Image Patch Similarity, Zhang et al. pointed out that methods based on utilization of deep features significantly outperform traditional algorithms on a wide range of distortions. LPIPS computes the distance between feature representations on multiple levels, similar to the Perceptual Score, building on the premise that different layers represent different structures of the image. To properly combine features, they are first unit-normalized to have similar scales and then summed with weights learned on the BAPPS [27] dataset to minimize error between model and human preference over two images. **PieAPP** [28], the Perceptual Image-Error Assessment through Pairwise Preference metric, uses a pairwise learning framework to predict the preference of one distorted image over the other. **PieAPP** gets features from the feature-extraction (FE) network and then computes the similarity using a score-computation (SC) network. Both FE and SC are trained from scratch targeting humans pairwise preference between images. **SWDN** [29], the Space Warping Difference Network, is specifically designed to handle geometric distortions. When comparing two image features, a new Space Warping Difference layer takes into account not only pixels on corresponding positions as in all pixel-wise methods, but also looks into a small range around them.

The successes of using deep-learning networks to compute image quality metrics gave rise to a number of other full-reference algorithms based on neural networks and relying on different deep features, pooling strategies, and pre-trained CNNs [29–31].

Metrics that use certain properties of human visual system can be divided into two distinct groups: **top-down** and **bottom-up**. Previously, we described bottom-up approaches to metrics design, which used properties of the human visual system to modify error-based MSE and PSNR to better reflect human perception. As an example, by adding adaptations to luminance, contrast sensitivity and contrast masking, more truthful metrics are derived.

On the contrary, approaches following top-down metrics design try to mimic human perception as a whole and do not model it by individual components, such as contrast sensitivity or other well-defined properties. Structural similarity is a perception-based model that considers image degradation as a perceived change in structural information, while also including luminance and contrast masking terms. These methods are based on the idea that pixels have strong inter-dependencies especially when they are spatially close.

SSIM [15], the structural similarity index measure, is the most popular top-down approach which has become a de-facto standard in the field of perceptual image processing (along with PSNR) and has inspired subsequent metrics based on feature similarity [16]. The main idea behind SSIM is to compute a number of sub-metrics that focusing on capturing structure perception in the human brain. More specifically, the comparison is done between luminance (average pixel intensity), contrast (standard deviation of the local image regions), and structure (cross-correlation values between two local image regions) components. This information is later combined into a single quality map by averaging the local quality scores, hence obtaining a global value for the image. The main disadvantage of the SSIM metric is that it takes into account only a single image scale, and thus can't adapt to different sets of viewing conditions.

MS-SSIM [6], the Multi-scale SSIM, measures SSIM on five different scales, computing contrast and similarity on all levels, while measuring luminance only at the final scale. The resulting scores at each scale are then combined through a weighted product, where the weights are obtained by a human perception study where the participants were asked to grade the effect of distortions at different scales. Although MS-SSIM is the most widely used extension, but many other exists which build on different methods of capturing contrast, structural and texture information.

IW-SSIM [18], the Information content Weighted SSIM, adds computation based on information theoretic principles. **ESSIM** [32], the Edge-based SSIM, uses first-order difference operators to compare information between image edges, as the authors observe that edges are the most important structure information for human perception. **GMSD** [33], the gradient magnitude similarity deviation, is focused on the computational efficiency of quality predictions. Based on the idea that global variation of image local quality degradation can reflect its overall quality, the authors proposed to compute the standard deviation of the pixel-wise gradient similarity map as an IQA index. This method is, however, problematic because an image with a large but constant local distortion yields a standard deviation of zero, indicating the best-predicted quality. **MS-GMSD** [34], the multi-scale GMSD, accounts for the variations in viewing conditions by measuring GMSD on five different scales, similar to what is done in MS-SSIM. The chromatic version, named MS-GMSDc, additionally measures chrominance dissimilarity on the last scale, motivated by the fact that humans have lower sensitivity to color distortions. **FSIM** [35], the feature similarity index measure, is built on the assumption that low-level features captured by the human visual system are used for understanding the image content. Two core features used in similarity computations are phase congruency, which is a contrast-invariant dimensionless measure of the local structure, and gradient magnitude maps obtained by the Scharr filter. During pooling, the phase congruency component serves as an adaptive local weighting factor to derive an overall visual quality score. The color-sensitive version of **FSIM**, is called **FSIMc** [35]. The method first converts RGB images into YIQ color space and then computes FSIM on luminance channel and chrominance similarity maps on I and Q channels.

VSI [36], the visual saliency induced quality index, states that the change of salience in degraded areas is the major predictor of image quality. The visual saliency index is computed using phase congruency and two simple priors (color temperature and center priors). Gradient magnitude maps are used as an additional feature and the pooling strategy mimics the FSIM measure. **VSI** shows a good correlation with human judgments on localized distortions, such as compression artifacts or local patch substitutions. The recently proposed **HaarPSI** [37], the Haar perceptual similarity index, decomposes both distorted and reference images into Haar wavelets and computes the structural similarity between magnitudes of high-frequency coefficients. The last level of the Haar wavelet is then used to weigh the importance of different regions. **MDSI** [38], mean deviation similarity index, combines the gradient similarity, chrominance similarity and deviation pooling into a single metric. MDSI contains seven configurable parameters that are jointly optimized on a mean opinion scores dataset to provide the best correlation results.

DISTS [31], the deep image structure and texture similarity measure, is a deep-learning-based algorithm that follows the LPIPS metric design principles. Image representations are extracted from the pre-trained convolutional neural network and combined using SSIM-like structure and texture similarity measurements. It is sensitive to structural distortions but at the same time robust to texture resampling and modest geometric transformations [39].

DSS, the DCT sub-bands similarity, aims to measure changes in structural information using sub-bands in the discrete cosine transform (DCT) domain. DSS extracts features in block-based DCT sub-bands and measures the distances between corresponding DCT sub-bands. The final quality index is computed by pooling those distances together with weights. The main motivation behind quality assessment in the DCT domain is the observation that the statistics of DCT coefficients change with the degree and type of image distortion.

Finally, the visual information fidelity (**VIF**) [40] uses a Gaussian scale mixture to statistically model the wavelet coefficients of a steerable pyramid decomposition of an image [41]. After that, it predicts the distorted image quality by quantifying the amount of preserved information from the reference image. Its spatial domain, named VIFp, computes fidelity on raw pixels.

As can be seen in this Section, a wide array of metrics has been introduced over the year in the full-reference group. In Section 5, we will provide an assessment of such metrics on a series of MRI use cases, and evaluate the different performances with respect to natural images, which was the focus of the research behind such metrics.

13.4.2 No reference assessment

The second group of metrics that we are going to discuss does not require a reference image which is considered the golden standard to compare against. To this end, no-reference metrics are developed to capture some intrinsic property of an image, rather than a notion of perceptual similarity compared to a second image. The properties of the images that the metrics capture may vary, but they tend to be more specialized than the metrics introduced in the previous section, hence the lower number of metrics present in the literature.

The Blind/Referenceless Image Spatial Quality Evaluator (**BRISQUE**) [42] operates on locally normalized luminance values in the spatial domain, called Mean Subtracted Contrast Normalized (MSCN) coefficients. Various features are extracted from the MSCN coefficients and their pairwise products, which are then used to estimate Generalized Gaussian Distribution (GGD) and Asymmetric GGD parameters. The SVR model is then used to learn to map the features to the quality score.

A different metric, **BIQI** (Blind Image Quality Index) [43], employs a two-stage process: feature extraction and quality prediction. In the feature extraction stage, BIQI derives features from natural images by modeling the statistical regularities observed in natural scenes. These features are then used to train a support vector machine (SVM) that acts as a quality predictor. The SVM generates a final quality score, reflecting the perceived quality of the input image.

BLIINDS-II (Blind Image Integrity Notator using DCT Statistics II) [44] is another no-reference IQA metric, focusing on assessing image integrity by modeling the statistical properties of discrete cosine transform (DCT) coefficients from natural images. BLIINDS-II first divides the input image into non-overlapping blocks and then computes the DCT of each block. Subsequently, the algorithm estimates relevant statistical features from the DCT coefficients, including mean, variance, and higher-order moments. These features are used as input for a trained regression model, which predicts the perceived image quality score.

NIQE (Naturalness Image Quality Evaluator) [45] metric based on the construction of a natural scene statistics model. NIQE quantifies the deviation of an input image's statistical properties from those of natural images. The algorithm focuses on the spatial domain, utilizing local mean subtracted contrast normalized (MSCN) coefficients to extract statistical features. By computing the mean and covariance of these coefficients, NIQE creates a feature vector representing the input image's properties. Finally, the distance between the feature vector of the input image and those of pristine natural images is calculated, resulting in an overall image quality score.

CLIP-IQA [46] represents a new family of so-called non-parametric metrics, gaining popularity due to their great generalization ability. Instead of training a new model on a relatively small and often biased datasets with mean-opinion-scores, the metric uses a holistic image understanding learned of a foundation model (CLIP [47]) learned by seeing hundreds of millions of image-text pairs. During this time-consuming and computationally demanding process, CLIP gains knowledge of various image properties, including their quality. CLIP is trained with the contrastive learning objective, mapping similar texts and images into similar feature vectors and dividing the opposite ones further apart. This property is utilized during inference by providing a pair of prompt antagonists (e.g., "Good image" and "Bad image") and checking the relative distance to each of them. It turns out that this simple prompting strategy results in relatively good and remarkably robust results across a series of benchmarks for quality assessment of natural images.

It can be seen that, with the exception of BRISQUE and CLIP-IQA, the other metrics present here are based on several assumptions on which features belong to a high-quality natural image. While this is not per se an issue, in introduces prior assumptions on what the characteristics that images must have in order to be considered of good quality. This can be an issue, as we will see in Section 13.5, as such characteristics do not directly correlate with predictors of image quality in MRI. This fact can also explain why such metrics remain relatively a niche in the MRI literature.

13.4.3 Distribution based

The final category we consider is distribution-based metrics. The idea for such an approach is to capture all the ideal high-quality images and consider them to form a probability distribution. Intuitively, we expect images, and MRI images in particular, to lay in close proximity of such distribution where the probability of an image is a direct representation of its quality. Moving away from such a distribution, meaning if an image lays ad the edges of it, such image will be considered of low quality. Overall, such metrics require the construction of the notion of distribution, which is hard given the intrinsic discrete nature of images, and the difficulty of collecting a large number of high-quality images describing the domain in its entirety. This is particularly challenging in the MRI context, where images are expensive to acquire compared to natural images, and even more difficult to share due to privacy and legal reasons.

Several metrics have been introduced in the computer vision literature. Freshet Inception Distance (**FID**) [48] considers embeddings of the two data distributions as a product of continuous multivariate Gaussian. The mean and covariance are estimated for both, and the Fréchet distance between these two Gaussians (a.k.a Wasserstein-2 distance) is then used to quantify the quality of the generated sample.

Kernel Inception Distance (**KID**) [49], sometimes referred to as Maximum Mean Discrepancy (MMD) - computes the dissimilarity between two probability distributions P and Q by measuring squared MMD for some fixed characteristic kernel function (e.g., Gaussian kernel).

Multi-Scale Intrinsic Distance (**MSID**) [50] develops an intrinsic and multi-scale method for characterizing and comparing data manifolds, using a lower-bound of the spectral Gromov-Wasserstein inter-manifold distance, which compares all data moments.

Geometry Score (**GS**) [51] constructs a performance measure by comparing the geometrical properties of the underlying data manifolds. In particular, topological approximation for computation of connected components ("holoes") in homology using witness complex [52] is introduced.

Inception score (**IS**) [53] uses the pre-trained InceptionNet model for feature extraction and captures properties of the generated sample to capture properties of generated samples: diversity with respect to class labels and high classifiability. An evolution, the Improved Inception score (**IS'**) [54] proposed a different way of feature aggregation, which improved both the calculation and interpretability of the Inception Score.

Due to the issues mentioned above, this category of metrics finds a hard application in the context of MRI image quality assessment, as it will also be reflected in the experiments presented in the next section.

13.5 Application of image-quality metrics in the context of MRI

Now that we have introduced a number of metrics to measure image quality, it is worth asking: "How do they perform in the context of magnetic resonance imaging?" This is not a trivial question to ask. Most of the metrics have been developed in the context of natural images by the computer vision community. It is safe to assume that some of such metrics may correlate with the notion of quality introduced in Section 13.2. However, it is unclear to which degree such correlation exists and which metrics better serve the purpose.

In the next sections, we will investigate such correlation. First, in Section 13.5.1, we introduce the fundamental differences between natural images and MR images. Then, in Section 13.5.2, we present our experimental framework, while in Section 13.5.3, we present the results and discuss the implications toward automated image quality assessment.

13.5.1 Fundamental difference between natural images and MR images

There are several fundamental differences between MRI images and natural images encountered in everyday life. These differences are primarily due to the context and purpose of these images, as well as the imaging techniques used.

The first and most obvious distinction arises from the difference in the acquisition process. MRI images are acquired using specialized imaging techniques designed to capture specific anatomical or physiological information. They employ complex equipment and imaging protocols, involving radio frequency signals, strong magnetic fields, and optionally, contrast agents. On the other hand, natural images are typically captured using cameras or mobile devices, which record visible light in everyday environments.

Second, these two imaging modalities differ in terms of information content. Medical images emphasize capturing detailed internal anatomical or physiological structures. They aim to visualize specific tissues, organs, or abnormalities, providing information for diagnosis and treatment. In contrast, natural images encompass a wide range of content, including landscapes, humans, animals, objects, and events, with a focus on visual aesthetics rather than specific biomedical information.

Noise and artifacts are another factor that makes the considered modalities significantly different. Medical images often suffer from inherent noise and artifacts due to the imaging process and limitations of the equipment. Various factors, such as patient motion, imperfections in image reconstruction algorithms, and distinct acquisition techniques, can introduce rather apparent noise and artifacts.

It would be erroneous to overlook the factor of image resolution: MRI imaging typically yields lower-resolution images, even though capturing small details is essential for accurate diagnosis. In contrast, modern optical cameras used to acquire natural images can exhibit a wide range of resolutions, depending on the capturing device, settings, and intended application. The widespread adoption of high-quality personal cameras in mobile devices has made 2K and even 4K imaging commonplace.

Another difference worth mentioning is the complex nature of MRI images. Although mainly the magnitude of the complex images is used for diagnosis, the image phase contains important information that can be used for secondary data analysis. Furthermore, computing IQA metrics based on the complex data may be advantageous since a linear reconstruction leads to zero mean Gaussian noise in the real and imaginary parts of the images.

From the image processing perspective, MRI scans are usually represented as stacks of gray scale images. Various visual properties and contrast adjustments are applied to highlight different tissue types, pathologies, or functional information. In contrast, natural images display a broader spectrum of visual appearances and color codings. They are influenced by factors such as lighting conditions, composition, and artistic choices, which are not applicable to medical data.

13.5.2 Evaluation of image quality metrics: an empirical study

We now present an empirical study investigating the correlation between image quality metrics and the perceived image quality by a group of radiologists [10].

This study specifically focuses on evaluating these metrics in the context of image processing algorithms and systems. For instance, when multiple image denoising and restoration algorithms are available to recover images corrupted by blur and noise, a perceptual objective metric should ideally identify the image that achieves the highest perceptual image quality post-restoration.

To reliably accomplish this task, the metrics must demonstrate a strong correlation with the perceptual quality estimates reported by human subjects for a given image processing algorithm.

The majority of modern benchmarks for full-reference metrics, such as the well-known TID2013 dataset [21] and KADID-10k dataset [55] for natural images, are derived from a limited set of distortion-free, real-world images. These datasets are created by applying a predefined set of augmentations, such as additive noise, artificial blurring, and color shift. However, the disparity between synthetic evaluations and real-world applications may lead to potentially misleading results: as an example, all augmentations based on color lose sense in an MRI context.

Furthermore, metrics trained on natural images aim to characterize the overall perception of the quality of an entire scene. In contrast, an MRI scan can be deemed high-quality when specific characteristics, crucial for the scan's diagnostic value, are considered adequate. These characteristics, identified as important components of radiographic image quality [56], encompass the perceived level of noise (signal-to-noise ratio, SNR), perceived soft tissue contrast (contrast-to-noise ratio, CNR), and the identification of artifacts, as discussed in Section 13.2.

Regrettably, none of the previous studies on metrics have taken these factors into consideration. Moreover, these specific quality criteria are interconnected. For instance, certain denoising algorithms may introduce additional blurring (resulting in the lower CNR) in exchange for increased SNR, and some motion correction approaches may introduce noticeable artifacts. Therefore, a more comprehensive evaluation of a metric's ability to express distinct MRI quality criteria is imperative.

Here, we adopt the paradigm of the **model-observer framework** [57]. In this framework, the quality of a medical image is defined by how well a clinical task, such as diagnostics, can be performed on it [58]. Ideally, the perfect MRI metric would be a task-based score, such as diagnostic accuracy, when the image is evaluated by radiologists. However, implementing such a metric is challenging due to the considerable diversity of diagnostic outcomes that radiologists encounter in practice. As a result, the convention is to utilize a **subjective estimation** of the overall diagnostic value instead [59].

Seven trained radiologists, each with 7–20 years of experience, participated in this study. The participants were tasked with scoring pairs of reconstructed-reference images based on three primary image quality criteria. The images were obtained by using an AI based reconstruction [60] on input data that was subject to three different kind of corruptions: 1) kspace undersampling, 2) simulated motion artifact, and 3) simulated noise. For each image pair and each criterion, radiologists assessed the perceived diagnostic quality of the reconstructed image in comparison to the ground truth. They used a four-point scale: not acceptable (1), weakly acceptable (2), rather acceptable (3), and fully acceptable (4). This four-point scale was chosen over the five-point Likert scale employed in a previous study [59] to reduce noise in data through lower ambiguity for expert labelers.

Here, we adapt the analysis of the scoring data proposed in Ref. [61] to the multiple IQ criteria. The voting scores for each scoring criteria are not analyzed in their raw format. Instead, they are converted to z-scores (averaged and re-scaled from 0 to 100 for each radiologist to account for their different scoring):

$$z_{nmk} = (D_{nmk} - \mu_{mk})/\sigma_{mk}, \tag{13.10}$$

where μ_{mk} and σ_{mk} are the mean and the standard deviation of the difference scores of the mth radiologist on the kth scoring criteria, and D_{nmk} are the difference scores for n^{th} degraded image defined as follows:

$$D_{nmk} = s_{mk,\text{ref}} - s_{nmk}. \tag{13.11}$$

In Eq. (13.11), $s_{mk,\text{ref}}$ is the raw score of the mth radiologist on the kth scoring criteria for the reference image corresponding to the nth degraded image, and s_{nmk} is the raw score of the mth radiologist on the nth degraded image on the kth scoring criteria. Note that in this study, the radiologists were asked to perform pair-wise comparison between degraded and reference images. Hence, it is possible to treat the raw labeling scores as the difference scores D_{nmk}.

After standardizing the expert votes by Eq. (13.10), their correlation statistics with each IQM were computed in the form of SRCC and KRCC coefficients, defined as follows:

$$\text{SRCC} = 1 - \frac{6\sum_{i=1}^{n} d_i^2}{n(n^2 - 1)}, \tag{13.12}$$

where d_i is the difference between the i-th image's ranks in the objective and the subjective ratings and n is the number of observations.

13.5.3 Evaluation results

The outcome of the study is shown in Fig. 13.1, capturing the correlation between radiologist evaluations based on three identified criteria for the MRI images, and for Natural images on the TID2013, KDADID-10k datasets. The four top-performing metrics are highlighted in bold.

From the figure, it is clear that the correlation observed in MRI images is, on average, lower than the notion of global quality on natural images captured in the two datasets. This suggests a gap in existing metrics concerning image quality assessment on MRI images. However, it is noteworthy that the best-performing metric in our study, DISTS, exhibits a correlation exceeding 0.7 across all three tasks, hinting at potential applications were the highest performance is not required.

The right side of the figure shows the metrics' ranking based on their performance. Particularly interesting is the observation that DISTS, while performing the best on MRI images, does not rank among the top four performers for natural images. This fact may indicate that there may be metrics that better capture the performance on MRI datasets but are less performant in a general case. Conversely, as the MRI community will look into computer vision literature for advancement in the field, it is a word of caution that "hidden gems," that is, well performing approaches for the MRI community, may not reside among the best performing approaches on generic images.

Another noteworthy observation is that, with the exception of HaarPSI, the best-performing metrics predominantly rely on deep-learning approaches employing pre-trained models for secondary tasks. The implications of this result in terms of robustness and trustworthiness will be discussed in Section 13.7. A more in-depth analysis of the results can be find in Kastryulin et al. [10].

13.6 Implementations and reproducibility consideration

The assessment of image quality, especially in the context of MRI images, holds significant importance when comparing various image formation and processing methods. For instance, when evaluating two distinct reconstruction algorithms, multiple metrics are commonly employed to gauge the advantages of one method over the other. Hence, it is essential to ensure the correct implementation of these metrics to enable a fair and meaningful comparison.

While it is relatively straightforward to implement basic metrics, as discussed in Section 13.1, more advanced metrics, as presented in Section 13.4, pose a challenge for proper implementation. Advanced metrics often rely on complex computations or necessitate the use of several parameters that influence their effectiveness. For

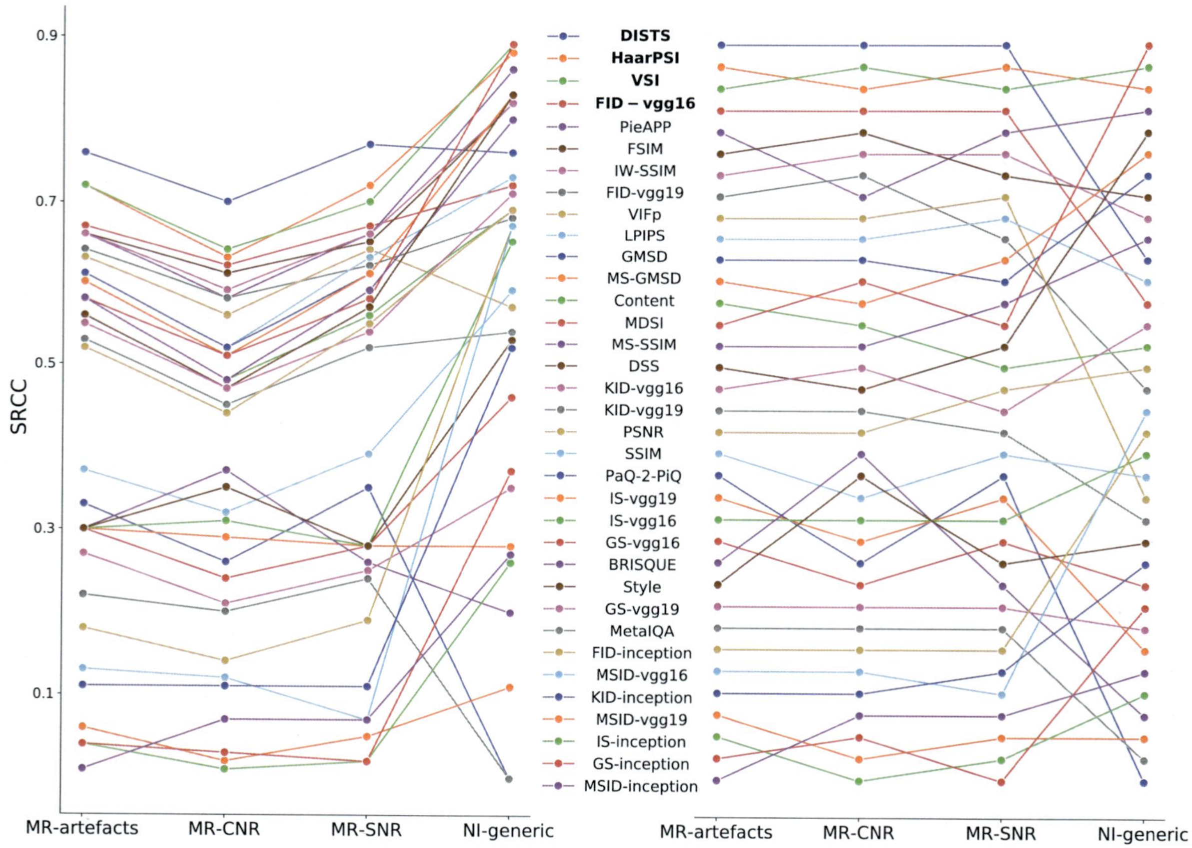

FIGURE 13.1

Performance of IQMs on different MRI tasks and on Natural Images (NIs), compared by their correlation with the expert votes (SRCC values, left) and sorted top-to-bottom by their rank (right). The ordering reflects the performance on the MRI data only. The same color-coding is used in both plots. *NI-generic* scores are the average between TID2013 [21] and KADID-10k [55] datasets. Note higher correlation of IQMs on NI and poor translation of ranking to MRI domain.

instance, the MS-SSIM metric involves creating an image pyramid, with different weightings assigned at each resolution based on a human study. Failure to use the appropriate weights can cast doubt on the reliability of the metrics and, consequently, the validity of the algorithm comparison.

While many libraries offer convenient access to image quality metrics through user-friendly packages, it's important to acknowledge their inherent limitations. Notably, the Blind Image Quality Toolbox and the Image Quality Assessment Toolbox stand out as verified MATLAB implementations of both no-reference (NR) and full-reference (FR) metrics. However, the evolving landscape of machine learning and scientific computing, with Python's increasing popularity, has led to these solutions becoming less relevant for a significant portion of research activities. Many image and vision computing experts now turn to Python due to its robust open-source repositories.

Fortunately, several Python libraries have emerged to bridge this gap, providing interfaces for the metrics discussed in this chapter. Among them, Kornia, PIQA, and IQA-PyTorch offer Python implementations of these metrics. Additionally, the IQA-optimization library provides these metrics in a format suitable for optimizing machine-learning models.

However, it's worth noting that these Python libraries may have a relatively limited selection of algorithms available, which restricts their utility as comprehensive solutions for all potential image quality metrics applicable in an MRI context.

Recently, Kastryulin et al. introduced the PyTorch Image Quality library [62], also known as PIQ, which provides majority of the metrics presented in this chapter in a well-tested and standardized format. The repository has gained prominence as one of the most widely recognized resources in this domain, adhering to straightforward guiding principles. First and foremost, PIQ focuses on user-friendliness, offering standardized interfaces that make it easy to utilize the metrics. Second, it prioritizes reliability by subjecting all metrics to extensive testing by both developers and the broader community. Finally, the library is developed with practical considerations in mind, striking a balance between computational efficiency and the applicability of the metrics.

Irrespective of one's choice of metrics and libraries, we strongly recommend that the research community move toward embracing well-tested and standardized metric implementations to enhance the reliability and reproducibility of their work.

13.7 AI-powered quality assessment: Robustness and trustworthiness

In this chapter, we have scrutinized several metrics that form the foundation for any automated quality control system. In addition, empirical studies have underscored how certain metrics, grounded in deep-learning models, closely align with the perceptual judgments made by radiologists when analyzing images.

This naturally prompts the question of the robustness of such metrics and their feasibility for large-scale deployment. Furthermore, given the era in which Large Language Models (LLM) have made a profound impact, showcasing remarkable capabilities at scale, it begs the question: is it still needed to retain a radiologist in the loop? Could we instead integrate the metrics presented in this chapter with a reasoning agent capable of directly generating a comprehensive report?

Addressing this question requires a comprehensive understanding of the scope of the application and the extent of automation that AI models can deliver. The AI-based metrics outlined in the preceding section can be broadly categorized into two groups when considering their robustness and trustworthiness:

1. Models trained to classify or regress a notion or level of quality.
2. Models initially trained for secondary tasks but subsequently repurposed as perception engines for quality assessment.

For the first category, the same considerations applies to training deep-learning models. Specifically, under specification [63] is a recognized issue that requires careful attention in the design and validation of models. An even more impactful concern is the well-known phenomenon of "garbage-in-garbage-out," signifying that a model is only as effective as the data used for its training. Given that image quality, as we discussed, is a loosely defined and often subjective concept, the creation of a well-curated and extensive dataset poses a significant challenge. It is unsurprising, considering these challenges, that metrics of this type do not emerge as top performers, as shown in the study presented in Section 13.5.3.

The second category presents a markedly different scenario. In this case, the model is initially trained for a secondary task, often starting with image classification using natural images. Our presented study indicates that such approaches consistently outperform others when compared to the perceived quality by a radiologist. What explains the efficacy of these methods, and how reliable are they?

Intuitively, we can attribute their performance to the perceptual qualities inherent in such methods. It is well-documented in the literature that deep-learning models trained for image classification, particularly in deeper layers, learn representations of complex objects using a notion of compositionality [26,64]. Studies have demonstrated that this behavior closely mirrors the functioning of the human visual system [65], providing an intuitive rationale for the effectiveness of these metrics.

However, in terms of robustness, it remains unclear whether models trained on natural images for specific tasks encompass complete feature representations in the deeper layers of the network. Further research is necessary to ascertain to what extent such representations translate to the medical imaging domain and, specifically, to MRI.

In conclusion, while the metrics discussed in the literature serve as excellent tools for image analysis, the question remains: are they sufficient to implement a fully automated workflow, potentially enabling the rejection of images based on poor diagnostic quality? Despite promising results documented in published studies [11,66–68], the current landscape suggests otherwise, given either suboptimal performance or limited-scale testing.

Recent advancements in foundation models for medical imaging [69], the evolution of multi-modal models, and the rise of Large Language Models (LLMs) may hold the key to achieving the next level of automation. Particularly noteworthy is the growing proficiency of LLMs in interpreting the content of images, paving the way for envisioning such capabilities being harnessed for the automation of MRI quality assessment.

13.8 Conclusions and outlook

In this chapter, we delved into the need for reliable automated image quality assessment and the ongoing efforts in developing corresponding metrics, creating a link between computer vision and MRI literature. While various metrics exhibit correlations with radiologists' assessments, a single metric capable of fully characterizing image quality remains elusive.

Despite notable advancements in the field, many challenges still confront the community with respect of image quality assessment metrics:

1. **Context-Dependent Nature:** The quality of MRI images cannot be considered independently of their imaging purpose. Defining absolute thresholds for parameters such as SNR, resolution, and uniformity, which universally determine diagnostic quality, proves, at this time, an insurmountable challenge.
2. **Multi-Faceted Image Quality:** Multiple factors contribute to image quality, making it difficult to distill them into a single numerical value. The complex interplay of elements adds a layer of difficulty to the assessment process.
3. **Impact of Reconstruction and Enhancement on Downstream Tasks:** Image reconstruction and enhancement techniques not only influence perceived image quality but may also disrupt existing post-processing pipelines. This becomes critical in applications such as quantitative parameter mapping, brain morphometry or tumor size measurements. In this settings current assessment approaches fall short in capturing the impact on downstream computations.
4. **Artifact Challenges:** Recognizing the significant impact of artifacts on image quality, their automatic detection and quantitative characterization prove challenging due to the variability in possible artifacts. This poses a barrier to the comprehensive and objective assessment of their impact in the resulting notion of quality.

We believe that the development of fully automated quality control for MRI images may be transformative for the field, not only to improve the acquisition workflow but potentially opening up new clinical applications driven by faster and more reliable scans. However, addressing the points above is critical, and we believe these are interesting challenges for the research community.

References

[1] Wülker C, et al. Digital reference objects for evaluating algorithm performance in MR image formation. *Magn Reson Imaging* 2024;**105**:67–74. https://doi.org/10.1016/j.mri.2023.10.008. https://www.sciencedirect.com/science/article/pii/S0730725X23001819.

[2] Kaufman L, et al. Measuring signal-to-noise ratios in MR imaging. *Radiology* 1989;**173**(1):265–7. https://doi.org/10.1148/radiology.173.1.2781018.

[3] Reeder SB, et al. Practical approaches to the evaluation of signal-to-noise ratio performance with parallel imaging: application with cardiac imaging and a 32-channel cardiac coil. *Magn Reson Med* 2005;**54**(3):748–54. https://doi.org/10.1002/mrm.20636.eprint. https://onlinelibrary.wiley.com/doi/abs/10.1002/mrm.20636.

[4] Scott Gazelle G, et al. A framework for assessing the value of diagnostic imaging in the era of comparative effectiveness research. *Radiology* 2011;**261**(3):692–8. https://doi.org/10.1148/radiol.11110155.

[5] Mannos J, Sakrison D. The effects of a visual fidelity criterion of the encoding of images. *IEEE Trans Inf Theor* 1974;**20**(4):525–36. https://doi.org/10.1109/TIT.1974.1055250.

[6] Wang Z, Simoncelli EP, Bovik AC. Multiscale structural similarity for image quality assessment. In: *The 37th Asilomar Conference on Signals, Systems & Computers. Vol. 2*; 2003. p. 1398–402.

[7] Ahumada AJ. Computational image quality metrics: a review. *SID Digest* 1993;**24**: 305–8.

[8] Kumar J, Ye P, Doermann D. A dataset for quality assessment of camera captured document images. In: *Lecture Notes in computer science 8357 LNCS*; 2013. p. 113–25. https://doi.org/10.1007/978-3-319-05167-3_9. https://link.springer.com/chapter/10.1007/978-3-319-05167-3_9.

[9] Yang H, Fang Y, Lin W. Perceptual quality assessment of screen content images. *IEEE TIP* 2015;**24**(11):4408–21.

[10] Kastryulin S, et al. Image quality assessment for magnetic resonance imaging. *IEEE Access* 2023;**11**:14154–68.

[11] Mortamet B, et al. Automatic quality assessment in structural brain magnetic resonance imaging. *Magn Reson Med* 2009;**62.2**:365–72.

[12] Pizarro RA, et al. Automated quality assessment of structural magnetic resonance brain images based on a supervised machine learning algorithm. *Front Neuroinf* 2016;**10**:52.

[13] Küstner T, et al. A machine-learning framework for automatic reference-free quality assessment in MRI. *Magn Reson Imaging* 2018;**53**:134–47.

[14] Wang Z, et al. Image quality assessment: from error visibility to structural similarity. *IEEE TIP* 2004;**13**(4):600–12.

[15] Wang Z, Bovik AC. Modern image quality assessment. *Synth Lect Image Video Multimedia Proc* 2006;**2**(1):1–156.

[16] Wang Z, Bovik AC. Mean squared error: love it or leave it? A new look at signal fidelity measures. *IEEE Signal Process Mag* 2009;**26.1**:98–117.

[17] Damera-Venkata N, et al. Image quality assessment based on a degradation model. *IEEE TMI* 2000;**9**(4):636–50. https://doi.org/10.1109/83.841940.

[18] Wang Z, Li Q. Information content weighting for perceptual image quality assessment. *IEEE TIP* 2010;**20**(5):1185–98.

[19] Cooper Larson E, Chandler DM. Most apparent distortion: full-reference image quality assessment and the role of strategy. *J Electron Imag* 2010;**19**(1)011006.

[20] Liang Y, et al. Image quality assessment using similar scene as reference. In: *European Conference on Computer Vision*; 2016. https://api.semanticscholar.org/CorpusID:11515951.

[21] Ponomarenko N, et al. Image database TID2013: peculiarities, results and perspectives. *Sig proces: Image com* 2015;**30**:57–77.

[22] Krizhevsky A, Sutskever I, Hinton GE. Imagenet classification with deep convolutional neural networks. In: *Advances in Neural Information Processing Systems*, vol 25; 2012.

[23] Simonyan K, Zisserman A. Very deep convolutional networks for large-scale image recognition. In: *ICLR*; 2015. http://arxiv.org/abs/1409.1556.

[24] F.N Iandola et al. "SqueezeNet: AlexNet-level accuracy with 50x fewer parameters and¡ 0.5 MB model size". In: arXiv preprint arXiv:1602.07360 (2016).

[25] Johnson J, Alahi A, Fei-Fei L. Perceptual losses for real-time style transfer and super-resolution. In: *ECCV*; 2016. p. 694–711.

[26] Mordvintsev A, et al. Differentiable image parameterizations. *Distill* 2018;**3**(7)e12.

[27] Zhang R, et al. The unreasonable effectiveness of deep features as a perceptual metric. In: *Proceedings of the IEEE Conference on CVPR*; 2018. p. 586–95.

[28] Prashnani E, et al. Pieapp: perceptual image-error assessment through pairwise preference. In: *Proceedings of the IEEE Conference on Computer Vision and Pattern Recognition*; 2018. p. 1808–17.

[29] J. Gu et al. "Image Quality Assessment for Perceptual Image Restoration: A New Dataset, Benchmark and Metric". In: arXiv preprint arXiv:2011.15002 (2020).

[30] Gu J, et al. PIPAL: a large-scale image quality assessment dataset for perceptual image restoration. *Lect Notes Comput Sci* 2020;**12356**(LNCS):633–51. https://doi.org/10.1007/978-3-030-58621-8_37. arXiv: 2007.12142.

[31] K. Ding et al. "Image quality assessment: unifying structure and texture similarity". In: arXiv:2004.07728 (2020).

[32] Chen G-H, et al. Edge-based structural similarity for image quality assessment. In: *2006 IEEE International Conference on Acoustics Speech and Signal Processing Proceedings. Vol 2*. IEEE; 2006. II.

[33] Xue W, et al. Gradient magnitude similarity deviation: a highly efficient perceptual image quality index. *IEEE TIP* 2013;**23**(2):684–95.

[34] Zhang B, Sander PV, Bermak A. Gradient magnitude similarity deviation on multiple scales for color image quality assessment. In: *2017 IEEE International Conference on Acoustics, Speech and Signal Processing (ICASSP)*. IEEE; 2017. p. 1253–7.

[35] Zhang L, et al. FSIM: a feature similarity index for image quality assessment. *IEEE TIP* 2011;**20**(8):2378–86.

[36] Zhang L, Shen Y, Li H. VSI: a visual saliency-induced index for perceptual image quality assessment. *IEEE TIP* 2014;**23**(10):4270–81.

[37] Reisenhofer R, et al. A Haar wavelet-based perceptual similarity index for image quality assessment. *Signal Process Image Commun* 2018;**61**:33–43.

[38] Ziaei Nafchi H, et al. Mean deviation similarity index: efficient and reliable full-reference image quality evaluator. *IEEE Access* 2016;**4**:5579–90.

[39] Ding K, et al. Comparison of full-reference image quality models for optimization of image processing systems. *Int J Comput Vis* 2021;**129**(4):1258–81. https://doi.org/10.1007/s11263-020-01419-7.

[40] Hamid RS, Bovik AC. *A visual information fidelity approach to video quality assessment. The First International Workshop on VPQM for Consumer Electronics*, **7**. sn; 2005. p. 2.

[41] Eero PS, et al. Shiftable multiscale transforms. *IEEE Trans Inf Theor* 1992;**38**(2):587–607.

[42] Mittal A, Krishna Moorthy A, Bovik AC. No-reference image quality assessment in the spatial domain. *IEEE Trans Image Process* 2012;**21.12**:4695–708.

[43] Saad MA, Bovik AC, Charrier C. A DCT statistics-based blind image quality index. *IEEE Signal Process Lett* 2010;**17**(6):583–6.
[44] Saad MA, Bovik AC, Charrier C. Blind image quality assessment: a natural scene statistics approach in the DCT domain. *IEEE Trans Image Process* 2012;**21**(8):3339–52.
[45] Cem Catalbas M. Modified VDSR-based single image super-resolution using naturalness image quality evaluator. *Signal Image Video Proces* 2022;**16**(3):661–8.
[46] J. Wang, K.C. K. Chan, and C.C. Loy. Exploring CLIP for assessing the look and feel of images. 2022. arXiv: 2207.12396 (cs.CV).
[47] A. Radford et al. Learning Transferable visual models from natural Language Supervision. 2021. arXiv: 2103.00020 (cs.CV).
[48] Heusel M, et al. Gans trained by a two time-scale update rule converge to a local nash equilibrium. *Adv Neural Inform Process Syst* 2017:6626–37.
[49] Bińkowski M, et al. Demystifying MMD GANs. In: *ICLR*; 2018. https://openreview.net/forum?id=r1lUOzWCW.
[50] Tsitsulin A, et al. The shape of data: intrinsic distance for data distributions. In: *ICLR*; 2020.
[51] Khrulkov V, Oseledets I. Geometry score: a method for comparing generative adversarial networks. In: *Proceedings of Machine Learning Research*, Vol 80. ICML; 2018. p. 2626–34.. In: *http://proceedings.mlr.press/v80/khrulkov18a.html.*
[52] De Silva V, Carlsson GE. Topological estimation using witness complexes. *SPBG* 2004; **4**:157–66.
[53] Salimans T, et al. Improved techniques for training gans. In: *Adv. Neural Inform. Process. Syst.*; 2016. p. 2234–42.
[54] S. Barratt and R. Sharma. "A note on the inception score". In: arXiv preprint arXiv: 1801.01973 (2018).
[55] Lin H, Hosu V, Saupe D. KADID-10k: a large-scale artificially distorted IQA database. In: *QoMEX*. IEEE; 2019. p. 1–3.
[56] Williams MB, et al. Digital radiography image quality: image acquisition. *J Am Coll Radiol* 2007;**4.6**:371–88.
[57] Harrison HB, et al. Model observers for assessment of image quality. *Proc Nat Acad Sci* 1993;**90**(21):9758–65. https://doi.org/10.1073/PNAS.90.21.9758. https://www.pnas.org/content/90/21/9758%20https://www.pnas.org/content/90/21/9758.abstract.
[58] He X, Park S. Model observers in medical imaging research. *Theranostics* 2013;**3**(10): 774. https://doi.org/10.7150/THNO.5138.
[59] Mason A, et al. Comparison of objective image quality metrics to expert radiologists' scoring of diagnostic quality of MR images. *IEEE TMI* 2020;**39**(4):1064–72. https://doi.org/10.1109/TMI.2019.2930338.
[60] Pezzotti N, et al. An adaptive intelligence algorithm for undersampled knee MRI reconstruction. *IEEE Access* 2020;**8**204825. https://doi.org/10.1109/ACCESS.2020.3034287.
[61] Hamid RS, Sabir MF, Bovik AC. A statistical evaluation of recent full reference image quality assessment algorithms. *IEEE TIP* 2006;**15**(11):3440–51. https://doi.org/10.1109/TIP.2006.881959.
[62] Kastryulin S, Zakirov D, Prokopenko D. PyTorch Image Quality: metrics and measure for image quality assessment. In: *Open-Source Software*; 2019. https://github.com/photosynthesisteam/piq.
[63] Amour AD', et al. Underspecification presents challenges for credibility in modern machine learning. *J Mach Learn Res* 2022;**23**(1):10237–97.

[64] Olah C, Mordvintsev A, Schubert L. Feature visualization. *Distill* 2017;**2**(11):e7.

[65] Dodge S, Karam L. A study and comparison of human and deep learning recognition performance under visual distortions. In: *2017 26th International Conference on Computer Communication and Networks (ICCCN)*. IEEE; 2017. p. 1–7.

[66] Davids M, et al. Fully-automated quality assurance in multi-center studies using MRI phantom measurements. *Magn Reson Imaging* 2014;**32**(6):771–80.

[67] Esses SJ, et al. Automated image quality evaluation of T2-weighted liver MRI utilizing deep learning architecture. *J Magn Reson Imag* 2018;**47**(3):723–8.

[68] Sujit SJ, et al. Automated image quality evaluation of structural brain magnetic resonance images using deep convolutional neural networks. In: *2018 9th Cairo International Biomedical Engineering Conference (CIBEC)*. IEEE; 2018. p. 33–6.

[69] Mazurowski MA, et al. Segment anything model for medical image analysis: an experimental study. In: *Medical Image Analysis*, vol 89; 2023102918.

CHAPTER

What is beyond the image? Machine learning for MR image analysis

14

Matthan W.A. Caan* and Mingshi Chen

Amsterdam UMC Location University of Amsterdam, Biomedical Engineering and Physics, Amsterdam, The Netherlands

**Corresponding author: E-mail: m.w.a.caan@amsterdamumc.nl*

14.1 Introduction

Magnetic resonance imaging (MRI), developed in the late 20th century, has revolutionized medical diagnostics by providing detailed images of the human body without the need for invasive procedures. Initially, radiologists depended on printed scans to diagnose and interpret medical conditions, meticulously reviewing each image to detect anomalies. This manual examination of static prints, though effective, was time-consuming and susceptible to human error. Over the years, the integration of digital technologies began enhancing this process, enabling more precise analysis and timely diagnosis. These analysis techniques have been developed or adopted from computer science. Techniques adopted from artificial intelligence (AI) allowed for (semi-)automated analysis. First rule-based approaches were used, such as in radiomics analysis, inspired by human logic or image analysis principles. Then, machine learning approaches appeared, learning from data. Representation learning allowed for exploiting the vast amount of information present in large imaging databases that have been built over the years, in hospitals and large scale population studies. This chapter introduces a number of these techniques and discusses their use in a range of medical applications.

14.2 Radiomics

14.2.1 Toward radiomics in MRI

Radiomics is a term that originates from combining the words "radio," referring to radiology, acquiring medical images for diagnosis and treatment of disease, and the suffix "omics," originally being the research on microscopic, molecular level, translated into the structure of an entire view of a system or organism [1,2]. While medical imaging operates at a resolution that is orders of magnitude higher than, e.g.,

Machine Learning in MRI. https://doi.org/10.1016/B978-0-443-14109-6.00014-6

genomics, proteomics or metabolomics, there is an abundance of information encapsulated in medical imaging data that is to be unraveled. The goal of radiomics is to convert images into high-dimensional mineable data, which involves the extraction of a large number of quantitative features. These can then be used for constructing models to predict disease prognosis, response to treatment, or to track disease noninvasively [3].

Radiomics has a high potential to be deployed in MRI, due to its conventionally non-quantitative nature. In contrast to CT, quantifying tissue density through radiation absorption in Hounsfield units, MRI is conventionally a relative imaging modality. The main clinical imaging modalities are weighted, including T1, T2, proton density, fluid attenuated inversion recovery (FLAIR), and diffusion weighting. These modalities provide relative contrast between tissue types. Quantifying MRI requires repeated acquisitions under varying sequence parameters, thus increasing the measurement time, which is not always feasible. This calls for a method to quantify information in weighted MRI data.

Originally, radiomics was developed for high-resolution radiation-based medical imaging techniques, including X-ray and CT. Thanks to developments in acquisition and reconstruction techniques, MRI can now also image the human body at high resolution. Thus, into today's practice, there is an abundance of information encapsulated in MRI data that is to be unraveled.

The first and main application of radiomics is in the domain of oncology. Beyond oncology, radiomics has been deployed in other areas such as neurology and cardiovascular medicine. Broader neuroscience applications have also become feasible. Within psychiatry, radiology can aid in identifying biological traits of observed phenomenological patterns, in a field termed *psychoradiology*. Below, we describe the methods of radiomics before detailing selected applications from different domains.

14.2.2 Study setup, data acquisition, and definition of target area

In order to answer the research question, clinical data collection may be performed in either a retrospective or prospective study design. Alternatively, a randomized clinical trial may be set up. Standardized protocols for capturing and reconstructing images are needed to ensure consistency and reliability across datasets. This is needed for radiomics to capture the relevant information present in the data, and to be invariant to possible noise or artifacts.

After data acquisition, minimal preprocessing is needed for radiomics. In contrast to voxel-based studies, no registration or spatial normalization to a standard space is needed. Rather, scans are processed individually in native space. Noise reduction may be achieved through Gaussian and median filtering, or edge-preserving smoothing through bilateral and anisotropic diffusion filters. The latter filters reduce noise while maintaining the sharpness of edges, which aids accurate feature extraction in radiomics studies. Furthermore, image enhancement techniques may be considered to prepare images for further analysis. These include the Laplacian of Gaussian and Sobel filters that enhance edges in an image, focusing on either coarse or fine textures. This can improve the detection of boundaries and structures within the image, which is crucial for accurate tumor segmentation and feature extraction in radiomics [4]. Finally, data intensity distributions need to be normalized such that these can be compared between scans.

In order to define the target area for analysis, a volume of interest (VOI) or single or multiple regions of interest (ROIs) are defined. This could include a tumor region, an organ or series of organs, or sub-structures within one organ to be studied. Manual, semi-automatic, or automatic segmentation of ROIs may be performed to isolate the specific areas of the image for detailed analysis. Quality assurance is needed to not affect the radiomics analysis by segmentation inaccuracies.

14.2.3 Radiomics features

Multiple feature banks with ranges of features have been described in the literature [5]. Most features defined below are in compliance with definitions as described by the Imaging Biomarker Standardization Initiative (IBSI), published elsewhere on conventional and convolution filters [6,7]. Fig. 14.1 displays and Table 14.1 summarizes different categories of features, categorized as first order statistics, shape-, texture- and transform-based features. These will be explained now in more detail.

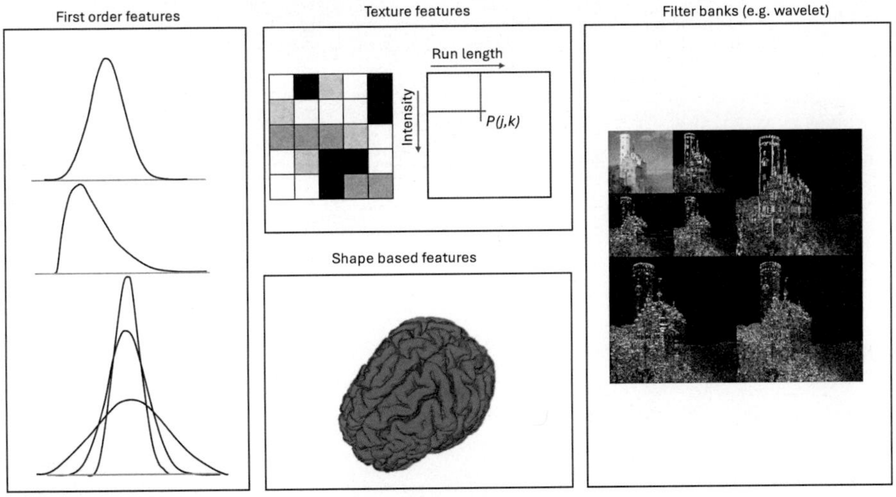

FIGURE 14.1

Different classes of radiomics features. **Left panel:** examples of first-order features, such as Gaussian distribution based second, third and fourth order features (mean, standard deviation, skewness, kurtosis). **Middle panel:** top: example of a texture feature, being the gray-level run length matrix (GLCM). Bottom: shape features can be extracted from three-dimensional isosurfaces. **Right panel:** transform-based features, for example, the wavelet filter bank. Wavelet filters are defined at different levels of resolution and orientation.

Images adopted from Refs. [8] and [9].

Table 14.1 Definition of feature classes, number of features (#) and features as implemented in the PyRadiomics toolbox [10].

Classes	#	Features
First order statistics	19	(Total) energy; entropy minimum, 10th/90th percentile, maximum, mean, median intensity; (interquartile) range; (robust) range; (robust) mean absolute deviation Root mean square; standard deviation; skewness; kurtosis; variance; Uniformity
Shape-based (3D)	17	Mesh/voxel volume; surface area (to volume ratio); sphericity; compactness 1/2; spherical disproportion; maximum diameter (2D slice, column, row; 3D); major/minor axis length; elongation; flatness
Shape-based (2D)	10	Mesh/pixel surface; perimeter (to surface ratio); sphericity; spherical disproportion; Maximum 2D diameter; major/minor axis length; elongation
Gray level Co-occurrence matrix	24	Autocorrelation; joint average; cluster prominence, shade, tendency; contrast; correlation; difference average/entropy/variance; joint energy/entropy; informational measure of correlation 1/2; inverse difference (moment, normalized); inverse variance; maximal correlation coefficient; maximum probability; sum average, entropy, sum of squares
Gray level run length matrix	16	Short/long run emphasis; gray/run level non-uniformity (normalized), run percentage, gray level variance, run variance/entropy (Short/long run) low/high gray level run emphasis
Gray level size zone matrix	16	Small/large area emphasis, gray level/size zone non-uniformity (normalized) Zone percentage, variance, entropy; gray level variance; small/large area low/high gray level (zone) emphasis
Neighboring gray tone difference matrix	5	Coarseness; contrast; busyness; complexity; strength
Gray level dependence matrix	14	The content provided seems to be technical in nature and contains terms related to image processing. There are a few punctuation errors that need to be corrected. Here is the revised version: "Small/large dependence emphasis; gray level/dependence non-uniformity (normalized); gray level variance; dependence variance/entropy; low/high gray level emphasis; small/large dependence low/high gray level emphasis"

First-order statistics features describe the distribution of voxel intensities within an image region without considering spatial relationships. They include metrics such as mean, median, standard deviation, skewness, and kurtosis. Usually, a histogram of the intensities is made, after which several first order statistics are extracted. Therefore, these features are commonly also referred to as histogram or intensity features. In clinical use, they are used for assessing tumor intensity characteristics which can help in distinguishing between benign and malignant lesions or identifying tissue types in various organs [11].

Shape-based features quantify the geometric properties of a region of interest, such as volume, surface area, sphericity, compactness, and elongation. Clinically, these are essential for characterizing the morphology of tumors, differentiating tumor subtypes, and assessing tumor growth or shrinkage over time [12].

Texture-based features seek to find a non-parametric representation of the variability in signal seen in a local environment. Different matrix-based metrics have been proposed, quantifying the relationships between adjacent groups of pixels, capturing texture information such as contrast, correlation, energy, and homogeneity. Moreover, they provide information about texture roughness, smoothness, coarseness or fineness, busyness, connectedness and complexity. Note that these terms are not mutually exclusive, rather the ensemble seeks to find a useful description for further analysis. Clinically, texture-based features are useful for identifying tissue heterogeneity, which can indicate the presence of tumors, tumor grading, and response to therapy [13]. Also, they help in distinguishing between different tissue types and assess tumor heterogeneity [14], while they are also important for evaluating tumor heterogeneity and predicting treatment outcomes [11].

Filtering is crucial in the radiomics workflow for several reasons, which is why transform-based features are proposed. Image filters may act as feature extractors, for representing tissue characteristics and enhancing specific structures. They may improve the robustness of features, making them less sensitive to variations in imaging protocols. Finally, some filters reduce dimensionality by emphasizing relevant features and suppressing less informative ones, simplifying analysis and improving computational efficiency.

A number of possible transform-based filters can be applied before extracting first-order statistics features [7]. The gradient magnitude is a transformation that may be applied to capture the degree or abruptness of gray-level intensity fluctuations across an image. Similar to histogram features, first-order statistics may subsequently be computed [15]. Wavelet filtering decomposes the images at various levels, allowing the extraction of texture features at multiple scales. Wavelet filters can help in identifying subtle textural patterns that are not visible at a single resolution. They are particularly useful in analyzing medical images for tumor characterization and prognosis prediction [4]. Gabor filters are designed to capture texture features at various orientations and scales, making them useful for texture analysis in radiomics. They can help in distinguishing between different tissue types based on their textural patterns, aiding in the differentiation of malignant and benign lesions. Further advanced model-based features have also been proposed, including a parametrization of the local texture, autoregressive models to estimate the local capturing of surrounding information, and fractal analysis as a measure of complexity when increasing the spatial scale.

In summary, these different classes of radiomic features collectively provide a comprehensive analysis of medical images, enabling improved diagnosis, prognosis, and treatment planning across various medical conditions. Each type of feature offers unique insights that contribute to a more detailed understanding of tissue characteristics and disease behavior.

14.2.3.1 Toolboxes

A commonly used toolbox to extract features is PyRadiomics as an open-source Python package [10]. The Radiomics Extensive Digital Interchangeable Classification Toolkit (PREDICT) is another toolbox, also available in Python. Although less frequently used, there are also toolboxes for radiomics in MATLAB.

14.2.4 Radiomics analysis

The extracted radiomic features capture various characteristics of tissues or lesions. Statistical and machine learning algorithms can be developed to form predictive models based on the radiomics features. These models can predict outcomes such as disease diagnosis, prognosis, and response to therapy. When analyzing the features, several aspects need to be specifically considered.

Features computed from data acquired in a multi-center setting need to be harmonized for differences in the data acquisition. The non-linear transformations applied when computing radiomic features challenge the harmonization procedure. It is now understood that first order correction is insufficient and that higher order modeling is required. In addition, all feature distributions can now be transformed to a chosen (external) reference, instead of the overall mean, providing more flexibility. Furthermore, bootstrap and Monte Carlo are added for improved robustness in the estimation. BM-ComBat integrates these aspects, reporting high performance in multi-center studies [16].

Computing a large number of radiomic features commonly leads to so-called *short-wide datasets*, with many features relative to the number of patients included in a study. Feature selection and extraction, as commonly performed in machine learning, must be planned with careful attention. Collinearity between features is common ground, and needs to be accounted for in feature selection. Non-Gaussian data distributions challenge the adoption of conventional feature extraction methods such as principal and independent component analysis. Tree-based networks and neural networks may be considered as promising learned dimension reduction approaches.

Testing the developed models on independent data is key to validate the effectiveness and reproducibility in different clinical settings. In a study into sleep deprivation, it was found that first order radiomics features were more reproducible than texture-based features were [17]. To promote reproducibility of studies, the

Radiomics Quality Score (RQS) was proposed [18]. Through a set of 16 criteria, in total 36 points can be achieved along the trajectory of design and acquisition to analysis and report. Interestingly, setting up a prospective study registered in a trial database is given the most points, since it provides the highest level of evidence for validity and use of a potential biomarker.

Interpreting the outcome of a radiomics study may be facilitated by the adoption of Shapley features [19]. They quantify the relevance of features in a given classification or regression task. Together with feature importance, this allows relating the performance back to the underlying data, at a level invisible to the naked eye in single scans. Visualized outliers on key performing features exemplify the underlying pathological process.

Eventually, radiomics data may be combined with clinical, demographic, genomic, and other biomarker data to enhance the predictive power and relevance of the models in a clinical context. Utilizing the models to predict clinical outcomes facilitates personalized treatment planning and management based on the predicted risks and benefits.

14.2.5 Clinical application

As mentioned in the introduction of this section, there is a wide range of clinical applications of radiomics with automated clinical information extraction from MRI dataset in the domain of cancer and beyond.

14.2.5.1 Cancer

Radiomics can enhance the diagnosis and characterization of tumors by extracting detailed quantitative features from medical images. These features can help distinguish between benign and malignant lesions and characterize the subtype of a tumor [20]. Furthermore, they can predict how well a tumor will respond to various treatments, such as chemotherapy, radiotherapy, or targeted therapies. This allows for personalized treatment planning and optimization based on the predicted response of the tumor to different therapeutic options [21]. The analysis of radiomic data can also provide prognostic information about patient outcomes, such as survival rates or risk of recurrence. Radiomics can identify patterns in the imaging data that correlate with more aggressive disease or poorer outcomes [22]. In surgical oncology, radiomics can assist in planning surgical interventions by providing detailed maps of tumor extent and the surrounding anatomy, helping surgeons to optimize surgical margins and reduce the risk of recurrence. In application to breast cancer, radiomics features extracted from PET and MRI have been used to characterize breast cancer phenotypes and prognosis, with certain features being discriminative of tumor grade and recurrence-free survival [23]. Finally, radiomics can help identify new biomarkers that are predictive of disease traits, therapeutic responses, or patient outcomes. These radiomic biomarkers can be integrated with genomic and proteomic data to further enhance the precision of medical diagnostics and personalized treatment strategies [24]. Longitudinal, non-invasive monitoring of disease progression can be done through serial imaging studies. This has been referred to as delta-radiomics [25]. Here, quantitative radiomic analysis can detect subtle changes in tumor characteristics over time, which may not be visible to the human eye [15].

14.2.5.2 Other domains

In stroke, radiomics provides a powerful paradigm for quantifying stroke lesion characteristics, with application in diagnosis of lesions, early outcome prediction and long-term prognosis evaluation [26]. A systematic review and meta-analysis into Alzheimer's disease and mild cognitive impairment concluded that MRI radiomics has promising performance and potential for serving as diagnosis and classification tool [27]. Furthermore, in Parkinson's disease, a differential diagnosis could be made based on features computed from multimodal MRI data [28]. A study into attention deficit hyperactivity disorder (ADHD) in the psychoradiology context found no differences in total brain, gray and white matter volume, but rather differences in shape of a small subset of brain regions. This demonstrates the added value of radiomics in identifying local cortical morphometric variation beyond volume alone [29].

These applications demonstrate the potential of radiomics to transform medical imaging from a primarily qualitative tool into a robust quantitative technique that supports decision-making in personalized medicine. In conclusion, MRI radiomics features hold significant potential as biomarkers for various clinical applications, including the prediction of disease characteristics and treatment outcomes. However, challenges such as feature reliability, methodological consistency, and robustness to image processing need to be addressed to facilitate the clinical implementation of radiomics.

14.3 Deep representation learning of features

For decades, the fields of computer vision and medical imaging have worked on designing features, filters and transformations for solving a range of problems in detection, diagnosis and prognosis problems. To further improve accuracy and performance, representation learning does not handcraft features, driven by human intelligence, but rather learn feature representations from data. Data were thought to be well represented when small variations in input space led to small perturbations in a learned embedding. Such an embedding can be decomposed into a non-linear combination of a layered series of basic transformation functions. This deeply layered embedding gradually grows in complexity, and increasingly gains capacity to learn abstract data representations. It is thus able to linearize highly non-linear problems in a high-dimensional manifold, where it may then compress to a low-dimensional efficient representation, as a common strategy in the field of deep learning today.

14.3.1 Invariant scattering transform

Medical Imaging data are subject to unwanted transformations when being acquired, which challenge the extraction of relevant features for subsequent learning. Among these transformations are translation, rotation, scaling, and deformation. Invariance of features to these transformations is generally required to arrive at a system that can operate on data in a real-life setting. The Invariant Scattering Transform (IST) was proposed as a method to extract features that meet this requirement [30], and may be seen as a hybrid method between conventional machine learning and representation learning.

The IST is designed as a set of wavelet coefficients that are subject to non-linear activation functions, thus rendering a set of scattering coefficients. The wavelet coefficients are also seen as Gabor filters. Through convolution with expanded filter banks, invariance to translation, rotation, and scaling is achieved. Multiple layers of transformations are generally defined and connected to arrive at a deep representation of the input data. This set aims at capturing local structure, symmetries, and global structure [31]. ISTs find their application in segmentation, classification, and registration domains. In the context of classification, ISTs may be connected to classifiers, such as Convolutional Neural Networks (CNNs) or Support Vector Machines.

Whereas ISTs are commonly defined in 2D, an extension to 3D was proposed, to support native analysis of medical imaging data [32]. Here, a major advantage of the IST was shown in the context of small datasets. 3D MRI scans comprise a huge number of features which makes classification a high-dimensional problem. Thanks to its use of predefined features, the ISTs is highly efficient in extracting discriminant information in the context of small datasets. In application to classifying Mild Cognitive Impairment, a high classification accuracy was feasible. Furthermore, it was shown that the ISTs is interpretable, by back-projecting the trained classifier to input space and visualizing the brain regions most relevant for classifying the disease.

The IST is limited in its predefined nature, such that it may not optimally represent the dataset at hand. It may serve as an initialization for CNNs, such that a tailored representation may be learned for a specific problem in the medical imaging domain, with the potential of faster convergence, compared to networks initialized at random.

14.3.2 Variational autoencoders

Autoencoders contain an encoding and decoding network and aim to learn efficient representations of the input data. The encoder aims to preserve relevant information in a low-dimensional manifold or latent space, while the decoder targets the reconstruction of the original input data. Non-informative information is disregarded by autoencoders, making them relevant in the context of image denoising. Autoencoders may be seen as an extension to Principal Component Analysis (PCA), maximizing the preserved variance in the subspace, or Independent Component Analysis

(ICA), which optimizes for independence or non-Gaussianity. Where PCA and ICA find orthogonal subspaces, autoencoders relax this constraint, and project data onto an arbitrarily orthonormal basis. While this allows for encoding with a high degree of non-linearity, no further constraints are imposed to ensure that the latent space is continuous. This prevents the use of the low-dimensional embedding as a continuous representation of the input data.

Variational autoencoders (VAEs) were introduced a decade ago as a way of unsupervised deep representation learning of features [33]. VAEs were proposed as an extension to conventional autoencoders. A VAE aims to learn a meaningful low-dimensional embedding in latent space, through Bayesian modeling. To this end, a VAE learns a latent posterior Gaussian distribution from which data are sampled that need to be close to the input data after decoding. The latent space is constrained to have a distribution close to the target Gaussian distribution.

Two applications of VAEs can be mentioned. First, a VAE acts as a generative model, from which new data instances can be obtained that are close to the input data. Since a VAE is trained in an unsupervised manner, these generated samples may be used for data augmentation in successive supervised training tasks.

Second, the low-dimensional latent space representation of high-dimensional input data may be used for classification tasks. While any classifier may be trained on the data in latent space, a conventional classifier such as the Support Vector Machine (SVM) may already suffice, assuming that the unsupervised dimension reduction is representative for the task at hand. Alternatively, a multi-layer perceptron (MLP) may be used.

In the context of stroke, such a classifier allowed for predicting functional outcome after treatment of acute ischemic stroke [34].

14.3.3 Zero-shot learning on imaging and report data

The first and major application of MR image analysis is in the domain of radiology, where radiologists read and diagnose a large volume of scans. They do so after a long period of training in a specific sub-discipline to become sensitive and specific enough to diagnose a broad range of diseases with high accuracy. The workload of radiologists can be quite demanding, as radiologists are nowadays required to review one image every 3 seconds to keep up with the workload [35]. This is partly driven by the increasing number of exams taking place and partly by technological developments leading to higher resolution data and thus more images to be reviewed.

Given recent developments in image and language recognition, there is now great potential in developing methods to automate the reading of radiological data. In the application to triaging breast MRI, it has been demonstrated that deep learning can dismiss almost 40% of MRI scans at 100% sensitivity, that is, without missing cancer on one of the scans [36]. However, in order to come to a written report and diagnosis in a broad range of diseases, further work is needed.

To train networks to perform such a task, a vast and rich source of information is enclosed in patient archiving and communication systems (PACS) in hospitals. Here, all imaging data is archived, together with written reports by radiologists, summarizing key observations and a diagnosis. These data are now being disclosed and made available for training.

Zero-shot learning has been considered as an approach to move forward. Here, a model is trained on a set of seen classes and then used to classify novel, previously unseen classes. This requires models to generalize during training through self-supervised tasks. In the radiological context, a text-vision framework is to be defined, integrating imaging and textual sources and information. From these, low-dimensional text and image representations need to be computed.

Textual representations are computed using advanced self-supervised models. The strategy is to build or adopt specific dictionaries used in certain domains. Here, domain adaptation is needed to adapt to the style of writing and to the specific language used in radiology reports. Also, similar to the training phase of specialisms in radiology, it may be advocated to learn networks on sub-domains in the field. For example, in neuroradiology, a concise terminology may be used to detect and diagnose disease in brain MRI scans. Masked language modeling (MLM) is an established way of doing so, where networks learn to infer masked words and thus exploit the natural redundancy in written text. MLM is a relevant training procedure for radiology applications. Other strategies, such as next sentence prediction (NSP), may be less informative, given the commonly punctuated style of reporting, with independent statements of diagnoses per sentence. A related method in this context is the deep bidirectional encoder representations from transformers (BERT) network [37]. In order to generate textual descriptions from clinical images, a pre-trained vision-language model was adopted, connecting a visual encoder to a large language model (LLM). In application to chest radiographs, prior knowledge was integrated through prompt learning for improved performance [38].

For encoding MRI scans, similar to textual encoding, a generic network infrastructure may be adopted and then fine tuned to the specific domain of interest. The DenseNet121 is a generic image encoding network and enables retrieving low-dimensional embeddings. Of interest is the option to capture non-disease specific morphological image variations, related to brain aging. It was therefore proposed to append image information by age [39]. This study focused on a specific set of neuroradiological diseases, and thereby a first step was taken on the path toward unsupervised anomaly detection. After training on a database of 82k hospital scans, the framework could, first, accurately classify abnormal from normal scans. Second, five diagnostic classifications could be made on stroke, multiple sclerosis, hemorrhage, meningioma and hydrocephalus. Saliency reports were added to explain the model's attention to specific MRI slices and regions.

To advance in the integration of medical imaging and LLMs, a robust validation practice needs to be followed. The validation process may be subdivided in three tiers, on general, task specific and clinical impact validation, respectively. When implemented in practice, the performance of such models needs to be continuously monitored, to avoid drift and bias [40].

14.4 Population statistics

14.4.1 Normative modeling

A conventional approach in studying disease patterns is the composition of two groups of patients or cases and otherwise healthy controls. These are composed of and seen as homogeneous and uniform entities, and assessed based on their overall differences. This "case-control" approach is well aligned with two-class classification in machine learning, and explains the popularity of the deployment of nearly non-adapted statistical classifiers in the field. Two main causes that limit the insights that may be learned from the data can be formulated. First, disease profiles are heterogeneous, and individual (subgroups of) cases, otherwise treated as outliers, may need to be studied separately to understand pathological processes. Second, natural demographic variation in the studied population, and variable measurement conditions, amongst other causes, introduce unwanted variation in the data. These processes may not accurately be accounted for based on the available dataset.

Normative modeling seeks to learn trends of variation of the data over parameters not of interest to the main outcome of a study. In large and independent normative databases, the distribution over the general population is learned, such that a normative range can be defined. Individual subjects are then projected onto the learned model distribution, as depicted in Fig. 14.2. They are subsequently statistically compared in terms of their degree of deviation of the reference data.

For brain imaging research, several large databases and population studies may be chosen to perform normative modeling. The Human Connectome Project (HCP)

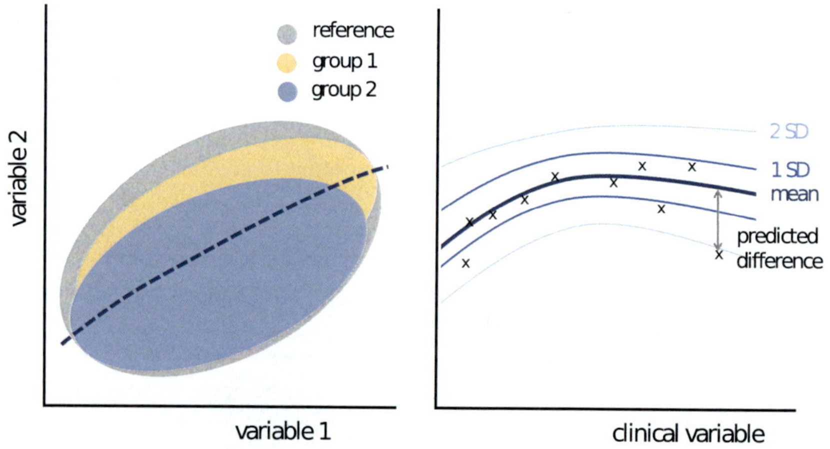

FIGURE 14.2

Principles of normative modeling and brain aging. Left: groups 1 and 2 largely overlap in a plot of two variables of interest. Partly, the variation in the data may be accounted for in relation to reference data. Right: new data is projected onto a normative model, and the predicted difference may be computed per individual data point and further considered as a signal of interest.

comprises a state-of-the-art MRI dataset in the decade from 20 to 30 years of age [41]. The Adolescent Brain Cognitive Development (ABCD) project included younger persons around 10 years of age [42]. The UK Biobank includes subjects above 45 years of age [43], while the German national cohort study held the entire adult age range as inclusion spectrum [44], These mega-cohorts are among the most popular population studies, which through their wealth of available information enable in depth study of how to build representative normative models.

In the field of psychiatry, the use of normative modeling is of high interest. Here, the mechanisms and criteria used for diagnosing disease are not directly rooted into the biology underneath. Henceforth, disease criteria may to a large amount overlap, and diagnostics may be non-dichotomous but rather include a spectrum of outcomes. Personalized assessment through normative modeling may stratify high risks, enable analysis of subtypes of diseases, and predict treatment outcome. Individual differences can be identified through normative modeling, whether strong effects at the group level exist, such as in schizophrenia, or these are absent, as seen in autism spectrum disorder [45].

In regression tasks as compared to classification tasks, normative modeling may appear to be of less added value. In predicting general cognitive ability, no significant improvement over linear regression was found [46]. This may be understood from the fact that normative modeling itself is a regression task. Sequential regression of non-informative nuisance parameters ahead of the analysis of the metric of interest may to a certain extent be superfluous. Essentially, this task would advocate pretraining of regression parameters of interest and embedding these in regression models. This however steps aside from the fact that nonlinear relations and interactions within the data may be present, e.g., through multivariate spatially correlated patterns. Studies into structural brain connectivity may shed new light upon this topic.

While trained normative models are currently made publicly available, they can gain from further generalization. The use of the term "normative" has been debated in this respect. The data on which normative models are trained, need to come from representative population cohorts, which match the demographics of the general population. Another important aspect is the need to normalize race or ethnicity as a step toward studying brain processes in the entire human population. The Centile-Brain project is an important step in this direction, by learning over the entire human life span in a cohort built with data from Europe, Australia, the USA, South Africa, and east Asia [47]. Future work would require global access to MRI scanners, and the development of accurate models of MRI data operating at lower field strengths, which can be made widely available against lower costs.

14.4.2 Brain aging

Brain aging seeks to find the deviation between cognitive and chronological age from brain MRI data. Much in line with parts of the methods used for normative modeling, it does so by learning a model to predict chronological age from MRI

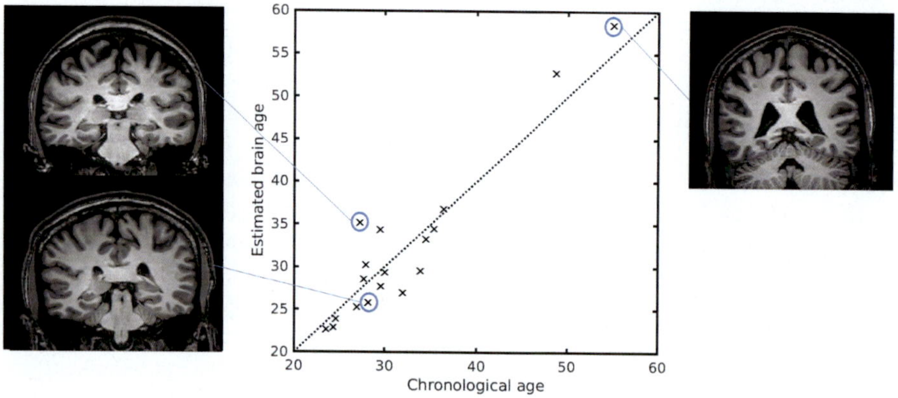

FIGURE 14.3

Estimation of brain age of a sample dataset of isotropic 1.0 mm 3D T1-weighted data from the author's institution, using a model distributed for routine clinical MRI examinations [48]. Three example coronal images are included of two subjects of younger age and a subject of older age. One subject of younger age shows slightly enlarged sulci (top left) and has a higher estimated brain age. The older subject (right) has enlarged ventricles.

data, and use the predicted age difference as a metric of accelerated or decelerated aging. Fig. 14.3 shows an example of a recent brain age model [48], applied to data owned by the authors, thus being external data with respect to the article's model. The brain age metric has been studied in a broad range of studies. For example, in persons with HIV, brains were marginally older but further aged at a similar rate as otherwise healthy controls [49]. Decelerated aging of 7.5 years was seen in meditation practitioners at 50 years of age [50]. Deep learning is capable of learning aging patterns from raw MRI data, and can nowadays be anatomically interpreted, as was demonstrated through domain-specific cognitive impairment [51].

Since aging refers to a developmental process over time, it is critical to model this process accurately. Conventionally, cross-sectional data are used to learn morphological variations over an age span. Thanks to repeated scanning efforts in ongoing population studies, the feasibility of modeling within-subject longitudinal aging can now be studied. Mapping the human brain charts longitudinally turned out to have approximately twice the effect size of cross-sectional analysis. The latter was thus interpreted to underestimate age-related changes. As expected, it was seen that factors unaccounted for in age modeling contribute to the error in predicting per-subject change over time [52]. Avoiding a type-II error in these studies is particularly pivotal. With recently developed methods and available reference big data, smaller studies may also benefit from these insights and should consider investing in periodically measuring a smaller cohort over cross-sectionally compiling a larger sample.

Beyond the human brain, deep learning has been used to estimate biological age from physical activity using LSTMs [53], and laboratory data [54]. Furthermore, neural networks were used to identify morphological indicators of aging on whole-body MRI [55]. Other work characterized environmental and genetic predictors of human cardiovascular aging [56]. The work expanded to multiple organ analyses, for predicting chronic disease and mortality [57], and biological age estimation in the UK Biobank using deep regression [58].

14.4.3 Learning from heterogeneous data sources

In the hospital setting, a patient is commonly subjected to a range of diagnostic instruments, which together can capture and describe the underlying pathological process. Aside from imaging data, clinical information can be included from laboratory measurements. Questionnaire data on cognition or self-reported health form another important data source.

In setting up a data analysis plan, previous work may be categorized into different levels of evidence. By defining different tiers with selections of subsets of features, analyses may be performed and compared on level 1 scientific evidence from meta-analyses and systematic reviews, or alternatively with weaker evidence from single studies. In addition, the entire feature set may be used in a tier.

XGBoost performs extreme gradient boosting on decision trees and is a broadly adapted model for tabular data. By randomly learning on subsets of data and features, it is naturally invariant to incomplete data sources. Given that partially missing data is frequently encountered in the medical domain, XGBoost forms a well-fitted method of choice. Other techniques consider the situation where the feature set is a priori known to be sparse, i.e., only a small subset of ideally non-correlated features is expected to be needed to accurately train a model. Lasso or l1-regularization is then adopted to steer the training into selecting a sparse subset of features. The repeated elastic net technique (RENT) for feature selection uses an ensemble of generalized linear models to this end [59].

Imaging data may be summarized in aggregate values over regions of interest (ROI) and thereby converted into tabular form. Alternatively, as alluded to earlier, information may be extracted through learned encoders or conventional feature extraction methods.

This raises the question how features originating from different modalities may be integrated. Naively, features may be concatenated to build one unordered feature vector. Alternatively, a hierarchical network infrastructure may be built, by stacking the output layers of networks trained per modality into an input layer of a higher order network [60]. Thereby, more complex non-linear interaction between data types may be captured. In certain occasions, weak prior labels are available on feature importance, either on the individual or group level. Bayesian methods have been proposed to incorporate soft constraints in modeling the prior distribution, guided by evidence from previous work [61].

14.4.4 Federated learning

Medical images may capture sensitive information that needs to be protected in order to respect the patient's privacy. Data processing techniques have been proposed to remove sensitive information from the data. This starts by anonymizing or pseudomizing the scans, on the level of metadata, and may be combined by image preprocessing technique to, for example, deface brain MRI scans through masking algorithms.

In many cases, data cannot be shared outside the hospital premises because informed consent has not been given, or legal restrictions prohibit the exchange of data, either on the national or international level. This challenges machine learning in scattered disease populations, where sufficient numbers can only be achieved by combining data from multiple sites.

Federated learning exploits data in cases where it cannot be shared outside the hospital premises. Rather than centralizing and merging data, followed by training a central network, federated learning reverts the procedure and brings the network to the individual data sources for local training of a network.

Care must be taken that the networks' data representation is indeed anonymous and cannot be traced back to specific data sites. A second challenge lies in accounting for site-to-site variations, both in local specific patient population statistics and differences in data representation due to variations in scanner hardware, sequence design and post-processing approaches.

One example application of federated learning is in a study into functional MRI, aiming to capture brain function in resting state or while performing a task in the scanner [62]. In application to autism spectrum disorder, a strategy was proposed to iteratively train local models, and share weights in a privacy-preserving manner with one global model. This was achieved through a mixture of expert domain adaptation. After adversarial domain alignment, individual subjects could no longer be mapped to the contributing site in latent space, demonstrating the feasibility of the approach. Federated learning led to improved performance in classifying autism as compared to classifying per individual site. This paves the way to harnessing the value of locally available patient data. Practically, enabling an infrastructure for federated learning is a significant effort, which is addressed through standardized central initiatives. The Medical Open Network for Artificial Intelligence (MONAI) framework provides software for federated learning, thus supporting a generic strategy in different application areas (https://docs.monai.io/en/stable/fl.html).

14.4.5 Challenges and future directions

The challenges that lie ahead in machine learning for MR image analysis are first related to dealing with small datasets. The use of foundation models may shed new light on improving performance, reliable estimation of uncertainty, and avoiding overfitting. Recently, a tabular foundation model reached performance levels beyond established machine learning models, supplemented with accurate estimates of uncertainty [63].

Much is to be expected from integrating image acquisition and image analysis in an end-to-end fashion. Hereby, learning on the clinical end goal may backpropagate to k-space, the image acquisition domain. Applications include joint reconstruction and segmentation, registration, or computation of clinical outcome metrics. Toolboxes to facilitate multitask medical imaging are appearing [64], to facilitate such work in an open science framework.

Ethical and legal aspects need to be carefully considered. The AI Act is regulation of the European Union is a framework being realized, aiming to guide initiatives in the process of entering the market. Globally, there has been an expansion of companies providing solutions to improve healthcare. Hospitals are making progress in organizing their activities with a manageable number of partners, to increase efficiency in the way of working, improving patient outcomes in Quality Adjusted Life Years. The availability of sufficient healthcare personnel, including radiographers and radiologists, is an important point of focus. By embedding AI in the training programs of these and other healthcare personnel, better ways of working can be realized in the future.

14.5 Conclusion

Over the past 5 decades, the world has seen a dramatic change in healthcare, where imaging based diagnosis has led to a huge improvement in quality of life. Machine learning has proven of significantly added value in analyzing the vast amount of MR imaging data. The future may show an increasing synergy between clinical experts and trained algorithms.

Conflict of interest

M.W.A. Caan is shareholder of Nico.lab International Ltd.

References

[1] Falk M, Hausmann M, Lukášová E, Biswas A, Hildenbrand G, Davídková M, Krasavin E, et al. Determining omics spatiotemporal dimensions using exciting new nanoscopy techniques to assess complex cell responses to DNA damage: Part A–radiomics. Crit Rev Eukaryot Gene Expr 2014;24(3):205–23.

[2] Avanzo M, Stancanello J, Naqa IEl. Beyond imaging: the promise of radiomics. Phys Med 2017;38(June):122–39.

[3] Rizzo S, Botta F, Raimondi S, Origgi D, Fanciullo C, Morganti AG, Bellomi M. Radiomics: the facts and the challenges of image analysis. Euro Radiol Experim 2018;2(1): 36.

[4] Aerts HJWL, Velazquez ER, Leijenaar RTH, Parmar C, Grossmann P, Carvalho S, Bussink J, et al. Decoding tumour phenotype by noninvasive imaging using a quantitative radiomics approach. Nat Commun 2014;5(June):4006.
[5] Ghandili S, Shayesteh S, Fouladi DF, Blanco A, Chu LC. Emerging imaging techniques for acute pancreatitis. Abdom Radiol 2020;45(5):1299–307.
[6] Zwanenburg A, Vallières M, Abdalah MA, Aerts HJWL, Andrearczyk V, Apte A, Ashrafinia S, et al. The image biomarker standardization initiative: standardized quantitative radiomics for high-throughput image-based phenotyping. Radiology 2020;295 (2):328–38.
[7] Whybra P, Zwanenburg A, Vincent A, Schaer R, Apte AP, Ayotte A, Baheti B, et al. The image biomarker standardization initiative: standardized convolutional filters for reproducible radiomics and enhanced clinical insights. Radiology 2024;310(2):e231319.
[8] wikipedia"https://en.wikipedia.org/wiki/Discrete_wavelet_transform.
[9] fieldtriptoolbox"https://www.fieldtriptoolbox.org/getting_started/seg3d/.
[10] Griethuysen JJ M van, Fedorov A, Parmar C, Ahmed H, Aucoin N, Narayan V, Beets-Tan RGH, Fillion-Robin J-C, Pieper S, Aerts HJWL. Computational radiomics system to decode the radiographic phenotype. Cancer Res 2017;77(21):e104–7.
[11] Altazi BA, Zhang GG, Fernandez DC, Montejo ME, Hunt D, Werner J, Biagioli MC, Moros EG. Reproducibility of F18-FDG PET radiomic features for different cervical tumor segmentation methods, gray-level discretization, and reconstruction algorithms. J Appl Clin Med Phys/Am Coll Med Phys 2017;18(6):32–48.
[12] Yang L, Yang J, Zhou X, Huang L, Zhao W, Wang T, Zhuang J, Tian J. Development of a radiomics nomogram based on the 2D and 3D CT features to predict the survival of non-small cell lung cancer patients. Eur Radiol 2019;29(5):2196–206.
[13] Ogbonnaya CN, Zhang X, Alsaedi BSO, Pratt N, Zhang Y, Johnston L, Nabi G. Prediction of clinically significant cancer using radiomics features of pre-biopsy of multiparametric MRI in men suspected of prostate cancer. Cancers 2021;13(24). https://doi.org/10.3390/cancers13246199.
[14] Park BW, Kim JK, Heo C, Park KJ. Reliability of CT radiomic features reflecting tumour heterogeneity according to image quality and image processing parameters. Sci Rep 2020;10(1):3852.
[15] Mayerhoefer ME, Materka A, Langs G, Häggström I, Szczypiński P, Gibbs P, Cook G. Introduction to radiomics. J Nucl Med 2020;61(4):488–95.
[16] Da-Ano R, Masson I, Lucia F, Doré M, Robin P, Alfieri J, Rousseau C, et al. Performance comparison of modified ComBat for harmonization of radiomic features for multicenter studies. Sci Rep 2020;10(1):10248.
[17] Poirot MG, Caan MWA, Ruhe HG, Bjørnerud A, Groote I, Reneman L, Marquering HA. Robustness of radiomics to variations in segmentation methods in multimodal brain MRI. Sci Rep 2022;12(1):16712.
[18] Lambin P, Leijenaar RTH, Deist TM, Peerlings J, de Jong EEC, van Timmeren J, Sanduleanu S, et al. Radiomics: the bridge between medical imaging and personalized medicine. Nat Rev Clin Oncol 2017;14(12):749–62.
[19] Lundberg, S., and S.-I. Lee. 2017. A unified approach to interpreting model predictions. arXiv (cs.AI). arXiv. http://arxiv.org/abs/1705.07874.
[20] Shur JD, Doran SJ, Kumar S, Dafydd Dap, Downey K, O'Connor JPB, Papanikolaou N, Messiou C, Koh D-M, Orton MR. Radiomics in oncology: a practical guide. Radiographics 2021. https://doi.org/10.1148/rg.2021210037.

[21] Nie K, Al-Hallaq H, Allen Li X, Benedict SH, Sohn JW, Moran JM, Fan Y, et al. NCTN assessment on current applications of radiomics in oncology. Int J Radiat Oncol Biol Phys 2019;104(2):302–15.

[22] Liu Z, Wang S, Dong Di, Wei J, Cheng F, Zhou X, Sun K, et al. The applications of radiomics in precision diagnosis and treatment of oncology: opportunities and challenges. Theranostics 2019;9(5):1303–22.

[23] Huang S-Y, Franc BL, Harnish RJ, Liu G, Mitra D, Copeland TP, Arasu VA, et al. Exploration of PET and MRI radiomic features for decoding breast cancer phenotypes and prognosis. NPJ Breast Cancer 2018;4(August):24.

[24] Lohmann P, Bundschuh RA, Miederer I, Mottaghy FM, Langen KJ, Galldiks N. Clinical applications of radiomics in nuclear medicine. Nuklearmedizin. Nuclear Med 2023;62 (6):354–60.

[25] Fave X, Zhang L, Yang J, Mackin D, Balter P, Gomez D, Followill D, et al. Delta-radiomics features for the prediction of patient outcomes in non-small cell lung cancer. Sci Rep 2017;7(1):588.

[26] Chen Q, Xia T, Zhang M, Xia N, Liu J, Yang Y. Radiomics in stroke neuroimaging: techniques, applications, and challenges. Aging Dis 2021;12(1):143–54.

[27] Shahidi R, Baradaran M, Ali A, Bagherieh S, Tajabadi Z, Farhadi A, Korani SS, et al. Diagnostic performance of MRI radiomics for classification of Alzheimer's disease, mild cognitive impairment, and normal subjects: a systematic review and meta-analysis. Aging Clin Exp Res 2023;35(11):2333–48.

[28] Bu S, Pang H, Li X, Zhao M, Wang J, Liu Y, Yu H. Multi-parametric radiomics of conventional T1 weighted and susceptibility-weighted imaging for differential diagnosis of idiopathic Parkinson's disease and multiple system atrophy. BMC Med Imag 2023;23 (1):204.

[29] Sun H, Chen Y, Huang Q, Lui S, Huang X, Shi Y, Xu X, Sweeney JA, Gong Q. Psychoradiologic utility of MR imaging for diagnosis of attention deficit hyperactivity disorder: a radiomics analysis. Radiology 2018;287(2):620–30.

[30] Bruna J, Mallat S. Invariant scattering convolution networks. IEEE Trans Pattern Anal Mach Intell 2013;35(8):1872–86.

[31] Ahsan Md M, Raman S, Siddique Z. Invariant scattering transform for medical imaging. In: Data driven approaches on medical imaging; 2023. p. 127–57.

[32] Adel T, Cohen T, Caan M, Welling M, AGEhIV study group and the Alzheimer's Disease Neuroimaging Initiative. 3D scattering transforms for disease classification in neuroimaging. NeuroImage Clin 2017;14(February):506–17.

[33] Kingma, D.P., and M. Welling. 2013. "Auto-encoding variational bayes. arXiv (stat. ML). arXiv. http://arxiv.org/abs/1312.6114v11.

[34] Tolhuisen ML, Hoving JW, Koopman MS, Kappelhof M, van Voorst H, Bruggeman AE, Demchuck AM, et al. Outcome prediction based on automatically extracted infarct core image features in patients with acute ischemic stroke. Diagnostics 2022;12(8). https://doi.org/10.3390/diagnostics12081786.

[35] Appliedradiology"https://appliedradiology.com/articles/the-Radiologist-S-Gerbil-Wheel-Interpreting-Images-Every-3-4-Seconds-Eight-Hours-a-Day-at-Mayo-Clinic.

[36] Verburg E, van Gils CH, van der Velden BHM, Bakker MF, Pijnappel RM, Veldhuis WB, Gilhuijs KGA. Deep learning for automated triaging of 4581 breast MRI examinations from the DENSE trial. Radiology 2022;302(1):29–36.

[37] Devlin J, Chang M-W, Lee K, Toutanova K. BERT: pre-training of deep bidirectional transformers for language understanding. In: Burstein J, Doran C, Solorio T, editors.

Proceedings of the 2019 Conference of the North AMerican Chapter of the Association for Computational Linguistics: Human Language Technologies, Volume 1 (Long and Short Papers). Minneapolis, Minnesota: Association for Computational Linguistics; 2019. p. 4171–86.

[38] Ji J, Hou Y, Chen X, Pan Y, Xiang Y. Vision-Language model for generating textual descriptions from clinical images: model development and validation study. JMIR Format Res 2024;8(February):e32690.

[39] Wood D, Guilhem E, Kafiabadi S, Busaidi AAl, Ayisha Al B, Ahmed H, Mansoor N, et al. Automated brain abnormality detection using a self-supervised text-vision framework. Authorea Inc.; 2024. https://doi.org/10.22541/au.171571444.45086524/v2.

[40] Hond Ade, Leeuwenberg T, Bartels R, van Buchem M, Kant I, Moons KG, van Smeden M. From text to treatment: the crucial role of validation for generative large language models in health care. Lancet Digital Health 2024;6(7):e441–3.

[41] Marcus DS, Harms MP, Snyder AZ, Jenkinson M, Anthony Wilson J, Glasser MF, Barch DM, et al. Human connectome project informatics: quality control, database services, and data visualization. Neuroimage 2013;80(October):202–19.

[42] Casey BJ, Cannonier T, Conley MI, Cohen AO, Barch DM, Heitzeg MM, Soules ME, et al. The adolescent brain cognitive development (ABCD) study: imaging acquisition across 21 sites. Develop Cogn Neurosci 2018;32(August):43–54.

[43] Miller KL, Alfaro-Almagro F, Bangerter NK, Thomas DL, Yacoub E, Xu J, Bartsch AJ, et al. Multimodal population brain imaging in the UK biobank prospective epidemiological study. Nat Neurosci 2016;19(11):1523–36.

[44] Peters A, German National Cohort (NAKO) Consortium, Peters A, Halina Greiser K, Göttlicher S, Ahrens W, Albrecht M, et al. Framework and baseline examination of the German national cohort (NAKO). Eur J Epidemiol 2022;37(10):1107–24.

[45] Rutherford S, Kia SM, Wolfers T, Fraza C, Zabihi M, Dinga R, Berthet P, et al. The normative modeling framework for computational psychiatry. Nat Protoc 2022;17(7):1711–34.

[46] Rutherford S, Barkema P, Tso IF, Sripada C, Beckmann CF, Ruhe HG, Marquand AF. Evidence for embracing normative modeling. eLife 2023;12(March). https://doi.org/10.7554/eLife.85082.

[47] Ge R, Yu Y, Qi YX, Fan Y-N, Chen S, Gao C, Haas SS, et al. Normative modelling of brain morphometry across the lifespan with CentileBrain: algorithm benchmarking and model optimisation. Lancet Digital Health 2024;6(3):e211–21.

[48] Wood DA, Kafiabadi S, Al Busaidi A, Guilhem E, Montvila A, Lynch J, Townend M, et al. Accurate brain-age models for routine clinical MRI examinations. Neuroimage 2022;249(April):118871.

[49] Cole JH, Underwood J, Caan MWA, De Francesco D, van Zoest RA, Leech R, Wit FWNM, et al. Increased brain-predicted aging in treated HIV disease. Neurology 2017;88(14):1349–57.

[50] Luders E, Cherbuin N, Gaser C. Estimating brain age using high-resolution pattern recognition: younger brains in long-term meditation practitioners. Neuroimage 2016;134(July):508–13.

[51] Yin C, Imms P, Cheng M, Amgalan A, Chowdhury NF, Massett RJ, Chaudhari NN, et al. Anatomically interpretable deep learning of brain age captures domain-specific cognitive impairment. Proc Natl Acad Sci USA 2023;120(2).

[52] Di Biase MA, Tian YE, Bethlehem RAI, Seidlitz J, Alexander-Bloch AF, Thomas Yeo BT, Zalesky A. Mapping human brain charts cross-sectionally and longitudinally. Proc Natl Acad Sci USA 2023;120(20).

[53] Rahman SA, Adjeroh DA. Deep learning using convolutional LSTM estimates biological age from physical activity. Sci Rep 2019;9(1):11425.

[54] Sagers L, Melas-Kyriazi L, Patel CJ, Manrai AK. Prediction of chronological and biological age from laboratory data. Aging 2020;12(9):7626–38.

[55] Langner T, Wikstrom J, Bjerner T, Ahlstrom H, Kullberg J. Identifying morphological indicators of aging with neural networks on large-scale whole-body MRI. IEEE Trans Med Imag 2020;39(5):1430–7.

[56] Shah M, de A Inácio MH, Lu C, Schiratti P-R, Zheng SL, Clement A, de Marvao A, et al. Environmental and genetic predictors of human cardiovascular ageing. Nat Commun 2023;14(1):4941.

[57] Tian YE, Cropley V, Maier AB, Lautenschlager NT, Breakspear M, Zalesky A. Heterogeneous aging across multiple organ systems and prediction of chronic disease and mortality. Nat Med 2023;29(5):1221–31.

[58] Ecker V, Früh M, Yang B, Gatidis S, Küstner T. Deep regression for biological age estimation in multiple organs: investigations on 40,000 subjects of the UK biobank. In: ICASSP 2024 - 2024 IEEE International Conference on Acoustics, Speech and Signal Processing (ICASSP). IEEE; 2024. p. 2255–9.

[59] Jenul A, Schrunner S, Hovde Liland K, Indahl UG, Futsæther CM, Tomic O. RENT—repeated elastic net technique for feature selection. IEEE Access 2021;9:152333–46.

[60] Jenul A, Schrunner S, Huynh BN, Helin R, Futsæther CM, Liland KH, Tomic O. Ranking feature-block importance in artificial multiblock neural networks. In: Artificial neural networks and machine learning – ICANN 2022. Switzerland: Springer Nature; 2022. p. 163–75.

[61] Jenul A, Schrunner S, Pilz J, Oliver T. A user-guided bayesian framework for ensemble feature selection in life science applications (UBayFS). Mach Learn 2022;111(10): 3897–923.

[62] Li X, Gu Y, Dvornek N, Staib LH, Ventola P, Duncan JS. Multi-site fMRI analysis using privacy-preserving federated learning and domain adaptation: ABIDE results. Med Image Anal 2020;65(October):101765.

[63] Hollmann N, Müller S, Purucker L, Krishnakumar A, Körfer M, Hoo SB, Schirrmeister RT, Hutter F. Accurate predictions on small data with a tabular foundation model. Nature 2025;637(8045):319–26.

[64] Karkalousos D, Išgum I, Marquering HA, Caan MWA. ATOMMIC: an advanced toolbox for multitask medical imaging consistency to facilitate artificial intelligence applications from acquisition to analysis in magnetic resonance imaging. Comput Methods Progr Biomed 2024;256(108377):108377.

CHAPTER

15 Give me that other image: Machine learning for image-to-image translation

Onat Dalmaz[1] and Tolga Çukur[2]

[1]*Stanford University, Department of Electrical Engineering, Stanford, CA, United States;* [2]*Bilkent University, Department of Electrical and Electronics Engineering, Ankara, Turkey*

15.1 Introduction

Medical imaging modalities can capture images of bodily anatomies under a diverse array of tissue contrasts, and this capability renders them an invaluable diagnostic tool in many clinical scenarios ranging from cancer to cardiovascular imaging. Unfortunately, prolonged duration of multi-modality medical imaging protocols restrict their in an effort to avoid excessive healthcare expenditures [1]. Even when resources admit prescription of comprehensive imaging protocols, prolonged scan durations inevitably increase susceptibility to artifacts from system limitations and uncooperative patients [2]. A fundamental framework to address these critical limitations is based on image-to-image translation for imputing unacquired images of target tissue contrasts from acquired images of source tissue contrasts [3,4]. This framework can be employed to synthesize medical images that have unacceptably poor quality or that are altogether absent, to present downstream analyses with a comprehensive array of images on the anatomy under investigation [5,6]. While promising, imputation of target-contrast images is a heavily ill-posed problem, given the high dimensionality of medical images, the absence of data samples from the target contrast, and the nonlinear disparities in tissue signals between source and target contrasts [7,8].

Paralleling their prowess in numerous medical imaging tasks with their ability to capture data-driven image priors of high representational capacity, machine learning approaches based on deep neural networks have recently enabled substantial leaps in image-to-image translation tasks [9–11]. Earlier efforts for medical image translation leveraged feed-forward convolutional neural networks (CNNs) to attain significant improvements over traditional approaches such as nonlinear regression models

Machine Learning in MRI. https://doi.org/10.1016/B978-0-443-14109-6.00015-8

[12,13]. These CNNs were originally trained with pixel-wise loss terms that are prone to loss of spatial details in synthesized images [14,15]. Later studies have adopted generative adversarial networks (GANs) that employ an additional discriminator subnetwork to express an adversarial loss term, offering demonstrated benefits in preserving structural details and attaining a degree of realism [10,16,17]. Quickly becoming a gold-standard in medical image translation following their introduction, many architectural improvements on CNN backbones [18–22] as well as more robust adversarial learning strategies [23–26] have been devised to push the envelope of translation performance.

Despite the prowess of CNN-based GAN models, recent developments in the field originating from advances in machine learning are fueling paradigm shifts in the de facto standard of building translation models. On the architectural front, attention-based networks, most prominently vision transformers [27], are on the rise as a substitute to CNNs based on local, non-adaptive filtering operations [28–30]. In the context of medical image translation, vision transformer backbones that leverage global, adaptive filtering operations have been demonstrated to improve capture of long-range spatial relationships among tissues, and improved reliability against atypical anatomy [31–33]. These attributes can enhance quality of synthesized images near pathological regions, thereby improving diagnostic evaluation and characterization in clinical applications [31].

While adversarial learning mechanisms in GAN models are commonly considered to offer state-of-the-art performance in terms of realism in synthesized images, they are also known to suffer from occasional instabilities that can compromise model training [34]. In the domain of generative models, an emergent framework substitutes adversarial learning with score-based techniques (e.g., as in diffusion models) to address this critical limitation [35]. GANs rely on a game-theoretic interplay to optimize indirect measures of the data likelihood (i.e., the likelihood of the measured target image given the source image and the translation model), which has been suggested as a primary reason for training instabilities [36]. Diffusion models instead provide an explicit characterization of the likelihood thorough a many-stage transformation from the source onto the target image [37]. Several recent studies have already demonstrated the great potential of diffusion modeling for enhanced sample fidelity in medical image translation [38,39].

The exploration of advanced machine learning models, particularly vision transformers and diffusion models, are paving the way for a new era in medical image translation, significantly enhancing the capability to synthesize high-fidelity, multi-modality images. This chapter aims to survey both the basics and recent developments in image-to-image translation techniques. The chapter begins by briefly introducing prominent applications of medical image translation. It then progresses through a background in fundamentals of GAN, transformer, and diffusion models; and showcases several successful adoptions of each model family from the literature. The chapter concludes by summarizing key points and suggesting directions for future studies.

15.2 Applications of medical image translation

Medical image translation encompasses a broad range of tasks aimed at generating synthetic representations of medical data to improve diagnostics, therapy planning, and clinical decision-making. Much of existing research in this area focuses on clinically relevant mappings including synthesis of missing contrasts from acquired contrasts in MRI, synthesis of contrast-enhanced images from non-invasive MRI scans, synthesis of motion-free images from motion-corrupted images, and synthesis of missing modalities to bridge between separate scanners [40]. Imputations of missing images in these tasks promise to improve accuracy, flexibility, and clinical utility in downstream assessments based on medical images.

A prime example of medical image translation, multi-contrast MRI translation involves mapping between different MRI sequences acquired without contrast agents. This task is fundamental in cases where complementary information is diagnostically valuable, such as combining information across T_1-weighted, T_2-weighted, and fluid-attenuated inversion recovery (FLAIR) images to improve lesion detection and characterization [10,23]. Traditional approaches rely on physics-based models or interpolation techniques to approximate missing contrasts, but these often struggle with maintaining fine structural details. Learning-based methods, by contrast, enable more precise synthesis of missing contrasts by leveraging patterns observed in training datasets, providing more accurate and consistent outputs that preserve anatomical features and pathological markers.

Another example task pertains to contrast-enhanced MRI images that improve visualization of vascular structures, tumors, and tissue perfusion by using gadolinium-based agents. However, concerns over safety, cost, and accessibility have motivated efforts to synthesize contrast-enhanced images from non-contrast acquisitions. Traditional approaches rely on heuristic models or statistical techniques that may fail to capture subtle patterns of enhancement. Learning-based methods, in contrast, enable data-driven modeling of enhancement patterns, improving fidelity and generalizability [19,41]. This approach provides a radiation-free and contrast-free alternative, particularly benefiting patients with renal impairment or other contraindications.

Medical image translation may also serve to mitigate artifacts corrupting images. For instance, patient motion during MRI acquisitions introduces artifacts that degrade image quality and can compromise clinical interpretations. Conventional motion-correction techniques, such as navigator echoes or retrospective reconstruction algorithms, are often limited by acquisition speed and sensitivity to severe motion. Learning-based methods address these challenges by leveraging data-driven priors to disentangle motion artifacts and restore spatial fidelity [42,43]. This enables the reconstruction of motion-free images with reduced acquisition complexity, making imaging more robust for pediatric, elderly, and uncooperative patients.

Going beyond the bounds of a single modality, one can also define cross-modality medical image translation tasks that synthesize anatomically aligned images across separate modalities. Traditional techniques often involve image registration combined with interpolation methods, which may introduce errors in alignment and intensity scaling. In contrast, learning-based approaches can offer direct mappings that preserve anatomical consistency and modality-specific features. These methods reduce the need for multiple acquisitions, minimize radiation exposure in CT [44,45]; and provide functional imaging information from PET without additional scanning [46]; supporting applications such as radiation therapy planning and metabolic studies.

In summary, medical image translation has expanded well beyond intra-modality mappings to address clinically significant tasks such as synthesizing contrast-enhanced, motion-free, and cross-modality images. Learning-based methods provide significant advantages over traditional techniques, including enhanced flexibility, improved accuracy, and reduced dependence on hardware modifications. These developments hold promise for improving diagnostic workflows, reducing patient burden, and expanding access to advanced imaging techniques. In the next chapter, we discuss the basics of adopting learning-based frameworks for medical image translation tasks, showcasing multi-contrast MRI translation as a prominent case.

15.3 Spotlight: Multi-contrast MRI translation

15.3.1 Physics of tissue contrast in MRI

The intrinsic mechanism that MRI relies on to image tissues in based on the magnetic spins of atoms with an odd number of protons, and hydrogen the most commonly imaged nucleus in MRI due to the abundance of water in the human body (see also Chapter 3). Each hydrogen atom contains a single proton that has a net magnetic moment and that nutates around the main moment axis. In the absence of any external magnetic field, the magnetic spins across a population of hydrogen atoms are oriented randomly. The incoherence in the spin directions causes destructive interference, resulting in zero net magnetization for the population. Yet, when the population is exerted in a strong, static magnetic field (denoted as B_0), the spins tend to align themselves in parallel or anti-parallel directions with B_0. There is a limited degree of polarization in the parallel direction (i.e., longitudinal direction), which causes a net magnetization along the external field.

To generate images, a brief radiofrequency (RF) pulse is applied on the tissues to tip their net magnetization onto the transverse plane (i.e., the plane whose normal points along B_0). Once the RF pulse stops, magnetic spins of tissues relax back to their equilibrium state by decaying the transverse component of their magnetization while growing back the longitudinal component. This relaxation process is governed by tissue-specific time constants, where T_2 denotes the constant for transverse decay, and T_1 denotes the constant for longitudinal recovery. Differences in relaxation parameters among tissue are a main mechanism to generate tissue contrast in MRI.

At a certain time point t where measurements are taken, a receiver coil is used to detect the magnetic flux emanating from the tissues due to the transverse component of their magnetization (denoted as $M(\mathbf{r})$ where $\mathbf{r} = [x, y, z]$ is the vector of Cartesian spatial coordinates). The measured signal can be expressed as:

$$s(t) = \int_{\mathbf{r}} M(\mathbf{r}) e^{-t/T_2(\mathbf{r})} \exp\left(-i\gamma \int_0^t \mathbf{G}(\tau) \cdot \mathbf{r}\, d\tau \right) d\mathbf{r} \tag{15.1}$$

where $\mathbf{G}$ is the vector of magnetic field gradients $\mathbf{G} = [G_x, G_y, G_z]$ that are used to encode spatial locations, and γ is the nucleus-specific gyromagnetic ratio. Since the time integral of gradients in the signal equation reflect encoding of specific spatial frequency components in k-space (i.e., the Fourier domain), the signal can also be expressed as:

$$s(t) = \int_{\mathbf{r}} M(\mathbf{r}) e^{-t/T_2(\mathbf{r})} e^{-i2\pi[\mathbf{k}_\mathbf{r}(t)]\cdot\mathbf{r}}\, d\mathbf{r} \tag{15.2}$$

where $\mathbf{k}_\mathbf{r} = [k_x, k_y, k_z]$ denotes the vector of spatial frequencies. Thus, an image of the tissue magnetization can be reconstructed through inverse Fourier transformation as $x(\mathbf{r}) = \mathscr{F}^{-1}\{s(t)\}$.

Eq. (15.2) describes the transient MRI signal at time t following excitation with a single RF pulse. Yet, MRI scans are commonly conducted through periodic repetitions of RF excitation and following data acquisition, where the MRI signal converges onto a steady-state distribution that depends on the repetition time (TR) and the echo time (TE) between the RF pulse and data acquisition in each period. In turn, these main sequence parameters interact with tissue relaxation parameters to elicit a rich set of tissue contrasts in MRI. For instance, the measured signal levels in an image acquired with a spin-echo sequence pervasively utilized in MRI exams is given as:

$$x(\mathbf{r}) = K\rho(\mathbf{r}) \left[1 - e^{-TR/T_1(\mathbf{r})}\right] e^{-TE/T_2(\mathbf{r})}, \tag{15.3}$$

where $\rho(\mathbf{r})$ is the proton density, and K is a scaling constant. By varying TE and TR, spin-echo sequences can be attuned to emphasize differences among tissues. For instance, T_1-weighted contrast can be generated through a relatively short TR, short TE sequence, T_2-weighted contrast can be generated through a long TR, long TE sequence, while a proton-density contrast can be generated through a long TR, short TE sequence. Therefore, sequences with different parameter ranges enable acquisition of images with distinct tissue contrasts for the same underlying anatomy, enhancing the diagnostic information captured in multi-contrast MRI.

15.3.2 Learning-based methods for MRI translation

Since their introduction to the domain of MRI contrast translation, GAN-based image-to-image translation models have emerged as a mainstream technique in the field [10]. For MRI translation, conditional variants of GAN models have been devised that can be trained to map source-contrast images onto target-contrast images of the same anatomy. Yet, unlike conventional deep learning models based on a single network, GANs offer improved realism in synthesized images by adopting an aggregate architecture comprising two subnetworks [47]: a generator G and a discriminator D (See Fig. 15.1). The generator receives the source image x^A of contrast A along with a random noise vector z to synthesize a target image of contrast B, $\hat{x}^B = G(z, x^A)$. Note that, for conditional image synthesis tasks, z is commonly dropped to avoid undesirable stochasticity in the trained models. Meanwhile, the discriminator receives the source image x^A along with either the synthesized target image $\hat{x}^B$ or the actual target image x^B in order to distinguish between synthetic and actual samples, $o = D(\hat{x}^B \text{ or } x^B, x^A)$ where o is a probability value between 0 and 1.

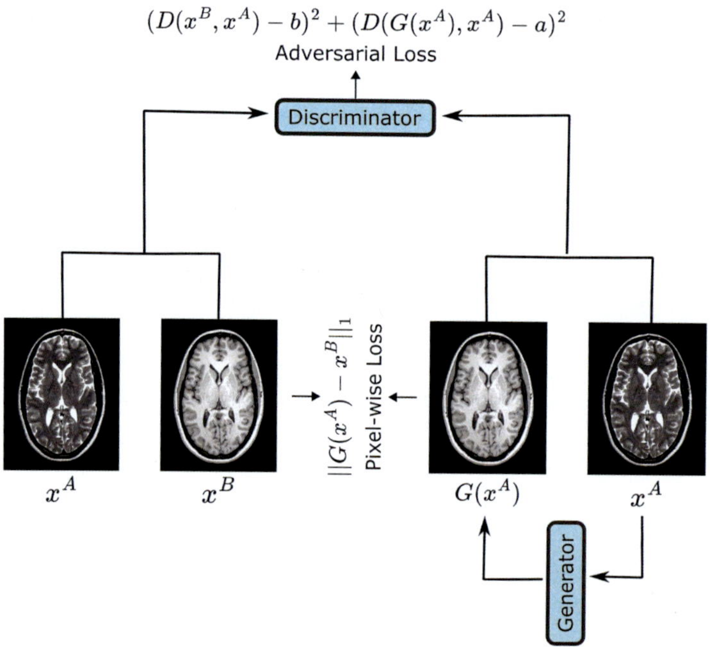

FIGURE 15.1

A supervised conditional GAN model for multi-contrast MRI translation. The model comprises a generator subnetwork that produces synthetic target-contrast images given as input actual source-contrast images of the same underlying anatomy. It also includes a discriminator that distinguishes between synthetic and actual target-contrast images. The interplay between the two subnetworks is set up through an adversarial loss term. In supervised GAN models, an additional pixel-wise loss term is commonly used to enforce pixel-level correspondence between synthetic and actual target-contrast images. Note that this requires source and target images of an anatomy that are co-registered in spatial domain and paired across subjects.

GAN models are commonly trained through a game theoretic interplay between the generator and the discriminator:

$$\min_G \max_D V(D,G) = \mathbb{E}_{x^B,x^A}\left[\log D\left(x^B,x^A\right)\right] + \mathbb{E}_{x^A}\left[\log\left(1 - D\left(G\left(x^A\right),x^A\right)\right)\right]. \tag{15.4}$$

In the mini-max game formulation above, the generator parameters are updated so as to minimize the second term in the optimization objective *V*, which reflects the expectation that the discriminator correctly classifies the synthetic target image. Meanwhile, the discriminator parameters are updated so as to maximize the sum of the two terms in *V*, which reflects the expectation that the discriminator correctly classifies both actual and synthetic target images. Since the mini-max formulation has been suggested to suffer from saturation problems, alternative optimization frameworks such as least-square GANs (LS-GAN) have received attention [48]. LS-GANs have been pervasively adopted in MRI contrast translation studies given their improved training behavior:

$$\min_G V^{(G)} = \mathbb{E}_{x^A}\left[\left(D\left(G\left(x^A\right),x^A\right) - c\right)^2\right], \tag{15.5}$$

$$\min_D V^{(D)} = \mathbb{E}_{x^b,x^A}\left[\left(D\left(x^B,x^A\right) - b\right)^2\right] + \mathbb{E}_{x^A}\left[\left(D\left(G\left(x^A\right),x^A\right) - a\right)^2\right]. \tag{15.6}$$

In the LS-GAN formulation, the objectives for the generator and discriminator are relatively decoupled to avoid saturation during training. The generator is trained to minimize the objective V^G, which reflects the expectation that the discriminator correctly classifies synthetic target images when $c = 1$. Meanwhile, the discriminator is trained to minimize the objective V^D, which reflects the expectation that the discriminator correctly classifies both actual and synthetic target images when $b = 0$ and $a = 1$.

Two separate families of conditional GAN models have been employed in the MRI literature based on the nature of training sets that they require. In supervised variants, it is assumed that the training set contains paired sets of source and target images from the same anatomical locations within the same subjects (See Fig. 15.1). Thus, pixel-wise loss (ℓ_1 or ℓ_2) terms that measure the pixel-level consistency between the synthetic and actual target-contrast images can be used jointly with adversarial loss in supervised variants. Commonly, an ℓ_1-norm based pixel-wise loss is used to enhance the ability to preserve edge sharpness and details in synthesized images:

$$L_{pix} = \mathbb{E}_{x^A,x^B}\left[\left\|G\left(x^A\right) - x^B\right\|_1\right]. \tag{15.7}$$

The combination of pixel-wise and adversarial losses is considered to be particularly beneficial in image translation tasks for maintaining visual fidelity and ensuring that translated images exhibit clear and distinct features.

While supervised variants have generally been reported to enable superior performance due to close guidance from ground-truth target images [10], they introduce stringent requirements on training data. In particular, prolonged exams are required to curate training data, where both source and target contrast images are collected in each subject [26]. Furthermore, images for the two contrasts must be spatially registered to compensate for potential motion in between the scans. The interpolation operations for motion correction can inevitably cause a degree of spatial smoothing, so registering onto target-contrast images may allow for better preservation of details in synthetic images [10].

In unsupervised variants, the requirements on the training set are relaxed to permit use of unpaired sets of source and target images [45]. This relaxation can allow for use of spatially unaligned sets of source-target images within the same subjects. Moreover, it can also allow for use of source and target contrast images from entirely different subjects. Unlike in supervised variants, pixel-wise loss terms cannot be employed with unsupervised variants due to lack of pixel-level correspondence between synthetic and actual target-contrast images. Instead, cycle-consistency losses are commonly adopted wherein two separate models are trained that respectively map source images onto target images, and target images onto source images [45]. In this way, an actual source image from a given subject can be used to generate a synthetic target image, and a reconstruction of the source image can then be obtained from the synthetic target image (See Fig. 15.2). Finally, a pixel-wise loss between actual and reconstructed source images can be used. A similar approach enables expression of a second pixel-wise loss between actual and reconstructed target images. These two pixel-wise losses collectively define the cycle-consistency loss:

$$L_{cyc} = \mathbb{E}_{x^A}\left[\left\|G_2\left(G_1\left(x^A\right)\right) - x^A\right\|_1\right] + \mathbb{E}_{x^B}\left[\left\|G_1\left(G_2\left(x^B\right)\right) - x^B\right\|_1\right], \tag{15.8}$$

where G_1 denotes the generator that maps from contrast A onto contrast B, and G_2 denotes the generator that maps from contrast B onto contrast A.

Whether supervised or unsupervised, once a conditional GAN model is trained for a specific MRI contrast translation task, the generator subnetwork is sufficient for inference on a test set. In particular, the acquired source-contrast image for a test subject can be input to the trained generator along with random noise variables (albeit these variables are often omitted to limit undesirable stochasticity in the output). A forward pass thorough the generator then produces synthetic target-contrast images for the same anatomy. Since only a single forward pass is sufficient to generate output, conditional GAN models have relatively fast inference times that improve practicality [39].

15.4 Advanced methods: Transformer-based translation

The mainstream GAN models for medical image translation have traditionally adopted CNN backbones in their generator and discriminator subnetworks. CNN architectures utilize compact, local filters to extract image features, and the filter

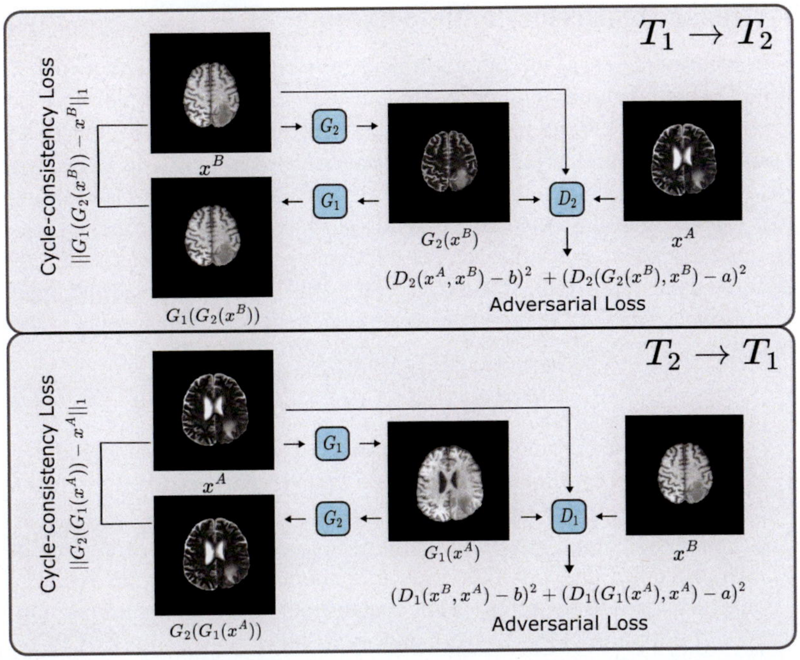

FIGURE 15.2

An unsupervised conditional GAN model for multi-contrast MRI translation. The model comprises a pair of generator subnetworks that perform mapping from source onto target contrast and from target onto source contrast, respectively. It also includes a pair of discriminators that distinguish between synthetic and actual images for target and for source contrasts, respectively. In addition to adversarial loss terms for the generator-discriminator pairs, cycle-consistency loss is often employed to enforce pixel-level correspondence between actual source images and their reconstructions, and between actual target images and their reconstructions. Note that this enables the model to be trained on source and target images from separate anatomies.

coefficients are kept static (i.e., they do not vary across input samples) and shift-invariant (i.e., they do not vary across different spatial locations within images). While these inductive biases promote high computational efficiency in CNN architectures by limiting the number of model parameters, they also limit the sensitivity of CNNs to long-range spatial interactions between distant image regions. Furthermore, the static, shift-invariant filters have limited capacity to adapt to subject-specific variations in anatomical features [30]. An emergent architectural framework to address these limitations is based on vision transformers [27]. In this section, we first overview the fundamental operators in vision transformers that permit non-local, adaptive image processing, and then showcase an early example of transformer-based models for MRI contrast translation.

15.4.1 Vision transformer architectures

Vision Transformers (ViTs) use attention-based filtering operations that span across long distances, and that utilize filter coefficients derived based on input images [27]. In ViT architectures, the input image is split into a sequence of tokens, where a token can be taken as an individual pixel or more commonly as a small, individual patch of pixels to attain lower computational complexity (see also Chapter 2). Each token is projected onto a high-dimensional embedding space prior to further processing. ViTs do not carry any locality bias, so the relative positions of tokens within the image are stored using additive positional encodings [27]. Assuming that the image x is split into N tokens where f^n denotes the nth token, these operations can be expressed as:

$$z_0 = \left[E(f^1); E(f^2); \ldots; E(f^N)\right] + P^{pos}. \tag{15.9}$$

Here, $E(\cdot)$ represents the projection operator that derives the token embeddings, and P^{pos} denotes positional encodings set based on known analytic functions such as sinusoids or taken as learnable variables [30].

Token embeddings supplemented with positional encodings are then passed through a transformer encoder that uses self-attention filtering to extract long-range contextual image features. The transformer encoder cascades multiple multi-head self-attention (MSA) layers and multi-layer perceptron (MLP) blocks, interleaved with layer normalization (LN) layers for stability [49]. Assuming that the input feature map to the l th stage of the cascade is z_{l-1}, the operations through the l th stage are given as:

$$z_l' = \mathbf{MSA}(\mathbf{LN}(z_{l-1})) + z_{l-1}, \tag{15.10}$$

$$z_l = \mathbf{MLP}(\mathbf{LN}(z_l')) + z_l'. \tag{15.11}$$

Here, z_l' denotes the compound output of the MSA and LN supplemented with the input feature map through a residual connection. Meanwhile, z_l denotes the output of the MLP and LN again supplemented with the input residually.

At the core of the ViT lies the MSA layer that leverages self-attention as a novel mixing operation for tokens in the input image [50]. To enable self-attention computations, the MSA layer first derives the "query" (q), "key" (k), and "value" (v) from the input feature map z through learnable linear transformations:

$$q = zW^Q, \tag{15.12}$$

$$k = zW^K, \tag{15.13}$$

$$v = zW^V, \tag{15.14}$$

where W^Q, W^K, and W^V are the learnable transformation matrices for query, key, and value, respectively. First, attention weights between each pair of tokens are computed based on the similarity between the "query" and "key" tokens, followed by a softmax operation for normalization:

$$A = \text{softmax}\left(\frac{q\,k^T}{\sqrt{N_D}}\right), \tag{15.15}$$

where N_D denotes the embedding dimensionality. These normalized attention weights are then used as filter coefficients where self-attention filtering is exercised by computing a weighted combination of the "value" tokens as $\boldsymbol{SA}(z) = Av$. Analogous to separate feature channels with distinct kernels in CNNs, separate attention heads are used that comprise a different set of filter coefficients in ViTs. Overall, the mapping through the MSA layer is given as:

$$\boldsymbol{MSA}(z) = [\boldsymbol{SA}_1(z); \boldsymbol{SA}_2(z); \ldots; \boldsymbol{SA}_H(z)]U_{msa}, \tag{15.16}$$

where $h \in \{1, 2, \ldots, H\}$ denotes the index of the attention head and U_{msa} is a projection matrix to compute the output feature maps.

15.4.2 Hybrid Transformer-CNN translation models

The ability of ViTs in capturing long-range spatial dependencies in MR images promises enhanced sensitivity to global context. Yet, this may come at the potential expense of reduced sensitivity to local context, if relatively large-sized patches are used to tokenize the images in an effort to limit computational complexity. In sharp contrast, the inherent design of CNNs forces them to focus primarily on local context while at the same time enjoying high computational efficiency. Thus, a common approach to combine the benefits of the two architectural families has been to devise hybrid architectures, where feature maps are processed with CNN modules at relatively high resolutions and with ViT modules at relatively low resolutions. An early example of this hybrid approach is residual vision transformers (ResViT), which introduces the first transformer-based generator for an adversarial MRI translation model in the literature [31].

ResViT employs a hybrid architecture that leverages the localized feature extraction capabilities of CNNs and the global contextual expressiveness afforded by ViTs. The generator within ResViT follows an encoder-information bottleneck-decoder pathway, where the bottleneck features a series of Aggregated Residual Transformer (ART) blocks (See Fig. 15.3). These blocks are designed to fuse local features extracted by CNN modules with contextual features derived from ViT modules, facilitated by external skip connections promoting a rich information flow. ART blocks in the information bottleneck distill task-relevant information through a synergistic interplay of convolutional and transformer operations in order maintain high local and global sensitivity. To mitigate risks of overfitting and improve computational efficiency, the weights for the ART blocks are shared across different stages in the information bottleneck. Representative synthetic target-contrast images recovered from source-contrast images depicted in Fig. 15.4 showcase the enhanced synthesis performance of ResViT over competing methods. In particular, ResViT achieves improved spatial acuity in depiction of both healthy and pathological tissue in synthetic images, and more accurate capture of the shape and signal intensity of lesions.

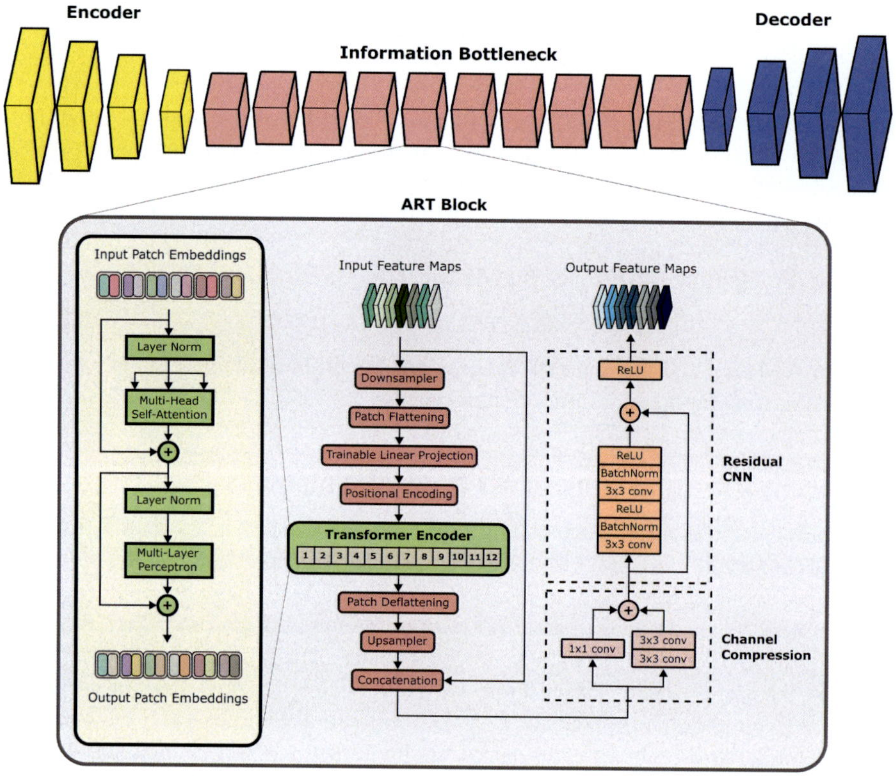

FIGURE 15.3

ResViT is a transformer-based GAN model for medical image translation applications including multi-contrast MRI translation [31]. ResViT's generator subnetwork follows an encoder—information bottleneck—decoder structure. While the encoder and decoder are composed of convolutional layers, the information bottleneck contains ART blocks, each equipped with a transformer encoder and a CNN encoder, residually connected to improve information flow. The transformer encoder improves capture of long-range spatial context, while the CNN encoder improves capture of local features.

The clinical utility of MR imaging is predicated on the ability to obtain a suite of different contrasts to extend the diagnostic information available on tissue characteristics and pathology. However, depending on many factors from patient motion to time limitations, the specific set of contrasts that are acquired can show undesirable variations across subjects. While a separate translation model could be trained in principle to perform the translation task for each source-target configuration, this can be impractical as it necessitates substantially higher run times for model training and it may compromise model performance as fewer training data will be available for each individual task.

To allow for imputation of an arbitrary target contrast within a multi-contrast MRI protocol, the encoder and decoder in recent transformer-based translation

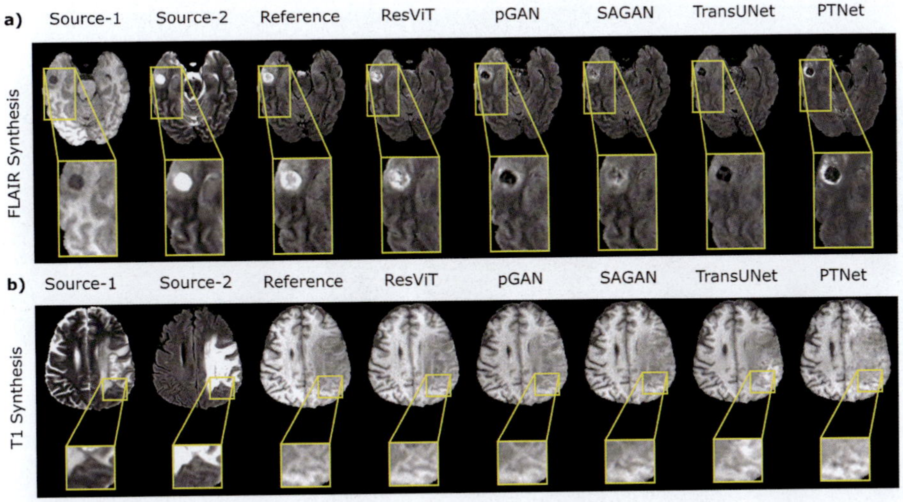

FIGURE 15.4

Representative synthetic FLAIR images (a) and synthetic T_1-weighted images (b) in Glioma patients. Synthetic target images recovered through ResViT, pGAN (a convolutional GAN model; [10], SAGAN (self-attention convolutional GAN model; [30], TransUNet (hybrid transformer-CNN model; [28], and PTNet (pure transformer model [51]; are displayed. Ground-truth target images (Reference) are also shown along with the two source contrasts (Source-1, Source-2) provided as input to the translation models.

models (e.g., ResViT [31]; MMT [32] incorporate multiple channels, one for each sequence within the protocol (See Fig. 15.5). On the encoder side, the missing sequences are masked through multiplication with zero and the acquired sequences are input without masking. On the decoder side, all sequences are reconstructed without any masking. Yet, the loss function is only expressed over output channels for the missing sequences. Model training can then be performed through this reconstruction loss added onto regular pixel-wise and adversarial losses. This unification approach improves versatility in practical utilization of MRI translation, since a single trained model can be deployed to handle various different types of missing contrasts within a desired MRI protocol [19,23].

15.5 Advanced methods: Diffusion-based translation

GAN models have served a pivotal role in enhancing the realism of synthetic images in medical image translation, significantly boosting their perceptual quality. However, despite their immense success and broad adoption in the field, several characteristic limitations can plague training of GAN models. Note that GANs employ a game theoretical interplay between the generator and discriminator subnetworks,

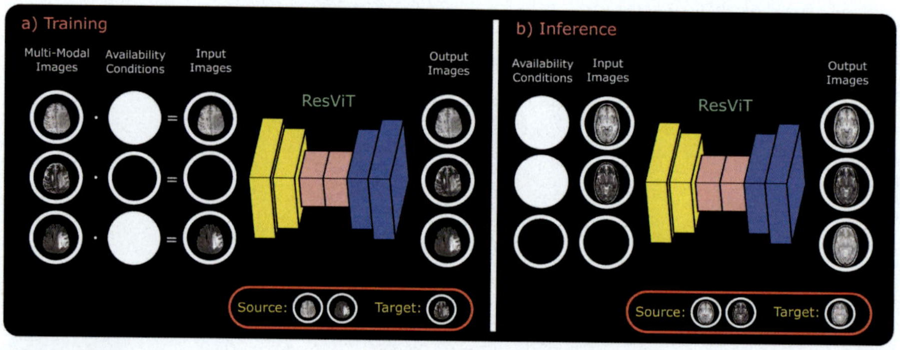

FIGURE 15.5

To allow for imputation of an arbitrary target contrast within a multi-contrast MRI protocol, the encoder and decoder in a translation model can be modified to include multiple channels, one for each sequence within the protocol. While missing sequences are masked on the encoder side, all sequences are reconstructed on the decoder side. A single unified model can be trained to cope with a diverse set of translation tasks, based on variable configurations for missing and acquired sequences. Afterward, an arbitrary missing sequence in the protocol can be imputed flexibly during inference using the trained model.

which seek antagonistic goals [36]. When the two subnetworks have reasonably well-matched and stable learning curves during training, the resultant GAN models excel in synthesizing diverse, high-quality images that accurately reflect the properties of actual target-contrast images [10]. Yet, the interplay between the two subnetworks is quite prone to instabilities that can cause premature convergence and mode collapse in GAN models [52,53]. When such instabilities are prominent, they can severely compromise image quality and diversity.

A recently emerging paradigm in generative modeling is based on diffusion probabilistic models that take an entirely different approach to learn the data distribution [37]. While GANs employ a one-shot synthesis process to map the source onto the target image, diffusion models use a very gradual process with thousands of steps to achieve a desired mapping. Furthermore, unlike GANs that only indirectly assess the likelihood of the data thorough an adversarial loss, diffusion models provide an explicit characterization of the likelihood in terms of the score function (i.e., gradient of the likelihood). Taken together, these attributes enable diffusion models to offer enhanced stability while maintaining similar quality to state-of-the-art GAN models in many image processing tasks [34]. In this section, we first overview the fundamentals of diffusion probabilistic models, and then showcase an early example of diffusion-based methods for MRI contrast translation.

15.5.1 Generative diffusion models

Diffusion models learn to synthesize images from a target distribution through a bidirectional, multi-step process between random Gaussian noise images on one end and actual image samples on the other end [37]. In the forward process, additive Gaussian noise is introduced on top a clean image x_0 progressively over time steps:

$$x_t = \sqrt{1-\beta_t} x_{t-1} + \sqrt{\beta_t}\epsilon, \epsilon \sim \mathcal{N}(0, I). \tag{15.17}$$

Here, x_t denotes the image sample at time step t, β_t represents the variance of the noise added at step t, ϵ denotes a random noise variable, $\mathcal{N}$ is the Gaussian distribution, and I stands for the identity matrix. Since the forward process forms a first-order Markov chain, the forward transition probability from x_{t-1} onto x_t can be expressed as:

$$q(x_t|x_{t-1}) = \mathcal{N}\left(x_t; \sqrt{1-\beta_t}\, x_{t-1}, \beta_t I\right). \tag{15.18}$$

While the forward process is analytically characterized based on Eq. (15.17), the reverse process starts on a pure noise sample x_T, and aims to gradually remove incremental noise added at each step to recover x_0. Since an analytical characterization of the denoising mappings is unavailable, diffusion models instead employ a neural network R to perform the denoising task. Note that probability theory dictates that for a sufficiently large number of time steps T, the reverse transition probability from x_t onto x_{t-1} can be approximated as:

$$q(x_{t-1}|x_t) = \mathcal{N}(x_{t-1}; \mu(x_t, t), \Sigma(x_t, t)), \tag{15.19}$$

where μ denotes the mean vector and Σ denotes the covariance matrix for the Gaussian distribution at time step t, both of which are unknown a priori. Yet, utilizing the network R as a means to estimate μ and Σ, one can draw samples from Eq. (15.20) to obtain denoised image samples at each time step. Assuming that the network-based estimates for these variables are μ_θ and Σ_θ, the network-based estimate for the reverse transition probability is given as:

$$p_\theta(x_{t-1}|x_t) = \mathcal{N}(x_{t-1}; \mu_\theta(x_t, t), \Sigma_\theta(x_t, t)), \tag{15.20}$$

where θ denotes the network parameters.

The training objective for the denoising network in diffusion models is commonly expressed as a variational lower bound on the log-likelihood of the data [34]:

$$L_{vb} = \mathbb{E}_{q(x_{0:T})}\left[\log \frac{p_\theta(x_{0:T})}{q(x_{1:T}|x_0)}\right] \le \mathbb{E}_{q(x_0)}[\log p_\theta(x_0)]. \tag{15.21}$$

Here, the objective comprises expectation over the distribution of clean images and the sequence of noisy image samples derived from those (i.e., $q(x_{0:T})$). The objective can be written as the sum of Kullback-Leibler (KL) divergences across all reverse transition steps, except for the final step T that is independent of θ:

$$L_{vb} = \log p_\theta\left(x_0|x_1\right) - \sum_{t=1}^{T} KL\left(q(x_{t-1}|x_t, x_0) \parallel p_\theta(x_{t-1}|x_t)\right). \tag{15.22}$$

In a simplified scenario where Σ is neglected, an estimate of μ can be derived based on an estimate of the added noise:

$$\mu_\theta(x_t, t) = \frac{1}{\sqrt{\alpha_t}}\left(x_t - \frac{\beta_t}{\sqrt{1-\overline{\alpha}_t}}\,\epsilon_\theta(x_t, t)\right), \tag{15.23}$$

with $\alpha_t = 1 - \beta_t$ and $\overline{\alpha}_t$ denotes the cumulative product of (α_t) from T to t. Therefore, the network can be trained to estimate the additive noise ϵ in between consecutive time steps:

$$L_{err} = \mathbb{E}_{t,x_0,\epsilon}\Big[\ \| \epsilon - \epsilon_\theta\Big(\sqrt{\overline{\alpha}_t}x_0 + \sqrt{1-\overline{\alpha}_t}\epsilon, t\Big)\|_2^2\Big]. \tag{15.24}$$

Note that a conventional approach to minimize the objective L_{err} is to perform Monte-Carlo simulations with samples drawn uniformly across time steps.

15.5.2 Hybrid diffusion-adversarial translation models

Machine learning studies have demonstrated benefits of diffusion models over conventional GAN-based methods in numerous applications. Yet, adapting diffusion models for MRI translation presents several non-trivial challenges. First, diffusion models are primarily devised to learn a mapping from Gaussian noise onto target images, so adopting them for contrast translation requires modifications to ensure that diffusion models receive guidance from a separate source image [54]. Second, diffusion models involve thousands of reverse diffusion steps for generating a single image sample, introducing substantial inefficiency during inference [55]. Third, diffusion models are typically trained with loss functions that assess pixel-level correspondence between synthetic and ground-truth target images, so they are inherently supervised [35].

Recently introduced to overcome these challenges, SynDiff is the first diffusion-based method in the literature for unsupervised medical image translation [39]. Demonstrated for both multi-contrast MRI and MRI-CT translation tasks, SynDiff uses a hybrid of diffusion and adversarial models to enjoy the benefits of both model families (See Fig. 15.6). In particular, SynDiff improves sampling efficiency by using only few large diffusion steps to transform random noise onto target images, and adversarial loss is used to maintain high accuracy over such large diffusion steps. To adapt this hybrid model for image-to-image translation, the denoising network in SynDiff employs an additional input channel that accepts the source-contrast image of an anatomy, and this source image is used as a constant reference while synthesizing the target-contrast image starting from random noise. Accordingly, the forward process in SynDiff that is reliant on only target-contrast images can be expressed as:

$$x_t^B = \sqrt{1-\gamma_t}x_{t-k}^B + \sqrt{\gamma_t}\epsilon, \tag{15.25}$$

$$q(x_t^B|x_{t-k}^B) = \mathcal{N}\Big(x_t^B; \sqrt{1-\gamma_t}\,x_{t-k}^B, \gamma_t I\Big), \tag{15.26}$$

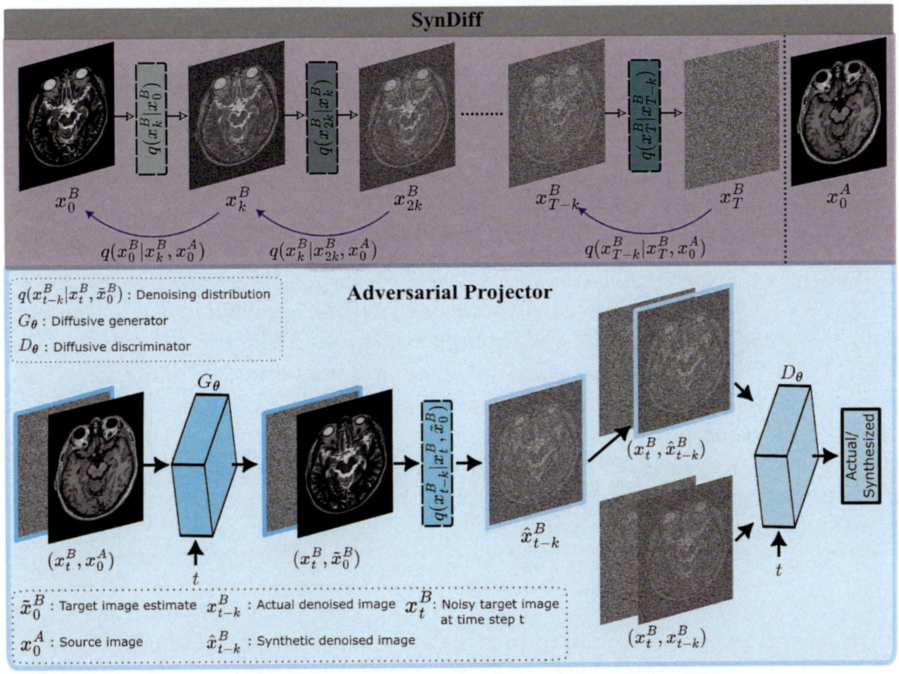

FIGURE 15.6

SynDiff is an adversarial diffusion model for unsupervised medical image translation applications, including multi-contrast MRI translation [39]. SynDiff uses a fast diffusion process to generate a synthetic target-contrast image starting from a random Gaussian noise image. This generation is guided through an acquired source-contrast image. While conventional diffusion models suffer from poor accuracy in reverse diffusion mapping required to generate synthetic images, SynDiff mitigates this limitation by adopting adversarial learning to capture the distribution of required reverse transition probabilities implicitly. As such, it is able to produce high-quality synthetic images with a fast sampling process.

where $k \leq 1$ denotes the number of time points covered in a single sampling step (i. e., larger steps taken across $[0\ T]$ for larger k), γ_t is the noise variance determined by an exponential schedule, and B denotes the target contrast. Meanwhile, a conditional reverse process is adopted to receive source-image guidance, and the reverse transition probability can be expressed using Bayes' rule as:

$$q\left(x^B_{t-k}|x^B_t, x^A_0\right) = q\left(x^B_t|x^B_{t-k}, x^A_0\right)\frac{q\left(x^B_{t-k}|x^A_0\right)}{q\left(x^B_t|x^A_0\right)}. \tag{15.27}$$

Note that clean images from the source and target contrasts depicting the same anatomy, that is, x^A_0 and x^B_0, are related through a non-linear albeit nearly deterministic relationship assuming relatively low noise levels in MRI acquisitions.

Assuming that this relationship is injective, then the reverse transition probability can be equivalently expressed as:

$$q(x_{t-k}^B|x_t^B, x_0^A) = q(x_t^B|x_{t-k}^B, x_0^B)\frac{q(x_{t-k}^B|x_0^B)}{q(x_t^B|x_0^B)}. \tag{15.28}$$

When small diffusion steps are employed, that is, $k/T \ll 1$, the reverse transition probability can be well approximated through a Gaussian distribution [37]:

$$q(x_{t-k}^B|x_t^B, x_0^A) = \mathcal{N}(x_{t-k}^B; \mu(x_t^B, x_0^B), \bar{\gamma}_t I), \tag{15.29}$$

In such cases, a denoising network can be used to estimate μ as in conventional diffusion models, and a denoised sample x_{t-k}^B can then be drawn based on Eq. (15.29). However, the normality assumption breaks down for large step sizes, that is, $k/T \gg 1$, and so an explicit characterization for reverse sampling lacks in such fast diffusion processes [39]. To improve accuracy in reverse diffusion mappings when $k/T \gg 1$, SynDiff uses adversarial learning to implicitly characterize the reverse transition probabilities. In particular, a generator subnetwork is used to parametrize the reverse transition probability as follows:

$$p_\theta(x_{t-k}^B|x_t^B, x_0^A) := q\left(x_{t-k}^B \middle| x_t^B, \tilde{x}_0^B = G_\theta(x_t^B, x_0^A, t)\right), \tag{15.30}$$

where the generator G_θ is used to produce an estimate of the clean target-contrast image $\tilde{x}_0^B$. A discriminator subnetwork, $D_\theta(\{\hat{x}_{t-k}^B \text{ or } x_{t-k}^B\}, x_t^B, t)$, is also employed to differentiate between image samples drawn from the parametrized and true reverse transition probabilities, $p_\theta(x_{t-k}^B|x_t^B, x_0^A)$ versus $q(x_{t-k}^B|x_t^B, x_0^A)$. Including an adversarial loss in the training objective ensures that the generator learns to mimic the actual reverse transition probabilities closely. Thus, SynDiff leverages a fast diffusion process to enhance sampling efficiency while benefiting from the stability of diffusive learning, and it mitigates performance losses due to large step sizes by implicitly capturing reverse transition probabilities through adversarial learning.

Finally, unsupervised learning is mediated by using an add-on cycle-consistent GAN architecture that receives source images to provide estimates for paired target images, and that receives target images to provide estimates for paired source images. These paired image estimates are then provided to the main diffusion architecture, and the entire architecture is trained end-to-end. When merged together, these building blocks enable SynDiff to train performant and computationally efficient medical image translation models on unpaired training datasets. Representative synthetic target-contrast images recovered from acquired source-contrast images as shown in Fig. 15.7 visually demonstrate the enhanced performance of SynDiff. Competing GAN and diffusion methods suffer from inaccurate estimation of signal intensity and shape particularly near pathology, and show a degree of spatial smoothing. In contrast, SynDiff achieves high spatial acuity, accurate intensity and shape estimation for lesions, and it generates synthetic images with significantly higher visual similarity to the reference images.

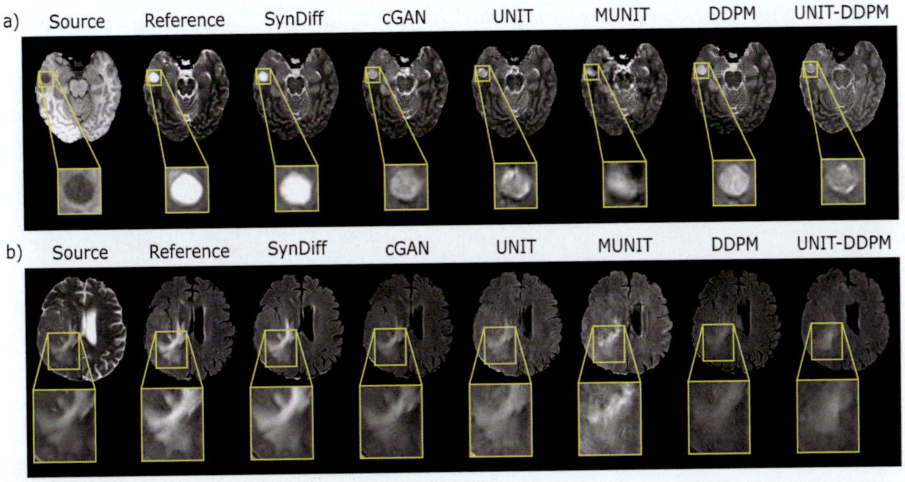

FIGURE 15.7

Representative synthetic T_2-weighted images (a) and synthetic FLAIR images (b) in Glioma patients. Synthetic target images recovered through SynDiff as well as cGAN, UNIT, MUNIT (GAN models [10,35]; and DDPM, UNIT-DDPM (diffusion models [35,37]; are displayed. Ground-truth target images (Reference) are also shown along with the source contrast (Source) provided as input to the translation models.

15.6 Summary and future directions

Medical image translation is a challenging inverse problem that involves prediction of nonlinear variations in tissue signals between source and target contrasts of the same anatomy. While traditional methods such as classical regression models have demonstrated that this task is approachable, their performance levels left much to be desired. Addressing this long-standing need, recent machine learning approaches based on deep neural networks have led a paradigm shift in medical image translation by providing regularization through data-driven image priors. P GAN methods have become a mainstream framework in this domain given their enhanced realism and overall image quality. Yet, canonical CNN-based generators in GAN models show limited capture of long-range context. Moreover, GAN models can also suffer from training instabilities that can compromise their reliability. Several emergent frameworks have recently been introduced that promise to address these limitations. On the architectural side, transformer models have emerged as a promising substitute for all-convolutional architectures, to improve sensitivity to long-range spatial relationships in medical images. On the side of learning paradigms, diffusion models have emerged in lieu of purely adversarial learning, in order to improve training stability without compromising sample quality. Recent studies have begun to demonstrate that, with a synergistic combination of these advanced approaches [56], it may be possible to attain more performance models for medical image translation.

Despite the promising results reported in the rapidly growing volume of studies on medical image translation, it is important to note that methodological efforts for translation tasks are still at a relatively early stage. Looking ahead, the field of medical image translation is poised to benefit from further innovations in learning-based modeling frameworks. On the architectural front, transformer backbones have set new standards on long-range contextual sensitivity, although this benefit comes at the expense of elevated computational complexity that yields inevitable trade-offs between capture of local versus global image features [27,57]. Development of alternative backbones that promise high sensitivity to long-range context in medical images while avoiding the memory and run time requirements of transformers can hence boost practicality of learning-based medical image translation for diverse deployment setups [58]. Parallel improvements are also viable for the learning paradigms, as diffusion models that offer enhanced training stability over adversarial approaches are not without limitation themselves. In particular, recently proposed denoising diffusion models use degradation operators based on Gaussian noise and hence recovery operators that intend to denoised noisy image samples. This results in a task-agnostic formulation as the learned task of image denoising in conventional diffusion models is irrelevant to the desired task of mapping between source and target modality images [37,59]. Development of alternative generative modeling approaches that instead inject a higher level of task-specific information into learned models such as emergent diffusion bridges can offer enhanced fidelity in medical image translation tasks [44].

Another major development milestone pertains to the successful translation of sophisticated machine-learning techniques into clinical settings. In this context, it will be essential to demonstrate that medical image translation models do not only perform well under within-domain settings where the scanner and sequence related parameters are similar between the training and test sets, but that they also generalize well under out-of-domain settings with substantial variability in scanner hardware and imaging protocols [60]. Out-of-domain generalization has been a long sought-after goal in many application domains of machine learning. Yet, robust solutions are yet to be devised to achieve this goal. This shortcoming is partly related to difficulties in curating training sets of sufficient size and diversity to densely cover the practically-relevant set of image features that can be encountered in broad subject populations. Collaborations between healthcare institutions that have access to different imaging hardware/software along with different subject pools can be a native solution to this paramount challenge [61,62]. Such multisite collaborations can be facilitated by overcoming regulatory and bureaucratic barriers through federated learning paradigms, which enable decentralized training of deep neural networks without requiring sites to transfer imaging data [63]. It is likely that the full potential of advanced translation models might be unraveled once the ability to flexibly train them on large-scale datasets emerges. Eventually, establishing a synergy between powerful models and learning paradigms might be the key to realize performant and generalizable medical image translation in the clinic.

References

[1] Bakas S, Sotiras A, Bilello M, Rozycki M, Kirby J, Freymann J, Farahani K, Davatzikos C. Advancing the cancer genome atlas glioma MRI collections with expert segmentation labels and radiomic features. Sci Data 2017;4.

[2] Krupa K, Bekiesińska-Figatowska M. Artifacts in magnetic resonance imaging. Pol J Radiol 2015;80:93–106.

[3] Iglesias JE, Konukoglu E, Zikic D, Glocker B, Van Leemput K, Fischl B. Is synthesizing MRI contrast useful for inter-modality analysis? MICCAI 2013:631–8.

[4] Roy S, Carass A, Prince J. A compressed sensing approach for MR tissue contrast synthesis. Inf Process Med Imaging 2011;22:371–83.

[5] Huang Y, Beltrachini L, Shao L, Frangi AF. Geometry regularized joint dictionary learning for cross-modality image synthesis in magnetic resonance imaging. In: Simulation and synthesis in medical imaging. Cham: Springer International Publishing; 2016. p. 118–26.

[6] Roy S, Carass A, Prince JL. Magnetic resonance image example-based contrast synthesis. IEEE Trans Med Imag 2013;32(12):2348–63.

[7] Huang Y, Shao L, Frangi AF. Cross-modality image synthesis via weakly-coupled and geometry co-regularized joint dictionary learning. IEEE Trans Med Imag 2018;37(3): 815–27.

[8] Jog A, Carass A, Roy S, Pham DL, Prince JL. Random forest regression for magnetic resonance image synthesis. Med Image Anal 2017;35:475–88.

[9] Chartsias A, Joyce T, Valerio Giuffrida M, Tsaftaris S. Multimodal MR synthesis via modality-invariant latent representation. IEEE Trans Med Imag 2018;37(3):803–14.

[10] Dar SUH, Yurt M, Karacan L, Erdem A, Erdem E, Çukur T. Image synthesis in multi-contrast MRI with conditional generative adversarial networks. IEEE Trans Med Imag 2019;38(10):2375–88.

[11] Van Nguyen H, Zhou K, Vemulapalli R. Cross-domain synthesis of medical images using efficient location-sensitive deep network. In: MICCAI. Cham: Springer International Publishing; 2015. p. 677–84.

[12] Joyce T, Chartsias A, Tsaftaris S. Robust multi-modal MR image synthesis. In: MICCAI; 2017. p. 347–55.

[13] Wei W, Poirion E, Bodini B, Durrleman S, Colliot O, Stankoff B, Ayache N. Fluid-attenuated inversion recovery MRI synthesis from multisequence MRI using three-dimensional fully convolutional networks for multiple sclerosis. J Med Imaging 2019;6:1.

[14] Liu F. SUSAN: segment unannotated image structure using adversarial network. Magn Reson Med 2018;81(5):3330–45.

[15] Yu B, Zhou L, Wang L, Fripp J, Bourgeat P. 3d cgan based cross-modality mr image synthesis for brain tumor segmentation. In: IEEE ISBI; 2018. p. 626–30.

[16] Armanious K, Jiang C, Fischer M, Kuestner T, Hepp T, Nikolaou K, Gatidis S, Yang B. MedGAN: medical image translation using GANs. Comput Med Imag Graph 2019b;79: 101684.

[17] Lee D, Moon W-J, Ye JC. Assessing the importance of magnetic resonance contrasts using collaborative generative adversarial networks. Nat Mach Intell 2020;2(1):34–42.

[18] Lan H, the Alzheimer Disease Neuroimaging Initiative, Toga AW, Sepehrband F. Three-dimensional self-attention conditional gan with spectral normalization for multimodal neuroimaging synthesis. Magn Reson Med 2021;86(3):1718–33.

[19] Lee D, Kim J, Moon W-J, Ye JC. Collagan: collaborative gan for missing image data imputation. In: IEEE CVPR; 2019. p. 2487–96.
[20] Yang H, Lu X, Wang S-H, Lu Z, Yao J, Jiang Y, Qian P. Synthesizing multi-contrast MR images via novel 3D conditional variational auto-encoding GAN. Mobile Network Appl 2021;26:1–10.
[21] Yurt M, Dar SU, Erdem A, Erdem E, Oğuz KK, Çukur T. mustGAN: multi-stream generative adversarial networks for MR image synthesis. Med Image Anal 2021;70: 101944.
[22] Zhou T, Fu H, Chen G, Shen J, Shao L. Hi-Net: hybrid-fusion network for multi-modal MR image synthesis. IEEE Trans Med Imag 2020;39(9):2772–81.
[23] Sharma A, Hamarneh G. Missing MRI pulse sequence synthesis using multi-modal generative adversarial network. IEEE Trans Med Imag 2020;39:1170–83.
[24] Wang G, Gong E, Banerjee S, Martin D, Tong E, Choi J, Chen H, Wintermark M, Pauly JM, Zaharchuk G. Synthesize high-quality multi-contrast magnetic resonance imaging from multi-echo acquisition using multi-task deep generative model. IEEE Trans Med Imag 2020;39(10):3089–99.
[25] Yu B, Zhou L, Wang L, Shi Y, Fripp J, Bourgeat P. Ea-GANs: edge-aware generative adversarial networks for cross-modality MR image synthesis. IEEE Trans Med Imag 2019;38(7):1750–62.
[26] Yurt M, Dalmaz O, Dar SUH, Ozbey M, Tinaz B, Oguz KK, Çukur T. Semi-supervised learning of MRI synthesis without fully-sampled ground truths. IEEE Trans Med Imag 2022;41(12):3895–906.
[27] Dosovitskiy, A., Beyer, L., Kolesnikov, A., Weissenborn, D., Zhai, X., Unterthiner, T., Dehghani, M., Minderer, M., Heigold, G., Gelly, S., Uszkoreit, J., and Houlsby, N. (2021). An image is worth 16x16 words: transformers for image recognition at scale. arXiv:2010.11929.
[28] Chen, J., Lu, Y., Yu, Q., Luo, X., Adeli, E., Wang, Y., Lu, L., Yuille, A. L., and Zhou, Y. (2021). Transunet: transformers make strong encoders for medical image segmentation. arXiv:2102.04306.
[29] He K, Gan C, Li Z, Rekik I, Yin Z, Ji W, Gao Y, Wang Q, Zhang J, Shen D. Transformers in medical image analysis. Intelli Med 2023;3(1):59–78.
[30] Korkmaz Y, Dar SUH, Yurt M, Ozbey M, Çukur T. Unsupervised mri reconstruction via zero-shot learned adversarial transformers. IEEE Trans Med Imag 2022;41(7):1747–63.
[31] Dalmaz O, Yurt M, Çukur T. ResViT: residual vision transformers for multi-modal medical image synthesis. IEEE Trans Med Imag 2022;44(10):2598–614.
[32] Liu J, Pasumarthi S, Duffy B, Gong E, Datta K, Zaharchuk G. One model to synthesize them all: multi-contrast multi-scale transformer for missing data imputation. IEEE Trans Med Imag 2023;42(9):2577–91.
[33] Yan S, Wang C, Chen W, Lyu J. Swin transformer-based gan for multi-modal medical image translation. Front Oncol 2022;12.
[34] Dhariwal P, Nichol A. Diffusion models beat gans on image synthesis. NeurIPS 2021; 34:8780–94.
[35] Sasaki, H., Willcocks, C. G., and Breckon, T. P. (2021). UNIT-DDPM: UNpaired image translation with denoising diffusion probabilistic models. arXiv:2104.05358.
[36] Isola P, Zhu J-Y, Zhou T, Efros AA. Image-to-image translation with conditional adversarial networks. In: IEEE CVPR; 2017. p. 1125–34.
[37] Ho J, Jain A, Abbeel P. Denoising diffusion probabilistic models. NeurIPS 2020;33: 6840–51.

[38] Meng, X., Gu, Y., Pan, Y., Wang, N., Xue, P., Lu, M., He, X., Zhan, Y., and Shen, D. (2022). A novel unified conditional score-based generative framework for multi-modal medical image completion. arXiv:2207.03430.

[39] Ozbey M, Dalmaz O, Dar SUH, Bedel HA, Ozturk S, Gungor A, Çukur T. Unsupervised medical image translation with adversarial diffusion models. IEEE Trans Med Imag 2023;42(12):3524–39.

[40] Armanious K, Jiang C, Abdulatif S, Kustner T, Gatidis S, Yang B. Unsupervised medical image translation using cycle-medgan. In: European Signal Processing Conference (EUSIPCO). IEEE; 2019a. p. 1–5.

[41] Gong E, Pauly JM, Wintermark M, Zaharchuk G. Deep learning enables reduced gadolinium dose for contrast-enhanced brain mri. J Magn Reson Imag 2018;48(2):330–40.

[42] Haskell MW, Cauley SF, Bilgic B, Hossbach J, Splitthoff DN, Pfeuffer J, Setsompop K, Wald LL. Network accelerated motion estimation and reduction (namer): convolutional neural network guided retrospective motion correction using a separable motion model. Magn Reson Med 2019;82(4):1452–61.

[43] Küstner T, Armanious K, Yang J, Yang B, Schick F, Gatidis S. Retrospective correction of motion-affected mr images using deep learning frameworks. Magn Reson Med 2019; 82(4):1527–40.

[44] Arslan, F., Kabas, B., Dalmaz, O., Ozbey, M., and Çukur, T. (2024). Self-consistent recursive diffusion bridge for medical image translation. arXiv:2405.06789.

[45] Yang H, Sun J, Carass A, Zhao C, Lee J, Xu Z, Prince J. Unpaired brain mr-to-ct synthesis using a structure-constrained cyclegan. In: Lecture Notes in Computer Science, volume 11045 LNCS. Springer Verlag; 2018. p. 174–82.

[46] Hu S, Lei B, Wang S, Wang Y, Feng Z, Shen Y. Bidirectional mapping generative adversarial networks for brain mr to pet synthesis. IEEE Trans Med Imag 2022;41(1):145–57.

[47] Goodfellow IJ, Pouget-Abadie J, Mirza M, Xu B, Warde-Farley D, Ozair S, Courville A, Bengio Y. Generative adversarial networks. NeurIPS 2014:2672–80.

[48] Mao X, Li Q, Xie H, Lau RY, Wang Z, Paul Smolley S. Least squares generative adversarial networks. In: IEEE CVPR; 2017. p. 2794–802.

[49] Ba JL, Kiros JR, Hinton GE. Layer normalization. 2016.

[50] Vaswani A, Shazeer N, Parmar N, Uszkoreit J, Jones L, Gomez AN, Kaiser Lu, Polosukhin I. In: Guyon I, Luxburg UV, Bengio S, Wallach H, Fergus R, Vishwanathan S, Garnett R, editors. Attention is all you need, vol 30. NeurIPS; 2017.

[51] Zhang X, He X, Guo J, Ettehadi N, Aw N, Semanek D, Posner J, Laine A, Wang Y. Ptnet3d: a 3d high-resolution longitudinal infant brain mri synthesizer based on transformers. IEEE Trans Med Imag 2022;41(10):2925–40.

[52] Gungor A, Dar SU, Ozturk S, Korkmaz Y, Elmas G, Ozbey M, Çukur T. Adaptive diffusion priors for accelerated mri reconstruction. Med Image Anal 2023;88:102872.

[53] Kodali, N., Abernethy, J., Hays, J., and Kira, Z. (2017). On convergence and stability of gans. arXiv:1705.07215.

[54] Xiao Z, Kreis K, Vahdat A. Tackling the generative learning trilemma with denoising diffusion GANs. In: ICLR; 2022.

[55] Bedel HA, Çukur T. DreaMR: diffusion-driven counterfactual explanation for functional MRI. IEEE Trans Med Imag 2024.

[56] Korkmaz Y, Cukur T, Patel VM. Self-supervised MRI reconstruction with unrolled diffusion models. In: MICCAI. Cham, Nature Switzerland: Springer; 2023. p. 491–501.

[57] Kabas, B., Arslan, F., Nezhad, V. A., Ozturk, S., Saritas, E. U., and Çukur, T. (2024). Physics-driven autoregressive state space models for medical image reconstruction. arXiv:2412.09331.

[58] Atli, O. F., Kabas, B., Arslan, F., Yurt, M., Dalmaz, O., and Çukur, T. (2024). I2I-Mamba: multi-modal medical image synthesis via selective state space modeling. arXiv:2405.14022.

[59] Mirza, M. U., Dalmaz, O., Bedel, H. A., Elmas, G., Korkmaz, Y., Gungor, A., Dar, S. U., and Çukur, T. (2023). Learning fourier-constrained diffusion bridges for mri reconstruction. arXiv:2308.01096.

[60] Rieke N, Hancox J, Li W, Milletarì F, Roth HR, Albarqouni S, Bakas S, Galtier MN, Landman BA, Maier-Hein K, Ourselin S, Sheller M, Summers RM, Trask A, Xu D, Baust M, Cardoso MJ. The future of digital health with federated learning. npj Digit Med 2020;3(1):119.

[61] Elmas G, Dar SUH, Korkmaz Y, Ceyani E, Susam B, Ozbey M, Avestimehr S, Çukur T. Federated learning of generative image priors for mri reconstruction. IEEE Trans Med Imag 2023;42(7):1996–2009.

[62] Guo P, Wang P, Zhou J, Jiang S, Patel VM. Multi-institutional collaborations for improving deep learning-based magnetic resonance image reconstruction using federated learning. CVPR; 2021. p. 2423–32.

[63] Dalmaz O, Mirza MU, Elmas G, Ozbey M, Dar SU, Ceyani E, Oguz KK, Avestimehr S, Çukur T. One model to unite them all: personalized federated learning of multi-contrast mri synthesis. Med Image Anal 2024:103121.

PART

Generalization and fairness

5

CHAPTER

16 The cause and effect of an MR image: Robustness and generalizability

Aasa Feragen[1], Eike Petersen[1] and Melanie Ganz-Benjaminsen[2, 3, *]

[1]*DTU Compute, Technical University of Denmark, Lyngby, Denmark;* [2]*Department of Computer Science, University of Copenhagen, Copenhagen, Denmark;* [3]*Neurobiology Research Unit, Rigshospitalet, Copenhagen University Hospital, Copenhagen, Denmark*

[*]*Corresponding author: E-mail: ganz@di.ku.dk*

16.1 Overview

As previous chapters have already described, artificial intelligence (AI) and machine learning (ML) have found various applications in MRI (Magnetic Resonance Imaging). AI and ML algorithms are utilized during raw data acquisition, image reconstruction, post-reconstruction for image enhancement, artifact detection and correction as well as post-acquisition for image segmentation, image registration and finally biomarker development and disease classification. In this chapter, we want to address the robustness and generalizability of AI and ML algorithms used in MRI and focus especially on pitfalls and what one can do to discover and/or avoid them.

16.2 Impact of domain shifts

Neural networks excel in diverse tasks in medical image analysis, such as image alignment and tumor detection. Training involves presenting the network with an input (e.g., an MRI scan) and a target output (e.g., an image highlighting a brain tumor). Through an optimization process, the network adjusts its parameters to progressively approach the desired output.

While neural networks in general have many successful applications, they also have limitations and often excel in only specific tasks. If trained on data with a certain domain, referring to a specific distribution or source of data, they struggle to generalize [1]. For instance, a network trained on real images of cats and dogs may fail when presented with drawings or paintings of these animals due to significant differences. In another example from within medical imaging, changes in imaging domain can be very subtle and even invisible to the human eye, but still be visible to a classification algorithm [2].

Machine Learning in MRI. https://doi.org/10.1016/B978-0-443-14109-6.00016-X

This challenge is pronounced in medical imaging, particularly with MRI scans. Variations in imaging distributions can impede a network's ability to generalize [3]. These shifts in image appearance can be due to population (e.g., disease state) or demographic differences (e.g., age, sex, or ethnicity) or stemming from factors such as different manufacturers or imaging techniques. They are known as distributional or domain shifts [1] (see also Chapter 17).

16.2.1 Causes of domain shifts in MRI

In MRI scans many different components can cause a domain shift. It is important to remember that compared to other medical imaging system, such as computed tomography scanners which measure density in absolute units, MRI scanners are not measurement devices [4] (for details see *IEC 60601-2-33:2022*). Therefore it is hard to exactly reproduce images with the same distribution of intensities, even on the same scanner of the same subject over time.

16.2.1.1 MRI hardware and software

We will start by assessing the influence of the scanner hardware and software on distributional shift in image intensities. Something as seemingly simple as measuring the T1 relaxation time, a quantitative measurement, is not straightforward. It depends on the given hardware, for example, scanner by different vendors, and software, e.g., different software versions or scanner settings. The ISMRM 2020 joint Reproducible Research and Quantitative MR study group's reproducibility challenge on phantom and human brain T1 mapping has shown that there are large differences in the distribution of T1 relaxation times measured on the same physical MRI phantom [5]. The differences stem on the one hand from difference in the hardware, for example, not every 3 T scanner actually has a 3 T B0 field. For example, Siemens "3T" scanners systematically operate at < 2.9T [6] as can also be seen in the DICOM tag (0018,0084).

But they also stem from differences in the software, since each MR scanner vendor uses proprietary sequences and hence not every pulse sequence is exactly the same. Even for applications with the same sequence parameters such as repetition time, echo time, flip angle, resolution, and the type of sequence used (e.g., spin echo or gradient echo) the resulting images and consequently their intensity distributions can be different, since the underlying pulse sequence is programmed slightly differently or image preparation is performed differently. Differences in RF amplifiers, receiver coils, and reconstruction software (filtering, etc.) can also contribute to a shift in intensities.

Another variation in signal intensity distribution over time is a well-known effect for magnetic resonance imaging scans, and is typically termed scanner drift. This can also lead to domain shifts in intensities over time that can even cancel out actual longitudinal brain volume changes [7].

Next, for many clinical but also research applications of AI algorithms in MRI not only a single MRI modality, but several, need to be combined. In general,

MRI protocols utilize very diverse sequences and modalities and the resulting variations in intensity distributions showcase a domain shift that algorithms need to be able to handle. Furthermore, there is a large difference between curated research-grade data consisting of high-resolution isotropically sampled 3D image acquisitions and clinical imaging protocols that typically are made up of anisotropic 2D acquisitions and also contain many more artifacts.

16.2.1.2 Experimental study setup

Besides the MRI hardware and software the choice of imaging subjects also can lead to domain shifts. Focusing first on the example of a single demographic, to obtain enough training data for an AI algorithm, data is often pooled across multiple studies. While this can help to broaden the overall training distribution [8,9], have shown that without harmonization it can be identified with high accuracy which neuroimaging study an image stems from. Data sets from different studies are subject to various types of biases. These include subject selection—the fact that subjects included in each study do not represent the overall population, acquisition method—where magnetic field strength, manufacturer, gradients, pulse sequences and image acquisition protocol cause variations in the images, and processing biases—which are in part related to acquisition bias.

Next, even within a single demographic the properties of the MRI can change drastically. Patient data changes over time and leads to differences in intensities beyond the scanner effects mentioned above that are of biological nature, for example, in aging [10]. In certain demographics these changes happens slowly and can be neglected, in others they occur rapidly. In infant neuroimaging changing tissue contrasts on structural MR images with age due to rapid myelination lead to data with inherent variations across the first year of life [11]. Hence, when trying to apply methods to different demographic groups, such as transferring computational tools developed on adults to children, toddlers and infants, these differences become exacerbated.

16.2.2 Addressing domain shifts

But how can we tackle these issues of domain shifts? One can either try and address domain shifts before they happen or remedy them afterward.

16.2.2.1 Addressing domain shifts during acquisition

Some of the ways to avoid domain shifts during MRI acquisition rely on using open source and scanner-independent imaging frameworks as well as standardization efforts as e.g., Ref. [12]. For example, one can work with open source sequence frameworks such as Pulseq [13] or QMRpullseq [14] to avoid proprietary vendor sequences. Next, there are also open-source MRI reconstruction engines such as Gadgetron [15] in order to avoid proprietary vendor image reconstruction software that uses in-built filtering or other settings one cannot control. Especially with the advent of deep learning reconstructions - which by now is also employed by some of the vendors - the reconstruction becomes a black box in itself that needs to be regarded with care [16]. In Chapter 24, more details regarding open-source frameworks are discussed. Machine learning code challenges to address differences in acquisition are also of importance and are discussed in Chapter 25. Scanner drift can be addressed by the use of phantoms. For example, the ADNI study has included the imaging of a geometric phantom for correcting differences due to drift in scanner performance and inter-scanner variability [17].

16.2.2.2 Algorithmic remedies

In order to address domain shift, first one needs to detect it; recent work [18] is trying to develop tools for this, but the field is still evolving and these tools need to be tested further. Often proper generalization tests can already be a good starting point to examine drops in performance. Once a domain shift is detected, multiple remedies exist to address performance drops resulting from a domain shift. One method is transfer learning as done in Ref. [19]. In transfer learning an algorithm is adapted often through feature extraction or fine-tuning; to the data it is supposed to be tested on. Other methods which avoid adapting the algorithm to each test domain anew can be categorized into two broad categories, domain harmonization, and domain generalization tools. Domain harmonization aims at removing domain-specific (e.g., scanner and type of image contrast) information from the images. Example studies using domain harmonization in MRI are [20–22]. They either utilize general adversarial networks or encoder-decoder-based frameworks for harmonization. Domain generalization aims to train neural networks in such a way that they work across images from multiple domains. Examples of this are given in [23–25]. This can be done by learning domain-independent features or by increasing data diversity through quantitative-map or image-based synthesis.

To conclude, detecting and handling domain shifts both before and after data acquisition is difficult. But it is of special importance, if one is aiming for clinical translation of newly developed AI tools.

16.3 Fair and robust image analysis

Besides domain shift another threat to the generalizability of AI tools on MRI data is bias. Bias in medical imaging has become a more prominent topic and a good introduction is given in [26,27]. A list of resources is distributed by the Fairness of AI in Medical Imaging (FAIMI) initiative (https://faimi-workshop.github.io/resources/). But what about the issue of bias, by which we here mean differences in AI performance, in MRI data?

16.3.1 Examples of bias in medical imaging

One of the very prominent examples in the general medical imaging literature where bias is an issue are AI models designed for analyzing dermatology images. They may exhibit bias based on skin tone [28]. Simply put, if these models are predominantly trained on images of individuals with lighter skin, they may not perform as effectively on those with darker skin, and vice versa. The distinct variation in skin tones between the two groups of subjects creates an evident and visible contrast in their data. Relating back to the previous section 3, we can think of this as a type of domain shift that is already present inside the training data, between subgroups of differently colored individuals. Another medical imaging example involves bias in sex-related classifications of chest X-ray models [29]. Here, the composition of the training data sets in certain applications has an influence on the test performance of the AI algorithms.

The same type of biases have also been detected in MRI applications and in AI applications where it was much less expected. Previous works [30,31] found statistically significant differences in segmentation performance between different racial groups in cardiac MR segmentation. This case is especially interesting for two reasons. First, it would be unlikely for a cardiologist to assert the ability to predict a patient's race from their MR scan, and most would likely be skeptical about the potential race bias in AI models for cardiac MR image analysis. Second, the bias was here not detected in a classification algorithm as in the other examples above, but in a segmentation algorithm. Nevertheless, such bias was identified without any clear attribution to confounding factors. Moreover, the anticipated source of bias, such as subject sex due to differences in heart size between men and women, turned out to be much less significant.

16.3.2 How to address bias

One of the first issues when trying to address bias is *detecting* it. Examples of quantitative studies that tried to analyze bias in medical imaging applications are [32–34]. Recent work [35] presents a framework to benchmark the fairness of machine learning models for medical imaging. Addressing the detected bias is then more complicated since it requires investigating and finding its *cause*. For instance, if a model tends to perform better for women compared to men, is that because of some intrinsic female property or because of some other factor that just happens to be correlated with biological sex or gender [36,37]? In Ref. [38], we have given an overview over different causes of disparate performance of machine learning models, suggested what the underlying causes could be and highlighted possible solutions. First, practitioners must examine where observed performance differences stem from, if e.g., factors such as label biases, selection biases, confounding factors, or intersectional effects are at play. Thorough subgroup performance analyses are crucial and an important tool here. But assessing label noise and biases is challenging and often also requires expert consultation [34]. Addressing it might necessitate the use of other target variables or a switch to bias-robust learning techniques.

To conclude, the FAIMI research field is still developing and new tools and techniques are popping up all the time. This can be discouraging and daring, but we suggest to start with low hanging fruit. If you have demographic data available, you can easily report your model's subgroup performance. This will boost transparency and showcase unbiased results. Finally, if bias is detected, it should not impede publication, but should be commented on in order to boost transparency and facilitate more research.

16.4 Shortcut learning

We next visit *shortcut learning*, or the *Clever Hans effect*, a generalizability problem known to appear with predictive machine learning models. Shortcut learning can happen when medical images contain features that correlate strongly with the predictive target at training time, but not necessarily at test time. While this phenomenon has been increasingly studied in medical imaging over the past few years [39–45], most of this work has considered other modalities than MRI. Thus, we first introduce the problem and state-of-the-art research from a general medical imaging viewpoint, before arguing why MRI based predictors are also at high risk of learning shortcuts.

16.4.1 Shortcut learning in medical imaging

We illustrate shortcut learning with a case from fetal ultrasound [41], where completed fetal ultrasound examinations from the clinic are used for training. Coming from a clinical workflow, these images contain text and measurement calipers annotated at acquisition time, see Fig. 16.1a. These annotations correlate strongly with the target anatomical plane, so a model that sorts images into head, abdomen, femur or other anatomical planes, can easily make predictions based on the annotations rather than anatomical features. That is, the text and calipers act as *shortcuts* for the model. When this happens, we risk that the model breaks down in real-life test scenarios, for example, if the model classifies images real-time during an examination, before images are annotated.

Dermatological images are, just as fetal ultrasound, often acquired and analyzed simultaneously. For this reason, dermatological images are often affected by the

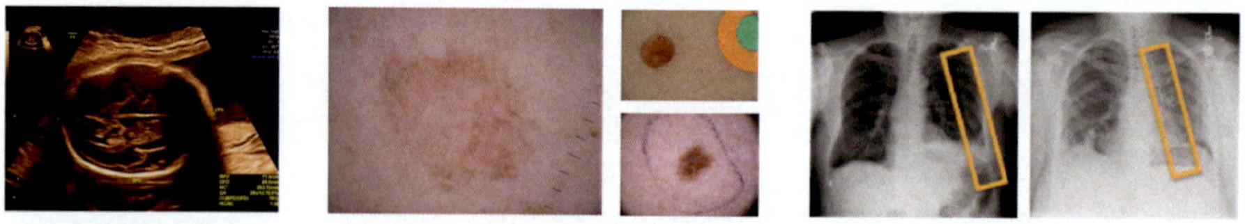

(a) (b) (c)

FIGURE 16.1

Example shortcuts learned by medical imaging classifiers: (a) Text and calipers in fetal ultrasound are shortcuts for the standard plane acquired. (b) Rulers, patches and ink act as a shortcut for skin lesion malignancy. (c) Support devices in chest X-ray are shortcuts for diagnosis.

clinicians' hypotheses: Clinicians tend to circle lesions of concern, and when photographing them, they tend to include a ruler in order to record the lesion size. Conversely, color calibration charts visible as colorful stickers are, at least in the publicly available ISIC database [46], primarily placed near benign lesions. See Fig. 16.1b for examples.

While this added information is clinically meaningful, it affects how machine learning models learn from the images: If it is easier to detect ink, rulers, or stickers than it is to actually diagnose the lesion, these features can become *shortcut features* for the model, in place of the signs of disease potentially visible in the image. Studies have confirmed shortcut learning for these datasets by demonstrating changes in predictions when shortcuts are synthetically removed or inserted [42,47].

In the two previous cases, the described shortcuts consisted of information added to the images by clinicians based on their suspicion of diagnosis during the clinical workflow. However, chest X-ray images used for diagnosing lung- and heart diseases are suspected of carrying shortcut information that is far more difficult to avoid, namely support devices such as pacemakers or chest drains [43,48], see Fig. 16.1c. These are hard to avoid, as treatment of acute respiratory or heart problems will necessarily precede image acquisition. Their suspected role as shortcuts for disease prediction is, however, supported by recent work showing that images with support devices tend to cluster in diagnostic feature spaces, and that the diagnosis confidence of the model varies systematically with the prevalence of support devices [43,48]. This is clearly harmful—image classifiers that only detect heart problems after identifying a pacemaker aren't clinically useful.

16.4.2 Shortcut learning in MRI

The increased emergence of cases like the three discussed above have spurred a frenzy of research on the existence, detection and mitigation of shortcut learning in medical imaging [37,39,42,43]. These examples of shortcuts seem both obvious and obviously noticeable, and they might also seem irrelevant from the point of view of MRI: As MR acquisition and analysis are typically performed separately; the images are not impacted by the clinicians' hypotheses. Support devices are rare in MRI images, in particular as anything metallic is impossible. Is there any reason to worry about shortcut learning for predictive models trained on MRI data? The short answer is yes! Shortcuts do not have to be image artifacts; they can also arise from correlation between the target and features that are naturally occurring, in ways that don't easily catch our eye.

Some diseases come with **symptoms** that affect MRI image appearance, most notably through movements. Examples include tremor associated with Parkinson's disease, but also decreased capability of following instructions of staying still, for example, due to diseases such as dementia [49]. It's well known that differences in motion between groups can lead to false clinical findings in both structural MRI and brain connectivity through fMRI [50,51]. While motion is traditionally known as a confounder for statistical analysis, the effects of motion might also act as shortcuts for machine learning models.

As discussed in Sec. 3, differences in **scanner hardware and MRI sequences** lead to unique image "styles" that enable machine learning models to recognize scanners from their images. While harmonization techniques are commonly used to reduce the effect of scanner bias on model training, experiments show after standard pipelines it is still possible to recognize the scanner, and that this can lead to overly optimistic estimation of performance [8]. That latter point, in particular, indicates that scanner bias acts as a shortcut.

Why does this happen? The key lies in the distribution of target variables across different types of scanners. If there are prevalence differences between scanners (e. g., as a result of prevalence differences between sites), the model may find it advantageous to use the site as a shortcut to recalibrate its prediction to match the scanner prevalence. Further examples of this effect are found in Ref. [52]. A different type of MRI shortcut learning is described in Ref. [53]; where the slice selected by a radiologist as containing a tumor, becomes a shortcut for the tumor. This is possible because particular types of tumors tend to appear in particular parts of the brain.

Finally, as any variable can in principle act as a shortcut, let's take a moment to recall **demographic features** that are often encoded in MRI images: Age can be predicted from brain MRI [54], where it, for example, manifests through atrophy; gender and race [55] have been shown to be encoded in MRI images; and comorbidities are often also naturally predictable from MRI. Just like with the scanner bias, models are vulnerable to any of these acting as shortcuts if the target classification- or risk prediction problem is hard, and the target variable is distributed differently for different values of the potential shortcut feature.

Consider a "toy" example where all the cases are elderly subjects, whereas all the controls are young, causing the model to learn age related features rather than disease specific features. A more realistic, and hence worrying, example interacts with algorithmic fairness: Assume a scenario where certain demographic groups are systematically underdiagnosed. This could, for instance, happen with mental or cognitive disorders where gender, age and culture affect the patients' eagerness to seek diagnosis and treatment. In machine learning terms, this would lead to a label bias, where an overweight of "healthy" labels assigned to a particular group are false. If the model is able to recognize the demographic group from the images, it will also have the opportunity—and this will be advantageous from the loss function's point of view—to recalibrate the model so that it makes fewer diagnoses on the underserved group. This would lead to continued underdiagnosis, and is extremely difficult to discover without access to true diagnostic labels. Such concerns lie at the heart of the recent discovery that race can be predicted from chest X-ray scans [56].

16.4.3 How can shortcut learning be detected and mitigated?

The most obvious way to detect shortcut learning is to observe the model fail at test time—and this can very well be how the suspicion of a shortcut first arises. However, once a concrete shortcut is suspected, the toolbox increases:

First, an observed drop in performance between test data with and without the suspected shortcut might indicate that the model is using the shortcut. However, the presence of the suspected shortcut might correlate with disease severity, as was the case with movement artifacts. Hence, a drop in performance *could* also just be the result of a more challenging prediction problem.

Having noticed this, one might instead try to remove the shortcut with image processing: Motion correction is one example of this. In the classical examples of ultrasound calipers or added features to dermatological images, the shortcuts can be removed with segmentation and inpainting, which can be done with high quality [41,42]. Again, a drop in performance after removing shortcuts from the test images would indicate shortcut learning—but could also be the result of insufficient quality in the image processing. For more complex shortcuts, however, such as age or gender that can be predicted from the images, it becomes quite challenging to remove the suspected shortcut without altering other aspects of the image. Some existing work employs generative image models to produce counterfactual images without the suspected shortcut, but again, it is hard to be certain whether a drop in performance—or no drop in performance—is caused by a shortcut or by inadequate counterfactual quality.

Another option for detecting shortcut learning is through explainable AI (XAI) [57,58]. Again, however, while XAI can provide supporting evidence, it is also known for its varying robustness [59], and again cannot be used on its own to reach certain conclusions.

Mitigation of shortcut learning can utilize similar tools as its detection: If it is possible to remove the shortcut from the images without altering the target-specific image content, then the shortcut learning can be mitigated by applying shortcut removal prior to training and testing. If adding the shortcut is easier than removing it, then artificially adding a shortcut might be applied as a form of augmentation to encourage robustness. However, in many scenarios this is unrealistic and impossible to validate: If gender, age or comorbidities act as shortcuts, it might not be possible to generate realistic counterfactuals with a different gender, age or comorbidities, as the disease manifestation might very well depend on either of these potential shortcut variables. Classical harmonization strategies fall under this family of mitigation techniques for scanner bias, but might be less prone to failure because they are usually less flexible.

The preprocessing and training of an algorithm might also affect how prone it is to shortcut learning. Indeed, standard normalization techniques can be seen as a form of harmonization. Recent work has shown that preprocessing and careful hyperparameter tuning can reduce demographic bias in predictive MRI models [60]; it is entirely possible that the observed difference is, in reality, a reduction of shortcut learning from, for example, scanner bias.

An alternative is given by applying a shortcut classifier to the data representation at the final layer of the model, and project all test data to its orthonormal complement in order to reduce the impact of the shortcut [61]. This, however, comes with the risk of losing predictive power for the true target.

16.5 Invariances and statistical independence

Many of the challenges discussed throughout this chapter so far—domain shifts, shortcut learning and some aspects of fairness—can be formalized in terms of *invariance* requirements. We might want the learned image representations and model predictions to be invariant with respect to scanner type, hospital, the recording physician, minor changes in body position, or a patient's biological sex or skin color. In practice, such invariance requirements are often formalized and enforced in terms of (conditional) statistical independence statements, as we will discuss below.

16.5.1 Notation

Before diving into further details, we will first introduce our notation. In the following, X will denote the inputs of a statistical model $f(X)$ (e.g., a deep neural network), such as raw or preprocessed volumes, volumetric characteristics extracted in a preprocessing step, or additional clinical or demographic input variables. The prediction targets, such as disease classifications, organ segmentations, or outcome predictions, will be denoted as Y, and the model's predictions of these targets as $\widehat{Y} = f(X)$. We will sometimes be interested in an intermediate *latent* representation Z of the inputs X derived by the model. Usually, this corresponds to the outputs of an intermediate layer $g(X)$ of a deep neural network, often the penultimate layer directly before a final classification layer $h(Z)$. Nuisance factors, with respect to which we might want the model $f(\cdot)$ to be invariant, are denoted as A. Finally, we will write $A \perp B$ and $A \perp B | C$ to denote marginal and conditional statistical independence, respectively.

16.5.2 Counterfactual invariance versus statistical independence

The strictest notion of invariance is *counterfactual* invariance [62–64]. This would mean, for example, that "the same image" from "the same patient" *except* for a different skin color should receive the same model predictions. There are some specific types of counterfactual invariances that can be formally enforced, such as rotational and translational invariances or *equivariances*, in the case of segmentation tasks [65–68]. In many cases, however, such counterfactual notions are difficult to formalize and enforce, because the true counterfactual images cannot be obtained and the counterfactual action with respect to which one would like to be invariant is challenging or impossible to formalize [69–71].

For this reason, most domain adaptation and generalization approaches resort to enforcing statistical independence statements of the form $Z \perp A$ instead, meaning that the learned image representations Z, and thus also the model predictions $\widehat{Y} = f(X) = h(g(X)) = h(Z)$, are forced to be statistically independent of some nuisance factor A, such as the scanner type or the patient's skin color. Such statistical independence statements are equivalent to stating that X and A share no *mutual information*, or that A cannot be predicted from X with higher-than-chance accuracy (and vice versa). Many of the studies reporting that, for example, patient race, sex, and age can be predicted with high accuracy from image analysis models' internal representations [56,72,73] can thus be understood as being concerned with such statistical (in)dependence relations: in such cases, the two variables are clearly *not* statistically independent, that is, $X \not\perp A$. As we will discuss below, while this appears problematic (because it opens the potential for shortcut learning to occur), it is not always avoidable, and also not *always* problematic.

Notably, such statistical independence requirements differ markedly from the above-mentioned counterfactual notions of invariance: they cannot provide guarantees about the treatment of *individual* recordings. If the overall distribution of image representations is the same in patients from two different scanners, this does *not* guarantee that individual patients would receive the same model predictions had they been recorded using a different scanner. In fact, naively enforcing such statistical independence statements can even *force* the model to treat patients scanned using one of the two scanner models differently. To see this, consider a simple binary disease classification scenario in which the disease prevalence differs between two patient cohorts recorded using two different scanner models, $A = a_1$ and $A = a_2$. Enforcing statistical independence $Z, \widehat{Y} \perp A$ in this setting would require the model to make equally many positive and negative predictions for the two cohorts. This would *require* the model to make wrong predictions for at least one of the cohorts, since prevalence differs and the same ratio of positive to negative predictions cannot be correct for both cohorts. Such an approach is thus in strong and direct conflict with the aim of accurate disease classification [74,75].

16.5.3 Which independence statements to enforce, and why?

The specific example outlined above, where the core problem stems from prevalence differences between the different patient groups or domains, can be addressed by either artificially balancing the disease prevalence in the two cohorts before training, or by applying a *conditional* independence statement of the form $X \perp A | Y$ instead [76]. The latter statement can be translated into two separate independence requirements in the disease-positive and disease-negative groups, respectively: in both disease-status groups, the distributions of internal model representations Z (and thus also the model predictions $\widehat{Y}$) should be identical between the two scanner cohorts. This is, however, only one specific instance of a larger question: which independence statements can or should be enforced in a given application, and why? Any answer to this question depends on (explicit or implicit) assumptions about the causal relationships between the different factors involved [77], and must thus be application-dependent. Once these assumptions have been agreed upon and formalized, typically in the shape of a causal graph, one can easily derive the appropriate (conditional) statistical independence statements to enforce. In the following section, we will lay out how to understand many of the previously discussed challenges in terms of causal relationships and causal graphs, and how to derive the correct statistical independence statements to enforce from such a graph.

16.6 Causal modeling

A causal graph is a graphical representation of assumptions about the causal relationships between a set of variables, such as MRI recordings, various patient properties, scanner type, hospital, and disease diagnoses [78–80]. In such a graph, nodes represent (measured or unobserved, latent) variables, and edges represent direct causal relationships. Figs. 16.2 and 16.3 show two illustrative examples of causal graphs describing some of the variables in typical medical image analysis scenarios; they will be discussed in detail below.

16.6.1 Deriving statistical independence statements from causal graphs

Two useful properties of causal graphs are that they

1. encode and visually represent domain-specific assumptions about the (directionality) of causal relationships in a given application scenario, and
2. enable easily deriving the statistical conditional independence statements that hold for any data conforming with the causal structure described by this graph.

The second property thus allows deriving from first principles the statistical independence properties that one might want to enforce when training a machine

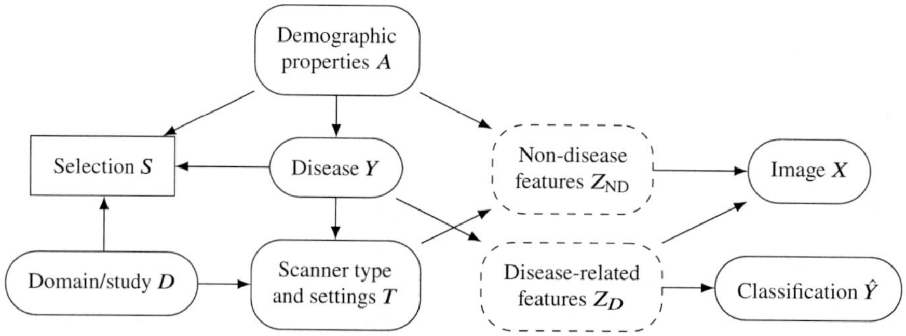

FIGURE 16.2

A causal graph describing a hypothetical disease classification scenario. The square node denotes a sample selection indicator S, dashed nodes indicate hidden (latent, unobserved) variables. In this graph, the conditional statistical independence statements $\hat{Y} \perp A, T, D | Y$ hold (among various other, less practically useful independence statements). Refer to the main text for further details.

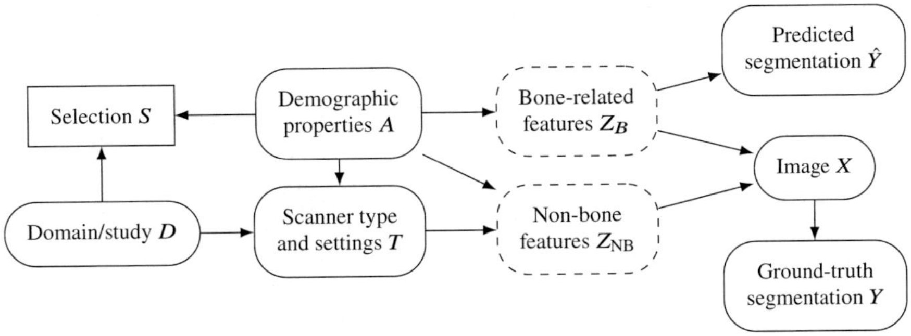

FIGURE 16.3

A causal graph describing a hypothetical bone segmentation scenario. The square node denotes a sample selection indicator S, dashed nodes indicate hidden (latent, unobserved) variables. In this graph, the conditional statistical independence statements $\widehat{Y} \perp T, D | A$ hold (among various other, less practically useful independence statements). Refer to the main text for further details.

learning model for a specific application. In this regard, it is important to note that it is the *absence* of edges (rather than their presence) in the graph that encodes prior knowledge and thus constrains the model: the most general (and least informative) causal graph is always the one with bidirectional causal links between all variables (The unnecessary causal effects could always be zero in that case). Dropping any such edge from the graph should thus always be well-justified from domain knowledge.

We will not discuss in detail here how exactly to derive the statistical conditional independence statements encoded by a graph. This is, however, not a difficult endeavor and can be done relatively easily "by eye" with some practice. The technical requirement for a conditional independence statement $A \perp B | C$ to hold in a causal graph is that C *d-separates* A and B in the graph[1]; refer to Ref. [80]; Def. 1.2.3 and Thm. 1.2.3), [78]; or [79] for details. Many tools and software packages exist that enable practitioners to easily and automatically derive the conditional independence statements that hold in a given graph, without having to understand the technicalities of d-separation themselves [81–84].

16.6.2 Two examples: Classification and segmentation

To illustrate how the conditional independence statements to enforce will differ between application scenarios, consider the two graphs shown in Figs. 16.2 and 16.3. The first shows a typical disease classification scenario, whereas the second

[1]Here, d-separation is a criterion for deciding, from a given a causal graph, whether a set A of variables is independent of another set B, given a third set C.

describes a hypothetical bone segmentation scenario. In both cases, the graphs include potential sample selection biases that may differ between studies or hospitals, as well as different scanner types. The scanner type may be influenced by whether a patient is sick or not, and the usage of different scanners may differ between studies. In both cases, we want to enforce model predictions to only depend on the (unknown and unobserved) relevant image features, and we can derive the necessary conditional independence statements to enforce from the specified causal graphs. The resulting conditional independence statements are given in the respective figure captions.

Our message here is not that these are the correct independence statements to enforce for all disease classification or structure segmentation tasks, but rather that the correct independencies to enforce will *differ between application scenarios*. Notice also that the first scenario describes an *anti-causal* prediction task (we are trying to estimate the presence or absence of disease, which is causing aspects of the image from which we are predicting), whereas the second scenario describes a *causal* prediction task (the ground-truth segmentation mask, which we are aiming to predict, is caused by the recorded image). We refer the reader to [77,85] for further discussions of causal modeling and shift robustness in medical imaging.

16.7 Summary and future direction

We have shown how different types of distribution shift can lead to problems such as shortcut learning, algorithmic bias, and other generalization challenges (see Fig. 16.4). We have discussed various forms of distribution shift that are common in MRI, such as scanner bias and demographic differences—indicating that the risk of generalization problems is real.

How does that affect best practices for robust AI within MRI analysis? The good news is that while some of the recently highlighted challenges are surprising, their underlying causes are well known: Generalization has always been at the heart of machine learning, and if we are able to understand *why* we have problems generalizing, we have a rich toolbox to mitigate the problem.

Hence, the main problems are reduced to realizing that one has a generalization problem, and understanding why. We therefore cannot emphasize enough the need for testing algorithms broadly and strategically to assess their potential failure modes: Identify subgroups at risk of generalization failure, and gather as much metadata as possible, in order to perform sub-group performance analyses. Evaluate on external/out-of-distribution datasets [86], design specific edge case stress tests [87], and use XAI tools or automatic slice detection methods for in-depth model inspection [88–92]. See also the surveys [93–95] for suggestions specific to healthcare.

Once you suspect that you indeed have a generalization problem—whether it is bias, shortcut learning, or of a different nature—design tests to uncover the causes of the problem. We have discussed how such tests could rely on causal inference or simple interventions such as artificially removing or adding suspected shortcuts.

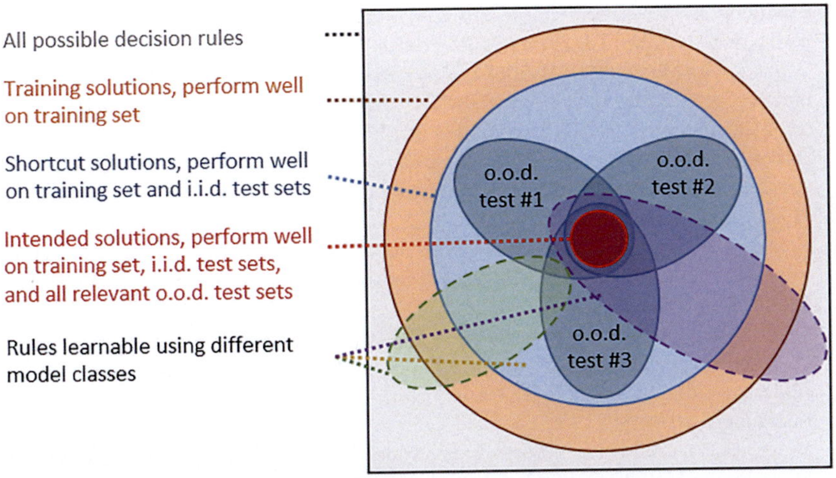

FIGURE 16.4

A schematic drawing showcasing how different types of domain shifts could look like—see the three out of distribution (o.o.d.) test sets in dark blue. It also schematically shows the space of shortcut models in light blue.

References

[1] Quinonero-Candela J, Sugiyama M, Schwaighofer A, Lawrence ND. Dataset Shift in Machine Learning. Mit Press; 2008.

[2] Burns JL, Zaiman Z, Vanschaik J, Luo G, Peng L, Price B. Ability of artificial intelligence to identify self-reported race in chest x-ray using pixel intensity counts. J Med Imag 2023;10(6):061106.

[3] Bashyam VM, Doshi J, Erus G, Srinivasan D, Abdulkadir A, Singh A. Deep generative medical image harmonization for improving cross-site generalization in deep learning predictors. J Magn Reson Imag 2022;55(3):908–16.

[4] Stikov N, Karakuzu A. The relaxometry hype cycle. Front Physiol 2023;14.

[5] Boudreau M, Karakuzu A, Cohen-Adad J, Bozkurt E, Carr M, Castellaro M. others. Results of the ISMRM 2020 Joint Reproducible Research and Quantitative MR Study Groups Reproducibility Challenge on Phantom and Human Brain T1 Mapping. 2023.

[6] Gracien R-M, Maiworm M, Brüche N, Shrestha M, Nöth U, Hattingen E, Deichmann R. How stable is quantitative MRI?–assessment of intra-and inter-scanner-model reproducibility using identical acquisition sequences and data analysis programs. Neuroimage 2020;207:116364.

[7] Takao H, Hayashi N, Ohtomo K. Effect of scanner in longitudinal studies of brain volume changes. J Magn Reson Imag 2011;34(2):438–44.

[8] Glocker B, Robinson R, Castro DC, Dou Q, Konukoglu E. Machine learning with multi-site imaging data: an empirical study on the impact of scanner effects. In: Medical Imaging Meets NeurIPS; 2019.

[9] Wachinger C, Becker BG, Rieckmann A, Pölsterl S. Quantifying confounding bias in neuroimaging datasets with causal inference. In: Medical Image Computing and

Computer Assisted Intervention–MICCAI 2019: 22nd International Conference, Shenzhen, China, October 13–17, 2019, Proceedings, Part Iv 22; 2019. p. 484–92.
[10] Fujita S, Mori S, Onda K, Hanaoka S, Nomura Y, Nakao T, Abe O. Characterization of brain volume changes in aging individuals with normal cognition using serial magnetic resonance imaging. JAMA Netw Open 2023;6(6):e2318153.
[11] Dubois J, Alison M, Counsell SJ, Hertz-Pannier L, Hüppi PS, Benders MJ. MRI of the neonatal brain: a review of methodological challenges and neuroscientific advances. J Magn Reson Imag 2021;53(5):1318–43.
[12] Karakuzu A, Appelhoff S, Auer T, Boudreau M, Feingold F, Khan AR. QMRI-BIDS: an extension to the brain imaging data structure for quantitative magnetic resonance imaging data. Sci Data 2022a;9(1):517.
[13] Layton KJ, Kroboth S, Jia F, Littin S, Yu H, Leupold J, Zaitsev M. PULSEQ: a rapid and hardware-independent pulse sequence prototyping framework. Magn Reson Med 2017; 77(4):1544–52.
[14] Karakuzu A, Biswas L, Cohen-Adad J, Stikov N. Vendor-neutral sequences and fully transparent workflows improve inter-vendor reproducibility of quantitative MRI. Magn Reson Med 2022b;88(3):1212–28.
[15] Hansen MS, Sørensen TS. Gadgetron: an open source framework for medical image reconstruction. Magn Reson Med 2013;69(6):1768–76.
[16] Kiryu S, Akai H, Yasaka K, Tajima T, Kunimatsu A, Yoshioka N, Ohtomo K. Clinical impact of deep learning reconstruction in MRI. Radiographics 2023;43(6):e220133.
[17] Jack CR, Bernstein MA, Fox NC, Thompson P, Alexander G, Harvey D. The alzheimer's disease neuroimaging initiative (adni): MRI methods. J Magn Reson Imag 2008;27(4):685–91.
[18] Kushol R, Wilman AH, Kalra S, Yang Y-H. DSMRI: domain shift analyzer for multicenter MRI datasets. Diagnostics 2023;13(18):2947.
[19] Ghafoorian M, Mehrtash A, Kapur T, Karssemeijer N, Marchiori E, Pesteie M. Transfer learning for domain adaptation in MRI: application in brain lesion segmentation. In: Medical Image Computing and Computer Assisted Intervention- MICCAI 2017: 20th International Conference, Quebec City, QC, Canada, September 11-13, 2017, Proceedings, Part iii 20; 2017. p. 516–24.
[20] Choi Y, Uh Y, Yoo J, Ha J-W. Stargan v2: diverse image synthesis for multiple domains. In: Proceedings of the IEEE/CVF Conference on Computer Vision and Pattern Recognition; 2020. p. 8188–97.
[21] Dewey BE, Zhao C, Reinhold JC, Carass A, Fitzgerald KC, Sotirchos ES. Deepharmony: a deep learning approach to contrast harmonization across scanner changes. Magn Reson Imaging 2019;64:160–70.
[22] Fatania K, Clark A, Frood R, Scarsbrook A, Al-Qaisieh B, Currie S, Nix M. Harmonisation of scanner-dependent contrast variations in magnetic resonance imaging for radiation oncology, using style-blind auto-encoders. Phys Imag Radiat Oncol 2022;22: 115–22.
[23] Guan H, Liu Y, Yang E, Yap P-T, Shen D, Liu M. Multi-site MRI harmonization via attention-guided deep domain adaptation for brain disorder identification. Med Image Anal 2021;71:102076.
[24] Iglesias JE, Billot B, Balbastre Y, Tabari A, Conklin J, González RG. Joint super-resolution and synthesis of 1 mm isotropic mp-rage volumes from clinical MRI exams with scans of different orientation, resolution and contrast. Neuroimage 2021;237:118206.

[25] Jog A, Hoopes A, Greve DN, Van Leemput K, Fischl B. Psacnn: pulse sequence adaptive fast whole brain segmentation. Neuroimage 2019;199:553–69.
[26] Chen RJ, Wang JJ, Williamson DF, Chen TY, Lipkova J, Lu MY, Mahmood F. Algorithmic fairness in artificial intelligence for medicine and healthcare. Nat Biomed Eng 2023;7(6):719–42.
[27] Mehrabi N, Morstatter F, Saxena N, Lerman K, Galstyan A. A survey on bias and fairness in machine learning. ACM Comput Surv 2021;54(6):1–35.
[28] Daneshjou R, Vodrahalli K, Novoa RA, Jenkins M, Liang W, Rotemberg V. Disparities in dermatology AI performance on a diverse, curated clinical image set. Sci Adv 2022;8(31):eabq6147.
[29] Larrazabal AJ, Nieto N, Peterson V, Milone DH, Ferrante E. Gender imbalance in medical imaging datasets produces biased classifiers for computer-aided diagnosis. Proc Natl Acad Sci 2020;117(23):12592–4.
[30] Puyol-Antón E, Ruijsink B, Mariscal Harana J, Piechnik SK, Neubauer S, Petersen SE, King AP. Fairness in cardiac magnetic resonance imaging: assessing sex and racial bias in deep learning-based segmentation. Front Cardiovasc Med 2022;9:859310.
[31] Puyol-Antón E, Ruijsink B, Piechnik SK, Neubauer S, Petersen SE, Razavi R, King AP. Fairness in cardiac MR image analysis: an investigation of bias due to data imbalance in deep learning based segmentation. In: Medical Image Computing and Computer Assisted Intervention–MICCAI 2021: 24th International Conference, Strasbourg, France, September 27–October 1, 2021, Proceedings, Part Iii 24; 2021. p. 413–23.
[32] Lee T, Puyol-Antón E, Ruijsink B, Aitcheson K, Shi M, King AP. An investigation into the impact of deep learning model choice on sex and race bias in cardiac MR segmentation. In: Workshop on clinical image-based procedures; 2023. p. 215–24.
[33] Seyyed-Kalantari L, Zhang H, McDermott MB, Chen IY, Ghassemi M. Underdiagnosis bias of artificial intelligence algorithms applied to chest radiographs in under-served patient populations. Nat Med 2021;27(12):2176–82.
[34] Zhang H, Dullerud N, Roth K, Oakden-Rayner L, Pfohl S, Ghassemi M. Improving the fairness of chest x-ray classifiers. In: Conference on health, inference, and learning; 2022. p. 204–33.
[35] Zong Y, Yang Y, Hospedales T. MEDFAIR: benchmarking fairness for medical imaging. In: Proceedings of ICLR 2023; 2023.
[36] Jiménez-Sánchez A, Juodelyte D, Chamberlain B, Cheplygina V. Detecting shortcuts in medical images – a case study in chest x-rays. In: International symposium on biomedical imaging (ISBI). IEEE; April 2023. https://doi.org/10.1109/isbi53787.2023.10230572.
[37] Weng N, Bigdeli S, Petersen E, Feragen A. Are sex-based physiological differences the cause of gender bias for chest x-ray diagnosis?. In: Lecture notes in computer science. Switzerland: Springer Nature; 2023. p. 142–52. https://doi.org/10.1007/978-3-031-45249-9_14.
[38] Petersen E, Holm S, Ganz M, Feragen A. The path toward equal performance in medical machine learning. Patterns 2023;4(7).
[39] Banerjee I, Bhattacharjee K, Burns JL, Trivedi H, Purkayastha S, Seyyed-Kalantari L, Gichoya J. "shortcuts" causing bias in radiology artificial intelligence: causes, evaluation and mitigation. J Am Coll Radiol 2023.
[40] Brown A, Tomasev N, Freyberg J, Liu Y, Karthikesalingam A, Schrouff J. Detecting shortcut learning for fair medical AI using shortcut testing. Nat Commun 2023;14(1):4314.

[41] Mikolaj, K., Lin, M., Bashir, Z., Svendsen, M. B. S., Tolsgaard, M., Nymark, A., & Feragen, A. (2023). Removing confounding information from fetal ultrasound images. arXiv preprint arXiv:2303.13918.

[42] Nauta M, Walsh R, Dubowski A, Seifert C. Uncovering and correcting shortcut learning in machine learning models for skin cancer diagnosis. Diagnostics 2021;12(1):40.

[43] Oakden-Rayner L, Dunnmon J, Carneiro G, Ré C. Hidden stratification causes clinically meaningful failures in machine learning for medical imaging. In: Proceedings of the ACM Conference on Health, Inference, and Learning; 2020. p. 151–9.

[44] Selby IA, Roberts M, Breger A, Rudd JH, Weir-McCall JR. Shortcut learning: reduced but not resolved. Radiology 2023;308(2):e230379.

[45] Weng N, Pegios P, Petersen E, Feragen A, Bigdeli S. Fast diffusion-based counterfactuals for shortcut removal and generation. In: European Conference on Computer Vision. Cham: Springer Nature Switzerland; 2024. p. 338–57.

[46] Codella, N., Rotemberg, V., Tschandl, P., Celebi, M. E., Dusza, S., Gutman, D., … others (2019). Skin lesion analysis toward melanoma detection 2018: a challenge hosted by the international skin imaging collaboration (isic). arXiv preprint arXiv: 1902.03368.

[47] Narla A, Kuprel B, Sarin K, Novoa R, Ko J. Automated classification of skin lesions: from pixels to practice. J Invest Dermatol 2018;138(10):2108–10.

[48] Juodelyte D, Jiménez-Sánchez A, Cheplygina V. Revisiting hidden representations in transfer learning for medical imaging. Trans Machine Learn Res 2023. Retrieved from, https://openreview.net/forum?id=ScrEUZLxPr.

[49] Godenschweger F, Kägebein U, Stucht D, Yarach U, Sciarra A, Yakupov R, Speck O. Motion correction in MRI of the brain. Phys Med Biol 2016;61(5):R32.

[50] Reuter M, Tisdall MD, Qureshi A, Buckner RL, van der Kouwe AJ, Fischl B. Head motion during MRI acquisition reduces gray matter volume and thickness estimates. Neuroimage 2015;107:107–15.

[51] Van Dijk KR, Sabuncu MR, Buckner RL. The influence of head motion on intrinsic functional connectivity MRI. Neuroimage 2012;59(1):431–8.

[52] Fay L, Cobos E, Yang B, Gatidis S, Küstner T. Avoiding shortcut-learning by mutual information minimization in deep learning-based image processing. IEEE Access 2023.

[53] Wallis D, Buvat I. Clever hans effect found in a widely used brain tumour MRI dataset. Med Image Anal 2022;77:102368.

[54] Cole JH, Poudel RP, Tsagkrasoulis D, Caan MW, Steves C, Spector TD, Montana G. Predicting brain age with deep learning from raw imaging data results in a reliable and heritable biomarker. Neuroimage 2017;163:115–24.

[55] Bi Y, Abrol A, Fu Z, Calhoun V. Deep learning prediction and visualization of gender related brain changes from longitudinal structural MRI data in the abcd study. In: 2022 44th Annual International Conference of the IEEE Engineering in Medicine and Biology Society (EMBC); 2022. p. 3814–7.

[56] Gichoya JW, Banerjee I, Bhimireddy AR, Burns JL, Celi LA, Chen L-C. AI recognition of patient race in medical imaging: a modelling study. Lancet Digit Health 2022;4(6): e406–14.

[57] Bach S, Binder A, Montavon G, Klauschen F, Müller K-R, Samek W. On pixel-wise explanations for non-linear classifier decisions by layer-wise relevance propagation. PLoS One 2015;10(7):e0130140.

[58] Saporta A, Gui X, Agrawal A, Pareek A, Truong SQ, Nguyen CD. Benchmarking saliency methods for chest x-ray interpretation. Nat Mach Intell 2022;4(10):867–78.

[59] Ghassemi M, Oakden-Rayner L, Beam AL. The false hope of current approaches to explainable artificial intelligence in health care. Lancet Digit Health 2021;3(11): e745–50.

[60] Wang R, Chaudhari P, Davatzikos C. Bias in machine learning models can be significantly mitigated by careful training: evidence from neuroimaging studies. Proc Natl Acad Sci 2023;120(6).

[61] Weber, T., Ingrisch, M., Bischl, B., & Rügamer, D. (2023). Unreading race: purging protected features from chest x-ray embeddings. arXiv preprint arXiv:2311.01349.

[62] Kusner MJ, Loftus J, Russell C, Silva R. Counterfactual fairness. In: Guyon I, et al., editors. Advances in neural information processing systems, vol 30. Curran Associates, Inc.; 2017. https://proceedings.neurips.cc/paper/2017/file/a486cd07e4ac3d270571622f4f316ec5-Paper.pdf.

[63] Mouli SC, Ribeiro B. Asymmetry learning for counterfactually-invariant classification in OOD tasks. In: International Conference on Learning Representations; 2022. Retrieved from, https://openreview.net/forum?id=avgclFZ221l.

[64] Veitch V, D'Amour A, Yadlowsky S, Eisenstein J. Counterfactual invariance to spurious correlations: why and how to pass stress tests. In: Beygelzimer A, Dauphin Y, Liang P, Vaughan JW, editors. Advances in neural information processing systems; 2021. Retrieved from, https://openreview.net/forum?id=BdKxQp0iBi8.

[65] Cohen T, Welling M. Group equivariant convolutional networks. In: International Conference on Machine Learning; 2016. p. 2990–9.

[66] Elaldi A, Gerig G, Dey N. E(3) x SO(3) - equivariant networks for spherical deconvolution in diffusion MRI. In: Medical imaging with deep learning; 2023. Retrieved from, https://openreview.net/forum?id=lri_iAbpn_r.

[67] Lafarge MW, Bekkers EJ, Pluim JP, Duits R, Veta M. Roto-translation equivariant convolutional networks: application to histopathology image analysis. Med Image Anal February 2021;68:101849. https://doi.org/10.1016/j.media.2020.101849.

[68] Pang S, Du A, Orgun MA, Wang Y, Sheng QZ, Wang S, Yu Z. Beyond CNNs: exploiting further inherent symmetries in medical image segmentation. IEEE Trans Cybern November 2023;53(11):6776–87. https://doi.org/10.1109/tcyb.2022.3195447.

[69] Hernán MA, Robins JM. Causal inference: what if. Boca Raton: Chapman and Hall/CRC; 2023.

[70] Kasirzadeh A, Smart A. The use and misuse of counterfactuals in ethical machine learning. In: Proceedings of the 2021 ACM Conference on Fairness, Accountability, and Transparency. ACM; March 2021. https://doi.org/10.1145/3442188.3445886.

[71] Robins JM, Greenland S. Causal inference without counterfactuals: comment. J Am Stat Assoc June 2000;95(450):431. https://doi.org/10.2307/2669381.

[72] Glocker B, Jones C, Bernhardt M, Winzeck S. Algorithmic encoding of protected characteristics in chest x-ray disease detection models. EBioMedicine March 2023;89: 104467. https://doi.org/10.1016/j.ebiom.2023.104467.

[73] Yi PH, Wei J, Kim TK, Shin J, Sair HI, Hui FK, Lin CT. Radiology "forensics": determination of age and sex from chest radiographs using deep learning. Emerg Radiol June 2021;28(5):949–54. https://doi.org/10.1007/s10140-021-01953-y.

[74] Zhao H, Dan C, Aragam B, Jaakkola TS, Gordon GJ, Ravikumar P. Fundamental limits and tradeoffs in invariant representation learning. J Mach Learn Res 2022;23(340): 1–49. Retrieved from, http://jmlr.org/papers/v23/21-1078.html.

[75] Zhao H, Gordon G. Inherent tradeoffs in learning fair representations. In: Wallach H, Larochelle H, Beygelzimer A, d'Alché-Buc F, Fox E, Garnett R, editors. Advances in

neural information processing systems, vol 32. Curran Associates, Inc.; 2019. Retrieved from, https://proceedings.neurips.cc/paper/2019/file/b4189d9de0fb2b9cce090bd1a15e3420-Paper.pdf.
[76] Li Y, Tian X, Gong M, Liu Y, Liu T, Zhang K, Tao D. Deep domain generalization via conditional invariant adversarial networks. In: Proceedings of the European Conference on Computer Vision (ECCV). Springer International Publishing; September 2018. p. 647–63. https://doi.org/10.1007/978-3-030-01267-0_38.
[77] Castro DC, Walker I, Glocker B. Causality matters in medical imaging. Nat Commun 2020;11(1). https://doi.org/10.1038/s41467-020-17478-w.
[78] Bishop CM. In: Jordan M, Kleinberg J, Schölkopf B, editors. Pattern recognition and machine learning. New York, NY: Springer; 2006.
[79] Koller D, Friedman N. Probabilistic graphical models. Cambridge, Mass: MIT Press; 2010 [u.a.].
[80] Pearl J. Causality. 2nd ed. ed. Cambridge University Press; 2009.
[81] Ankan A, Panda A. pgmpy: probabilistic graphical models using python. Proceedings of the 14th Python in Science Conference (SCIPY 2015), vol 10; 2015.
[82] Hagberg AA, Schult DA, Swart PJ. Exploring network structure, dynamics, and function using NetworkX. In: Varoquaux G, Vaught T, Millman J, editors. Proceedings of the 7th Python in Science Conference; 2008. p. 11–5. Pasadena, CA USA.
[83] Textor J, van der Zander B, Gilthorpe MS, Liśkiewicz M, Ellison GT. Robust causal inference using directed acyclic graphs: the r package ‘dagitty. Int J Epidemiol January 2017;dyw341. https://doi.org/10.1093/ije/dyw341.
[84] Tikka S, Karvanen J. Identifying causal effects with the R package causaleffect. J Stat Software 2017;76(12). https://doi.org/10.18637/jss.v076.i12.
[85] Subbaswamy A, Saria S. From development to deployment: dataset shift, causality, and shift-stable models in health AI. Biostatistics 2019. https://doi.org/10.1093/biostatistics/kxz041.
[86] Yu AC, Mohajer B, Eng J. External validation of deep learning algorithms for radiologic diagnosis: a systematic review. Radiology: Artif Intell May 2022;4(3). https://doi.org/10.1148/ryai.210064.
[87] D'Amour A, Heller K, Moldovan D, Adlam B, Alipanahi B, Beutel A, Sculley D. Underspecification presents challenges for credibility in modern machine learning. J Mach Learn Res 2022;23(1).
[88] d'Eon G, d'Eon J, Wright JR, Leyton-Brown K. The spotlight: a general method for discovering systematic errors in deep learning models. In: 2022 ACM Conference on Fairness, Accountability, and Transparency. ACM; June 2022. https://doi.org/10.1145/3531146.3533240.
[89] Eyuboglu S, Varma M, Saab K, Delbrouck J-B, Lee-Messer C, Dunnmon J, Ré C. Domino: discovering systematic errors with cross-modal embeddings. ICLR March 2022. https://doi.org/10.48550/ARXIV.2203.14960.
[90] Jain S, Lawrence H, Moitra A, Madry A. Distilling model failures as directions in latent space. In: The Eleventh International Conference on Learning Representations; 2023. Retrieved from, https://openreview.net/forum?id=99RpBVpLiX.
[91] Olesen V, Weng N, Feragen A, Petersen E. Slicing through bias: explaining performance gaps in medical image analysis using slice discovery methods. In: Ethics and fairness in medical imaging. Switzerland: Springer Nature; October 2024. p. 3–13. https://doi.org/10.1007/978-3-031-72787-0_1.

[92] Plumb G, Johnson N, Cabrera A, Talwalkar A. Towards a more rigorous science of blindspot discovery in image classification models. Trans Machine Learn Res 2023. Retrieved from, https://openreview.net/forum?id=MaDvbLaBiF.

[93] Alderman JE, Palmer J, Laws E, McCradden MD, Ordish J, Ghassemi M, Liu X. Tackling algorithmic bias and promoting transparency in health datasets: the STANDING Together consensus recommendations. Lancet Digit Health January 2025;7(1):e64–88. https://doi.org/10.1016/s2589-7500(24)00224-3.

[94] Lekadir K, Feragen A, Fofanah AJ, Frangi AF, Buyx A, Emelie A, Starmans MPA. FUTURE-AI: international consensus guideline for trustworthy and deployable artificial intelligence in healthcare. August 2023. https://www.bmj.com/content/388/bmj-2024-081554.abstract.

[95] Liu X, Glocker B, McCradden MM, Ghassemi M, Denniston AK, Oakden-Rayner L. The medical algorithmic audit. Lancet Digit Health May 2022;4(5):e384–97. https://doi.org/10.1016/s2589-7500(22)00003-6.

CHAPTER

17 Scale it up: Large-scale MR data processing

Taro Langner
Antaros Medical, BioVenture Hub, Mölndal, Sweden
Corresponding author: E-mail: taro.langner@tensorlabbet.com

17.1 Introduction

Deep learning has faced widespread adoption for medical image analysis tasks ever since the breakthrough of convolutional neural networks for image recognition in 2012 [1,2], and especially the publication of the U-Net [3] for biomedical image segmentation in 2015 [4]. As a technology, it has since also become substantially more accessible due to increased availability of information, abstraction through evolving software frameworks, and ease of access to hardware acceleration with consumer-grade graphics cards or cloud computing resources. Together with open-access repositories for medical imaging data (see also Chapter 24), this has put practitioners, students, and enthusiasts in a position to implement and train deep learning models, within a span of hours, that have the potential to reach human expert performance for a growing number of specific image analysis tasks [4].

There are several key challenges involved, however, that lead to only a small minority of such projects ever achieving an impact beyond experimental or research settings. For the practitioner, acceptance management and workflow integration can pose the perhaps greatest challenges for applications in medicine. Regulatory constraints and considerations of data governance furthermore impose restrictions on access to medical data and can prevent technologically sound approaches from receiving approval for deployment. This chapter more closely examines the methodological challenge of generalization across different imaging cohorts and the special case of inference at scale on large, homogeneous imaging datasets such as clinical trials and cohort studies. It also explores how transfer learning can provide benefits from training on comparable large datasets by reusing the acquired information on downstream tasks with potentially more limited access to training data. With such data often being heavily access-restricted, federated learning is furthermore discussed as a way of providing safe access to sensitive data siloed at hospitals and research institutions without infringing on patient rights or privacy concerns. Finally, a brief overview of representation learning is provided for overcoming missing input

Machine Learning in MRI. https://doi.org/10.1016/B978-0-443-14109-6.00017-1

data and establishing common links between abstract concepts encoded in multimodal input such as images, text and more.

17.2 Large-scale data processing on clinical or epidemiological cohorts

Large-scale medical imaging studies can examine thousands of volunteers with standardized imaging protocols that are replicated across multiple sites. UK Biobank [5], for example, has collected extensive medical data of more than half a million volunteers, including imaging data for a subgroup of 100,000 participants of whom about 70,000 are furthermore planned to undergo follow-up imaging at a later point in time [6]. Similarly, the German National Cohort (NAKO) [7] aims to acquire MRI of 30,000 volunteers. For neuroimaging, further studies of note include the Human Connectome Project (HCP) with multimodal MRI of 1100 participants [8], the Adolescent Brain Cognitive Development (ABCD) study with about 10,000 participants [9] and the HEALthy Brain and Child Development (HBCD) dataset which follows 7500 pregnant volunteers [10].

With time, ever more large-scale MRI datasets are becoming available [11]. With multiple imaging protocols, a need for quality control, and a plethora of research hypotheses that link the imaging data to extensive metadata ranging from lifestyle questionnaires over biochemical assays to genetic information, the limitations of manual effort for solving these tasks quickly become evident. At these scales, spending just 1 second of analysis time per image can accumulate to an entire week of full-time work or more.

Many potentially interesting analyses, such as semantic segmentation of organs, muscle, and tissues, require far more than a single second, ranging from minutes to hours or even days when performed with no automation by a trained expert [12]. Whereas a single expert annotator would consequently be required to spend several years or decades on such a task, spreading the work across a wider team poses a challenge to repeatability, with both options incurring significant cost and labor. Whereas these approaches are nonetheless still pursued in the industry, they are often augmented with increasingly powerful semi-automated image analysis techniques [13,14].

Beyond manual and semi-automated analysis, the high degree of standardization in these studies in terms of imaging devices, protocols, and data formatting poses almost ideal conditions for machine learning methodology. Under these controlled conditions, major sources of variability are eliminated, enabling the collection of training data that are directly comparable to those cases that will be encountered by a resulting model when deployed for inference. At the same time, it becomes possible to scale up the quantity of training data, utilizing from dozens [15,16], over hundreds [17,18], and thousands [19,20] to tens of thousands [21] of volumetric scans. A quickly growing body of literature has been dedicated to these tasks that could in all likelihood be covered in an entire book of its own. Rather than attempting an exhaustive review here, only a few examples will be examined in the following sections.

17.2.1 Large-scale applications of machine learning

A common application of machine learning in this setting is the partial or complete automation of tasks that would otherwise be prohibitive in cost and effort due to sheer scale, even if they could be performed manually at least in principle. But beyond providing what is predominantly an efficiency boost or time savings that merely replace manual effort, machine learning techniques can also be used in this scenario to augment human abilities and perform tasks that are conceptually infeasible for reasons other than simple scalability. One example for this consists of age estimation from brain MRI with neural networks for image-based regression, which not only reached high accuracy (see also Chapter 19) but also provided brain-predicted age as a biomarker, with systematic overestimation being associated with premature aging related to injury, degeneration, and mortality [22]. Later work also linked this biomarker to genetic information based on genome-wide association studies [23]. Other work utilized saliency analysis [24] for visualizing relevant image features for a given prediction among a larger group of different subjects. This approach has also been combined with image-based regression for properties such as age, highlighting age-related anatomical structures with aggregation across entire subgroups of a given study [21,25].

17.2.2 Quality control of predictions

Once a model has been deployed for inference on a large-scale study, it can generate thousands or even tens of thousands of predictions. At this scale, anomalies and outlier cases must be expected, along with flawed or entirely failed predictions. For automatically generated segmentation masks, quality control could be performed manually by visual inspection of every single two-dimensional image slice, together with its proposed segmentation mask. In the literature, quality control of this type found nine out of 10 images to be error-free, requiring no further corrections [26]. However, identifying the remaining failure cases and providing objective quality metrics for the model performance during deployment accordingly represents an important factor in making the generated results reliable as a whole (see also Chapter 13). The paradigm of Machine Learning Operations, or MLOps, combines several best practices for this under the concept of a monitoring phase [27]. The scientific literature has likewise devoted increasing interest to methodological tools for addressing this challenge, providing automated approaches not only for image quality assessment but also for evaluations of machine learning prediction results. For medical image analysis, this typically involves a major overlap with image quality control as a first step, which has already been described in more depth in a previous chapter. In the following, the focus is instead of evaluating the actual predictions themselves.

For semantic segmentation (see also Chapter 12), a common way of identifying failed predictions consists in statistical evaluation of the resulting segmentation masks and underlying image properties. This type of evaluation can also utilize a combination of statistical image features relating to aspects such as texture and expected intensities in combination with a machine learning classifier such as random forests trained to distinguish between correct and incorrect segmentations [28]. Another approach consists of training a separate convolutional neural network to provide, from an entire given image together with its previously predicted segmentation, an estimate of the Dice score for each predicted class [29]. As an alternative, the concept of Reverse Classification Accuracy (RCA) was proposed to assign a quality metric to each test case by using it, together with its prediction, as a minimal training set for a separate, smaller model referred to as RCA classifier [30]. This RCA classifier is evaluated against the original ground training truth data. Its performance metrics on the ground truth data are then used to assign a quality score to the test case itself, under the assumption that only a high-quality prediction would enable for it to learn the segmentation task. For image segmentation, this RCA classifier can be implemented as an atlas segmentation [31].

Other approaches use the model itself to generate not only a prediction, but also a measure of its quality with the help of uncertainty quantification. As one such technique, Monte-Carlo dropout [32] can provide a measure of empirical variance by utilizing dropout layers in a model not only during training but also during testing. By keeping the dropout layers active, a random subset of neuron activations will be suppressed, causing multiple repeated predictions to produce slightly different results. This empirical variance can be interpreted as a measure of uncertainty, as high variability for a given prediction may imply low confidence of the model. This approach has also been proposed as a tool for quality control in large-scale MRI segmentation [33]. Whereas this methodology models a distribution with an empirical variance by repeated inference, other techniques can treat each prediction directly as a distribution with a corresponding specific, or heteroscedastic, uncertainty of its own [34,35]. This approach has also been utilized to provide automated quality estimates or prediction intervals on large-scale MRI [25,36].

17.2.3 Real-world impact

These methodologies represent only a small subset of proposed approaches and published work, with ever more sophisticated and powerful methods being developed. Although the literature is speeding ahead of the adoption in the industry and endorsement in the official data catalogs of the underlying large-scale studies, results from machine learning in this space have nonetheless already left an impact on medical research, producing measurements and biomarkers that have formed the basis for further hypotheses and experiments. At the time of writing, UK Biobank in particular (see also Chapter 14) has increasingly already admitted into the official data catalog, for sharing with international researchers, a range of derived data fields based on techniques like those discussed in this section [16,17,37].

Playing to the strengths of machine learning, these approaches were often able to analyze the entire imaged cohort with a single model for a given endpoint. It would be natural to expect that these models, trained on unusually large and well-curated medical imaging datasets, would now represent an ideal, fully-automated solution for analysis of images from any other study in the future. However, this expectation can typically not be met and these models tend to generalize only to a narrow focus of specific images that closely resemble the original training data. Although they may provide excellent performance on the domain that they were trained for, even minor changes in imaging protocol, device type, or patient demographics may radically change their behavior during inference. The capabilities of common models to generalize (see also Chapter 16) can be negatively impacted even when moving between whole-body MRI of studies as superficially similar as UK Biobank and the German National Cohort study [38] due to the phenomenon of distribution shifts, which will be examined more closely in the following section.

17.3 Distribution shifts and correction between imaging cohorts

A fundamental assumption of supervised learning is the ability of a machine learning model to learn a given task from a limited set of training data with available ground truth annotations. Once training is complete, the model is expected to generalize and provide reliable predictions for new, previously unseen data when deployed for inference. To avoid simple memorization of the training samples, this ability is typically evaluated on validation or test data that was not included in the training process. However, even excellent evaluation results on validation or test data are no guarantee for robust generalization during deployment. In medical image analysis, it is therefore not uncommon to observe human expert-level performance on one study with a model that nonetheless fails to produce any plausible predictions on images from another dataset, based on subtle differences in image characteristics that may not be evident to the human eye [39].

17.3.1 Types of distribution shifts

Covariate shifts occur when the features associated with a desired label change during deployment in ways not represented among the training data. In MR imaging, this is a common occurrence due to the variable contrast and qualitative nature of image intensities. Variations in imaging protocols, magnetic field strength, coil placement, and imaging device manufacturers can contribute to these shifts. Likewise, changing fields of view, but also medically relevant characteristics of the imaged anatomy such as subject demographics and anomalies like implants or pathologies can play a role [40].

Label shifts refer to changes in the distribution of the property to be predicted in a given domain relative to its distribution among the training data. A typical example consists of changing class distributions, where a model may have been trained on a balanced dataset before being applied to highly imbalanced data.

Concept shift relates to a definitional change of desired labels. In medicine, this can be a common occurrence as diagnoses or even anatomical delineations can vary between experts of the field, with consensus decisions trying to strike a compromise between different subjective interpretations. A model for semantic segmentation of the liver may, for example, fail to meet expectations when the desired behavior on partial volume, blood vessels, or cysts diverges from the ground truth style provided as training data. Even with just a single operator creating manual reference annotations, variability and style drifts can occur over time [41].

17.3.2 The challenges of generalization in CT and MRI

At first glance, covariate shift in particular could be addressed by aggressive expansion of the training domain and aggregation of extensive datasets capturing the widest-possible variability of data for any given task. Efforts in this direction have enabled substantial progress in the field of natural language processing [42] and also produced models for general-purpose image segmentation [43]. Despite training on millions of samples, the latter have nonetheless been reported to perform worse on medical image data than models trained for this specific domain [44]. Until recently, a common curation step for training data in medical image analysis nonetheless consists in the isolation of imaging modalities, with benchmark challenges reporting superior top performances for modality-specific models as compared to those trained on combined data of, for example, both CT and MRI with the goal of being able to accurately segment both image types with a single model [45]. Ongoing research is continually pushing the boundaries in this area and may well produce more universally viable datasets and approaches, some of which will be examined later on in this chapter.

For CT imaging in particular, the quantitative nature of Hounsfield units can eliminate one major aspect of variability between images. This has enabled models such as Totalsegmentator [46] to achieve robust performance, out-of-the-box, on CT data from various sources, leading to widespread sharing of its freely available implementation for segmentation of over 100 anatomical structures. MR images, in contrast, tend to exhibit greater variability due to the qualitative nature of image intensities and variety of available protocols, field strengths, and scanner settings [40], which makes the training of general-purpose models more challenging (although more recently, a newer iteration of Totalsegmentator also showed favorable results for MRI from multiple sources [47]).

When it comes to the reliability of a machine learning model, it would be desirable to provide a formal verification of an expected minimum accuracy across future data, especially from the perspective of acceptance management and regulatory requirements. However, the complex interplay between modern neural network architectures, with potentially millions of trainable parameters, and volumetric images with tens of thousands of voxel intensity values makes it difficult to provide any universal proof for robust generalization under domain shifts. Even though techniques have been proposed for quantifying domain shifts in medical image analysis [48], it is typically infeasible to provide any guarantees without applying the model to gather empirical results on the actual data in question.

17.3.3 Correction and training for cross-study analysis

Whereas distribution shifts between different data sources can affect the performance of a given model, they can also affect medical measurements obtained from source-specific models. Demographic factors such as age, sex, and body weight can have a direct impact on image-derived phenotypes such as measures of organ volumes and body composition, which may moreover be affected by the appearance of the underlying image data. For cross-study analysis, this has prompted the development of data harmonization techniques with model-based approaches that can calibrated one study to the properties of another [38]. Data transformation has been proposed as another technique to make the target domain more similar to the samples encountered in training, which can involve transformations of the feature space utilized by a model, for example, by principal component analysis [49].

Apart from transforming the data and image-derived measurements, other methodologies aim to improve robustness to domain shifts during training. Here, many common techniques that were designed for improved generalization can also provide immediate benefits for the robustness of a model. This can range from using a suitable pre-processing strategy with regularization and data augmentation in training to techniques such as ensembling and test-time augmentation during inference. The highly successful nnU-Net [50], for example, resamples images to a common resolution to be more robust against variability in image spacing. It also converts the image intensities of modalities such as MRI to z-scores, making images from different scanners and protocols more comparable. Pre-processing like this can eliminate several aspects of variability in the data before it even reaches the underlying neural network itself. During training, its augmentation strategy with random scaling, elastic deformations and brightness changes can likewise expose the network to samples resembling the results of different scanners or patient demographics. Fundamental techniques such as these, which are primarily intended for improved generalization in the target domain, can often help overcome minor domain shifts when trying to perform the same task on a slightly different domain.

Data synthesis is yet another approach that has been proposed as a tool for overcoming distribution shifts [51]. One of its applications is the generation of new samples and augmentation during training, which has been proposed for tasks with imbalanced class labels [52]. It has also been proposed for the imputation of missing data or entire additional modalities with cross-domain image generation, for example, by performing image-to-image translation with generative adversarial networks [53,54].

With these methodologies, different approaches have been laid out for training techniques that aim to improve generalization or modify the input or output data of a machine learning model to improve its robustness when moving from one domain to another. A common property of the techniques examined here is that the training is entirely constrained to the original domain. A different strategy extends beyond this domain border with the help of transfer learning that utilizes information from one domain to more effectively train a model on another.

17.4 Transfer learning

The availability of training data is often a limiting factor for machine learning in medicine, especially when ground truth annotations for specific tasks are required. One common strategy to address this challenge consists of transfer learning, which can re-purpose knowledge from related upstream tasks to more effectively address a given downstream task. A wide range of techniques have been developed under this concept [55], some of which have also been proposed for medical image analysis [56,57] and MRI in particular [58].

One application of transfer learning is domain adaptation, where a model is applied to a downstream task of another domain with the goal of overcoming distribution shifts [39]. This can involve single-shot and few-shot techniques, in which either one or a small number of labeled training samples are available from the downstream task [59]. In the extreme case of zero-shot learning, a model is applied to a new task that is posed with no additional downstream training data [60]. Addressing these challenges tends to require models that incorporate vast amounts of information from upstream tasks with extensive training data, prompting the search for ever more comprehensive foundation models that can provide a starting point for transfer learning approaches.

17.4.1 Using pre-trained models

As ever more training data are being collected and curated, neural network models can draw on an increasing wealth of information. Whereas in 2012, about 1.2 million training samples of ImageNet [61,62] enabled AlexNet [1] to achieve breakthrough success in image recognition, more recent work in 2024 showcased Large Language Models (LLMs) such as Llama3 trained on as many as 15 trillion samples [63].

For convolutional neural networks, supervised pre-training on large-scale visual recognition challenge datasets such as ImageNet remains a popular technique. Whereas this type of model can be trained without prior knowledge by initializing its trainable parameters, or weights, with a randomized distribution of values, copying the parameters from a model that has already converged on another upstream task can enable potentially faster convergence and improved generalization during inference in the given downstream task. This can occur even when the upstream and downstream tasks are substantially different. As such, supervised pre-training by image classification on millions of images of cars, fruits, and animals can provide advantages even for tasks in medical image analysis [56,57,64].

Although models trained on such seemingly unrelated upstream tasks may have never learned to process medical images, the configuration of trainable parameters that enabled their training to converge on another task may still outperform a model initialization that is entirely random. In the case of convolutional neural networks trained on ImageNet, the first network layers have been observed to learn general structural image patterns resembling Gabor filters or color blobs, whereas the final layers tend to be more task-specific [65]. Learning to extract such general-purpose features may accordingly provide a better starting point for further optimization on a new downstream task than a random weight initialization. On medical image data, these features have been reported to evolve into less generic patterns during continued training or never arise at all when training from scratch [57]. Structural benefits can arise even from pre-training on an upstream task for which labels themselves have been generated at random [66].

For the model to be initialized with pre-trained weights, differences between the upstream and downstream task can require changes to the neural network architecture, so that not all pre-trained weights can be copied directly. As an example, a convolutional neural network trained on the ImageNet [62] visual recognition challenge to predict 1000 different classes may require for its last fully-connected layer to be replaced if the new downstream task only requires two class labels [57]. Similarly, its first convolutional layer with three channel dimensions for RGB input may have to be pruned or replaced if the goal is to train on a downstream task with only one input channel. When replacing model components in this manner, the affected weights can either be reused in part of replaced entirely with random initialization.

Due to the benefits of supervised pre-training, various approaches have been developed to pose a given downstream task to resemble the original upstream task as closely as possible. For medical image analysis, this can involve utilizing a model in a way that closely resembles the processing of natural RGB images as encountered in ImageNet [62]. Conceptually, posing the downstream task in this way aims to keep the distribution shift to the upstream task as small as possible. One common step for this consists in pre-processing to scale or limit the image intensities to a specific range. Often more challenging is the disparity in the image shape, which for medical data can often be volumetric with potentially more or fewer than just three RGB color channels. Approaches for bridging this gap can involve 2.5D representations that combine different slices of volumetric images into a single, two-dimensional image with three channels. These slices can be orthogonal patches that provide different views on the same anatomical point in multi-view fusion [67], three consecutive slices of the same imaging plane [50], or combinations of different imaging modalities [68].

Once a model has been successfully initialized with pre-trained weights, different strategies have been proposed for subsequent use on the downstream task [64]. One approach consists of merely applying the model as a feature extractor without performing any further training. The resulting "off-the-shelf" features can then be utilized by other means and potentially be combined with non-imaging data as proposed for radiomics applications [69]. Other approaches limit the training to only a subset of model weights. By keeping the remainder of the model unchanged, or "frozen," a complex model can be trained to adapt to a more limited pool of downstream data without greater risk of overfitting. Variations of this approach can range from freezing all but the final layer for fine-tuning to using different learning rates that inhibit the adaptation of some layers [64]. Perhaps most commonly, the entire model can also be trained with no further restrictions, so that the initialization with pre-trained weights merely serves as a deliberate starting point. Other work has more recently started to utilize weights of ever more comprehensive upstream tasks involving multiple modalities, such as images and text, which will be examined more closely later on in this chapter under the aspect of representation learning.

In modern software frameworks, supervised pre-training can also be performed implicitly by loading various existing model weights as an initialization option rather than requiring for any computational resources to be spent on actually training on an upstream task. Whereas these options can cover a wide range of popular benchmark challenge datasets that received extensive attention in research, they have historically favored neural network architectures for natural, two-dimensional image data with color channels. In turn, more specialized architectures with volumetric filter kernels that have achieved substantial success for voxel-based image data in medicine are not as commonly represented. Ongoing research efforts are aiming to change this by collecting more extensive volumetric medical imaging data for the purpose of pre-training suitable foundation models for medical imaging data [70,71].

17.4.2 Positive and negative transfer

Positive transfer is said to occur when transfer learning benefits the downstream task. However, it is worth noting that transfer learning is not always a required prerequisite for good performance. There is no universal guarantee for transfer learning to provide performance benefits, and it can indeed cause negative transfer to occur in which the downstream task is solved worse than it would have been with a random weight initialization [57]. At the time of writing, volumetric, convolutional architectures such as nnU-Net [50] have established state-of-the-art results in a wide range of medical benchmark challenges even without being able to reap the full benefits of pre-training. Likewise, even transformer models such as UNETR reported no benefit from using pre-trained weights for medical data [72], despite the underlying Vision Transformer (ViT) [73] architecture having been proposed with heavy reliance on pre-training for natural images.

Beyond transfer learning, multiple related methodologies have been proposed that do not strictly focus on one given target domain. In multi-task learning, for example, one model is trained to concurrently perform several tasks in parallel [74]. This can involve training the same model on different sub-tasks and potentially using different, task-specific, output layers, whereas the remaining model weights are shared between tasks. The model is thereby required to learn features that are relevant to all included tasks, which can have a regularization effect that has been reported as beneficial especially at smaller sample sizes [49].

Many more techniques have been proposed in this context, which can help to overcome limited training data in the target domain [58]. Beyond moving information between different domains, other machine learning methodologies have been proposed that can utilize training data from several sources. As this type of aggregation can pose considerable challenges in the field of medicine, the next section provides an overview of how federated learning can achieve this without any requirement for transmitting, pooling, or otherwise exposing any sensitive data directly.

17.5 Federated learning

Many institutions around the world have exclusive access to medical imaging data, such as hospitals, research organizations, and companies. Despite the wealth of contained information, sharing this data for research is largely restricted by regulatory constraints and concerns relating to data privacy and security. These challenges of data governance mean that only a small minority of the images acquired by thousands of MRI devices around the world in daily use are accessible for analysis with machine learning [75].

Federated learning is a technique with the potential for bridging this gap [76–78]. It allows for the distributed training of models without pooling or exposing the underlying images or records, so that the respective owners can retain full data sovereignty. Instead of transmitting the data itself, this methodology can share the parameters of a global consensus model that is refined locally and aggregated with different compute plans and infrastructure topologies to optimize a global loss function [79].

In each learning round, the participating entities of the federation can perform one or more iterative optimization steps of the global loss function by training on the locally available data and applying the resulting parameter updates to their local copy of the model. To conclude the round, the aggregated parameter updates are then shared with the other participants. When a high number of local optimization steps is performed as part of a given round, the overarching training process as a whole tends to increasingly diverge from a conventional approach of training on pooled data and special care must be taken to avoid potential performance penalties. In turn, sharing more granular updates can pose a risk of exposing more directly identifiable elements that could be vulnerable to exploitation or information leakage [80]. Ongoing research toward privacy-preserving federated learning is concerned with addressing these limitations with concepts such as local differential privacy [81].

The participating entities can be arranged as nodes into different topologies, such as decentralized architectures, in which they communicate directly with each other and the overarching model is updated in parallel or centralized designs in which all nodes are connected to one aggregation server. In the latter, the participating institutions do not necessarily have to be aware of each other and can remain anonymous or be blinded to one another. Hybrid approaches can furthermore combine both of these structures in hierarchical architectures that can be composed of nested subfederations. The compute plan determines how the global model is updated based on the local processing results. For example, the update can be performed on a peer-to-peer basis, over the aggregation server or in sequential order between nodes [75,82]. In practice, the latter is typically avoided due to the risk of catastrophic forgetting, where a model may no longer be able to correctly process training samples that only occurred early on in the training process [83].

Semantic segmentation of brain MRI has received special attention for applications of federal learning, with early work reporting competitive results for brain lesion segmentation when simulating different institutions collaborating over a centralized aggregation server [84]. Another approach proposed decentralized, peer-to-peer learning for whole brain segmentation on data of different institutions, enabling shared training on ground truth segmentations that may require up to 1 week of manual work by expert annotators per case [85]. Other work proposed decentralized federated learning as a privacy-preserving approach for combining training data from fMRI of the brain and explored domain adaptation techniques for overcoming domain shifts between different sites [86]. Further work explored federated learning for MRI reconstruction [87,88] and synthesis [89].

Ongoing work is increasing the scope of comparable initiatives, with the goal of aggregating ever more data for training of machine learning models [75]. Beyond segmentation, classification, and reconstruction, there is increasing interest in also linking the underlying images to multi-modal metadata such as textual descriptions and reports that could be explored with more recent methodologies for representation learning.

17.6 Representation learning

One of the primary reasons for the popularity of deep learning has been the ability of neural networks to learn relevant features for a given task from data through mathematical optimization [90]. In medical image analysis, these automatically learned features have been found to not only outperform hand-crafted features in a variety of downstream tasks, but also to hold potential as generic "off-the-shelf" representations for image data [64]. Research has since been exploring further ways to form, manipulate, and utilize these feature representations for various applications including MR image processing.

Among these applications have been attempts to overcome domain shifts by finding representations of MR image features that are invariant to the exact type of scanner and protocol that they were acquired with. Domain-adversarial networks have been proposed as one such approach that can form domain-invariant feature representations by using an adversarial domain discriminator [91]. In this adversarial setting, this discriminator is trained to predict the given domain from intermediate feature representations of a given network. The training is successfully concluded when these feature representations no longer provide any indication to the discriminator as to which input domain they were extracted from.

Related work similarly proposed strategies for disentangling feature representations into factors that could differentiate between characteristics of the imaging modality, such as CT or MRI, and an underlying spatial and anatomical structure that could then be processed independently of which modality it originates from Ref. [92]. This concept has also been leveraged to process multi-modal MRI in which one or more modalities are missing by embedding the input modalities into a shared modality-invariant latent space [93,94]. This not only enabled segmentation with one or more missing input modalities, but could also be used in a generative strategy, for example, to create images with synthetic lesions [95].

17.6.1 Language–image pre-training

The concept of multi-modal input has received increasing attention and can be taken beyond merely combining different types of images toward concurrent processing of entirely different formats such as images, video, and text [96]. Following earlier work that examined X-ray images paired with text data [97,98], Contrastive Language–Image Pre-training (CLIP) [99] was a proposed and received substantial attention as a method that could represent both text and images in a multi-modal embedding space. This approach can be pretrained on batches of images, each of which is paired with one corresponding text. The contrastive optimization aims for the similarity between the image embedding as created by an image encoder and the text embedding as created by a text encoder to be maximized for each pair, whereas the similarity between incorrect pairings is minimized. Once pretraining on 400 million input pairs is concluded, the model has thereby been trained to relate visual and textual concepts to each other by converting both input formats into a similar feature embedding [99].

The ability of models like CLIP to learn representations that link text and images together has since been applied for a growing range of applications outside of medical image analysis. One such example consists of image classification in a zero-shot setting, where no training is conducted on the target domain. This task was posed by creating one textual caption for each potential class and comparing its text embedding to the image embedding of a given input image. The class that scored the highest similarity between its embedded caption and the embedded image was then assigned, with an accuracy rivaling specialized architectures trained for the specific task [99]. In later work, CLIP-guided synthesis has furthermore been used to control the process of image generation with textual prompts in methods such as Stable Diffusion [100].

Whereas many of these approaches examined non-medical settings, similar techniques have since been proposed for medical image analysis as well. For organ segmentation and tumor detection in CT, a CLIP-driven universal model was proposed that trained on data of several pooled benchmark challenge datasets with mutually inconsistent labels and could isolate specific targets based on textual prompts [101]. This approach was subsequently expanded to CT, MRI and PET data in a continued push toward a universal segmentation model [71].

17.6.2 Universal segmentation models

Another effort toward a universal segmentation model has been building on the Segment Anything Model (SAM) [43]. This approach represents two-dimensional images with embeddings that are then combined with a prompt consisting of either an input mask, points, boxes or free text to produce multiple output segmentation masks. Whereas the original system was trained on 11 million diverse natural images, its performance on medical data was reported with mixed results [44]. More recent work proposed MedSAM as a closely related approach for medical images that included over 1 million paired images and segmentations from 15 modalities in training [102]. Ongoing work is continually widening the scope of inputs and can benefit from ever more data being available from research organizations and benchmark challenges that has been released for shared use.

17.7 Current challenges and future directions

Many of the techniques discussed in this chapter were designed to address limited access to high-quality training data. Whereas this emphasis on data-efficiency can solve specific tasks well with just dozens of labeled training samples, it poses a stark contrast to other disciplines. Natural language processing, where training data are far more readily available, has instead leveraged scaling laws [103] that showed conclusive mutual benefits between larger datasets, compute and model sizes for increasingly powerful Large Language Models (LLMs). Their multi-modal capabilities and ability to process prompts with image data have been met with interest even in radiology, but hallucinations and privacy concerns remain open challenges [104]. Nonetheless, access to more domain-specific, medical data holds great potential and techniques like federated learning are likely to play a major role in the future.

17.8 Summary

By learning more universal and sophisticated feature representations, recent machine learning methodologies have become increasingly powerful in extracting underlying concepts of human knowledge from diverse types of data and relating them to each other. In the search of universally viable embeddings, these systems go beyond memorization of specific shapes and patterns toward progressively more abstract relationships to capture information that, while difficult to formalize or encode explicitly, is of essential interest and intuitive use to a human observer regardless of the particular medium or format at hand. At the same time, these techniques explore new ways to open up additional interfaces for human interaction, such as text prompts, seed points, bounding boxes and more types of input that can inform a given task. There are more ways in which this relationship can be manifested and the following chapter examines more closely how processing of MRI can benefit from human guidance.

References

[1] Krizhevsky A, Sutskever I, Hinton GE. Imagenet classification with deep convolutional neural networks. Adv Neural Inf Process Syst 2012;25.

[2] Dan C, Ueli M, Jonathan M, Jürgen S. Multi-column deep neural network for traffic sign classification. Neural Netw 2012;32(1):333–8.

[3] Ronneberger O, Fischer P, Brox T. U-net: convolutional networks for biomedical image segmentation. In: Medical image computing and computer-assisted intervention–MICCAI 2015: 18th international conference, Munich, Germany, October 5-9, 2015, Proceedings, Part III 18. Springer; 2015. p. 234–41.

[4] Litjens G, Kooi T, Ehteshami Bejnordi B, Setio AAA, Ciompi F, Ghafoorian M, Awm Van Der Laak J, Van Ginneken B, Sánchez CI. A survey on deep learning in medical image analysis. Med Image Anal 2017;42:60–88.

[5] Sudlow C, Gallacher J, Allen N, Beral V, Burton P, Danesh J, Downey P, Elliott P, Green J, Landray M, et al. UK biobank: an open access resource for identifying the causes of a wide range of complex diseases of middle and old age. PLoS Med 2015;12(3):e1001779.

[6] Littlejohns TJ, Holliday J, Gibson LM, Garratt S, Oesingmann N, Alfaro-Almagro F, Bell JD, Boultwood C, Collins R, Conroy MC, et al. The UK biobank imaging enhancement of 100,000 participants: rationale, data collection, management and future directions. Nat Commun 2020;11(1):2624.

[7] Bamberg F, Kauczor H-U, Weckbach S, Schlett CL, Forsting M, Ladd SC, Greiser KH, Weber M-A, Schulz-Menger J, Niendorf T, et al. Whole-body mr imaging in the German national cohort: rationale, design, and technical background. Radiology 2015;277(1):206–20.

[8] Stine Elam J, Glasser MF, Harms MP, Sotiropoulos SN, Andersson JLR, Burgess GC, Curtiss SW, Oostenveld R, Larson-Prior LJ, Schoffelen J-M, et al. The human connectome project: a retrospective. Neuroimage 2021;244:118543.

[9] Casey BJ, Cannonier T, Conley MI, Cohen AO, Barch DM, Heitzeg MM, Soules ME, Teslovich T, Dellarco DV, Garavan H, et al. The adolescent brain cognitive development (abcd) study: imaging acquisition across 21 sites. Dev Cogn Neurosci 2018; 32:43–54.

[10] Dean III DC, Dylan Tisdall M, Wisnowski JL, Feczko E, Gagoski B, Alexander AL, Edden RAE, Gao W, Hendrickson TJ, Howell BR, et al. Quantifying brain development in the healthy brain and child development (hbcd) study: the magnetic resonance imaging and spectroscopy protocol. Dev Cogn Neurosci 2024;70:101452.

[11] Selvikvåg Lundervold A, Lundervold A. An overview of deep learning in medical imaging focusing on mri. Z Med Phys 2019;29(2):102–27.

[12] Fischl B, Salat DH, Busa E, Albert M, Dieterich M, Haselgrove C, Van Der Kouwe A, Killiany R, Kennedy D, Klaveness S, et al. Whole brain segmentation: automated labeling of neuroanatomical structures in the human brain. Neuron 2002;33(3):341–55.

[13] Wilman HR, Kelly M, Garratt S, Matthews PM, Milanesi M, Herlihy A, Gyngell M, Neubauer S, Bell JD, Banerjee R, et al. Characterisation of liver fat in the UK biobank cohort. PLoS One 2017;12(2):e0172921.

[14] Linge J, Borga M, West J, Tuthill T, Miller MR, Dumitriu A, Thomas EL, Romu T, Tunón P, Bell JD, et al. Body composition profiling in the UK biobank imaging study. Obesity 2018;26(11):1785–95.

[15] Basty N, Liu Y, Cule M, Thomas EL, Bell JD, Whitcher B. Automated measurement of pancreatic fat and iron concentration using multi-echo and t1-weighted mri data. In: 2020 IEEE 17th international symposium on biomedical imaging (ISBI). IEEE; 2020. p. 345–8.

[16] Langner T, Östling A, Maldonis L, Karlsson A, Olmo D, Lindgren D, Wallin A, Lowe L, Strand R, Ahlström H, et al. Kidney segmentation in neck-to-knee body mri of 40,000 UK biobank participants. Sci Rep 2020;10(1):20963.

[17] Liu Y, Basty N, Whitcher B, Bell JD, Sorokin EP, van Bruggen N, Louise Thomas E, Cule M. Genetic architecture of 11 organ traits derived from abdominal mri using deep learning. eLife 2021;10:e65554.

[18] Kart T, Fischer M, Küstner T, Hepp T, Bamberg F, Winzeck S, Glocker B, Rueckert D, Gatidis S. Deep learning-based automated abdominal organ segmentation in the UK biobank and German national cohort magnetic resonance imaging studies. Investig Radiol 2021;56(6):401–8.

[19] Bai W, Sinclair M, Tarroni G, Oktay O, Rajchl M, Vaillant G, Lee AM, Aung N, Lukaschuk E, Sanghvi MM, et al. Automated cardiovascular magnetic resonance image analysis with fully convolutional networks. J Cardiovasc Magn Reson 2018;20(1): 1–12.

[20] Haas ME, Pirruccello JP, Friedman SN, Wang M, Emdin CA, Ajmera VH, Simon TG, Homburger JR, Guo X, Budoff M, et al. Machine learning enables new insights into genetic contributions to liver fat accumulation. Cell Genom 2021;1(3).

[21] Langner T, Strand R, Ahlström H, Kullberg J. Large-scale biometry with interpretable neural network regression on UK biobank body mri. Sci Rep 2020;10(1):17752.

[22] Cole JH, Ritchie SJ, Bastin ME, Hernández V, Maniega SM, Royle N, Corley J, Pattie A, Harris SE, Zhang Q, et al. Brain age predicts mortality. Mol Psychiatr 2018;23(5):1385–92.

[23] Atli Jónsson B, Bjornsdottir G, Thorgeirsson TE, Ellingsen LM, Walters GB, Gudbjartsson DF, Stefansson H, Stefansson K, Ulfarsson MO. Brain age prediction

using deep learning uncovers associated sequence variants. Nat Commun 2019;10(1): 5409.

[24] Selvaraju RR, Cogswell M, Das A, Vedantam R, Devi P, Batra D. Grad-cam: visual explanations from deep networks via gradient-based localization. In: Proceedings of the IEEE international conference on computer vision; 2017. p. 618–26.

[25] Hepp T, Blum D, Karim A, Schoelkopf B, Stern D, Yang B, Gatidis S. Uncertainty estimation and explainability in deep learning-based age estimation of the human brain: results from the German national cohort mri study. Comput Med Imaging Graph 2021:101967.

[26] Kart T, Fischer M, Winzeck S, Glocker B, Bai W, Bülow R, Emmel C, Friedrich L, Hans-Ulrich K, Keil T, et al. Automated imaging-based abdominal organ segmentation and quality control in 20,000 participants of the UK biobank and German national cohort studies. Sci Rep 2022;12(1):18733.

[27] Mäkinen S, Skogström H, Laaksonen E, Mikkonen T. Who needs mlops: what data scientists seek to accomplish and how can mlops help?. In: 2021 IEEE/ACM 1st workshop on AI engineering-software engineering for AI (WAIN). IEEE; 2021. p. 109–12.

[28] Alba X, Lekadir K, Pereanez M, Medrano-Gracia P, Young AA, Frangi AF. Automatic initialization and quality control of large-scale cardiac mri segmentations. Med Image Anal 2018;43:129–41.

[29] Robinson R, Oktay O, Bai W, Valindria VV, Sanghvi MM, Aung N, Paiva JM, Zemrak F, Fung K, Lukaschuk E, et al. Real-time prediction of segmentation quality. In: Medical image computing and computer assisted intervention–MICCAI 2018: 21st international conference, Granada, Spain, September 16-20, 2018, proceedings, Part IV 11. Springer; 2018. p. 578–85.

[30] Valindria VV, Lavdas I, Bai W, Kamnitsas K, Aboagye EO, Rockall AG, Rueckert D, Glocker B. Reverse classification accuracy: predicting segmentation performance in the absence of ground truth. IEEE Trans Med Imag 2017;36(8):1597–606.

[31] Robinson R, Valindria VV, Bai W, Oktay O, Kainz B, Suzuki H, Sanghvi MM, Aung N, Paiva JM, Zemrak F, et al. Automated quality control in image segmentation: application to the UK biobank cardiovascular magnetic resonance imaging study. J Cardiovasc Magn Reson 2019;21(1):1–14.

[32] Gal Y, Ghahramani Z. Dropout as a bayesian approximation: Representing model uncertainty in deep learning. International conference on machine learning. PMLR; 2016. p. 1050–9.

[33] Bard A, Raisi-Estabragh Z, Ardissino M, Lee AM, Pugliese F, Dey D, Sarkar S, Munroe PB, Neubauer S, Harvey NC, et al. Automated quality-controlled cardiovascular magnetic resonance pericardial fat quantification using a convolutional neural network in the UK biobank. Front Cardiovasc Med 2021;8:677574.

[34] Kendall A, Gal Y. What uncertainties do we need in bayesian deep learning for computer vision? Adv Neural Inf Process Syst 2017;30.

[35] Lakshminarayanan B, Alexander P, Blundell C. Simple and scalable predictive uncertainty estimation using deep ensembles. Adv Neural Inf Process Syst 2017;30.

[36] Langner T, Gustafsson FK, Avelin B, Strand R, Ahlström H, Kullberg J. Uncertainty-aware body composition analysis with deep regression ensembles on UK biobank mri. Comput Med Imag Graph 2021;93:101994.

[37] Suinesiaputra A, Sanghvi MM, Aung N, Paiva JM, Zemrak F, Fung K, Lukaschuk E, Lee AM, Carapella V, Kim YJ, et al. Fully-automated left ventricular mass and volume

mri analysis in the UK biobank population cohort: evaluation of initial results. Int J Cardiovasc Imag 2018;34:281–91.

[38] Gatidis S, Kart T, Fischer M, Winzeck S, Glocker B, Bai W, Bülow R, Emmel C, Friedrich L, Hans-Ulrich K, et al. Better together: data harmonization and cross-study analysis of abdominal mri data from UK biobank and the German national cohort. Investig Radiol 2023;58(5):346.

[39] Guan H, Liu M. Domain adaptation for medical image analysis: a survey. IEEE (Inst Electr Electron Eng) Trans Biomed Eng 2021;69(3):1173–85.

[40] Hagiwara A, Fujita S, Ohno Y, Aoki S. Variability and standardization of quantitative imaging: monoparametric to multiparametric quantification, radiomics, and artificial intelligence. Investig Radiol 2020;55(9):601.

[41] Withey DJ, Koles ZJ. Medical image segmentation: methods and software. In: 2007 joint meeting of the 6th international symposium on noninvasive functional source imaging of the brain and heart and the international conference on functional biomedical imaging. IEEE; 2007. p. 140–3.

[42] Devlin J, Chang M-W, Lee K, Toutanova K. Bert: pre-training of deep bidirectional transformers for language understanding. arXiv preprint arXiv:1810.04805, 2018.

[43] Kirillov A, Mintun E, Ravi N, Mao H, Rolland C, Gustafson L, Xiao T, Whitehead S, Berg AC, Lo W-Y, et al. Segment anything. arXiv preprint arXiv:2304.02643, 2023.

[44] Roy S, Wald T, Koehler G, Rokuss MR, Disch N, Holzschuh J, Zimmerer D, Maier-Hein KH. Sam. md: zero-shot medical image segmentation capabilities of the segment anything model. arXiv preprint arXiv:2304.05396, 2023.

[45] Emre Kavur A, Sinem Gezer N, Barıs M, Aslan S, Conze P-H, Groza V, Pham DD, Chatterjee S, Ernst P, Özkan S, et al. Chaos challenge-combined (ct-mr) healthy abdominal organ segmentation. Med Image Anal 2021;69:101950.

[46] Wasserthal J, Breit H-C, Meyer MT, Pradella M, Hinck D, Sauter AW, Heye T, Boll DT, Cyriac J, Yang S, et al. Totalsegmentator: robust segmentation of 104 anatomic structures in ct images. Radiol Artif Intell 2023;5(5).

[47] D'Antonoli TD, Berger LK, Indrakanti AK, Vishwanathan N, Weiß J, Jung M, Berkarda Z, Rau A, Reisert M, Küstner T, et al. Totalsegmentator mri: sequence-independent segmentation of 59 anatomical structures in mr images. arXiv preprint arXiv:2405.19492, 2024.

[48] Stacke K, Eilertsen G, Unger J, Lundström C. Measuring domain shift for deep learning in histopathology. IEEE J Biomed Health Inform 2020;25(2):325–36.

[49] Cheplygina V, de Bruijne M, Pluim JPW. Not-so-supervised: a survey of semi-supervised, multi-instance, and transfer learning in medical image analysis. Med Image Anal 2019;54:280–96.

[50] Isensee F, Jaeger PF, Kohl SAA, Petersen J, Maier-Hein KH. nnu-net: a self-configuring method for deep learning-based biomedical image segmentation. Nat Methods 2021;18(2):203–11.

[51] Yi X, Walia E, Babyn P. Generative adversarial network in medical imaging: a review. Med Image Anal 2019;58:101552.

[52] Shin H-C, Tenenholtz NA, Rogers JK, Schwarz CG, Senjem ML, Gunter JL, Andriole KP, Michalski M. Medical image synthesis for data augmentation and anonymization using generative adversarial networks. In: Simulation and synthesis in medical imaging: third international workshop, SASHIMI 2018, held in conjunction with MICCAI 2018, Granada, Spain, September 16, 2018, proceedings 3. Springer; 2018. p. 1–11.

[53] Armanious K, Jiang C, Fischer M, Küstner T, Hepp T, Nikolaou K, Gatidis S, Yang B. Medgan: medical image translation using gans. Comput Med Imaging Graph 2020;79: 101684.

[54] Ben-Cohen A, Klang E, Raskin SP, Soffer S, Ben-Haim S, Konen E, Marianne Amitai M, Greenspan H. Cross-modality synthesis from ct to pet using fcn and gan networks for improved automated lesion detection. Eng Appl Artif Intell 2019;78: 186–94.

[55] Zhuang F, Qi Z, Duan K, Xi D, Zhu Y, Zhu H, Xiong H, He Q. A comprehensive survey on transfer learning. Proc IEEE 2020;109(1):43–76.

[56] Van Opbroek A, Ikram MA, Vernooij MW, De Bruijne M. Transfer learning improves supervised image segmentation across imaging protocols. IEEE Trans Med Imag 2014;34(5):1018–30.

[57] Raghu M, Zhang C, Kleinberg J, Bengio S. Transfusion: understanding transfer learning for medical imaging. Adv Neural Inf Process Syst 2019;32.

[58] Valverde JM, Imani V, Abdollahzadeh A, De Feo R, Prakash M, Ciszek R, Tohka J. Transfer learning in magnetic resonance brain imaging: a systematic review. J Imaging 2021;7(4):66.

[59] Wang Y, Yao Q, Kwok JT, Ni LM. Generalizing from a few examples: a survey on few-shot learning. ACM Comput Surv 2020;53(3):1–34.

[60] Palatucci M, Pomerleau D, Hinton GE, Mitchell TM. Zero-shot learning with semantic output codes. Adv Neural Inf Process Syst 2009;22.

[61] Kolesnikov A, Beyer L, Zhai X, Puigcerver J, Yung J, Gelly S, Houlsby N. Big transfer (bit): general visual representation learning. In: Computer vision–ECCV 2020: 16th European conference, Glasgow, UK, August 23–28, 2020, Proceedings, Part V 16. Springer; 2020. p. 491–507.

[62] Russakovsky O, Jia D, Su H, Krause J, Satheesh S, Ma S, Huang Z, Karpathy A, Khosla A, Bernstein M, et al. Imagenet large scale visual recognition challenge. Int J Comput Vis 2015;115:211–52.

[63] Dubey A, Jauhri A, Pandey A, Kadian A, Al-Dahle A, Letman A, Mathur A, Schelten A, Yang A, Fan A, et al. The llama 3 herd of models. arXiv preprint arXiv:2407.21783 2024.

[64] Shin H-C, Roth HR, Gao M, Lu Le, Xu Z, Nogues I, Yao J, Mollura D, Summers RM. Deep convolutional neural networks for computer-aided detection: Cnn architectures, dataset characteristics and transfer learning. IEEE Trans Med Imag 2016;35(5): 1285–98.

[65] Yosinski J, Clune J, Bengio Y, Lipson H. How transferable are features in deep neural networks? Adv Neural Inf Process Syst 2014;27.

[66] Maennel H, Alabdulmohsin IM, Tolstikhin IO, Baldock R, Bousquet O, Gelly S, Keysers D. What do neural networks learn when trained with random labels? Adv Neural Inf Process Syst 2020;33:19693–704.

[67] Roth HR, Lu Le, Liu J, Yao J, Seff A, Cherry K, Kim L, Summers RM. Improving computer-aided detection using convolutional neural networks and random view aggregation. IEEE Trans Med Imag 2015;35(5):1170–81.

[68] Hesam Hesamian M, Jia W, He X, Kennedy P. Deep learning techniques for medical image segmentation: achievements and challenges. J Digit Imag 2019;32: 582–96.

[69] Avanzo M, Wei L, Stancanello J, Vallieres M, Rao A, Morin O, Mattonen SA, El Naqa I. Machine and deep learning methods for radiomics. Med Phys 2020;47(5): e185–202.

[70] Chen S, Ma K, Zheng Y. Med3d: transfer learning for 3d medical image analysis. arXiv preprint arXiv:1904.00625, 2019.

[71] Ye Y, Xie Y, Zhang J, Chen Z, Xia Y. Uniseg: a prompt-driven universal segmentation model as well as a strong representation learner. arXiv preprint arXiv:2304.03493, 2023.

[72] Hatamizadeh Ali, Tang Y, Nath V, Yang D, Myronenko A, Landman B, Roth HR, Xu D. Unetr: transformers for 3d medical image segmentation. In: Proceedings of the IEEE/CVF winter conference on applications of computer vision; 2022. p. 574–84.

[73] Dosovitskiy A, Beyer L, Kolesnikov A, Weissenborn D, Zhai X, Unterthiner T, Dehghani M, Minderer M, Heigold G, Gelly S, et al. An image is worth 16x16 words: transformers for image recognition at scale. arXiv preprint arXiv:2010.11929 2020.

[74] Zhang Y, Yang Q. A survey on multi-task learning. IEEE Trans Knowl Data Eng 2021; 34(12):5586–609.

[75] Rieke N, Hancox J, Li W, Milletari F, Roth HR, Albarqouni S, Bakas S, Galtier MN, Landman BA, Maier-Hein K, et al. The future of digital health with federated learning. npj Digit Med 2020;3(1):119.

[76] McMahan B, Moore E, Ramage D, Hampson S, Arcas BAy. Communication-efficient learning of deep networks from decentralized data. In: Artificial intelligence and statistics. PMLR; 2017. p. 1273–82.

[77] Xu J, Glicksberg BS, Su C, Walker P, Bian J, Wang F. Federated learning for healthcare informatics. Journal of Healthcare Informatics Research 2021;5:1–19.

[78] Nguyen DC, Pham Q-V, Pathirana PN, Ding M, Seneviratne A, Lin Z, Dobre O, Hwang W-J. Federated learning for smart healthcare: a survey. ACM Comput Surv 2022;55(3):1–37.

[79] Konečnỳ J, Brendan McMahan H, Ramage D, Richtárik P. Federated optimization: distributed machine learning for on-device intelligence. arXiv preprint arXiv: 1610.02527 2016.

[80] Wang Z, Song M, Zhang Z, Song Y, Wang Q, Qi H. Beyond inferring class representatives: User-level privacy leakage from federated learning. In: IEEE INFOCOM 2019-IEEE conference on computer communications. IEEE; 2019. p. 2512–20.

[81] Sun L, Qian J, Chen X. Ldp-fl: practical private aggregation in federated learning with local differential privacy. arXiv preprint arXiv:2007.15789 2020.

[82] Kairouz P, Brendan McMahan H, Avent B, Bellet A, Bennis M, Nitin Bhagoji A, Bonawitz K, Charles Z, Cormode G, Cummings R, et al. Advances and open problems in federated learning. Foundations and Trends® in Machine Learning 2021;14(1–2): 1–210.

[83] French RM. Catastrophic forgetting in connectionist networks. Trends Cognit Sci 1999;3(4):128–35.

[84] Sheller MJ, Reina GA, Edwards B, Martin J, Bakas S. Multi-institutional deep learning modeling without sharing patient data: a feasibility study on brain tumor segmentation. In: Brainlesion: Glioma, multiple sclerosis, stroke and traumatic brain injuries: 4th international workshop, BrainLes 2018, held in conjunction with MICCAI 2018, Granada, Spain, September 16, 2018, revised selected papers, Part I 4. Springer; 2019. p. 92–104.

[85] Roy AG, Siddiqui S, Pölsterl S, Navab N, Braintorrent CW. A peer-to-peer environment for decentralized federated learning. arXiv preprint arXiv:1905.06731 2019.

[86] Li X, Gu Y, Dvornek N, Staib LH, Ventola P, Duncan JS. Multi-site fmri analysis using privacy-preserving federated learning and domain adaptation: Abide results. Med Image Anal 2020;65:101765.

[87] Guo P, Wang P, Zhou J, Jiang S, Patel VM. Multi-institutional collaborations for improving deep learning-based magnetic resonance image reconstruction using federated learning. In: Proceedings of the IEEE/CVF conference on computer vision and pattern recognition; 2021. p. 2423–32.

[88] Elmas G, Dar SUH, Korkmaz Y, Ceyani E, Susam B, Ozbey M, Avestimehr S, Çukur T. Federated learning of generative image priors for mri reconstruction. IEEE Trans Med Imag 2022.

[89] Dalmaz O, Mirza U, Elmas G, Özbey M, Uh Dar S, Ceyani E, Avestimehr S, Çukur T. One model to unite them all: personalized federated learning of multi-contrast mri synthesis. arXiv preprint arXiv:2207.06509 2022.

[90] Bengio Y, Courville A, Vincent P. Representation learning: a review and new perspectives. IEEE Trans Pattern Anal Mach Intell 2013;35(8):1798–828.

[91] Kamnitsas K, Baumgartner C, Ledig C, Newcombe V, Simpson J, Kane A, Menon D, Nori A, Criminisi A, Rueckert D, et al. Unsupervised domain adaptation in brain lesion segmentation with adversarial networks. In: Information processing in medical imaging: 25th international conference, IPMI 2017, Boone, NC, USA, June 25-30, 2017, proceedings 25. Springer; 2017. p. 597–609.

[92] Chartsias A, Joyce T, Papanastasiou G, Semple S, Williams M, Newby DE, Dharmakumar R, Tsaftaris SA. Disentangled representation learning in cardiac image analysis. Med Image Anal 2019;58:101535.

[93] Havaei M, Guizard N, Chapados N, Bengio Y. Hemis: hetero-modal image segmentation. In: Medical image computing and computer-assisted intervention–MICCAI 2016: 19th international conference, Athens, Greece, October 17-21, 2016, proceedings, Part II 19. Springer; 2016. p. 469–77.

[94] Zhou T, Canu S, Vera P, Su R. Latent correlation representation learning for brain tumor segmentation with missing mri modalities. IEEE Trans Image Process 2021;30: 4263–74.

[95] Chartsias A, Joyce T, Giuffrida MV, Tsaftaris SA. Multimodal mr synthesis via modality-invariant latent representation. IEEE Trans Med Imag 2017;37(3): 803–14.

[96] Baltrušaitis T, Ahuja C, Morency L-P. Multimodal machine learning: a survey and taxonomy. IEEE Trans Pattern Anal Mach Intell 2018;41(2):423–43.

[97] Huang S-C, Shen L, Lungren MP, Yeung S. Gloria: a multimodal global-local representation learning framework for label-efficient medical image recognition. In: Proceedings of the IEEE/CVF international conference on computer vision; 2021. p. 3942–51.

[98] Zhang Y, Jiang H, Miura Y, Manning CD, Langlotz CP. Contrastive learning of medical visual representations from paired images and text. In: Machine learning for healthcare conference. PMLR; 2022. p. 2–25.

[99] Radford A, Kim JW, Hallacy C, Ramesh A, Goh G, Agarwal S, Sastry G, Askell A, Mishkin P, Clark J, et al. Learning transferable visual models from natural language supervision. In: International conference on machine learning. PMLR; 2021. p. 8748–63.

[100] Rombach R, Blattmann A, Lorenz D, Esser P, Ommer B. High-resolution image synthesis with latent diffusion models. In: Proceedings of the IEEE/CVF conference on computer vision and pattern recognition; 2022. p. 10684–95.

[101] Liu J, Zhang Y, Chen J-N, Xiao J, Lu Y, Landman BA, Yuan Y, Yuille A, Tang Y, Zhou Z. Clip-driven universal model for organ segmentation and tumor detection. In: Proceedings of the IEEE/CVF international conference on computer vision; 2023. p. 21152–64.

[102] Ma J, Wang B. Segment anything in medical images. arXiv preprint arXiv:2304.12306 2023.

[103] Kaplan J, McCandlish S, Henighan T, Brown TB, Chess B, Child R, Gray S, Radford A, Wu J, Amodei D. Scaling laws for neural language models. arXiv preprint arXiv:2001.08361 2020.

[104] Bhayana R. Chatbots and large language models in radiology: a practical primer for clinical and research applications. Radiology 2024;310(1):e232756.

CHAPTER

Human in the loop: Integration of experts in MR data processing

18

Gustavo Carneiro

Centre for Vision, Speech and Signal Processing (CVSSP), University of Surrey, Surrey, United Kingdom

Corresponding author: E-mail: g.carneiro@surrey.ac.uk

18.1 Introduction

Magnetic resonance (MR) technology is a noninvasive imaging technique that uses powerful magnets and radio waves to generate detailed images of the body's internal structures [1]. Unlike X-rays or CT scans, MR imaging (MRI) does not use ionizing radiation, making it a safer option for repeated imaging studies. MRI is particularly effective for soft tissue visualization, providing high-resolution images of the brain, spinal cord, muscles, and joints, as well as detecting abnormalities like tumors, inflammation, and injuries. The versatility and precision of MRI make it a crucial tool in diagnostic medicine, enabling clinicians to diagnose and monitor a wide range of medical conditions.

Artificial intelligence (AI) has increasingly become an integral part of the automated analysis of MRI data [2]. AI algorithms can process and analyze vast amounts of imaging data, identifying patterns and anomalies quite efficiently. In MRI, AI applications range from automating the segmentation of anatomical structures (see also Chapter 12) and detecting pathologies, to predicting disease outcomes and personalizing treatment plans (see also Chapter 14). By leveraging large datasets, AI models can learn to differentiate between normal and abnormal tissues, classify disease types, and even assist in image reconstruction, enhancing the clarity and quality of MR images.

Even though such AI models show promising results, they tend to be vulnerable to biases and errors present in datasets, particularly when the models are trained on limited or nonrepresentative datasets [3]. In fact, AI models are susceptible to many types of typo of biases, including [4]:

- selection bias: data concentrated in particular demographics groups,
- label bias: noise in the labels,
- measurement bias: noise in the data due to specific variations in imaging equipment or protocols,

Machine Learning in MRI. https://doi.org/10.1016/B978-0-443-14109-6.00018-3

- sampling bias: imbalanced distribution of class samples,
- confirmation bias: data collection influenced by preexisting beliefs or hypotheses about samples, and
- recall bias: case-enriched sample distribution.

Such biases can lead to overfitting during training, resulting in poor generalization of the AI model (see also Chapter 16).

A particularly successful way to mitigate these biases consists of leveraging human knowledge into the training and testing of AI models, forming human-in-the-loop (HITL) systems. In HITL systems, clinicians and radiologists play a crucial role in mitigating these biases by providing critical oversight to identify and correct them [5]. Such HITL systems require careful analysis of the complex and variable interactions between human expertise and MR data for effective development [6]. Radiologists and medical professionals bring essential domain expertise and clinical experience, allowing them to accurately interpret MR images by discerning subtle patterns and anomalies that AI models might overlook or misinterpret. This expertise is especially valuable in cases of ambiguous data or rare conditions that the models may not have frequently encountered during training. By integrating human input, AI models can be trained and tested more effectively, leading to higher diagnostic accuracy and reliability, which can increase the clinical relevance of the AI models and improve their impact on patient care. Furthermore, HITL systems facilitate the interpretability of AI models and real-time interaction, fostering greater transparency and trust in AI-driven MR data analysis [7]. This synergy between human expertise and AI is essential for the successful and ethical deployment of these advanced AI tools in healthcare.

This critical oversight to mitigate biases can occur both during AI model training and testing [5], as shown in Fig. 18.1. During training, the human oversight is explored to improve training convergence and generalization, where the trained model becomes more robust across diverse patient populations and imaging conditions [5]. Consequently, the accuracy, reliability, and transparency of automated

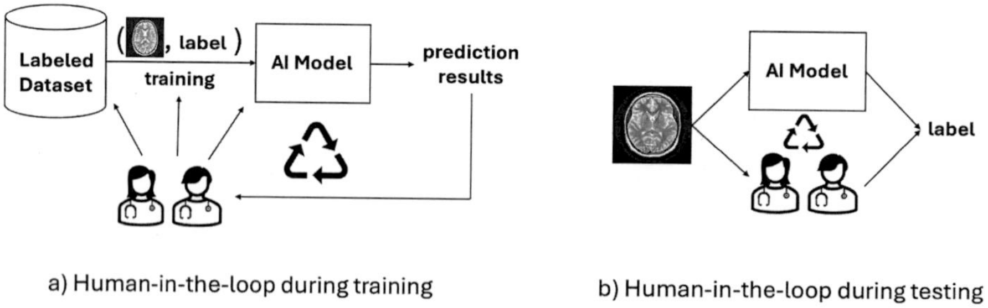

FIGURE 18.1

Human in the loop during training (a) and testing (b). During training, humans can be involved in data collection, preprocessing, and annotation, and in the design of training algorithms or AI models. On the contrary, during testing, humans and AI can cooperate to reach joint predictions, and to influence each other's predictions in cooperative or collaborative approaches.

diagnoses can be significantly improved, increasing the likelihood of successful AI system translation into clinical practice. During testing, AI models can cooperate or collaborate with human experts, forming human-AI (HAI) cooperative or collaborative teams [8]. It is important to emphasize that human-AI cooperation focuses on working alongside each other with clearly defined roles, whereas collaboration entails a deeper, ongoing interaction where both parties actively influence and shape the outcome together. This cooperation or collaboration aims to optimize accuracy while minimizing the cost of involving experts in the analysis of test images. These HAI cooperative or collaborative models tend to be more accurate than isolated AI models and clinicians, leading to better patient outcomes. In addition, the cooperation or collaboration with humans can also address ethical and legal concerns associated with autonomous AI systems in healthcare, as it may allow humans to make the final decision [9].

In this chapter, we first describe *human-in-the-loop (HITL)* [10] *training* techniques. We start by describing how datasets are labeled to be used for training AI models in medical image analysis, and in particular in MR data analysis. For instance, we discuss basic dataset annotation techniques that produce training sets used by supervised learning methods [11–14]. Then, we explain more sophisticated HITL training methods, such as active learning [15], crowdsourcing [16], multirater learning [17], self-supervised learning [18], and reinforcement learning [19]. Next, we discuss *HITL testing* techniques, where we focus on the following techniques: learning to defer [20–23], learning to complement [24–28], learning to assist, and interactive refinement, adjustments, and interpretation. We conclude the chapter with a discussion on practical considerations, limitations and future work, and final considerations.

18.2 Human in the loop during training

The training of AI models for analyzing medical images, particularly MR data, depends on the involvement of clinical experts. These experts are essential for collecting data and producing high-quality annotations that are critical for training robust AI systems. However, the process of gathering data and annotating large-scale, rich datasets is often time-consuming and expensive, posing challenges in terms of project budget and resource allocation. This section explores the integration of human expertise in the training phase, emphasizing the importance of balancing quality with costs. By leveraging the strengths of both AI and human experts, the discussed methods aim to achieve efficient and accurate model development, ultimately improving the clinical applicability of AI-driven solutions in healthcare.

18.2.1 Supervised learning

Dataset collection and annotation are an essential part of the supervised learning of medical imaging data, where human experts provide labeled data to train and validate AI models. In supervised learning, the quality of the training data and respective labels is crucial for model performance, so the expertise, decisions, and potential errors of the human annotators have a direct influence on the model's learning process. This interaction between human annotators and the machine learning model is the simplest example of a human-in-the-loop (HITL) system, where human input is essential for guiding and improving the model's performance. In HITL scenarios, humans may not only label data but also provide feedback during model development, correct model errors, and refine the model iteratively, making the process collaborative between humans and AI.

The HITL process of building a supervised learning dataset generally involves a data collection stage, subsequently followed by an annotation stage, as explained below [11–14].

Data Collection: (1) *Patient selection*—the process begins with selecting a representative sample of patients whose MR images will be included in the dataset. This selection must consider various factors, such as the diversity of conditions, demographics, and imaging protocols, to ensure the dataset is comprehensive and generalizable. (2) *Image acquisition*—MR images are acquired using standardized imaging protocols. This step often involves multiple imaging sequences to capture different tissue contrasts and anatomical details. The consistency of imaging parameters across different machines and sites is crucial to reduce variability that could affect the model's performance. (3) *Data preprocessing*—before annotation, the MR data typically undergoes preprocessing steps such as normalization, noise reduction, and artifact correction. This preprocessing ensures that the images are in a uniform format, enhancing the reliability of subsequent annotations.

The output from this data collection step is a set of unlabeled data, denoted by $\mathcal{U} = \{(\mathbf{x}_i)\}_{i=1}^{M}$, where $\mathbf{x} \in \mathscr{X}$ denotes the image data. This data collection stage is followed by the data annotation stage, where annotators provide labels to the data samples in $\mathcal{U}$, as explained in the steps below.

Annotation: (1) *Annotation guidelines development*—clear and detailed annotation guidelines are developed to ensure consistency across annotators. These guidelines define what structures or abnormalities need to be labeled, the specific labeling conventions to be used, and how ambiguous cases should be handled. (2) *Multiexpert labeling*—to improve the accuracy and reliability of annotations, the same MR data are often labeled by multiple experts. This approach helps capture different perspectives and reduces the likelihood of individual bias. Multiple annotations can be crucial in cases where the data are complex or ambiguous. When multiple experts provide annotations, there are often discrepancies between their labels. To resolve these conflicts, a consensus annotation can be generated in many ways [29], such as majority voting (the most common label among the experts is selected as the consensus annotation), averaging (labels are averaged to produce a soft label) [30], deliberation or adjudication (experts may discuss and review the conflicting annotations to reach an agreement on the most accurate label) [14], and weighted consensus (the expertise level of each annotator is considered, giving more weight to the labels provided by more experienced experts) [31]. This consensus annotations may then undergo a final review by a senior expert or a committee to ensure that the labels are accurate and consistent with clinical standards [32]. (3) *Labeling of multiple structures or conditions*—in many cases, each MR image may contain several structures or conditions of interest, requiring multiple labels per image. For instance, in brain MRIs, different regions such as the cortex, white matter, and lesions may all be annotated separately [33]. This comprehensive labeling provides richer information for the AI model to learn from.

This data collection and annotation also involve the documentation of metadata about the patient (e.g., demographics and imaging protocols) and annotators (e.g., accuracy and seniority). This metadata is crucial for understanding the dataset's characteristics and for interpreting the AI model's outputs. The last step in data collection usually involves the documentation of a detailed description of all steps above, which is essential for the reproducibility of research and for other researchers or clinicians who may use the dataset in the future.

In addition to providing initial labeled datasets, human experts are also critical in the validation and iterative improvement of the AI models [4]. As the model processes new data, human experts review its predictions, comparing them against their own interpretations and the original labels. They can identify discrepancies, errors, or areas where the model's performance may be lacking. This feedback is used to adjust and refine the model, a process known as iterative learning.

The output from the annotation stage forms is a labeled dataset represented by $\mathcal{D} = \{(\mathbf{x}_i, \mathbf{y}_i)\}_{i=1}^{N}$, where $\mathbf{x} \in \mathscr{X}$ is the image, and $\mathbf{y} \in \mathscr{Y}$ denotes the label, with $\mathscr{Y} \subset \{0,1\}^{|\mathscr{Y}|}$. This dataset is then used to train the AI model $f_\theta : \mathscr{X} \rightarrow [0,1]^{|\mathscr{Y}|}$, parameterized by $\theta \in \Theta$, with an optimization method that minimizes the following loss function

$$\theta^* = \arg\min_{\theta \in \Theta} \frac{1}{N} \sum_{(\mathbf{x}_i, \mathbf{y}_i) \in \mathscr{D}} \ell(\mathbf{y}_i, f_\theta(\mathbf{x}_i)), \tag{18.1}$$

where $\ell(.)$ is the cost function that penalizes incorrect predictions made by the AI model $f_\theta(.)$.

Overall, the integration of human expertise in the supervised learning process ensures that the resulting AI models are both accurate and clinically relevant. By providing high-quality labeled data and continuous validation, human experts enhance the AI model's ability to interpret MR images accurately, ultimately leading to accurate diagnostic tools and improved patient outcomes.

18.2.1.1 Active learning

Ideally, in MR data analysis, AI models should be trained using a large and accurately labeled dataset $\mathscr{D}$. However, collecting and annotating such dataset is challenging due to many reasons. First, only highly trained radiologists and specialists can provide accurate labels for MR data, making the annotation process time-consuming and costly. Second, representing the large range of variation typically shown by MR data is challenging because of the multitude of imaging settings and patient demographics. Active

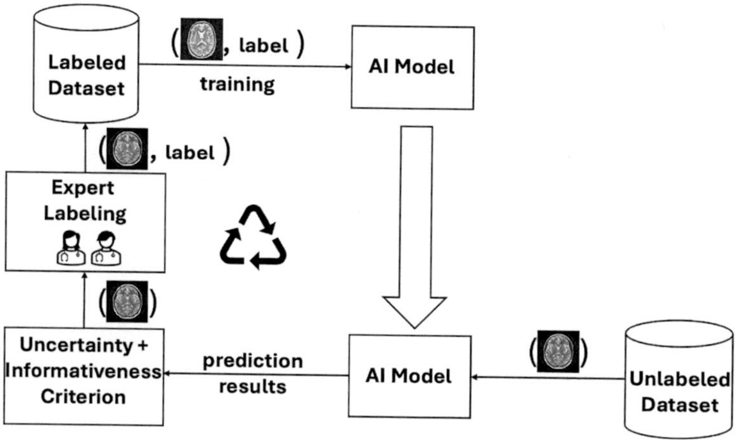

FIGURE 18.2

In active learning, after each training iteration of the AI model, we take its prediction for unlabeled data and use an uncertainty or informativeness criterion to select the unlabeled samples to be annotated by experts. Such newly labeled samples are removed from the unlabeled dataset and included in the labeled dataset for the next training iteration of the AI model.

learning [34,35] addresses these challenges by prioritizing the labeling of the most uncertain or informative cases, mitigating the challenges involved in the collection of data and its annotation by human experts, as depicted in Fig. 18.2.

There are three methods to sample data for labeling in active learning strategies [34], where we assume to have a small set of labeled training samples, denoted by $\mathcal{D} = \{(\mathbf{x}_i, \mathbf{y}_i)\}_{i=1}^{N}$, alongside another set of unlabeled data, represented by $\mathcal{U} = \{(\mathbf{x}_i)\}_{i=1}^{M}$. One type of sampling is called stream-based selective (SBS) sampling [36], where initially $\mathcal{U} = \emptyset$ and gets populated during the training with unlabeled data. Every time a new sample $\widehat{\mathbf{x}} \in \mathscr{X}$ is added to $\mathcal{U}$, we test its uncertainty or informativeness with

$$\mathrm{h}(f_\theta(\widehat{\mathbf{x}})) > \tau, \tag{18.2}$$

where $\mathrm{h} : [0, 1]^{|\mathscr{Y}|} \rightarrow \mathbb{R}$ is a function that measures the uncertainty or informativeness in the classification of the unlabeled sample $\widehat{\mathbf{x}}$ by the AI model $f_\theta(.)$, and $\tau \in \mathbb{R}$ is a threshold. If the condition in Eq. (18.2) is true, then we request a human expert to label $\widehat{\mathbf{x}}$ with $\widehat{\mathbf{y}} \in \mathscr{Y}$, and update the labeled dataset with $\widehat{\mathcal{D}} = \mathcal{D} \cup \{(\widehat{\mathbf{x}}, \widehat{\mathbf{y}})\}$. The model is retrained with Eq. (18.1) by setting $\mathcal{D} = \widehat{\mathcal{D}}$ [34,35]. Even though simple, its sample-wise decision process that depends on a predefined threshold τ is nonoptimal. Alternatively, membership query synthesis (MQS) generates samples that need to be labeled [37] with a generative function $\widehat{\mathbf{x}} \sim \mathbb{P}_{\mathscr{X}}$, which are then labeled with $\widehat{\mathbf{y}}$ by a human expert. Although interesting, MQS approaches tend to struggle to generate samples that are from unpopulated regions of the data distribution.

Arguably, the most common sampling strategy is the pool-based sampling active learning [38] that assumes large pool of unlabeled data, represented by the set $\mathcal{U}$. During each training iteration, the algorithm selects the most uncertain or informative unlabeled sample with

$$\widehat{\mathbf{x}} = \arg\max_{\mathbf{x}\in\mathcal{U}} \mathrm{h}(f_\theta(\mathbf{x})), \tag{18.3}$$

where h(.) is defined in Eq. (18.2). The sample $\widehat{\mathbf{x}}$ selected from Eq. (18.3) is then sent to an expert, who annotates it with label $\widehat{\mathbf{y}}$. This sample is subsequently added to the labeled dataset to form $\widehat{\mathcal{D}} = \mathcal{D} \cup \{(\widehat{\mathbf{x}}, \widehat{\mathbf{y}})\}$ and removed from the unlabeled dataset $\widehat{\mathcal{U}} = \mathcal{U} \backslash \{(\widehat{\mathbf{x}})\}$. Then, the model is retrained with Eq. (18.1) by setting $\mathcal{D} = \widehat{\mathcal{D}}$ [34,35], where the unlabeled set is updated with $\mathcal{U} = \widehat{\mathcal{U}}$. This process repeats iteratively, with the model progressively refining its performance until it reaches the desired level of accuracy or convergence.

The function $\mathrm{h}(f_\theta(\mathbf{x}))$ measures the uncertainty or informativeness of the sample $\mathbf{x}$ by measuring how hard it is for the model $f_\theta(\mathbf{x})$ to classify between different classes. This can be done in many ways [35], such as with entropy, where $\mathrm{h}(f_\theta(\mathbf{x})) = -f_\theta(\mathbf{x})^\top \log(f_\theta(\mathbf{x}))$, assuming that $f_\theta(\mathbf{x})$ outputs a probability distribution in the space of $|\mathcal{Y}|$ classes. Another common strategy for h(.), called query-by-committee, relies on a committee of models defined by $\{f_{\theta_i}(\mathbf{x})\}_{i=1}^{C}$, and unlabeled samples where these models disagree are prioritized for human annotation. This can be measured by the entropy of the ensemble classification with $\mathrm{h}(\{f_{\theta_i}(\mathbf{x})\}_{i=1}^{C}) = -\overline{f}_\theta(\mathbf{x})^\top \log(\overline{f}_\theta(\mathbf{x}))$, where $\overline{f}_\theta(\mathbf{x}) = \frac{1}{C}\sum_{i=1}^{C} f_{\theta_i}(\mathbf{x})$. This strategy is useful in MR image analysis to explore different interpretations of ambiguous or complex images, ensuring that the final model benefits from a broad range of expert insights. Active learning can also explore the expected model change strategy [35], which selects images that, if labeled, would most significantly impact the model's performance. This can be achieved with $\mathrm{h}(f_\theta(\widehat{\mathbf{x}})) = \sum_{\widehat{\mathbf{y}}\in\mathcal{Y}} f_\theta^{\widehat{\mathbf{y}}}(\widehat{\mathbf{x}}) \left\| \sum_{(\mathbf{x},\mathbf{y})\in\mathcal{D}\cup(\widehat{\mathbf{x}},\widehat{\mathbf{y}})} \nabla_\theta \ell(\mathbf{y}, f_\theta(\mathbf{x})) \right\|$, where $f_\theta^{\widehat{\mathbf{y}}}(\mathbf{x})$ denotes the probability of classifying the class set in $\widehat{\mathbf{y}}$. The images selected with the functions above represent rare conditions or edge cases, where additional data could greatly improve the model's ability to generalize to new, unseen cases. The selection of a diverse set of images is another strategy that ensures that the training data covers a wide range of conditions and imaging scenarios [35]. This is critical in medical imaging to avoid overfitting and ensure that the model performs well across different patient populations and imaging settings.

Active learning offers many advantages over the original supervised learning methodology [35]. It fosters close collaboration between AI and human experts, optimizing the annotation process and improving model accuracy. Active learning creates a dynamic feedback loop where human expertise is continuously leveraged to address the model's weaknesses. For instance, if the model consistently misclassifies a particular type of tumor, experts can provide targeted annotations for similar cases, helping the model learn more effectively. By focusing on the most informative cases, active learning significantly reduces the number of labeled examples needed to train the model, making the process more cost-effective. This approach is particularly valuable in MR image analysis, where expert time is costly, as highlighted by the significant attention active learning has received in MR tasks [39–43]. Despite these advantages, active learning also needs to address some challenges. The continuous selection of hard cases for experts to label can lead to annotation fatigue, potentially affecting the quality of the labels. It is important to manage this by balancing the workload, providing support tools to assist experts in the annotation process, and by selecting samples for annotation that will truly enhance model performance without overwhelming the experts.

Active learning is a critical human-in-the-loop supervised training strategy, offering a way to efficiently leverage human expertise to train AI models with high accuracy and reliability. By focusing on the most informative cases, active learning not only enhances model performance but also ensures that these models are more generalizable and better suited for real-world clinical applications.

18.2.1.2 Crowdsourcing

Supervised learning for medical tasks using MR images typically relies on highly trained experts like radiologists, but crowdsourcing introduces an approach to managing the vast amounts of MR data. Crowdsourcing, displayed in Fig. 18.3, distributes large sets of MR images to a group of nonexperts or partially trained individuals for labeling, offering a scalable solution to the challenges of annotation [44–50]. Hence, crowdsourcing methods aim to efficiently obtain a large number of annotations from a broad pool of annotators with the objective of quickly and cost-effectively producing relatively reliable data annotation.

In crowdsourcing, we start with a set of unlabeled data samples, denoted by $\mathscr{U} = \{(\mathbf{x}_i)\}_{i=1}^{M}$, where each sample is sent to K nonexpert annotators to label it, producing $\{\widehat{\mathbf{y}}_{ij}\}_{j=1}^{K}$ for each $\mathbf{x}_i \in \mathscr{U}$. This forms the set $\mathscr{R} = \{(\mathbf{x}_i, \{\widehat{\mathbf{y}}_{ij}\}_{j=1}^{K})\}_{i=1}^{N}$. The next step consists of aggregating these labels to produce a final label for the data sample $\mathbf{x}_i$ with

$$\mathbf{y}_i = \mathrm{a}\left(\left\{\widehat{\mathbf{y}}_{ij}\right\}_{j=1}^{K}\right), \tag{18.4}$$

where $\mathrm{a} : \mathscr{Y}^K \rightarrow \mathscr{Y}$ represents an label aggregator function.

Crowdsourcing, despite the variability in accuracy and reliability of annotations from unskilled or partially skilled contributors, aims to train classifiers using the

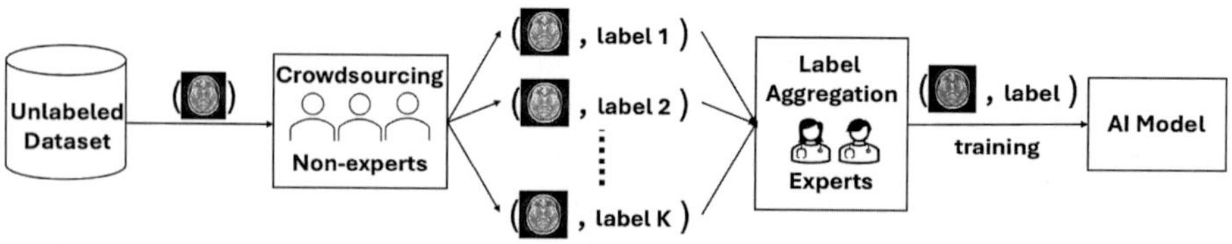

FIGURE 18.3

In crowdsourcing, unlabeled data are provided to nonexperts to provide a set of K labels, which are then aggregated, usually with the support of experts, to produce a single label per data sample for training the AI model.

potentially noisy labels obtained from Eq. (18.4). The primary challenge is to extract "true" labels from these noisy annotations, which have been typically achieved via redundancy, gold standard tasks, or expert reviews.

In redundancy-based quality control, multiple crowd participants annotate the same image, and the final label is estimated by consensus methods like majority voting [51,52], where $a(\{\widehat{\mathbf{y}}_{ij}\}_{j=1}^{K}) = \text{mode}(\{\widehat{\mathbf{y}}_{ij}\}_{j=1}^{K})$ returns the most frequent label in the set or averaging [53], where $a(\{\widehat{\mathbf{y}}_{ij}\}_{j=1}^{K}) = \frac{1}{K}\sum_{j=1}^{K}\widehat{\mathbf{y}}_{ij}$. Alternatively, to estimate the "true" label, gold standard tasks intersperse known reference tasks within the workload, allowing the accuracy and reliability of participants' contributions to be assessed [54]. Then, participants with higher accuracy on gold standard tasks can have their contributions weighted more heavily in final consensus-building. Also, participants with low accuracy can be excluded from further tasks to prevent poor-quality data from impacting results, or they can alternatively get further training to improve their performance. Another way to estimate the "true" label is with expert review that leverages medical experts to validate a subset of the crowd-generated annotations to ensure high-level quality assurance [55]. Such expert-validated labels can then be used for training the classifier. After estimating a "true" label per data sample, then we can form the labeled dataset $\mathscr{D} = \{(\mathbf{x}_i, \mathbf{y}_i)\}_{i=1}^{N}$ and train the model $f_\theta(.)$ with Eq. (18.1).

Crowdsourcing has many advantages over regular data annotation for supervised learning methods [44–50]. It can enable the annotation of a vast amount of data quickly, which is particularly useful in large-scale datasets for AI training. Compared with hiring experts, like radiologists, for large-scale annotation, crowdsourcing can be more cost-effective, especially for less complex labeling tasks. Crowdsourcing also has the potential to reveal insights that might be overlooked by a single expert or a small group of specialists. Nevertheless, there are many challenges associated with crowdsourcing. The main issue is ensuring the accuracy of the annotations. Nonexperts may lack the knowledge to correctly annotate complex MR data, leading to label noise. The need for user-friendly interfaces and effective instructional materials adds complexity to crowdsourcing projects. Ensuring that crowdsourcing platforms comply with legal and ethical guidelines is essential, particularly concerning patient data protection. Not all medical image analysis tasks are suitable for crowdsourcing. Tasks like outlining anatomical structures, labeling regions of interest, or identifying clear patterns in images (e.g., presence of tumors or fractures) can be effectively crowdsourced, but highly specialized tasks that require deep clinical knowledge are typically beyond the capability of a general crowd and still require expert involvement. Crowdsourcing is increasingly more integrated with AI, with AI models first performing an initial analysis, followed by crowdsourcing to validate or refine the results.

18.2.1.3 Multirater learning

Similarly to crowdsourcing presented in Section 18.2.1.2, multirater learning (MRL) is a methodology that involves multiple annotators or raters to label data to be used in the training of AI methods. However, they differ in their primary objectives, methodologies, applications, and underlying assumptions, which are explained in this section. While crowdsourcing leverages the collective wisdom from large, diverse, and often nonexpert crowds of people for data labeling, MRL relies on a small set of experts, potentially with varying expertise levels, enabling an accurate estimation of the "true" labels of data points, while simultaneously modeling the reliability and biases of each annotator, as shown in Fig. 18.4. Hence, MRL tends to be more expensive and more reliable than crowsourcing given the focus on inferring true labels and understanding annotator behavior (e.g., bias and error rates).

MRL methods incorporate statistical models (e.g., confusion or label transition matrices) to estimate each annotator's reliability and error patterns, where latent variables are introduced to learn the "true" labels to be estimated alongside the annotators' confusion matrices. One of the most explored MRL methods is the one based on the expectation-maximization (EM) algorithm [44,50,56]. In such EM algorithms [49], we assume that each annotator provides a noisy version of $\mathbf{y}_i$ defined by $p(\widehat{\mathbf{y}}_{ij}|\mathbf{y}_i, \Pi_j) = \text{multinomial}(\widehat{\mathbf{y}}_{ij}|\pi^j_{\mathbf{y}_i})$, where $\Pi_j = [\pi_j(1,:), \ldots, \pi_j(|\mathcal{Y}|,:)] \in [0,1]^{|\mathcal{Y}|\times|\mathcal{Y}|}$ is the annotator's confusion matrix showing the probability of classifying the class represented by the column index given that the true class is denoted by the row index, and $p(E|C)$ computes the probability of the event E conditioned on event C (note that when C is absent, then this returns the unconditional probability of E). Assuming independence between labelers, the expected value of the data log-likelihood under the estimate of the posterior distribution of the "true" labels $q(\mathbf{y}_i)$ is defined by

$$\mathbb{E}_{q(\mathbf{y}_i)}\left[p\left(\mathcal{R}, \{\mathbf{y}_i\}_{i=1}^N \middle| \{\Pi_j\}_{j=1}^K\right)\right] = \sum_{i=1}^N \sum_{\mathbf{y}_i\in\mathcal{Y}} q(\mathbf{y}_i)\log\left(p(\mathbf{y}_i)\prod_{j=1}^K p\left(\widehat{\mathbf{y}}_{ij}|\mathbf{y}_i,\Pi_j\right)\right). \tag{18.5}$$

To estimate the hidden "true" labels and the confusion matrices we can use the EM algorithm, where in the E-step, we estimate the posterior $q(\mathbf{y}_i)$ with

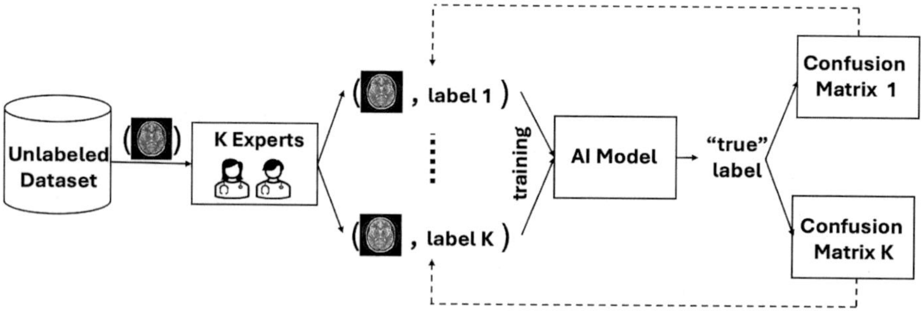

FIGURE 18.4

In multirater learning, each training sample is labeled by K experts, which are used to train the AI model to produce the "true" label, which is used by each of the K trainable annotator-specific confusion matrices to recover each of the K experts' labels, where these K matrices represent the labeling biases of the experts.

$$q(\mathbf{y}_i) \propto p(\mathbf{y}_i) \prod_{j=1}^{K} p\left(\widehat{\mathbf{y}}_{ij} | \mathbf{y}_i, \Pi_j\right). \tag{18.6}$$

In the M-step, we update the annotators' confusion matrices based on the current estimate of the "true" label $\mathbf{y}_i$ with

$$\pi_j(c,l) = \frac{\sum_{i=1}^{N} q(\mathbf{y}_i = c)\mathbb{I}\left(\widehat{\mathbf{y}}_{ij} = l\right)}{\sum_{i=1}^{N} q(\mathbf{y}_i = c)}, \tag{18.7}$$

where $\mathbb{I}(.)$ denotes an indicator function. These E and M steps are iteratively repeated until the parameters converge. At convergence, the algorithm provides both an estimate of the true labels and the confusion matrix for each annotator. A more recent redundancy-based quality control approach is the CrowdLab method [57], which achieves state-of-the-art results by leveraging multiple noisy-label samples and predictions provided by a model trained directly using the noisy labels to estimate a consensus label (with a confidence score) per training sample, and a performance assessment for each annotator. This allows the elimination of the labels produced by low-ranked annotators and requests the relabeling of low-confidence labels.

Instead of estimating the "true" label per data sample, crowdsourcing algorithms directly use the multirater dataset $\mathscr{R} = \{(\mathbf{x}_i, \{\widehat{\mathbf{y}}_{ij}\}_{j=1}^{K})\}_{i=1}^{N}$ to train a classifier, frequently relying on the EM algorithm [47,48,58]. For instance, Raykar et al. [47], extends the redundancy-based quality control methods described above in Eqs. (18.6) and (18.7) to learn not only the "true" label and annotators' confusion matrices, but also the parameters θ of the classifier $f_\theta : \mathscr{X} \to [0,1]^{|\mathscr{Y}|}$. More specifically, the estimate of the posterior $q(\mathbf{y}_i)$ is calculated using the current estimates of the annotators' confusion matrices and the classifier, with

$$q(\mathbf{y}_i) \propto p(\mathbf{y}_i|\mathbf{x}_i,\theta) \prod_{j=1}^{K} p\left(\widehat{\mathbf{y}}_{ij} | \mathbf{y}_i, \Pi_j\right). \tag{18.8}$$

where $p(\mathbf{y}_i|\mathbf{x}_i,\theta) = f_\theta(\mathbf{x}_i)$. In the M-step, we update the confusion matrix for each annotator in the same way as described in Eq. (18.7), and the classifier is trained with

$$\theta^{(t+1)} = \arg\min_{\theta} - \sum_{i} p(\mathbf{y}_i = 1|\mathbf{x}_i, \theta^{(t)}) \log(p(\mathbf{y}_i = 1|\mathbf{x}_i, \theta)), \tag{18.9}$$

where t denotes the training iteration index. These E and M steps are iteratively repeated until the parameters converge, producing an estimate of the true labels, confusion matrices for the annotators, and classifier parameters.

Such EM algorithms for joint "true" label estimation and classifier training have been superseded by gradient-based learning methods. For example, CrowdLayer [49] incorporates crowdsourced labels into deep neural networks by modeling annotator-specific transition matrices. Techniques such as WDN [59] and UnionNet-B [60] further refine classification by aggregating outputs or coordinating responses from multiple annotators. The multiannotator deep learning (MaDL) framework [45] captures potential correlations among annotators through probabilistic learning of their embeddings. Methods like CrowdCons [61] focus on creating noise-tolerant representations, while CrowdAR [62] predicts annotation reliability, using this information as a soft gold standard in training. In addition, ADMoE [63] employs a Mixture of Experts architecture to enhance learning from noisy sources, particularly for anomaly detection. In general, the approaches above formulate a model that contains a classifier $f_\theta : \mathscr{X} \rightarrow [0,1]^{|\mathscr{Y}|}$ that predicts the "true" label, and a set of K confusion matrices to be learned for the annotators, which transforms the "true" label to the jth annotator's label, denoted by $\{\mathbf{T}_j\}_{j=1}^{K}$, with $\mathbf{T}_j \in [0,1]^{|\mathscr{Y}| \times |\mathscr{Y}|}$. The training is achieved through an optimization that minimizes the following loss function

$$\theta^*, \left\{\mathbf{T}_j^*\right\}_{j=1}^{K} = \arg \min_{\theta, \left\{\mathbf{T}_j\right\}_{j=1}^{K}} -\sum_{i=1}^{N} \sum_{j=1}^{K} \ell\left(\widehat{\mathbf{y}}_{ij}, \mathbf{T}_j \times f_\theta(\mathbf{x}_i)\right), \tag{18.10}$$

where $\ell(.)$ is a classification loss function.

In summary, MRL within the HITL framework enhances the model's ability to handle complex medical image analysis tasks by leveraging the diversity of expert opinions, modeling uncertainty, and continuously incorporating expert feedback. While MRL and Crowdsourcing both deal with multiple annotators, they serve different primary purposes and employ distinct methodologies. MRL is more focused on accurately inferring true labels by modeling annotator reliability, making it suitable for scenarios requiring high label quality and understanding of annotator behavior. On the contrary, Crowdsourcing emphasizes scalability and efficiency in obtaining large volumes of annotations, often leveraging the diversity and sheer number of annotators to achieve acceptable label quality through aggregation techniques. Therefore, MRL usually leads to more accurate, reliable, and clinically useful AI systems than crowdsourcing, ultimately improving patient outcomes.

18.2.2 Self-supervised learning

Self-supervised learning (SSL) relies on the autonomous generation of labels for unlabeled data to enable the pretraining of AI models, which can then be fine-tuned using relatively small labeled training sets [64–71]. As presented in Fig. 18.5, SSL data labels are generated to solve pretext tasks related to the main AI task for subsequent fine-tuning. The success of SSL depends on effective pretext tasks, especially when combined with human-in-the-loop (HITL) methods, where human experts design tasks like image reconstruction or segmentation, allowing the model to learn features directly from the data without labels.

Therefore, in medical image analysis problems, such as classification from MRI data, expert guidance is critically important to ensure that the pretext labels are clinically relevant. For example, human experts can design pretext tasks focused on clinically significant areas, such as tumor margins or regions susceptible to specific

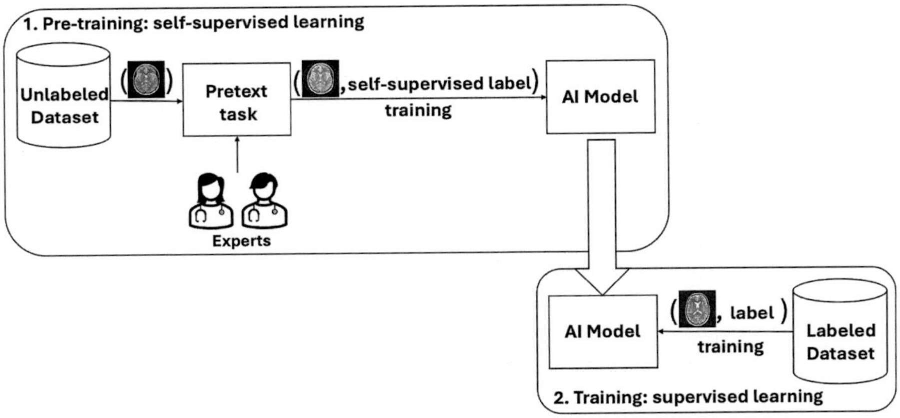

FIGURE 18.5

Self-supervised learning methods consist of two stages. The first stage (pretraining: self-supervised learning) takes unlabeled data and use information provided by domain experts (i.e., clinicians) to produce a pretext task that will provide self-supervised labels to pretrain the AI model. The second stage (training: supervised learning) consists of a regular training process to fine tune the AI model using the labeled dataset.

diseases, guaranteeing that the auto-generated data is both accurate and relevant to clinical needs. Such collaborative synergy between human expertise and SSL algorithms can improve the model's performance for downstream tasks. This integration ultimately leads to better diagnostic tools, enhanced patient outcomes, and greater confidence in training advanced AI model in healthcare. The exploration of SSL in medical image analysis started with the direct use of computer vision-designed pretext tasks [71], which, even though unrelated to the task to be learned by the AI model, showed promising results.

More recently, SSL methods have relied on human expertise for the design of pretext tasks. For instance, unsupervised anomaly detection methods are trained to detect diseases using a training set containing exclusively normal images [18]. However, relying solely on normal images can lead to learning suboptimal representations that struggle to detect diseases of various sizes, appearances, and shapes. In Tian et al. [18], SSL pretext tasks are designed to generate multiple types of images containing simulated lesions from images of healthy subjects. Similarly, to reduce the labor and time required for annotating video object bounding boxes, Le et al. [72] propose an efficient interactive self-annotation framework based on cyclic self-supervised learning. The framework combines automatic model learning with interactive processes, where the automatic learning phase accelerates model training and enhances the interaction speed. During interactive recursive annotation, the model receives feedback from human annotators to refine the annotation process. To further save time, they introduce a novel correction module that leverages AI models to reduce the distance between annotated frames, improving the use of neighboring frames. Building on this, Adhikari and Huttunen [73] refine the framework into a one-stage process, where the human's primary role is correcting errors rather than performing full annotations, further improving the whole process.

The incorporation of SSL into HITL pretraining offers a promising approach for enhancing model performance, especially in domains with limited labeled data. SSL allows models to learn meaningful representations from data without requiring extensive manual annotation, while the HITL process ensures that the tasks designed for pretraining are relevant and aligned with domain-specific knowledge. This synergy between SSL and human expertise not only reduces the reliance on labeled data but also produces more robust models capable of handling complex tasks, such as medical image analysis, by capturing important features that are clinically relevant. As SSL methods evolve, their integration with HITL frameworks will continue to drive advancements in machine learning, enabling more accurate and efficient AI systems across various fields.

18.2.3 Reinforcement learning

Reinforcement learning (RL) in the context of human-in-the-loop involves an iterative process where an AI model, referred to as AI agent in the context of RL, learns to make decisions based on feedback from human experts [74–81]. RL approaches are particularly useful in medical imaging tasks that require complex decision-making, such as diagnosis, treatment planning, or anomaly detection in MR images.

In a typical RL framework, the AI agent interacts with the environment, which provides the medical image. Given this image, the AI agent takes actions, such as making predictions or segmenting the image, and receives feedback from the environment in the form of rewards or penalties based on the accuracy or clinical relevance of those actions [82]. State-of-the-art RL approaches are able to finding optimal decision-making policies, but they require a large amount of time and data because of the complexity of the task. In such a scenario, human experts play a critical role in this feedback loop by providing guidance, corrections, or additional information that helps the agent improve its decision-making process more efficiently.

Leveraging RL in a human-in-the-loop (HITL) context has not been heavily studied in medical image analysis [75,83,84], but it is important to discuss it here given its potential in the area, particularly for tasks involving MRI data. This methodology has been explored to increase the speed of the RL agent training [75,77,85–89]. Specifically, following the formulation in Arakawa et al. [85], the RL setting, as depicted in Fig. 18.6, consists of an AI agent that interacts with the environment $\mathscr{E}$ through a sequence of observations, actions, and rewards. At training step t, the agent receives from $\mathscr{E}$ the observation $\mathbf{s}_t \in \mathscr{S}$, takes action $\mathbf{a}_t \in \mathscr{A}$ and receives reward $\mathbf{r}_t \in \mathscr{R}$. The AI agent has a trainable policy $\pi(\mathbf{a}|\mathbf{s})$ and the objective of RL is to find the optimal policy that maximizes the total reward expectation $R_t = \sum_{k=0}^{\infty} \gamma^k \mathbf{r}_{t+k}$ at each state $\mathbf{s}_t$, with γ representing a discount factor for future rewards.

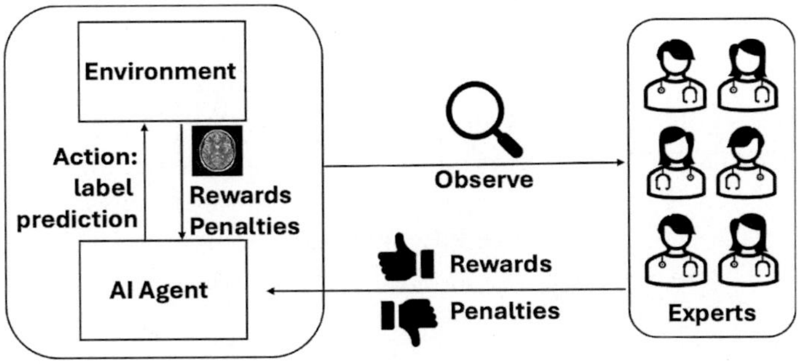

FIGURE 18.6

In human-in-the-loop reinforcement learning, the AI agent interacts with the environment by providing actions (predictions) given the images provided by the environment, which then returns rewards (or penalties) based on the accuracy or clinical relevance of the actions. Simultaneously, human experts observe the images and actions and return rewards and penalties.

HITL RL methodologies [77,80,85] generally rely on the Q-learning algorithm to estimate the optimal action value function $Q_\pi : \mathscr{S} \times \mathscr{A} \rightarrow \mathscr{R}$, which returns the expected total reward for a given state and action when following policy π, defined by

$$Q(\mathbf{a}, \mathbf{s}) = \mathbb{E}\left[\sum_{t=0}^{\infty} \gamma^t \mathbf{r}_{t+1} \middle| \mathbf{s}_0 = \mathbf{s}, \mathbf{a}_0 = \mathbf{a}, \pi\right], \tag{18.11}$$

where $\mathbf{r}_t$ is the reward at time step t, and γ is the discount factor. The optimal action value is defined by

$$Q^*(\mathbf{s}, \mathbf{a}) = \max_\pi Q_\pi(\mathbf{s}, \mathbf{a}). \tag{18.12}$$

When dealing with images, this function $Q(.)$ is approximated with a deep neural network, represented by $f_\theta^Q : \mathscr{S} \times \mathscr{A} \rightarrow \mathscr{R}$, enabling the handling of high-dimensional state spaces, such as MRI images. This network takes as input the MRI image as the state $\mathbf{s}$ and outputs a Q-value for each possible action $\mathbf{a}$ in that state. The training of the network consists of minimizing the following mean square error loss:

$$\theta^* = \arg\min_\theta \sum_{(\mathbf{s},\mathbf{a})} \| f_\theta^Q(\mathbf{s}, \mathbf{a}) - Q(\mathbf{s}, \mathbf{a}) \|^2. \tag{18.13}$$

After training, the AI agent's policy is defined by $\pi^Q(\mathbf{s}) = \text{argmax}_{\mathbf{a}} f_\theta^Q(\mathbf{s}, \mathbf{a})$.

The HITL RL setting introduces a human who assesses $\mathbf{a}$ at the state $\mathbf{s}$ to provide a feedback $H : \mathscr{S} \times \mathscr{A} \rightarrow \mathscr{R}$ for state-action pair to the AI agent [80,85]. An example of a HITL RL is the deep TAMER method [80] that aims to approximate $H(.)$ with a deep learning model $f_\theta^H : \mathscr{S} \times \mathscr{A} \rightarrow \mathscr{R}$, where training minimizes the loss:

$$\theta^* = \underset{\theta}{\text{argmin}} \sum_{(\mathbf{s},\mathbf{a})} \| f_\theta^H(\mathbf{s},\mathbf{a}) - H(\mathbf{s},\mathbf{a}) \|^2. \quad (18.14)$$

After training, the AI agent's policy is $\pi^H(\mathbf{s}) = \text{argmax}_{\mathbf{a}} f_\theta^H(\mathbf{s},\mathbf{a})$. Arakawa et al. [85] introduced DQN-TAMER by exploring not only the human rewards, denoted by $H(\mathbf{s},\mathbf{a})$ in Eq. (18.13), but also the environmental rewards represented by $Q(\mathbf{s},\mathbf{a})$ from Eq. (18.14). This is achieved by combining the training algorithms in Eqs. (18.13) and (18.14), producing the AI agent's policy is $\pi^{DQN-TAMER}(\mathbf{s}) = \text{argmax}_{\mathbf{a}} \alpha^Q f_\theta^Q(\mathbf{s},\mathbf{a}) + \alpha^H f_\theta^H(\mathbf{s},\mathbf{a})$, where α^Q, α^H weight the agent's reliance on the reward from the environment and from a human.

HITL RL offers a promising approach to complex medical imaging tasks, such as diagnosis and anomaly detection, by integrating expert feedback into the learning process. By allowing AI agents to interact with medical data and receive feedback from clinicians, RL can accelerate the development of accurate and clinically relevant decision-making policies. Although traditional RL methods can be slow and data-intensive, human guidance helps overcome these challenges by providing corrections and domain-specific knowledge, ultimately improving the efficiency and effectiveness of AI systems in healthcare applications. This synergy between human expertise and AI learning is likely to enhance the quality of medical decision-making.

18.3 Human in the loop during testing

The traditional view that automation reduces human involvement is being reconsidered [90,91], as the uncertainties inherent in automation often demand greater human participation, leading to new strategies for human-AI cooperation or collaboration [92]. Despite this shift, most AI systems in machine learning have been developed to be tested in isolation, neglecting the dynamics of human-AI cooperative or collaborative classification [93–95]. Although this isolated development is scientifically sound, it fails to account for the practical implications of how AI decisions influence human behavior—a crucial and unpredictable factor. Recent research by Chiou and Lee [96] highlights that AI's trustworthiness is closely tied to model confidence [97,98] and explainability [99,100], underscoring the importance of integrating human-AI collaboration and cooperation into AI system design.

The relationship between humans and AI is bidirectional; just as AI influences human decision-making, human input can also affect AI outcomes. The understanding of this relationship is a key focus of Human-AI Collaborative or Cooperative Classification (HAI-CCC) research [24,28,101–103]. When classifying an image, HAI-CCC systems typically consider the following possible approaches:

I AI makes the prediction independently,
II AI defers the decision to human experts,
III AI complements human experts' predictions to support joint decision-making;

IV AI assists human experts in their decision process; and
V AI interacts with humans to provide a progressive corrections, refinements, and updates.

Predictions from independent AI models are made without human involvement during the testing phase, leaving the training process unchanged. However, when AI defers to, complements, assists, or interacts with humans, both training and testing must be redesigned, as discussed in the following sections.

18.3.1 Learning to defer

Learning to defer (L2D) methods consist of HAI-CCC approaches, where the decision-making process depends on either a "single" prediction by the AI model or the human expert. These methods aim to develop a classifier and a rejector to determine when a human expert's prediction should override the AI's prediction [20–23]. This approach, known as *rejection learning* [104], has been expanded to incorporate human expertise into the decision process [21].

The L2D formulation [21,104] involves a data point (e.g., an MR image) $\mathbf{x} \in \mathcal{X}$, a label $\mathbf{y} \in \mathcal{Y}$, the AI model $f_\theta : \mathcal{X} \to \Delta^{|\mathcal{Y}|-1}$, the human expert's decision $h : \mathcal{X} \to \mathcal{Y}$, and the deferral decision function $d : \mathcal{X} \to \{0, 1\}$, with 0 representing the AI should predict, and 1 meaning deferral to expert (see Fig. 18.7). The dataset to train L2D methods is denoted by $\mathcal{D}_{L2D} = \{(\mathbf{x}_i, \mathbf{y}_i, h_i(\mathbf{x}))\}_{i=1}^N$, and the labeling of a test image $\mathbf{x}$ is obtained with:

$$\widehat{\mathbf{y}} = \begin{cases} \text{argmax}_{\mathbf{y} \in \mathcal{Y}} f_\theta^{\mathbf{y}}(\mathbf{x}) & , \text{if } d(\mathbf{x}) = 0 \\ h(\mathbf{x}) & , \text{if } d(\mathbf{x}) = 1 \end{cases}, \tag{18.15}$$

where $f_\theta^{\mathbf{y}}(\mathbf{x}) \in [0, 1]$ denotes the AI model's prediction probability associated with class $\mathbf{y}$. The goal of learning to defer is to minimize a loss function that combines the cost of both the AI and human predictions, taking into account accuracy and the cost of deferring to the human. Such a loss is defined as:

$$\ell_{L2D}(\mathbf{x}, \mathbf{y}, f_\theta, h, d) = d(\mathbf{x})(\ell(\mathbf{y}, h(\mathbf{x})) + \lambda) + (1 - d(\mathbf{x}))\ell(\mathbf{y}, f_\theta(\mathbf{x})), \tag{18.16}$$

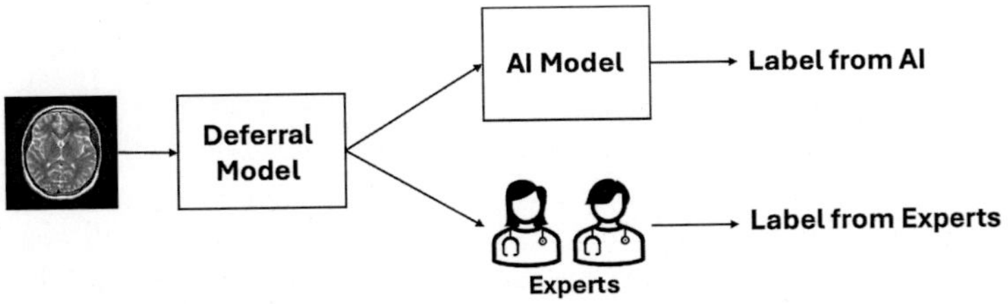

FIGURE 18.7

In learning to defer, the image is first analyzed by a deferral model that decides if the image will be classified by the AI model or an expert.

where λ is the cost to defer to a human, and the deferral decision function can either be defined by an a separate classifier or by thresholding the AI model with

$$d(\mathbf{x}) = \begin{cases} 1 & ,\text{if } \max_{\mathbf{y}}\left(f_\theta^{\mathbf{y}}(\mathbf{x})\right) < \tau \\ 0 & ,\text{otherwise} \end{cases}, \tag{18.17}$$

where $\tau \in [0, 1]$ is a classification threshold.

Subsequent research on L2D methods has focused on creating various surrogate loss functions [23,105–115]. The true L2D loss function typically involves discrete choices, as shown in Eqs. (18.16) and (18.17). Either the AI model makes a prediction or defers to a human, making it difficult to train directly. Surrogate losses provide a smooth, differentiable approximation that can be optimized using gradient-based methods. For example, the surrogate loss from Mozannar and Sontag [111] consists of unifying the classifier and deferral decision with an augmented label space defined as $\mathcal{Y}^{\perp} = \mathcal{Y} \cup \{\perp\}$, where $\perp$ represents the deferral class. The model is defined by $|\mathcal{Y}| + 1$ classifiers, where the multiclass classifiers are represented by $\{g_{\theta_k} : \mathcal{X} \rightarrow \mathbb{R}\}_{k=1}^{|\mathcal{Y}|}$, and the deferral classifier is denoted by $g_{\theta_\perp} : \mathcal{X} \rightarrow \mathbb{R}$. The loss function is then defined as:

$$\begin{aligned} \ell_{L2D_{sur}}\left(\mathbf{x}, \mathbf{y}, \{\theta_k\}_{k=1}^{|\mathcal{Y}|}, \theta_\perp, h(\mathbf{x})\right) = & -\log\left(\frac{\exp\left(g_{\theta_\mathbf{y}}(\mathbf{x})\right)}{\sum_{\mathbf{y}' \in \mathcal{Y}^\perp} \exp\left(g_{\theta_{\mathbf{y}'}}(\mathbf{x})\right)}\right) \\ & - \mathbb{I}(\mathbf{y} = h(\mathbf{x}))\log\left(\frac{\exp\left(g_{\theta_\perp}(\mathbf{x})\right)}{\sum_{\mathbf{y}' \in \mathcal{Y}^\perp} \exp\left(g_{\theta_{\mathbf{y}'}}(\mathbf{x})\right)}\right), \end{aligned} \tag{18.18}$$

where $\mathbb{I}(.)$ represents the indicator function. In the loss function in Eq. (18.18), the first term optimizes the function $g_{\theta_\mathbf{y}}$ associated with the true label $\mathbf{y}$, while the second term optimizes the deferral function $g_{\theta_\perp}(.)$ if the expert's prediction $h(\mathbf{x})$ is correct. The training is achieved with the following optimization:

$$\{\theta_k^*\}_{k=1}^{|\mathcal{Y}|}, \theta_\perp^* = \arg \max_{\{\theta_k\}_{k=1}^{|\mathcal{Y}|}, \theta_\perp} \sum_{(\mathbf{x},\mathbf{y},h(\mathbf{x})) \in \mathcal{D}_{L2D}} \ell_{L2D_{sur}}\left(\mathbf{x}, \mathbf{y}, \{\theta_k\}_{k=1}^{|\mathcal{Y}|}, \theta_\perp, h(\mathbf{x})\right). \tag{18.19}$$

During testing, the classification is obtained by

$$\widehat{\mathbf{y}} = \begin{cases} \operatorname{argmax}_{\mathbf{y} \in \mathcal{Y}} g_{\theta_\mathbf{y}}(\mathbf{x}) & ,\text{if } g_{\theta_\perp}(\mathbf{x}) < g_{\theta_{\widehat{\mathbf{y}}}}(\mathbf{x}) \\ h(\mathbf{x}) & ,\text{otherwise} \end{cases}. \tag{18.20}$$

Alternatively, score-based triage systems optimize the allocation of instances to algorithms [113], while differentiable triage applies deterministic thresholds based on error differences between the model and human experts on a per-instance basis [112]. In addition, Verma and Nalisnick [115] introduce a one-vs-all classifier to address underfitting, and Mozannar and Sontag [111] propose a cost-sensitive softmax cross-entropy loss, later extending it to a realizable-consistent surrogate loss for learning with deferral using halfspaces [109]. Cao et al. [105] further developed a statistically consistent asymmetric softmax-based surrogate loss, addressing issues of unboundedness.

A common limitation of L2D methods is their reliance on a single-expert setting, which fails to consider more complex environments involving multiple experts. As a result, recent L2D research has shifted toward multiexpert settings [20,22,116–123]. Verma et al. [122] proposed an L2D method that can defer to one of multiple users, while Mao et al. [22] introduced a two-stage surrogate loss for learning to defer to multiple experts. Tailor et al. [120] advance this by incorporating meta-learning into L2D, allowing for deferral to an average of experts, even when encountering new experts at test time.

Despite these advancements, current L2D methods do not support joint decision-making between AI models and human experts. To bridge this gap, learning to complement methods have been introduced, as discussed below.

18.3.2 Learning to complement

Learning to complement (L2C) methods target HAI-CCC systems that focus on optimizing the cooperation between human experts and AI models to maximize the expected utility of human-AI decisions [24–28,124], as displayed in Fig. 18.8. The learning to complement (L2C) formulation in Kerrigan et al. [25] suggested combining human and AI predictions. Following the same nomenclature as the L2D methods in Section 18.3.1, we have a data point $\mathbf{x} \in \mathscr{X}$, a label $\mathbf{y} \in \mathscr{Y}$, the AI model $f_\theta : \mathscr{X} \rightarrow \Delta^{|\mathscr{Y}|-1}$, and the human expert's decision $h : \mathscr{X} \rightarrow \mathscr{Y}$. The dataset to train L2C methods is represented by $\mathscr{D}_{L2C} = \{(\mathbf{x}_i, \mathbf{y}_i, h_i(\mathbf{x}))\}_{i=1}^{N}$. The classification from the combination of the human and AI model's predictions for a test image $\mathbf{x}$ is achieved with

$$\hat{\mathbf{y}} = \operatorname*{argmax}_{\mathbf{y} \in \mathscr{Y}} p(\mathbf{y}|h(\mathbf{x}), f_\theta(\mathbf{x})), \tag{18.21}$$

where the conditional probability is defined by

$$p(\mathbf{y}|h(\mathbf{x}), f_\theta(\mathbf{x})) \propto p(h(\mathbf{x})|\mathbf{y}) \times p(\mathbf{y}|f_\theta(\mathbf{x})), \tag{18.22}$$

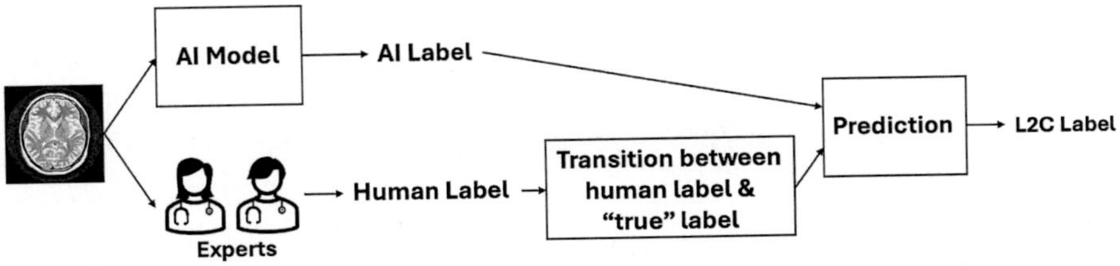

FIGURE 18.8

Learning to complement (L2C) methods contain the AI model, trained to predict the labeled data, and the predictor of the transition between human and "true" labels, trained to represent the experts' labeling confusion, which produce the AI model's and experts' labels that are combined to produce the L2C prediction.

which assumes that the human label $h(\mathbf{x})$ is conditionally independent from the AI mode's prediction $f_\theta(\mathbf{x})$ given $\mathbf{y}$. In Eq. (18.22), $p(h(\mathbf{x})|\mathbf{y})$ denotes the probability that the human labeler confuses "true" label $\mathbf{y}$ with $h(\mathbf{x})$—such probability is represented by a confusion matrix $\mathbf{T} \in [0,1]^{|\mathscr{Y}|\times|\mathscr{Y}|}$, where $\mathbf{T}_{\tilde{\mathbf{y}},\mathbf{y}} = p(h(\mathbf{x}) = \tilde{\mathbf{y}}|\mathbf{y})$. Also in Eq. (18.22), $p(\mathbf{y}|f_\theta(\mathbf{x}))$ represents the AI model's prediction probabilities. Therefore, the predicted probability for label $\mathbf{y}$, given the human prediction $\tilde{\mathbf{y}}$ and the AI model prediction $f_\theta(\mathbf{x})$, is defined by

$$p\left(\mathbf{y}\middle|h(\mathbf{x}) = \tilde{\mathbf{y}}, f_\theta(\mathbf{x})\right) = \frac{\mathbf{T}_{\tilde{\mathbf{y}},\mathbf{y}} \times f_\theta^{\mathbf{y}}(\mathbf{x})}{\sum_{\mathbf{y}' \in \mathscr{Y}} \mathbf{T}_{\tilde{\mathbf{y}},\mathbf{y}'} \times f_\theta^{\mathbf{y}'}(\mathbf{x})}, \tag{18.23}$$

where $f_\theta^{\mathbf{y}}(\mathbf{x})$ denotes the AI model's prediction associated with label $\mathbf{y}$. The training of this L2C method [25] is achieved by first training the AI model with

$$\theta^* = \arg\min_\theta \sum_{(\mathbf{x},\mathbf{y},h(\mathbf{x}))\in\mathscr{D}_{L2C}} \ell(\mathbf{y}, f_\theta(\mathbf{x})). \tag{18.24}$$

Next, we train the confusion matrix with

$$\mathbf{T}^* = \arg\min_{\mathbf{T}} \sum_{(\mathbf{x},\mathbf{y},h(\mathbf{x}))\in\mathscr{D}_{L2C}} \ell(h(\mathbf{x}), \mathbf{T} \times \mathbf{y}), \tag{18.25}$$

which in Kerrigan et al. [25] is formulated as a Bayesian training. Other L2C approaches introduced similar methods, such as [27], which proposed a Bayesian framework that integrates predictions and various confidence scores from both humans and machines, demonstrating that a hybrid approach outperforms standalone human or machine predictions. More recently, Liu et al. [26] leverage perceptual differences between humans and AI to enhance system performance, while Hemmer et al. [124] proposed an ensemble model combining AI and human predictions, albeit without optimizing cooperation costs.

A common limitation across existing HAI-CCC methods, including L2D- and L2C-based approaches, is the assumption of clean labels, which is often unrealistic since real-world data typically involves multiple noisy labels per sample. In contrast, the LECOMH [125] and LECODU [123] approaches are designed to handle multiple noisy labels in the training set by leveraging multirater learning methods, explained in Section 18.2.1.3. Another interesting limitation of HAI-CCC methods is that previous methods are either L2D or L2C, but not both at the same time. This has been recently addressed by LECODU [123] that can simultaneously defer and complement with multiple users.

18.3.3 Learning to assist

Learning to assist (L2A) approaches focus on enhancing human decision-making by providing AI-generated predictions as a supportive tool rather than a definitive answer [126] — see Fig. 18.9. L2A methods are particularly prevalent in critical domains such as law [127] and medicine [128], where the ultimate decision-making

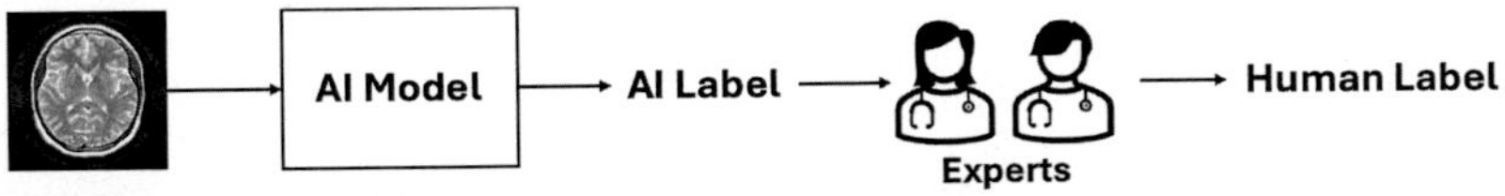

FIGURE 18.9

Learning to assist (L2A) approaches comprise an AI model that returns a prediction to the expert, who will make the final prediction.

authority remains with humans. In these settings, the AI model acts as an assistant, offering predictions or insights that help human experts make more informed decisions.

A significant area of research within L2A methods involves improving the explainability and transparency of AI models [129]. This is crucial because, in high-stakes environments like healthcare and legal systems, the ability of the AI to explain its predictions can build trust and facilitate better collaboration between humans and AI. Enhanced transparency ensures that human experts understand the rationale behind AI suggestions, which is vital for making informed, accurate final decisions.

The development of L2A methods also involves addressing challenges related to out-of-distribution (OOD) detection [127], where the AI model must recognize when it encounters data that is significantly different from what it was trained on. This ability is critical in preventing misleading predictions that could negatively influence human decision-making.

As these approaches evolve, ongoing research continues to focus on making AI models more interpretable and reliable, ensuring they serve as effective aids in decision-making processes rather than replacements for human judgment.

18.3.4 Interactive refinement and adjustments

In HAI-CCC methods, interactive refinement and adjustments play a critical role in enhancing the synergy between human expertise and AI capabilities. These methods allow human experts to fine-tune AI outputs and correct errors in real-time, leading to more accurate and reliable results, as displayed in Fig. 18.10. In this section, we explore interactive refinement and adjustments in HAI-CCC systems, particularly for medical image analysis and MR data processing tasks.

Several interactive segmentation methods have been developed to refine medical image analysis through human-AI collaboration. Amrehn et al. [130] introduce UI-Net, allowing users to provide scribbles to iteratively refine segmentations, trained using simulated user interactions. Conditional Random Fields (CRFs) and graph cuts have also been integrated with Convolutional Neural Networks (CNN) [131,132], where user feedback refines initial segmentations. Bredell et al. [133] introduce interCNN, a two-network approach that improves segmentation over multiple iterations of user input. Kurzendorfer et al. [134] use a 2D-to-3D method for

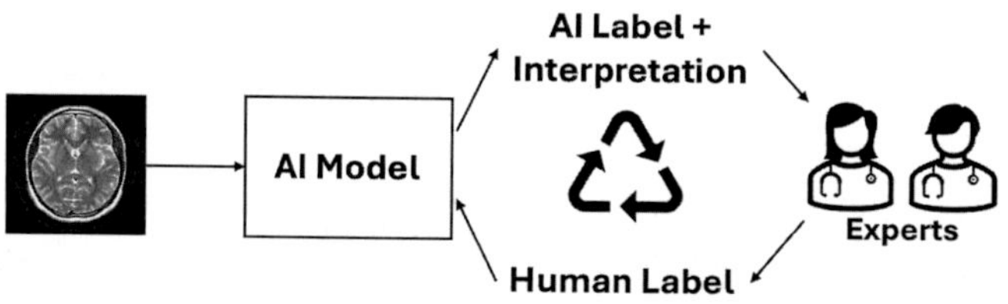

FIGURE 18.10

In interactive refinement and adjustments and interative interpretation, methods usually consist of an iterative process where the AI model produces AI label + interpretation, and the expert can interact with the model by correcting the AI models' outputs. Such correction is then used by the AI model to refine its label and interpretation, which will be again used by the experts in an iterative process.

generating segmentation surfaces, while Jang and Kim [135] present a scheme to generalize any segmentation model with user interaction, improving mis-segmented regions. Interactive refinement has also been formulated as a Markov Decision Process, using multiagent reinforcement learning to capture voxel dependencies and achieve more precise segmentations with fewer interactions [136]. Zhou et al. [137] propose the Volumetric Memory Network (VMN) that enables interactive 3D segmentation by incorporating user input, propagating segmentation across slices, and suggesting further refinements through a quality assessment module. A more effective attention to object details has been the target of the FocusCut method [138], which prioritizes user repair clicks to focus on local object details by adaptively cropping click-centered patches for progressive refinement, achieving state-of-the-art results. In Luo et al. [139], the proposed interactive segmentation method requires minimal user input to efficiently segment objects by encoding user-provided interior points using exponentialized geodesic distance and refining the segmentation with a few additional clicks. The Segment Anything Model (SAM) [140], trained on over 1 billion annotations for natural images, demonstrates impressive zero-shot segmentation performance on some medical imaging datasets but performs moderately to poorly on others, with better results for well-circumscribed objects and box prompts, and slightly improved performance with iterative multi-point prompts [141].

The problem of interactive refinement for classification and regression tasks in medical image analysis has also been addressed, as shown by Lian et al. [142]; who introduce a hierarchical fully convolutional CNN (H-FCN) for Alzheimer's diagnosis, which identifies discriminative brain regions from structural MRI (sMRI). Their approach allows for user interaction to refine patch selection and guide the classifier toward more relevant regions. Similarly, Liu et al. [143] propose a landmark-based deep multiinstance learning (LDMIL) framework for brain disease diagnosis. By detecting anatomical landmarks in MRIs, users can intervene to refine the classifier by updating which regions are used for analysis, improving overall diagnostic accuracy. These methods highlight the potential for human-in-the-loop refinement in classification tasks.

18.3.5 Interactive interpretation

While Section 18.3.4 discusses how users communicate with the AI model, it is also important to discuss how the AI model can communicate with the user [34]. The information that the AI model provides back to users is designed to help users make informed decisions about the system's predictions [34] or to inform users about the system's uncertainty [144,145]—a simplified diagram of an interactive interpretation is shown in Fig. 18.10.

Interactive interpretation methods, depicted in Fig. 18.10, is associated with the field of explainable AI (XAI) [146], which studies AI methods that can provide explanations for the predictions to be used by clinicians for informed decision support. The main difference is that XAI does not explore the idea of interacting with the user—the XAI model provides an explanation, and that is the end of the process. The methods below assume that the user can interact with the system. Real-time systems like the RL-based ultrasound imaging proposed by Milletari et al. [147] allow users to interact with model feedback. The interaction between users and models is increasingly vital, and research in interpretability, uncertainty, and human-computer interaction is converging, as shown in works like [146,148–151], particularly within active learning systems in clinical applications.

18.4 Practical considerations, limitations and future work

In this section, we discuss challenges, limitations and opportunities in the development and deployment of human-in-the-loop methods. In particular, we highlight the primary practical research areas, and suggest potential directions for future exploration.

Human-in-the-loop training methods, discussed in Section 18.2.1, can rely on expert or nonexpert annotators. In general, experts are expensive, but can provide accurate training data, while nonexperts are cheap, but can provide noisy labels, which can negatively impact the accuracy of medical image data annotations. Kuo et al. [152] proposed a cost-sensitive active learning approach for hemorrhage detection that optimizes annotation time investment, while Schlemper et al. [153] developed a budget-based framework combining strong and weak annotations to minimize costs while maintaining performance. When relying exclusively on nonexperts, it is important to add some quality assurance mechanisms, like the ones below. Zhang and Chaudhuri [154] propose active learning methods that mitigate this by assessing annotator quality and assigning challenging cases to stronger annotators. Li et al. [155] model crowd expertise and label correlations to improve annotation quality, extending this approach to active learning. Cheplygina et al. [53] found that nonexpert annotations via Amazon MTurk were interpretable but often unusable due to complex instructions, although aggregated novice annotations correlated well with expert data. Rodrigues and Pereira [49] further address annotator reliability through a crowd layer in deep models. While models can mitigate low-quality annotations, providing task-specific instructions remains essential for effective nonexpert annotation. Future research should focus on how the annotator's expertise, annotation type, and cost interact to influence data curation.

Human-in-the-loop methods in medical image annotation often overlook the interface used by human annotators for interacting with images, despite its potential impact on annotation accuracy. Nalisnik et al. [156] introduced an online framework for classifying whole-slide images, improving glioma quantification through active learning. Khosravan et al. [157] developed an interface using eye-tracking to initiate seed points for segmentation, showing high performance, while Stember et al. [158] demonstrated that eye-tracked annotations performed comparably to hand annotations. Tinati et al. [159] explored the gamification of neuron segmentation with EyeWire, where over 500,000 players contributed. While this was task-specific, it showed that nonmonetary motivations like scientific contribution can drive crowdsourced annotations. As interfaces evolve, they will play a crucial role in the success of medical image analysis applications.

Continual learning (CL) in medical image analysis is a rapidly evolving field focused on enabling AI models to learn from new data while retaining knowledge from previous tasks, without requiring retraining on all prior data [160,161]. This approach addresses the challenge of "catastrophic forgetting," where models forget previously learned information when updated with new data. In medical image analysis, CL is particularly valuable because of the evolving nature of medical knowledge, data, and imaging techniques. By incorporating continual learning, AI systems can adapt to new diseases, imaging modalities, and annotations while maintaining accuracy on prior tasks, offering a more flexible and sustainable solution for clinical applications. Introducing human in the loop to CL offers a solution to not only refining results based on user input, but also retraining the AI model to improve task performance on both current and future images while preserving knowledge from previous tasks [162,163].

Foundation models are large-scale deep learning models that have been trained on a wide range of multimodal data to be applied across different tasks, such as few-shot learning, zero-shot learning, or prompt engineering [164]. Given that it is possible to continuously interact with foundation models through prompts, they become natural models to explore for human-in-the-loop tasks.

An important open research question is the combined application of a set of the techniques presented in this chapter. For example, we can consider the application of many of the human-in-the-loop training and testing techniques for continual learning using foundation models. Hence, human-in-the-loop methods are likely to become the focus in deep learning training and testing given their potential to reduce annotation costs and increase model performance.

18.5 Conclusions

In this chapter, we discussed the integration of human expertise and artificial intelligence (AI) in the analysis of magnetic resonance (MR) data, with a particular focus on human-in-the-loop (HITL) systems in medical imaging. These systems leverage the domain knowledge of clinicians during both the training and testing phases of AI models to mitigate biases inherent in datasets and improve model performance. During training, humans provide less uncertain labels or annotate samples the model struggles with, enhancing training convergence and generalization. During testing, human-AI collaboration can take various forms, such as combining the predictions by experts and the AI model or allowing mutual influence on decision-making. This process improves the accuracy, transparency, and reliability of MR imaging analysis, ensuring the clinical relevance of AI models. As research in this area progresses, different techniques, such as active learning and collaborative models, will converge to create systems that support continual learning, where humans continually influence AI results and model training.

References

[1] Berger A. How does it work?: magnetic resonance imaging. Br Med J 2002;324(7328):35.

[2] Lundervold AS, Lundervold A. An overview of deep learning in medical imaging focusing on MRI. Z Med Phys 2019;29(2):102–27.

[3] Norori N, Hu Q, Aellen FM, Faraci FD, Tzovara A. Addressing bias in big data and ai for health care: a call for open science. Patterns 2021;2(10).

[4] Liang W, Tadesse GA, Ho D, Fei-Fei L, Zaharia M, Zhang C, Zou J. Advances, challenges and opportunities in creating data for trustworthy ai. Nat Mach Intell 2022;4(8):669–77.

[5] Schwartz R, Schwartz R, Vassilev A, Greene K, Perine L, Burt A, Hall P. Towards a standard for identifying and managing bias in artificial intelligence, 3. US Department of Commerce, National Institute of Standards and Technology; 2022.

[6] Mosqueira-Rey E, Hernández-Pereira E, Bobes-Bascarán J, Alonso-Ríos D, Pérez-Sánchez A, Fernández-Leal Á, Moret-Bonillo V, Vidal-Ínsua Y, Vázquez-Rivera F. Addressing the data bottleneck in medical deep learning models using a human-in-the-loop machine learning approach. Neural Comput Appl 2024;36(5):2597–616.

[7] Kumar S, Datta S, Singh V, Datta D, Singh SK, Sharma R. Applications, challenges, and future directions of human-in-the-loop learning. IEEE Access 2024.

[8] Mozannar H. Training human-AI teams. Ph.D. Thesis, Massachusetts Institute of Technology; 2024.

[9] He H, Gray J, Cangelosi A, Meng Q, McGinnity TM, Mehnen J. The challenges and opportunities of human-centered ai for trustworthy robots and autonomous systems. IEEE Transact Cognit Dev Syst 2021;14(4):1398–412.

[10] Mosqueira-Rey E, Hernández-Pereira E, Alonso-Ríos D, Bobes-Bascarán J, Fernández-Leal Á. Human-in-the-loop machine learning: a state of the art. Artif Intell Rev 2023;56(4):3005–54.

[11] Fu H, Li F, Xu Y, Liao J, Xiong J, Shen J, Liu J, Zhang X, iChallenge-GON study group. A retrospective comparison of deep learning to manual annotations for optic disc and optic cup segmentation in fundus photos. medRxiv 2020. 2020–05.

[12] Joskowicz L, Cohen D, Caplan N, Sosna J. Inter-observer variability of manual contour delineation of structures in ct. Eur Radiol 2019;29:1391–9.

[13] Kumar S, Giubilato A, Morgan W, Jitskaia L, Barry C, Bulsara M, Constable IJ, Yogesan K. Glaucoma screening: analysis of conventional and telemedicine-friendly devices. Clin Exp Ophthalmol 2007;35(3):237–43.

[14] Schaekermann M, Beaton G, Habib M, Lim A, Larson K, Law E. Understanding expert disagreement in medical data analysis through structured adjudication. Proc ACM Human-Comput Interact 2019;3(CSCW):1–23.

[15] Monarch RM. Human-in-the-loop machine learning: active learning and annotation for human-centered AI. Simon and Schuster; 2021.

[16] Sheng VS, Zhang J. Machine learning with crowdsourcing: a brief summary of the past research and future directions. In: Proceedings of the AAAI Conference on Artificial Intelligence. 33; 2019. p. 9837–43.

[17] Baumgartner CF, Tezcan KC, Chaitanya K, Hötker AM, Muehlematter UJ, Schawkat K, Becker AS, Donati O, Konukoglu E. Phiseg: capturing uncertainty in medical image segmentation. In: Medical Image Computing and Computer Assisted Intervention–MICCAI 2019: 22nd International Conference, Shenzhen, China, October 13–17, 2019, Proceedings, Part II 22. Springer; 2019. p. 119–27.

[18] Tian Y, Liu F, Pang G, Chen Y, Liu Y, Verjans JW, Singh R, Carneiro G. Self-supervised pseudo multi-class pre-training for unsupervised anomaly detection and segmentation in medical images. Med Image Anal 2023;90:102930.

[19] Retzlaff CO, Das S, Wayllace C, Mousavi P, Afshari M, Yang T, Saranti A, Angerschmid A, Taylor ME, Holzinger A. Human-in-the-loop reinforcement learning: a survey and position on requirements, challenges, and opportunities. J Artif Intell Res 2024;79:359–415.

[20] Keswani V, Lease M, Kenthapadi K. Towards unbiased and accurate deferral to multiple experts. In: AAAI/ACM Conference on AI, Ethics, and Society; 2021. p. 154–65.

[21] Madras D, Pitassi T, Zemel R. Predict responsibly: improving fairness and accuracy by learning to defer. In: Advances in neural information processing systems. 31; 2018.

[22] Mao A, Mohri C, Mohri M, Zhong Y. Two-stage learning to defer with multiple experts. In: Advances in neural information processing systems; 2023.

[23] Narasimhan H, Jitkrittum W, Menon AK, Rawat A, Kumar S. Post-hoc estimators for learning to defer to an expert. In: Advances in neural information processing systems. 35; 2022. p. 29292–304.

[24] Bansal G, Nushi B, Kamar E, Horvitz E, Weld DS. Is the most accurate AI the best teammate? Optimizing AI for teamwork. In: AAAI Conference on Artificial Intelligence. 35; 2021. p. 11405–14.

[25] Kerrigan G, Smyth P, Steyvers M. Combining human predictions with model probabilities via confusion matrices and calibration. In: Advances in neural information processing systems. 34; 2021. p. 4421–34.

[26] Liu M, Wei J, Liu Y, Davis J. Do humans and machines have the same eyes? Human-machine perceptual differences on image classification. arXiv preprint arXiv: 2304.08733 2023.

[27] Steyvers M, Tejeda H, Kerrigan G, Smyth P. Bayesian modeling of human–AI complementarity. Natl Acad Sci 2022;119(11):e2111547119.

[28] Wilder B, Horvitz E, Kamar E. Learning to complement humans. In: Proceedings of the Twenty-Ninth International Joint Conference on Artificial Intelligence. 20. IJCAI'; 2021.

[29] Davani AM, Díaz M, Prabhakaran V. Dealing with disagreements: looking beyond the majority vote in subjective annotations. Transact Assoc Comput Linguistics 2022;10: 92–110.

[30] Sabou M, Bontcheva K, Derczynski L, Scharl A. Corpus annotation through crowdsourcing: towards best practice guidelines. In: LREC. Citeseer; 2014. p. 859–66.

[31] Zhang J, Sheng VS, Li Q, Wu J, Wu X. Consensus algorithms for biased labeling in crowdsourcing. Inf Sci 2017;382:254–73.

[32] Sylolypavan A, Sleeman D, Wu H, Sim M. The impact of inconsistent human annotations on ai driven clinical decision making. NPJ Digit Med 2023;6(1):26.

[33] Menze BH, Jakab A, Bauer S, Kalpathy-Cramer J, Farahani K, Kirby J, Burren Y, Porz N, Slotboom J, Wiest R, et al. The multimodal brain tumor image segmentation benchmark (brats). IEEE Trans Med Imag 2014;34(10):1993–2024.

[34] Budd S, Robinson EC, Kainz B. A survey on active learning and human-in-the-loop deep learning for medical image analysis. Med Image Anal 2021;71:102062.

[35] Settles B. In: Active learning literature survey; 2009.

[36] Atlas L, Cohn D, Ladner R. Training connectionist networks with queries and selective sampling. Adv Neural Inf Process Syst 1989;2.

[37] Angluin D. Queries and concept learning. Mach Learn 1988;2:319–42.

[38] Lewis DD, Catlett J. Heterogeneous uncertainty sampling for supervised learning. In: Machine Learning Proceedings. Elsevier; 1994. p. 148–56.

[39] Boehringer AS, Sanaat A, Arabi H, Zaidi H. An active learning approach to train a deep learning algorithm for tumor segmentation from brain MR images. Insights Imaging 2023;14(1):141.

[40] Küstner T, Gatidis S, Liebgott A, Schwartz M, Mauch L, Martirosian P, Schmidt H, Schwenzer NF, Nikolaou K, Bamberg F, et al. A machine-learning framework for automatic reference-free quality assessment in MRI. Magnet Reson Imaging 2018;53: 134–47.

[41] Mahapatra D, Vos FM, Buhmann JM. Active learning based segmentation of Crohns disease from abdominal MRI. Comput Methods Progr Biomed 2016;128:75–85.

[42] Wang H, Jin Q, Li S, Liu S, Wang M, Song Z. A comprehensive survey on deep active learning in medical image analysis. Med Image Anal 2024:103201.

[43] Zhang Z, Li J, Tian C, Zhong Z, Jiao Z, Gao X. Quality-driven deep active learning method for 3d brain MRI segmentation. Neurocomputing 2021;446:106–17.

[44] Dawid AP, Skene AM. Maximum likelihood estimation of observer error-rates using the em algorithm. J Roy Stat Soc: Series C (Appl Stat) 1979;28(1):20–8.

[45] Herde M, Huseljic D, Sick B. Multi-annotator deep learning: a probabilistic framework for classification. Transact Machine Learning Res 2023.

[46] Ørting S, Doyle A, van Hilten A, Hirth M, Inel O, Madan CR, Mavridis P, Spiers H, Cheplygina V. A survey of crowdsourcing in medical image analysis. arXiv preprint arXiv:1902.09159 2019.

[47] Raykar VC, Yu S, Zhao LH, Valadez GH, Florin C, Bogoni L, Moy L. Learning from crowds. J Mach Learn Res 2010;11(4).

[48] Rodrigues F, Lourenco M, Ribeiro B, Pereira FC. Learning supervised topic models for classification and regression from crowds. IEEE Trans Pattern Anal Mach Intell 2017; 39(12):2409–22.

[49] Rodrigues F, Pereira F. Deep learning from crowds. In: Proceedings of the AAAI conference on artificial intelligence. 32; 2018.

[50] Whitehill J, Wu T-f, Bergsma J, Movellan J, Ruvolo P. Whose vote should count more: optimal integration of labels from labelers of unknown expertise. In: Advances in neural information processing systems. 22; 2009.
[51] Gurari D, Theriault D, Sameki M, Isenberg B, Pham TA, Purwada A, Solski P, Walker M, Zhang C, Wong JY, et al. How to collect segmentations for biomedical images? a benchmark evaluating the performance of experts, crowdsourced non-experts, and algorithms. In: 2015 IEEE Winter Conference on Applications of Computer Vision. IEEE; 2015. p. 1169–76.
[52] Zhou Z-H. Ensemble methods: foundations and algorithms. CRC Press; 2012.
[53] Cheplygina V, Perez-Rovira A, Kuo W, Tiddens HA, De Bruijne M. Early experiences with crowdsourcing airway annotations in chest ct. In: Deep Learning and Data Labeling for Medical Applications: First International Workshop, LABELS 2016, and Second International Workshop, DLMIA 2016, Held in Conjunction with MICCAI 2016, Athens, Greece, October 21, 2016, Proceedings 1. Springer; 2016. p. 209–18.
[54] Foncubierta Rodríguez A, Müller H. Ground truth generation in medical imaging: a crowdsourcing-based iterative approach. In: Proceedings of the ACM Multimedia 2012 Workshop on Crowdsourcing for Multimedia; 2012. p. 9–14.
[55] Khare R, Good BM, Leaman R, Su AI, Lu Z. Crowdsourcing in biomedicine: challenges and opportunities. Briefings Bioinf 2016;17(1):23–32.
[56] Warfield SK, Zou KH, Wells WM. Simultaneous truth and performance level estimation (staple): an algorithm for the validation of image segmentation. IEEE Trans Med Imag 2004;23(7):903–21.
[57] Goh HW, Tkachenko U, Mueller J. CROWDLAB: supervised learning to infer consensus labels and quality scores for data with multiple annotators. In: NeurIPS human in the loop learning workshop; 2022.
[58] Rodrigues F, Pereira F, Ribeiro B. Gaussian process classification and active learning with multiple annotators. In: International Conference on Machine Learning. PMLR; 2014. p. 433–41.
[59] Guan M, Gulshan V, Dai A, Hinton G. Who said what: modeling individual labelers improves classification. In: AAAI Conference on Artificial Intelligence. 32; 2018.
[60] Wei H, Xie R, Feng L, Han B, An B. Deep learning from multiple noisy annotators as a union. IEEE Transact Neural Networks Learn Syst 2022.
[61] Yang H, Li X, Pedrycz W. Learning from crowds with contrastive representation. IEEE Access 2023.
[62] Cao Z, Chen E, Huang Y, Shen S, Huang Z. Learning from crowds with annotation reliability. In: International Conference on Research and Development in Information Retrieval; 2023. p. 2103–7.
[63] Zhao Y, Zheng G, Mukherjee S, McCann R, Awadallah A. Admoe: anomaly detection with mixture-of-experts from noisy labels. In: AAAI Conference on Artificial Intelligence. 37; 2023. p. 4937–45.
[64] Bergman L, Hoshen Y. Classification-based anomaly detection for general data. arXiv preprint arXiv:2005.02359 2020.
[65] Chen T, Kornblith S, Norouzi M, Hinton G. A simple framework for contrastive learning of visual representations. In: International Conference on Machine Learning. PMLR; 2020. p. 1597–607.
[66] Cho J, Kang I, Park J. Self-supervised 3d out-of-distribution detection via pseudoanomaly generation. In: International Conference on Medical Image Computing and Computer-assisted Intervention. Springer; 2021. p. 95–103.

[67] He K, Fan H, Wu Y, Xie S, Girshick R. Momentum contrast for unsupervised visual representation learning. In: Proceedings of the IEEE/CVF Conference on Computer Vision and Pattern Recognition; 2020. p. 9729–38.

[68] Hendrycks D, Mazeika M, Kadavath S, Song D. Using self-supervised learning can improve model robustness and uncertainty. Adv Neural Inf Process Syst 2019;32.

[69] Liu F, Jonmohamadi Y, Maicas G, Pandey AK, Carneiro G. Self-supervised depth estimation to regularise semantic segmentation in knee arthroscopy. In: Medical Image Computing and Computer Assisted Intervention–MICCAI 2020: 23rd International Conference, Lima, Peru, October 4–8, 2020, Proceedings, Part I 23. Springer; 2020. p. 594–603.

[70] Schlüter HM, Tan J, Hou B, Kainz B. Natural synthetic anomalies for self-supervised anomaly detection and localization. In: European Conference on Computer Vision. Springer; 2022. p. 474–89.

[71] Tian Y, Pang G, Liu F, Chen Y, Shin SH, Verjans JW, Singh R, Carneiro G. Constrained contrastive distribution learning for unsupervised anomaly detection and localisation in medical images. In: Medical Image Computing and Computer Assisted Intervention–MICCAI 2021: 24th International Conference, Strasbourg, France, September 27–October 1, 2021, Proceedings, Part V 24. Springer; 2021. p. 128–40.

[72] Le T-N, Sugimoto A, Ono S, Kawasaki H. Toward interactive self-annotation for video object bounding box: Recurrent self-learning and hierarchical annotation based framework. In: Proceedings of the IEEE/CVF Winter Conference on Applications of Computer Vision; 2020. p. 3231–40.

[73] Adhikari B, Huttunen H. Iterative bounding box annotation for object detection. In: 2020 25th International Conference on Pattern Recognition (ICPR). IEEE; 2021. p. 4040–6.

[74] Christiano PF, Leike J, Brown T, Martic M, Legg S, Amodei D. Deep reinforcement learning from human preferences. Adv Neural Inf Process Syst 2017;30.

[75] da Cruz LC, Sierra-Franco CA, Silva-Calpa GFM, Raposo AB. Enabling autonomous medical image data annotation: a human-in-the-loop reinforcement learning approach. In: 2021 16th Conference on Computer Science and Intelligence Systems (FedCSIS). IEEE; 2021. p. 271–9.

[76] Griffith S, Subramanian K, Scholz J, Isbell CL, Thomaz AL. Policy shaping: integrating human feedback with reinforcement learning. Adv Neural Inf Process Syst 2013;26.

[77] Knox WB, Stone P. Tamer: training an agent manually via evaluative reinforcement. In: 2008 7th IEEE International Conference on Development and Learning. IEEE; 2008. p. 292–7.

[78] Ross S, Gordon G, Bagnell D. A reduction of imitation learning and structured prediction to no-regret online learning. In: Proceedings of the Fourteenth International Conference on Artificial Intelligence and Statistics. JMLR Workshop and Conference Proceedings; 2011. p. 627–35.

[79] Saunders W, Sastry G, Stuhlmueller A, Evans O. Trial without error: towards safe reinforcement learning via human intervention. arXiv preprint arXiv:1707.05173 2017.

[80] Warnell G, Waytowich N, Lawhern V, Stone P. Deep tamer: interactive agent shaping in high-dimensional state spaces. In: Proceedings of the AAAI Conference on Artificial Intelligence. 32; 2018.

[81] Zhu Y, Mottaghi R, Kolve E, Lim JJ, Gupta A, Fei-Fei L, Farhadi A. Target-driven visual navigation in indoor scenes using deep reinforcement learning. In: 2017 IEEE International Conference on Robotics and Automation (ICRA). IEEE; 2017. p. 3357–64.
[82] Hu M, Zhang J, Matkovic L, Liu T, Yang X. Reinforcement learning in medical image analysis: concepts, applications, challenges, and future directions. J Appl Clin Med Phys 2023;24(2):e13898.
[83] Saripalli VR, Pati D, Potter M, Avinash G, Anderson CW. Ai-assisted annotator using reinforcement learning. SN Comput Sci 2020;1(6):327.
[84] Wang J, Yan Y, Zhang Y, Cao G, Yang M, Ng MK. Deep reinforcement active learning for medical image classification. In: Medical Image Computing and Computer Assisted Intervention–MICCAI 2020: 23rd International Conference, Lima, Peru, October 4–8, 2020, Proceedings, Part I 23. Springer; 2020. p. 33–42.
[85] Arakawa R, Kobayashi S, Unno Y, Tsuboi Y, Maeda S-i. Dqn-tamer: human-in-the-loop reinforcement learning with intractable feedback. arXiv preprint arXiv:1810.11748 2018.
[86] Krening S. In: Humans teaching intelligent agents with verbal instruction; 2019.
[87] Liang H, Yang L, Cheng H, Tu W, Xu M. Human-in-the-loop reinforcement learning. In: 2017 Chinese Automation Congress (CAC). IEEE; 2017. p. 4511–8.
[88] Lin Z, Harrison B, Keech A, Riedl MO. Explore, exploit or listen: combining human feedback and policy model to speed up deep reinforcement learning in 3d worlds. arXiv preprint arXiv:1709.03969 2017.
[89] Torrey L, Taylor M. Teaching on a budget: agents advising agents in reinforcement learning. In: Proceedings of the 2013 International Conference on Autonomous Agents and Multi-agent Systems; 2013. p. 1053–60.
[90] Committee O-RADO. In: Taxonomy and definitions for terms related to on-road motor vehicle automated driving systems. SAE International; 2014.
[91] Parasuraman R, Sheridan TB, Wickens CD. A model for types and levels of human interaction with automation. IEEE Trans Syst Man Cybern Syst Hum May 2000;30(3):286–97.
[92] Strauch B. Ironies of automation: still unresolved after all these years. IEEE Transact Human-Machine Syst 2018;48(5):419–33.
[93] Kamar E, Hacker S, Horvitz E. Combining human and machine intelligence in large-scale crowdsourcing. In: International Conference on Autonomous Agents and Multiagent Systems. 12; 2012. p. 467–74.
[94] Rosenfeld A, Solbach MD, Tsotsos JK. Totally looks like-how humans compare, compared to machines. In: CVPRW; 2018. p. 1961–4.
[95] Serre T. Deep learning: the good, the bad, and the ugly. Ann Rev Vision Sci 2019;5:399–426.
[96] Chiou EK, Lee JD. Trusting automation: Designing for responsivity and resilience. Hum Factors 2023;65(1):137–65.
[97] Lu Z, Yin M. Human reliance on machine learning models when performance feedback is limited: heuristics and risks. In: CHI Conference on Human Factors in Computing Systems; 2021. p. 1–16.
[98] Yin M, Wortman Vaughan J, Wallach H. Understanding the effect of accuracy on trust in machine learning models. In: CHI Conference on Human Factors in Computing Systems; 2019. p. 1–12.
[99] Shin D. The effects of explainability and causability on perception, trust, and acceptance: implications for explainable AI. Int J Hum Comput Stud 2021;146:102551.

[100] Weitz K, Schiller D, Schlagowski R, Huber T, André E. "do you trust me?" increasing user-trust by integrating virtual agents in explainable ai interaction design. In: ACM International Conference on Intelligent Virtual Agents; 2019. p. 7–9.
[101] Agarwal N, Moehring A, Rajpurkar P, Salz T. Combining human expertise with artificial intelligence: experimental evidence from radiology. Tech Rep, National Bureau of Economic Research 2023.
[102] Vodrahalli K, Daneshjou R, Gerstenberg T, Zou J. Do humans trust advice more if it comes from AI? an analysis of human-AI interactions. In: AAAI/ACM Conference on AI, Ethics, and Society; 2022. p. 763–77.
[103] Wu X, Xiao L, Sun Y, Zhang J, Ma T, He L. A survey of human-in-the-loop for machine learning. Future Gener Comput Syst October 2022;135(C):364–81. https://doi.org/10.1016/j.future.2022.05.014.
[104] Cortes C, DeSalvo G, Mohri M. Learning with rejection. In: International Conference on Algorithmic Learning Theory. Springer; 2016. p. 67–82.
[105] Cao Y, Mozannar H, Feng L, Wei H, An B. In defense of softmax parametrization for calibrated and consistent learning to defer. In: Advances in neural information processing systems. 36; 2024.
[106] Charoenphakdee N, Cui Z, Zhang Y, Sugiyama M. Classification with rejection based on cost-sensitive classification. In: International Conference on Machine Learning. PMLR; 2021. p. 1507–17.
[107] Charusaie M-A, Mozannar H, Sontag D, Samadi S. Sample efficient learning of predictors that complement humans. In: International Conference on Machine Learning. PMLR; 2022. p. 2972–3005.
[108] Liu S, Cao Y, Zhang Q, Feng L, An B. Mitigating underfitting in learning to defer with consistent losses. In: International Conference on Artificial Intelligence and Statistics; 2024.
[109] Mozannar H, Lang H, Wei D, Sattigeri P, Das S, Sontag D. Who should predict? exact algorithms for learning to defer to humans. In: Ruiz F, Dy J, van de Meent J-W, editors. Proceedings of the 26th International Conference on Artificial Intelligence and Statistics. Vol. 206 of Proceedings of Machine Learning Research. PMLR; 25–27 April 2023. p. 10520–45. https://proceedings.mlr.press/v206/mozannar23a.html.
[110] Mozannar H, Satyanarayan A, Sontag D. Teaching humans when to defer to a classifier via exemplars. In: AAAI Conference on Artificial Intelligence. 36; 2022. p. 5323–31.
[111] Mozannar H, Sontag D. Consistent estimators for learning to defer to an expert. In: International Conference on Machine Learning. PMLR; 2020. p. 7076–87.
[112] Okati N, De A, Rodriguez M. Differentiable learning under triage. In: Advances in neural information processing systems. 34; 2021. p. 9140–51.
[113] Raghu M, Blumer K, Corrado G, Kleinberg J, Obermeyer Z, Mullainathan S. The algorithmic automation problem: prediction, triage, and human effort. arXiv preprint arXiv:1903.12220 2019.
[114] Straitouri E, Wang L, Okati N, Rodriguez MG. Improving expert predictions with conformal prediction. In: International Conference on Machine Learning. PMLR; 2023. p. 32633–53.
[115] Verma R, Nalisnick E. Calibrated learning to defer with one-vs-all classifiers. In: International Conference on Machine Learning. PMLR; 2022. p. 22184–202.
[116] Babbar V, Bhatt U, Weller A. On the utility of prediction sets in human-AI teams. In: International Joint Conference on Artificial Intelligence; 2022.

[117] Hemmer P, Thede L, Vössing M, Jakubik J, Kühl N. Learning to defer with limited expert predictions. In: AAAI Conference on Artificial Intelligence. 37; 2023. p. 6002–11.
[118] Leitão D, Saleiro P, Figueiredo MA, Bizarro P. Human-AI collaboration in decision-making: beyond learning to defer. In: ICML Workshop on Human-Machine Collaboration and Teaming; 2022.
[119] Mao A, Mohri M, Zhong Y. Principled approaches for learning to defer with multiple experts. In: International symposium on artificial intelligence and mathematics; 2024.
[120] Tailor D, Patra A, Verma R, Manggala P, Nalisnick E. Learning to defer to a population: a meta-learning approach. In: International Conference on Artificial Intelligence and Statistics; 2024.
[121] Verma R, Barrejón D, Nalisnick E. On the calibration of learning to defer to multiple experts. In: ICML workshop on human-machine collaboration and teaming; 2022.
[122] Verma R, Barrejon D, Nalisnick E. Learning to defer to multiple experts: consistent surrogate losses, confidence calibration, and conformal ensembles. In: International Conference on Artificial Intelligence and Statistics. PMLR; April 25–27, 2023. p. 11415–34.
[123] Zhang Z, Ai W, Wells K, Rosewarne D, Do T-T, Carneiro G. Learning to complement and to defer to multiple users. In: European Conference on Computer Vision. Springer; 2024.
[124] Hemmer P, Schellhammer S, Vössing M, Jakubik J, Satzger G. Forming effective human-AI teams: building machine learning models that complement the capabilities of multiple experts. In: International Joint Conference on Artificial Intelligence; 2022. p. 2478–84.
[125] Zhang Z, Wells K, Carneiro G. Learning to complement with multiple humans (LECOMH): integrating multi-rater and noisy-label learning into human-AI collaboration. arXiv preprint arXiv:2311.13172 2023.
[126] Straitouri E, Wang L, Okati N, Rodriguez MG. Improving expert predictions with prediction sets. arXiv preprint arXiv:2201.12006 2023.
[127] Liu H, Lai V, Tan C. Understanding the effect of out-of-distribution examples and interactive explanations on human-AI decision making. Proc ACM Hum-Comput Interact October 2021;5(CSCW2). https://doi.org/10.1145/3479552.
[128] Levy A, Agrawal M, Satyanarayan A, Sontag D. Assessing the impact of automated suggestions on decision making: domain experts mediate model errors but take less initiative. In: Proceedings of the 2021 CHI Conference on Human Factors in Computing Systems. CHI '21. New York, NY, USA: Association for Computing Machinery; 2021. https://doi.org/10.1145/3411764.3445522.
[129] Tjoa E, Guan C. A survey on explainable artificial intelligence (xai): toward medical xai. IEEE Transact Neural Networks Learn Syst 2021;32(11):4793–813.
[130] Amrehn M, Gaube S, Unberath M, Schebesch F, Horz T, Strumia M, Steidl S, Kowarschik M, Maier A. Ui-net: interactive artificial neural networks for iterative image segmentation based on a user model. arXiv preprint arXiv:1709.03450 2017.
[131] Wang G, Li W, Zuluaga MA, Pratt R, Patel PA, Aertsen M, Doel T, David AL, Deprest J, Ourselin S, et al. Interactive medical image segmentation using deep learning with image-specific fine tuning. IEEE Trans Med Imag 2018;37(7):1562–73.
[132] Wang G, Zuluaga MA, Li W, Pratt R, Patel PA, Aertsen M, Doel T, David AL, Deprest J, Ourselin S, et al. Deepigeos: a deep interactive geodesic framework for

medical image segmentation. IEEE Trans Pattern Anal Mach Intell 2018;41(7): 1559–72.

[133] Bredell G, Tanner C, Konukoglu E. Iterative interaction training for segmentation editing networks. In: Machine Learning in Medical Imaging: 9th International Workshop, MLMI 2018, Held in Conjunction with MICCAI 2018, Granada, Spain, September 16, 2018, Proceedings 9. Springer; 2018. p. 363–70.

[134] Kurzendorfer T, Fischer P, Mirshahzadeh N, Pohl T, Brost A, Steidl S, Maier A. Rapid interactive and intuitive segmentation of 3d medical images using radial basis function interpolation. J Imaging 2017;3(4):56.

[135] Jang W-D, Kim C-S. Interactive image segmentation via backpropagating refinement scheme. In: Proceedings of the IEEE/CVF Conference on Computer Vision and Pattern Recognition; 2019. p. 5297–306.

[136] Liao X, Li W, Xu Q, Wang X, Jin B, Zhang X, Wang Y, Zhang Y. Iteratively-refined interactive 3d medical image segmentation with multi-agent reinforcement learning. In: Proceedings of the IEEE/CVF Conference on Computer Vision and Pattern Recognition; 2020. p. 9394–402.

[137] Zhou T, Li L, Bredell G, Li J, Unkelbach J, Konukoglu E. Volumetric memory network for interactive medical image segmentation. Med Image Anal 2023;83:102599.

[138] Lin Z, Duan Z-P, Zhang Z, Guo C-L, Cheng M-M. Focuscut: diving into a focus view in interactive segmentation. In: Proceedings of the IEEE/CVF Conference on Computer Vision and Pattern Recognition; 2022. p. 2637–46.

[139] Luo X, Wang G, Song T, Zhang J, Aertsen M, Deprest J, Ourselin S, Vercauteren T, Zhang S. Mideepseg: minimally interactive segmentation of unseen objects from medical images using deep learning. Med Image Anal 2021;72:102102.

[140] Kirillov A, Mintun E, Ravi N, Mao H, Rolland C, Gustafson L, Xiao T, Whitehead S, Berg AC, Lo W-Y, et al. Segment anything. In: Proceedings of the IEEE/CVF International Conference on Computer Vision; 2023. p. 4015–26.

[141] Mazurowski MA, Dong H, Gu H, Yang J, Konz N, Zhang Y. Segment anything model for medical image analysis: an experimental study. Med Image Anal 2023;89:102918.

[142] Lian C, Liu M, Zhang J, Shen D. Hierarchical fully convolutional network for joint atrophy localization and Alzheimer's disease diagnosis using structural MRI. IEEE Trans Pattern Anal Mach Intell 2018;42(4):880–93.

[143] Liu M, Zhang J, Adeli E, Shen D. Landmark-based deep multi-instance learning for brain disease diagnosis. Med Image Anal 2018;43:157–68.

[144] Kompa B, Snoek J, Beam AL. Second opinion needed: communicating uncertainty in medical machine learning. NPJ Digit Med 2021;4(1):4.

[145] Oktay O, Schlemper J, Folgoc LL, Lee M, Heinrich M, Misawa K, Mori K, McDonagh S, Hammerla NY, Kainz B, et al. Attention u-net: learning where to look for the pancreas. arXiv preprint arXiv:1804.03999 2018.

[146] Jin W, Li X, Fatehi M, Hamarneh G. Guidelines and evaluation of clinical explainable ai in medical image analysis. Med Image Anal 2023;84:102684.

[147] Milletari F, Birodkar V, Sofka M. Straight to the point: reinforcement learning for user guidance in ultrasound. In: Smart Ultrasound Imaging and Perinatal, Preterm and Paediatric Image Analysis: First International Workshop, SUSI 2019, and 4th International Workshop, PIPPI 2019, Held in Conjunction with MICCAI 2019, Shenzhen, China, October 13 and 17, 2019, Proceedings 4. Springer; 2019. p. 3–10.

[148] Amrehn M, Steidl S, Kortekaas R, Strumia M, Weingarten M, Kowarschik M, Maier A. A semi-automated usability evaluation framework for interactive image segmentation systems. Int J Biomed Imag 2019;2019(1):1464592.

[149] Beede E, Baylor E, Hersch F, Iurchenko A, Wilcox L, Ruamviboonsuk P, Vardoulakis LM. A human-centered evaluation of a deep learning system deployed in clinics for the detection of diabetic retinopathy. In: Proceedings of the 2020 CHI Conference on Human Factors in Computing Systems; 2020. p. 1–12.
[150] Capel T, Brereton M. What is human-centered about human-centered ai? a map of the research landscape. In: Proceedings of the 2023 CHI Conference on Human Factors in Computing Systems; 2023. p. 1–23.
[151] Lim WX, Chen Z, Ahmed A. The adoption of deep learning interpretability techniques on diabetic retinopathy analysis: a review. Med Biol Eng Comput 2022;60(3):633–42.
[152] Kuo W, Häne C, Yuh E, Mukherjee P, Malik J. Cost-sensitive active learning for intracranial hemorrhage detection. In: Medical Image Computing and Computer Assisted Intervention–MICCAI 2018: 21st International Conference, Granada, Spain, September 16-20, 2018, Proceedings, Part III 11. Springer; 2018. p. 715–23.
[153] Schlemper J, Oktay O, Schaap M, Heinrich M, Kainz B, Glocker B, Rueckert D. Attention gated networks: learning to leverage salient regions in medical images. Med Image Anal 2019;53:197–207.
[154] Zhang C, Chaudhuri K. Active learning from weak and strong labelers. Adv Neural Inf Process Syst 2015;28.
[155] Li S-Y, Jiang Y, Chawla NV, Zhou Z-H. Multi-label learning from crowds. IEEE Trans Knowl Data Eng 2018;31(7):1369–82.
[156] Nalisnik M, Gutman DA, Kong J, Cooper LA. An interactive learning framework for scalable classification of pathology images. In: 2015 IEEE International Conference on Big Data (Big Data). IEEE; 2015. p. 928–35.
[157] Khosravan N, Celik H, Turkbey B, Cheng R, McCreedy E, McAuliffe M, Bednarova S, Jones E, Chen X, Choyke P, et al. Gaze2segment: a pilot study for integrating eye-tracking technology into medical image segmentation. In: Medical Computer Vision and Bayesian and Graphical Models for Biomedical Imaging: MICCAI 2016 International Workshops, MCV and BAMBI, Athens, Greece, October 21, 2016, Revised Selected Papers 8. Springer; 2017. p. 94–104.
[158] Stember JN, Celik H, Krupinski E, Chang PD, Mutasa S, Wood BJ, Lignelli A, Moonis G, Schwartz LH, Jambawalikar S, et al. Eye tracking for deep learning segmentation using convolutional neural networks. J Digit Imag 2019;32:597–604.
[159] Tinati R, Luczak-Roesch M, Simperl E, Hall W. An investigation of player motivations in eyewire, a gamified citizen science project. Comput Hum Behav 2017;73:527–40.
[160] Parisi GI, Kemker R, Part JL, Kanan C, Wermter S. Continual lifelong learning with neural networks: a review. Neural Netw 2019;113:54–71.
[161] Verma T, Jin L, Zhou J, Huang J, Tan M, Choong BCM, Tan TF, Gao F, Xu X, Ting DS, et al. Privacy-preserving continual learning methods for medical image classification: a comparative analysis. Front Med 2023;10:1227515.
[162] Baweja C, Glocker B, Kamnitsas K. Towards continual learning in medical imaging. arxiv, Cornell University, website 2022.
[163] Zheng E, Yu Q, Li R, Shi P, Haake A. A continual learning framework for uncertainty-aware interactive image segmentation. In: Proceedings of the AAAI Conference on Artificial Intelligence. 35; 2021. p. 6030–8.
[164] Zhou C, Li Q, Li C, Yu J, Liu Y, Wang G, Zhang K, Ji C, Yan Q, He L, et al. A comprehensive survey on pretrained foundation models: a history from bert to chatgpt. arXiv preprint arXiv:2302.09419 2023.

PART

Clinical application

6

CHAPTER

Machine learning applications in neuroradiology

19

Kyle Tegtmeyer and Long Tu*

Department of Radiology and Biomedical Imaging, Yale University School of Medicine, New Haven, CT, United States

**Corresponding author: E-mail: long.tu@yale.edu*

19.1 Synopsis

MRIs are widely used throughout all subdomains of neuroradiology, favored for the superior detail it provides for both anatomic and pathologic structures. Within brain imaging, CT remains the most commonly used modality for initial assessment, particularly for assessment of urgent/emergent neurologic presentations. Abnormalities detected on CT (or suspicion of abnormality not detected by CT) will often prompt further evaluation with MRI. Analysis of imaging trends from 2000 to 2016 demonstrate an increased use of MRI imaging of the brain across all age ranges, with use of brain MRI nearing that of brain CT within the pediatric population. MRI has become the preferred method for imaging of the spine among all age ranges (particularly in adults) between 2000 and 2016 [1].

Machine learning (ML) models can utilize the information extracted from MRI scans to guide diagnosis, differentiation, molecular subtyping, prognostication, survival prediction, treatment planning, therapy response prediction, and disease surveillance across a range of conditions affecting brain, neck, and spine. In this chapter, we will cover ML applications of brain MRI related to various neurological conditions, including cancer and neoplasms, dementia imaging, stroke, and demyelinating disorders. Regarding neck MRI, we primarily focus on ML applications in differentiation and treatment planning of soft tissue and pharyngeal cancer. Finally, we will review ML applications in spine MRI with focus on cancers, and neoplasms, and demyelinating disease. We will also discuss some of the potential pitfalls and limitations of ML applications within neuroradiology MRI that must be addressed before more widespread clinical adoption.

Machine Learning in MRI. https://doi.org/10.1016/B978-0-443-14109-6.00019-5

19.2 Brain imaging

Imaging of primary pathologies within the brain is the area of greatest interest within the current ML research investigations. MRI of the brain is one of the most commonly performed imaging exams within neuroradiology, due to its superior anatomic and pathologic detail, and its utility for evaluation of a wide range of pathologies, including acute stroke, neoplasia, hemorrhage, white matter disease, and meningeal disease.

ML applications utilizing brain MRI have thus far focused on neoplastic conditions, stroke, dementia, and demyelinating diseases [2]. Within these categories, a range of ML algorithms have been developed for tasks, including segmentation, risk stratification, classification and molecular subtyping, diagnosis, and treatment planning and prediction of treatment response. These algorithms use various ML strategies and methodologies which are further detailed in prior chapters, including conventional ML methodologies such as support vector machines (SVM) and stratification models such as random forest, to more complex neural network structures, such as convolutional neural networks (CNNs), recurrent neural networks (RNNs), and vision transformers [3–5].

19.2.1 Cancer and neoplasia

MRI is the most sensitive and specific imaging modality for the evaluation of brain cancers and neoplasms within the brain. Many studies have sought to apply ML approaches to the diagnosis and analysis of tumors within the brain. As mentioned above, these studies typically fall within one of the following few major categories, based on analysis task each study is attempting to perform: segmentation of the tumor, classification and molecular subtyping, differential diagnosis, risk stratification, treatment planning, and assessment of therapy response. Many studies have been published within each sub-task of interest in tumor imaging; while a conclusive review of all studies performed is beyond the scope of this text, representative examples of studies within each major category are described below.

19.2.1.1 Segmentation

A major area of interest within the imaging of brain cancers and neoplasms is segmentation, or the use of ML to determine the exact boundaries and volumes of tumor versus surrounding normal tissues. Many different ML approaches have been utilized in the literature for tumor segmentation. A relatively early study by Amin et al. created a computer vision method employing a random forest classifier approach for tumor segmentation, which helps identify growing regions of tumors based on these characteristics [3]. SVM-based approaches have demonstrated 100% sensitivity and specificity for segmentation and differentiation of enhancing and non-enhancing regions of tumor within longitudinal assessments of treatment response within high-grade gliomas [4].

K-means clustering algorithms have also been used for brain tumor segmentation, largely within the field of radiomics [6]. Many studies have used CNNs for tumor segmentation, which can show strong performance in detected and differentiating imaging features, such as edges, shapes, and textures [7]. One such study demonstrated strong performance when segmenting gliomas, with relatively lower sensitivity in successfully segmenting meningiomas [8].

Much of the recent work on tumor segmentation is built on U-net, an early CNN architecture that demonstrated strong performance in an cell tracking image segmentation challenge in 2015, demonstrating an overall intersection over union (a measure of segmentation accuracy) of 92% versus 83% in the competition runner up model [9]. While this model was originally applied outside of radiologic image analysis, numerous subsequent derivative models demonstrated strong performance in segmentation of a variety of intracranial neoplasms [10–12]. An ongoing annual tumor segmentation challenge for brain neoplasms (Brain Tumor Segmentation Challenge, or BraTS) has run since 2012 [13], with winning models often building on the u-net architecture, such as through the inclusion of an automated hyperparameter optimization and a transformer-based computer vision model in 2021 [14].

19.2.1.2 Classification and molecular subtyping

ML efforts have been applied to classification of brain neoplasms based on data provided by MRI examinations, including both initial diagnosis (determining broad tumor class, such as glioma) and molecular subtyping (narrowing to specific subtype or even molecular signatures). Multiple groups have demonstrated radiomics methods for characterization of brain lesions, including radiomic histogram analysis and various machine learning approaches for determination of tumor histology, WHO grade, and molecular characteristics (Figs. 19.1 and 19.2) [5,15–18]. A separate study developed a CNN that differentiated meningiomas, pituitary lesions, and gliomas with an overall accuracy of 99.3% [19]. A transformer-based approach was utilized to classify H3K27M alteration status in diffuse midline gliomas with an overall accuracy of 85.1% [20]. Differentiating molecular subtypes of various primary brain neoplasms may prove helpful for prognostication and treatment as research efforts further elucidate the importance of these mutations.

Metastatic lesions to the brain are far more common than primary brain tumors; therefore, analysis of metastatic lesions within the brain is another area of research that may prove clinically valuable. One study utilizing CNN analysis of 3D T1-weighted acquisitions demonstrated an overall accuracy of 85% in differentiating metastatic lesions of breast and lung origin [21]. Another study used a random forest classification model of radiomic features extracted from brain MRI images of patients with non-small cell lung cancer brain metastases to ascertain PD-L1 status, and demonstrated an overall AUC of 0.84 [22]. Such models that can detect treatable mutations within central nervous system metastases may allow more targeted therapy within the appropriate patient subgroups, and may help to improve survival within these patients.

19.2.1.3 Treatment planning

ML approaches for treatment planning in intracranial neoplasms with MRI imaging is mainly used within radiation oncology for contouring and radiation planning. Treatment planning broadly encompasses selecting, designing, and delivering the

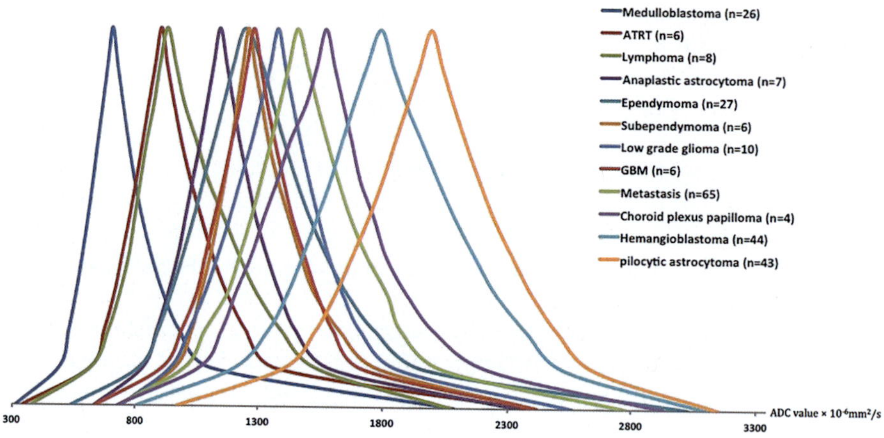

FIGURE 19.1

Representative averaged MRI diffusion coefficient (ADC) histogram distributions of various posterior fossa brain tumors. Hyper-cellular tumors such as medulloblastomas, atypical teratoid/rhabdoid tumors (ATRT), and lymphomas exhibited the lowest ADC percentile values, reflecting their densely cellular nature. In contrast, pilocytic astrocytomas and hemangioblastomas demonstrated the highest ADC percentile values, consistent with their lower tumor cell density, prominent cystic or microcystic architecture, and abundant extracellular or interstitial space. Of note, for each tumor type, average ADC percentile values were calculated, and representative histograms were normalized so that their median values align at the same vertical position, facilitating direct visual comparison of distribution shapes. Abbreviation: *GBM*, glioblastoma multiforme.

From Payabvash et al. [5]

most appropriate therapeutic interventions for a patient. ML can help optimize treatment planning using MRI features to determine the intervention's optimal type, dose, and timing. For example, ML can help plan the optimal radiation therapy for patients with gliomas using MRI features to estimate the tumor volume, shape, and growth rate and to tailor the radiation dose and field to the tumor and the surrounding healthy tissue [23]. Some groups have demonstrated that CNNs are able to determine effective radiotherapy treatment contours achieving dice similarity coefficients of nearly 90% [15].

Machine learning methods may help improve efficiency of the radiation therapy workflow through decreasing the need for additional correlative imaging modalities. Electron density images are typically needed for radiation therapy dosimetry, which are typically provided by CT imaging. One study utilized a CNN to construct a pseudo-CT image stack from 3D MRI brain images, with dice coefficients of 0.95, 0.94, and 0.85 when compared to CT images of air, soft tissue, and bone, respectively; no significant difference between planning target volume (PTV) was observed between CT and pseudo-CT [24]. Further studies have expanded on pseudo-CT generation, and demonstrated no significant degradation in dosimetry performance with T1-weighted only or T1-weighted post-contrast only starting image sets [25].

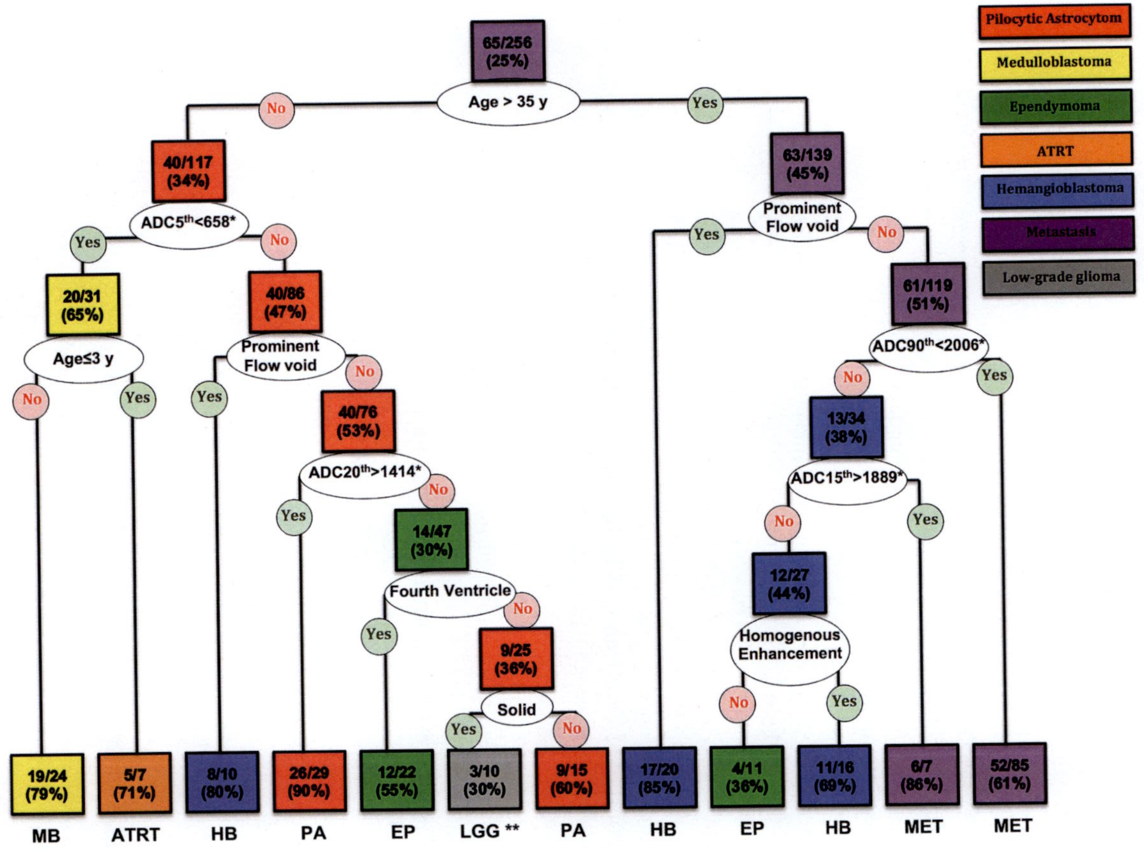

FIGURE 19.2

A Classification and Regression Tree (CART) machine learning model developed to distinguish among seven histopathologic types of posterior fossa brain tumors based on apparent diffusion coefficient (ADC) histogram metrics of solid tumor component, morphological findings on structural MRI, and patient's age. Each node in the decision tree displays the number of correctly classified cases (positive predictive value) as a ratio of the most common tumor type to the total number of cases meeting the criteria at that node. Color coding denotes the predicted tumor type, while color intensity corresponds to classification accuracy – darker shades indicate higher predictive certainty.

Abbreviations: *ATRT*, atypical teratoid/rhabdoid tumor; *EP*, ependymoma; *HB*, hemangioblastoma; *LGG*, low-grade glioma/astrocytoma; *MET*, metastasis; *MB*, medulloblastoma; *PA*, pilocytic astrocytoma.

From Payabvash et al. [5]

19.2.1.4 Assessment of treatment response

Therapy response prediction estimates a treatment's likely outcome and effectiveness for a patient based on their baseline and follow-up MRI features. ML can help predict the therapy response using MRI to assess the degree of tumor shrinkage, necrosis, or recurrence and to evaluate the patient's survival and quality of life. The ability to determine efficacy of a current treatment course early in one's treatment course theoretically allows treating physicians to continue effective treatments or pivot to other treatment options if current regimen is not demonstrating appropriate efficacy. Early work in this field utilized SVM-based approaches to differentiate recurrent metastatic disease from treated lesions. For example, Larroza et al. demonstrated an AUC >0.9 through SVM feature extraction for differentiating active brain metastases from radiation necrosis [26]. Another study used ML techniques to predict the response to chemotherapy or immunotherapy for patients with gliomas using MRI features to measure the tumor response criteria, such as the modified response assessment in neuro-oncology (mRANO) or the immune-related response criteria (irRC) [27].

Prognostication through MRI image analysis is another area of interest within the literature. Efforts are underway to use MRI images to not only diagnose a patient's malignancy, but to also assess overall prognosis and risk of disease recurrence after treatment. One study used a radiomic analysis of 801 patients with IDH wild-type glioblastoma and developed a scoring system using 20 radiomic features, and subsequently demonstrated significance as an independent prognostic factor. This score demonstrated the highest prognostic accuracy when combined with a standard clinical prognosis model [28].

In addition to treatment response, a similar area of interest involves the prediction of early features suggestive of recurrent or metastatic disease, which would allow earlier intervention. One study comparing pre- and post-treatment diffusion weighted imaging (DWI) of brain tumors demonstrated an accuracy of up to 85.8% in detecting early metastatic disease on DWI before manifestations on T1--weighted images, with random forest and gradient boosting methods demonstrating the strongest performance [29]. Another study demonstrates a model that combines temporal input (in the form of prior imaging) with adjustable parameters that allow one to favor sensitivity or positive predictive value (PPV) in the detection of brain metastases, achieving increases in sensitivity from 85.3% to 97.5% and PPV from 69.1% to 98.7% over the base model, with respective parametric adjustments [30].

19.2.2 Stroke imaging

Another major area of neuroimaging is the evaluation of cerebrovascular accidents (CVA), more commonly referred to as stroke. Stroke is a major public health concern, representing the second leading cause of death and a leading cause of long-term disability, costing the global economy roughly $721 billion USD per year [31,32]. Approximately 87% of strokes are ischemic strokes (due to occlusive blood clots within the intracranial vasculature), with the remaining 13%

representing hemorrhagic strokes (bleeding within the brain) [32]. Non-contrast CT imaging of the head is typically performed as an initial screening modality for strokes, and demonstrates high sensitivity for hemorrhagic strokes. CT angiogram or CT perfusion studies may subsequently be performed using intravenous contrast to assess for vascular occlusions (ischemic stroke) and delineate the ischemic core and tissue at risk for infarct. MRI examinations may be helpful for both troubleshooting cases not detected by CT (such as detection of acute stroke with DWI imaging), or follow-up imaging to assess for degree of brain infarction or for monitoring secondary effects, such edema or mass effect. ML can be helpful within many facets of stroke imaging, including diagnosis, differentiation of stroke subtypes, and prognostication.

In stroke imaging, the phrase "time is brain" captures the urgency of medical care—the longer before detection and treatment, the more brain volume (and thus function) will be lost. A strong emphasis both within the literature and in clinical practice is placed on reducing the time to diagnosis. Much work has been placed on detection and segmentation of stroke volumes. One study demonstrated a model that segmented regions of acute stroke on DWI sequences with Dice coefficients of 0.61 for small lesions and 0.83 for larger lesions, with an overall stroke detection rate of 94% [33].

Studies have demonstrated that thrombectomy for large vessel occlusions (LVO) even up to 24 h after time of last known normal may lead to significantly improved patient outcomes [34]; understanding the timeline/chronicity of one's stroke can be helpful in determining treatment possibilities. For cases where exact timing of stroke symptoms is unclear, one study trained a model to predict timing based on mismatch in abnormalities on DWI (near-immediate impact) and FLAIR (abnormalities apparent within ~4 h) sequences, with accuracy matching that of radiologist raters [35]. Another study investigated ML models for estimating time since stroke onset using radiomic feature extraction from DWI sequences and demonstrated an overall accuracy of 78.8% [36].

In diagnosing acute stroke, one key diagnostic dilemma for radiologists involves the differentiation of acute stroke from a variety of conditions that mimic stroke, both in clinical and imaging presentations. Many possible mimicking conditions exist and can be considered, including seizures, hypoglycemia, migraine, peripheral vestibular disease, and brain tumors [37]. To date, no studies have been reported utilizing AI to broadly compare diagnostic evaluation of stroke and stroke mimics based on clinical findings or imaging data, with limited evidence investigating subsets within the broad category of stroke mimics. One such example is evaluation of dizziness, which remains a diagnostic dilemma as there is often a wide overlap between symptoms of benign causes of dizziness with posterior circulation strokes, leading to frequent imaging evaluations for these patients in the emergency department setting [38]. The previous work has demonstrated CNN approaches to lesion segmentation after posterior circulation strokes based on non-contrast head CT imaging [39]. Studies have demonstrated that specially designed MRI protocols for posterior fossa stroke evaluation can demonstrate high sensitivity for posterior circulation strokes in an ED setting [40].

Fewer studies to date investigate the role for AI evaluation of MRI imaging in posterior circulation strokes. One such example trained existing CNN models to classify acute strokes based on involved vascular territory from DWI images [41].

Prognostication in stroke patient care is important, as understanding the expected degree of disability from one's stroke may help with treatment plan development and continued care coordination. ML may help in the assessment of patient prognosis after stroke. One study demonstrated a combined model utilizing clinical parameters and a CNN using acute-period DWI sequences trained to predict modified Rankin Scale (mRS) at 90 days post-stroke and achieved an AUC of 0.92, with 79% accuracy within +/− 1 mRS category [42].

Another area of interest in the ML literature is synthetic image generation. One study demonstrated the ability to create synthetic time-of-flight (TOF) magnetic resonance angiogram (MRA) from time-resolved MRA sequences; while image quality was worse in synthetic images for distal middle cerebral, basilar and posterior cerebral artery circulations, radiologists were not able to distinguish synthetic from true TOF images [43]. Synthetic MRI sequence production with machine learning-based reconstruction of image quality can accelerate acquisition times and improve image quality, with applications in stroke and other domains of neuroimaging [44].

Prior work demonstrates that rapid brain MRI protocols are preferable to CT for pediatric stroke evaluations [45], which is now standard of care within many institutions. Despite a strong clinical preference for pediatric stroke evaluation with MRI, it remains an area that has not received much attention in the literature [46]. One such study investigates a range of multimodal biomarkers, including MRI imaging, fMRI and tractography to predict functional outcomes after perinatal stroke, finding that tractographic evaluation of brain connectivity can help to longer-term predict clinical outcomes [47]. To date, no further literature is available investigating applications of AI analysis methods for diagnosis or prognostication of pediatric or neonatal stroke care.

More recently, numerous multimodal AI models have become commercially available for real-time use in PACS systems for detection and prognostication of stroke cases. This includes prognostic evaluations such as automatic ASPECTS score evaluation to detect degree of brain involvement in middle cerebral artery (MCA) strokes, detection of LVO from CTA or MRI datasets, automation of CT perfusion data, and detection of intracranial hemorrhage [46]. Multiple studies have compared performance of commercially available automated CT Perfusion algorithms, demonstrating evolving performance over time as models are updated [48–50]. Another study compares ASPECTS scoring algorithms from two commercial vendors to junior and senior radiology staff, finding overall similar performance of both commercial products, with near equivalent performance to junior radiologists and slightly inferior performance relative to more senior radiologists [51]. Ongoing work is needed to understand the clinical impact of these commercially available systems, including comparison data and evaluations of long-clinical impact of such models.

19.2.3 Dementia imaging

Much like the analysis of brain neoplasms, the analysis of dementia imaging relies on a range of processes, including segmentation of normal brain structures, volumetric analysis, and detection/differentiation of different dementia sub-types, with some additional value provided in prognostic and treatment response evaluation. ML demonstrates promise for multiple facets of dementia imaging, including both volumetric analysis and differentiation of dementia types. An early study utilized an SVM-based approach demonstrated an overall accuracy of 94.5% (sensitivity of 91.5% and specificity of 96.6%) in differentiating patients with Alzheimer's dementia from normal elderly control patients based on MRI imaging of the brain [52]. A CNN-based pipeline approach subsequently demonstrated an AUC of up to 99.6% in differentiating patients with Alzheimer's dementia from healthy controls across three previously published MRI image datasets [53]. A separate study combines image analysis with a CNN and longitudinal analysis of multiple T1-weighted MRI examinations over time through a recurrent neural network (RNN), and demonstrates an accuracy of 91.3% for differentiating Alzheimer's dementia from normal controls and 71.7% accuracy for differentiating stable from progressive mild cognitive impairment [54]. An SVM-based weight map for differentiating multiple forms of dementia was shown to improve accuracy of radiologist classification of dementia types, particularly among junior radiologists [55].

Machine learning methods have demonstrated utility in prognostication from MRI images. CNN approaches have also been used in earlier-stage patients to evaluate imaging features that predict progression of cognitive impairment, with accuracy between 95.4% and 97.0% in evaluating risk of disease progression [56]. Additional work with neural networks has demonstrated improving accuracy and specificity with the addition of additional clinical data, such as cognitive testing results, biomarkers, or genetic data [57].

Another area of interest in the literature concerns brain aging, or the analysis of imaging findings of the brain as patient's age that may allow for early prediction and intervention in cases where dementia may likely develop. Early work used an ensemble of 21 neural networks to assess a range of serum biomarkers and assign a biologic age, demonstrating an accuracy coefficient of 90% for matching to chronological age within 10 years [35,58]. Neuroimaging provides a strong variable in the assessment of biologic age, mainly in the form of brain MRI examinations. One study used CNNs to evaluate biological age of a patient's brain based on MRI imaging, and demonstrated the ability to predict developing cases of Alzheimer dementia based on advancing biologic age of patient's brains relative to chronologic age [59]. Another study combined brain MRI segmentation analysis with DNA methylation data from epigenetic analyses through deep learning approaches, demonstrating a strong correlation between biologic age (assessed through imaging and epigenetic data) and future mortality risk [60]. Others have demonstrated strong performance in assessment of brain age through deep learning analyses using white matter tract morphology on MRI as a measure of biological aging [61]. As advances in treatment modalities for early Alzheimer dementia lead to clinical improvement [62], such models may play a critical role in identifying patients for whom such interventions may improve clinical outcomes. Clinical applications of such models for determining patient in need of intervention remains an area in need of further investigation.

19.2.4 Demyelinating diseases

ML assessment of demyelinating diseases is another major area of research focus, with the majority of studies focusing on applications within evaluation of multiple sclerosis (MS). A study utilizing a random forest approach demonstrated strong performance in differentiating healthy controls and clinically isolated syndrome (CIS - the first clinical instance within the MS spectrum) from relapsing remitting (RR) and secondary progressive (SP) forms of MS (AUC of healthy controls vs. RR 0.82 and vs. SP 0.99; CIS vs. RR 0.83 and vs. SP 0.95), with relatively poorer performance of healthy controls versus CIS (AUC 0.79) [63]. An unsupervised learning study classified MRI images of MS into three distinct imaging appearances (cortex-led, normal-appearing white matter-led, and lesion-led), with patients demonstrating the lesion led morphology showing the highest rates of disability progression, as well as clinical improvement in some clinical trials [64]. Additional studies have assessed deep learning approaches for plaque segmentation (including number and volume of lesions) in MS for disease monitoring and surveillance of disease progression [65].

While multiple sclerosis is the most common form of chronic demyelinating disease, other disease processes may demonstrate similar clinical presentations and imaging appearances, hindering diagnosis, and treatment. A CNN-based approach for differentiating MS from neuromyelitis optica spectrum disorders (NMOSD) demonstrated an AUC of 0.859 with an overall accuracy of 81.1% [66]. Similarly, a radiomics model differentiated MS from neuropsychiatric systemic lupus erythematosus with an AUC of 0.967 and an accuracy of 89.2% [67].

19.3 Head and neck imaging

Relatively less research has developed ML applications for MRI imaging of the head and neck, which may be in part due to the lower volume of MRI imaging for the head and neck compared to the central nervous system. Many studies demonstrating ML approaches for imaging of the head and neck focus on the evaluation of head and neck malignancies, most often squamous cell carcinoma of the head and neck. Studies have often followed similar approaches to evaluation of the many facets of cancer imaging and evaluation relative to brain neoplasms as outlined above. A few demonstrative studies are summarized below.

Much like with intracranial neoplasms, neural networks have shown promise for the segmentation and evaluation of head and neck cancers. For example, a neural network-based approach evaluating recurrence of nasopharyngeal carcinoma demonstrated as overall accuracy of 84.1% compared to expert radiologists' accuracy of 85.9%, with radiologists achieving improved accuracy 87.7% with access to neural network model outputs [68]. A study comparing CNN (DeepLab V3+ and U-net) tumor segmentation of hypopharyngeal tumors to manual segmentation demonstrated high correlation, with a higher dice similarity coefficient with the DeepLab V3+ model, particularly with smaller volume tumors (0.74 vs. 0.70) [69]. Similarly, a segmentation study using a variety of ML approaches (including SVM, K-nearest neighbors, and ensemble classifiers) predicted pathologic stage of oral tongue squamous cell carcinoma with an overall accuracy of 0.86 [70].

Numerous studies have reported on radiomic analysis for the evaluation of imaging features of head and neck cancers. A study investigating radiomic features of nasopharyngeal carcinoma recurrence demonstrated an AUC of 0.727–0.835 for a set of eight radiomic features used to differentiate locally recurrent versus non-progressive disease, with a neural network approach demonstrating an accuracy of 0.812 in classifying progressive disease (Fig. 19.3) [71]. While radiomic feature extraction may not translate directly to image interpretation by radiologists, such models may demonstrate clinical utility in cancer evaluation and prognostication settings.

19.4 Spinal imaging

Numerous machine learning approaches have been explored to assist with the imaging evaluation of spinal pathology. Many of the published ML approaches for spinal imaging mirror the most common indication for spinal MRI: evaluation of degenerative disease and disc pathologies. For example, a model called BianqueNet was developed using a deep learning approach (based on DeepLab V3+) demonstrated strong performance in evaluation and quantification of intervertebral degenerative disease. A two-part CNN-based approach including tumor segmentation and classification models demonstrated strong agreement with expert radiologists on severity ratings for central canal, lateral recess, and neuroforaminal stenosis, with kappa statistic $\geq$0.89 across all stenosis types and severity categories [72]. A subsequent study using this model demonstrated a decrease in interpretation time of spinal MRI studies by 76 s for MSK radiologists and 156 s for general radiologists, with an improved kappa statistic for studies interpreted with model assistance (0.71 vs. 0.39 without model) [73]. A meta-analysis investigating a variety of ML approaches for the evaluation of disc pathologies, including disc herniation, degeneration, and Modic changes, demonstrated similar overall performance of deep learning models compared to other methodologies, including SVM, K-nearest neighbors and random forest [74].

Additional machine learning studies have been performed for evaluation of spinal cord injuries. A model trained to evaluated cervical spinal cord compression demonstrated an area under the curve of 0.94 (sensitivity of 0.88 and specificity of 0.89) [75]. A review of articles using ML to assess spinal cord injuries in 2022 summarized 13 prior studies utilizing a range of ML approaches (including random forest, SVM, gradient boosting, K-nearest neighbors, and CNNs), including diagnostic and prognostication/management models, and found the highest performance among SVM models [76].

Finally, multiple studies have investigated the use of deep learning upscaling approaches for image completion using limited k-space sampling (shorter image acquisition times). One study demonstrated overall improved image quality with similar radiologist interpretations for deep learning reconstructed 3D T2-weigthed fast spin echo (T2W FSE) images of the lumbar spine compared to conventionally acquired 3D and 2D T2W FSE sequences, with significantly reduced acquisition time [77]. An additional study demonstrated an overall reduction in scan time for lumbar spine MRI exams of 32.3% through ML-acceleration, with improved signal-to-noise ratio for T1 and T2 weighted images and improved contrast-to-noise ratio for T1 weighted images compared to a conventional acquisition protocol [78]. A deep learning-accelerated lumbar spine MRI protocol including sagittal T1-, T2-weighted, Short TI inversion recovery (STIR), and axial T2-weighted images was performed within 1 min and 53 s versus 12 min and 31 s (conventional MRI protocol), and demonstrated equivalent findings for all major spinal pathologies except foraminal stenosis and disc contour irregularities [79].

19.5 Challenges and pitfalls

ML in MRI has shown promising results and potential benefits for improving MRI's quality, efficiency, and effectiveness in clinical settings. However, ML in MRI also faces some challenges and pitfalls that must be addressed before it can be widely adopted in clinical practice.

19.5.1 Data quality and ethics

The performance of ML models depends mainly on the quality, quantity, and diversity of the data used for training, validation, and testing. Poor quality data may lead to inaccurate or biased predictions or decisions by the ML model. For example, if the data contains noise, artifacts, or missing values, the ML model may learn incorrect patterns or relationships from the images used for interpretation. More balanced data must be used to avoid overfitting or underfitting by the ML model. Lack of data diversity may lead to poor transferability of the ML model across different scanners, protocols, or populations [80].

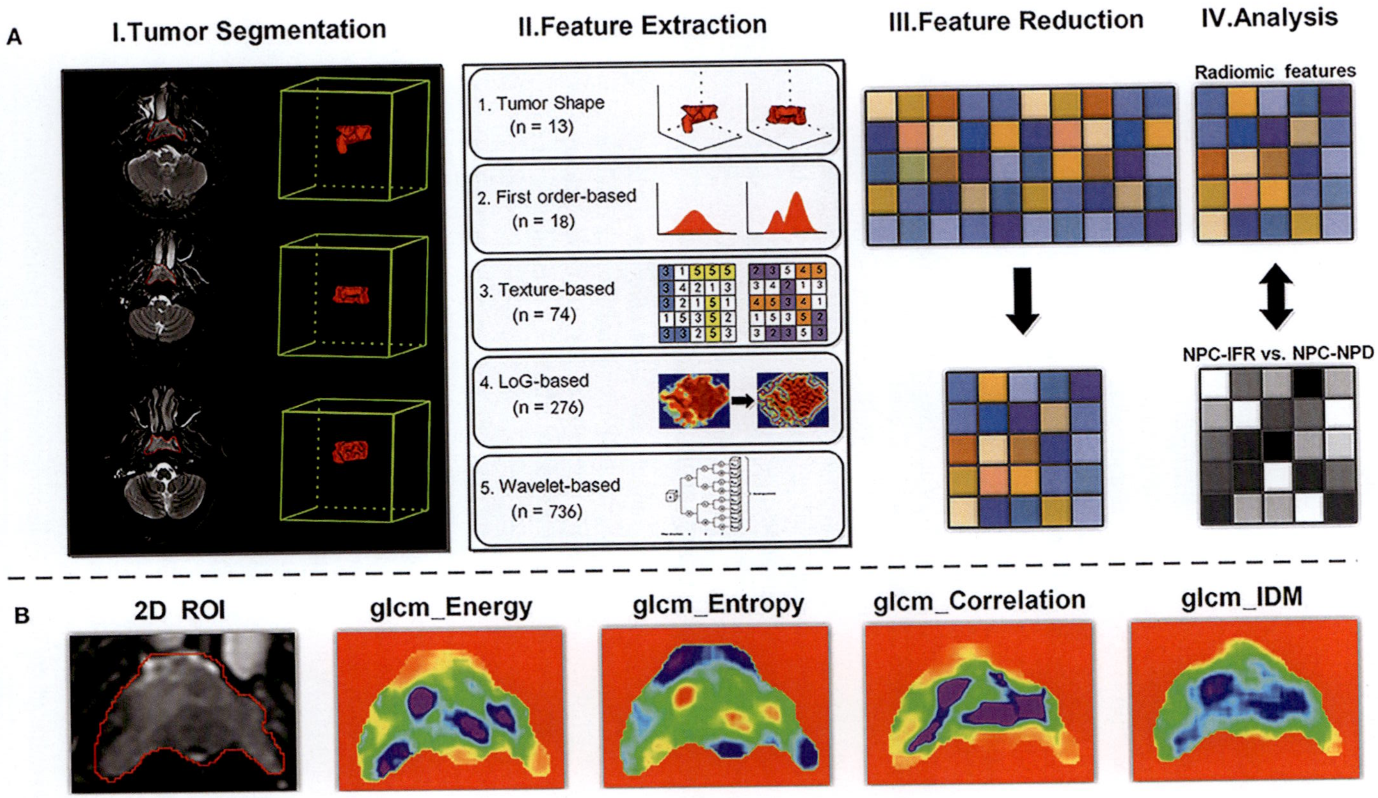

FIGURE 19.3

An overview of workflow for machine learning MRI radiomics model to predict nasopharyngeal carcinoma recurrence after intensity-modulated radiotherapy. (A) The radiomics model differentiates in-field recurrence (NPC-IFR) from cases of non-progression disease (NPC-NPD). The pipeline includes (I) manual tumor segmentation was performed on T2-weighted MRI, (II) quantitative radiomic features—including shape, first-order intensity, texture, Laplacian of Gaussian (LoG), and wavelet-based features—were extraction from the segmented tumor regions, (III) a Principal Component Analysis (PCA) for dimensionality reduction, followed by (IV) supervised machine learning algorithms—that is, Artificial neural network (ANN), k-nearest neighbor (KNN), and support vector machine (SVM)—to classify cases as NPC-IFR or NPC-NPD. (B) Representative radiomic feature maps are shown, derived from two-dimensional tumor regions using the Gray-Level Co-occurrence Matrix (GLCM) method. Examples include maps of Energy, Entropy, Correlation, and Inverse Difference Moment (IDM), which illustrate the spatial texture patterns quantified during the analysis.

From Li et al. [71]

Therefore, it is essential to ensure that the data used for ML are reliable, representative, and relevant. Data preprocessing, augmentation, normalization, and standardization are some techniques that can improve data quality. Data preprocessing is the process of cleaning, transforming, and integrating the data before feeding it to the ML model. For example, data preprocessing can involve removing noise, artifacts, or outliers [81].

Data augmentation creates new data from existing data by applying transformations, such as flipping, rotating, cropping, or adding noise. Data augmentation can increase the quantity and diversity of the data and reduce overfitting [82]. Data normalization is scaling the data to a standard range, between 0 and 1 or -1 and 1. Data normalization can reduce the influence of different units and scales on the ML model. Data standardization is transforming the data to have zero mean and unit variance. Data standardization can reduce the influence of outliers and skewed distributions on the ML model [83].

Data privacy, security, and ethics are additional issues that must be considered when dealing with sensitive medical data. Patients must be aware and in control of how their data is collected, used, and shared. Data must be protected from unauthorized access, use, or disclosure. It is essential to respect the rights and interests of data subjects, ensure the trust and confidence of data users, and comply with applicable laws and regulations of data governance.

19.5.2 Making the "black box" transparent

Many ML models, especially deep learning models, are often considered "black boxes" that are difficult or impossible to interpret by patients and clinicians alike. Their internal workings are unknown or hidden from the user. Despite high accuracy, poor transparency of ML applications can limit trustworthiness, acceptability, and accountability in clinical settings.

If a deep learning model makes a diagnosis or recommendation based on MRI data, the clinicians and patients may want to know how and why the model made that decision and whether they can trust or accept it. Therefore, developing methods to enhance ML models' interpretability is vital. Attention mechanisms, feature attribution, and counterfactual analysis can improve model interpretability. Similarly, radiologists may benefit from "heat maps" provided by ML models to demonstrate the regions of a given image most associated with the result it calls, such as highlighting the area the model believes to represent a brain tumor.

Attention mechanisms are a neural network layer that can learn to focus on or attend to specific parts of the data or the model. Attention mechanisms can help users understand what the model is paying attention to or ignoring when making predictions or decisions [84]. For example, attention mechanisms can show which parts of the MRI image are more influential or informative for the model.

Feature attribution is the technique of assigning or attributing a score or value to each feature or input variable of the model, indicating how much it contributes to or influences the output or prediction of the model. Feature attribution can help users understand which features or input variables are more important or relevant for the model [85]. For example, feature attribution can show which MRI image pixels or regions positively or negatively impact the model's diagnosis or recommendation.

19.5.3 Clinical implementation

While there are many studies investigating possible ML applications within neuroimaging, few of these models ever reach clinical practice. Many ML models are currently available clinically, though many more never leave the research stage. This may be in part secondary to regulatory approval requirements and the complexity of developing a robust system that is broadly applicable within a range of radiology informatics systems. Many of these models are also computationally expensive, and likely require resources not available in most practice settings; implementation would require bolstering of *on-prem* server computational headroom or sending data to third-party servers (which many departments limit due to data security concerns). In addition, model implementation requires significant efforts and expertise on the part of a practice's information technology staff, making implementation nearly insurmountable for many institutions without the requisite technical capabilities. In parallel with the groundbreaking research work described above, significant work is needed to facilitate the process of clinical implementation of said work before patients may experience the true benefits of ML within neuroradiology.

References

[1] Smith-Bindman R, et al. Trends in use of medical imaging in US health care systems and in Ontario, Canada, 2000-2016. JAMA 2019;322(9):843–56.

[2] Duong MT, Rauschecker AM, Mohan S. Diverse applications of artificial intelligence in neuroradiology. Neuroimaging Clin N Am 2020;30(4):505–16.

[3] Amin J, et al. Detection of brain tumor based on features fusion and machine learning. J Ambient Intell Hum Comput 2018.

[4] Blumenthal DT, et al. Classification of high-grade glioma into tumor and nontumor components using support vector machine. AJNR Am J Neuroradiol 2017;38(5): 908–14.

[5] Payabvash S, et al. Machine learning decision tree models for differentiation of posterior fossa tumors using diffusion histogram analysis and structural MRI findings. Front Oncol 2020;10:71. https://doi.org/10.3389/fonc.2020.00071.

[6] Biratu ES, et al. A survey of brain tumor segmentation and classification algorithms. J Imaging 2021;7(9).

[7] Paul JS, et al. Deep learning for brain tumor classification. In: Medical Imaging 2017: Biomedical Applications in Molecular, Structural, and Functional Imaging. SPIE; 2017.

[8] Gao P, et al. Development and validation of a deep learning model for brain tumor diagnosis and classification using magnetic resonance imaging. JAMA Netw Open 2022;5 (8):e2225608.

[9] Ronneberger O, Fischer P, Brox T. U-net: convolutional networks for biomedical image segmentation. Cham: Springer International Publishing; 2015.

[10] Zhang Y, Han Y, Zhang J. MAU-Net: mixed attention U-Net for MRI brain tumor segmentation. Math Biosci Eng 2023;20(12):20510–27.

[11] Ghosh S, Chaki A, Santosh KC. Improved U-Net architecture with VGG-16 for brain tumor segmentation. Phys Eng Sci Med 2021;44(3):703–12.

[12] Li P, et al. mResU-Net: multi-scale residual U-Net-based brain tumor segmentation from multimodal MRI. Med Biol Eng Comput 2023.

[13] Baid U, et al. The rsna-asnr-miccai brats 2021 benchmark on brain tumor segmentation and radiogenomic classification. arXiv preprint arXiv:2107.02314 2021.

[14] Futrega M, et al. Optimized U-Net for brain tumor segmentation. In: International MICCAI Brainlesion Workshop. Springer; 2021.

[15] Kocher M, et al. Applications of radiomics and machine learning for radiotherapy of malignant brain tumors. Strahlenther Onkol 2020;196(10):856–67.

[16] Merkaj S, et al. Machine learning tools for image-based glioma grading and the quality of their reporting: challenges and opportunities. Cancers (Basel) 2022;14(11).

[17] Payabvash S, Tihan T, Cha S. Volumetric voxelwise apparent diffusion coefficient histogram analysis for differentiation of the fourth ventricular tumors. NeuroRadiol J 2018;31(6):554–64.

[18] Payabvash S, Tihan T, Cha S. Differentiation of cerebellar hemisphere tumors: combining apparent diffusion coefficient histogram analysis and structural MRI features. J Neuroimaging 2018;28(6):656–65.

[19] Ullah N, et al. TumorDetNet: a unified deep learning model for brain tumor detection and classification. PLoS One 2023;18(9):e0291200.

[20] Huang B, et al. Deep learning-based prediction of H3K27M alteration in diffuse midline gliomas based on whole-brain MRI. Cancer Med 2023;12(16):17139–48.

[21] Gultekin MA, et al. Differentiation of lung and breast cancer brain metastases: comparison of texture analysis and deep convolutional neural networks. J Clin Ultrasound 2023;51(9):1579–86.

[22] Meißner AK, et al. Radiomics for the non-invasive prediction of PD-L1 expression in patients with brain metastases secondary to non-small cell lung cancer. J Neuro Oncol 2023;163(3):597–605.

[23] Peeken JC, et al. Deep learning derived tumor infiltration maps for personalized target definition in Glioblastoma radiotherapy. Radiother Oncol 2019;138:166–72.

[24] Liu F, et al. MR-based treatment planning in radiation therapy using a deep learning approach. J Appl Clin Med Phys 2019;20(3):105–14.

[25] Alvarez Andres E, et al. Dosimetry-driven quality measure of brain pseudo computed tomography generated from deep learning for MRI-only radiation therapy treatment planning. Int J Radiat Oncol Biol Phys 2020;108(3):813–23.

[26] Larroza A, et al. Support vector machine classification of brain metastasis and radiation necrosis based on texture analysis in MRI. J Magn Reson Imaging 2015;42(5):1362–8.

[27] Zhang H, et al. Machine learning-based tumor-infiltrating immune cell-associated lncRNAs for predicting prognosis and immunotherapy response in patients with glioblastoma. Briefings Bioinf 2022;23(6).

[28] Guan F, et al. Biological underpinnings of radiomic magnetic resonance imaging phenotypes for risk stratification in IDH wild-type glioblastoma. J Transl Med 2023;21(1):841.

[29] Madamesila J, et al. Can machine learning models improve early detection of brain metastases using diffusion weighted imaging-based radiomics? Quant Imaging Med Surg 2023;13(12):7706–18.

[30] Huang Y, et al. Deep learning for brain metastasis detection and segmentation in longitudinal MRI data. Med Phys 2022;49(9):5773–86.

[31] Feigin VL, et al. World stroke organization (WSO): global stroke fact sheet 2022. Int J Stroke 2022;17(1):18–29.

[32] Tsao CW, et al. Heart disease and stroke statistics-2023 update: a report from the American Heart Association. Circulation 2023;147(8):e93–621.

[33] Chen L, Bentley P, Rueckert D. Fully automatic acute ischemic lesion segmentation in DWI using convolutional neural networks. Neuroimage Clin 2017;15:633–43.

[34] Jadhav AP, Desai SM, Jovin TG. Indications for mechanical thrombectomy for acute ischemic stroke: current guidelines and beyond. Neurology 2021;97(20 Suppl. 2): S126–s136.

[35] Akay EMZ, et al. A deep learning analysis of stroke onset time prediction and comparison to DWI-FLAIR mismatch. Neuroimage Clin 2023;40:103544.

[36] Zhang YQ, et al. MRI radiomic features-based machine learning approach to classify ischemic stroke onset time. J Neurol 2022;269(1):350–60.

[37] Pohl M, et al. Ischemic stroke mimics: a comprehensive review. J Clin Neurosci 2021; 93:174–82.

[38] Tu LH, et al. Cost-effectiveness of CT, CTA, MRI, and specialized MRI for evaluation of patients presenting to the emergency department with dizziness. AJR Am J Roentgenol 2024;222(2):e2330060.

[39] Zoetmulder R, et al. Automated final lesion segmentation in posterior circulation acute ischemic stroke using deep learning. Diagnostics 2021;11(9).

[40] Tu LH, et al. Pilot MRI-based strategies to improve the detection of stroke in patients with dizziness/vertigo. Clin Imaging 2022;82:234–6.

[41] Koska IO, et al. Deep learning classification of ischemic stroke territory on diffusion-weighted MRI: added value of augmenting the input with image transformations. J Imaging Inform Med 2024.

[42] Liu Y, et al. Functional outcome prediction in acute ischemic stroke using a fused imaging and clinical deep learning model. Stroke 2023;54(9):2316–27.

[43] You SH, et al. Deep learning-based synthetic TOF-MRA generation using time-resolved MRA in fast stroke imaging. AJNR Am J Neuroradiol 2023;44(12):1391–8.

[44] Tanenbaum LN, et al. Synthetic MRI for clinical neuroimaging: results of the magnetic resonance image compilation (MAGiC) prospective, multicenter, multireader trial. Am J Neuroradiol 2017;38(6):1103–10.

[45] Christy A, Murchison C, Wilson JL. Quick brain magnetic resonance imaging with diffusion-weighted imaging as a first imaging modality in pediatric stroke. Pediatr Neurol 2018;78:55–60.

[46] Pacchiano F, et al. Artificial intelligence applied in acute ischemic stroke: from child to elderly. Radiol Med 2024;129(1):83–92.

[47] Carlson HL, et al. Structural and functional connectivity of motor circuits after perinatal stroke: a machine learning study. Neuroimage Clin 2020;28:102508.

[48] Xiong Y, et al. Comparison of automated CT perfusion softwares in evaluation of acute ischemic stroke. J Stroke Cerebrovasc Dis 2019;28(12):104392.

[49] Koopman MS, et al. Comparison of three commonly used CT perfusion software packages in patients with acute ischemic stroke. J Neurointerv Surg 2019;11(12):1249–56.

[50] Mallon DH, et al. Comparison of automated ASPECTS, large vessel occlusion detection and CTP analysis provided by Brainomix and RapidAI in patients with suspected ischaemic stroke. J Stroke Cerebrovasc Dis 2022;31(10):106702.

[51] Li X, et al. Automated ASPECTS in acute ischemic stroke: comparison of the overall scores and Hounsfield unit values of two software packages and radiologists with different levels of experience. Acta Radiol 2023;64(1):328–35.

[52] Magnin B, et al. Support vector machine-based classification of Alzheimer's disease from whole-brain anatomical MRI. Neuroradiology 2009;51(2):73–83.

[53] Zhang Q, Du Q, Liu G. A whole-process interpretable and multi-modal deep reinforcement learning for diagnosis and analysis of Alzheimer's disease(*). J Neural Eng 2021; 18(6).

[54] Cui R, Liu M. RNN-based longitudinal analysis for diagnosis of Alzheimer's disease. Comput Med Imaging Graph 2019;73:1–10.

[55] Chagué P, et al. Radiological classification of dementia from anatomical MRI assisted by machine learning-derived maps. J Neuroradiol 2021;48(6):412–8.

[56] Wu C, et al. Discrimination and conversion prediction of mild cognitive impairment using convolutional neural networks. Quant Imaging Med Surg 2018;8(10):992–1003.

[57] Ning K, et al. Classifying Alzheimer's disease with brain imaging and genetic data using a neural network framework. Neurobiol Aging 2018;68:151–8.

[58] Putin E, et al. Deep biomarkers of human aging: application of deep neural networks to biomarker development. Aging (Albany NY) 2016;8(5):1021–33.

[59] Armanious K, et al. Age-net: an MRI-based iterative framework for brain biological age estimation. IEEE Trans Med Imag 2021;40(7):1778–91.

[60] Cole JH, et al. Brain age predicts mortality. Mol Psychiatr 2018;23(5):1385–92.

[61] Salih A, et al. Brain age estimation at tract group level and its association with daily life measures, cardiac risk factors and genetic variants. Sci Rep 2021;11(1):20563.

[62] Cohen S, et al. Lecanemab clarity AD: quality-of-life results from a randomized, double-blind phase 3 trial in early alzheimer's disease. J Prev Alzheimers Dis 2023; 10(4):771–7.

[63] Ricciardi A, et al. Patterns of inflammation, microstructural alterations, and sodium accumulation define multiple sclerosis subtypes after 15 years from onset. Front Neuroinform 2023;17:1060511.

[64] Eshaghi A, et al. Identifying multiple sclerosis subtypes using unsupervised machine learning and MRI data. Nat Commun 2021;12(1).

[65] Moazami F, et al. Machine learning approaches in study of multiple sclerosis disease through magnetic resonance images. Front Immunol 2021;12:700582.

[66] Hagiwara A, et al. Differentiation between multiple sclerosis and neuromyelitis optica spectrum disorders by multiparametric quantitative MRI using convolutional neural network. J Clin Neurosci 2021;87:55–8.

[67] Luo X, et al. Multi-lesion radiomics model for discrimination of relapsing-remitting multiple sclerosis and neuropsychiatric systemic lupus erythematosus. Eur Radiol 2022;32(8):5700–10.

[68] Deng Y, et al. Deep learning-based recurrence detector on magnetic resonance scans in nasopharyngeal carcinoma: a multicenter study. Eur J Radiol 2023;168:111084.

[69] Lin YC, et al. Fully automated segmentation and radiomics feature extraction of hypopharyngeal cancer on MRI using deep learning. Eur Radiol 2023;33(9):6548–56.

[70] Vidiri A, et al. Magnetic resonance imaging-based prediction models for tumor stage and cervical lymph node metastasis of tongue squamous cell carcinoma. Comput Struct Biotechnol J 2023;21:4277–87.

[71] Li S, et al. Use of radiomics combined with machine learning method in the recurrence patterns after intensity-modulated radiotherapy for nasopharyngeal carcinoma: a preliminary study. Front Oncol 2018;8:648. https://doi.org/10.3389/fonc.2018.00648.

[72] Hallinan J, et al. Deep learning model for automated detection and classification of central canal, lateral recess, and neural foraminal stenosis at lumbar spine MRI. Radiology 2021;300(1):130–8.

[73] Lim DSW, et al. Improved productivity using deep learning-assisted reporting for lumbar spine MRI. Radiology 2022;305(1):160–6.

[74] Compte R, et al. Are current machine learning applications comparable to radiologist classification of degenerate and herniated discs and Modic change? A systematic review and meta-analysis. Eur Spine J 2023;32(11):3764–87.

[75] Merali Z, et al. A deep learning model for detection of cervical spinal cord compression in MRI scans. Sci Rep 2021;11(1):10473.

[76] Dietz N, et al. Machine learning in clinical diagnosis, prognostication, and management of acute traumatic spinal cord injury (SCI): a systematic review. J Clin Orthop Trauma 2022;35:102046.

[77] Sun S, et al. Evaluation of deep learning reconstructed high-resolution 3D lumbar spine MRI. Eur Radiol 2022;32(9):6167–77.

[78] Yoo H, et al. Deep learning-based reconstruction for acceleration of lumbar spine MRI: a prospective comparison with standard MRI. Eur Radiol 2023;33(12):8656–68.

[79] Fujiwara M, et al. Ultrafast lumbar spine MRI protocol using deep learning–based reconstruction: diagnostic equivalence to a conventional protocol. Skelet Radiol 2023;52(2):233–41.

[80] Meyer MI, et al. A contrast augmentation approach to improve multi-scanner generalization in MRI. Front Neurosci 2021;15:708196.

[81] Chaki J. Diagnosis of neurological disorders based on deep learning techniques. CRC Press; 2023.

[82] Mikołajczyk A, Grochowski M. Data augmentation for improving deep learning in image classification problem. In: 2018 International Interdisciplinary PhD Workshop. IIPhDW); 2018.

[83] Muhammad Ali P, Faraj R. Data normalization and standardization: a technical report. 2014.

[84] Soydaner D. Attention mechanism in neural networks: where it comes and where it goes. Neural Comput Appl 2022;34(16):13371–85.

[85] Zhou Y, et al. Do feature attribution methods correctly attribute features?. In: Proceedings of the AAAI Conference on Artificial Intelligence; 2022.

CHAPTER

AI in cardiovascular MRI 20

O. Burak Demirel[1,2], Kelly K. Horst[1] and Tim Leiner[1,*]

[1]*Department of Radiology, Mayo Clinic Rochester, Rochester, MN, United States;* [2]*Philips Healthcare, MR Clinical Science, Philips North America, Cambridge, MA, United States*

[*]*Corresponding author: E-mail: Tim@mayo.edu*

20.1 Introduction

Cardiovascular magnetic resonance (CMR) imaging is a cornerstone of modern cardiology, offering unparalleled insights into cardiac structure, function, and tissue characteristics [1,2]. Its applications range from assessing ventricular function and myocardial viability to evaluating ischemia and fibrosis [3]. Despite its clinical importance, CMR is often time-intensive, operator-dependent, and challenging to standardize, particularly in resource-limited settings. Integrating Artificial Intelligence (AI) into CMR workflows promises to overcome these limitations, transforming how clinicians acquire, process, and interpret cardiac imaging data.

AI enables automated, efficient, and accurate analysis of complex imaging datasets. From reconstructing high-quality images from undersampled data to automating segmentation and extracting quantitative biomarkers, AI enhances every step of the CMR pipeline [4,5]. Leveraging architectures such as convolutional neural networks (CNNs), AI accelerates routine tasks and provides new opportunities for precision diagnostics and personalized treatment planning.

This review explores the fundamentals of AI, its architectures, and learning strategies and delves into its transformative applications in CMR. Special attention is given to image acquisition, reconstruction, segmentation, and quantitative analysis advancements. While the potential of AI in CMR is immense, challenges such as data availability, standardization, and clinical validation remain barriers to widespread adoption. By addressing these hurdles, AI has the potential to redefine the role of CMR in cardiovascular care, bridging gaps in accessibility and improving patient outcomes.

20.2 Fundamentals of AI

Artificial intelligence (AI) empowers machines to learn from experience, adapt to new inputs, and execute tasks that mimic human intelligence. This process involves

Machine Learning in MRI. https://doi.org/10.1016/B978-0-443-14109-6.00020-1

using large datasets to identify patterns or relationships of interest and generalizing these relationships to new, unseen data [6]. For example, determining the left ventricular (LV) ejection fraction (LVEF) typically requires manual contouring of the LV cavity and myocardium across multiple slices and cardiac phases. AI leverages a database of these pre-drawn "ground truth" segmentations to optimize a mathematical model, enabling it to predict segmentations for future acquisitions.

While AI broadly refers to machines performing tasks characteristic of human intelligence—such as reasoning, deduction, or induction—machine learning (ML) is a subset of AI focused on using mathematical models and training data to make predictions [7]. Within ML, neural networks serve as a foundational technique, and deep learning expands on this by utilizing complex neural network architectures to train models on large-scale data.

An AI system typically consists of two primary components: the architecture and the learning strategy. The architecture determines the type of deep learning techniques used, such as convolutional neural networks (CNNs), which are commonly applied to image-based tasks. The learning strategy, on the other hand, dictates how the AI system is trained. For instance, supervised learning involves using labeled data, where each input is paired with a corresponding ground truth, to guide the training process. A schematic of a deep learning training process is shown in Fig. 20.1.

20.2.1 Deep learning architectures

Several deep learning architectures are utilized in cardiac MRI, including convolutional neural networks (CNNs), recurrent neural networks (RNNs), generative

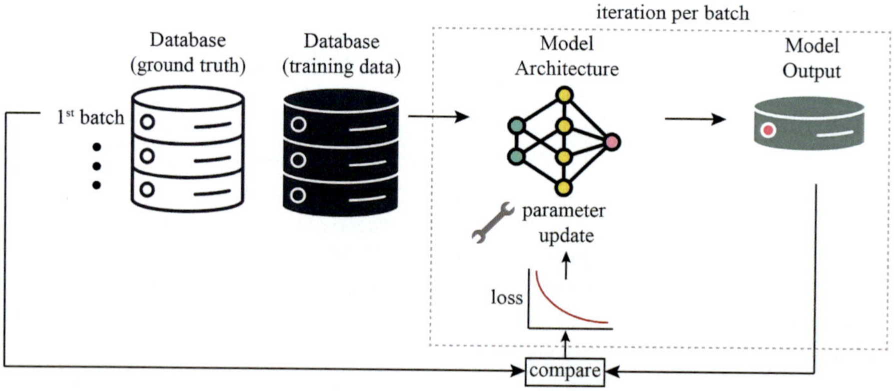

FIGURE 20.1

Deep learning training schematic. The process begins with a dataset containing ground truth and training data divided into batches for processing. Each batch is fed into the model architecture, where predictions (model outputs) are generated. These outputs are compared to the ground truth, and a loss value is computed to quantify the error. This loss guides the parameter updates (e.g., weights in the network) through an optimization algorithm, iteratively improving the model's performance with each batch. The process continues until the model achieves the desired accuracy or a predefined stopping criterion is met.

adversarial networks (GANs), transformers, and diffusion models. The choice of architecture depends on various factors such as the specific task (e.g., image reconstruction, segmentation, or predictive modeling), the size and structure of the dataset (e.g., static images such as late gadolinium enhancement (LGE) or time series such as cine imaging), computational requirements, and the ease of integration into clinical workflows. Different deep-learning architectures are depicted in Fig. 20.2.

20.2.1.1 Convolutional neural networks (CNNs)

Convolutional neural networks (CNNs) are among the most widely used architectures in cardiac MRI, excelling at tasks such as segmentation, classification, and image reconstruction [8]. They employ convolutional layers to extract features, pooling layers to reduce spatial dimensions, and fully connected layers to produce outputs for classification or regression. Common architectures include ResNet, which uses skip connections to improve training in deep networks; U-Net, designed for medical image segmentation with encoder-decoder paths; and DenseNet [9], which connects layers densely to improve gradient flow and feature reuse. In cardiac MRI, CNNs automate the segmentation of structures such as ventricles and myocardium, assist in disease classification for conditions such as myocardial infarction, and reconstruct high-quality images from undersampled data. In addition, CNNs enable multi-modality integration, combining MRI features with data like ECG for comprehensive analysis [10].

3D CNNs extend traditional 2D CNNs by simultaneously processing spatial and temporal data, making them particularly effective for volumetric segmentation and dynamic analysis in cine imaging or 3D late gadolinium enhancement (LGE) datasets [11]. However, 3D CNNs demand more computational power due to their complexity.

20.2.1.2 Recurrent neural networks (RNNs)

Recurrent neural networks (RNNs) specialize in sequential or time-series data, making them ideal for analyzing cine MRI or physiological signals like ECG [12,13]. Unlike CNNs, which excel at spatial data, RNNs capture temporal patterns by maintaining the memory of previous inputs through recurrent connections and shared weights, allowing efficient modeling of sequential relationships. They analyze cardiac motion dynamics, extract features such as myocardial strain, and integrate time-series data with imaging. RNNs are also valuable in prediction tasks, such as forecasting cardiac states or detecting arrhythmias [14,15].

20.2.1.3 Generative adversarial networks (GANs)

Generative adversarial networks (GANs) create realistic data representations using a generator-discriminator framework, where the generator produces data resembling real samples, and the discriminator evaluates their authenticity by distinguishing between real and synthetic data [16]. This adversarial training process allows GANs to excel in tasks such as data augmentation, denoising, and cross-modality synthesis. In cardiac MRI, GANs are widely utilized to expand datasets by creating additional realistic images, addressing challenges such as limited data availability or class

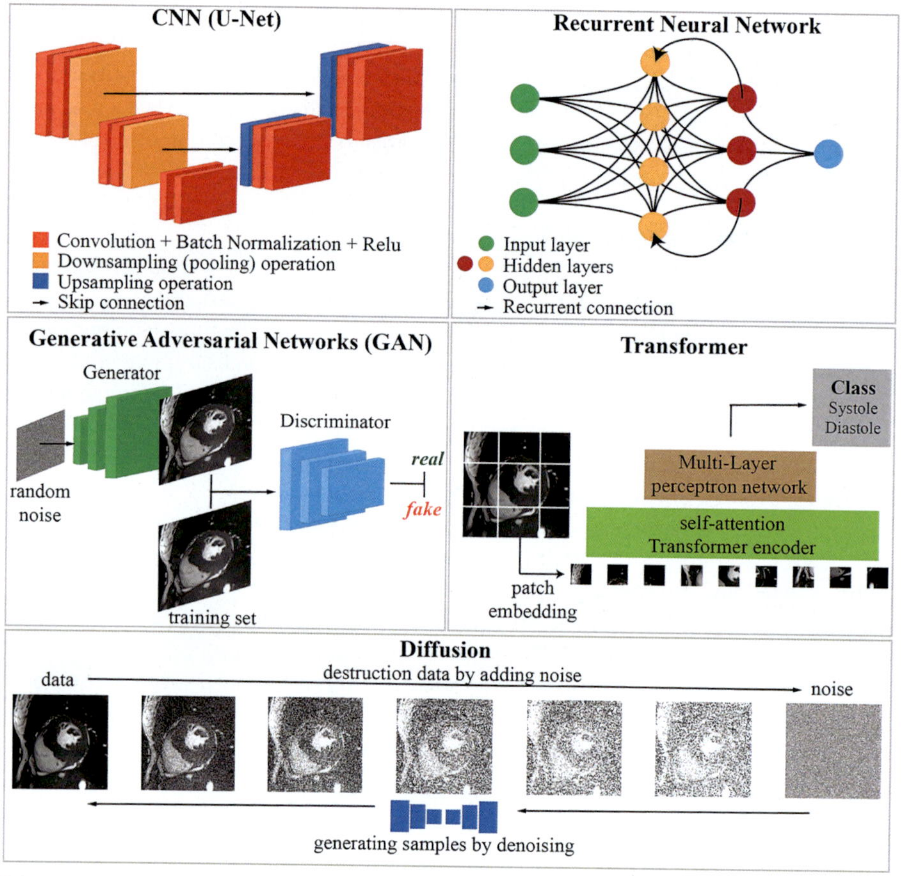

FIGURE 20.2

Various deep learning architectures commonly used in machine learning tasks, particularly in medical imaging. The CNN (U-Net) is designed for image segmentation, featuring layers for downsampling, upsampling, and skip connections to preserve spatial information. The Recurrent Neural Network (RNN) is ideal for sequential data, utilizing recurrent connections to pass information across layers, often applied in time-series analysis. Generative Adversarial Networks (GAN) consist of a generator, which creates synthetic data, and a discriminator, which distinguishes real from fake data, making them widely used for image generation. The Transformer employs self-attention mechanisms and transformer encoders, frequently used in natural language processing and image classification. Finally, Diffusion Models generate samples by progressively denoising data corrupted by noise, increasingly applied for high-quality image synthesis.

imbalance. They also enhance low-quality or noisy scans by removing artifacts and reconstructing high-resolution images from undersampled k-space data. Moreover, GANs contribute to segmentation workflows by refining outputs or producing pre-segmented images, reducing the reliance on manual annotation [17].

20.2.1.4 Transformers

Transformers leverage self-attention mechanisms to model relationships between different parts of input data, capturing long-range dependencies and global context [18]. Originally developed for natural language processing [19] transformers are increasingly applied to tasks in the medical imaging domain, such as cardiac MRI segmentation, reconstruction, temporal modeling, and disease prediction tasks. Positional encoding is used to preserve the order of sequential data, such as cardiac phases in cine MRI, while parallel processing enables efficient handling of large datasets. Vision Transformers (ViTs) and hybrid models combining CNNs with transformers have demonstrated state-of-the-art performance in segmentation and classification tasks. Transformers also integrate imaging data with clinical information to predict conditions such as cardiomyopathies or ischemia. In addition, transformers can perform cross-modality synthesis, generating synthetic images across modalities, such as CT to MRI [20–22].

20.2.1.5 Diffusion models

Diffusion models are generative architectures that progressively transform noise into structured outputs through iterative denoising, making them highly effective for image reconstruction, denoising, and super-resolution imaging [23]. In cardiac MRI, they are used to reconstruct high-quality images from undersampled k-space data, enhance noisy or low-resolution scans, and generate synthetic data for augmentation [24]. Diffusion models also enable cross-modality translation, such as converting CT or PET to MRI, facilitating multi-modality analysis [25].

While deep learning models have transformed cardiac MRI, they face significant challenges. CNNs require large, labeled datasets and high computational resources, and their 2D models struggle with inter-slice context, necessitating 3D CNNs for volumetric data. RNNs suffer from vanishing gradients, computational intensity, and limited applicability to high-dimensional spatial data. GANs are difficult to train due to instability, require extensive datasets, and lack interpretability. Transformers demand significant computational power and large datasets, and their clinical adoption remains limited. Diffusion models, although promising, are computationally expensive and still in the early stages of use in cardiac imaging. These limitations highlight the need for ongoing innovation to improve efficiency, accessibility, and clinical utility in deep learning for cardiac MRI.

20.2.2 Learning strategies

Deep learning models use different learning strategies to optimize performance and ensure accurate predictions. A fundamental step in the training process is dividing datasets into three subsets: training, validation, and testing. The training set, typically the largest portion, is used to teach the model patterns and relationships from the data. The validation set, a smaller subset, is used during training to evaluate the model's performance on unseen data, helping to fine-tune hyperparameters and prevent overfitting. The testing set, reserved for final evaluation, provides an unbiased assessment of the model's ability to generalize to new data. This separation ensures the model performs well in real-world scenarios by detecting overfitting or underfitting and confirming its robustness in unseen cases.

Overfitting and underfitting are two challenges in machine learning that impact a model's ability to generalize effectively to new data [26]. Overfitting occurs when a model learns the training data too thoroughly, capturing not only the essential patterns but also the noise and minor variations that are not representative of the broader dataset. This results in excellent performance on training data but poor performance on unseen data, as the model struggles to adapt to new inputs. Conversely, underfitting happens when a model fails to capture the underlying patterns in the training data, often due to being too simple, insufficiently trained, or working with inadequately prepared data. An underfitted model performs poorly on both training and unseen data because it has not learned the relationships necessary for accurate predictions. Balancing a model to avoid both overfitting and underfitting is critical, requiring appropriate model complexity, sufficient training, and effective data preprocessing.

Various learning strategies complement this dataset preparation. Supervised learning trains models on labeled data, using ground truth labels to guide the learning process, making it effective for tasks such as classification and segmentation [27]. On the other hand, unsupervised learning relies on unlabeled data to identify inherent structures or patterns, often used for clustering or dimensionality reduction [28]. Semi-supervised learning bridges the gap by combining a small amount of labeled data with a larger set of unlabeled data, reducing dependency on manual labeling [29]. Self-supervised learning, an emerging approach, creates its own labels through pretext tasks, enabling models to leverage large unlabeled datasets, which is particularly valuable in fields such as medical imaging [30]. Transfer learning uses pre-trained models from similar domains, fine-tuning them for specific tasks to address challenges such as limited labeled data [30]. Reinforcement learning, where an agent learns through interactions with an environment based on rewards or penalties, has potential applications in areas such as optimizing imaging protocols or treatment planning [31].

By integrating these strategies, deep learning models can maximize their effectiveness. For instance, a model may use self-supervised learning to pretrain on unlabeled data, followed by supervised fine-tuning, or employ transfer learning to enhance generalization. These approaches, tailored to the task and dataset, ensure robust and accurate results in cardiac MRI and other medical imaging applications.

20.3 Applications of AI in CMR

20.3.1 Image acquisition

20.3.1.1 Self-driving scanning and user-independent CMR planning

Traditional CMR imaging has relied on manual planning to acquire cardiac short- and long-axis views, a time-consuming and operator-dependent process. Major MRI system manufacturers have developed AI-driven solutions to address these challenges, introducing user-independent CMR planning tools that streamline workflows and enhance image consistency.

Siemens Healthineers offers the Cardiac Dot Engine [32], an AI-powered tool that automates cardiac plane planning, reducing the need for manual intervention. GE HealthCare provides AIR x [33], an AI-based system designed to automate slice positioning in brain imaging, which can be adapted for cardiac applications. Philips has developed MR Smart Workflow [34–36],which includes AI-driven features for automatic exam planning and patient positioning, determining landmarks for desired anatomical structures, and positioning them within the magnet isocenter with a single command [35,36].

These AI-driven tools represent a significant advancement in CMR imaging, reducing the reliance on manual planning and positioning. Automating these processes improves workflow efficiency and enhances image quality and consistency, allowing MRI technologists to focus more on quality control and patient care.

20.3.2 Image reconstruction

Image reconstruction is critical in cardiac MRI, converting raw k-space data into clinically useful images. The objective is to achieve high-quality imaging while minimizing scan times and overcoming challenges such as noise, artifacts, and undersampled data. Early reconstruction methods relied on linear techniques such as Generalized Autocalibrating Partially Parallel Acquisitions (GRAPPA) [37] and Sensitivity Encoding (SENSE) [38], which leveraged parallel imaging to reduce scan times but were limited by noise amplification and artifacts in highly undersampled data. Undersampling k-space data violates the Nyquist sampling criterion, leading to aliasing artifacts where multiple signals overlap, making accurate image reconstruction challenging. Compressed sensing [39] further improved reconstruction by introducing sparse representations and regularization, which mitigated some of these issues but often relied on handcrafted parameters and priors, restricting flexibility and efficiency. Sparse representation represents the image reconstructed using only the most essential data components, much like summarizing a song with just a few keynotes. On the other hand, regularization ensures that those notes are arranged to create a smooth and harmonious melody, avoiding discordant or jarring sounds.

Recent advancements in AI-driven methods have revolutionized image reconstruction, incorporating both purely data-driven and physics-driven approaches. These modern techniques address the challenges of undersampled data by integrating deep learning with imaging physics, allowing for more accurate reconstruction while minimizing artifacts and noise. By leveraging these methods, it is possible to accelerate imaging, maintain high-quality image reconstruction, and enable advanced quantitative analysis. Four key areas illustrate how deep learning is transforming cardiac MRI reconstruction.

20.3.2.1 Image enhancement: image-to-image networks

Image-to-image networks are a data-driven approach to image enhancement, designed to address artifacts such as aliasing caused by undersampled k-space data. These networks are trained using pairs of target images (fully sampled acquisitions) and accelerated images (synthesized by undersampling). The network learns to map the accelerated images to high-quality, artifact-free outputs. Image-to-image networks significantly improve image clarity and diagnostic utility by eliminating aliasing and wraparound artifacts, particularly in cases involving undersampled data from Cartesian and non-Cartesian scans, such as radial or spiral trajectories [40–45].

Image-to-image networks are also versatile in handling various undersampling patterns, including truncated sampling, incoherent sampling, and equispaced sampling. Truncated sampling involves acquiring only a subset of k-space data, typically focusing on the central region where most signal energy resides. Incoherent sampling, often used in compressed sensing, spreads the sampling points randomly across k-space to make aliasing artifacts more recoverable. Equispaced sampling, in contrast, collects data at uniform intervals but is more prone to coherent aliasing.

However, a limitation of image-to-image networks is their lack of integration with the underlying physics of MRI acquisition, which can lead to inconsistencies in multi-scale k-space data. This issue becomes particularly critical when imaging complex pathologies, as the absence of physical modeling may compromise reconstruction quality [46]. Despite this limitation, image-to-image networks provide a fast and effective solution for enhancing image quality in both routine clinical settings and research environments, making them a valuable tool in modern cardiac MRI.

20.3.2.2 Iterative image reconstruction: unrolled networks

Unrolled networks combine physics-driven models with deep learning to iteratively reconstruct images. This approach incorporates a forward imaging model that represents the physics of MRI acquisition and a learned image regularizer, enhancing the reconstructed images' quality. Unrolled networks enforce data consistency by iterating between these two components while leveraging deep learning for regularization [47–49].

Compared to purely data-driven methods, unrolled networks consistently achieve superior image quality by directly integrating coil sensitivities and k-space data into the reconstruction process [50]. These networks are particularly effective for accelerated imaging, where undersampled data often results in severe artifacts. By learning the regularization parameters and image priors from data, unrolled networks address the limitations of handcrafted priors in traditional compressed sensing methods. As a result, they are highly effective for both routine imaging and complex cases requiring high fidelity, such as cardiac cine imaging [40,51–57].

20.3.2.3 Denoising

Denoising is critical in cardiac MRI to improve signal-to-noise ratio (SNR) while preserving anatomical and pathological details [58] Traditional denoising techniques, such as filtering, often cause image blurring, whereas AI-based methods, such as CNNs and GANs, achieve superior results by learning complex noise patterns directly from data [59,60]. These models are trained on synthetic noisy and noise-free image pairs, as acquiring noise-free MRI scans *in vivo* is not feasible.

AI-based denoising is particularly valuable for low-SNR scenarios, such as low-field imaging, accelerated protocols, or noisy sequences such as late gadolinium enhancement (LGE) [61]. It enhances myocardial tissue characterization, recovers SNR lost in accelerated scans, and reduces the need for time-consuming signal averaging in cardiac diffusion imaging. Low-field, cost-effective, but inherently noisy systems benefit from AI-denoising techniques, producing sharper, high-quality images.

20.3.2.4 Quantitative myocardial tissue characterization

AI-driven image reconstruction has advanced quantitative myocardial tissue characterization, a critical aspect of cardiac MRI. To estimate tissue parameters, techniques such as T_1 and T_2 mapping involve acquiring multiple images with varying contrast across heartbeats. These methods are traditionally time-intensive, but AI models reduce scan times by estimating quantitative values from fewer images, improving efficiency without compromising accuracy [62–64]. Cardiac MR fingerprinting, which estimates multiple tissue properties simultaneously, further benefits from AI through accelerated dictionary generation, matching, and reconstruction [65].

Deep learning models enhance the precision and reproducibility of tissue parameter estimation by leveraging spatial information in images. AI can replace conventional numerical fitting methods, such as those based on the Bloch equation, with neural networks trained to estimate tissue properties directly. These approaches also mitigate motion artifacts and enhance the detection of subtle abnormalities such as fibrosis or edema, enabling earlier, and more reliable diagnosis of cardiac diseases.

Deep learning models for myocardial tissue characterization have also been extended to novel techniques such as radiomics [66] and texture analysis (TA) [67], which extract high-dimensional quantitative imaging features from medical images. Radiomics involves converting digital images into mineable data, enabling the analysis of morphologic, intensity-based, fractal-based, and texture features.

These features provide insights into underlying tissue textures, offering a more objective alternative to purely visual image interpretation.

For example, texture features have been applied to late gadolinium enhancement (LGE) imaging to enhance the segmentation and analysis of scarred myocardium in myocardial infarction [66,68]. Studies have also demonstrated the utility of TA in differentiating acute from chronic infarction and distinguishing between conditions such as hypertrophic cardiomyopathy (HCM), amyloidosis, and aortic stenosis [67,69,70].

20.3.3 Image segmentation

Image segmentation is the process of delineating anatomical structures and functional regions within cardiac MRI scans, enabling precise quantification of parameters such as ventricular volumes, myocardial thickness, and scar tissue [71]. It is a critical step in cardiac imaging workflows, forming the foundation for diagnosis, treatment planning, and disease monitoring. Segmentation is required across various cardiac MRI sequences, such as cine imaging for functional assessments, T_1 and T_2 mapping for tissue characterization, and late gadolinium enhancement (LGE) imaging for scar visualization [72–74].

Traditionally, segmentation has been performed manually, requiring substantial time and expertise from clinicians. This approach is labor-intensive and prone to inter- and intra-observer variability, which can impact clinical decision-making. Automated segmentation using AI has emerged as a solution to these challenges, offering faster, more consistent, and scalable methods for analyzing cardiac MRI data.

Commercial software solutions like cvi42 by circle cardiovascular imaging [75], Medis Suite by Medis Medical Imaging [76], and IntelliSpace Portal by Philips [77] have incorporated AI-based segmentation features to enhance the analysis of cardiac MRI data. These tools provide automated functional and volumetric measurements, often able to validate and adjust contours for improved accuracy. Such software is widely used in clinical and research settings because it can streamline workflows and improve consistency. Additionally, these platforms often integrate with Picture Archiving and Communication System (PACS) systems, enabling seamless data analysis, and reporting.

AI-based segmentation, particularly through deep learning architectures like U-Net, has revolutionized the field. These models learn to identify and segment structures directly from training datasets, reducing the need for manual intervention. Automated segmentation enables rapid analysis of large datasets and is increasingly being adopted across various cardiac MRI applications, enhancing both clinical and research workflows.

Different deep learning techniques applied to cardiovascular magnetic resonance (CMR) imaging are shown in Fig. 20.3.

20.4 Integration of AI in clinical practice

Artificial intelligence has significantly advanced MR for cardiac imaging by improving image acquisition, reconstruction, and analysis. Many of these techniques

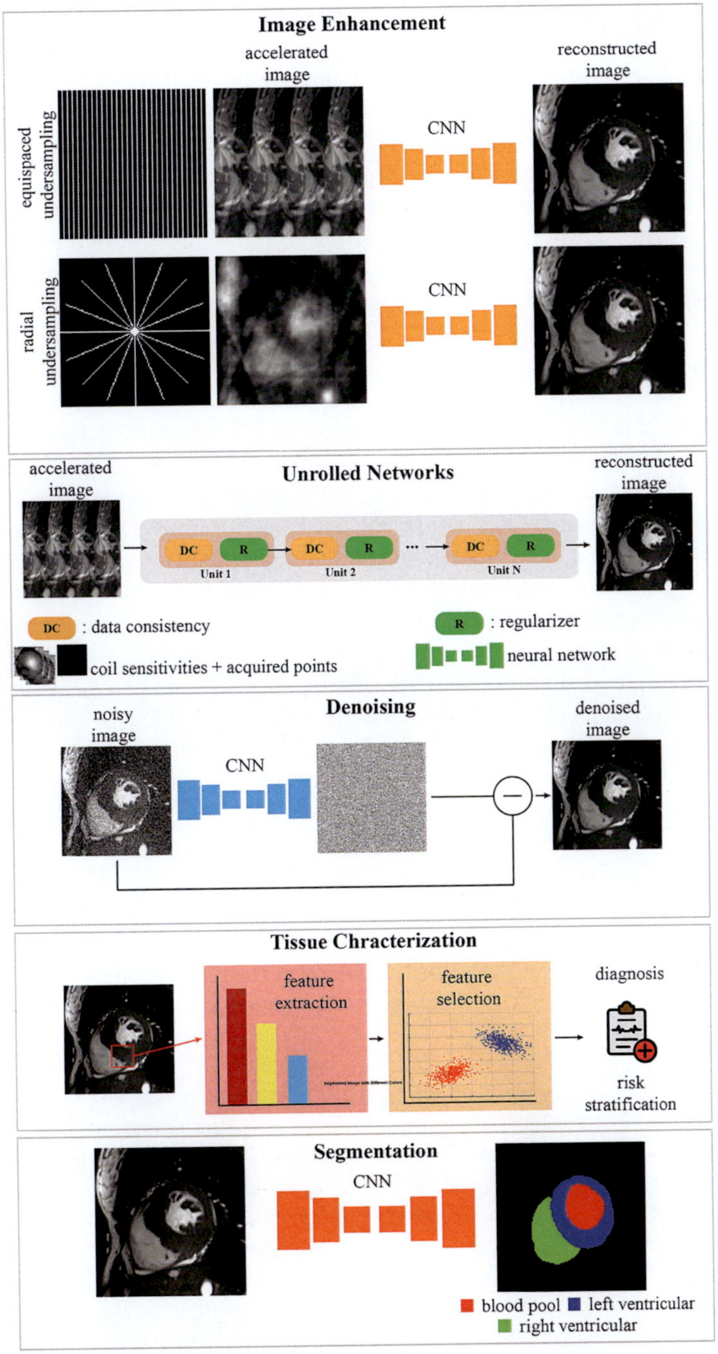

Image Enhancement
accelerated image
reconstructed image
equispaced undersampling
CNN
radial undersampling
CNN
Unrolled Networks
accelerated image
reconstructed image
DC
R
Unit 1
Unit 2
Unit N
DC : data consistency
R : regularizer
coil sensitivities + acquired points
neural network
Denoising
noisy image
CNN
denoised image
Tissue Chracterization
feature extraction
feature selection
diagnosis
risk stratification
Segmentation
CNN
blood pool
left ventricular
right ventricular

FIGURE 20.3

are broadly applicable to different series types and clinical applications. In general, AI methods are used to maximize the utilization of under-sampled k-space signal data to reduce scan times. Algorithms also improve the quality of reconstructed images through denoising techniques, super-resolution, and the reduction of artifacts, including aliasing and motion. This, in turn increases the accuracy and precision of quantitative measurements. The auto-segmentation of anatomy and calculation of functional cardiac parameters is yet another incorporation of AI that streamlines workflows, increases inter-reader and intra-reader reliability. These factors combine with the extraction of imaging biomarkers to potentially increase diagnostic accuracy. Examples of specific AI applications are further discussed below.

20.4.1 Cine imaging

Cine imaging in cardiac MR provides detailed, high-resolution imaging of cardiac anatomy and function throughout the cardiac cycle, allowing for the simultaneous assessment of cardiac structure, function, and motion. Multiple studies have been published that showcase the automated analysis of cine cardiac MR images [78–84]. Most utilize a U-Net like architecture for automatic chamber segmentation. U-Net is a type of convolutional neural network consisting of an encoder-decoder [8]. It may be adapted for 2D or 3D images and remains the backbone for biomedical imaging segmentation [85]. Chamber segmentation allows for the calculation of functional cardiac endpoints such as ventricular volume, stroke volume, and LVEF. Some have emphasized the need for specialized models for pathologies such as single ventricle morphology and tetralogy of Fallot [86,87]. In complex cases, multiple networks are employed for specific subtasks such as shape-modeling and multi-view integration.

20.4.2 Flow

AI applications in assessing blood flow in cardiac MR include comprehensive hemodynamic analysis. Blood flow quantification is performed through phase-contrast (PC) MRI. As with cine MR techniques discussed above, deep learning-based

Deep learning techniques applied to cardiovascular magnetic resonance (CMR) imaging. In image enhancement, convolutional neural networks (CNNs) are employed to reconstruct high-quality images from accelerated undersampled data, such as equispaced or radial undersampling, enhancing the usability of CMR images. Unrolled networks integrate data consistency layers with learned regularization through iterative units to provide superior image reconstruction from incomplete or accelerated datasets. Denoising techniques use CNNs to reduce noise in CMR images, restoring high-quality images. In tissue characterization, features are extracted and selected from segmented regions, aiding in cardiac abnormalities' diagnosis and risk stratification. Finally, segmentation employs CNNs to delineate key cardiac structures such as the blood pool, left ventricle, and right ventricle, enabling detailed cardiac function and anatomy analysis.

compressed sensing significantly reduces scan times while maintaining spatial and temporal resolution [88,89]. One AI application that particularly enhances blood flow analysis is the application of denoising, which improves image quality where there is low signal-to-noise ratio, effectively increasing the reliability of blood flow measurements [90,91]. This is because noise can mimic flow-related signals, creating artifacts or false flow patterns. The ability to remove this noise allows for the use of lower resolution scans that can be acquired more quickly.

As with cine MR, AI-based auto-segmentation is used with MR flow to segment anatomy relevant to blood flow analysis. This typically involves the segmentation of vessels, shunts, and across valves in addition to cardiac chamber volumes in order to quantify parameters such as regurgitant fraction, pressures, and peak flow velocity [92–95]. The repeated application of these AI models to quantify blood flow over time increases inter-reader and intra-reader reliability and the consistency of measurement [96,97].

4D flow is a subtype of flow MR that includes large datasets and complex flow dynamics such as vortices, shear stress, and turbulence. AI models may be used to efficiently process the large, complex datasets as well as predict clinical endpoints in 4D flow [91,96,98].

20.4.3 T_1/T_2 mapping

In cardiac MRI, T_1 and T_2 mapping are techniques used for measuring cardiac signal on specific MR series to quantify tissue characteristics. Convolutional neural networks have been used to simultaneously acquire data for T_1 and T_2 relaxation mapping and cine imaging, reducing the need for separate acquisitions [84,99]. Latent transformer neural network modules have also been used to process compressed data representations that model complex spatial and temporal dependencies in T_1 and T_2 mapping, a technique that circumvents the need for image reconstruction for segmentation to increase efficiencies [100]. Auto-segmentation can further help to localize signal changes within the endocardium, myocardium, and/or epicardium for incorporation into imaging biomarkers.

Machine learning algorithms have been used to extract imaging features from T_1 and T_2 mapping and combine them with data and MR series to predict a wide range of pathologies. Algorithms have incorporated both segmentation and signal quantification in T_2 mapping to predict acute myocardial infarction [101]. T_1 mapping of radiomic features has been used to distinguish between subtypes of hypertrophic cardiomyopathy [102]. Predictive models have utilized a combination of T_1 mapping and late gadolinium enhancement patterns to predict acute and chronic myocardial infarction, hypertrophic cardiomyopathy, dilated cardiomyopathy, myocarditis, and cardiac amyloidosis [103].

20.4.4 Strain imaging

Cardiac strain imaging assesses myocardial shape and deformity during MR cine imaging and global myocardial strain has been shown to detect cardiac dysfunction earlier than LVEF [103]. In recent years, deep learning frameworks have been used to fully automate myocardial strain estimation [104,105]. Recurrent and convolutional neural network frameworks have been combined to analyze myocardial strain from short axis images, distinguishing patterns related to diabetes, hypertension, previous myocardial infarction, and arrhythmogenic right ventricular cardiomyopathy [105,106].

An important feature of neural networks applied to cardiac strain assessment is the incorporation of domain knowledge into the loss function, which assumes the relative incompressibility of cardiac tissue and penalizes unrealistic cardiac deformations [106,107]. Implicit neural representations (INRs) have been applied to a combination of long and short-axis views to automatically assess cardiac motion at varied time points throughout the cardiac cycle in order to quantify the global and segmental cardiac strain. This represents an application of unsupervised deep learning with super-resolution to distinguish disease-specific patterns of altered cardiac motion [106]. Likewise, WarpPINN is an open-source neural network specifically developed for estimating cardiac motion that builds parameters into the loss function for unrealistic cardiac motion. This drives the model toward realistic predictions of cardiac strain.

20.4.5 Perfusion

MR perfusion imaging is used to estimate myocardial blood flow and quantify the presence of ischemic and infarcted tissue. AI applications in MR perfusion rely on auto-segmentation of the left ventricle to calculate myocardial blood flow (MBF) and myocardial perfusion reserve (MPR).

Predictive algorithms have used CNNs to automate quantification of mean stress MBF and MPR and outperformed conventional metrics to predict adverse outcomes [108]. Automated pixel-wise quantitative myocardial perfusion mapping has been shown to effectively detect coronary artery disease and coronary microvascular dysfunction compared with invasive fractional flow reserve (FFR) and coronary flow reserve (CFR), and outperforms manual, subjective calculations [109]. Fully-automated pipelines have been proposed and some validated against manual segmentations to quantify cardiac MR perfusion in clinical practice [110–113].

20.4.6 LGE

Late gadolinium enhancement (LGE) is used to detect myocardial scar, where gadolinium accumulates in the relatively increased extracellular space of scar compared to healthy tissue. Thus, myocardial scar demonstrates increased signal on LGE, ECV imaging, and T_1 mapping, all of which can be used as inputs to predict various pathologies [114–117].

AI models have successfully identified and quantified scar, as well as characterized and stratified the patterns of scar according to etiology and severity. There is also evidence that AI models can detect more accurate imaging biomarkers than conventional LGE findings in order to detect undiagnosed myocardial infarction [117].

The application of artificial intelligence has done much to contribute to standardizing LGE imaging techniques and validating models across datasets and vendors. Convolutional neural networks have been used to optimize the inversion time for LGE [118–121]. Combined convolutional and recurrent neural networks have been used to identify a vendor-neutral LGE inversion time [121]. Unsupervised CycleGAN-based domain adaptation, a generative technique that adapts models trained on one dataset to work effectively on a target dataset without new labels, has been used to automate the quantification of myocardial scar across protocols from multiple institutions while mitigating domain shift [122]. AI-driven cardiac MR techniques have even been designed to imitate the results of LGE without the need for intravenous contrast, a concept that alters the fundamental imaging technique of "LGE" altogether [116]. However, more research is needed to investigate whether native techniques lead to similar outcomes versus LGE and can be used for clinical decision making regarding the presence, extent and transmurality of myocardial scarring.

20.4.7 AI-driven CMR reporting

Natural language processing can leverage the above applications to translate the model outputs into radiology report text, known as automatic radiology report generation (ARRG). This process of automation streamlines workflow and standardizes radiology practices. Overall, the aim is to use these models to increase diagnostic precision.

The standards for cardiac MR reporting are outlined by the Society for Cardiovascular Magnetic Resonance [123,124]. Clinical parameters generated by algorithms above can auto-populate reports. The future is in image-to-text vision-language models (VLMs). Although not specific to cardiac MR, models have been developed with multimodal transformer architecture, which combine image encoders that extract imaging features with language models for text generation. One is mapped to another through a projection layer which maps the imaging to text [125]. This conceptualized framework has been successfully applied to both 2D and 3D radiology images [126], and it is likely that these techniques will soon be applied to CMR.

Another key aspect of automatic radiology report generation is its ability to incorporate clinical recommendations based on the findings. This aim would have to be carefully curated. Extending this concept even further is the incorporation of clinical information into the imaging interpretation model and the possibility that end-to-end multimodal models will be used for the interpreting each instance of cardiac MR imaging at the patient level, generating an individualized, and comprehensive report.

20.5 Challenges and limitations

Although it is difficult to predict the exact shape of future outcomes, the pace of progress in medical AI continues to accelerate. The availability of quality data has been a primary limitation; many of the most innovative steps have come from work with public datasets which, through a process of carefully curated patient consent and deidentification, has anonymized what is otherwise highly confidential and personal data to a world of innovators [127].

Another limitation has been the time-consuming process of expert labeling. However, this may be somewhat circumvented with the continued development of semi-supervised and unsupervised training techniques. Building models to perform accurate inference on datasets acquired from diverse populations with varied protocols is technically challenging, as is achieving inference to accurately characterize edge cases. Models should be also prospectively validated and monitored for model drift, particularly as imaging technology and protocols change over time.

One limitation of CMR when integrated with AI tools is the variability of radiomic feature extraction, particularly in texture analysis, due to preprocessing steps such as denoising. Texture features derived from myocardial radiomics can be sensitive to imaging parameters [128], where factors such as spatial resolution and noise levels were shown to influence feature reproducibility. Denoising algorithms, while effective in reducing noise and enhancing image quality, can inadvertently alter subtle image textures that are critical for radiomic analysis, potentially compromising feature stability, and diagnostic accuracy. The importance of reproducibility in segmentation-based radiomic workflows notes that preprocessing variability could impact feature consistency across datasets [129]. Thus, balancing noise reduction with preserving critical texture details remains a significant challenge in leveraging AI for robust and reproducible CMR radiomics.

Clinician and patient model adoption must go beyond data for widespread adoption. Model transparency and explainability are also important components of widespread acceptance. This is particularly important for those who will adopt liability for decisions implemented with model results. There are also significant concerns for workflow disruptions that may result in retraining and a shifting workforce. All of these factors require investment of resources for innovation, computational resources, and trustworthy, accountable leadership.

20.6 Conclusion and future perspectives

Over the past several years, AI has been integrated in every step of the cardiac imaging pipeline, and advances continue to accelerate and facilitate faster and more robust CMR. Previously labor-intensive tasks are in the process of being fully automated, so operators and readers can devote more attention to high level, complex tasks. Many of the algorithms discussed in this chapter are already commercially available through CMR hardware vendors and post processing software vendors, and are generating measurable improvements in patient throughput, image quality, and reproducible image analysis and, hopefully soon, improved outcomes.

References

[1] Rajiah PS, François CJ, Leiner T. Cardiac MRI: state of the art. Radiology 2023;307 (3):e223008.

[2] Guo R, et al. Emerging techniques in cardiac magnetic resonance imaging. J Magn Reson Imag 2022;55(4):1043–59.

[3] Emrich T, Halfmann M, Schoepf UJ, Kreitner K-F. CMR for myocardial characterization in ischemic heart disease: state-of-the-art and future developments. Eur Radiol Exp 2021;5(1):14.

[4] Gore JC. Artificial intelligence in medical imaging. Elsevier; 2020.

[5] Panayides AS, et al. AI in medical imaging informatics: current challenges and future directions. IEEE J Biomed Health Inform 2020;24(7):1837–57.

[6] Rajpurkar P, Chen E, Banerjee O, Topol EJ. AI in health and medicine. Nat Med 2022; 28(1):31–8.

[7] Janiesch C, Zschech P, Heinrich K. Machine learning and deep learning. Electron Mark 2021;31(3):685–95.

[8] Ronneberger O, Fischer P, Brox T. U-net: convolutional networks for biomedical image segmentation. In: Medical image computing and computer-assisted intervention–MICCAI 2015: 18th international conference, Munich, Germany, October 5-9, 2015, proceedings, part III 18. Springer; 2015. p. 234–41.

[9] Huang G, Liu Z, Van Der Maaten L, Weinberger KQ. Densely connected convolutional networks. In: Proceedings of the IEEE conference on computer vision and pattern recognition; 2017. p. 4700–8.

[10] Ahmad Z, Tabassum A, Guan L, Khan NM. ECG heartbeat classification using multimodal fusion. IEEE Access 2021;9:100615–26.

[11] Baumgartner CF, Koch LM, Pollefeys M, Konukoglu E. An exploration of 2D and 3D deep learning techniques for cardiac MR image segmentation. In: Statistical atlases and computational models of the heart. ACDC and MMWHS challenges: 8th international workshop, STACOM 2017, held in conjunction with MICCAI 2017, Quebec city, Canada, September 10-14, 2017, revised selected papers 8. Springer; 2018. p. 111–9.

[12] Sherstinsky A. Fundamentals of recurrent neural network (RNN) and long short-term memory (LSTM) network. Physica D 2020;404:132306.

[13] Lyu Q, et al. Cine cardiac MRI motion artifact reduction using a recurrent neural network. IEEE Trans Med Imaging 2021;40(8):2170–81.

[14] Ansari Y, Mourad O, Qaraqe K, Serpedin E. Deep learning for ECG Arrhythmia detection and classification: an overview of progress for period 2017–2023. Front Physiol 2023;14:1246746.

[15] Ferdiana E, et al. Fully automated myocardial strain estimation from CMR tagged images using a deep learning framework in the UK Biobank. Radiology 2020;2(1): e190032.

[16] Goodfellow I, et al. Generative adversarial networks. Commun ACM 2020;63(11): 139–44.

[17] Ali M, Ali M, Hussain M, Koundal D. Generative adversarial networks (GANs) for medical image processing: recent advancements. Arch Comput Methods Eng 2024: 1–14.

[18] Dosovitskiy A. An image is worth 16x16 words: transformers for image recognition at scale. arXiv preprint arXiv:2010.11929, 2020.

[19] Vaswani A. Attention is all you need. Adv Neural Inf Process Syst 2017;30.
[20] Li Y, Han N, Qin Y, Zhang J, Su J. Trans-cGAN: transformer-Unet-based generative adversarial networks for cross-modality magnetic resonance image synthesis. Stat Comput 2023;33(5):113.
[21] Khader F, et al. Medical transformer for multimodal survival prediction in intensive care: integration of imaging and non-imaging data. Sci Rep 2023;13(1):10666.
[22] Fan Y, Song J, Yuan L, Jia Y. HCT-Unet: multi-target medical image segmentation via a hybrid CNN-transformer Unet incorporating multi-axis gated multi-layer perceptron. Vis Comput 2024:1–16.
[23] Ho J, Jain A, Abbeel P. Denoising diffusion probabilistic models. Adv Neural Inf Process Syst 2020;33:6840–51.
[24] Chung H, Lee ES, Ye JC. MR image denoising and super-resolution using regularized reverse diffusion. IEEE Trans Med Imaging 2022;42(4):922–34.
[25] Li Y, Yakushev I, Hedderich DM, Wachinger C. PASTA: pathology-A ware MRI to PET Cross-modal translation with diffusion models. In: International conference on medical image computing and computer-assisted intervention. Springer; 2024. p. 529–40.
[26] Alpaydin E. Introduction to machine learning. MIT press; 2020.
[27] Bishop CM. Pattern recognition and machine learning, vol. 2. Springer google schola; 2006. p. 1122–8.
[28] Ziegel ER. The elements of statistical learning. Taylor & Francis; 2003.
[29] S.-S. Learning, Semi-supervised learning, CSZ2006.html, vol. 5, p. 2, 2006.
[30] LeCun Y, Bengio Y, Hinton G. Deep learning. Nature 2015;521(7553):436–44.
[31] Sutton RS, Barto AG. Reinforcement learning: an introduction. MIT press; 2018.
[32] J. C. Pueyo et al., "Cardiac dot engine: significant time reduction at cardiac magnetic resonance imaging".
[33] Peters RD, Harris H, Lawson S. The clinical benefits of AIR™ Recon DL for MR image reconstruction. GE Healthcare; 2020.
[34] MR Smart workflow, https://www.usa.philips.com/healthcare/resources/landing/smart workflow.
[35] Leiner T, et al. Machine learning in cardiovascular magnetic resonance: basic concepts and applications. J Cardiovasc Magn Reson 2019;21(1):61.
[36] Villegas-Martinez M, de Villedon de Naide V, Muthurangu V, Bustin A. The beating heart: artificial intelligence for cardiovascular application in the clinic. Magn Reson Mater Phys, BiolMed 2024;37(3):369–82.
[37] Griswold MA, et al. Generalized autocalibrating partially parallel acquisitions (GRAPPA). Magn Reson Med 2002;47(6):1202–10.
[38] Pruessmann KP, Weiger M, Scheidegger MB, Boesiger P. SENSE: sensitivity encoding for fast MRI. Magn Reson Med 1999;42(5):952–62.
[39] Lustig M, Donoho DL, Santos JM, Pauly JM. Compressed sensing MRI. IEEE Signal Process Mag 2008;25(2):72–82.
[40] Hauptmann A, Arridge S, Lucka F, Muthurangu V, Steeden JA. Real-time cardiovascular MR with spatio-temporal artifact suppression using deep learning–proof of concept in congenital heart disease. Magn Reson Med 2019;81(2):1143–56.
[41] Wang J, Weller DS, Kramer CM, Salerno M. DEep learning-based rapid spiral image REconstruction (DESIRE) for high-resolution spiral first-pass myocardial perfusion imaging. NMR Biomed 2022;35(5):e4661.

[42] Yoon S, et al. Accelerated cardiac MRI cine with use of resolution enhancement generative adversarial inline neural network. Radiology 2023;307(5):e222878.

[43] Shen D, et al. Rapid reconstruction of highly undersampled, non-Cartesian real-time cine k-space data using a perceptual complex neural network (PCNN). NMR Biomed 2021;34(1):e4405.

[44] Morales MA, et al. An inline deep learning based free-breathing ECG-free cine for exercise cardiovascular magnetic resonance. J Cardiovasc Magn Reson 2022;24(1):47.

[45] Fan L, et al. Rapid dealiasing of undersampled, non-Cartesian cardiac perfusion images using U-net. NMR Biomed 2020;33(5):e4239.

[46] Oscanoa JA, et al. Deep learning-based reconstruction for cardiac MRI: a review. Bioengineering 2023;10(3):334.

[47] Hammernik K, et al. Learning a variational network for reconstruction of accelerated MRI data. Magn Reson Med 2018;79(6):3055–71.

[48] Aggarwal HK, Mani MP, Jacob M. MoDL: model-based deep learning architecture for inverse problems. IEEE Trans Med Imaging 2018;38(2):394–405.

[49] Schlemper J, Caballero J, V Hajnal J, Price A, Rueckert D. A deep cascade of convolutional neural networks for MR image reconstruction. In: Information processing in medical imaging: 25th international conference, IPMI 2017, Boone, NC, USA, June 25-30, 2017, proceedings 25. Springer; 2017. p. 647–58.

[50] Muckley MJ, et al. Results of the 2020 fastMRI challenge for machine learning MR image reconstruction. IEEE Trans Med Imaging 2021;40(9):2306–17.

[51] Küstner T, et al. CINENet: deep learning-based 3D cardiac CINE MRI reconstruction with multi-coil complex-valued 4D spatio-temporal convolutions. Sci Rep 2020;10(1): 13710.

[52] Qin C, et al. Complementary time-frequency domain networks for dynamic parallel MR image reconstruction. Magn Reson Med 2021;86(6):3274–91.

[53] Qin C, Schlemper J, Caballero J, Price AN, V Hajnal J, Rueckert D. Convolutional recurrent neural networks for dynamic MR image reconstruction. IEEE Trans Med Imaging 2018;38(1):280–90.

[54] Pednekar A, et al. Accelerated cine cardiac MRI using deep learning-based reconstruction: a systematic evaluation. J Magn Reson Imaging 2024;60(2):640–50.

[55] Demirel OB, et al. Accelerated cardiac cine with spatio-coil regularized deep learning reconstruction. Magn Reson Med 2025;93(3):1132–48.

[56] Huang W, et al. Deep low-rank plus sparse network for dynamic MR imaging. Med Image Anal 2021;73:102190.

[57] Kofler A, Dewey M, Schaeffter T, Wald C, Kolbitsch C. Spatio-temporal deep learning-based undersampling artefact reduction for 2D radial cine MRI with limited training data. IEEE Trans Med Imaging 2019;39(3):703–17.

[58] Tian C, Fei L, Zheng W, Xu Y, Zuo W, Lin C-W. Deep learning on image denoising: an overview. Neural Netw 2020;131:251–75.

[59] Ran M, et al. Denoising of 3D magnetic resonance images using a residual encoder–decoder Wasserstein generative adversarial network. Med Image Anal 2019;55: 165–80.

[60] Phipps K, et al. Accelerated in vivo cardiac diffusion-tensor MRI using residual deep learning–based denoising in participants with obesity. Radiol Cardiothorac Imaging 2021;3(3):e200580.

[61] Muscogiuri G, et al. Feasibility of late gadolinium enhancement (LGE) in ischemic cardiomyopathy using 2D-multisegment LGE combined with artificial intelligence

reconstruction deep learning noise reduction algorithm. Int J Cardiol 2021;343: 164–70.
[62] V Le J, et al. Accelerated cardiac T1 mapping with recurrent networks and cyclic, model-based loss. Med Phys 2022;49(11):6986–7000.
[63] Guo R, et al. Accelerated cardiac T1 mapping in four heartbeats with inline MyoMap-Net: a deep learning-based T1 estimation approach. J Cardiovasc Magn Reson 2022;24 (1):6.
[64] Amyar A, et al. Scanner-independent MyoMapNet for accelerated cardiac MRI T1 mapping across vendors and field strengths. J Magn Reson Imag 2024;59(1):179–89.
[65] Velasco C, Fletcher TJ, Botnar RM, Prieto C. Artificial intelligence in cardiac magnetic resonance fingerprinting. Front Cardiovasc Med 2022;9:1009131.
[66] Di Noto T, et al. Radiomics for distinguishing myocardial infarction from myocarditis at late gadolinium enhancement at MRI: comparison with subjective visual analysis. Radiol Cardiothorac Imaging 2019;1(5):e180026.
[67] Baessler B, et al. Cardiac MRI and texture analysis of myocardial T1 and T2 maps in myocarditis with acute versus chronic symptoms of heart failure. Radiology 2019;292 (3):608–17.
[68] Durmaz ES, et al. Radiomics-based machine learning models in STEMI: a promising tool for the prediction of major adverse cardiac events. Eur Radiol 2023;33(7): 4611–20.
[69] Chen B-H, et al. Myocardial extracellular volume fraction radiomics analysis for differentiation of reversible versus irreversible myocardial damage and prediction of left ventricular adverse remodeling after ST-elevation myocardial infarction. Eur Radiol 2021;31:504–14.
[70] Fahmy AS, Rowin EJ, Arafati A, Al-Otaibi T, Maron MS, Nezafat R. Radiomics and deep learning for myocardial scar screening in hypertrophic cardiomyopathy. J Cardiovasc Magn Reson 2022;24(1):40.
[71] Chen C, et al. Deep learning for cardiac image segmentation: a review. Front Cardiovasc Med 2020;7:25.
[72] Campello VM, et al. Multi-centre, multi-vendor and multi-disease cardiac segmentation: the M&Ms challenge. IEEE Trans Med Imaging 2021;40(12):3543–54.
[73] Taha AA, Hanbury A. Metrics for evaluating 3D medical image segmentation: analysis, selection, and tool. BMC Med Imaging 2015;15:1–28.
[74] Isensee F, Jaeger PF, Kohl SAA, Petersen J, Maier-Hein KH. nnU-Net: a self-configuring method for deep learning-based biomedical image segmentation. Nat Methods 2021;18(2):203–11.
[75] cvi42, https://www.circlecvi.com/cardiac-mr.
[76] Medis medical imaging, https://medisimaging.com/software-solutions/medis-suite-ct/.
[77] Philips IntelliSpace, https://www.usa.philips.com/healthcare/product/HC881072/intellispace-portal-advanced-visualization-solution.
[78] Ruijsink B, et al. Fully automated, quality-controlled cardiac analysis from CMR: validation and large-scale application to characterize cardiac function. Cardiovasc Imaging 2020;13(3):684–95.
[79] Vergani V, Razavi R, Puyol-Antón E, Ruijsink B. Deep learning for classification and selection of cine CMR images to achieve fully automated quality-controlled CMR analysis from scanner to report. Front Cardiovasc Med 2021;8:742640.
[80] Tao Q, Lelieveldt BPF, van der Geest RJ. Deep learning for quantitative cardiac MRI. Am J Roentgenol 2020;214(3):529–35.

[81] Fadil H, et al. A deep learning pipeline for automatic analysis of multi-scan cardiovascular magnetic resonance. J Cardiovasc Magn Reson 2021;23(1):47.

[82] Mariscal-Harana J, et al. An artificial intelligence tool for automated analysis of large-scale unstructured clinical cine cardiac magnetic resonance databases. Eur Heart J Digi Health 2023;4(5):370–83.

[83] Wang S, et al. AI based CMR assessment of biventricular function: clinical significance of intervendor variability and measurement errors. Cardiovasc Imaging 2022; 15(3):413–27.

[84] Lyu J, et al. The state-of-the-art in cardiac mri reconstruction: results of the cmrxrecon challenge in miccai 2023, arXiv preprint arXiv:2404.01082, 2024.

[85] Xie X, Yang M. USCT-UNet: rethinking the semantic gap in U-net network from U-shaped skip connections with multichannel fusion transformer. IEEE Trans Neural Syst Rehabil Eng 2024;32:3782–93. https://doi.org/10.1109/TNSRE.2024.3468339.

[86] Govil S, et al. A deep learning approach for fully automated cardiac shape modeling in tetralogy of Fallot. J Cardiovasc Magn Reson 2023;25(1):15.

[87] Yao T, et al. A deep learning pipeline for assessing ventricular volumes from a cardiac MRI registry of patients with single ventricle physiology. Radiol Artif Intell 2023;6(1): e230132.

[88] Bustin A, Fuin N, Botnar RM, Prieto C. From compressed-sensing to artificial intelligence-based cardiac MRI reconstruction. Front Cardiovasc Med 2020;7:501814.

[89] Lin L, et al. Free-breathing cardiac cine MRI with compressed sensing real-time imaging and retrospective motion correction: clinical feasibility and validation. Eur Radiol 2023;33(4):2289–300.

[90] Ferdian E, et al. 4DFlowNet: super-resolution 4D flow MRI using deep learning and computational fluid dynamics. Front Phys 2020;8:138.

[91] Shit S, et al. SRflow: deep learning based super-resolution of 4D-flow MRI data. Front Artif Intell 2022;5:928181.

[92] Chan E, et al. Automated quality controlled analysis of 2d phase contrast cardiovascular magnetic resonance imaging. In: International workshop on statistical atlases and computational models of the heart. Springer; 2022. p. 101–11.

[93] Frueh M, Kuestner T, Nachbar M, Thorwarth D, Schilling A, Gatidis S. Self-supervised learning for automated anatomical tracking in medical image data with minimal human labeling effort. Comput Methods Progr Biomed 2022;225:107085.

[94] Montalt-Tordera J, et al. Automatic segmentation of the great arteries for computational hemodynamic assessment. J Cardiovasc Magn Reson 2022;24(1):57.

[95] Pajaziti E, et al. Shape-driven deep neural networks for fast acquisition of aortic 3D pressure and velocity flow fields. PLoS Comput Biol 2023;19(4):e1011055.

[96] Trauzeddel RF, et al. The influence of post-processing software on quantitative results in 4D flow cardiovascular magnetic resonance examinations. Front Cardiovasc Med 2024;11:1465554.

[97] Oechtering TH, et al. Repeatability and reproducibility of various 4D Flow MRI post-processing software programs in a multi-software and multi-vendor cross-over comparison study. J Cardiovasc Magn Reson 2023;25(1):22.

[98] Bustamante M, Viola F, Engvall J, Carlhäll C, Ebbers T. Automatic time-resolved cardiovascular segmentation of 4D flow MRI using deep learning. J Magn Reson Imag 2023;57(1):191–203.

[99] Guo R, et al. MyoFold: rapid Myocardial tissue and movement quantification via a highly Folded sequence, arXiv preprint arXiv:2307.08062, 2023.

[100] Wang F, et al. T1/T2 relaxation temporal modelling from accelerated acquisitions using a Latent Transformer, arXiv preprint arXiv:2309.16853, 2023.
[101] Zhou Y, et al. Simultaneous deep learning of myocardium segmentation and T2 quantification for acute myocardial infarction MRI, arXiv preprint arXiv:2405.10570, 2024.
[102] Antonopoulos AS, et al. Machine learning of native T1 mapping radiomics for classification of hypertrophic cardiomyopathy phenotypes. Sci Rep 2021;11(1):23596.
[103] Brady B, King G, Murphy RT, Walsh D. Myocardial strain: a clinical review. Ir J Med Sci 2023;192(4):1649–56.
[104] Backhaus SJ, et al. Artificial intelligence fully automated myocardial strain quantification for risk stratification following acute myocardial infarction. Sci Rep 2022;12 (1):12220.
[105] Ferdian E, et al. Fully automated myocardial strain estimation from cardiovascular MRI–tagged images using a deep learning framework in the UK Biobank. Radiol Cardiothorac Imaging 2020;2(1):e190032.
[106] Alvarez-Florez L, Sander J, Bourfiss M, Tjong FVY, Velthuis BK, Išgum I. Deep learning for automatic strain quantification in arrhythmogenic right ventricular cardiomyopathy. In: International workshop on statistical atlases and computational models of the heart. Springer; 2023. p. 25–34.
[107] López PA, Mella H, Uribe S, Hurtado DE, Costabal FS. WarpPINN: cine-MR image registration with physics-informed neural networks. Med Image Anal 2023;89: 102925.
[108] Knott KD, et al. The prognostic significance of quantitative myocardial perfusion: an artificial intelligence–based approach using perfusion mapping. Circulation 2020;141 (16):1282–91.
[109] Kotecha T, et al. Automated pixel-wise quantitative myocardial perfusion mapping by CMR to detect obstructive coronary artery disease and coronary microvascular dysfunction: validation against invasive coronary physiology. JACC Cardiovasc Imaging 2019;12(10):1958–69.
[110] Engblom H, et al. Fully quantitative cardiovascular magnetic resonance myocardial perfusion ready for clinical use: a comparison between cardiovascular magnetic resonance imaging and positron emission tomography. J Cardiovasc Magn Reson 2017;19:1–9.
[111] Bernard O, et al. Deep learning techniques for automatic MRI cardiac multi-structures segmentation and diagnosis: is the problem solved? IEEE Trans Med Imaging 2018;37 (11):2514–25.
[112] Jacobs M, et al. Automated segmental analysis of fully quantitative myocardial blood flow maps by first-pass perfusion cardiovascular magnetic resonance. IEEE Access 2021;9:52796–811.
[113] Xue H, et al. Automated inline analysis of myocardial perfusion MRI with deep learning. Radiol Artif Intell 2020;2(6):e200009.
[114] Moccia S, et al. Development and testing of a deep learning-based strategy for scar segmentation on CMR-LGE images. Magn Reson Mater Phys Biol Med 2019;32: 187–95.
[115] Paciorek AM, et al. Automated assessment of cardiac pathologies on cardiac MRI using T1-mapping and late gadolinium phase sensitive inversion recovery sequences with deep learning. BMC Med Imaging 2024;24(1):43.
[116] Zhang Q, et al. Toward replacing late gadolinium enhancement with artificial intelligence virtual native enhancement for gadolinium-free cardiovascular magnetic

resonance tissue characterization in hypertrophic cardiomyopathy. Circulation 2021; 144(8):589–99.

[117] Lu X, et al. Evaluation of deep learning-based reconstruction late gadolinium enhancement images for identifying patients with clinically unrecognized myocardial infarction. BMC Med Imaging 2024;24(1):127.

[118] Bahrami N, Retson T, Blansit K, Wang K, Hsiao A. Automated selection of myocardial inversion time with a convolutional neural network: spatial temporal ensemble myocardium inversion network (STEMI-NET). Magn Reson Med 2019;81(5): 3283–91.

[119] Zhuang X, et al. Cardiac segmentation on late gadolinium enhancement MRI: a benchmark study from multi-sequence cardiac MR segmentation challenge. Med Image Anal 2022;81:102528.

[120] Ohta Y, et al. Optimization of null point in Look-Locker images for myocardial late gadolinium enhancement imaging using deep learning and a smartphone. Eur Radiol 2023;33(7):4688–97.

[121] Xie C, et al. Automated inversion time selection for late gadolinium–enhanced cardiac magnetic resonance imaging. Eur Radiol 2024:1–13.

[122] Crawley R, et al. Automated cardiovascular MR myocardial scar quantification with unsupervised domain adaptation. Eur Radiol Exp 2024;8(1):93.

[123] Hundley WG, et al. Society for Cardiovascular Magnetic Resonance (SCMR) guidelines for reporting cardiovascular magnetic resonance examinations. J Cardiovasc Magn Reson 2022;24(1):29.

[124] Grosse-Wortmann L, et al. Society for Cardiovascular Magnetic Resonance guidelines for reporting cardiovascular magnetic resonance examinations in patients with congenital heart disease. J Cardiovasc Magn Reson 2024;26(2):101062.

[125] Bannur S, et al. Maira-2: grounded radiology report generation, arXiv preprint arXiv: 2406.04449, 2024.

[126] Luo L, Tang B, Chen X, Han R, Chen T. VividMed: vision language model with versatile visual grounding for medicine, arXiv preprint arXiv:2410.12694, 2024.

[127] Zhang Q, et al. Improving the efficiency and accuracy of cardiovascular magnetic resonance with artificial intelligence—review of evidence and proposition of a roadmap to clinical translation. J Cardiovasc Magn Reson 2024;26(2):101051.

[128] Jang J, et al. Sensitivity of myocardial radiomic features to imaging parameters in cardiac MR imaging. J Magn Reson Imag 2021;54(3):787–94.

[129] Jang J, et al. Reproducibility of segmentation-based myocardial radiomic features with cardiac MRI. Radiol Cardiothorac Imaging 2020;2(3):e190216.

CHAPTER

Machine learning in body MRI 21

Pouria Yazdian Anari[1, 2], Jovitha Nelson[3] and Ashkan A. Malayeri[2, *]

[1]*Radiology Department, Creighton University/CHI, Omaha, NE, United States;* [2]*Radiology Department, Clinical Center, NIH, Bethesda, MD, United States;* [3]*Interventional Radiology Department, Case Western University Hospitals Cleveland Medical Center, Cleveland, OH, United States*

[*]*Corresponding author: E-mail: ashkan.malayeri@gmail.com*

21.1 Introduction

Body magnetic resonance imaging (MRI) has emerged as an essential technique for diagnosing and monitoring various diseases due to its noninvasive approach and the yield of high-resolution images without exposure to ionizing radiation [1]. With the recent accelerated advancements in machine learning (ML) and artificial intelligence (AI), an opportunity to revolutionize clinical practices by adopting these techniques into body MRI procedures presents itself [2].

Current applications of AI in body MRI span from image acquisition/reconstruction innovations to postprocessing developments with advanced analytical tools. The realm of medical imaging has significantly advanced with the integration of ML algorithms for enhancing the efficiency and accuracy of the body MRI [3]. These advancements provide opportunities for early disease diagnosis and precise treatments through image segmentation, lesion detection, and tissue characterization [3,4]. The incorporation of ML assists in optimizing MRI imaging protocols by reducing scan times and improving image quality by removing artifacts and other limitations associated with conventional methods. Furthermore, implementing ML-based quantitative imaging biomarkers facilitates the evaluation of disease severity, the prediction of treatment responses, and personalized patient management [5].

Effectively incorporating ML in body MRI requires the close cooperation of specialists from various disciplines, including radiologists, medical physicists, computer scientists, and engineers. Radiologists provide clinical insights, define problem scopes, and validate results. Engineers and medical physicists contribute to designing and improving MRI acquisition protocols, image reconstruction techniques, and hardware design. Computer scientists are responsible for creating and refining ML algorithms that cater to specific clinical needs. Collaboration among the interdisciplinary team is critical for developing pioneering solutions and

Machine Learning in MRI. https://doi.org/10.1016/B978-0-443-14109-6.00021-3

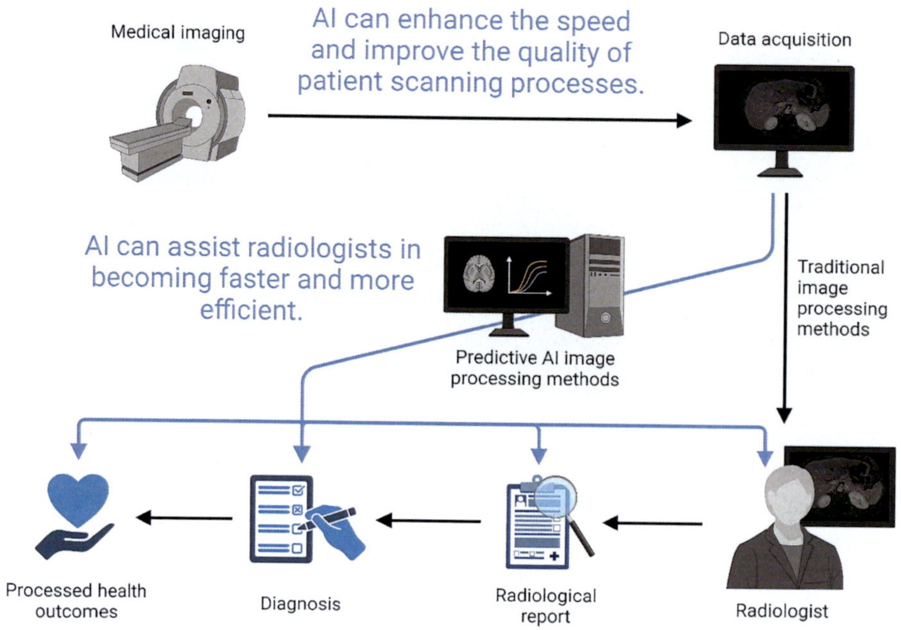

FIGURE 21.1

Illustration of the role of artificial intelligence (AI) in medical imaging workflows. AI enhances the speed and quality of patient scanning processes, optimizing data acquisition. Predictive AI image processing methods assist radiologists by improving efficiency and accuracy in image interpretation, diagnosis, and reporting. The integration of AI into radiological practice contributes to faster diagnoses, improved health outcomes, and streamlined workflows.

translating them into practical use. Implementing ML tools into clinical workflows for body MRI requires the consideration of regulatory requirements, ethical implications, and practical challenges. Safe and effective integration can be achieved through collaboration between researchers and clinicians from different disciplines, improving patient outcomes (Fig. 21.1).

This chapter provides a comprehensive overview of the current status of AI in body MRI, its applications in clinical care and oncological imaging, anticipated difficulties, and promising future directions.

21.1.1 Liver and abdominal imaging

21.1.1.1 Fibrosis and cirrhosis assessment

ML has shown promising results in the noninvasive assessment of liver stiffness, a surrogate marker for liver fibrosis and cirrhosis. Most of the models for evaluation of the liver stiffness have been developed to categorically classify liver stiffness measurements derived from Magnetic Resonance Elastography (MRE), using clinical data and MRI radiomic features. Clinical data may include demographics, medical history, and laboratory features. MRI radiomic features encompass geometric, histogram-based, and textural characteristics of liver tissue.

ML, particularly deep learning model convolutional neural networks (CNNs), and algorithms such as Random Forest (RF), Support Vector Machine (SVM), Naïve Bayes, K-Nearest Neighbor (KNN), and Extreme Gradient Boosting (XGBoost), have been employed to significantly enhance the diagnosis and staging of hepatic fibrosis and cirrhosis, at times outperforming traditional methods such as liver biopsy and serological testing [1–6].

Chen et al. found that among the ML algorithms tested for analyzing real-time tissue elastography (RTE) images for hepatic fibrosis, RF achieved the highest accuracy. Specifically, the RF classifier demonstrated an accuracy of 82.87% when differentiating between stage S0 and stage S1-4 of hepatic fibrosis. In addition, when comparing medium-level fibrosis stages S0–2 and high-level fibrosis stages S3–4, the RF classifier showed an impressive accuracy of 88.09%. These results highlight the potential of ML, notably the RF algorithm, in accurately staging hepatic fibrosis, which is crucial for the effective clinical management of liver diseases [7].

A study by Hong-Ye Peng et al. validated five novel ML models for predicting nonalcoholic fatty liver disease (NAFLD). Notably, the XGBoost model displayed the best predictive performance, with an area under the curve (AUC) increased to 0.938 after parameter tuning. This indicates that XGBoost is a reliable reference for early identification of high-risk patients with NAFLD in clinical practice [8].

Other than ML-based algorithms, others have studied the role of deep learning algorithms in evaluation of liver fibrosis. For example, Hectors and colleagues developed a cross-contrast neural network (CCNN) that integrates the benefits of statistical analysis and CNNs, fitting the problem of medical image analysis with a relatively small dataset. The CCNN achieved high accuracy scores ranging from 83.3% to 98.33%, particularly in binary classification tasks related to liver fibrosis, which demonstrates the method's efficiency [9]. Furthermore, in Hectors et al.'s study in 2020, a VGG16 CNN model was utilized for staging liver fibrosis, achieving AUROCs ranging from 0.77 to 0.91 for different fibrosis stages. This demonstrates the model's capability to distinguish between various stages of liver fibrosis with considerable accuracy [10].

The combined use of clinical features and radiomic features has resulted in an improved diagnostic performance when compared to the separate use of these features. In an internal validation study, liver stiffness was predicted with ML using a combination of clinical and radiomic features (e.g., demographics, medical history, diagnoses including NAFLD, and laboratory testing including alanine aminotransferase level, aspartate aminotransferase level, bilirubin level, albumin level, aspartate aminotransferase-to-platelet ratio index score, and fibrosis-4 score). The SVM model used in the study achieved an accuracy of 81.8%, a sensitivity of 72.2%, and a specificity of 87.0%. This performance was reaffirmed in an external validation study, where the model maintained a strong accuracy of 75.0%. By accurately identifying patients with normal or near-normal liver stiffness, this type of model has the potential to reduce the need for invasive and costly procedures like MRE [15].

Interestingly, when considering different liver stiffness cutoff values, it was noted that the standard cutoff of less than 3 kPa provided the best balance between sensitivity and specificity [11]. Moreover, the diagnostic performance of SVM models was affected by fatty liver disease, which can confound liver stiffness measurements. For patients without fatty liver disease, the model's accuracy and AUC improved slightly [1].

Despite the promising results, limitations such as the smaller sample size in the external cohort and the heterogeneity of the study population need to be addressed in future research to ensure the generalizability of the findings. Models that can distinguish between simple fatty liver and steatohepatitis and between early-stage and late-stage NAFLD, which have significantly different clinical prognoses, need to be developed.

21.1.1.2 Lesion segmentation and classification

Another important application of ML and deep learning is the automatic segmentation of the entire liver. Recent studies in liver MRI segmentation have showcased significant advancements in AI techniques, mainly using deep learning models. Winther et al. utilized a 3D U-Net for segmenting liver from Gd-EOB-DTPA-enhanced MRI images for volume calculations, achieving a DICE score of 0.96 $\pm$ 1.9. This study underlines the growing accuracy and reliability of AI in liver segmentation, which is crucial for various clinical applications, including surgical planning, and liver volume assessment [12].

The use of AI in segmenting the liver is valuable for a wide range of diagnostic assessments and is already in use in many clinical platforms. One of these studies was done by Liu et al. focused on segmenting the liver for automated liver iron quantification using a 2D U-Net on T2 maps, achieving a DICE score of 0.86 $\pm$ 0.01. This highlights the model's effectiveness in specific quantitative assessments [8]. Another study by Wang et al. employed a 2D U-Net to segment the liver across multiple imaging modalities, including T1- and T2-weighted images. As a result, the DICE scores of 0.95 $\pm$ 0.03 for T1-weighted images and 0.92 $\pm$ 0.05 for T2--weighted images were obtained [13].

AI segmentation methods also have profound implications in treatment and therapy planning. The study by Chen et al. uses a 2D Dense U-Net to segment multiple organs in abdominal scans to aid radiotherapy planning. The researchers achieved a notable DICE score of 0.96 $\pm$ 0.009 in the liver [14]. The 2D Dense U-Net model in this study reflects the potential for AI use in precise delineation of liver and tumor boundaries, which is essential for effective radiotherapy planning.

Other than whole liver segmentation studies, there is an evolving role for not only segmenting the outline of the liver but also for more detailed segmentation of liver structures, including the vasculature and the biliary tree. Recent studies employing sophisticated methods like 4D k-mean clustering and 2D U-Nets demonstrate the progression toward more nuanced and comprehensive segmentation approaches

[15,16]. These methods have shown promising results in segmenting the biliary tree and liver vasculature with a more accurate assessment of liver parenchymal volume, an important predictor for success of liver interventions.

The application of AI in segmenting liver lesions has also been investigated in several studies; one of these studies by Bousabarah et al. utilized a 2D U-Net for segmenting and delineating hepatocellular carcinomas (HCCs) in Gadoxetic acid-enhanced MRI, achieving the DICE scores of 0.91 ± 0.01 for the liver and 0.68 $\pm$ 0.03 for tumors. These scores indicate the growing proficiency of AI models in segmenting complex liver lesions, which is critical for accurate tumor characterization and subsequent treatment strategies [17].

In addition, the application of AI in liver MRI extends to tumor characterization. In an interesting study, Wu et al. developed an AlexNet CNN model to classify liver tumors according to their LI-RADS grade, achieving an impressive accuracy of 90%, showcasing the potential role of AI in decision-making support [18]. In similar studies, Messaoudi et al. and Hamm et al. utilized deep learning models for classifying liver tumors into HCC or non-HCC, demonstrating the versatility and effectiveness of AI in differentiating various types of liver tumors [19,20]. In another study, Liu et al. used SVM with radiomic features for classifying combined hepatocellular cholangiocarcinoma (cHCC-CC) versus non-cHCC-CC, achieving a mean AUROC of 0.77 [21]. There have also been recent advances in screening for liver masses; for instance, CNN has shown to be highly accurate and specific in classifying tumors into type and LI-RADS-derived classes [1,20,22].

Despite the notable advancements in AI technology for liver disease classification and segmentation, several challenges remain, necessitating areas for improvement to enhance diagnostic accuracy and efficiency. Key among these challenges is the need to develop AI models that are robust to variations in imaging equipment and protocols, ensuring consistent performance across different clinical settings. Moreover, fostering ongoing collaboration between AI researchers, radiologists, and hepatologists is vital for refining AI models to meet clinical needs better, encompassing both technical aspects and the practical challenges of liver disease diagnosis and treatment (Fig. 21.2).

21.1.1.3 Renal function assessment

Integrating ML techniques in MRI can aid in renal function assessment, offering noninvasive methods for early disease detection and monitoring. Several studies have delved into this domain, applying various ML models to enhance the diagnostic accuracy and predictability of renal diseases, particularly chronic kidney disease (CKD).

Diabetes presents significant challenges in managing renal health, prompting researchers to explore innovative methods for assessing renal function. Yi Deng and colleagues lead an approach utilizing radiomics texture in diffusion tensor imaging (DTI) to identify early kidney damage in individuals with diabetes. Their method extracted texture from specific kidney areas using MaZda software. This method achieved an area under the receiver operating characteristic (ROC) curve (AUC) of 0.882, indicating promising diagnostic performance [23]. Expanding upon these

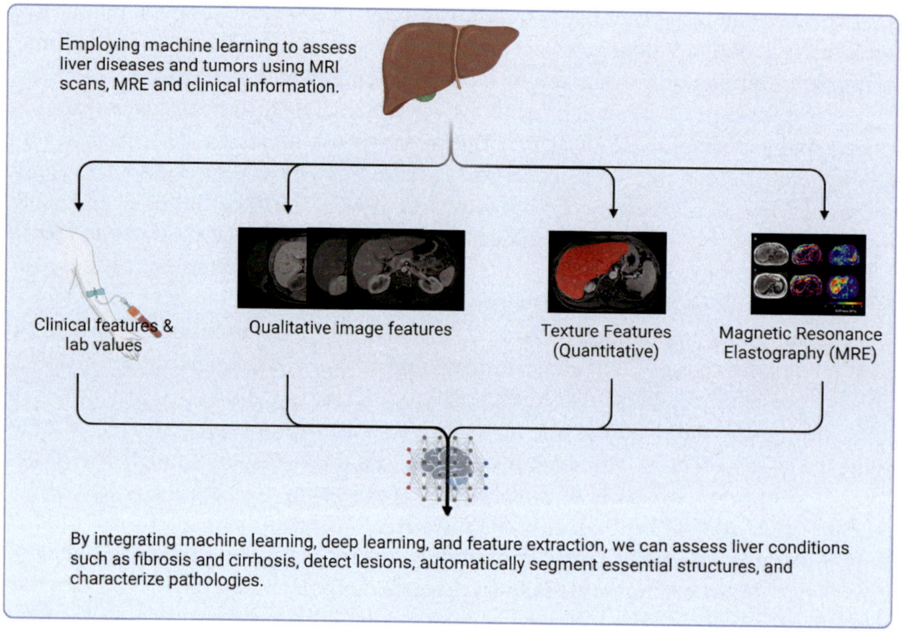

FIGURE 21.2

Machine learning-driven assessment of liver diseases and tumors using MRI, Magnetic Resonance Elastography (MRE), and clinical information. This approach integrates clinical features, qualitative and quantitative image features, and MRE-derived data to improve the detection, segmentation, and characterization of liver pathologies such as fibrosis, cirrhosis, and tumors.

methods, Xiaokai Mo and colleagues developed MRI texture-based ML models to noninvasively assess renal function in diabetic patients, employing various kidney segmentation methods on T2-weighted images. This study, involving 174 patients, indicated that models based on the largest coronal slices and all coronal slices achieved high accuracy. The optimal model yielded AUCs of 0.938, 0.919, and 0.959 for classifying normal function, mild or moderate impairment, and severe impairment, respectively [24].

Continuing the exploration to produce models that can predict renal function in diabetes patients, we can find several other studies. For instance, in a study by Wenbo Chen et al., the team utilized ML-based multimodal MRI texture analysis (mMRI-TA) to assess renal function and fibrosis in diabetic nephropathy. Employing the speeded-up robust features (SURF) algorithm for texture feature extraction from T2-weighted MRI and incorporating quantitative values from blood oxygenation level-dependent (BOLD) and diffusion-weighted imaging (DWI) sequences, their mMRI-TA model demonstrated exceptional performance. The model proficiently classified the severe renal function impairment (sRI) group, nonsevere renal function impairment (non-sRI) group, and normal renal function (normal-RF) group, yielding AUCs of 0.961 (95% CI: 0.853, 1.000), 0.809 (95% CI: 0.600, 0.980), and 0.850 (95% CI: 0.638, 0.988), respectively. This study underlines the substantial potential of integrating various MRI modalities through ML to significantly enhance diagnostic accuracy for DN [25].

Building on the foundation of innovative diagnostic methods for renal health, the examination of renal function in transplant patients introduces a crucial dimension to the utilization of ML in MRI analyses. Patients with renal transplants are at risk of kidney function deterioration, necessitating continuous monitoring to evaluate the transplanted organ's health. In this context, a study by Małgorzata Grzywinska et al. explored the use of texture analysis on T2-weighted MRI images to assess renal function after kidney transplantation. The study found a correlation between texture parameters and estimated glomerular filtration rate, underscoring the potential of texture analysis as a tool for evaluating transplanted kidney function [26].

Other MRI acquisition techniques have been unitized for texture analysis; for example, Jiule Ding et al. explored the value of texture analysis based on DWI, -BOLD, and susceptibility-weighted imaging (SWI) in evaluating renal dysfunction. The study included 72 patients who were classified based on their estimated glomerular filtration rate (eGFR) into either severe renal function impairment (sRI, eGFR <30 mL/min/1.73 m^2), nonsevere renal function impairment (non-sRI, eGFR ≥ 30 mL/min/1.73 m^2 and <80 mL/min/1.73 m^2), or control (CG, eGFR ≥ 80 mL/min/1.73 m^2) groups. Thirteen texture features were extracted and analyzed. Six features emerged as the most valuable for assessing renal dysfunction: 0.25QuantileDWI, 0.5QuantileDWI, HomogeneityDWI, EntropyBOLD, SkewnessSWI, and CorrelationSWI. The study found that features derived from DWI were smaller in sRI than in non-sRI; EntropyBOLD and CorrelationSWI were smaller in non-sRI than in CG, all with $P < .05$. This study concluded that texture analysis based on DWI, BOLD, and SWI can assist in assessing renal dysfunction, especially in the early stages, demonstrating that texture features based on BOLD and SWI may be suitable for assessing renal function impairment during these stages [27]. The comprehensive body of research underscores the transformative impact of ML on enhancing MRI's diagnostic ability for kidney function analysis. By leveraging the capabilities of ML, the medical community can be equipped with more reliable and efficient tools for early disease detection, enabling timely interventions and potentially improving patient outcomes.

21.1.2 Renal mass segmentation, detection, and characterization

Despite the potential of deep learning, particularly CNN, in revolutionizing the segmentation, detection, and characterization of renal masses such as clear cell renal cell carcinomas (ccRCC) and associated cysts, research in this area remains relatively sparse. This scarcity of studies underscores a significant opportunity for further exploration and development within this field. The application of CNNs has shown promise in enhancing diagnostic precision, yet it faces limitations, including the need for large, annotated datasets and the challenge of model generalization across diverse imaging protocols. Future research directions could focus on addressing these challenges by developing strategies for data augmentation, transfer learning, and the creation of more robust models capable of handling variability in imaging data. In addition, interdisciplinary collaboration could open opportunities for novel approaches in data collection and annotation [28].

In evaluating 2D and 3D U-Net models, findings reveal differences in their performance for segmenting renal masses, offering insights into the potential and limitations of each model in clinical application. A recent study demonstrated the automated segmentation of ccRCC, cysts, and the surrounding normal kidney parenchyma in VHL patients using CNNs on MRI images. The study, leveraging a database of 115 VHL patients with 173 pathologically confirmed ccRCCs, employed U-Net architectures for both 2D and 3D CNN models. These models were trained on a combination of MRI phases, including precontrast, corticomedullary, nephrographic, and excretory phases, with manual segmentation of lesions serving as ground truth.

The 2D U-Net model showcased an average lesion detection AUC of 0.88, with Dice Similarity Coefficient (DSC) scores of 0.78 for kidneys, 0.40 for cysts, and 0.46 for tumors. Meanwhile, the 3D U-Net model exhibited slightly lower performance with an average ccRCC lesion detection AUC of 0.79 and DSC scores of 0.67 for kidney, 0.32 for cysts, and 0.34 for tumors. These findings indicate the efficacy of CNN-based models, particularly the 2D U-Net, in accurately segmenting and detecting renal masses and cysts, thus facilitating the quantification of disease burden in VHL patients [28].

In addition to the techniques previously discussed, recent advancements have further refined the capabilities of ML in MRI for renal mass analysis, specifically through the development of a differentiable decision forest, known as Hinge Forest (HF), tailored for segmenting hereditary ccRCC in MRI scans. This innovative approach, applied to a dataset encompassing 115 patients with 504 ccRCCs and 1171 cystic lesions, demonstrated notable segmentation accuracy. HF showcased superior performance in tumor segmentation compared with traditional U-Net architectures, achieving an average DSC of 0.53 for tumors, thus outperforming U-Net in this aspect. However, U-Net demonstrated slightly better results in segmenting kidney parenchyma, highlighting the trade-offs between different ML models. The HF model's capability to provide interpretable insights into its decision-making process, through the generation of leaf maps, represents a significant step toward more transparent and understandable AI-driven diagnostics in renal disease management [29].

AI and ML can also help with detection of renal masses. In a recent study, utilizing YOLOv7 (You Only Look Once) on excretory phase MRI scans across a large patient cohort with renal cell carcinoma (RCC), employing both 2D and innovative 2.5D approaches. the best model performance in 2.5D showed even highest efficacy with a PPV of 0.75, sensitivity of 0.69, and F1 score of 0.72 [30].

Exploring the integration of ML with MRI in renal health represents a frontier teeming with potential, yet it navigates a landscape punctuated by challenges. Future directions in this burgeoning field must address issues such as data heterogeneity, where variability in MRI equipment and protocols across institutions can obstruct the development of universally applicable models. In addition, the imperative of enhancing model interpretability; as these algorithms become increasingly integral to diagnostic processes, understanding the "why" behind their predictions is crucial for clinician trust and decision-making. As researchers and practitioners chart the course forward, addressing these limitations will be paramount in fully unleashing the potential of ML-enhanced MRI for renal disease detection and monitoring.

21.1.3 Pelvic imaging

21.1.3.1 Prostate cancer detection and staging

Prostate cancer (PCa) is the second most common cancer and the fifth leading cause of cancer-related death among men. Its incidence rates are higher in developed countries (37.5 per 100,000) compared with developing countries (11.3 per 100,000). The mortality rates follow a similar trend with 8.1 per 100,000 in developed countries and 5.9 per 100,000 in developing countries. In 2020, approximately 1.4 million new PCa cases were reported globally, resulting in around 375,000 deaths [31].

MRI has become a crucial tool in the detection and staging of prostate cancer and is recommended as the first-line screening method for patients with a clinical suspicion of PCa. The Prostate Imaging-Reporting and Data System (PI-RADS) provides guidelines to stratify the probability of clinically significant prostate cancer on MRI. These guidelines have undergone numerous revisions to improve their effectiveness, but differentiating cancer from noncancerous pathology on MRI remains a challenge [32–34].

Recent advancements in deep learning-based methods have shown promise in improving the detection and staging of PCa. Deep CNNs have been developed for prostate segmentation and demonstrate higher accuracy than conventional methods. These CNNs are capable of learning appropriate features of input images and incorporating data augmentation methods. Several studies have utilized deep learning for detecting, grading, and segmenting PCa and have achieved high sensitivity and specificity values [35–37].

A study conducted by Wang et al. focused on the application of AI in enhancing the accuracy of prostate cancer detection on MRI. This retrospective study utilized a deep learning algorithm trained on a large dataset of prostate MRI scans. The AI model was designed to differentiate between benign and malignant lesions, with a specific focus on clinically significant prostate cancer (csPCa). The study found that the AI model achieved a sensitivity of 90%, a specificity of 85%, and an AUC of 0.93. These results indicate that AI, particularly deep learning models, may significantly improve the precision of prostate cancer diagnosis, potentially leading to improved patient outcomes [38].

In the context of radiotherapy for PCa, MRI plays a vital role as it aids in localizing tumors, establishing radiotherapy treatment plans, and assessing radiotherapy efficacy. Precision therapy techniques like intensity-modulated radiotherapy and 3D conformal radiation therapy have been developed to improve the accuracy of radiotherapy. However, challenges remain in manual segmentation of the prostate gland for accurate mapping of the planned target volume and organs at risk. To address these challenges, MRI-only radiotherapy has been developed, combining MRI and CT simulations to enhance radiotherapy planning.

Furthermore, MRI plays a significant role in monitoring PCa prognosis and reducing mortality, particularly through the assessment of biochemical recurrence (BCR) postradical prostatectomy. A study using deep learning for BCR prediction achieved maximum Area Under Curve (AUC) values of 0.85 and 0.88 for BCR-free survival prediction at 3 and 5 years, respectively. MRI-based techniques are also significant in detecting PCa metastasis, with recent studies demonstrating accurate detection and segmentation of pelvic bone metastases [39].

Prostate and prostate tumor segmentation using MRI is a critical process in the diagnosis and treatment planning of PCa. The segmentation of the prostate and its tumors in MRI images allows for precise localization, size estimation, and characterization of the lesions. This process is essential for determining the appropriate course of treatment. Multiparametric MRI (mpMRI) is the primary imaging technique used, offering high-resolution images that combine anatomical and functional data. The mpMRI typically includes T2-weighted imaging for anatomical details, DWI for cell density, and dynamic contrast-enhanced (DCE) imaging for blood flow patterns. These diverse data sets from mpMRI provide a comprehensive view of the prostate gland and any potential tumors [40].

Deep learning methods, particularly CNNs, have significantly advanced the field of prostate and tumor segmentation. These methods automate segmentations, which were traditionally done manually, leading to increased precision and accuracy of prostate tumor identification. The deep learning models are trained on large datasets of annotated MRI images to identify the patterns and features characteristic of prostate tumors [41,42].

Studies have shown that deep learning-based segmentation can achieve high accuracy. For example, a study using a 3D deep learning model for prostate segmentation achieved a DSC of up to 0.95, indicating a high level of agreement between the automated segmentation and expert radiologists' annotations. Another study utilizing a deep CNN reported a sensitivity of over 90% in detecting clinically significant prostate tumors, demonstrating the effectiveness of these methods in identifying tumors that require treatment [41].

In addition to improving diagnostic accuracy, automated prostate and tumor segmentation plays a vital role in treatment planning, especially in radiotherapy. Accurate segmentation ensures that the radiation dose can be precisely targeted to the tumor while minimizing exposure to healthy tissue. This precision is vital for effective treatment and the reduction in side effects. Overall, the integration of advanced imaging techniques like mpMRI with deep learning algorithms has revolutionized the approach to prostate and prostate tumor segmentation. It has enhanced the accuracy, efficiency, and consistency of the diagnosis and staging of prostate cancer, leading to more personalized and effective treatment strategies [43].

21.1.3.2 Female pelvic imaging: Cervical, uterine, and ovarian

The application of ML, especially using multiomics data, is revolutionizing our approach to female reproductive health. Models created from this approach use a broad range of variables, including clinical, metabolic, hormonal, and genetic data, and have been validated in different studies. For example, researchers like Günakan et al. have developed predictive models for lymph node metastasis in endometrial cancer using ML, achieving an accuracy of approximately 85% [44].

MRI plays a crucial part in the management of endometrial carcinoma, the most common gynecological tumor in developed countries. Radiomics analysis of MRI data can predict molecular and clinicopathological prognostic factors in patients with endometrial carcinoma. Studies involving shown MRI's effectiveness in predicting high-grade endometrial carcinoma and other key factors, indicating its potential to improve the accuracy of staging and treatment planning. For instance, Dong et al. developed a deep learning model based on 4896 MRI slices from 72 patients with stage I endometrial carcinoma to predict deep muscle invasion, achieving an accuracy rate of 75%. The accuracy of the model in contrast-enhanced T1--weighted images was found to be higher (79.2%) compared with T2-weighted images (70.8%), indicating a differential performance based on the imaging technique [45]. Similarly, Chen et al. used MR images from 530 patients to develop a deep learning-based model that predicted myometrial invasion depth with an accuracy of 84%, a sensitivity of 66.7%, and a specificity of 87.5%. These studies exemplify the potential of deep learning in enhancing diagnostic accuracy and treatment planning in endometrial cancer [46,47].

In ovarian tumor assessment, a study by Wang et al. evaluated a CNN for ovarian tumor differentiation based on MRI, analyzing 545 lesions from 451 patients. This model, compared with junior radiologists, achieved a higher test accuracy (0.87 vs. 0.64) and specificity (0.92 vs. 0.64) with comparable sensitivity, showcasing AI's capability to surpass traditional radiological assessments and support radiologists, particularly less experienced ones, in improving their diagnostic performance. These findings collectively underscore the promising role of AI and radiomics in advancing ovarian tumor diagnostics.

Dong et al. developed and validated a deep-learning radiomics model with a cohort of 226 patients in cervical cancer treatment outcomes. This model predicted lymph node status in operable cervical cancer with an accuracy of 97.16% and an AUC of 0.99 in internal validation. It maintained high performance in independent external validation with an AUC of 0.90 and accuracy of 92.00%. The results highlight the model's utility in enhancing presurgical planning and decision-making.

In the same realm, a study by Takada et al. aimed to refine prognostic predictions in cervical cancer treated with definitive radiotherapy. By incorporating peritumoral regions into the volume of interest (VOI) for MRI radiomics analysis, they enhanced the model's predictive capacity. The expanded VOI models achieved AUC-ROCs up to 0.86, outperforming the predictions based solely on the tumor VOI.

Wu et al. employed ML to predict lymph node metastasis in cervical cancer, leveraging MRI-based radiomics. This study, while not providing specific numerical details initially, utilized a cohort of 115 patients to analyze the association between texture features derived from DWI images and apparent diffusion coefficient (ADC) maps with the extent of local invasion and nodal involvement. Through their rigorous ML approach, they identified independent predictors of local invasion and nodal involvement, with their models achieving AUCs of 0.793 and 0.802, respectively. The sensitivity and specificity for predicting nodal involvement were 80.77% and 68.25%, respectively, illustrating the significant potential of radiomics in accurately identifying patients at risk of metastasis.

In the area of applying ML to the assessment of female pelvic lesions, significant challenges persist, notably in the areas of data diversity and algorithmic bias. Future research must prioritize collecting and utilizing diverse datasets that accurately reflect the broad spectrum of patient demographics encountered in gynecological oncology. This diversity is crucial for training models that are equitable and effective across different populations. In addition, there is a pressing need to address algorithmic bias, which can lead to disparities in diagnostic accuracy and treatment recommendations. Efforts should be directed toward developing methodologies for detecting and correcting biases within ML models. Moreover, interdisciplinary research is essential to bridge the gap between ML advancements and their clinical application, ensuring that these technologies are developed with a clear understanding of the clinical context. This approach will not only enhance the reliability and applicability of ML models but also ensure they contribute effectively to patient care in the field of gynecologic oncology.

21.1.4 Gastrointestinal imaging

21.1.4.1 Colorectal cancer detection and staging

The integration of AI with MRI is setting new benchmarks in managing colorectal cancer (CRC). Advanced AI technologies not only aid in predicting the risk of metastases with enhanced precision but also significantly lower the miss rates for colorectal neoplasia through Computer-Aided Detection (CAD) systems.

Radiomics is at the forefront of transforming CRC screening. The study by Ma et al. serves as an example, where an MRI-based radiomics model utilizing high-resolution T2-weighted images (T2WIs) was developed to predict the pathological features of rectal cancer. In this retrospective analysis of 152 patients, the application of ML classifiers to 1029 extracted radiomics features highlighted the predictive capabilities of these models. Specifically, the Support Vector Machine (SVM) classifier demonstrated a superior performance in predicting the degree of differentiation, achieving an AUC of 0.862, with sensitivity and specificity rates of 83.3% and 85.0%, respectively. Furthermore, the RF classifier showed promise in N stage prediction, with an AUC of 0.746, sensitivity of 79.3%, and specificity of 72.2%. These findings illustrate the potential of T2WI-based radiomics as effective biomarkers for rectal cancer [48]. Building upon this foundation, the research by Lu et al., focused on assessing local invasion of rectal cancer through texture analysis from sagittal fat-suppression and transverse T2-weighted MRI. Their study of 174 rectal cancer cases aimed to distinguish between high (T3/4) and low (T1/2) T-stage tumors using univariate and multivariate analyses. The multivariate analysis, incorporating texture features, achieved an AUC of 0.789, indicating a moderate correlation with T-stage [49].

Moreover, another investigation sought to improve preoperative evaluation by analyzing texture features from DWI and ADC maps, aiming to associate these features with stages of rectal cancer, particularly local invasion (T1-2 vs. T3-4) and nodal involvement (N0 vs. N1-2). Involving 115 patients, the research assessed ADC values and a range of texture features, employing statistical and multivariate logistic regression analyses to develop predictive models. The models for predicting local invasion and nodal involvement demonstrated promising predictive performance, with AUCs of 0.793 and 0.802, sensitivities of 78.57% and 80.77%, and specificities of 74.19% and 68.25%, respectively [50].

Advancements in imaging and ML can be use in colorectal cancer treatment, enabling precise predictions of responses to neoadjuvant therapy in patients with locally advanced rectal cancer (LARC). This progress is critical as accurately foreseeing treatment outcomes can significantly optimize therapeutic strategies, improving patient prognosis and management. In an approach to predict treatment response to neoadjuvant chemotherapy for LARC, Li et al., in their study, developed a multimodal radiomics model combining computed tomography (CT) and MRI radiomics features. This retrospective analysis included 118 patients, showing that the high-resolution T2-weighted imaging (HR-T2WI) model outperformed the CT, dynamic contrast-enhanced T1 images (DCE-T1), and ADC models in validation, with superior AUC and accuracy (ACC). Specifically, the HR-T2WI model achieved an AUC of 0.859 and an ACC of 0.896, while the multimodal radiomics model demonstrated even higher efficacy with an AUC of 0.93 and an ACC of 0.875 in the validation set [51].

Parallelly, Fu et al., in their research, explored the potential of deep learning D-based radiomic features extracted from pretreatment DWIs for predicting the response to neoadjuvant chemoradiation treatment (nCRT) in LARC patients. Including 43 patients categorized into responders and nonresponders based on post-nCRT assessments, they discovered that DL-based features significantly outperformed human extracted features in predictive performance. The DL-based model not only achieved a notable mean AUC of 0.73 but also demonstrated an improvement in classification performance, with a model accuracy of up to 87.69% for distinguishing between T1/T2 and T3/T4 cancer stages [52].

The use of deep learning in auto-segmentation of colorectal cancer on MRI represents a significant advance in the treatment of LARC. This technology facilitates precise tumor delineation, essential for effective treatment planning and producing models for detection and characterization of tumors.

One pivotal study by Trebeschi et al. explores the application of CNNs for the automatic localization and segmentation of rectal cancers on multiparametric MRI scans. Involving 140 patients, this study demonstrates the CNN's proficiency in differentiating tumor from nontumor tissues with high segmentation accuracy. The CNN achieved DSC scores of 0.68 and 0.70 for two independent readers and outstanding AUC values of 0.99, showcasing the potential of deep learning to expedite and refine MRI-based rectum segmentations [53].

Concurrently, a study conducted by Soomro et al. introduces a cutting-edge method for automatically segmenting colorectal cancer from 3D MRI images, employing a synergistic blend of 3D fully convolutional neural networks (3D-FCNNs) and a 3D level-set algorithm. This innovative combination aims to enhance the training phase and refine output during testing by incorporating smoothing functions and prior knowledge in a postprocessing step. The effectiveness of this method was validated through comparison with standalone 3D-FCNNs, demonstrating superior performance. Specifically, the integration of the 3D level-set algorithm resulted in higher DSC scores on both training and testing datasets, with scores of (0.91813 vs. 0.8568) and (0.9378 vs. 0.86238), respectively. These findings underscore the proposed method's ability to achieve more precise segmentation outcomes for 3D colorectal MRI data compared with using 3D-FCNNs alone, marking a significant advancement in the field [54]. In conjunction with the innovative method proposed by Soomro et al., the study by Jian et al. delves into the use of fully convolutional networks (FCNs) to segment colorectal tumors on T2-weighted MRI. Facing the challenge of blurred lesion boundaries. The methodology is distinct for using side-output blocks from VGG-16's convolutional layers, enabling the capture of multiscale features and enhancing tumor boundary detection. A comprehensive evaluation using 2772 manually segmented colorectal tumor images from T2-weighted MRIs revealed that the proposed method outperforms conventional U-net approaches, with significant metrics including an average Dice similarity coefficient of 83.56, positive predictive value of 82.67, specificity of 96.75, and sensitivity of 87.85. These results underscore the FCN-based method's superiority in accurately segmenting colorectal tumors, offering a promising alternative to the labor-intensive and subjective manual segmentation processes currently in practice [55].

21.2 Discussion and conclusion

Despite these advancements, the integration of AI with MRI in oncology faces challenges, including the need for large, annotated datasets to train models, potential biases, and the requirement for extensive validation in diverse clinical settings. In addition, the interpretability of AI models remains a concern, as the complex algorithms can act as "black boxes," making it difficult for clinicians to understand the rationale behind specific predictions or recommendations. Addressing these limitations is crucial for ensuring the ethical and effective application of AI in clinical practice, and ongoing research is focused on enhancing the transparency, reliability, and inclusivity of AI models in healthcare.

List of abbreviations

ADC	Apparent Diffusion Coefficient
AI	Artificial Intelligence
AUC	Area Under the Curve
BCR	Biochemical Recurrence
BOLD	Blood Oxygenation Level-Dependent
CAD	Computer-Aided Detection
CCNN	Cross-Contrast Neural Network
ccRCC	Clear Cell Renal Cell Carcinoma
CKD	Chronic Kidney Disease
CNNs	Convolutional Neural Networks
CRC	Colorectal Cancer
csPCa	Clinically Significant Prostate Cancer
CT	Computed Tomography
DCE	Dynamic Contrast-Enhanced
DSC	Dice Similarity Coefficient
DTI	Diffusion Tensor Imaging
DWI	Diffusion-Weighted Imaging
eGFR	Estimated Glomerular Filtration Rate
FCNNs	Fully Convolutional Neural Networks
Gd-EOB-DTPA	Gadolinium-ethoxybenzyl-diethylenetriamine pentaacetic acid
HCC	Hepatocellular Carcinoma
HF	Hinge Forest
KNN	K-Nearest Neighbor
LARC	Locally Advanced Rectal Cancer
ML	Machine Learning
mpMRI	Multiparametric MRI
MRE	Magnetic Resonance Elastography
MRI	Magnetic Resonance Imaging
NAFLD	Nonalcoholic Fatty Liver Disease
nCRT	Neoadjuvant Chemoradiation Treatment
PCa	Prostate Cancer
PI-RADS	Prostate Imaging-Reporting and Data System
RCC	Renal Cell Carcinoma
RF	Random Forest
ROC	Receiver Operating Characteristic
RTE	Real-Time Elastography
SURF	Speeded-Up Robust Features
SVM	Support Vector Machine

SWI	Susceptibility-Weighted Imaging
VOI	Volume of Interest
XGBoost	Extreme Gradient Boosting
YOLOv7	You Only Look Once version 7

References

[1] He L, et al. Machine learning prediction of liver stiffness using clinical and T2-weighted MRI radiomic data. AJR Am J Roentgenol 2019;213(3):592–601.
[2] Yu Q, et al. An imaging-based artificial intelligence model for non-invasive grading of hepatic venous pressure gradient in cirrhotic portal hypertension. Cell Rep Med 2022;3 (3):100563.
[3] Wang K, et al. Deep learning radiomics of shear wave elastography significantly improved diagnostic performance for assessing liver fibrosis in chronic hepatitis B: a prospective multicentre study. Gut 2019;68(4):729–41.
[4] Yasaka K, et al. Deep learning for staging liver fibrosis on CT: a pilot study. Eur Radiol 2018;28(11):4578–85.
[5] Ahmed Y, et al. Detecting liver fibrosis using a machine learning-based approach to the quantification of the heart-induced deformation in tagged MR images. NMR Biomed 2020;33(1):e4215.
[6] Yasaka K, et al. Liver fibrosis: deep convolutional neural network for staging by using gadoxetic acid-enhanced hepatobiliary phase MR images. Radiology 2018;287(1): 146–55.
[7] Chen Y, et al. Machine-learning-based classification of real-time tissue elastography for hepatic fibrosis in patients with chronic hepatitis B. Comput Biol Med 2017;89:18–23.
[8] Peng HY, et al. Development and validation of machine learning models for nonalcoholic fatty liver disease. Hepatobiliary Pancreat Dis Int 2023;22(6):615–21.
[9] Yudan H, et al. A liver fibrosis staging method using cross-contrast network. Expert Syst Appl 2019;130:124–31.
[10] Hectors SJ, et al. Fully automated prediction of liver fibrosis using deep learning analysis of gadoxetic acid-enhanced MRI. Eur Radiol 2021;31(6):3805–14.
[11] Michelotti FC, et al. An empirical approach to derive water T from multiparametric MR images using an automated pipeline and comparison with liver stiffness. J Magn Reson Imag 2024;59(4):1193–203.
[12] Winther H, et al. A 3D deep neural network for liver volumetry in 3T contrast-enhanced MRI. Röfo 2021;193(3):305–14.
[13] Wang K, et al. Automated CT and MRI liver segmentation and biometry using a generalized convolutional neural network. Radiol Artif Intell 2019;1(2).
[14] Chen Y, et al. Fully automated multiorgan segmentation in abdominal magnetic resonance imaging with deep neural networks. Med Phys 2020;47(10):4971–82.
[15] Ivashchenko OV, et al. A workflow for automated segmentation of the liver surface, hepatic vasculature and biliary tree anatomy from multiphase MR images. Magn Reson Imaging 2020;68:53–65.
[16] Irving B, et al. Deep quantitative liver segmentation and vessel exclusion to assist in liver assessment. In: Medical Image Understanding and Analysis: 21st Annual Conference, MIUA 2017, Edinburgh, UK, July 11–13, 2017, Proceedings 21. Springer; 2017.

[17] Bousabarah K, et al. Automated detection and delineation of hepatocellular carcinoma on multiphasic contrast-enhanced MRI using deep learning. Abdom Radiol (NY) 2021; 46(1):216–25.
[18] Wu Y, et al. Deep learning LI-RADS grading system based on contrast enhanced multiphase MRI for differentiation between LR-3 and LR-4/LR-5 liver tumors. Ann Transl Med 2020;8(11):701.
[19] Messaoudi R, et al. A novel deep learning approach for liver MRI classification and HCC detection. In: International Conference on Pattern Recognition and Artificial Intelligence. Springer; 2020.
[20] Hamm CA, et al. Deep learning for liver tumor diagnosis part I: development of a convolutional neural network classifier for multi-phasic MRI. Eur Radiol 2019;29(7): 3338–47.
[21] Liu X, et al. Can machine learning radiomics provide pre-operative differentiation of combined hepatocellular cholangiocarcinoma from hepatocellular carcinoma and cholangiocarcinoma to inform optimal treatment planning? Eur Radiol 2021;31(1):244–55.
[22] Trivizakis E, et al. Extending 2-D convolutional neural networks to 3-D for advancing deep learning cancer classification with application to MRI liver tumor differentiation. IEEE J Biomed Health Inform 2018;23(3):923–30.
[23] Deng Y, et al. DTI-based radiomics signature for the detection of early diabetic kidney damage. Abdom Radiol (NY) 2020;45(8):2526–31.
[24] Mo X, et al. MRI texture-based machine learning models for the evaluation of renal function on different segmentations: a proof-of-concept study. Insights Imaging 2023; 14(1):28.
[25] Chen W, et al. Machine learning-based multimodal MRI texture analysis for assessing renal function and fibrosis in diabetic nephropathy: a retrospective study. Front Endocrinol 2023;14:1050078.
[26] Grzywińska M, et al. Computation of the texture features on T2-weighted images as a novel method to assess the function of the transplanted kidney: primary research. Transplant Proc 2020;52(7):2062–6.
[27] Ding J, et al. Evaluation of renal dysfunction using texture analysis based on DWI, BOLD, and susceptibility-weighted imaging. Eur Radiol 2019;29(5):2293–301.
[28] Anari PY, et al. Automatic segmentation of clear cell renal cell tumors, kidney, and cysts in patients with von Hippel-Lindau syndrome using U-net architecture on magnetic resonance images. ArXiv 2023.
[29] Lay N, et al. Deep learning-based decision forest for hereditary clear cell renal cell carcinoma segmentation on MRI. Med Phys 2023;50(8):5020–9.
[30] Anari PY, et al. Deep learning algorithm (YOLOv7) for automated renal mass detection on contrast-enhanced MRI: a 2D and 2.5D evaluation of results. Abdom Radiol (NY) 2024;49(4):1194–201.
[31] Sung H, et al. Global cancer statistics 2020: GLOBOCAN estimates of incidence and mortality worldwide for 36 cancers in 185 countries. CA Cancer J Clin 2021;71(3):209–49.
[32] Mottet N, et al. EAU-ESTRO-SIOG guidelines on prostate cancer. Part 1: screening, diagnosis, and local treatment with curative intent. Eur Urol 2017;71(4):618–29.
[33] Barentsz JO, et al. Synopsis of the PI-RADS v2 guidelines for multiparametric prostate magnetic resonance imaging and recommendations for use. Eur Urol 2016;69(1):41–9.
[34] Yilmaz EC, et al. Prospective evaluation of PI-RADS version 2.1 for prostate cancer detection and investigation of multiparametric MRI-derived markers. Radiology 2023;307(4):e221309.

[35] Zhao L, et al. Detecting adverse pathology of prostate cancer with a deep learning approach based on a 3D swin-transformer model and biparametric MRI: a multicenter retrospective study. J Magn Reson Imag 2024.

[36] Li-Yin Y, et al. Medical image diagnosis of prostate tumor based on PSP-Net+VGG16 deep learning network. Comput Methods Progr Biomed 2022;221:106770.

[37] Romera E, et al. ERFNet: efficient residual factorized ConvNet for real-time semantic segmentation. IEEE Trans Intell Transport Syst 2018;19(1):263–72.

[38] Wang B, et al. Deeply supervised 3D fully convolutional networks with group dilated convolution for automatic MRI prostate segmentation. Med Phys 2019;46(4):1707–18.

[39] Hensel J, Thalmann GN. Biology of bone metastases in prostate cancer. Urology 2016; 92:6–13.

[40] Gavade AB, et al. Enhancing prostate cancer diagnosis with deep learning: a study using mpMRI segmentation and classification. arXiv preprint arXiv:2310.05371 2023.

[41] Lai C-C, et al. Autosegmentation of prostate zones and cancer regions from biparametric magnetic resonance images by using deep-learning-based neural networks. Sensors 2021;21(8):2709.

[42] Kawula M, et al. Patient-specific transfer learning for auto-segmentation in adaptive 0.35 T MRgRT of prostate cancer: a bi-centric evaluation. Med Phys 2023;50(3): 1573–85.

[43] Kawula M, et al. Prior knowledge based deep learning auto-segmentation in magnetic resonance imaging-guided radiotherapy of prostate cancer. Phys Imaging Radiat Oncol 2023;28:100498.

[44] Emre G, et al. A novel prediction method for lymph node involvement in endometrial cancer: machine learning. Int J Gynecol Cancer 2019;29(2):320.

[45] Dong HC, et al. Using deep learning with convolutional neural network approach to identify the invasion depth of endometrial cancer in myometrium using MR images: a pilot study. Int J Environ Res Public Health 2020;17(16).

[46] Dong T, et al. Development and validation of a deep learning radiomics model predicting lymph node status in operable cervical cancer. Front Oncol 2020;10.

[47] Chen X, et al. Deep learning for the determination of myometrial invasion depth and automatic lesion identification in endometrial cancer MR imaging: a preliminary study in a single institution. Eur Radiol 2020;30(9):4985–94.

[48] Ma X, et al. MRI-based radiomics of rectal cancer: preoperative assessment of the pathological features. BMC Med Imaging 2019;19(1):86.

[49] Lu HC, Wang F, Yin JD. Texture analysis based on sagittal fat-suppression and transverse T2-weighted magnetic resonance imaging for determining local invasion of rectal cancer. Front Oncol 2020;10:1476.

[50] Yin JD, et al. Prediction of different stages of rectal cancer: texture analysis based on diffusion-weighted images and apparent diffusion coefficient maps. World J Gastroenterol 2020;26(17):2082–96.

[51] Li ZY, et al. Multi-modal radiomics model to predict treatment response to neoadjuvant chemotherapy for locally advanced rectal cancer. World J Gastroenterol 2020;26(19): 2388–402.

[52] Fu J, et al. Deep learning-based radiomic features for improving neoadjuvant chemoradiation response prediction in locally advanced rectal cancer. Phys Med Biol 2020;65 (7):075001.

[53] Trebeschi S, et al. Deep learning for fully-automated localization and segmentation of rectal cancer on multiparametric MR. Sci Rep 2017;7(1):5301.

[54] Soomro MH, et al. Automated segmentation of colorectal tumor in 3D MRI using 3D multiscale densely connected convolutional neural network. J Healthc Eng 2019;2019: 1075434.
[55] Jian J, et al. Fully convolutional networks (FCNs)-based segmentation method for colorectal tumors on T2-weighted magnetic resonance images. Australas Phys Eng Sci Med 2018;41(2):393–401.

CHAPTER

Clinical applications of machine learning in breast MRI

22

Sara Mozayan[1] and Maryam Etesami[2,*]

[1]*UCLA Health (University of California), Los Angeles, CA, United States;* [2]*University of Toronto Temerty Medicine, Department of Medical Imaging, Toronto, ON, Canada*

[*]*Corresponding author: E-mail: maryametesami@gmail.com*

22.1 Introduction

Breast cancer is a significant health concern and ranks as the second leading cause of cancer-related mortality among women. Over the last 3 decades, due to increasing focus on early detection and improvements in cancer treatments, breast cancer mortality rates have decreased by 43% [1]. This progress has been bolstered by technological advancements in imaging modalities. Breast magnetic resonance imaging (MRI) has emerged as a valuable complement to mammography and ultrasound, offering the highest sensitivity for breast cancer detection among imaging modalities.

The main indications for a contrast-enhanced breast MRI examination include screening for breast cancer in high-risk women, staging biopsy-proven breast cancer, and evaluating response to neoadjuvant chemotherapy. The dynamic T1-weighted contrast-enhanced sequence serves as the foundation of the breast MRI protocol, with T2-weighted and diffusion-weighted imaging providing additional support in lesion characterization. With the advancement of MRI techniques, multiparametric breast MRI has become widely adopted.

However, despite the high sensitivity of breast MRI, challenges persist. Radiologist assessment of breast cancer on MRI can be limited by factors such as variations in technique and image quality, reader variability, and interpretive errors. Breast background parenchymal enhancement can also obscure or mimic lesions on MRI, potentially leading to oversights. These challenges are further exacerbated by the growing demand for breast MRI services amidst a shortage of radiologists, resulting in higher workloads. Moreover, long scan time, high costs, and limited availability of breast MRI pose barriers to widespread patient access and clinical implementation. In this context, machine learning demonstrates promise as a transformative tool to aid in image interpretation and improve the diagnostic accuracy of breast MRI, shifting from subjective visual interpretations to objective, data-driven

Machine Learning in MRI. https://doi.org/10.1016/B978-0-443-14109-6.00030-4

analyses. This, in turn, may streamline scan protocols, enhancing efficiency and cost-effectiveness.

Recently, significant progress has been made in applying machine learning models to breast MRI for various clinical task, such as in determining diagnosis, prognosis, and response to therapy. In this chapter, we review some of the current uses of machine learning in breast MRI, including breast segmentation, lesion classification, cancer molecular subtyping, prediction of invasive histopathology, lymph node metastasis, and response to neoadjuvant chemotherapy. In addition, we review some of the limitations of the current machine learning efforts.

22.2 Components of the multiparametric breast MRI Protocol

A multiparametric breast MRI combines anatomic details from T1 and T2 weighted sequences with functional insights from dynamic contrast enhanced (DCE) and diffusion-weighted sequences. A precontrast T1-weighted acquisition and postcontrast T1-weighted acquisition at multiple timepoints are obtained to assess background parenchymal enhancement as well as identify and characterize abnormal enhancement, including the morphology and kinetic profile of lesions. The angiogenic properties of tumors, which include increased vascularity and vessel permeability, result in rapid uptake and washout of contrast material. This kinetic phenomenon is leveraged to accentuate areas of enhancement that are concerning for malignancy. Postprocessing techniques may be used on DCE sequences to aid in lesion identification, including subtraction analysis, creation of maximum intensity projections, and generation of time-signal intensity curves and color map overlays.

T2-weighted sequences are utilized to identify structures with high water content, which is often an indicator of benign conditions, like cysts, lymph nodes, and fibroadenomas. Malignant considerations, however, include entities such as necrotic cancers, mucinous carcinoma, and metaplastic carcinoma. Diffusion-weighted sequences depict the properties of random movement of water molecules, which can help highlight highly cellular lesions such as tumors [2].

The multiparametric breast MRI provides a rich dataset which requires a systematic approach for interpretation. Background parenchymal enhancement, lesion analysis, and assessment of associated findings should be described according to the Breast Imaging Reporting and Data System (BI-RADS) lexicon. The final assessment and management of a breast MRI examination is determined using the BI-RADS categories as well. While currently reliant on human expertise and manual analysis which can be difficult, time-consuming, and error-prone, there is growing evidence that machine learning, particularly deep learning, can enhance the interpretation and reporting of breast MRI, with several areas of research underway.

22.3 Diagnosis and differentiation of breast lesions

22.3.1 Breast segmentation

Segmenting pertinent areas of interest from extensive imaging datasets is a crucial step in the quantitative evaluation of the multiparametric breast MRI. This segmentation process can delineate regions of interest, tissue volumes, and tumor dimensions, providing valuable information for diagnosis and treatment. Furthermore, the segmentation of large, complex datasets allows for improved computational efficiency and enables the extraction of features and patterns for further analysis. Traditional, manual segmentation is labor-intensive, dependent on radiologist expertise, and susceptible to reader variability and errors. However, automated image segmentation utilizing neural networks has shown high accuracy, which could streamline the process, expedite reporting, and ensure more reproducible results.

Breast boundary segmentation detects the breast-air and breast-chest wall edges, allowing for exclusion of clinically inessential information such as the lung, mediastinum, and background air [3]. Fibroglandular tissue segmentation can delineate fat and fibroglandular tissue in the breast, thus providing volumetric measurements of breast density, which has a proven association with breast cancer risk [4]. Based on fibroglandular tissue enhancement following injection of a contrast agent, background parenchymal enhancement can then be quantified, which also is a predictor of breast cancer risk [5].

Automated anatomic breast segmentation using deep learning approach has demonstrated an increasing trend in achieving high levels of accuracy comparable to manual segmentation by experts. For example, Dalmis et al. found that convolutional neural networks (CNN) such as U-Net performed better than conventional computer vision segmentation methods such as atlas-based techniques on a dataset of 66 breast MRIs. The study reported a Dice similarity coefficient of 0.93 and 0.94 for breast segmentation by two different U-net methods and 0.86 by atlas-based methods relative to manual segmentation. [6]. Ma et al. implemented a generative adversarial network that improved upon the U-Net architecture, resulting in a high Dice similarity coefficient for fibroglandular tissue segmentation (0.87) compared with classical U-Net methods (0.81) on a dataset of 100 breast MRIs [7]. Muller-Franzes et al. demonstrated that a transformer-based neural network outperformed the established nnU-Net based methods for fibroglandular tissue segmentation on a multiinstitutional dataset (Fig. 22.1) [8]. The growing effectiveness of deep learning-based segmentation methods holds promise as a tool that can be used to enhance the quantification of breast density and background parenchymal enhancement in breast MRI, thereby improving breast cancer risk assessment and ultimately patient care.

In addition to anatomic breast segmentation, deep learning has been studied for lesion detection and segmentation in DCE breast MRI. U-Net architectures and its variants have been prominently used for this task. In a notable study by Hirsch et al., a 3D U-Net was trained on a substantial dataset of 38,229 breast MRI exams

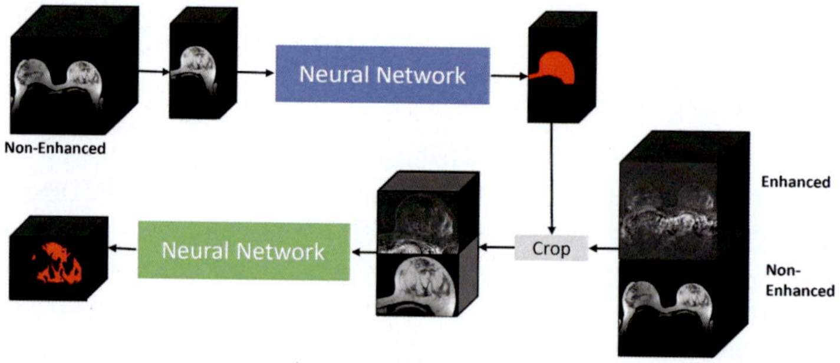

FIGURE 22.1

Deep Learning-Based Segmentation Framework for Breast and Fibroglandular Tissue. Müller-Franzes et al. reported a two-stage segmentation deep learning framework for breast fibroglandular tissue trained on the internal dataset (UKA). In the first stage, a neural network was used to segment the whole breast using noncontrast, nonfat-suppressed T1- and T2-weighted MRI sequences. The resulting whole-breast segmentation mask was then applied to crop the subtraction image from the contrast-enhanced T1-weighted sequence, as well as the nonenhanced sequences. In the second stage, a separate neural network used the cropped images to generate segmentation masks of the fibroglandular tissue (FGT). During inference on the DUKE dataset, only nonfat-suppressed T1-weighted sequences were used as input for whole breast segmentation, as T2-weighted images were available only in the UKA dataset. Both neural networks processed all input data as 3D volumes [8].

From Muller-Franzes et al. [8].

and tested on a dataset of 250 manually segmented breast cancers by four radiologists. The study found a median Dice similarity coefficient of 0.77, showcasing that the 3D U-NET achieved comparable performance in 2D segmentation of breast cancers to that of radiologists [9]. A study by Dalmis et al. employed a U-Net architecture using only the early-phase contrast-enhanced sequence and found that it had a higher sensitivity for lesion detection when compared to traditional machine learning methods applied to the full dynamic breast MRI protocol [10]. This study highlights the promising application of deep learning in the context of abbreviated MRI protocols, which are of particular interest due to their time and cost efficiency.

22.3.2 Lesion classification

While breast MRI offers the highest sensitivity among all breast imaging techniques, the specificity is moderate, ranging from 30% to 80% [11]. As both benign and malignant lesions can exhibit similar enhancement, avoiding a false-positive diagnosis requires recognizing normal enhancing structures and interpreting the shape, size, enhancement patterns, and kinetics of breast lesions according to the BI-RADS lexicon. Convolutional neural networks have shown diagnostic accuracy in lesion classification, achieving high sensitivities, specificities, and area under the receiver operating characteristic curve (AUC).

Most pretrained convolutional neural networks require a reduction in the large four-dimensional dataset generated with breast MR imaging. Despite the volumetric and temporal data reduction, the application of preprocessing techniques followed by CNN models has shown robust performance in lesion classification. In one study by Hu et al., using a retrospective dataset of 1990 lesions, researchers collapsed information in the 4D DCE breast MR into a 2D maximum intensity projection (MIP) utilizing two different approaches: one involving a MIP of the images and one involving a MIP of features within the neural network. The VGG19 deep learning model was then applied to extract imaging features which were then classified as benign or malignant using support vector machines. The classification results were compared with the ground truth based on histopathology. The study demonstrated a high AUC, with a higher AUC achieved for the feature MIP (AUC 0.93, 94% sensitivity, 72% specificity) compared with the image MIP (AUC 0.91, 90% sensitivity, 73% specificity) [12].

Additional deep learning architectures have also been applied for lesion characterization. On a dataset of 278,685 DCE-MRI image slices from 438 patients, Liu et al. used a weakly supervised approach based on the ResNet-101 architecture to classify benign from malignant lesions, reporting an AUC of 0.92, with a 74% sensitivity and 95% specificity [13]. Adachi et al. employed the RetinaNet architecture on a dataset of MIP images from DCE-breast MRI. The study compared the performance of the deep learning-based model with that of radiologists, finding superior performance by the deep learning system (AUC 0.92 for the deep learning system, 0.88 for radiologists without deep learning assistance, and 0.90 for radiologists with deep learning assistance) [14].

Lesion classification has been improved by the inclusion of multiparametric MR images in addition to the DCE-MR images. For example, Hu et al. found that using a pretrained CNN to extract features from DCE and T2 weighted sequences which were then fused to create a single ensemble of features and classified using a support vector machine resulted in a higher AUC compared with DCE alone. Specifically, the AUC for the combined features was 0.87, statistically significantly outperforming DCE alone ($P < .001$), which had an AUC of 0.85 (Fig. 22.2) [15]. Accuracies up to 93% have been reported using models that combine diffusion-weighted and DCE sequences for lesion classification [16].

22.3.3 Cancer molecular subtyping

Breast cancer is a complex disease with distinct molecular subtypes based on gene expression patterns that influence tumor behavior, response to treatment, and clinical outcomes. The four main molecular subtypes of breast cancer are luminal A, luminal B, human epidermal growth factor receptor-2 (HER2) enriched, and basal-like/triple-negative. These subtypes have significant prognostic implications, with the luminal A subtype associated with the most favorable prognosis and the basal-

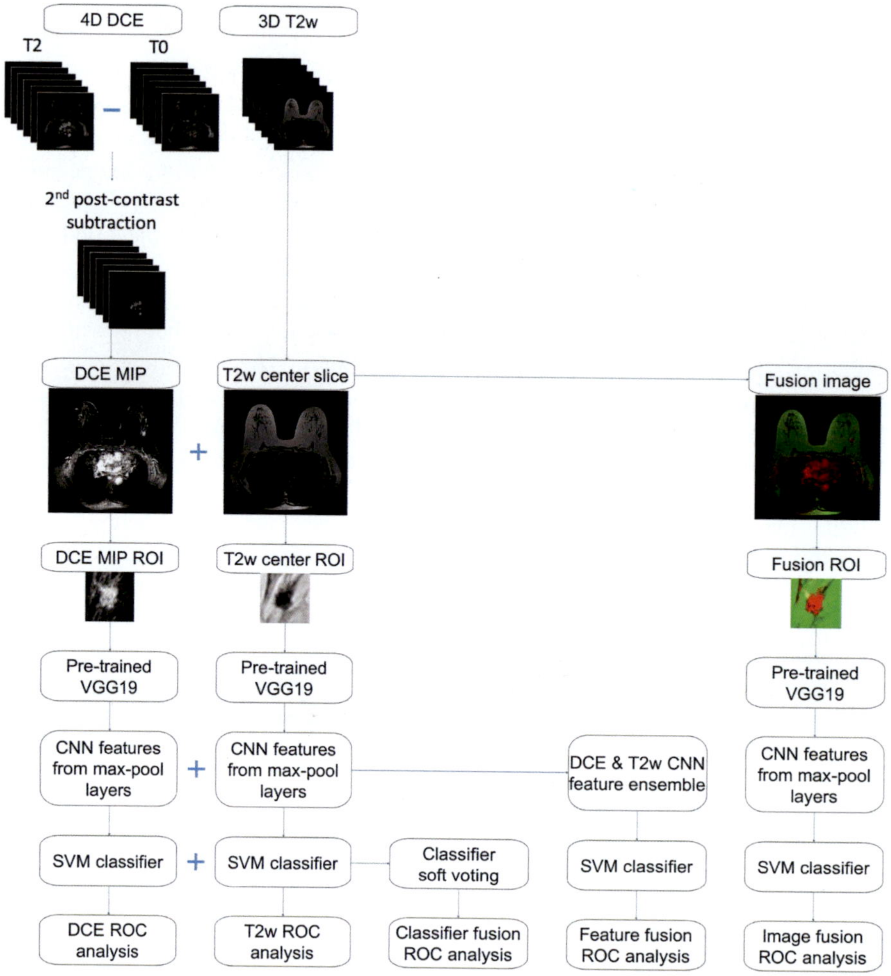

FIGURE 22.2

Lesion Classification Pipeline Incorporating Multiparametric MRI Data. Hu et al. reported a lesion classification framework based on multiparametric MRI, utilizing both dynamic contrast enhanced (DCE) and T2-weighted (T2w) sequences. Information from these modalities is integrated through three complementary strategies: (1) Image fusion—DCE and T2w images are combined to form a composite RGB image, enabling joint spatial representation; (2) Feature fusion—deep features extracted from each modality using convolutional neural networks (CNNs) are merged and used as input to a support vector machine (SVM) classifier; and (3) Classifier fusion—the malignancy probabilities predicted independently from DCE and T2w classifiers are aggregated using a soft voting approach to enhance overall performance. Abbreviations: MIP, maximum intensity projection; ROI, region of interest; ROC, receiver operating characteristic.

From Hu et al. [15].

like/triple-negative subtype associated with the worst prognosis [17]. Knowledge of the molecular subtype also aids in tailoring treatment, especially the use of targeted therapy. As immunohistochemistry for molecular subtype classification is invasive, variable, and costly, there is growing interest in using imaging features as a noninvasive, alternative approach for evaluation of breast cancer [18].

Machine learning techniques have been explored for automatically extracting and analyzing features from MRI to classify molecular subtypes. For example, one study using a database of 216 patients found that a custom convolutional neural network trained on postcontrast MR images was able to predict the four molecular subtypes with an overall accuracy of 70% on a testing set of 40 patients [19]. Other studies have also demonstrated promising results using different machine learning models, including transfer learning and ensemble learning methods such as random forest and eXtreme Gradient Boosting (XGBoost) [20–22].

22.4 Staging

22.4.1 Prediction of occult invasive disease

The differentiation between invasive cancer and ductal carcinoma in situ (DCIS) is crucial for prognosis and treatment planning. However, approximately one in four cases of DCIS diagnosed at core needle biopsy is upstaged to invasive disease at the time of surgery [23]. Several studies have explored using MR imaging characteristics and machine learning to predict occult invasive disease in DCIS. In a study evaluating semiautomatically extracted MR imaging features of tumor morphology, texture, and enhancement, specific texture features demonstrated a high predictive value for invasive disease, yielding an AUC of 0.72 [24]. Zhu et al. employed a pretrained deep learning network (GoogleNet) to extract imaging features from DCE-MRI data and a support vector machine to predict the upstaging of DCIS based on these features. Their method achieved encouraging results, with an AUC of 0.70 [25]. While the research in this area is limited, these studies suggest machine learning may capture MRI features that are valuable for detecting occult invasive disease.

22.4.2 Prediction of lymph node metastasis

Accurately predicting lymph node metastasis is a critical aspect in evaluating patients with breast cancer as it has significant prognostic and therapeutic implications. Sentinel lymph node biopsy or axillary lymph node dissection can stage the axilla; however, these procedures are invasive, costly, and carry risk of complications and morbidity. Therefore, a noninvasive evaluation of axillary lymph node status based on preoperative imaging is desirable. Machine learning models, particularly deep learning, hold potential in accomplishing this task, extracting features and patterns from imaging to preoperatively stage lymph node involvement.

Several studies have used machine learning models to predict lymph node metastasis based on imaging features of the primary tumor, imaging features of the axillary lymph node, or a combination of both. Nguyen et al. used a 4D CNN (3 spatial dimensions + time) model that used imaging features from the primary tumor on DCE-MRI and nonimaging clinical information for 357 patients with breast cancer, achieving an AUC of 0.71% and 72% accuracy [26]. A study by Ren et al. used a convolutional neural network on a dataset of 259 nodes to detect axillary lymph node metastasis on breast MRI with an AUC of 0.91 and accuracy of 85%, which outperformed an expert radiologist who achieved an AUC of 0.80 and accuracy of 78% [27]. Similarly, Ha et al. used a convolutional neural network on a dataset of 275 nodes and demonstrated a comparable accuracy of 84% [28]. Gao et al. found that a convolutional block attention model that could focus on the tumor and axillary lymph node region on DCE-MRI achieved an AUC of 0.91 and accuracy of 83% on a dataset of 941 patients [29]. These models show promise in utilizing machine learning models and dynamic MR imaging to noninvasively determine lymph node status.

22.5 Treatment planning

22.5.1 Response to neoadjuvant therapy

Neoadjuvant therapy is the standard of care in the systemic treatment of locally advanced and inflammatory breast cancers. It has also been increasingly used with triple negative or HER2-positive cancers as well as in patients with an unfavorable breast-to-tumor ratio or clinically node-positive disease [30]. The main benefits of neoadjuvant therapy are to reduce tumor size to enable less extensive surgery, to guide subsequent treatments by assessing in vivo disease response, and to provide prognostic information based on the pathologic complete response (pCR) rate which is associated with better overall survival and decreased recurrence [31]. The response rate to neoadjuvant chemotherapy varies depending on the breast cancer molecular subtype, with a reported overall pCR rate of 19% [32]. Disease response and residual tumor are evaluated through clinical examination and imaging, with MRI having the most accuracy in accomplishing this task [31]. Furthermore, MRI has shown superiority in predicting final pathologic response during early treatment, which enables improved patient care through the ability to modify the treatment and surgical strategy accordingly [33].

Although MRI is the most accurate modality to predict neoadjuvant treatment response, it remains a difficult task as MRI can overestimate or underestimate the true tumor burden [34]. Machine learning has been studied to improve the predictive power of MR imaging, especially in applying these models to pretreatment or early treatment imaging when there is the potential to alter therapy to increase rates of pathologic response and treatment tolerability. For example, Braman et al. used radiomic textural features of intratumoral and peritumoral areas on pretreatment

DCE-MRI to train machine learning classifiers to predict pCR for a given patient. They reported a maximum AUC of 0.78 using a diagonal linear discriminant analysis classifier [35]. Ravichandran et al. developed a CNN to predict pCR from pretreatment DCE-MRI with an AUC of 0.77 and overall accuracy of 82%, with the inclusion of clinical variables further improving the ability to predict response (AUC 0.85, accuracy 85%) [36]. Even greater accuracy was achieved by Ha et al., who used a VGG16 model on pretreatment DCE-MRI to classify predicted response into complete, partial, and no response/progression, achieving an overall accuracy of 88% [37]. Using DCE-MRI data from exams prior to and following administration of neoadjuvant chemotherapy, Qu et al. achieved great performance with their CNN model for pCR prediction, reporting an AUC of 0.97 [38]. These studies illustrate the potential of using MRI and machine learning to improve prediction of neoadjuvant treatment response, which can have significant implications for enabling a personalized approach to treatment and improving patient outcomes.

22.6 Challenges

Machine learning in breast MRI is an evolving research field that has demonstrated great potential to improve the diagnosis and treatment of breast cancer. However, several challenges remain before machine learning models can be applied in a clinical practice.

A major limitation of current machine learning efforts in MRI analysis is the lack of large-scale, diverse, and high-quality datasets. The performance of machine and deep learning models improves when trained on extensive and varied data. However, data availability remains a challenge due to various factors, such as the lack of data standardization, limitations in sharing data across sites, the cost and time burden of curating large volumes of annotated data, and the ethical and legal issues related to data privacy and security. When models are trained on small-scale, single-center datasets without external validation, they tend to suffer from poor generalization and overfit the training data, resulting in reduced accuracy and reliability.

Much of the published research also utilized deep learning networks such as neural network which remain a challenge to apply in the clinical setting due to their lack of interpretability. The term black box is often used, where the input features and output predictions of these models are known, but the decision-making process remains opaque. Balancing model complexity with model interpretability is crucial for clinical adoption.

22.7 Conclusion

Given the impressive progress in machine learning for breast MRI, challenges remain in data availability, standardization, and interpretability. However, with continued research and collaborative efforts across institutions, these models are expected to evolve into powerful diagnostic tools that complement radiologists' expertise. By integrating advanced algorithms with emerging imaging technologies, we can enhance the precision, efficiency, and patient-centered outcomes in breast cancer diagnosis, treatment planning, and cost-effectiveness.

References

[1] Giaquinto AN, et al. Breast cancer statistics, 2022. CA Cancer J Clin 2022;72(6): 524–41.
[2] Mann RM, Cho N, Moy L. Breast MRI: state of the art. Radiology 2019;292(3):520–36.
[3] Piantadosi G, et al. Multi-planar 3D breast segmentation in MRI via deep convolutional neural networks. Artif Intell Med 2020;103:101781.
[4] Zhao X, et al. Clinical applications of deep learning in breast MRI. Biochim Biophys Acta Rev Cancer 2023;1878(2):188864.
[5] You C, et al. Association between background parenchymal enhancement and pathologic complete remission throughout the neoadjuvant chemotherapy in breast cancer patients. Transl Oncol 2017;10(5):786–92.
[6] Dalmis MU, et al. Using deep learning to segment breast and fibroglandular tissue in MRI volumes. Med Phys 2017;44(2):533–46.
[7] Ma X, et al. Automated fibroglandular tissue segmentation in breast MRI using generative adversarial networks. Phys Med Biol 2020;65(10):105006.
[8] Muller-Franzes G, et al. Fibroglandular tissue segmentation in breast MRI using vision transformers: a multi-institutional evaluation. Sci Rep 2023;13(1):14207.
[9] Hirsch L, et al. Radiologist-level performance by using deep learning for segmentation of breast cancers on MRI scans. Radiol Artif Intell 2022;4(1):e200231.
[10] Dalmis MU, et al. Fully automated detection of breast cancer in screening MRI using convolutional neural networks. J Med Imaging 2018;5(1):014502.
[11] Bluemke DA, et al. Magnetic resonance imaging of the breast prior to biopsy. JAMA 2004;292(22):2735–42.
[12] Hu Q, et al. Improved classification of benign and malignant breast lesions using deep feature maximum intensity projection MRI in breast cancer diagnosis using dynamic contrast-enhanced MRI. Radiol Artif Intell 2021;3(3):e200159.
[13] Liu MZ, et al. Weakly supervised deep learning approach to breast MRI assessment. Acad Radiol 2022;29(Suppl. 1):S166–72.
[14] Adachi M, et al. Detection and diagnosis of breast cancer using artificial intelligence based assessment of maximum intensity projection dynamic contrast-enhanced magnetic resonance images. Diagnostics 2020;10(5).
[15] Hu Q, Whitney HM, Giger ML. A deep learning methodology for improved breast cancer diagnosis using multiparametric MRI. Sci Rep 2020;10(1):10536.
[16] Reig B, et al. Machine learning in breast MRI. J Magn Reson Imaging 2020;52(4): 998–1018.
[17] Carey LA, et al. Race, breast cancer subtypes, and survival in the carolina breast cancer study. JAMA 2006;295(21):2492–502.
[18] Hylton N. Dynamic contrast-enhanced magnetic resonance imaging as an imaging biomarker. J Clin Oncol 2006;24(20):3293–8.
[19] Ha R, et al. Predicting breast cancer molecular subtype with MRI dataset utilizing convolutional neural network algorithm. J Digit Imaging 2019;32(2):276–82.

[20] Sheng W, et al. Invasive ductal breast cancer molecular subtype prediction by MRI radiomic and clinical features based on machine learning. Front Oncol 2022;12:964605.

[21] Zhang Y, et al. Prediction of breast cancer molecular subtypes on DCE-MRI using convolutional neural network with transfer learning between two centers. Eur Radiol 2021; 31(4):2559–67.

[22] Liu HQ, et al. Machine learning on MRI radiomic features: identification of molecular subtype alteration in breast cancer after neoadjuvant therapy. Eur Radiol 2023;33(4): 2965–74.

[23] Brennan ME, et al. Ductal carcinoma in situ at core-needle biopsy: meta-analysis of underestimation and predictors of invasive breast cancer. Radiology 2011;260(1):119–28.

[24] Harowicz MR, et al. Can algorithmically assessed MRI features predict which patients with a preoperative diagnosis of ductal carcinoma in situ are upstaged to invasive breast cancer? J Magn Reson Imaging 2017;46(5):1332–40.

[25] Zhu Z, et al. Deep learning analysis of breast MRIs for prediction of occult invasive disease in ductal carcinoma in situ. Comput Biol Med 2019;115:103498.

[26] Nguyen S, et al. Preoperative prediction of lymph node metastasis from clinical DCE MRI of the primary breast tumor using a 4D CNN. Med Image Comput Comput Assist Interv 2020;12262:326–34.

[27] Ren T, et al. Convolutional neural network detection of axillary lymph node metastasis using standard clinical breast MRI. Clin Breast Cancer 2020;20(3):e301–8.

[28] Ha R, et al. Axillary lymph node evaluation utilizing convolutional neural networks using MRI dataset. J Digit Imaging 2018;31(6):851–6.

[29] Gao J, et al. Attention-based deep learning for the preoperative differentiation of axillary lymph node metastasis in breast cancer on DCE-MRI. J Magn Reson Imaging 2023;57(6):1842–53.

[30] Korde LA, et al. Neoadjuvant chemotherapy, endocrine therapy, and targeted therapy for breast cancer: ASCO guideline. J Clin Oncol 2021;39(13):1485–505.

[31] Reig B, et al. Breast MRI for evaluation of response to neoadjuvant therapy. Radiographics 2021;41(3):665–79.

[32] Haque W, et al. Response rates and pathologic complete response by breast cancer molecular subtype following neoadjuvant chemotherapy. Breast Cancer Res Treat 2018; 170(3):559–67.

[33] Hylton NM, et al. Locally advanced breast cancer: MR imaging for prediction of response to neoadjuvant chemotherapy–results from ACRIN 6657/I-SPY TRIAL. Radiology 2012;263(3):663–72.

[34] Fowler AM, Mankoff DA, Joe BN. Imaging neoadjuvant therapy response in breast cancer. Radiology 2017;285(2):358–75.

[35] Braman NM, et al. Intratumoral and peritumoral radiomics for the pretreatment prediction of pathological complete response to neoadjuvant chemotherapy based on breast DCE-MRI. Breast Cancer Res 2017;19(1):57.

[36] Ravichandran K, et al. A deep learning classifier for prediction of pathological complete response to neoadjuvant chemotherapy from baseline breast DCE-MRI. Medical Imaging 2018: Computer-Aided Diagnosis 2018:10575.

[37] Ha R, et al. Prior to initiation of chemotherapy, can we predict breast tumor response? Deep learning convolutional neural networks approach using a breast MRI tumor dataset. J Digit Imaging 2019;32(5):693–701.

[38] Qu YH, et al. Prediction of pathological complete response to neoadjuvant chemotherapy in breast cancer using a deep learning (DL) method. Thorac Cancer 2020;11 (3):651–8.

CHAPTER

Clinical applications of machine learning in musculoskeletal MRI

23

Delaram Shakoor[1] and Iman Khodarahmi[2,*]

[1]*Hospital for Special Surgery, New York, NY, United States;* [2]*New York University School of Medicine, New York, NY, United States*

[*]*Corresponding author: E-mail: Iman.khodarahmi@nyulangone.org*

23.1 Introduction

In the field of machine learning (ML) in medical imaging, numerous algorithms have been applied to various diagnostic radiology tasks, drawing substantial interest among radiology societies [1]. Notably, deep learning (DL) algorithms, renowned for their autonomous learning capabilities, have emerged as key players, demonstrating proficiency in different tasks essential to musculoskeletal radiology. The allure of DL algorithms in musculoskeletal radiology lies in their potential to tackle two significant challenges: enhancing diagnostic performance by reducing the human error rate and accelerating the interpretation of medical images [1,2]. Despite the high accuracy of subspecialized radiologists in musculoskeletal MRI interpretations, achieving proficiency typically requires extensive training over several years. Disease-detecting DL algorithms offer the prospect of providing expert-level interpretations for readers with less expertise and contributing to the education of radiology trainees [1]. In this chapter, we intend to describe various ML applications that have been developed for various aspects of musculoskeletal MR imaging.

23.2 Spine imaging

23.2.1 Disease diagnosis/differentiation

ML applications have been developed for various aspects of spine imaging [3–7]. However, ML algorithms for automated interpretation of lumbar spine MRI have drawn significant attention, due to growing demand on radiological departments [3]. Several studies have described DL algorithms for the detection of Modic changes and disc degeneration, as well as evaluation of central canal, neural foramina and lateral recesses [8–18]. A recent meta-analysis by Compte et al. reviewed 27 studies which utilized ML algorithms for assessing lumbar degeneration disease [3]. The

Machine Learning in MRI. https://doi.org/10.1016/B978-0-443-14109-6.00023-7

authors mentioned that the majority of studies used DL applications, however no statistically significant difference was found in the performance of DL, compared with other ML applications [3]. They observed while SpineNet dominated studies on DL applications [10,18,19], none of the current ML applications has been used as a clinical decision-making tool [3]. Additional studies have also described using ML to evaluate central canal, lateral recess and neural foramina stenosis [8,9,11,13]. A few studies also suggested that using DL improved radiologists' efficiency and objectivity in reporting cervical and lumbar MRIs [17,20]. Interestingly, a study by Shemesh et al. described a novel DL algorithm to detect the ossification of posterior longitudinal ligaments on MRI images, obviating the need for additional CT [21], which could have potential to aid in the diagnosis of several spinal disorders.

While the focus of ML applications has been on the quantifying and standardizing degenerative disease of lumbar spine, few studies have assessed using ML for differentiating inflammatory and infectious spondylarthritis. For instance, Kim et al. described a DL scoring system, which differentiated between pyogenic and tuberculous spondylitis, comparable to the performance of skilled radiologists [22]. In another study by Keep et al., a quantitative ML application supported by texture analysis (TA) was proposed which outperformed qualitative analysis of radiologist in correctly differentiation degenerative from inflammatory changes of sacroiliac joints [23].

Several studies have also developed and validated ML applications for spine oncologic evaluation. Promising results have been reported for using both ML algorithms and MRI based TA to differentiate malignant from benign vertebral lesions as well as detecting bone marrow infiltration in hematological malignancy [6,24–27]. Gitto et al. developed a support vector machine based on radiomics features obtained from T2-weighted and DWI images, which could differentiate benign from malignant osseous lesions with an accuracy of 76% [25]. Another support vector machine texture classifier was reported to have comparable performance to radiologists for detecting bone marrow infiltration [27]. Further attempts have been made to develop and validate models which are able to discriminate benign from malignant vertebral fractures [28–30]. Prior reports indicate that a newly developed Gaussian naïve Bayes model and a Two-Stream Compare and Contrast Network model had higher diagnostic performance compared to radiologist for detecting malignant vertebral fracture [28,30]. However, in a study by Yeh et al., ResNet50 algorithm was shown to improve the sensitivity and accuracy of less experienced readers, while showing less accuracy comparing to the senior readers [29]. There is limited data available in the current literature on evaluation of cord tumors and spinal soft tissue neoplasms. Ito et al. reported an accuracy of 93.5% for a DL based system for automatic detection of spinal schwannomas based on T1 and T2 weighted images [31].

23.2.2 Prediction and prognostication

Few studies have evaluated the use of DL applications in predicting cervical spine myelopathy and postoperative changes of the spine. Jin et al. explored different artificial intelligence (AI) algorithms to analyze diffusion tensor imaging data for cervical myelopathy prognostication in patients with clinical symptoms who underwent surgical intervention [32]. In addition, Hopkins et al. developed and validated an ML model based on the conventional cervical spine MRI which could diagnose and predict cervical spine myelopathy with a median accuracy of 90% [33]. Goedmakers et al. described a DL algorithm for predicting adjacent segment disease using pre-operative cervical MRI of patients undergoing single level ACDF [5]. In a study by Jiang et al., an MRI based random survival forest model was described which could predict new vertebral fracture within 2 years following a vertebral augmentation [34]. Another study developed and validated models based on radiomic features of lumbar T2 weighted images along with clinical risk factors with reported AUC of 0.93 to predict imminent new vertebral fracture following vertebral augmentation [35]. Takahashi et al. also developed an ML based model using MRI features to predict nonunion of osteoporotic vertebral fracture with reported AUC up to 86% [36]. In addition, Poullain et al. described using texture indices obtained from opportunistic lumbar MRI along with age and Hounsfield Unit Bone density from body CT to assess risk of vertebral fracture [37].

23.3 Knee imaging

23.3.1 Osteoarthritis diagnosis and differentiation

Osteoarthritis (OA) is a multifactorial disorder marked by biochemical and morphological changes in the articular cartilage, underlying osseous structures, and the surrounding supportive tissues of the knee [38]. Magnetic resonance imaging (MRI) is instrumental in the diagnosis and research of osteoarthritis (OA) [39]. In recent years, it has provided insights into disease etiology, identified potential treatment pathways, and contributed to the prediction of long-term outcomes [39]. Machine learning (ML) has also played a pivotal role in advancing the field of OA imaging, particularly in the field of MRI [40]. Its extensive utilization has significantly contributed to the enhancement of diagnostic capabilities, enabling more accurate and efficient assessments of osteoarthritic conditions [1]. Through sophisticated algorithms and pattern recognition techniques, machine learning applications have demonstrated the capacity to discern subtle abnormalities, contributing to a more comprehensive understanding of musculoskeletal disorders [1].

As the initial step, attempts were made to develop and validate various DL algorithms including convolutional neural networks (CNNs) for automatic segmentation of MR images to evaluate their applicability in multiple aspects of osteoarthritis research [41–43]. In addition, DL applications such as conditional Generative Adversarial Networks (cGANs) have been used to generate accurate automated semantic maps of knee joint tissues [44,45]. Various ML applications, such as Deformable Model-based, Graph-based, and Atlas-based methods, have been developed for bone segmentation with the objective of generating the bone-cartilage interface

[45–50]. These applications aim to extract cartilage tissue from the bone surface, quantify and monitor changes in bone shape and surface related to structural deformations, and compute bone models to investigate the impact of biomechanical stress at different localized knee sites [51]. Cheng et al. developed a fully automated DL framework, utilizing holistically nested networks (HNN) architecture, which demonstrated exceptional accuracy in segmenting the patella and distal femur, providing a promising tool for comprehensive investigations in both adults and actively growing adolescents, even when growth plates are present [47].

Since the hallmark of osteoarthritis (OA) is the loss of articular cartilage, the segmentation of MR images of articular cartilage serves as a valuable tool for both detection of OA presence and determining the severity of disease by grading OA [52,53]. However, manual segmentation of such images is a time-consuming task that is susceptible to high reader variability, making it prone to errors [52]. Therefore, numerous studies have concentrated on the segmentation of articular cartilage and automated techniques utilizing graph-based methods, active shape modeling and atlas databases have been extensively reported for cartilage segmentation [53–62]. These efforts have been undertaken as an initial step toward comprehending normal articular cartilage thickness and morphology, establishing a reference for future evaluations [62,63]. Shah et al., by applying a CNN model to 3,910 MRIs of healthy knees, reported normal variations in the thickness of articular cartilage within different compartments of the knee and found that female gender was associated with thinner cartilage [63]. Other DL algorithms have been developed for simultaneous automatic segmentation of bone and cartilage tissues of the knee. A modified version of CNN, iMaskRCNN, was recently developed for bone and cartilage segmentation [64]. The authors reported high dice scores up to 98% for bone segmentation and improved detection of effusion, showcasing its reliability for automated OA assessment [64].

Furthermore, ML applications have been employed to identify abnormal articular cartilage and to detect cartilage defects in MRI images [65–67]. The weighted neighbor distance using a compound hierarchy of algorithms representing morphology (WND-CHRM), initially introduced for X-ray imaging, has been successfully extended to MR images. In this innovative ML application, the algorithm has demonstrated a remarkable ability to identify abnormal cartilage, achieving an impressive accuracy of 86% [65]. In a study by Liu et al., a fully automated DL system was developed to detect various cartilage lesions within the knee MR images using CNNs [66]. The system demonstrated high diagnostic performance, with sensitivity ranging from 80.5% to 84.1%, specificity ranging from 85.2% to 87.9%, and good interobserver agreement [66]. In addition, ML algorithms based on MRI radiomics analysis were developed and assessed for Knee OA diagnosis [68]. These models, employing logistic regression, K-nearest neighbor, and support vector machine classifiers, demonstrated promising performance, with the final model achieving high accuracy (0.968) and area under the ROC curve (0.983) in the validation cohort, showcasing the potential of MRI radiomics for KOA diagnosis across all three compartments of knee joints [68].

Prior advancements in quantitative MRI techniques, such as the quantification of T1ρ and T2 relaxation times, enable the characterization of the biochemical composition of articular cartilage [69,70]. Few fully automated femoral cartilage segmentation models for measuring T2 relaxation values were developed and validated [71–73]. One of these models demonstrated strong agreement with expert segmentations, achieving a Dice score of 0.85 ± 0.03, and accurately estimated T2 values and 4-year changes in individual subregions, showing potential for accelerating osteoarthritis research [72].

23.3.2 Osteoarthritis prediction and prognostication

Several ML models have been proposed which can incorporate the clinical and imaging data to predict the outcome of knee OA and identify individuals who will benefit from early surgical intervention such as total knee arthroplasty [74]. These models are beyond the scope of this chapter, and we only discuss ML models which use imaging data only for morphological phenotyping and quantitative analysis of cartilage thickness loss [75,76]. These models have also demonstrated significant promise in their application for investigating the structural progression of knee OA and predicting the outcome of knee OA [77–80]. In a recent study by Jamshidi et al., authors, using baseline OAI MRI images, employed machine learning methods for the classification of structural knee OA progressor. The findings underscored the predictive capacity of baseline X-ray and MRI-based features in identifying early OA knee progressor, with Multi-Layer Perceptron (MLP) emerging as the most effective classification method across various outcomes, including cartilage volume loss and joint space narrowing [78]. Eckstein et al. investigated the longitudinal performance characteristics of an automated CNN based cartilage segmentation method employing MR images [79]. They reported favorable outcomes, indicating the method's high sensitivity to changes in cartilage thickness loss and its efficiency in distinguishing rates of cartilage loss between distinct strata, such as progressor knees versus non-progressor knees [79]. WND-CHRM was also applied to T2 map images and was reported to predict progression to symptomatic osteoarthritis with an accuracy of 75%, which could add a valuable dimension to its clinical utility [81].

Statistical shape modeling (SSM), a form of supervised machine learning, utilizes principal component analysis to simplify intricate 3D geometric shapes into a single metric value [82]. Through SSM, a distinctive OA 3D bone shape, called B score has been proposed, encompassing features such as osteophyte ridge formation and alterations in the width and flattening of the articular surfaces [83]. Bowes and colleagues validated this B score, by testing on OA Initiative database (OAI) and reported that B-score offers a reader-independent quantification at a single timepoint, delivering OA status with well-defined clinical risks throughout the entire spectrum of the disease [83]. In addition, this B score was applied to Applied Public-Privat Research enabling Osteoarthritis Clinical Headway, APPROACH data base and the authors reported high correlation and minimal bias in B-score for prediction of clinical knee OA and suggest that it can be reliably measured using both CT or MRI scans [84].

23.3.3 Internal derangements of the knee

To identify injuries of the internal structures of the knee, attempts were initially made for segmentation of cruciate ligaments and menisci [85,86], and these efforts were expanded to include post-surgical changes including detection of reconstructed anterior cruciate ligament (ACL) [87]. The majority of the studies have focused on developing DL algorithms which can detect ACL injuries and they reported favorable outcomes which can be applied in different settings and MR scanners [2,88–94]. For instance, in a study by Germann et al., the clinical validation of a Deep Convolutional Neural Network (DCNN) was investigated in 512 patients with surgically proven ACL tear, encompassing diverse magnetic resonance imaging (MRI) examinations from multiple institutions and varying protocols [2]. The DCNN demonstrated a commendable sensitivity of 96.1%, comparable to expert radiologists, although with a significantly lower specificity of 93.1%, with no significant difference between 1.5 and 3 T scanners [2]. In another study, a dual-modules deep learning model, known as double-linear layers U-Net (DCLU-Net), was developed as a semi-supervised framework for the diagnosis and grading of ACL tears on MRI [89]. The model, incorporating radiomic features, demonstrated high accuracies of 90%, 82%, and 92% for classifying intact, partially torn, and fully ruptured ACL, respectively. Further studies have also explored using DL application for detection of meniscal tears with reported area under the curves (AUCs) of our model ranging between 0.889–0.924 for medial and lateral meniscal tears [95–98].

23.4 Skeletal maturity

This area is relatively underexplored, with a few studies attempting to assess skeletal maturity based on MR images [99,100], as the majority of prior research has primarily focused on radiographic evaluation [101]. Zech et al. reported using a deep learning approach employing a long short-term memory (LSTM) model to predict chronological age [99]. The model's ability to infer skeletal maturity from knee MR sequences was comparable to that of radiology residents, showcasing potential for efficient and automated evaluation of lower extremity skeletal maturity in a fraction of the time required by human experts.

23.5 Shoulder imaging

23.5.1 Rotator cuff injuries

Rotator cuff tendon (RCT) injury is a common cause of shoulder pain and instability, necessitating MR evaluation for accurate assessment [102]. However, the accurate interpretation of MRI for diagnosing RCTs presents challenges for individuals lacking musculoskeletal training [103]. Moreover, this process is time-consuming and demanding, requiring the scrutiny of numerous MRI sequences [103]. Therefore, the majority of ML applications in the field of shoulder imaging have focused on the segmentation of RCT and its associated musculature to facilitate detection of RCT injuries [103–111]. Lin and colleagues developed and validated a DL model using a sophisticated ensemble algorithm combined with fluid-sensitive sequences in three planes and sagittal oblique T1-weighted sequences obtained from 11,925 noncontrast shoulder MRI for detection and classification of RCT injuries [107]. The authors reported excellent diagnostic accuracy, with AUC values ranging from 0.89 to 0.93 [107] (Fig. 23.1). Particularly impressive was the model's ability to excel in detecting full-thickness tears, reaching an exceptional AUC of 0.99 [107]. Furthermore, a CNN model was developed to select a specific shoulder sagittal MR image (Y-view), which could automatically segmenting rotator cuff (RC) muscles [108]. The developed CNN method demonstrated exceptional performance in both tasks, achieving accuracy exceeding 98% for Y-view selection and accurate segmentation of rotator cuff (RC) muscles with mean Dice scores surpassing 0.93 [108]. Another study employed a DL framework to analyze the occupation ratio and an automated region-based Otsu thresholding technique to assess fatty infiltration in the supraspinatus (SS) muscle [110]. The results demonstrated exceptional agreement between clinician assessment and the deep neural network, with high values for the Dice similarity coefficient, accuracy, sensitivity, specificity, and minimal relative area difference for both the supraspinatus fossa and muscle [110]. In addition, the Otsu thresholding method effectively differentiated fatty infiltration levels among Goutallier grades, revealing a significant correlation with the occupation ratio. This comprehensive analysis, conducted on 240 patients using shoulder magnetic resonance imaging highlights the potential of CNN for efficient and objective quantification of shoulder MRI parameters.

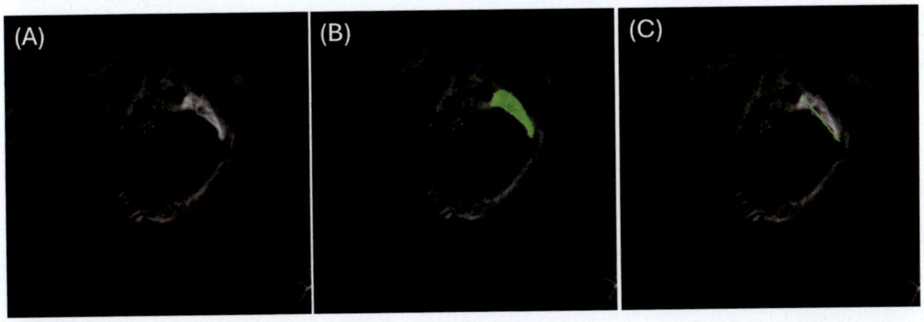

FIGURE 23.1

Deep learning–based automated rotator cuff tear measurement on shoulder MRI. The tear in (A) is first segmented in three dimensions, as denoted by the green area in (B). Later, the in-plane size of the tear is measured (C).

Images courtesy of Dana Lin, MD. NYU Langone Health, New York, NY.

23.5.2 Biceps tendinitis

Since biceps tendon (BT) pathologies including, tear or tendinitis are another common cause of shoulder pain and instability, few studies have attempted to develop DL models to evaluate BT disorders [105,112]. A recent study utilized radiomic features extracted from preoperative MRI to Biceps tendinitis, employing machine learning classifiers such as random forest (RF) and support vector machine (SVM) [112]. Among the classifiers, the random forest model, incorporating 20 selected radiomic features, exhibited the highest sensitivity, specificity, accuracy, and AUC in recognizing Biceps tendinitis, emphasizing its potential for predictive accuracy in this context.

23.6 Imaging of other joints

Despite the potential of ML in musculoskeletal imaging, its application in other joints, such as hip, ankle, and wrist imaging, remain limited. Few scattered studies have investigated ML's role in understanding and diagnosing the most prevalent pathologies affecting these joints, shedding light on the challenges and opportunities in this evolving field. Using hip MRI of 1016 patients, Ni and colleagues developed a sophisticated deep learning model, integrating CNN-6, Mask R–CNN, and LeNet-5, to enhance the efficiency and accuracy of diagnosing and classifying labral injuries on MR images [113]. The authors reported high AUC values for segmentation and detection of labral injuries, highlighting the potential of deep learning as a valuable tool in aiding radiological interpretations of hip joint pathologies [113]. In another study, Klontzas et al. developed a robust approach, utilizing MRI-based radiomics, and machine learning, particularly XGboost, to differentiate avascular necrosis from transient osteoporosis of the hip [114]. The XGboost classifier, leveraging 38 relevant radiomics features, surpassed the diagnostic performance of a general radiologist, emphasizing its efficacy in aiding the differentiation between transient osteoporosis and avascular necrosis of the hip [114].

In the realm of ankle injuries, particularly those involving the Anterior Talofibular Ligament (ATFL), a recent study endeavored to enhance diagnostic precision through computer vision analysis on MRI [115]. Utilizing feature extraction algorithms and diverse classifiers, including Multi-Layer Perceptron (MLP), SVM, k-Nearest Neighbors (kNN), and RF, the method achieved a remarkable 85.03% hit rate, surpassing human expert-based analyses by 22% in cases lacking distinct morphologies [115].

In the intricate landscape of wrist injuries, where clinical diagnosis faces challenges due to the complexity of ligaments and joint capsules, a recent study attempted to enhance diagnostic accuracy [116]. Employing the Canny algorithm for edge feature extraction from 160 MR images and the particle swarm optimization-support vector machine (PSO-SVM) algorithm for lesion segmentation, the study evaluated the algorithm's image processing efficacy using indicators such as peak signal-to-noise ratio (PSNR), mean square error (MSE), figure of merit (FOM), and structural similarity (SSIM) [116]. The PSO-SVM algorithm showed stable edge stability, with diagnostic accuracy, and Dice similarity coefficient reaching 0.9413 and 0.8715, respectively, outperforming other algorithms and holding promise for clinical auxiliary diagnosis of wrist joint injuries [116].

23.7 Oncology imaging

23.7.1 Disease diagnosis and differentiation

The conventional MRI sequences combined with dynamic enhanced MRI techniques currently detect potential malignant neoplasms with acceptable reliability [117]. Nevertheless, certain imaging features of benign and malignant lesions can exhibit similarities, posing a challenge in formulating a clear-cut differential diagnosis. ML applications using radiomics features obtained from MR images have shown promising results in enhancing diagnostic precision by differentiating various malignant osseous and soft tissue neoplasms from benign lesions with similar imaging features [117–134]. Prior studies have developed radiomics models using MR images which could predict the histopathological grades of soft tissue sarcomas preoperatively with high level of accuracy [131,132]. Additional study also reported that DL applications could enhance diagnostic accuracy of radiologists interpreting preoperative MR images to the level comparable to presurgical biopsies [135].

23.7.2 Prediction and prognostication

ML applications have also been employed to assess the treatment response of soft tissue sarcomas and predict treatment outcomes based on pretreatment MR features [136–140]. This approach leverages the power of ML algorithms to analyze imaging data, providing valuable insights into treatment efficacy and prognosis for patients with soft tissue sarcomas. Notably, radiomics models based on MR images have emerged, demonstrating the capability to predict the recurrence in soft tissue sarcomas following surgical or oncological interventions [139,140].

23.8 Bone marrow signal and infectious work up

While the majority of studies have concentrated on assessing bone marrow (BM) signal in whole-body MR images for detection of systemic disease such as multiple myeloma [141], only a few have explored MR images as a means to facilitate the detection of signal abnormalities in patients with infectious etiologies [142]. A prior study investigated using a 2D CNN to develop a model for automated segmentation of bone marrow signal in knee images from both healthy individuals and those with chronic non-bacterial osteomyelitis among children and adolescents [142]. The results demonstrate promising feasibility, with the developed model exhibiting acceptable performance across various signal intensity levels, paving the way for potential advancements in automated assessment of bone marrow signals. Furthermore, a recent study explored the potential of ML application in detecting septic total hip arthroplasty (THA) failure using 173 preoperative MRIs [143]. The study used SVM to predict THA infection which showed strong performance, achieving 92% sensitivity, and 82% accuracy, in predicting THA infection.

23.9 Challenges, limitations, and future directions

Despite the promising applications of ML and DL in musculoskeletal imaging, several challenges hinder their widespread clinical adoption. The need for large, high-quality, and well-annotated datasets, along with variability in imaging protocols and scanner types, limits model generalizability. In addition, the "black box" nature of DL models raises concerns about interpretability and clinician trust. Regulatory hurdles and the lack of extensive external validation further delay clinical integration. Future research should focus on improving model explainability, developing standardized datasets, and conducting large-scale validation studies across diverse populations. Collaborative efforts between radiologists, data scientists, and regulatory bodies will be crucial in ensuring ML's safe and effective implementation, ultimately enhancing diagnostic accuracy, efficiency, and patient outcomes.

23.10 Conclusion

Machine learning and deep learning have emerged as transformative tools in musculoskeletal imaging, offering enhanced diagnostic accuracy, efficiency, and prognostic capabilities. From detecting rotator cuff injuries and knee derangements to differentiating benign from malignant musculoskeletal tumors, ML applications continue to expand across various clinical domains. While promising advancements have been made in automating image interpretation, segmenting structures, and predicting disease progression, most ML models remain in the research phase, requiring further validation and regulatory approval for clinical integration. Despite these challenges, ML's potential to revolutionize musculoskeletal radiology is undeniable, offering radiologists valuable decision-support tools that improve diagnostic precision, streamline workflows, and ultimately enhance patient care. Continued research and collaboration are essential to fully realize ML's clinical impact.

References

[1] Fritz B, Fritz J. Artificial intelligence for MRI diagnosis of joints: a scoping review of the current state-of-the-art of deep learning-based approaches. Skelet Radiol 2022;51 (2):315–29.

[2] Germann C, Marbach G, Civardi F, Fucentese SF, Fritz J, Sutter R, et al. Deep convolutional neural network-based diagnosis of anterior cruciate ligament tears: performance comparison of homogenous versus heterogeneous knee MRI cohorts with different pulse sequence protocols and 1.5-T and 3-T magnetic field strengths. Investig Radiol 2020;55(8):499–506.

[3] Compte R, Granville Smith I, Isaac A, Danckert N, McSweeney T, Liantis P, et al. Are current machine learning applications comparable to radiologist classification of degenerate and herniated discs and Modic change? A systematic review and meta-analysis. Eur Spine J 2023;32(11):3764–87.

[4] Bharadwaj UU, Christine M, Li S, Chou D, Pedoia V, Link TM, et al. Deep learning for automated, interpretable classification of lumbar spinal stenosis and facet arthropathy from axial MRI. Eur Radiol 2023;33(5):3435–43.

[5] Goedmakers CMW, Lak AM, Duey AH, Senko AW, Arnaout O, Groff MW, et al. Deep learning for adjacent segment disease at preoperative MRI for cervical radiculopathy. Radiology 2021;301(3):664–71.

[6] Chianca V, Cuocolo R, Gitto S, Albano D, Merli I, Badalyan J, et al. Radiomic machine learning classifiers in spine bone tumors: a multi-software, multi-scanner study. Eur J Radiol 2021;137:109586.

[7] Hess M, Allaire B, Gao KT, Tibrewala R, Inamdar G, Bharadwaj U, et al. Deep learning for multi-tissue segmentation and fully automatic personalized biomechanical models from BACPAC clinical lumbar spine MRI. Pain Med 2023;24(Suppl. 1): S139–48.

[8] Gaonkar B, Beckett J, Villaroman D, Ahn C, Edwards M, Moran S, et al. Quantitative analysis of neural foramina in the lumbar spine: an imaging informatics and machine learning study. Radiol Artif Intell 2019;1(2):180037.

[9] Gaonkar B, Villaroman D, Beckett J, Ahn C, Attiah M, Babayan D, et al. Quantitative analysis of spinal canal areas in the lumbar spine: an imaging informatics and machine learning study. AJNR Am J Neuroradiol 2019;40(9):1586–91.

[10] Grob A, Loibl M, Jamaludin A, Winklhofer S, Fairbank JCT, Fekete T, et al. External validation of the deep learning system "SpineNet" for grading radiological features of degeneration on MRIs of the lumbar spine. Eur Spine J 2022;31(8):2137–48.

[11] Hallinan J, Zhu L, Yang K, Makmur A, Algazwi DAR, Thian YL, et al. Deep learning model for automated detection and classification of central canal, lateral recess, and neural foraminal stenosis at lumbar spine MRI. Radiology 2021;300(1):130–8.

[12] Han M, Liu L, Hu M, Liu G, Li P. Medical expert and machine learning analysis of lumbar disc herniation based on magnetic resonance imaging. Comput Methods Programs Biomed 2022;213:106498.

[13] Huber FA, Stutz S, Vittoria de Martini I, Mannil M, Becker AS, Winklhofer S, et al. Qualitative versus quantitative lumbar spinal stenosis grading by machine learning supported texture analysis-Experience from the LSOS study cohort. Eur J Radiol 2019;114:45–50.

[14] Ishimoto Y, Jamaludin A, Cooper C, Walker-Bone K, Yamada H, Hashizume H, et al. Could automated machine-learned MRI grading aid epidemiological studies of lumbar spinal stenosis? Validation within the Wakayama spine study. BMC Musculoskelet Disord 2020;21(1):158.

[15] Athertya JS, Saravana Kumar G. Classification of certain vertebral degenerations using MRI image features. Biomed Phys Eng Express 2021;7(4).

[16] Lagerstrand K, Hebelka H, Brisby H. Identification of potentially painful disc fissures in magnetic resonance images using machine-learning modelling. Eur Spine J 2022;31 (8):1992–9.
[17] Lim DSW, Makmur A, Zhu L, Zhang W, Cheng AJL, Sia DSY, et al. Improved productivity using deep learning-assisted reporting for lumbar spine MRI. Radiology 2022;305(1):160–6.
[18] McSweeney TP, Tiulpin A, Saarakkala S, Niinimäki J, Windsor R, Jamaludin A, et al. External validation of SpineNet, an open-source deep learning model for grading lumbar disk degeneration MRI features, using the northern Finland birth cohort 1966. Spine (Phila Pa 1976) 2023;48(7):484–91.
[19] Windsor RJA, Kadir T, Zisserman A. SpineNetV2: automated detection, labelling and radiological grading of clinical MR scans. arXiv e-prints May 01, 2022.
[20] Merali Z, Wang JZ, Badhiwala JH, Witiw CD, Wilson JR, Fehlings MG. A deep learning model for detection of cervical spinal cord compression in MRI scans. Sci Rep 2021;11(1):10473.
[21] Shemesh S, Kimchi G, Yaniv G, Harel R. MRI-based detection of cervical ossification of the posterior longitudinal ligament using a novel automated machine learning diagnostic tool. Neurosurg Focus 2023;54(6):E11.
[22] Kim K, Kim S, Lee YH, Lee SH, Lee HS, Kim S. Performance of the deep convolutional neural network based magnetic resonance image scoring algorithm for differentiating between tuberculous and pyogenic spondylitis. Sci Rep 2018;8(1):13124.
[23] Kepp FH, Huber FA, Wurnig MC, Mannil M, Kaniewska M, Guglielmi R, et al. Differentiation of inflammatory from degenerative changes in the sacroiliac joints by machine learning supported texture analysis. Eur J Radiol 2021;140:109755.
[24] Filograna L, Lenkowicz J, Cellini F, Dinapoli N, Manfrida S, Magarelli N, et al. Identification of the most significant magnetic resonance imaging (MRI) radiomic features in oncological patients with vertebral bone marrow metastatic disease: a feasibility study. Radiol Med 2019;124(1):50–7.
[25] Gitto S, Bologna M, Corino VDA, Emili I, Albano D, Messina C, et al. Diffusion-weighted MRI radiomics of spine bone tumors: feature stability and machine learning-based classification performance. Radiol Med 2022;127(5):518–25.
[26] Larhmam SM MA, Drisis S, Benjelloun M. A texture analysis approach for spine metastasis classification in T1 and T2 MRI. Bioinforma Biomed 2018;10814:198–211.
[27] Hwang EJ, Jung JY, Lee SK, Lee SE, Jee WH. Machine learning for diagnosis of hematologic diseases in magnetic resonance imaging of lumbar spines. Sci Rep 2019;9 (1):6046.
[28] Liu B, Jin Y, Feng S, Yu H, Zhang Y, Li Y. Benign vs malignant vertebral compression fractures with MRI: a comparison between automatic deep learning network and radiologist's assessment. Eur Radiol 2023;33(7):5060–8.
[29] Yeh LR, Zhang Y, Chen JH, Liu YL, Wang AC, Yang JY, et al. A deep learning-based method for the diagnosis of vertebral fractures on spine MRI: retrospective training and validation of ResNet. Eur Spine J 2022;31(8):2022–30.
[30] Zhang H, Yuan G, Wang C, Zhao H, Zhu K, Guo J, et al. Differentiation of benign versus malignant indistinguishable vertebral compression fractures by different machine learning with MRI-based radiomic features. Eur Radiol 2023;33(7):5069–76.
[31] Ito S, Ando K, Kobayashi K, Nakashima H, Oda M, Machino M, et al. Automated detection of spinal schwannomas utilizing deep learning based on object detection from magnetic resonance imaging. Spine (Phila Pa 1976) 2021;46(2):95–100.

[32] Jin R, Luk KD, Cheung JPY, Hu Y. Prognosis of cervical myelopathy based on diffusion tensor imaging with artificial intelligence methods. NMR Biomed 2019;32(8):e4114.

[33] Hopkins BS, Weber 2nd KA, Kesavabhotla K, Paliwal M, Cantrell DR, Smith ZA. Machine learning for the prediction of cervical spondylotic myelopathy: a post hoc pilot study of 28 participants. World Neurosurg 2019;127:e436–42.

[34] Jiang Y, Cai J, Zeng Y, Ye H, Yang T, Liu Z, et al. Development and validation of a machine learning model to predict imminent new vertebral fractures after vertebral augmentation. BMC Musculoskelet Disord 2023;24(1):472.

[35] Cai J, Shen C, Yang T, Jiang Y, Ye H, Ruan Y, et al. MRI-based radiomics assessment of the imminent new vertebral fracture after vertebral augmentation. Eur Spine J 2023;32(11):3892–905.

[36] Takahashi S, Terai H, Hoshino M, Tsujio T, Kato M, Toyoda H, et al. Machine-learning-based approach for nonunion prediction following osteoporotic vertebral fractures. Eur Spine J 2023;32(11):3788–96.

[37] Poullain F, Champsaur P, Pauly V, Knoepflin P, Le Corroller T, Creze M, et al. Vertebral trabecular bone texture analysis in opportunistic MRI and CT scan can distinguish patients with and without osteoporotic vertebral fracture: a preliminary study. Eur J Radiol 2023;158:110642.

[38] Pedoia V, Haefeli J, Morioka K, Teng HL, Nardo L, Souza RB, et al. MRI and biomechanics multidimensional data analysis reveals R(2) -R(1ρ) as an early predictor of cartilage lesion progression in knee osteoarthritis. J Magn Reson Imaging 2018;47(1):78–90.

[39] Joseph GB, Hou SW, Nardo L, Heilmeier U, Nevitt MC, McCulloch CE, et al. MRI findings associated with development of incident knee pain over 48 months: data from the osteoarthritis initiative. Skelet Radiol 2016;45(5):653–60.

[40] Yeoh PSQ, Lai KW, Goh SL, Hasikin K, Hum YC, Tee YK, et al. Emergence of deep learning in knee osteoarthritis diagnosis. Comput Intell Neurosci 2021;2021:4931437.

[41] Burton 2nd W, Myers C, Rullkoetter P. Semi-supervised learning for automatic segmentation of the knee from MRI with convolutional neural networks. Comput Methods Programs Biomed 2020;189:105328.

[42] Pedoia V, Majumdar S, Link TM. Segmentation of joint and musculoskeletal tissue in the study of arthritis. Magma 2016;29(2):207–21.

[43] Tang X, Guo D, Liu A, Wu D, Liu J, Xu N, et al. Fully automatic knee joint segmentation and quantitative analysis for osteoarthritis from magnetic resonance (MR) images using a deep learning model. Med Sci Monit 2022;28:e936733.

[44] Kessler DA, MacKay JW, Crowe VA, Henson FMD, Graves MJ, Gilbert FJ, et al. The optimisation of deep neural networks for segmenting multiple knee joint tissues from MRIs. Comput Med Imaging Graph 2020;86:101793.

[45] Zhou Z, Zhao G, Kijowski R, Liu F. Deep convolutional neural network for segmentation of knee joint anatomy. Magn Reson Med 2018;80(6):2759–70.

[46] Gan H-S, Ramlee MH, Wahab AA, Lee Y-S, Shimizu A. From classical to deep learning: review on cartilage and bone segmentation techniques in knee osteoarthritis research. Artif Intell Rev 2021;54(4):2445–94.

[47] Cheng R, Alexandridi NA, Smith RM, Shen A, Gandler W, McCreedy E, et al. Fully automated patellofemoral MRI segmentation using holistically nested networks: implications for evaluating patellofemoral osteoarthritis, pain, injury, pathology, and adolescent development. Magn Reson Med 2020;83(1):139–53.

[48] Ambellan F, Tack A, Ehlke M, Zachow S. Automated segmentation of knee bone and cartilage combining statistical shape knowledge and convolutional neural networks: data from the Osteoarthritis Initiative. Med Image Anal 2019;52:109–18.
[49] Khalid H, Hussain M, Al Ghamdi MA, Khalid T, Khalid K, Khan MA, et al. A comparative systematic literature review on knee bone reports from MRI, X-rays and CT scans using deep learning and machine learning methodologies. Diagnostics 2020;10(8).
[50] Kim-Wang SY, Bradley PX, Cutcliffe HC, Collins AT, Crook BS, Paranjape CS, et al. Auto-segmentation of the tibia and femur from knee MR images via deep learning and its application to cartilage strain and recovery. J Biomech 2023;149:111473.
[51] Marques J, Genant HK, Lillholm M, Dam EB. Diagnosis of osteoarthritis and prognosis of tibial cartilage loss by quantification of tibia trabecular bone from MRI. Magn Reson Med 2013;70(2):568–75.
[52] Rodriguez-Vila B, Gonzalez-Hospital V, Puertas E, Beunza JJ, Pierce DM. Democratization of deep learning for segmenting cartilage from MRIs of human knees: application to data from the osteoarthritis initiative. J Orthop Res 2023;41(8):1754–66.
[53] Ebrahimkhani S, Jaward MH, Cicuttini FM, Dharmaratne A, Wang Y, de Herrera AGS. A review on segmentation of knee articular cartilage: from conventional methods towards deep learning. Artif Intell Med 2020;106:101851.
[54] Kijowski R, Fritz J, Deniz CM. Deep learning applications in osteoarthritis imaging. Skelet Radiol 2023;52(11):2225–38.
[55] Fripp J, Crozier S, Warfield SK, Ourselin S. Automatic segmentation and quantitative analysis of the articular cartilages from magnetic resonance images of the knee. IEEE Trans Med Imaging 2010;29(1):55–64.
[56] Tamez-Peña JG, Farber J, González PC, Schreyer E, Schneider E, Totterman S. Unsupervised segmentation and quantification of anatomical knee features: data from the Osteoarthritis Initiative. IEEE Trans Biomed Eng 2012;59(4):1177–86.
[57] Chadoulos CG, Tsaopoulos DE, Moustakidis S, Tsakiridis NL, Theocharis JB. A novel multi-atlas segmentation approach under the semi-supervised learning framework: application to knee cartilage segmentation. Comput Methods Programs Biomed 2022;227:107208.
[58] Panfilov E, Tiulpin A, Nieminen MT, Saarakkala S, Casula V. Deep learning-based segmentation of knee MRI for fully automatic subregional morphological assessment of cartilage tissues: data from the Osteoarthritis Initiative. J Orthop Res 2022;40(5): 1113–24.
[59] Prasoon A, Igel C, Loog M, Lauze F, Dam EB, Nielsen M. Femoral cartilage segmentation in knee MRI scans using two stage voxel classification. Annu Int Conf IEEE Eng Med Biol Soc 2013;2013:5469–72.
[60] Zhang Q, Geng J, Zhang M, Kan T, Wang L, Ai S, et al. Cartilage morphometry and magnetic susceptibility measurement for knee osteoarthritis with automatic cartilage segmentation. Quant Imaging Med Surg 2023;13(6):3508–21.
[61] Zhang K, Lu W, Marziliano P. Automatic knee cartilage segmentation from multi-contrast MR images using support vector machine classification with spatial dependencies. Magn Reson Imaging 2013;31(10):1731–43.
[62] Schmidt AM, Desai AD, Watkins LE, Crowder HA, Black MS, Mazzoli V, et al. Generalizability of deep learning segmentation algorithms for automated assessment of cartilage morphology and MRI relaxometry. J Magn Reson Imaging 2023;57(4): 1029–39.

[63] Shah RF, Martinez AM, Pedoia V, Majumdar S, Vail TP, Bini SA. Variation in the thickness of knee cartilage. The use of a novel machine learning algorithm for cartilage segmentation of magnetic resonance images. J Arthroplast 2019;34(10):2210–5.

[64] Felfeliyan B, Hareendranathan A, Kuntze G, Jaremko JL, Ronsky JL. Improved-Mask R-CNN: towards an accurate generic MSK MRI instance segmentation platform (data from the Osteoarthritis Initiative). Comput Med Imaging Graph 2022;97:102056.

[65] Ashinsky BG, Coletta CE, Bouhrara M, Lukas VA, Boyle JM, Reiter DA, et al. Machine learning classification of OARSI-scored human articular cartilage using magnetic resonance imaging. Osteoarthr Cartil 2015;23(10):1704–12.

[66] Liu F, Zhou Z, Samsonov A, Blankenbaker D, Larison W, Kanarek A, et al. Deep learning approach for evaluating knee MR images: achieving high diagnostic performance for cartilage lesion detection. Radiology 2018;289(1):160–9.

[67] Huo J, Ouyang X, Si L, Xuan K, Wang S, Yao W, et al. Automatic grading assessments for knee MRI cartilage defects via self-ensembling semi-supervised learning with dual-consistency. Med Image Anal 2022;80:102508.

[68] Cui T, Liu R, Jing Y, Fu J, Chen J. Development of machine learning models aiming at knee osteoarthritis diagnosing: an MRI radiomics analysis. J Orthop Surg Res 2023;18 (1):375.

[69] Dunn TC, Lu Y, Jin H, Ries MD, Majumdar S. T2 relaxation time of cartilage at MR imaging: comparison with severity of knee osteoarthritis. Radiology 2004;232(2): 592–8.

[70] Linka K, Thüring J, Rieppo L, Aydin RC, Cyron CJ, Kuhl C, et al. Machine learning-augmented and microspectroscopy-informed multiparametric MRI for the non-invasive prediction of articular cartilage composition. Osteoarthr Cartil 2021;29(4): 592–602.

[71] Peuna A, Thevenot J, Saarakkala S, Nieminen MT, Lammentausta E. Machine learning classification on texture analyzed T2 maps of osteoarthritic cartilage: oulu knee osteoarthritis study. Osteoarthr Cartil 2021;29(6):859–69.

[72] Thomas KA, Krzemiński D, Kidziński Ł, Paul R, Rubin EB, Halilaj E, et al. Open source software for automatic subregional assessment of knee cartilage degradation using quantitative T2 relaxometry and deep learning. Cartilage 2021;13(1_suppl): 747S–56S.

[73] Pedoia V, Lee J, Norman B, Link TM, Majumdar S. Diagnosing osteoarthritis from T (2) maps using deep learning: an analysis of the entire Osteoarthritis Initiative baseline cohort. Osteoarthr Cartil 2019;27(7):1002–10.

[74] Tolpadi AA, Lee JJ, Pedoia V, Majumdar S. Deep learning predicts total knee replacement from magnetic resonance images. Sci Rep 2020;10(1):6371.

[75] Namiri NK, Lee J, Astuto B, Liu F, Shah R, Majumdar S, et al. Deep learning for large scale MRI-based morphological phenotyping of osteoarthritis. Sci Rep 2021;11(1): 10915.

[76] Nelson AE, Fang F, Arbeeva L, Cleveland RJ, Schwartz TA, Callahan LF, et al. A machine learning approach to knee osteoarthritis phenotyping: data from the FNIH Biomarkers Consortium. Osteoarthr Cartil 2019;27(7):994–1001.

[77] Du Y, Almajalid R, Shan J, Zhang M. A novel method to predict knee osteoarthritis progression on MRI using machine learning methods. IEEE Trans NanoBioscience 2018;17(3):228–36.

[78] Jamshidi A, Leclercq M, Labbe A, Pelletier JP, Abram F, Droit A, et al. Identification of the most important features of knee osteoarthritis structural progressors using machine learning methods. Ther Adv Musculoskelet Dis 2020;12. 1759720X20933468.
[79] Eckstein F, Chaudhari AS, Fuerst D, Gaisberger M, Kemnitz J, Baumgartner CF, et al. Detection of differences in longitudinal cartilage thickness loss using a deep-learning automated segmentation algorithm: data from the foundation for the National Institutes of Health biomarkers study of the osteoarthritis initiative. Arthritis Care Res 2022;74 (6):929–36.
[80] Jørgensen DR, Dam EB, Lillholm M. Predicting knee cartilage loss using adaptive partitioning of cartilage thickness maps. Comput Biol Med 2013;43(8):1045–52.
[81] Ashinsky BG, Bouhrara M, Coletta CE, Lehallier B, Urish KL, Lin PC, et al. Predicting early symptomatic osteoarthritis in the human knee using machine learning classification of magnetic resonance images from the osteoarthritis initiative. J Orthop Res 2017;35(10):2243–50.
[82] Cootes TF, Taylor CJ, Cooper DH, Graham J. Active shape models-their training and application. Comput Vis Image Understand 1995;61(1):38–59.
[83] Bowes MA, Kacena K, Alabas OA, Brett AD, Dube B, Bodick N, et al. Machine-learning, MRI bone shape and important clinical outcomes in osteoarthritis: data from the Osteoarthritis Initiative. Ann Rheum Dis 2021;80(4):502–8.
[84] Brett A, Bowes MA, Conaghan PG. Comparison of 3D quantitative osteoarthritis imaging biomarkers from paired CT and MR images: data from the IMI-APPROACH study. BMC Musculoskelet Disord 2023;24(1):76.
[85] Byra M, Wu M, Zhang X, Jang H, Ma YJ, Chang EY, et al. Knee menisci segmentation and relaxometry of 3D ultrashort echo time cones MR imaging using attention U-Net with transfer learning. Magn Reson Med 2020;83(3):1109–22.
[86] Flannery SW, Kiapour AM, Edgar DJ, Murray MM, Fleming BC. Automated magnetic resonance image segmentation of the anterior cruciate ligament. J Orthop Res 2021;39 (4):831–40.
[87] Flannery SW, Kiapour AM, Edgar DJ, Murray MM, Beveridge JE, Fleming BC. A transfer learning approach for automatic segmentation of the surgically treated anterior cruciate ligament. J Orthop Res 2022;40(1):277–84.
[88] Chang PD, Wong TT, Rasiej MJ. Deep learning for detection of complete anterior cruciate ligament tear. J Digit Imaging 2019;32(6):980–6.
[89] Dung NT, Thuan NH, Van Dung T, Van Nho L, Tri NM, Vy VPT, et al. End-to-end deep learning model for segmentation and severity staging of anterior cruciate ligament injuries from MRI. Diagn Interv Imaging 2023;104(3):133–41.
[90] Minamoto Y, Akagi R, Maki S, Shiko Y, Tozawa R, Kimura S, et al. Automated detection of anterior cruciate ligament tears using a deep convolutional neural network. BMC Musculoskelet Disord 2022;23(1):577.
[91] Siouras A, Moustakidis S, Giannakidis A, Chalatsis G, Liampas I, Vlychou M, et al. Knee injury detection using deep learning on MRI studies: a systematic review. Diagnostics 2022;12(2).
[92] Štajduhar I, Mamula M, Miletić D, Ünal G. Semi-automated detection of anterior cruciate ligament injury from MRI. Comput Methods Programs Biomed 2017;140: 151–64.
[93] Tran A, Lassalle L, Zille P, Guillin R, Pluot E, Adam C, et al. Deep learning to detect anterior cruciate ligament tear on knee MRI: multi-continental external validation. Eur Radiol 2022;32(12):8394–403.

[94] Zhang L, Li M, Zhou Y, Lu G, Zhou Q. Deep learning approach for anterior cruciate ligament lesion detection: evaluation of diagnostic performance using arthroscopy as the reference standard. J Magn Reson Imaging 2020;52(6):1745–52.

[95] Shin H, Choi GS, Shon OJ, Kim GB, Chang MC. Development of convolutional neural network model for diagnosing meniscus tear using magnetic resonance image. BMC Musculoskelet Disord 2022;23(1):510.

[96] Rizk B, Brat H, Zille P, Guillin R, Pouchy C, Adam C, et al. Meniscal lesion detection and characterization in adult knee MRI: a deep learning model approach with external validation. Phys Med 2021;83:64–71.

[97] Fritz B, Marbach G, Civardi F, Fucentese SF, Pfirrmann CWA. Deep convolutional neural network-based detection of meniscus tears: comparison with radiologists and surgery as standard of reference. Skelet Radiol 2020;49(8):1207–17.

[98] Pedoia V, Norman B, Mehany SN, Bucknor MD, Link TM, Majumdar S. 3D convolutional neural networks for detection and severity staging of meniscus and PFJ cartilage morphological degenerative changes in osteoarthritis and anterior cruciate ligament subjects. J Magn Reson Imaging 2019;49(2):400–10.

[99] Zech JR, Carotenuto G, Jaramillo D. Inferring pediatric knee skeletal maturity from MRI using deep learning. Skelet Radiol 2022;51(8):1671–7.

[100] Mutasa S, Chang PD, Ruzal-Shapiro C, Ayyala R. MABAL: a novel deep-learning architecture for machine-assisted bone age labeling. J Digit Imaging 2018;31(4):513–9.

[101] Larson DB, Chen MC, Lungren MP, Halabi SS, Stence NV, Langlotz CP. Performance of a deep-learning neural network model in assessing skeletal maturity on pediatric hand radiographs. Radiology 2018;287(1):313–22.

[102] Sambandam SN, Khanna V, Gul A, Mounasamy V. Rotator cuff tears: an evidence based approach. World J Orthop 2015;6(11):902–18.

[103] Lee SH, Lee J, Oh KS, Yoon JP, Seo A, Jeong Y, et al. Automated 3-dimensional MRI segmentation for the posterosuperior rotator cuff tear lesion using deep learning algorithm. PLoS One 2023;18(5):e0284111.

[104] Kim H, Shin K, Kim H, Lee ES, Chung SW, Koh KH, et al. Can deep learning reduce the time and effort required for manual segmentation in 3D reconstruction of MRI in rotator cuff tears? PLoS One 2022;17(10):e0274075.

[105] Key S, Demir S, Gurger M, Yilmaz E, Barua PD, Dogan S, et al. ViVGG19: novel exemplar deep feature extraction-based shoulder rotator cuff tear and biceps tendinosis detection using magnetic resonance images. Med Eng Phys 2022;110:103864.

[106] Guo D, Liu X, Wang D, Tang X, Qin Y. Development and clinical validation of deep learning for auto-diagnosis of supraspinatus tears. J Orthop Surg Res 2023;18(1):426.

[107] Lin DJ, Schwier M, Geiger B, Raithel E, von Busch H, Fritz J, et al. Deep learning diagnosis and classification of rotator cuff tears on shoulder MRI. Investig Radiol 2023;58(6):405–12.

[108] Medina G, Buckless CG, Thomasson E, Oh LS, Torriani M. Deep learning method for segmentation of rotator cuff muscles on MR images. Skelet Radiol 2021;50(4):683–92.

[109] Oeding JF, Pareek A, Nieboer MJ, Rhodes NG, Tiegs-Heiden CA, Camp CL, et al. A machine learning model demonstrates excellent performance in predicting subscapularis tears based on pre-operative imaging parameters alone. Arthroscopy 2023.

[110] Ro K, Kim JY, Park H, Cho BH, Kim IY, Shim SB, et al. Deep-learning framework and computer assisted fatty infiltration analysis for the supraspinatus muscle in MRI. Sci Rep 2021;11(1):15065.

[111] Yao J, Chepelev L, Nisha Y, Sathiadoss P, Rybicki FJ, Sheikh AM. Evaluation of a deep learning method for the automated detection of supraspinatus tears on MRI. Skelet Radiol 2022;51(9):1765–75.
[112] Yin L, Kong Y, Guo M, Zhang X, Yan W, Zhang H. A preliminary attempt to use radiomic features in the diagnosis of extra-articular long head biceps tendinitis. Magma 2023;36(4):651–8.
[113] Ni M, Wen X, Chen W, Zhao Y, Yuan Y, Zeng P, et al. A deep learning approach for MRI in the diagnosis of labral injuries of the hip joint. J Magn Reson Imaging 2022;56(2):625–34.
[114] Klontzas ME, Manikis GC, Nikiforaki K, Vassalou EE, Spanakis K, Stathis I, et al. Radiomics and machine learning can differentiate transient osteoporosis from avascular necrosis of the hip. Diagnostics 2021;11(9).
[115] Astolfi RS, da Silva DS, Guedes IS, Nascimento CS, Damaševičius R, Jagatheesaperumal SK, et al. Computer-aided ankle ligament injury diagnosis from magnetic resonance images using machine learning techniques. Sensors (Basel) 2023;23(3).
[116] Li Z, Ji X. Magnetic resonance imaging image segmentation under edge detection intelligent algorithm in diagnosis of surgical wrist joint injuries. Contrast Media Mol Imaging 2021;2021:6891120.
[117] Ong W, Zhu L, Tan YL, Teo EC, Tan JH, Kumar N, et al. Application of machine learning for differentiating bone malignancy on imaging: a systematic review. Cancers (Basel) 2023;15(6).
[118] Georgeanu VA, Mămuleanu M, Ghiea S, Selisteanu D. Malignant bone tumors diagnosis using magnetic resonance imaging based on deep learning algorithms. Medicina (Kaunas) 2022;58(5).
[119] Gitto S, Cuocolo R, Albano D, Chianca V, Messina C, Gambino A, et al. MRI radiomics-based machine-learning classification of bone chondrosarcoma. Eur J Radiol 2020;128:109043.
[120] Crombé A, Fadli D, Buy X, Italiano A, Saut O, Kind M. High-grade soft-tissue sarcomas: can optimizing dynamic contrast-enhanced MRI postprocessing improve prognostic radiomics models? J Magn Reson Imaging 2020;52(1):282–97.
[121] Fields BKK, Demirjian NL, Hwang DH, Varghese BA, Cen SY, Lei X, et al. Whole-tumor 3D volumetric MRI-based radiomics approach for distinguishing between benign and malignant soft tissue tumors. Eur Radiol 2021;31(11):8522–35.
[122] Foreman SC, Llorián-Salvador O, David DE, Rösner VKN, Rischewski JF, Feuerriegel GC, et al. Development and evaluation of MR-based radiogenomic models to differentiate atypical lipomatous tumors from lipomas. Cancers (Basel) 2023;15(7).
[123] Fradet G, Ayde R, Bottois H, El Harchaoui M, Khaled W, Drapé JL, et al. Prediction of lipomatous soft tissue malignancy on MRI: comparison between machine learning applied to radiomics and deep learning. Eur Radiol Exp 2022;6(1):41.
[124] Gitto S, Interlenghi M, Cuocolo R, Salvatore C, Giannetta V, Badalyan J, et al. MRI radiomics-based machine learning for classification of deep-seated lipoma and atypical lipomatous tumor of the extremities. Radiol Med 2023;128(8):989–98.
[125] Juntu J, Sijbers J, De Backer S, Rajan J, Van Dyck D. Machine learning study of several classifiers trained with texture analysis features to differentiate benign from malignant soft-tissue tumors in T1-MRI images. J Magn Reson Imaging 2010;31(3):680–9.

[126] Lee S, Lee SY, Jung JY, Nam Y, Jeon HJ, Jung CK, et al. Ensemble learning-based radiomics with multi-sequence magnetic resonance imaging for benign and malignant soft tissue tumor differentiation. PLoS One 2023;18(5):e0286417.

[127] Leporq B, Bouhamama A, Pilleul F, Lame F, Bihane C, Sdika M, et al. MRI-based radiomics to predict lipomatous soft tissue tumors malignancy: a pilot study. Cancer Imaging 2020;20(1):78.

[128] Nakagawa M, Nakaura T, Yoshida N, Azuma M, Uetani H, Nagayama Y, et al. Performance of machine learning methods based on multi-sequence textural parameters using magnetic resonance imaging and clinical information to differentiate malignant and benign soft tissue tumors. Acad Radiol 2023;30(1):83–92.

[129] Navarro F, Dapper H, Asadpour R, Knebel C, Spraker MB, Schwarze V, et al. Development and external validation of deep-learning-based tumor grading models in soft-tissue sarcoma patients using MR imaging. Cancers (Basel) 2021;13(12).

[130] Peeken JC, Neumann J, Asadpour R, Leonhardt Y, Moreira JR, Hippe DS, et al. Prognostic assessment in high-grade soft-tissue sarcoma patients: a comparison of semantic image analysis and radiomics. Cancers (Basel) 2021;13(8).

[131] Zhang Y, Zhu Y, Shi X, Tao J, Cui J, Dai Y, et al. Soft tissue sarcomas: preoperative predictive histopathological grading based on radiomics of MRI. Acad Radiol 2019;26 (9):1262–8.

[132] Xu W, Hao D, Hou F, Zhang D, Wang H. Soft tissue sarcoma: preoperative MRI-based radiomics and machine learning may Be accurate predictors of histopathologic grade. AJR Am J Roentgenol 2020;215(4):963–9.

[133] Yang Y, Zhou Y, Zhou C, Zhang X, Ma X. MRI-based computer-aided diagnostic model to predict tumor grading and clinical outcomes in patients with soft tissue sarcoma. J Magn Reson Imaging 2022;56(6):1733–45.

[134] Gitto S, Cuocolo R, Albano D, Morelli F, Pescatori LC, Messina C, et al. CT and MRI radiomics of bone and soft-tissue sarcomas: a systematic review of reproducibility and validation strategies. Insights Imaging 2021;12(1):68.

[135] Zhao K, Zhu X, Zhang M, Xie Z, Yan X, Wu S, et al. Radiologists with assistance of deep learning can achieve overall accuracy of benign-malignant differentiation of musculoskeletal tumors comparable with that of pre-surgical biopsies in the literature. Int J Comput Assist Radiol Surg 2023;18(8):1451–8.

[136] Blackledge MD, Winfield JM, Miah A, Strauss D, Thway K, Morgan VA, et al. Supervised machine-learning enables segmentation and evaluation of heterogeneous post-treatment changes in multi-parametric MRI of soft-tissue sarcoma. Front Oncol 2019;9:941.

[137] Fields BKK, Demirjian NL, Cen SY, Varghese BA, Hwang DH, Lei X, et al. Predicting soft tissue sarcoma response to neoadjuvant chemotherapy using an MRI-based delta-radiomics approach. Mol Imaging Biol 2023;25(4):776–87.

[138] Gao Y, Kalbasi A, Hsu W, Ruan D, Fu J, Shao J, et al. Treatment effect prediction for sarcoma patients treated with preoperative radiotherapy using radiomics features from longitudinal diffusion-weighted MRIs. Phys Med Biol 2020;65(17):175006.

[139] Liu S, Sun W, Yang S, Duan L, Huang C, Xu J, et al. Deep learning radiomic nomogram to predict recurrence in soft tissue sarcoma: a multi-institutional study. Eur Radiol 2022;32(2):793–805.

[140] Peeken JC, Asadpour R, Specht K, Chen EY, Klymenko O, Akinkuoroye V, et al. MRI-based delta-radiomics predicts pathologic complete response in high-grade

soft-tissue sarcoma patients treated with neoadjuvant therapy. Radiother Oncol 2021; 164:73–82.

[141] Wennmann M, Neher P, Stanczyk N, Kahl KC, Kächele J, Weru V, et al. Deep learning for automatic bone marrow apparent diffusion coefficient measurements from whole-body magnetic resonance imaging in patients with multiple myeloma: a retrospective multicenter study. Investig Radiol 2023;58(4):273–82.

[142] von Brandis E, Jenssen HB, Avenarius DFM, Bjørnerud A, Flatø B, Tomterstad AH, et al. Automated segmentation of magnetic resonance bone marrow signal: a feasibility study. Pediatr Radiol 2022;52(6):1104–14.

[143] Albano D, Gitto S, Messina C, Serpi F, Salvatore C, Castiglioni I, et al. MRI-based artificial intelligence to predict infection following total hip arthroplasty failure. Radiol Med 2023;128(3):340–6.

PART

Reproducibility

7

CHAPTER

24 Let's share: Open-source frameworks and public databases

Fernando Pérez-García[1, *], Benjamin Murray[2], Eric Kerfoot[2], Sebastien Ourselin[2], Marc Modat[2]

[1]Microsoft Research Cambridge, Cambridge, United Kingdom; [2]School of Biomedical Engineering & Imaging Sciences, King's College London, London, United Kingdom

[*]Corresponding-author: E-mail: fperezgarcia@microsoft.com

24.1 Introduction

Open-source software is any software artifact, from a small piece of code to whole libraries or whole applications, whose source code is publicly accessible under license terms permitting a broad use of the code and running software derived from it. Typically for modern software, this source code is available online through code repositories for anyone to read and download, and released under licenses that protect authorship and certain commercial rights but otherwise allow any use of the code. Anyone can get this code, compile it (if necessary), run it on their computers, modify it, and distribute it to others.

Open source evolved from the Free Software Movement,[1] which advocated for software use with few legal or commercial constraints, in contrast to proprietary and closed-source commercial software. Broadly, free software under this definition and open-source software are equivalent categories, albeit with some differences in philosophy[2] compared with the Open Source Initiative (OSI).[3]

Many open-source licenses serve different purposes within the open-source community. The OSI, set up as a steward of the Open Source Definition,[4] approves licenses as being open source or not through their license review process. Although it is possible to write a bespoke license for open-source code, most open-source projects choose from an existing open-source approved license.

Adjacent to the open-source movement are source-available licenses, often adopted by corporations who see the benefits of open source in terms of code quality and

[1]https://www.fsf.org.
[2]https://www.gnu.org/philosophy/free-sw.html.
[3]https://opensource.org/osd.
[4]https://opensource.org/definition-annotated.

Machine Learning in MRI. https://doi.org/10.1016/B978-0-443-14109-6.00024-9

contributor community, but, for different reasons, still prefer that the project is distributed under commercial licensing agreements, for some or all of the potential users.

These concepts of free availability with few constraints of use can be extended to any other area of intellectual property, such as databases, designs and schematics, literature, and other media formats.

Within the context of this chapter, we can consider open-source tools to be software systems, used by other applications or directly by users, whose licensing conditions allow the tool to be used for data analysis in almost any context. This includes noncommercial usage, such as research, and commercial usage, provided certain limitations about derivative works are adhered to. Similarly, open datasets containing imaging and other medical data freely available with no restrictions on research use or constraining legal agreements can be considered open source. Furthermore, in the era of deep learning, the weights (the parameters of a neural network that are the result of the training process) of a neural network can also be made open source (or, rather, open-access).

24.2 Why open source?

So, why has the open-source movement become so important for both research and software development? The answer is a confluence of beneficial incentives, both in terms of using and contributing to open-source software. Open-source software brings communities together. It gives would-be developers the chance to contribute to something and raise their own skill set in the process. It provides software that, defying its early critics, can beat proprietary software in terms of security, transparency, and consistency.

Open-access datasets, too, provide quality and consistency in research through their critical role as benchmarks [1], providing common reference points against which cutting-edge methods can be evaluated.

24.2.1 Community

Open-source communities can act as hubs of expertise and best practices. The broad open-source ecosystem that thrives around the Python programming language is the perfect exemplar of the benefits of open-source communities. The scientific programming ecosystem of Python, in particular, produces open-source libraries such as NumPy [2], SciPy [3], and scikit-learn [4], and data analytics tools such as Pandas [5] that underpin a great deal of computational science work in both research and industry.

Python has been a natural choice for these open-source projects because another of its ecosystems is responsible for producing open-source tools such as PyPI[5] and virtualenv[6] that have significantly lowered the level of difficulty involved in publishing, accessing, and using libraries written in Python and aid in producing reproducible research.

The Python Software Foundation is one of the driving forces behind Python and its open-source focus and philosophy. It is primarily concerned with maintaining and developing the language and providing some of the critical open-source infrastructure. It also organizes PyCon,[7] which brings open-source developers from across the Python ecosystem together.

24.2.2 Benefits of using open source

Beyond the concentration of specialist domain knowledge and best practices in key open-source libraries, open-source software offers several operational advantages. Its transparency has proven effective in security-critical applications, with many projects developing and adopting robust practices to mitigate risks from malicious contributors.

24.2.2.1 Security

The notion that sharing code, particularly code for which security is important, may seem counterintuitive. The ability to see code means that malicious actors can take advantage of vulnerabilities in the code. The open-source movement, however, considers that open-sourcing software means that a broad array of experts can find and fix defects in the code that lead to security vulnerabilities.

It is hard to quantify whether open-source software is more secure than its closed-source counterparts, but some attempts at quantification have been made, such as fuzz testing. Miller et al. generated a proxy for security whereby a suite of proprietary and open-source software was fed random streams of data while being monitored for crashes or hangs [6]. The principle is that software application programming interfaces (APIs) should be tolerant to malformed and random inputs, with such tolerance serving as a proxy for security. Over 5 years, Miller tested proprietary Unix and analogous open-source subsystems and found that, while all tested systems improved their rate of crashes and hangs, the two open-source variants produced the best reliability rates.

This is not to say that open-source systems are inherently more secure than proprietary systems, but the open-source movement's emphasis on exposing of vulnerabilities as a best practice is difficult to dispute. However, for open source to be most effective, it must come with a deep knowledge among key maintainers of the

[5]https://pypi.org.
[6]https://virtualenv.pypa.io/en/latest/.
[7]https://pycon.org/.

software in question, as well as a commitment to finding and resolving security flaws timely. As a result, prominent open-source projects are successfully used in some of the most security-critical environments. This is due to both the fundamental dynamics of open-source projects and the measures that open-source projects put in place to ensure the continued good reputation of the project.

24.2.2.2 Transparency

Open-source code is transparent by its very nature. The full source code is available for anyone to inspect and so all potential vulnerabilities are in the open. The active contributor and user bases of open-source projects have an interest, reputationally and/or professionally, in the software not being the cause of security breaches. There is, therefore, a vested interest in finding and reporting the vulnerabilities in the codebase, and, as such, vulnerabilities tend to be fixed more quickly than in proprietary codebases.

24.2.2.3 Auditing

Open-source projects, particularly those where security is a significant aspect of the project, typically have regular audits from both within the body of contributors and from independent researchers and security experts.

24.2.3 Contributing to open-source projects

For users of open-source software, there are many benefits to becoming contributors to open-source software projects. As a researcher looking to improve their programming skill set, there are strong pedagogical and reputational benefits to involving oneself in established or up-and-coming open-source libraries.

24.2.3.1 Pedagogical benefits

Contributing to open-source projects is an opportunity to take part in a larger-scale development process than developers of small-scale projects may typically be exposed to. This process requires that the contributor be able to make the case for a given feature change or addition to be made, to negotiate with other stakeholders the appropriate implementation for the change in question, and add documentation for and testing of the change.

24.2.3.2 Reputational benefits

Contributing to open-source projects is increasingly seen as a critical part of a software developer's experience and skill set, alongside vocational experience. It also provides opportunities to demonstrate expertise in a particular area of software engineering outside of vocational experience.

While bad actors have sometimes attempted to deliberately introduce vulnerabilities into open-source projects, in practice, changes to security-critical elements of open-source projects tend to follow stringent review procedures to detect and prevent such occurrences. This must be set in contrast to security vulnerabilities in proprietary software, which often relies on "security by obscurity."

24.2.4 Private sector and open source

Several prominent open-source libraries have been created by corporations or started life as proprietary libraries that subsequently became open-source. The field of deep learning is dominated by libraries that were released as open-source projects by major corporations. TensorFlow [7], initially released by Google under the Apache 2.0 license, and PyTorch [8], initially released by Meta under the BSD 3-Clause license, are two of the dominant deep learning libraries for both research and commercial applications using deep learning.

When done well, this practice is highly beneficial to both the corporation releasing the library and the community that uses and contributes to it.

24.2.4.1 Controversies around corporate open source

Both Google and Meta include Contributor License Agreements that link contributors to their respective libraries. These are quite common for corporate open-source projects but often contain intellectual property transfer from contributors to the open-source project owner, i.e., the corporation in question.

Some consider these agreements controversial as they enable the library owner to change the licensing agreement from open-source to non–open-source without any legal recourse available to contributors.

Even with slightly less restrictive Contributor License Agreements, this has happened a number of times with prominent open source libraries, including MongoDB[8] and ElasticSearch,[9] although usually with considerable reputational damage to the entity doing so. In these cases, the Supplier Security and Privacy Assurance (SSPA) license (not approved as open source by the OSI) was adopted to limit the ability of cloud providers to freely use, adapt, and monetize their libraries without contributing to them in terms of functionality or monetarily.

In the case of ElasticSearch, the repository was forked by the open-source community OpenSearch,[10] which has continued to develop a free library with similar functionality to ElasticSearch's library. Notably, ElasticSearch has since moved to relicense their library under the GNU Affero General Public License (AGPL),[11] which is an OSI-approved license.

24.3 Requirements of open source and public databases

Ideally, the funding, time, and effort that goes into the creation of research datasets and research code are not wasted in the medium term. The goal is that research software and datasets have broad utility across multiple research efforts and in a

[8]https://www.mongodb.com/.
[9]https://www.elastic.co/elasticsearch.
[10]https://opensearch.org/.
[11]https://www.gnu.org/licenses/agpl-3.0.en.html.

sustainable fashion, and many principles have been discovered and refined into best practices that help to achieve these goals. Many of these principles and best practices have been encoded into findability, accessibility, interoperability, and reusability (FAIR) [9], a set of principles published in 2016 for data management and stewardship, and adapted as FAIR Principles for Research Software (FAIR4RS) [10] in 2022. FAIR is widely endorsed and has been progressively adopted within the scientific community.

In this section, we discuss how each of these aspects applies to open-source projects and public databases in the context of machine learning for medical image computing.

24.3.1 Findability (and citability)

Findability refers to making data easily discoverable through indexing, metadata, and persistent identifiers. There is an emphasis on this being done in such a way that allows automated discovery by data and code repositories and aggregators.

24.3.1.1 Code repositories

Source code should be preferably hosted on a cloud-based platform such as GitHub[12] or GitLab.[13]

24.3.1.2 Data repositories

Data repositories provide persistent digital object identifiers (DOIs) for data and code, and allow for the creation of a citable record. Discipline-specific data repositories are generally recommended.[14] For example, PhysioNet[15] is "a repository of freely available medical research data," and OpenNeuro[16] is "a free and open platform for validating and sharing Brain Imaging Data Structure (BIDS)-compliant MRI, PET, MEG, EEG, and iEEG data." If a suitable discipline-specific data repository does not exist, authors may upload their data to generic data repositories such as Zenodo[17] and Figshare[18]. Academic institutions may also provide data repositories for their members and authors, respectively. Some examples include the University College London (UCL)[19] and King's College London (KCL)[20] research data repositories.

[12]https://github.com.
[13]https://gitlab.com.
[14]https://www.nature.com/sdata/policies/repositories.
[15]https://physionet.org.
[16]https://openneuro.org.
[17]https://zenodo.org.
[18]https://figshare.com.
[19]https://rdr.ucl.ac.uk.
[20]https://kcl.figshare.com.

24.3.1.3 Versioning

Version control, particularly through sequence-based identifiers, plays an important role in managing both software development and data management. These identifiers offer a chronological record of changes, providing a clear historical timeline of modifications. A commonly used convention for versioning is Semantic Versioning (SemVer),[21] which defines a set of rules and requirements that dictate how version numbers are assigned and incremented.

In the context of data management, sequence-based identifiers in version control systems allow for precise tracking of dataset alterations, and new releases can be assigned a new DOI to ensure precise citability.

When a software library is written for a specific publication, it is encouraged to release a new version and upload a snapshot of the code to a data repository such as Zenodo, which will generate a DOI for the release that can be cited in the publication [11].

24.3.1.4 Publishing

Publications describing the software or data may be submitted to peer-reviewed journals such as the Journal of Open Source Software (JOSS)[22] or the Journal of Open Research Software (JORS).[23] Research software articles are also often published in discipline-specific journals such as Computer Methods and Programs in Biomedicine (CMPB)[24] or Frontiers in Neuroinformatics.[25]

`CITATION.cff`[26] is a human- and machine-readable format for describing how to cite software and data. It is recommended to include a `CITATION.cff` file in the root directory of the repository, which can be used by reference managers such as Zotero[27] to automatically generate citations.

24.3.2 Accessibility

Sharing code on public repositories allows one to easily get a copy of (clone) the codebase using standard tools such as `git`.

Unlike code, different levels of accessibility may be required for data. Sometimes, the data is freely available under certain requirements, such as accepting terms and conditions, registering on a website, or requesting access. For example, the authors of the Information eXtraction from Images (IXI) dataset[28] made the data available under the CC BY-SA 3.0 license and only required that the authors

[21] https://semver.org.
[22] https://joss.theoj.org.
[23] https://openresearchsoftware.metajnl.com.
[24] https://www.journals.elsevier.com/computer-methods-and-programs-in-biomedicine.
[25] https://www.frontiersin.org/journals/neuroinformatics.
[26] https://citation-file-format.github.io.
[27] https://www.zotero.org.
[28] https://brain-development.org/ixi-dataset.

of publications using the data cite the IXI website. To download the MIMIC-CXR dataset [12], users must register on the PhysioNet website and perform a training course on the use of the data. The authors of RoentGen [13] asked users to fill out a form requesting the weights of their generative model trained on the MIMIC-CXR dataset. To publish manuscripts using data from the Alzheimer's Disease Neuroimaging Initiative (ADNI),[29] authors must include the phrase "for the Alzheimer's Disease Neuroimaging Initiative*," with the asterisk referring to a footnote containing a statement of a complete list of ADNI investigators.

The level of accessibility is a balance between the needs of the community and the requirements of the data owners, which often include privacy and ethical considerations.

24.3.2.1 Packaging

Effective packaging of software tools and libraries enhances their accessibility. Packaging refers to the process of bundling software so that it can be easily distributed and installed by end-users. This aspect is particularly important in a research environment where ease of installation and configuration can significantly impact the adoption and use of software.

Python's open-source software ecosystem provides standardized packaging formats such as Python wheels or Conda packages, which may be published on cloud-based platforms such as Python Package Index (PyPI)[30] or conda-forge [14], which ensure compatibility across different environments and platforms. These packaging formats also support dependency management, making it easier for users to install software with all its necessary components.

24.3.3 Interoperability

Interoperability refers to the ability of different systems, platforms, or applications to work together and understand each other's data.

Sharing data stored in commonly used formats for medical images such as Digital Imaging and Communications in Medicine (DICOM) ensures operability with a wide range of software tools in the field. Friendlier formats for research include Neuroimaging Informatics Technology Initiative (NIfTI)[31] and BIDS [15] (both typically used for neuroimaging data), and, more generally, Nearly Raw Raster Data (NRRD).[32]

Employing software tools and frameworks commonly adopted by the community, such as the Insight Toolkit (ITK) [16] and Visualization Toolkit (VTK) [17], also plays a significant role in promoting interoperability, easing learning curves

[29] http://adni.loni.usc.edu.
[30] https://pypi.org.
[31] https://nifti.nimh.nih.gov.
[32] https://teem.sourceforge.net/nrrd/format.html.

and lowering the entry barrier for new users who are likely to already be familiar with these tools.

In the context of machine learning, interoperability is also promoted by the use of common frameworks such as Scikit-learn [18], PyTorch [8], and (less often) TensorFlow [7], which allow for the development of models that can be easily deployed and used by others.

24.3.4 Reusability

24.3.4.1 Licensing

Adding a license file to a code repository allows others to use, modify, and share the code. The license file should be placed in the root directory of the repository and named `LICENSE`. Different licenses have different requirements, and it is important to choose a license that is compatible with the intended use of the code. For example, the MIT license[33] is a permissive license that allows others to use, modify, and share the code, as long as the original license is included in derivative works. An easy way to choose a license is to use GitHub's *Choose a License* website.[34]

For data, Creative Commons[35] licenses are commonly used. For example, the CC BY-NC-SA (Attribution-NonCommercial-ShareAlike) "lets others remix, tweak, and build upon your work noncommercially, as long as they credit you and license their new creations under the identical terms." A good resource for choosing a license is the *Creative Commons License Chooser.*[36]

24.3.4.2 Continuous integration and deployment

Leveraging automated systems such as continuous integration (CI) and continuous deployment (CD) pipelines helps in maintaining up-to-date and stable versions of the software. Tools such as GitHub Actions can automate the testing and deployment process, ensuring that the latest versions are always accessible to the community. For example, every time new code is pushed to the online repository, the CI pipeline can automatically run a series of tests to ensure that the code is working as expected. Similarly, when a new release is created, the CD pipeline can automatically deploy the code to different specialized platforms (see Section 3.2.1).

24.3.4.3 Documentation

Comprehensive documentation, including installation guides, API references, usage examples, and tutorials, is an essential component of reusable software. If the Julia, Python, or R programming languages are used, interactive tutorials in Jupyter

[33] https://opensource.org/licenses/MIT.
[34] https://choosealicense.com.
[35] https://creativecommons.org.
[36] https://creativecommons.org/choose.

Notebooks can be particularly effective in demonstrating the use of software in real-world scenarios.

Documentation may be built automatically using tools such as Sphinx[37] or MkDocs[38] (CI), and hosted on cloud-based platforms such as Read the Docs[39] or GitHub Pages[40] (CD).

24.4 Examples of successful open-source projects and public datasets

This section discusses some examples of open-source software projects and open-source data repositories commonly used with biomedical machine learning applications.

24.4.1 Software projects

Machine learning tools and applications are typically built on top of software frameworks, which themselves rely on other frameworks and projects. While the open-source ecosystem for machine learning is vast, in the following we highlight a few essential projects within the Python ecosystem.

24.4.1.1 NumPy

Numerical Python or NumPy[41] [2] is a Python library for defining large and efficient numerical arrays and the algorithms to operate on them. Python represents numbers natively as individual objects, which becomes very inefficient when working with large-scale numerical data. NumPy instead represents large number arrays as contiguous memory, much like compiled languages (i.e., Fortran or C), which exchanges the type flexibility of Python for speed and memory efficiency. Many algorithms are defined for common mathematical operations on these arrays, which are themselves defined in compiled code, and presented through an API with deliberate similarities to the MATLAB environment to aid user migration.

Many libraries use NumPy as the foundation for handling large-scale data such as 3D biomedical images, building on its multidimensional array type. This leverages NumPy as a common platform and avoids reimplementing core functionality for loading, processing, and saving array data.

[37]https://www.sphinx-doc.org.
[38]https://www.mkdocs.org.
[39]https://readthedocs.org.
[40]https://pages.github.com.
[41]https://numpy.org.

24.4.1.2 Scikit-learn

Scikit-learn[42] [18] provides implementations of many machine learning algorithms and techniques, though mostly not neural network–based deep learning techniques. From classification, clustering, and regression, NumPy is used extensively to define data representation and numeric operations.

Scikit-learn is used widely for applications where deep learning would be less suitable. Within the biomedical application sphere, it can be applied effectively to simple regression of medical signal data, dimensionality reduction of nonimage data, and many other ancillary tasks as part of a larger data analysis system.

24.4.1.3 Matplotlib

Matplotlib[43] [19] implements many data visualization routines and facilities, and integrates these with stand-alone interfaces or with other tools such as Jupyter notebooks. The Python interface explicitly presents routines and behavior similar to that of MATLAB to help user migration. Common functionalities include the ability to graph simple values, scatter plots, bar plots, render 2D image data, combine such plots in a larger view, and a limited ability to render data in 3D.

When dealing with biomedical data, especially imaging data, a fast and flexible plotting toolkit, which integrates with existing tools, is extremely useful for validating and monitoring where this data is being used. When developing or using biomedical applications, Matplotlib can be used to inspect data before being used for training, monitoring, and displaying the output of the machine learning system during training, and displaying the results of data analysis after inference. Visually validating data in all stages of machine learning software pipelines is a critical technique for the researcher, developer, and user, as a check against corrupted inputs or nonsensical results. Many other libraries, such as Seaborn [20] or Pandas [5], build on top of the base functionality Matplotlib provides to provide a greater range of visual styles or integrate with other features.

24.4.1.4 PyTorch

In contrast to Scikit-learn, PyTorch[44] [8] focuses exclusively on deep learning to the exclusion of other machine learning techniques. Its primary data structure is a tensor representing multidimensional data arrays very similar to NumPy's array type, but one that can reside in the main system memory or the memory of a graphics processing unit (GPU). Structures and routines build upon the tensor type to define the common components of neural networks, training algorithms, data processing types, and other facilities for developing deep learning models.

PyTorch provides an accessible interface for GPU computation within the Python language and so almost entirely eliminates the need to program directly for

[42]https://scikit-learn.org.
[43]https://matplotlib.org.
[44]https://pytorch.org.

such hardware. GPUs have become essential to deep learning model development and deployment due to being significantly faster than central processing units (CPUs) for large-scale array operations. Making this hardware accessible through Python greatly expands the pool of researchers and practitioners who can develop and use deep learning models, and so accelerates the development of this area of artificial intelligence (AI).

24.4.1.5 TorchIO

TorchIO [21] is a Python library for efficient loading, preprocessing, augmentation, and patch-based sampling of 3D medical images in deep learning, following the design of PyTorch and TorchVision. It presents an intuitive interface for users without much experience with 3D medical images while providing easy-to-use powerful preprocessing and data augmentation tools, such as MRI-specific augmentation including simulation of bias field inhomogeneity, ghosting artifacts, and motion artifacts.

TorchIO is part of the official PyTorch ecosystem,[45] promoting standardization and good practices in the development of deep learning applications for medical image computing.

24.4.1.6 MONAI

Project Medical Open Network for Artificial Intelligence (MONAI) [22] builds on PyTorch, NumPy, and other libraries to define a framework for deep learning applications in healthcare. The core library provides many common components for data handling and network definitions particular to biomedical and clinical applications, which is built upon by other parts of the project for annotating new data (MONAI Label [23]) and facilitating the deployment of applications in the clinic (MONAI Deploy). The project's objective is to support the whole process of developing deep learning systems and their associated training datasets, getting finalized products into actual use by healthcare professionals, and thus getting the benefit of biomedical AI research to patients sooner.

MONAI was developed as a collaboration primarily between King's College London and Nvidia but has expanded to include many other commercial and academic partners and contributors. The initial impetus was the clear need to accelerate the pathway from research to clinic by providing implementations of common techniques and models for researchers to use, while also defining deployment infrastructure as well as readily usable pretrained models. Common sets of tools and definitions help users avoid reinventing certain software wheels as well as provide a common standard of scientific results for new research to be compared against.

[45]https://pytorch.org/ecosystem/.

24.4.1.7 nnU-Net

nnU-Net [24] is a self-configuring method for deep learning–based biomedical image segmentation, based on the U-Net architecture [25]. Its release as an out-of-the-box tool and its numerous successes in segmentation challenges have made it a popular choice for many researchers and developers aiming to train a segmentation models without the need to write much code.

24.4.2 Support organizations

NumPy, Matplotlib, Scikit-learn, and many others are sponsored by the NumFOCUS Foundation,[46] a nonprofit setup to support these and other open source projects with funding and expertise. PyTorch has its own foundation[47] to provide the same sort of support as well as governance, and is itself part of the Linux Foundation.

Many other projects are supported and sponsored by such foundations, charities, and consortia. These provide governance through various bodies to determine the development path of the software projects, collect and distribute donated funds and other resources, coordinate with other projects, organize events, and coordinate activities and support from corporate and government sponsors. This pattern of software projects creating or joining larger collective institutions has become more common as the size of the open source space and the projects themselves have expanded rapidly. The coordination among many more contributors with differing objectives and expertise, and the need to attract greater financial and other support, requires more time and organization compared with the small-scale projects of the past. Many projects such as PyTorch are the basis for significant commercial products, and therefore many industrial stakeholders are very keen to see such projects succeed and be sustainable, and see the value in their own developers contributing back.

Medical imaging software projects are often supported and gathered together by online platforms such as the Neuroimaging Informatics Tools and Resources Clearinghouse (NITRC) [26] (which is also an image database), funded by the National Institutes of Health (NIH), and the MR-Hub[48], managed by the Reproducible Research Study Group of the International Society for Magnetic Resonance in Medicine (ISMRM) and focusing on MRI reconstruction. Some examples of software projects related to MRI reconstruction are MERLIN [27], the BART Toolbox [28], SigPy [29], TensorFlow MRI [30], or DIRECT [31].

24.4.3 Data

Open datasets are becoming more critical to machine learning research and practice as the scale of data needed for training models increases, and as the challenges of

---

[46]https://numfocus.org.

[47]https://pytorch.org/foundation.

[48]https://ismrm.github.io/mrhub/.

using complex models in practice become better understood. Models require large amounts of data to produce any result, but very often require much more beyond this to be generalizable in practice and so safe to be used in a clinical context. With a training dataset that is too small or lacks variation, a model will learn only the distribution of that limited data—just a narrow subset of the true real-world distribution.

For example, imaging models might be trained on data from a single source with thorough curation for quality and content, and will then fail to produce results when given images from unseen sources which have far wider variations in quality and context than can be found in the original dataset. Scanner model, acquisition protocol, presence of artifact, presence of foreign objects, demographic variation, anatomical variation, and many other factors impact the raw appearance of medical images even when other factors are controlled. The ideal dataset would contain enough images with variances in such properties to represent the real-world distribution, but this is rarely possible.

Many solutions exist to help models generalize better, i.e., learn the data features relevant to the prediction through the noise and variation found in the input data. Data augmentation is one strategy that seeks to create more variation by manipulating images such that the important features are still present, but the task is made harder. Consequently, models must learn to detect such features through this added noise, and so the noise and variation found in images not seen during training represent a similar challenge the model is now robust against. Other strategies, such as network dropout, attempt to make the training process harder in a similar manner, with the expectation that models will be forced to learn the relevant features in their inputs, rather than rely on ancillary features or even overfit to the dataset and essentially memorize the training set.

However, large datasets will always be a critical component in generalization and robust training. Many variations in image content and properties cannot be created through augmentation, and so the distribution of training sets can only be expanded by including real data. Open datasets with the scale to train generalizable models are thus key tools making machine learning research and development accessible to many more people than would be possible with closed, proprietary datasets.

Challenge platforms such as Kaggle[49] and Grand Challenge[50] have become popular for hosting open datasets and challenges to encourage research and development in specific areas. These platforms provide a common interface for researchers to submit their work, and for others to access the datasets and results. The datasets can then be leveraged to train models that generalize across different modalities and anatomies. Some recent examples of models leveraging a large range of datasets are TotalSegmentator [32], nnInteractive [33], or MedSAM [34].

[49]https://www.kaggle.com.
[50]https://grand-challenge.org.

24.4.3.1 Medical Segmentation Decathlon

Organized as a challenge at MICCAI 2019, the Medical Segmentation Decathlon [1] is a dataset of MR and CT images designed to challenge segmentation methodologies. Each of the 10 tasks involves segmenting different organs in different modalities, and so the dataset itself [35] is composed of 10 subsets of varying size as outlined in Table 24.1.

As the challenge website accepts submissions long after the initial conference is over, this dataset continues to serve as a benchmark for other segmentation methodologies. Scientifically, it enables comparison of new approaches and also provides a training set that can serve as a starting point for new models. However, despite its size, the diversity of images within each task remains far narrower than that of real-world scenarios, limiting model generalization.

24.4.3.2 BraTS

The Brain Tumor Segmentation (BraTS) datasets consist of MR image volumes of the brain captured with four separate protocols: T_1, postcontrast T_1-weighted, T_2-weighted, and T_2 fluid attenuation inversion recovery. The BraTS challenges were presented as a number of instances at the MICCAI conference and involved the multiclass segmentation of brain tumors in these images. Previous iterations of the challenge have had a range of dataset sizes, from 65 clinical volume images in the 2012 iteration [36] to 2200 in the most recent 2024 iteration [37].

The challenges themselves are clearly valuable events for encouraging research and exchange of ideas in the area of brain tumor segmentation; however, beyond this

Table 24.1 Contents of the Medical Segmentation Decathlon dataset.

Task name	Modality	Num. images	Annotations
Brain tumors	MRI	750	Glioma (necrotic/active tumor), edema
Cardiac	MRI	30	Left atrium
Liver and tumors	CT	201	Liver, tumors
Hippocampus	MRI	394	Hippocampus, head and body
Prostate	MRI	48	Prostate central gland, peripheral zone
Lung tumors	CT	96	Lung tumors
Pancreas	CT	420	Pancreas, masses (cyst or tumor)
Hepatic vessels	CT	443	Liver vessels
Spleen	CT	61	Spleen
Colon tumors	CT	190	Colon cancer

scope, the datasets are very useful training tools for a range of other machine learning applications. One drawback is that the BraTS datasets contain only pathological images. To train a machine learning application for a different task, the dataset would have to be combined with images of subjects with other conditions or normal presentation; otherwise, the application will recognize only images with tumors present.

24.4.3.3 ADNI

The ADNI [38] dataset consists of longitudinal biomarker data for the study of Alzheimer's disease (AD), including MRI imaging of the brain acquired on a range of scanners over a period of two decades, for 2600 subjects. The imaging data covers structural and functional MR sequences captured at regular intervals (approximately 6 months apart, depending on the study), and is accompanied by biographical information, biomarkers, and genetic data. This data has been acquired across over 60 sites for the study of AD and other related conditions. As a result of over 26,000 investigators being given access to this dataset, over 4400 publications[51] have been written citing ADNI.

24.4.3.4 TCIA

The Cancer Imaging Archive (TCIA) [39] contains a wide range of imaging for many forms of cancer. These include all imaging modalities, including histopathology and other forms of data. Many of the subdatasets hosted on TCIA focus on specific cancers and were acquired as part of individual studies or are collected datasets from specific sites. TCIA is composed of submitted datasets rather than containing the results of funded studies or initiatives, so it relies on submissions from researchers who have relevant new data to provide or analysis of existing subdatasets they want to contribute back. Animal and phantom imaging data are also included in many subdatasets with sizes ranging from single subject to over 26,000 subjects.

In many cases, analyses by the submitting study or by subsequent research represent the input data suitable for training machine learning models [40]. This information includes segmentation and classification data for images, allowing applications to be trained to perform these tasks. Data is provided in standard formats such as DICOM Seg or RTSTRUCT, and TCIA provides Python code and examples for efficiently accessing items from the database.[52]

24.5 Remaining challenges in open source

Open-source projects are subject to the same management and organizational requirements as any other software project. First, requirements and features must be

[51]https://adni.loni.usc.edu/news-publications/publications.
[52]https://wiki.cancerimagingarchive.net/pages/viewpage.action?pageId=59310412.

determined with clarity, an architecture must be defined to meet these requirements, and software is written to implement the architecture, which is tested during and after development. Then, upon deployment, the long-term management and maintenance of the software is performed. Whether the project began as a personal, small-scale project where these steps are very informal and experimental, or a commercial project sponsored by a large entity able to devote professional resources to it, these activities in one form or another cannot be avoided and so require effort and ability to be managed well.

For all projects relying on community engagement and contribution, rather than those developed by a single organization and made available without expectation of external developers contributing, the challenge of building and maintaining this community is ever present. Why developers volunteer their time to contribute to a project, if not paid by their employers to do so, and why they choose to eventually cease involvement, is a critical question that garners no small amount of research [41,42]. The human aspect of keeping a community going, much as it is for many other endeavors, often proves to be the difficult activity rather than the expected but well-understood technical dimension.

These challenges are ever present: how to gather interest from those who would benefit from the project, how to encourage new people to begin contributing to the project, how to manage the interactions and collaborations of team members effectively, and how to maintain the momentum and interest over the life cycle of the project. Contributors will begin interacting with a project when they have a personal or professional need for the software, but may drift away from active involvement as time or professional pressures change [42]. Conversely, engagement may fall as the project reaches "completion," has less technical relevance, or accumulates too much technical debt to maintain. Managing personal and personnel issues within projects becomes a significant effort as the project size grows, which can surprise and overwhelm team members.

Open-source projects are meant to provide resources to users who, in turn, will have feedback, feature requests, bug reports, and other communications with the developers. Many hosting platforms, such as GitHub or Gitlab, provide means for users to raise issues, start discussions in forums, and propose code changes directly to the code base. To be an effective project which provides utility to its users, team members will have to manage and respond to this feedback to ensure users get the help they need and ensure that the project is developed to better meet their requirements. This "customer service" requirement can also become a significant and unexpected resource burden; however, maintaining good community interaction is often key to the long-term success of a project.

Often, projects can be deemed "dead" if their communities have little activity and code contributions or releases become rare. The perception here, even if the current code is useful and of good quality, is that in the future, there won't be maintainers who will update the code and keep it fit for purpose. Being directly responsive to requirements, by providing requested features or fixing bugs promptly, encourages new people to engage with the community and avoid this fate.

Many interactions between open-source software developers and users are questions from the latter about how to do some particular task or explanations of why an error was encountered. Although this is just providing debugging or troubleshooting help rather than discussing the development or direction of the project, these interactions are valuable to users, who see an active community they can engage with. Over time, the accumulated questions and answers form a large amount of documentation that users can turn to if they have similar issues, as well as help identify bugs or new requirements. Continued interactions of this sort show the community is active, and so the project is sustainable over the long term.

Many core contributors also come from academic environments and so have a vested interest in education itself. Building this body of knowledge produces a significant source of educational materials that students and researchers can exploit and avoid duplication of effort.

Managing the core developers, those who have been the principal contributors to the project, is another important and unexpected area of community management. Each developer has their own professional or personal reason for wanting to develop certain aspects of the project, with only limited time outside their professional roles to do so. Differences of opinion on direction or development, lack of firm agreement on aspects such as deadlines, professional duties temporarily making contribution impossible, and many other factors must be managed. Open-source projects are often defined as a community project without a top-down decision-making structure, and so the contributors, through their teams and the advisory board, must resolve these issues and come to an agreement on development direction.

In short, the human organizational and social factors are significant for open-source projects, perhaps to a surprising degree compared with the technical factors. Considering instead the technological and scientific challenges, many projects face the challenges of maintaining the technical value over time while continuing to be manageable by their communities. This value can be reduced if the technologies the project relies on change significantly or become obsolete, porting the system to new platforms may be difficult or even impossible, the accumulated technical debt of the system may eventually make further changes intractable tasks, or the very purpose of the project itself may evolve or disappear altogether as new solutions or techniques arise.

All these characteristics are not specific to open-source projects, as any piece of software can run the risk of becoming obsolete, unmanageable technically, or simply unneeded. The developers, whether as a closed team or a communal effort, must ensure these issues are addressed to maintain that value. MONAI, for example, relies on the regular contribution of biomedical researchers, clinicians, and commercial users to keep the project on track and continue to provide features with clinical and scientific value. The core system must be continually updated to maintain compatibility with its major dependencies as they change. This is especially critical in the deep learning field, which sees rapid change and the desire to use the most current version of libraries such as PyTorch. There are a wide range of skill levels and experience among the contributors; therefore, maintaining code quality is a continual challenge addressed through core developer oversight and the enforcement of standards through automated testing frameworks.

24.6 What's next

The trends of future development in deep learning will significantly define those in biomedical machine learning as well, with the focus on new model architectures, bigger models, and bigger data.

Frameworks such as TorchIO or MONAI will continue to grow and incorporate support for new methodologies through community contributions, while open datasets of biomedical data such as TCIA will hopefully also continue to grow.

As the size of models grows to meet the need for greater capability, and the necessity to train models to generalize well enough to be reliable in clinical practice, the need for very large accessible datasets will become critical for researchers to develop such large and reliable machine learning systems. Given the extraordinary amounts of computing power required to train large models from scratch, fine-tuning is used by many researchers to make experiments feasible within a given compute budget. There is therefore an ongoing and even expanding need for cutting-edge foundation models whose weights (and preferably also source) are also open-sourced.

Foundation models [43] are large models (in terms of number of parameters) trained on very large datasets for accomplishing specific tasks, but which excel at being adapted to other tasks or are easily fine-tuned. The advantage of these models is to make the value of the training done on the large dataset available to users and researchers without needing to reproduce the training process. Without making the dataset available, the value of the data and the information the networks learned from it can be shared and built upon, such that many users need little or no further training to effectively repurpose this value. Many who cannot access or construct such large datasets, and who lack the resources to train large models with it, can nonetheless access and exploit the results of such an effort.

The earliest foundation models are large language models such as the GPT series [44]. Many other large models suitable for adaptation to other tasks have been developed for areas other than language processing, hence the need for the "foundation model" term. Of concern for this chapter are those models for the understanding and processing of biomedical text and image data.

24.7 Conclusion

Open-source code and open-access databases have become a critical part of the development of machine learning tools. They enable reproducibility, foster collaboration, and accelerate innovation in the field. Ubiquitous software frameworks such as TorchIO or MONAI, and databases such as TCIA or BraTS, are examples of successful projects that have been adopted by a large community of researchers and developers. These projects demonstrate the value of open source and open access to drive progress in the field of machine learning in healthcare.

References

[1] Antonelli M, Reinke A, Bakas S, Farahani K, Kopp-Schneider A, Landman BA, Litjens G, Menze B, Ronneberger O, Summers RM, van Ginneken B, Bilello M, Bilic P, Christ PF, Do RKG, Gollub MJ, Heckers SH, Huisman H, Jarnagin WR, McHugo MK, Napel S, Pernicka JSG, Rhode K, Tobon-Gomez C, Vorontsov E, Meakin JA, Ourselin S, Wiesenfarth M, Arbeláez P, Bae B, Chen S, Daza L, Feng J, He B, Isensee F, Ji Y, Jia F, Kim I, Maier-Hein K, Merhof D, Pai A, Park B, Perslev M, Rezaiifar R, Rippel O, Sarasua I, Shen W, Son J, Wachinger C, Wang L, Wang Y, Xia Y, Xu D, Xu Z, Zheng Y, Simpson AL, Maier-Hein L, Cardoso MJ. The medical segmentation decathlon. *Nat Commun* 2022;**13**(4128).

[2] Harris CR, Millman KJ, van der Walt SJ, Gommers R, Virtanen P, Cournapeau D, Wieser E, Taylor J, Berg S, Smith NJ, Kern R, Picus M, Hoyer S, van Kerkwijk MH, Brett M, Haldane A, del Río JF, Wiebe M, Peterson P, Gérard-Marchant P, Sheppard K, Reddy T, Weckesser W, Abbasi H, Gohlke C, Oliphant TE. Array programming with NumPy. *Nature* Sept. 2020;**585**:357–62.

[3] Virtanen P, Gommers R, Oliphant TE, Haberland M, Reddy T, Cournapeau D, Burovski E, Peterson P, Weckesser W, Bright J, van der Walt SJ, Brett M, Wilson J, Millman KJ, Mayorov N, Nelson ARJ, Jones E, Kern R, Larson E, Carey CJ, Polat İ, Feng Y, Moore EW, VanderPlas J, Laxalde D, Perktold J, Cimrman R, Henriksen I, Quintero EA, Harris CR, Archibald AM, Ribeiro AH, Pedregosa F, van Mulbregt P, SciPy 1.0 Contributors. SciPy 1.0: fundamental algorithms for scientific computing in Python. *Nat Methods* 2020;**17**:261–72.

[4] Pedregosa F, Varoquaux G, Gramfort A, Michel V, Thirion B, Grisel O, Blondel M, Prettenhofer P, Weiss R, Dubourg V, Vanderplas J, Passos A, Cournapeau D, Brucher M, Perrot M, Duchesnay E. Scikit-learn: machine learning in Python. *J Mach Learn Res* 2011;**12**:2825–30.

[5] McKinney W. Data structures for statistical computing in Python. In: *Proceedings of the 9th Python in science conference (Stéfan van der Walt and Jarrod Millman*; 2010. p. 56–61.

[6] Miller BP, Koski D, Lee CP, Maganty V, Murthy R, Natarajan A, Steidl J. Fuzz revisited: a re-examination of the reliability of unix utilities and services. In: *Tech Rep*. Univ. of Wisconsin, Computer Science Dept; 1995.

[7] Abadi M, Barham P, Chen J, Chen Z, Davis A, Dean J, Devin M, Ghemawat S, Irving G, Isard M, Kudlur M, Levenberg J, Monga R, Moore S, Murray DG, Steiner B, Tucker P, Vasudevan V, Warden P, Wicke M, Yu Y, Zheng X. TensorFlow: a system for large-scale machine learning. In: *Proceedings of the 12th USENIX conference on operating systems design and implementation, OSDI'16*. USA: USENIX Association; Nov. 2016. p. 265–83.

[8] Paszke A, Gross S, Massa F, Lerer A, Bradbury J, Chanan G, Killeen T, Lin Z, Gimelshein N, Antiga L, Desmaison A, Kopf A, Yang E, DeVito Z, Raison M,

Tejani A, Chilamkurthy S, Steiner B, Fang L, Bai J, Chintala S. PyTorch: an imperative style, high-performance deep learning library. In: Wallach H, Larochelle H, Beygelzimer A, Alché-Buc Fd, Fox E, Garnett R, editors. *Advances in neural information processing systems 32*. Curran Associates, Inc.; 2019. p. 8026–37.

[9] Wilkinson MD, Dumontier M, Aalbersberg IJ, Appleton G, Axton M, Baak A, Blomberg N, Boiten J-W, da Silva Santos LB, Bourne PE, Bouwman J, Brookes AJ, Clark T, Crosas M, Dillo I, Dumon O, Edmunds S, Evelo CT, Finkers R, Gonzalez-Beltran A, Gray AJG, Groth P, Goble C, Grethe JS, Heringa J, 't Hoen PAC, Hooft R, Kuhn T, Kok R, Kok J, Lusher SJ, Martone ME, Mons A, Packer AL, Persson B, Rocca-Serra P, Roos M, van Schaik R, Sansone S-A, Schultes E, Sengstag T, Slater T, Strawn G, Swertz MA, Thompson M, van der Lei J, van Mulligen E, Velterop J, Waagmeester A, Wittenburg P, Wolstencroft K, Zhao J, Mons B. The FAIR Guiding Principles for scientific data management and stewardship. *Sci Data* Mar. 2016;**3**160018.

[10] Barker M, Chue Hong NP, Katz DS, Lamprecht A-L, Martinez-Ortiz C, Psomopoulos F, Harrow J, Castro LJ, Gruenpeter M, Martinez PA, Honeyman T. Introducing the FAIR Principles for research software. *Sci Data* Oct. 2022;**9**:622.

[11] Katz DS, Chue Hong NP, Clark T, Muench A, Stall S, Bouquin D, Cannon M, Edmunds S, Faez T, Feeney P, Fenner M, Friedman M, Grenier G, Harrison M, Heber J, Leary A, MacCallum C, Murray H, Pastrana E, Perry K, Schuster D, Stockhause M, Yeston J. Recognizing the value of software: a software citation guide. *F1000Research* Jan. 2021;**9**:1257.

[12] Johnson AEW, Pollard TJ, Berkowitz SJ, Greenbaum NR, Lungren MP, Deng C-y, Mark RG, Horng S. MIMIC-CXR, a de-identified publicly available database of chest radiographs with free-text reports. *Sci Data* Dec. 2019;**6**:317.

[13] Chambon P, Bluethgen C, Delbrouck J-B, Van der Sluijs R, Połacin M, Chaves JMZ, Abraham TM, Purohit S, Langlotz CP, Chaudhari A. In: *RoentGen: vision-language foundation model for chest X-ray generation*; 2022.

[14] conda-forge community. *The conda-forge project: community-based software distribution built on the conda package format and ecosystem*. July 2015.

[15] Gorgolewski KJ, Auer T, Calhoun VD, Craddock RC, Das S, Duff EP, Flandin G, Ghosh SS, Glatard T, Halchenko YO, Handwerker DA, Hanke M, Keator D, Li X, Michael Z, Maumet C, Nichols BN, Nichols TE, Pellman J, Poline J-B, Rokem A, Schaefer G, Sochat V, Triplett W, Turner JA, Varoquaux G, Poldrack RA. The brain imaging data structure, a format for organizing and describing outputs of neuroimaging experiments. *Sci Data* June 2016;**3**160044.

[16] McCormick MM, Liu X, Ibanez L, Jomier J, Marion C. ITK: enabling reproducible research and open science. *Front Neuroinf* 2014;**8**.

[17] Schroeder W, Martin K, Lorensen B. *The visualization toolkit: an object-oriented approach to 3D graphics*. Kitware; 2006.

[18] Pedregosa F, Varoquaux G, Gramfort A, Michel V, Thirion B, Grisel O, Blondel M, Prettenhofer P, Weiss R, Dubourg V, Vanderplas J, Passos A, Cournapeau D, Brucher M, Perrot M, Duchesnay E. Scikit-learn: machine learning in Python. *J Mach Learn Res* 2011;**12**(85):2825–30.

[19] Hunter JD. Matplotlib: a 2d graphics environment. *Comput Sci Eng* 2007;**9**(3):90–5.

[20] Waskom ML. seaborn: statistical data visualization. *J Open Source Softw* 2021;**6**(60): 3021.

[21] Pérez-García F, Sparks R, Ourselin S. TorchIO: a Python library for efficient loading, preprocessing, augmentation and patch-based sampling of medical images in deep learning. *Comput Methods Progr Biomed* Sept. 2021;**208**106236.

[22] Cardoso MJ, Li W, Brown R, Ma N, Kerfoot E, Wang Y, Murrey B, Myronenko A, Zhao C, Yang D, Nath V, He Y, Xu Z, Hatamizadeh A, Myronenko A, Zhu W, Liu Y, Zheng M, Tang Y, Yang I, Zephyr M, Hashemian B, Alle S, Darestani MZ, Budd C, Modat M, Vercauteren T, Wang G, Li Y, Hu Y, Fu Y, Gorman B, Johnson H, Genereaux B, Erdal BS, Gupta V, Diaz-Pinto A, Dourson A, Maier-Hein L, Jaeger PF, Baumgartner M, Kalpathy-Cramer J, Flores M, Kirby J, Cooper LAD, Roth HR, Xu D, Bericat D, Floca R, Zhou SK, Shuaib H, Farahani K, Maier-Hein KH, Aylward S, Dogra P, Ourselin S, Feng A. Monai: an open-source framework for deep learning in healthcare. *arXiv:2211.02701* 2022.

[23] Diaz-Pinto A, Alle S, Nath V, Tang Y, Ihsani A, Asad M, Pérez-García F, Mehta P, Li W, Flores M, Roth HR, Vercauteren T, Xu D, Dogra P, Ourselin S, Feng A, Cardoso MJ. Monai label: a framework for ai-assisted interactive labeling of 3d medical images. *Med Image Anal* 2024;**95**103207.

[24] Isensee F, Jaeger PF, Kohl SAA, Petersen J, Maier-Hein KH. nnU-Net: a self-configuring method for deep learning-based biomedical image segmentation. *Nat Methods* Feb. 2021;**18**:203–11.

[25] Ronneberger O, Fischer P, Brox T. U-net: convolutional networks for biomedical image segmentation. In Navab, N., Hornegger, J., Wells, W., Frangi, A. (eds) Medical Image Computing and Computer-Assisted Intervention – MICCAI 2015. MICCAI 2015. Lecture Notes in Computer Science, vol 9351. Springer, Cham. https://doi.org/10.1007/978-3-319-24574-4_28 May 2015.

[26] Kennedy DN, Haselgrove C, Riehl J, Preuss N, Buccigrossi R. The nitrc image repository. *Neuroimage* Jan. 2016;**124**:1069–73.

[27] Hammernik K, Küstner T. Machine enhanced reconstruction learning and interpretation networks (merlin). In: *Proceedings of the international society for magnetic resonance in medicine*. ISMRM); 2022.

[28] Blumenthal M, Heide M, Holme C, Juschitz M, Rapp B, Schaten P, Scholand N, Tamir J, Tönnes C, Uecker M. *mrirecon/bart: version 0.9.00*. 2023.

[29] Ong F, Martin J, Grissom W, Srinivasan S, Johnson KM, Huynh C, Desai A, -Li Z, Tamir J, Sandino C, EfratShimron, Zeng D, Mickevicius N. *mikgroup/sigpy: minor release to trigger zenodo for*. 2022.

[30] Montalt Tordera J. *Tensorflow mri*. 2022.

[31] Yiasemis G, Moriakov N, Karkalousos D, Caan M, Teuwen J. Direct: deep image reconstruction toolkit. *J Open Source Softw* 2022;**7**(73):4278.

[32] Wasserthal J, Breit H-C, Meyer MT, Pradella M, Hinck D, Sauter AW, Heye T, Boll DT, Cyriac J, Yang S, Bach M, Segeroth M. Totalsegmentator: robust segmentation of 104 anatomic structures in ct images. *Radiology: Artif Intell* Sept. 2023;**5**.

[33] Isensee F, Rokuss M, Krämer L, Dinkelacker S, Ravindran A, Stritzke F, Hamm B, Wald T, Langenberg M, Ulrich C, Deissler J, Floca R, Maier-Hein K. nninteractive: redefining 3d promptable segmentation. *arXiv:2503.08373* 2025.

[34] Ma J, He Y, Li F, Han L, You C, Wang B. Segment anything in medical images. *Nat Commun* Jan. 2024;**15**.

[35] Simpson AL, Antonelli M, Bakas S, Bilello M, Farahani K, van Ginneken B, Kopp-Schneider A, Landman BA, Litjens G, Menze B, Ronneberger O, Summers RM, Bilic P, Christ PF, Do RKG, Gollub M, Golia-Pernicka J, Heckers SH, Jarnagin WR,

McHugo MK, Napel S, Vorontsov E, Maier-Hein L, Cardoso MJ. A large annotated medical image dataset for the development and evaluation of segmentation algorithms. *arXiv:1902.09063* 2019.

[36] Menze BH, Jakab A, Bauer S, Kalpathy-Cramer J, Farahani K, Kirby J, Burren Y, Porz N, Slotboom J, Wiest R, Lanczi L, Gerstner E, Weber M-A, Arbel T, Avants BB, Ayache N, Buendia P, Collins DL, Cordier N, Corso JJ, Criminisi A, Das T, Delingette H, Demiralp C, Durst CR, Dojat M, Doyle S, Festa J, Forbes F, Geremia E, Glocker B, Golland P, Guo X, Hamamci A, Iftekharuddin KM, Jena R, John NM, Konukoglu E, Lashkari D, Mariz JA, Meier R, Pereira S, Precup D, Price SJ, Raviv TR, Reza SMS, Ryan M, Sarikaya D, Schwartz L, Shin H-C, Shotton J, Silva CA, Sousa N, Subbanna NK, Szekely G, Taylor TJ, Thomas OM, Tustison NJ, Unal G, Vasseur F, Wintermark M, Ye DH, Zhao L, Zhao B, Zikic D, Prastawa M, Reyes M, Van Leemput K. The multimodal brain tumor image segmentation benchmark (brats). *IEEE Trans Med Imag* 2015;**34**(10):1993–2024.

[37] de Verdier MC, Saluja R, Gagnon L, LaBella D, Baid U, Tahon NH, Foltyn-Dumitru M, Zhang J, Alafif M, Baig S, Chang K, D'Anna G, Deptula L, Gupta D, Haider MA, Hussain A, Iv M, Kontzialis M, Manning P, Moodi F, Nunes T, Simon A, Sollmann N, Vu D, Adewole M, Albrecht J, Anazodo U, Chai R, Chung V, Faghani S, Farahani K, Kazerooni AF, Iglesias E, Kofler F, Li H, Linguraru MG, Menze B, Moawad AW, Velichko Y, Wiestler B, Altes T, Basavasagar P, Bendszus M, Brugnara G, Cho J, Dhemesh Y, Fields BKK, Garrett F, Gass J, Hadjiiski L, Hattangadi-Gluth J, Hess C, Houk JL, Isufi E, Layfield LJ, Mastorakos G, Mongan J, Nedelec P, Nguyen U, Oliva S, Pease MW, Rastogi A, Sinclair J, Smith RX, Sugrue LP, Thacker J, Vidic I, Villanueva-Meyer J, White NS, Aboian M, Conte GM, Dale A, Sabuncu MR, Seibert TM, Weinberg B, Abayazeed A, Huang R, Turk S, Rauschecker AM, Farid N, Vollmuth P, Nada A, Bakas S, Calabrese E, Rudie JD. The 2024 brain tumor segmentation (brats) challenge: glioma segmentation on post-treatment mri. *arXiv:2405.18368* 2024.

[38] Petersen R, Aisen P, Beckett L, Donohue M, Gamst A, Harvey D, Jack C, Jagust W, Shaw L, Toga A, Trojanowski J, Weiner M. Alzheimer's disease neuroimaging initiative (adni): clinical characterization. *Neurology* Jan. 2010;**74**:201–9.

[39] Clark K, Vendt B, Smith K, Freymann J, Kirby J, Koppel P, Moore S, Phillips S, Maffitt D, Pringle M, Tarbox L, Prior F. The cancer imaging archive (tcia): maintaining and operating a public information repository. *J Digit Imag* 2013;**26**(07).

[40] Fedorov A, Hancock M, Clunie D, Brochhausen M, Bona J, Kirby J, Freymann J, Pieper S, Aerts HJWL, Kikinis R, Prior F. Dicom re-encoding of volumetrically annotated lung imaging database consortium (lidc) nodules. *Med Phys* Sept. 2020;**47**: 5953–65.

[41] Miller C, Widder DG, Kästner C, Vasilescu B. Why do people give up flossing? a study of contributor disengagement in open source. In: Bordeleau F, Sillitti A, Meirelles P, Lenarduzzi V, editors. *Open source systems*. Cham: Springer International Publishing; 2019. p. 116–29.

[42] Coelho J, Valente MT. Why modern open source projects fail. In: *Proceedings of the 2017 11th Joint meeting on foundations of software engineering, ESEC/FSE 2017*. New York, NY, USA: Association for Computing Machinery; 2017. p. 186–96.

[43] Bommasani R, Hudson DA, Adeli E, Altman R, Arora S, von Arx S, Bernstein MS, Bohg J, Bosselut A, Brunskill E, Brynjolfsson E, Buch S, Card D, Castellon R, Chatterji N, Chen A, Creel K, Davis JQ, Demszky D, Donahue C, Doumbouya M,

Durmus E, Ermon S, Etchemendy J, Ethayarajh K, Fei-Fei L, Finn C, Gale T, Gillespie L, Goel K, Goodman N, Grossman S, Guha N, Hashimoto T, Henderson P, Hewitt J, Ho DE, Hong J, Hsu K, Huang J, Icard T, Jain S, Jurafsky D, Kalluri P, Karamcheti S, Keeling G, Khani F, Khattab O, Koh PW, Krass M, Krishna R, Kuditipudi R, Kumar A, Ladhak F, Lee M, Lee T, Leskovec J, Levent I, Li XL, Li X, Ma T, Malik A, Manning CD, Mirchandani S, Mitchell E, Munyikwa Z, Nair S, Narayan A, Narayanan D, Newman B, Nie A, Niebles JC, Nilforoshan H, Nyarko J, Ogut G, Orr L, Papadimitriou I, Park JS, Piech C, Portelance E, Potts C, Raghunathan A, Reich R, Ren H, Rong F, Roohani Y, Ruiz C, Ryan J, Ré C, Sadigh D, Sagawa S, Santhanam K, Shih A, Srinivasan K, Tamkin A, Taori R, Thomas AW, Tramèr F, Wang RE, Wang W, Wu B, Wu J, Wu Y, Xie SM, Yasunaga M, You J, Zaharia M, Zhang M, Zhang T, Zhang X, Zhang Y, Zheng L, Zhou K, Liang P. On the opportunities and risks of foundation models. *arXiv: 2108.07258* 2022.

[44] Brown T, Mann B, Ryder N, Subbiah M, Kaplan JD, Dhariwal P, Neelakantan A, Shyam P, Sastry G, Askell A, Agarwal S, Herbert-Voss A, Krueger G, Henighan T, Child R, Ramesh A, Ziegler D, Wu J, Winter C, Hesse C, Chen M, Sigler E, Litwin M, Gray S, Chess B, Clark J, Berner C, McCandlish S, Radford A, Sutskever I, Amodei D. Language models are few-shot learners. In: Larochelle H, Ranzato M, Hadsell R, Balcan M, Lin H, editors. *Advances in neural information processing systems*, vol 33. Curran Associates, Inc.; 2020. p. 1877–901.

CHAPTER

System under test: Challenges for algorithm benchmarking

25

Minu D. Tizabi[1, 2] and Annika Reinke[1, 3, *]

[1]German Cancer Research Center (DKFZ) Heidelberg, Division of Intelligent Medical Systems, Heidelberg, Germany; [2]National Center for Tumor Diseases (NCT), NCT Heidelberg, a partnership between DKFZ and University Medical Center, Heidelberg, Germany; [3]German Cancer Research Center (DKFZ) Heidelberg, HI Helmholtz Imaging, Heidelberg, Germany

[*]Corresponding author: E-mail: a.reinke@dkfz-heidelberg.de

25.1 Introduction

Machine learning (ML) has emerged as a key disruptive technology in biomedical image analysis. Modern algorithms have the potential to facilitate clinical routine by performing a variety of different image processing tasks. One application area, in which ML presents particularly promising solutions to long-standing challenges, is magnetic resonance imaging (MRI). MRI challenges that can be solved through the use of ML methods range from the overcoming of slow imaging speed by the reconstruction of MRI images from an under-sampled k-space to the segmentation of tiny brain lesions that may easily be overlooked by radiologists.

However, while massive methodological progress can be observed, identifying the best-working solution for a specific biomedical research problem remains challenging. In the area of ML-based MRI research alone, the annual publication of over a 1000 articles poses a significant challenge to researchers in accurately discerning algorithms that genuinely surpass the state of the art. This difficulty is compounded by the common, but rarely conclusively proven assertion in these publications that the respective presented methodology outperforms all others. Given that each work uses its own conditions, including dataset (split), computational infrastructure, and validation method, a fair, and direct comparison of methods is nearly impossible.

To address this issue, the ML community began conducting international comparative studies of algorithms. The first such study was organized in 1997 and compared retrospective intermodality brain image registration methods using volumes from MRI, computed tomography (CT), and positron emission tomography (PET) [1]. From 2004 onward, comparative studies became an integral part of biomedical image analysis research in the shape of large-scale international

Machine Learning in MRI. https://doi.org/10.1016/B978-0-443-14109-6.00025-0

benchmarking competitions, also termed *challenges*. With their number and impact continuously on the rise, challenges have over the past 20 years evolved into the undisputed state-of-the-art method to compare ML algorithms in a fair manner and thus validate them.

Challenges address a specific research question and are typically organized at international conferences or on standard challenge platforms (discussed in Section 25.2) to attract a broad diversity of researchers from all over the world. They can be either used to explore a new research problem, examine whether current methods are capable of handling it and/or identify open research questions, or to determine the best-working method for a specific question [2].

Challenges have a tremendous impact on the research community. Winning methods often rise to prominence and are considered the new state of the art for a specific research question, garnering many citations and serving as a basis for future algorithm development in the field. As shown by Varoquaux and Cheplygina [3], challenges substantially drive scientific research in two ways: First, organization of a challenge often results in publishing a new benchmarking dataset which is used in many future publications. Second, the comparison of algorithm performance to the state of the art itself drives algorithm development.

Numerous competitions provide financial incentives to entice participants. However, according to research conducted by Eisenmann et al. [4], monetary rewards were considered the primary motivation for participation by only 16% of challenge participants in 2021. A substantial majority valued knowledge exchange (70%), the opportunity to compare methodologies to others (65%), access to the challenge data (52%), prospective co-authorship in forthcoming challenge publications (50%), and winning the competition (42%) as more compelling incentives [4].

Given the importance of challenges in the comparative validation of ML algorithms, it is of particular importance to ensure that they are conducted in the highest quality possible. Neglecting the appropriate validation of ML algorithms may lead to their success in a challenge merely by chance or the dissemination of algorithms that are entirely impractical for clinical use. This could be a contributing factor to the frequent failure of ML algorithm advancements to transition into clinical practice.

Accordingly, over the past years, several initiatives have sought to increase the general quality of challenges. In the following, we discuss different aspects to be taken into account in the context of MRI challenges.

25.2 Recent MRI challenges

In the period of 2020–2023, a total of 87 challenges using MRI data were organized, with the number increasing throughout the years; for example, 31% of the MRI challenges in this period were organized in 2023 and 21% in 2020, as shown in Fig. 25.1a. Most challenges (82%) were hosted at the International Conference on Medical Image Computing and Computer Assisted Intervention (MICCAI), which

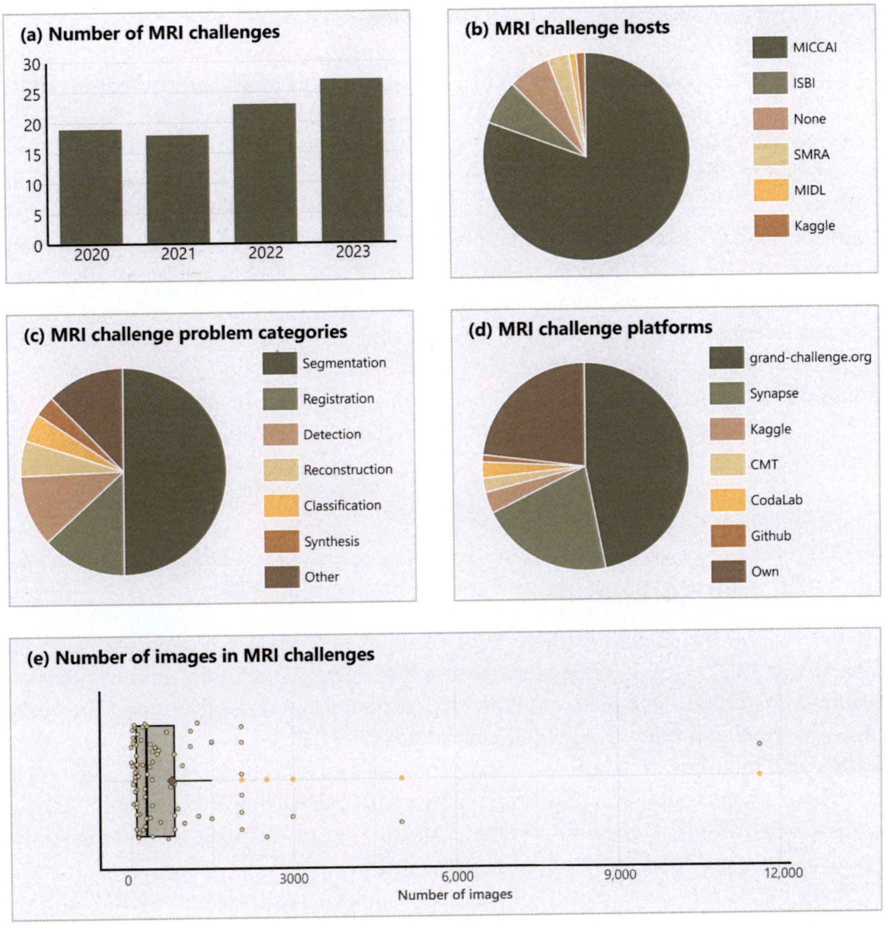

FIGURE 25.1

Overview of recent MRI challenges in the period of 2020–2023. (a) Number of MRI challenges from 2020 to 2023. (b) MRI challenge hosts. (c) MRI challenge problem categories. (d) MRI challenge platforms. (e) Number of images in MRI challenges. The median is shown as a thick black vertical line, the mean is shown as a rhombus, outliers are indicated by yellow dots. The individual numbers of images are shown by small jittered dots.

represents the largest conference in the field of biomedical image analysis. Other hosts include the IEEE International Symposium on Biomedical Imaging (ISBI; 7%), the annual meeting of the Society for Magnetic Resonance Angiography (SMRA; 3%), the Conference on Medical Imaging with Deep Learning (MIDL, 1%), and the online platform Kaggle (1%). Another 7% of MRI challenges were hosted without being affiliated to any specific platform or conference (Fig. 25.1b).

Half of the recent MRI challenges dealt with segmentation problems, followed by registration (13%), detection (11%), and reconstruction (6%) tasks. Less common challenges addressed tasks such as classification, image synthesis, regression, and prediction (Fig. 25.1c). MRI is one of the most commonly used imaging techniques in challenges. For example, of all 2020–23 challenges at MICCAI, the largest challenge host, 48% dealt with MRI data.

Some MRI challenges are repeated every year and have become a solid anchor in the refinement of methods tackling a specific research problem. The most prominent example is the Brain Tumor Segmentation (BraTS) challenge [5], which was first organized in 2011 and repeated annually since then. It addresses the segmentation of heterogeneous brain tumors (gliomas). In 2021, the challenge was jointly organized by the Radiological Society of North America (RSNA) and MICCAI for its 10th anniversary and was extended to multi-institutional multi-parametric MRI (mpMRI) scans. The challenge typically organizes multiple sub-tasks including segmentation of adult glioma, meningioma, brain metastases, or pediatric tumors, as well as survival and pseudoprogression prediction or modality synthesis of MRI volumes.

25.3 Challenge platforms

Whether or not organizers affiliate their challenge to a large conference such as MICCAI or ISBI, they have to decide on a platform that actually allows them to host their challenge. This platform typically offers a general description of the challenge objective and main design choices, such as the dataset description, validation metrics, and rankings. In addition, it often contains (possibly as a separate website) a submission and/or validation platform, where challenge participants can upload their results or algorithms. Depending on the platform, the results may be automatically validated for a pre-final and/or final leaderboard.

Many organizers choose to use their own websites, platforms, or clouds to host their challenges. But there also exist a number of internationally well-known, partly open-source platforms that can be used. The most commonly known platform is *grand-challenge.org*, "a platform for end-to-end development of machine learning solutions in biomedical imaging."[1] The platform helps researchers to easily create a challenge website and offers an overview of all previously and currently organized challenges on the platform with filter options based on modality, anatomical region, conference, and similar. In the past few years, the platform has added many functionalities, for example the automatic validation of submissions on secret test data. It further offers users a challenge-specific discussion forum, options to form teams, support, and a live leaderboard. In addition, grand-challenge.org also provides a data management system, allows for annotation generation, deploying the submitted algorithms and finding certified solutions that successfully competed in challenges.

[1] https://grand-challenge.org/about/

While most of the features are free of cost, the platform charges a fee for its advanced options.

Previously recognized within the realm of biological machine learning, *Synapse by Sage Bionetworks* is another open-source platform for challenges, general collaborative research, and biomedical data sharing. It similarly facilitates creating challenge documentation websites and discussion forums, uploading challenge data, both public and private, as well as validation of algorithm submissions. It further educates users in data privacy policies and requires them to pass a short exam to make sure that they follow the data sharing principles. Similar to grand-challenge.org, Synapse offers free basic and paid advanced options.

Kaggle constitutes a widely known platform for hosting datasets and data challenges. Kaggle is not specifically dedicated to biomedical applications but can rather be used for any general data science-related problem. Moreover, as opposed to grand-challenge.org and Synapse, Kaggle's target audience often goes beyond academic research and the platform thus typically offers much higher prize money. Compared to the other platforms, the free options are also more restricted; for example, custom validation metrics can only be chosen in a paid plan. In addition, to our knowledge, Kaggle does not feature a strict quality control mechanism. It thus comes as no surprise that the majority of biomedical imaging challenges are hosted on other platforms.

For the 87 MRI challenges conducted between 2020 and 2023, grand-challenge.org was the most commonly used hosting platform (47%). While grand-challenge.org used to be the standard platform, other platforms such as Synapse are gaining more ground (21%). Kaggle was only used in 3% of the challenges. About 23% of all MRI challenges used their own platform or cloud for hosting (Fig. 25.1d).

25.4 Challenge organization: Common and best practice

Every challenge is individual in its goal, tackled research question, and design. However, common patterns in challenge organization can be observed. Challenge organizers typically first come up with the research problem to tackle and curate a dataset for the challenge. This challenge dataset is often split into training, validation, and test sets. The *training dataset* is used to train ML algorithms and typically includes images and labels. Depending on challenge design, this dataset may be complemented by a dedicated *validation dataset*, or participants are free to split the training set into train and validation, depending on preference. The training phase typically lasts several weeks or even months and serves to develop a working solution for the challenge's research problem. To guarantee later fair comparability, challenge organizers are required to carefully define the rules for training algorithms: is it allowed to complement the training dataset provided by the organizers with publicly available or private data? Are pre-trained models on other datasets eligible for participation?

After tuning their methods, participants are sometimes allowed to test their performance on another data subset. If this *pre-validation* step is chosen by the organizers, it can be handled in various ways. For example, challenge participants may submit their results or algorithms and are informed about their metric scores. Other organizers implement a leaderboard, which depicts participants' pre-validation performance in comparison to others. This strategy should be carefully considered. Some challenges allow an infinite number of resubmissions, making them vulnerable to deception by overfitting the algorithms to the data. This phenomenon is also called "leaderboard climbing" [2]. Nowadays, organizers typically prevent overfitting by defining a limited number of resubmissions per day or in total.

Once method development is complete, the final algorithm can be submitted to the challenge organizers or a specified platform to be tested on a (usually withheld) *testing dataset*. This step, too, can be conducted in various ways. In the past, challenge participants often uploaded the results of their methods directly based on the test images, which were published without labels. Nowadays, it has become more common to submit the full algorithm rather than its output to overcome data privacy issues and to prevent cheating. The latter is of importance as in a situation where only algorithm outputs are submitted, challenge organizers can never be certain whether participants manipulated their results. In addition, participants could have chosen to systematically submit only the most plausible results while omitting those outputs with low performance. If these missing values are ignored in the final ranking calculation, algorithms submitting only on a fraction of the images may be scored higher compared to those who submitted all results. As shown by Reinke et al. [6], this strategy could completely change rankings in theory and was indeed exploited for cheating by some participants in major credible challenges in the past 10 years [7].

The challenge organizers validate the results of the submitted solutions with one or multiple validation metrics. These performance metrics are often aggregated into a ranking to determine the challenge winner. The winning submission is typically considered the best-performing method, often regarded as the new state of the art in tackling the particular research question, and awarded a substantial prize money. Finally, the challenge results may be published by the organizers in the shape of a peer-reviewed publication, which is in many cases highly cited.

Challenge organizers have high responsibility in guaranteeing the quality and integrity of their challenge design. This necessitates careful planning of the competition well in advance, with the primary objectives being the precise formulation of the research question and problem domains that algorithms are to target, the curation of high-quality datasets, and the establishment of a validation strategy that aligns with the nature of the problem and the characteristics of the dataset.

To preserve representativeness of the dataset, it is essential to procure the dataset from diverse sources encompassing various sites and devices. The dataset should capture the inherent diversity present in imaged objects, and particular attention must be given to ensuring that the test set accurately reflects the general population [8]. In addition to comprehensive data acquisition, the quality of annotations holds paramount importance. Annotators should be provided with explicit and detailed labeling instructions, accompanied by a robust set of images illustrating potential corner cases [9]. Furthermore, employing multiple annotators per image is advised to compensate for inter-rater variability [10]. Ideally, the annotators should either be trained domain experts or employed in professional labeling companies.

In recent MRI challenges, dataset sizes varied massively. While the mean dataset size was 758, the median size was 290, showing a large prevalence of outliers. The minimum dataset size was 10, the maximum size was 11,607 (Fig. 25.1e). It should be noted that creating a large dataset is not always possible due to several restrictions, including patient privacy and long acquisition times. Thus, some challenges also employ synthetic data.

To ensure high quality challenge designs, the MICCAI society, and subsequently ISBI and MIDL, recently introduced an in-depth peer-review process for challenges to be conducted at their conferences. To be accepted, organizers nowadays have to prepare a comprehensive challenge design document by filling in all items of the BIAS reporting guideline for biomedical image analysis challenges [11] (see Section 25.7 for details). The proposal document is then reviewed and rated by independent reviewers, similarly to a standard publication peer-review process. The design documents of accepted MICCAI challenges are registered, that is, uploaded and fully visible to the research community to ensure full transparency to challenge participants.[2] Challenge organizers are encouraged to link their registered design documents on their websites. Any changes to the challenge design have to be reported and reviewed before they can be implemented.

25.5 Validation metrics in challenges: Common practice and pitfalls

Numerous validation metrics have been developed in the past decades, each designed to assess a specific aspect for a particular type of problem. Given the large number of available metric candidates, researchers are faced with the challenge of deciding which are most suitable for their research problem. Reinke et al. [12] found that when searching for information on a specific metric, the median number of search results across about 30 validation metrics was 22,100 hits—a number of publications impossible to be screened by any individual researcher. It was further shown that 24% of recent challenges justified their choice of metrics by their popularity rather than their clinical suitability [7].

Relying on common metrics while disregarding their suitability for the problem at hand may have severe consequences. For example, in MR-based brain tumor segmentation, the standard validation metric is the Dice Similarity Coefficient (DSC)

[2]http://www.miccai.org/special-interest-groups/challenges/miccai-registered-challenges/

on voxel level, which is additionally by far the most common metric used in challenges [8]. The DSC measures the voxel overlap between a reference segmentation and the predicted segmentation. While it may be an appropriate metric for many imaging problems, it was not designed to handle multiple objects. In the example of brain tumor segmentation, this means that an algorithm might obtain a high DSC score and thus appear deceptively promising although it completely misses small lesions, as shown in Fig. 25.2.

The adequate choice of validation metrics is of particularly great importance in the context of challenges. As challenges are typically seen as a general benchmark generating new baseline methods, many researchers copy the metrics used in a challenge when testing their own algorithm on the challenge dataset. The use of poorly chosen metrics may thus be propagated through numerous publications, over time becoming an unquestioned standard validation procedure for a specific problem despite being entirely unsuited and thus affecting the entire research field. It thus becomes imperative for challenge organizers to lead the way by choosing clinically suitable and relevant metrics.

A number of pitfalls has to be avoided in the selection of appropriate validation metrics. To categorize potential pitfalls, Reinke et al. [12] introduced a taxonomy of metric-related pitfalls in image analysis. Based on this taxonomy, metric pitfalls can

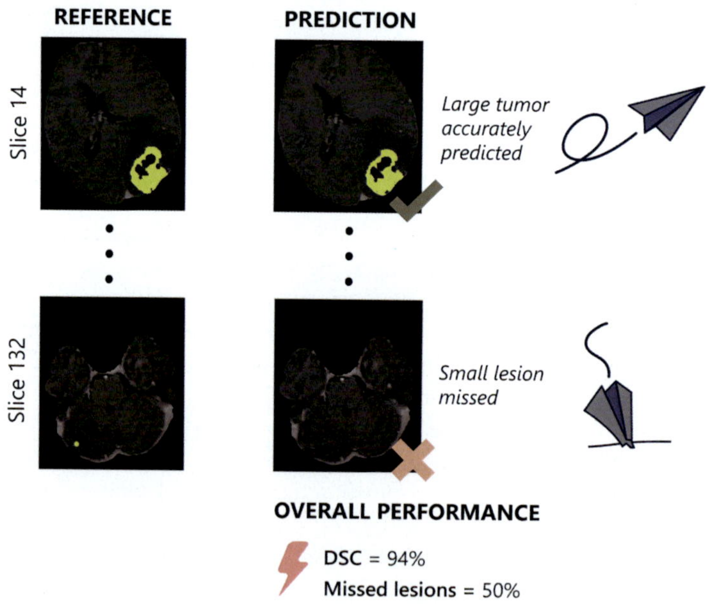

FIGURE 25.2

Assessment of brain tumor segmentation algorithms. The algorithm shows good performance in segmenting large tumors, but fails to detect small lesions. This is overlooked by standard metrics like the Dice Similarity Coefficient (DSC).

generally be grouped into three categories: inadequate phrasing of the problem, inadequate metric selection, and inadequate metric application [12,13]. The first category comprises the use of the wrong problem category to define the underlying research question. Object detection tasks, for example, are concerned with roughly localizing objects, such as tumors in MR scans, and classifying them. For validation, many practitioners tend to use a segmentation metric, such as the DSC or the closely related metric Intersection over Union (IoU), which are, as explained above, not capable of handling multiple objects and thus unsuitable for the task at hand. Challenge organizers should thus carefully define their problems according to the research question and only choose metrics suitable for the type of task.

The second category encompasses the inadequate selection of metrics while neglecting their pitfalls. Examples of pitfalls include choosing accuracy in the presence of class imbalance, choosing DSC when interested in structure boundaries, or prevalence-dependent metrics such as the positive predictive value (PPV) to compare methods across datasets with different prevalences. While a comprehensive overview of metric pitfalls is out of scope for this chapter, interested readers may find a comprehensive list of pitfalls for classification, detection, and segmentation problems in Reinke et al. [12,14]. For tracking tasks, we refer readers to Luiten et al. [15], and for calibration metrics to Refs. [16,17].

The final category of metric pitfalls is related to the application of chosen metrics, such as their aggregation or implementation. When aggregating metrics, one should keep in mind that different aggregation schemes may yield different rankings (see Section 25.6). Moreover, in biomedical research, data often has a hierarchical structure, being commonly acquired from different hospitals, scanners, or patients. This means that the data are not necessarily independent, and hierarchical structures must be taken into account when aggregating results. For example, challenge results for test data from two hospitals and five different patients should first be aggregated per patient and subsequently per hospital.

When implementing the validation pipeline for a challenge, organizers should further be aware that different libraries may yield different metric scores depending on how specific parts of a metric were implemented. A prominent example would be the boundary calculation for distance-based metrics such as the Hausdorff Distance (HD) or the strategy to handle identical confidence scores in object detection tasks for the average precision (AP). A standardized implementation for classification, detection, and segmentation metrics was recently included within the Medical Open Network for Artificial Intelligence (MONAI) framework.[3]

We generally recommend challenge organizers to choose multiple metrics for a comprehensive performance assessment of the submitted algorithms. Using more than a single metric comes with the advantage of avoiding pitfalls related to that metric. For example, when segmenting tumors in MRI, for which it outlining them accurately is important, the DSC would be capable of measuring the pixel

[3]https://github.com/Project-MONAI/MetricsReloaded/

overlap of predicted and reference tumor while it is not capable of testing the accuracy of the boundaries of a segmentation. However, it could be complemented by a distance-based metric (e.g., Normalized Surface Distance (NSD)), which focuses on that specific aspect. Of note, the selection of more than one metric only makes sense if the metrics measure complementary properties. While DSC and IoU are very often reported together, both measure the same properties of the algorithm, and one can be directly computed from the other [13]. Thus, reporting both metrics does not add any useful information for the reader and may even negatively impact the validation because overlap between prediction and reference may be overemphasized compared to other aspects, such as boundary accuracy.

In the context of classification, detection, and segmentation, a comprehensive metric recommendation framework guiding researchers toward choosing suitable metrics in a problem-aware manner was recently developed [13]. We advise challenge organizers to use this framework to select appropriate metrics for their challenge.[4]

25.6 Rankings: Calculation, uncertainties, and interpretation

Rankings not only determine the winner of a competition, but are also in use in general algorithm benchmarking in comparison to other baseline methods. A standardized ranking scheme used across challenges, however, does not exist, as ranking is instead performed depending on the underlying research question. Most challenges employ a ranking based on aggregating all metric scores per participant with the mean or other aggregation operators such as the median [8]. The ranks are then determined by sorting the aggregated scores. One could also choose to calculate a ranking per image and aggregate the per-image ranks per participant. Another option is to base the ranking on calculating statistical tests and comparing the number of statistically significant results.

The ranking scheme chosen should always be suitable for the problem at hand. If, for example, the worst-case performance of algorithms should be assessed, organizers may choose a low percentile (e.g., fifth percentile) instead of the mean for calculation. Maier-Hein et al. [8] demonstrated that the choice of ranking scheme has a huge impact on the final ranking. Rankings were shown to be sensitive to the chosen metrics, aggregation scheme and operator, but also to which annotator generated the reference standard.

To be fair and comprehensive and avoid a submission being ranked the winner solely by chance, challenge organizers should analyze their rankings in depth. Wiesenfarth et al. [18] presented an extensive overview of various analysis and

[4]https://metrics-reloaded.dkfz.de/

visualization techniques for challenge results. With challenge reports often only based on ranking tables without any further information, advanced visualization methods are shown to facilitate the interpretation of results.

The authors emphasize the importance of also visualizing raw challenge results, that is, visualizing the individual metric scores per challenge participants. Such visualization directly provides evidence on whether an algorithm produced many low-quality outputs or outliers, whether algorithms performed similarly enough to be ranked together, and, depending on the type of visualization, may highlight influential test images. The easiest visualizations are box plots overlaid with the individual metric values as jittered dots or violin plots illustrating the distribution of scores.

To assess ranking variability and robustness against small perturbations, Wiesenfarth et al. [18] suggest to bootstrap the challenge dataset and recalculate rankings for the bootstrapped data. Quantitatively, the ranking robustness can be assessed by calculating ranking correlation coefficients between the original and the individual bootstrapped rankings. Descriptive statistics over correlation coefficients or plotting of values can be helpful for interpretation. The authors also suggest specific plots to visualize the frequencies of ranks achieved by each participating team over all bootstrap rankings. Using those techniques enables easy assessment of ranking robustness.

In the case of unstable rankings, especially those with substantially varying challenge winners, challenge organizers may consider not naming a single algorithm the winner, but rather defining a group of winners, which may further boost collaboration between researchers. Further analyses beyond challenge rankings are possible and can allow advanced insights. Such analyses may include combining algorithms of different participants through ensembling to calculate inter-algorithm variability, or the analysis of common problems or biases [11]. Roß et al. [19] propose a statistical strength-weakness analysis of challenge participants for identifying image characteristics which reduce or increase algorithm performance. Such an analysis helps identify common strengths of algorithms as well as failure cases, which can be used for future algorithm development.

25.7 Reporting: Transparency and reproducibility

Challenges are validation studies, for which interpretability and reproducibility of results should always be ensured. Transparent reporting of a challenge is relevant both for participants while the challenge is ongoing and for researchers after its publication.

Participants should always be able to understand the full challenge design. They need to know the data, the task, submission methods, and rules as well as the planned validation strategy to prevent misunderstandings. Organizers are thus advised to publish a comprehensive challenge website containing all necessary information. It is necessary to keep in mind that while some information may seem obvious to an organizer, it may not be for a first-time challenge participant. Example images on the website could help even first-time participants understand the data and challenge structure. Moreover, a high number of participants use the challenge's validation metrics and ranking schemes in their method development to optimize algorithms for those properties that are relevant to the challenge organizers, implying that they should be prominently stated on the website.

Not stating relevant information may prevent potential participants from submitting results. Even worse, changing details in the challenge design without notifying participants early enough may be interpreted as cheating by the organizers. As shown in previous research, ranking schemes were often not published before the challenge took place [8]. Reinke et al. [6] showed that this could be easily exploited by challenge organizers changing the ranking to more desirable outcomes. More generally, Maier-Hein et al. [8] found that many important details are never published by organizers, neither before nor after the challenge. To combat this, the MICCAI society both implemented the aforementioned structured submission process (see Section 25.4) to ensure all required challenge details were determined, as well as took it upon itself to publish the design documents of accepted challenges on its own website for all participants to see (so-called *challenge registration*), after it was observed that many organizers still failed to publish the details.

Another substantial effort to elevate the standards of challenge reporting was the development of the "Biomedical Image Analysis ChallengeS" (BIAS) guideline, the first full reporting guideline for biomedical image analysis challenges [11]. The BIAS guideline advises organizers how to report the challenge in full detail, beginning with an informative introduction (e.g., introduction from both technical and biomedical points of view). The guideline proceeds with the challenge methods, giving detailed descriptions of the overall organization including rules and submission methods, the mission of the challenge, in-depth description of the dataset including acquisition and annotation, and assessment methods, including the full validation procedure. For many parameters such as data splits or metrics, the guideline also requires authors to provide a justification of their choice. These justifications were included in the guideline to ensure parameters being chosen for their suitability to the underlying problem rather than their popularity. Finally, the guideline helps authors to structure the results of their challenge, provides tips for further analyses and interpretation, as well as advises points to be included in the discussion and conclusion sections. In this manner, the guideline ensures transparent reporting of challenges by organizers. In a publication context, it also helps reviewers identify whether a manuscript reports all relevant parameters as well as facilitates their retrieval. In this capacity, the guideline is already used in the journal Medical Image Analysis for challenge reports.

Of note, many items of the guideline should also be reported in non-challenge related algorithm publications. For an algorithm to be interpretable and reproducible, it is necessary to describe all relevant details of the method, but also the dataset, hardware, and validation schemes. Here, too, important details such as how baseline methods were optimized or patients in the dataset were selected are frequently missing in literature. In this regard, the "Enhancing the QUAlity and Transparency Of health Research" (EQUATOR) network [20], an umbrella organization for transparent health research, offers several guidelines for reporting ML algorithms in different applications, including specialized guidelines for MRI research. For example, Nichols et al. [21] describe best practices in data analysis and sharing in neuroimaging using MRI.

25.8 Challenge participation: Strategies in MRI challenges

Throughout this chapter, we have focused on giving insights on challenges from the perspective of their organizers. In this section, we will turn the spotlight on the participant perspective.

In a recent study, Eisenmann et al. [4,22] investigated the current participation practice as well as winning strategies in all 80 biomedical image analysis challenges conducted within MICCAI and ISBI 2021. In this following, we will focus on the 18 MRI tasks of this study and give insights on both general participation in MRI challenges as well as strategies of MRI challenge winners.

MRI challenge participants typically did not work alone, but formed teams with a median of three members. The lead of the participating team had taken part in a median of two challenges before the current one. The vast majority of participants developed their methods in Python and used PyTorch for method development. The teams spent a median of 48 hours for method development, which is almost half the time observed in the general study of Eisenmann et al. [4,22]. 40% of participants claimed to not have had enough time to prepare their submission.

88% of participants employed deep learning algorithms, 6% less compared to the general study. A median of 10 hyperparameter combinations were explored, while 64% of MRI challenge participants took the metrics into account while searching for hyperparameters. Only 10% used additional data and only 8% used pre-trained models. A median of 6.3×10^6 trainable parameters were used, which is 1.5×10^6 less compared to the general study results. Most of the participants did not use early stopping (66%). 48% post-processed the challenge data. The majority of MRI challenge participants worked on a Graphics Processing Unit (GPU) cluster and the total training time was a median of 300 GPU hours, 30 hours more compared to the general study.

38%, which is 12% less than all challenges in that study, submitted one single model, while the remaining used ensembling composed of a median of five models. Ensembling was either applied for multiple identical models of either different initialization (5%), different training subsets (20%), different models trained on the whole dataset (8%), or randomly drawn training data subsets (4%). Most time was spent on choosing the right network architecture for the given task and choosing appropriate data augmentation, which was used by the majority of algorithms.

Winners of MRI challenges rated multi-task learning as the most important design decision related to the training paradigm and deciding on the augmentation method as the most important design choice for data usage. For model initialization, both supervised pre-training on the domain data and self-supervised or unsupervised pre-training were seen as crucially important. When asked specifically to identify strategies for winning a challenge, MRI challenge winners attributed knowing the state of the art in the field and investing in an experiment pipeline that allows rapid iteration of experiments as crucially important. They further identified reflecting the metrics in the method design, analyzing and handling failure cases, optimizing hyperparameters systematically, optimizing the augmentation method, leveraging external data from the same domain, having domain knowledge, and incorporating domain experts in the challenge team as very important for winning a challenge. Working in a large team and ensembling heterogeneous models were seen as barely important for winning an MRI challenge.

Of note, many challenge organizers provide their own baseline models, on the one hand to determine how much value the newly developed methods add compared to an already existing algorithm, and on the other hand to provide a starting point for challenge participants. In addition, many challenges publish the winning method's codes and frameworks after a challenge has ended. For challenge participants, it is therefore beneficial to investigate and build upon existing methods, such as the renowned nnU-Net, an open-source, self-configuring framework for image segmentation, which revolutionized segmentation challenges [23] (see Chapter 24).

25.9 Conclusion and outlook: The future of challenges in MRI analysis

In this chapter, we demonstrated the relevance of biomedical challenges for scientific progress and the benchmarking of algorithms in MR image analysis. We discussed that challenges are powerful tools which, however, can be easily misused or lead to false conclusions if not conducted thoughtfully and correctly. We pointed out several potential pitfalls in organizing, conducting and reporting challenges, and algorithm benchmarking to the reader and presented solutions on how to overcome these issues. To summarize the solutions, we advise organizers to carefully define their challenge objective, adjust the full challenge design according to the underlying (clinical) research question, with a particular focus on using adequate metrics and rankings and transparently report all relevant parameters to both the participants and in the final challenge publication or report.

We showed that MRI data plays a crucial role in current biomedical image analysis challenges. With the annual number of dedicated MRI-related challenges on the rise and a general trend toward larger, multi-centered datasets, we expect a further increase in the use of MRI data as well as an expansion of the already diverse range of MRI-related tasks in biomedical challenges. MRI is a complex imaging modality that offers numerous opportunities for ML-based research in the following years. To exemplify this potential, we examined the work of Eisenmann et al. [4], who investigated the current participation practice as well as winning strategies in all 80 biomedical image analysis challenges conducted within MICCAI and ISBI 2021, and specifically analyzed the answers of those survey participants who organized MRI challenges in 2021. For only two of a total of 18 tasks conducted within MRI challenges, the problem stated by the challenge organizers was seen as solved. Half of the challenge organizers indicated that the problem was not solved but substantial progress has been made, and 39% only reported minor progress in solving the problem at hand. Compared to all challenges in the study, more MRI challenges made substantial (4% higher) or minor progress (6% higher).

Moreover, our analysis of the results of Eisenmann et al. [4] revealed that only 28% of MRI challenge organizers think that the winning algorithm(s) exceed the current state of the art. Most methodological progress was seen in new architectures or the combination of architectures (39%) and new phrasing of the optimization problem, such as the loss or labels (33%). About 22% of organizers did not see any substantial progress. Common failure cases were seen in imaging conditions (e.g., motion blur, lighting, contrast, artifacts; 39%), specific shapes/sizes/arrangement that were present during training (33%), domain transfer/generalization (new device, new patient, shape/size/object arrangement which was not present during training; 28%), and specific class(es) that perform below average/aggregation (17%).

Several open research questions were highlighted by the 2021 MRI challenge organizers who took part in the survey. For instance, MRI reconstruction algorithms were based on the challenge performance, found to be currently not yet able to handle realistically degraded brain MRI images. Large deformations, large variations in voxel size (before pre-processing), and contrast were found to be difficult as well for the current state-of-the-art algorithms. When organizing a challenge, dataset representativeness was an open problem, due to, for example, sparsity of small elements of interest with potentially large variability in shape and intensity signature. When progressing toward more advanced challenge setups, federated learning is currently seen as an important step [24], as conducted in the Federated Tumor Segmentation (FeTS) challenge. However, the organizers indicated that the federated setup limited the analysis of failure cases because they could not directly be accessed.

In conclusion, many open research questions and unsolved problems remain in the benchmarking of MRI algorithms, particularly MRI-related challenges. When organized thoroughly, challenges help foster scientific progress and define new state-of-the-art methods. They offer an opportunity to comparatively test the performance of new image analysis algorithms under fair and transparent conditions that ideally reflect the underlying clinical problem. In MRI research as in other imaging modalities, adequate benchmarking is the ultimate prerequisite to translation of algorithms into clinical practice. Novel image analysis methods can only ever be as good in practice as the validation they have undergone. Rigorous validation and challenge design is thus of utmost importance for ML algorithms to be successfully applied in real-life clinical practice and for future algorithm development.

References

[1] West J, Fitzpatrick JM, Wang MY, Dawant BM, Maurer Jr CR, Kessler RM, Woods RP. Comparison and evaluation of retrospective intermodality brain image registration techniques. J Comput Assist Tomogr 1997;21(4):554–68.

[2] Mendrik AM, Aylward SR. A framework for challenge design: insight and deployment challenges to address medical image analysis problems. arXiv preprint arXiv: 1911.08531 2019.

[3] Varoquaux G, Cheplygina V. Machine learning for medical imaging: methodological failures and recommendations for the future. NPJ Digit Med 2022;5(1):48.

[4] Eisenmann M, Reinke A, Weru V, Tizabi MD, Isensee F, Adler TJ, Maier-Hein L. Why is the winner the best?. In: Proceedings of the IEEE/CVF Conference on Computer Vision and Pattern Recognition; 2023. p. 19955–66.

[5] Menze BH, Jakab A, Bauer S, Kalpathy-Cramer J, Farahani K, Kirby J, Van Leemput K. The multimodal brain tumor image segmentation benchmark (BRATS). IEEE Trans Med Imag 2014;34(10):1993–2024.

[6] Reinke A, Eisenmann M, Onogur S, Stankovic M, Scholz P, Full PM, Maier-Hein L. How to exploit weaknesses in biomedical challenge design and organization. In: Medical Image Computing and Computer Assisted Intervention–MICCAI 2018: 21st International Conference, Granada, Spain, September 16-20, 2018, Proceedings, Part IV, 11. Springer International Publishing; 2018. p. 388–95.

[7] Reinke A. Eliminating flaws in biomedical image analysis validation (Doctoral dissertation). 2023.

[8] Maier-Hein L, Eisenmann M, Reinke A, Onogur S, Stankovic M, Scholz P, Kopp-Schneider A. Why rankings of biomedical image analysis competitions should be interpreted with care. Nat Commun 2018;9(1):5217.

[9] Rädsch T, Reinke A, Weru V, Tizabi MD, Schreck N, Kavur AE, Maier-Hein L. Labelling instructions matter in biomedical image analysis. Nat Mach Intell 2023;5(3): 273–83.

[10] Joskowicz L, Cohen D, Caplan N, Sosna J. Inter-observer variability of manual contour delineation of structures in CT. Eur Radiol 2019;29:1391–9.

[11] Maier-Hein L, Reinke A, Kozubek M, Martel AL, Arbel T, Eisenmann M, Landman BA. BIAS: transparent reporting of biomedical image analysis challenges. Med Image Anal 2020;66:101796.

[12] Reinke A, Tizabi MD, Baumgartner M, Eisenmann M, Heckmann-Nötzel D, Kavur AE, Maier-Hein L. Understanding metric-related pitfalls in image analysis validation. Nat Meth 2024;21(2):182–94.

[13] Maier-Hein L, Reinke A, Christodoulou E, Glocker B, Godau P, Isensee F, Jäger PF. Metrics reloaded: pitfalls and recommendations for image analysis validation. Nat Meth 2024;21(2):195–212.

[14] Reinke A, Tizabi MD, Sudre CH, Eisenmann M, Rädsch T, Baumgartner M, Maier-Hein L. Common limitations of image processing metrics: a picture story. arXiv preprint arXiv:2104.05642 2021.

[15] Luiten J, Osep A, Dendorfer P, Torr P, Geiger A, Leal-Taixé L, Leibe B. Hota: a higher order metric for evaluating multi-object tracking. Int J Comput Vis 2021;129:548–78.
[16] Gruber S, Buettner F. Better uncertainty calibration via proper scores for classification and beyond. Adv Neural Inf Process Syst 2022;35:8618–32.
[17] Guo C, Pleiss G, Sun Y, Weinberger KQ. On calibration of modern neural networks. In: International Conference on Machine Learning. PMLR; July 2017. p. 1321–30.
[18] Wiesenfarth M, Reinke A, Landman BA, Eisenmann M, Saiz LA, Cardoso MJ, Kopp-Schneider A. Methods and open-source toolkit for analyzing and visualizing challenge results. Sci Rep 2021;11(1):2369.
[19] Roß T, Bruno P, Reinke A, Wiesenfarth M, Koeppel L, Full PM, Maier-Hein L. Beyond rankings: learning (more) from algorithm validation. Med Image Anal 2023;86:102765.
[20] Simera I, Moher D, Hirst A, Hoey J, Schulz KF, Altman DG. Transparent and accurate reporting increases reliability, utility, and impact of your research: reporting guidelines and the EQUATOR Network. BMC Med 2010;8(1):1–6.
[21] Nichols TE, Das S, Eickhoff SB, Evans AC, Glatard T, Hanke M, Yeo BT. Best practices in data analysis and sharing in neuroimaging using MRI. Nat Neurosci 2017;20(3): 299–303.
[22] Eisenmann M, Reinke A, Weru V, Tizabi MD, Isensee F, Adler TJ, Finzel R. Biomedical image analysis competitions: the state of current participation practice. arXiv preprint arXiv:2212.08568 2022.
[23] Isensee F, Jaeger PF, Kohl SA, Petersen J, Maier-Hein KH. nnU-Net: a self-configuring method for deep learning-based biomedical image segmentation. Nat Methods 2021;18 (2):203–11.
[24] Rieke N, Hancox J, Li W, Milletari F, Roth HR, Albarqouni S, Cardoso MJ. The future of digital health with federated learning. NPJ Digit Med 2020;3(1):119.

PART

Conclusion

8

CHAPTER

Current solutions, future challenges, and directions 26

Thomas Küstner[1,*] and Sergios Gatidis[2]

[1]*University Hospital Tübingen, Medical Image and Data Analysis, Department of Radiology, Tübingen, Germany;* [2]*Department of Radiology, Stanford University, Stanford, CA, United States*

[*]*Corresponding author: E-mail: thomas.kuestner@med.uni-tuebingen.de*

26.1 Introduction

The integration of machine learning (ML) into magnetic resonance imaging (MRI) has revolutionized the field, offering promising advancements along the whole imaging pipeline from data acquisition, over image reconstruction, to postprocessing, and for various clinical applications. Once realized, this intelligent imaging revolution will enable faster acquisition times, lower clinician workloads, reduced healthcare costs, and more personalized treatment decisions for patients.

With the increasing availability of data and computational resources, numerous ML solutions which are primarily based on deep learning methods, have been proposed in recent years, and these methods are increasingly translated into clinical practice. ML algorithms have been explored for various aspects of the MR imaging workflow, including scan planning, accelerated acquisition and reconstruction, and image analysis. While current research often focuses on just one segment of this comprehensive imaging pipeline, ML offers broader potential to enhance the entire workflow, from image acquisition to analysis and diagnosis. Future ML applications will further refine the selection of exams based on real-time physiological parameters, such as heart rate, or patient data gathered during planning. Moreover, ML techniques will continue to accelerate scan times, enabling real-time interventional MRI.

A notable trend is the integration of different elements of the imaging pipeline, such as acquisition, reconstruction, postprocessing, analysis, and diagnosis, into unified models, often in the form of multitask and multimodal networks. These models capitalize on the increasing data availability and will shape the future of ML-driven MR imaging. However, for ML algorithms to be reliable in routine clinical practice, challenges surrounding usability, robustness, reproducibility, data quality, and performance must be addressed. Rigorous validation across diverse and heterogeneous patient populations will be critical to ensure the effectiveness and reliability of these technologies.

Machine Learning in MRI. https://doi.org/10.1016/B978-0-443-14109-6.00026-2

The future of ML in MRI is centered around making these technologies more accessible, transparent, robust, clinically relevant, and integrated into routine workflows. This chapter summarizes the key challenges facing the field of ML in MRI and discusses future directions for overcoming these barriers, based on insights from the discussions in the chapters of this book. The chapter discusses these findings for the respective sections of this book addressing MR image acquisition, reconstruction, postprocessing, and their clinical applications, while highlighting critical aspects of generalization, fairness, and reproducibility.

26.2 MR image acquisition

Magnetic Resonance (MR) image acquisition involves capturing detailed internal body structures using strong magnetic fields and radio waves. It begins with selecting the appropriate imaging sequence based on the clinical or research objective, such as T1-weighted or T2-weighted imaging, to obtain the desired contrast, functional and/or quantitative information. Advanced imaging techniques like functional MRI (fMRI), magnetic resonance spectroscopy (MRS), or diffusion tensor imaging (DTI) provide more specific information on tissue function and pathology.

MR imaging offers a substantial opportunity for ML, to explore and exploit the high dimensionality, redundancy, and/or complementary nature of the data. One of the primary challenges in this context is understanding exactly what ML models are learning from the data, as highlighted in Chapter 4: "We still need to understand what else the machine is learning as beneficial that we have not seen yet." This is especially true when such insights are not immediately apparent, have not been observed through conventional imaging methods, or when human observation and intuition may fall short. It underscores the need for deeper understanding and greater transparency in how ML algorithms interpret and digest MRI data. In addition, this raises the question which sequence, contrast, sampling, and so on are most informative for the ML method and vice versa how ML could be informative or supportive for the imaging process. If adequately addressed, ML holds the potential to unlock new diagnostic features or hidden patterns in imaging that were previously unexplored.

Chapters 4–6 discuss various aspects and potentials of ML along the MR imaging process for improving the sampling efficiency (Chapter 4), supporting "low-field MRI in a high-field world" (Chapter 5), and acquiring spectroscopic information (Chapter 6). While MRI is usually prone to long acquisition times, over the past decades different acceleration strategies have been proposed and established [1]. Long acquisition times are not tolerated by a substantial proportion of patients and are associated with other downsides such as decreased image quality due to motion artifacts, increased cost, reduced patient throughput and decreased access, particularly in regions with limited availability of MRI scanners. ML has thus the potential to address challenges related to acquisition time, signal-to-noise ratio (SNR), spatial and temporal resolution, acquisition of multicontrast information, and mitigation of artifacts. Addressing these challenges, Chapter 4 sheds light on recent learning-based developments for optimizing MR sampling for efficient extraction of underlying information. These methods could personalize scanning protocols to individual patients, improving both efficiency and image quality. It is argued that the use of generative models and data augmentation techniques may refine training data quality and mitigate the challenges posed by undersampling. The chapter stresses the importance of transparency in ML algorithms so that clinicians can trust and interpret the results produced by these systems. Chapter 5 shifts focus to the challenges posed by low-field MRI systems, which are widely used in resource-constrained environments but can suffer from lower SNR and resolution compared with high-field MRI systems. The chapter explores the potential for ML to enhance the performance of low-field MRI by improving image quality, reducing noise and artifacts, and addressing the resolution limitations that have traditionally hindered their clinical utility. Future developments need to focus on improving the generalizability of ML so that it can be applied across both low- and high-field MRI systems. Chapter 6 introduces a proactive approach to MR image acquisition, emphasizing the importance of addressing potential issues before they escalate into significant diagnostic errors. This concept, encapsulated in the phrase "Fix it before it's broken" highlights the value of using ML to predict and correct problems during the image acquisition process. Challenges, such as patient motion, scanner malfunctioning, and suboptimal scanning parameters can compromise image quality and diagnostic accuracy. ML models, however, could help identify these issues in real time, allowing for adjustments and corrections before they affect the scan. By integrating ML into the imaging workflow, the process could become more efficient, reducing the need for repeated scans, improving diagnostic reliability, and enhancing patient care. Future efforts remain to focus on refining these models to enhance their clinical applicability.

26.3 MR image reconstruction

MR image reconstruction involves solving the inverse problem of the MR encoding process which links the acquired k-space to the reconstructed image. This relationship is characterized by the Bloch equations and depends on a number of parameters, such as motion (Chapter 9) or MR relaxation effects (Chapter 10). An analytical expression can be derived for the MR signal model with the encoding operator incorporating such relevant physical effects (Chapter 3). ML has shown great promise in improving MR image reconstruction especially in the context of accelerated/sub-Nyquist-sampled imaging (Chapter 7). Chapter 7 introduces the different aspects and variations of ML algorithms and architectures. It connects these approaches to practical applications such as motion correction, low-field MR image reconstruction, and quantitative imaging, which are further detailed in the subsequent Chapters

9 and 10. In Chapter 8, concepts and architectures for improving spatial resolution are presented and super-resolution methods in the context of neuroimaging are highlighted. Chapter 11 addresses the trustworthiness and robustness of ML models in the safety-critical environment of MR imaging.

Motion artifacts caused by physiological and patient movement, as well as variability in MRI data due to scanner differences, long acquisition times, and calibration issues are major obstacles to clinical implementation. While ML models have demonstrated an ability to mitigate some artifacts, they are not immune to generating new artifacts or hallucinations, especially in more challenging or out-of-domain cases. As emphasized in Chapter 11, "We stress that while DL reconstruction output may provide high-quality images upon initial inspection, the reconstructions may also include hallucinations or omit small structural elements, limiting its diagnostic value. Moreover, it is important to ensure the robustness of MR image reconstruction models, given the sensitive nature of radiological data." ML models often perform well on training datasets but fail to generalize across different scanners, protocols, or patient populations, limiting their clinical applicability. In addition, the black-box nature of ML models raises concerns about their transparency, equity, and trustworthiness in clinical practice. Addressing these issues requires robust evaluation, clinical validation, and improved generalization methods, alongside the publication of open-access datasets and development of reproducible research practices to enhance trust and accuracy in these applications. Future investigations should thus focus on more robust ML models and data augmentation techniques that can prevent or reduce the introduction of new errors. Incorporating the physical model into ML reconstruction algorithms, known as physics-informed ML, may help to reduce artifacts and improve model reliability. To enhance model generalization across diverse clinical settings, transfer learning and multicenter studies need to be performed. Lastly, incorporating explainable AI techniques or the ability to deploy personalized models would foster clinicians to trust in these models.

While early ML developments helped to reduce computation times for reconstruction, especially in contrast to iterative reconstruction approaches, novel models show a trend toward higher computational demand, making their deployment into clinical workflows challenging. Model compression and cloud-based solutions may reduce the computational demands of ML-based reconstructions, making them again more feasible in clinical practice.

Bias is another concern, arising from nonrepresentative or misused datasets, which can result in inequities in healthcare by delivering less accurate diagnoses for underrepresented populations. The misuse or inappropriate documentation of datasets—referred to as "data crimes"—can further distort model performance, as preprocessing steps may not be accounted for, providing misleading results. In addition, ML models are vulnerable to adversarial attacks, where subtle changes in data can drastically alter outputs. Instabilities in models, particularly in noisy environments, can lead to poor or unpredictable reconstructions, and models may struggle with generalization when there are substantial differences between training and test data. On the contrary, there is also potential for ML to harmonize and standardize processing and to provide more objective measures in changing domains and environments. This in turn requires trustworthy and robust models as discussed above. As noted in Chapter 11, "It is of course easy to find a provably stable, non-hallucinating DNN: the zero network is perfectly stable and non-hallucinating, yet wholly ineffective in practice," it is crucial to find a balance between stability, trustworthiness, and diagnostic effectiveness.

The future of ML for MR image reconstruction lies in addressing the key challenges of data quality, processing speed, and generalization. Physics-informed and model-based techniques will play crucial roles in improving image quality and enhancing the clinical applicability of these technologies. Ensuring that ML models are efficient, accurate, and capable of generalizing across diverse clinical settings is essential for their successful integration into clinical workflows and for impacting patient care. The development of standardized benchmarks, open-source tools, and data-sharing platforms will facilitate collaborative efforts across institutions, enabling faster progress and innovation in the field.

26.4 MR image postprocessing

Once MRI images are acquired and reconstructed, postprocessing steps such as segmentation, quality control, and advanced analysis become necessary to extract clinically relevant information. Moreover, it is desirable to analyze MRI findings in conjunction with information from other imaging modalities, laboratory data, or further nonimaging data, like electronic healthcare records. However, managing, linking, and analyzing diverse data from within the MRI domain (e.g., different sequences, scanners, and so on) or between MRI and another domain still presents unique challenges.

Image segmentation is one of the most fundamental tasks in MRI postprocessing, but segmenting complex anatomical structures across diverse patient populations remains challenging. One key issue is the lack of large, annotated datasets for training ML models, particularly for rare conditions or unusual anatomies. Future directions include the use of self-supervised learning and multitask learning models. Such frameworks can learn from unannotated or weakly labeled data and perform multiple tasks (e.g., reconstruction and segmentation) simultaneously. Cross-modality approaches that integrate MRI with other imaging modalities, such as CT or PET, also help to overcome the limitations of isolated MRI datasets. The combination of tasks and data show a stronger resilience to domain shifts, thus improving generalization.

With increasing automation, data complexity, and quantity, embedded image quality control becomes critical to ensure accurate image processing. However, the subjective nature of quality assessments and the diversity of MRI protocols, applications, and artifacts render it a challenging task. "What precisely constitutes image quality? Is it merely the attainment of clear, noise-free images, or does it extend beyond – encompassing the avoidance of motion artifacts, irregularities, and other distortions? The pursuit of quality in MRI, in the end, encompasses the synthesis of images that provide clear diagnostic value, enabling radiologists to provide the correct and appropriate diagnosis" (Chapter 13). The responsibility of quality control is thus to be extended beyond mere quantitative measures and to provide meaningful information or guidance for downstream tasks. ML models capable of real-time, automated feedback can improve the efficiency and consistency of quality measures, reducing the reliance on manual assessments. "Is it still needed to retain a radiologist in the loop? Could we instead integrate the metrics [...] with a reasoning agent capable of directly generating a comprehensive report?" (Chapter 13). While human experts remain an important factor, their role in this loop will gradually change. Future systems will focus on objective and automated assessment to support humans in downstream tasks (diagnosis, treatment monitoring, and so on). Hence, research will need to explore robust methods, allowing quality control tools to work effectively in these settings and across various protocols and patient populations. Quality control measures will thus become an integral part of any postprocessing method, especially when it comes to advanced image analysis and diagnostics. Future research will refine the quality metrics and interaction with their downstream tasks and upstream data so that they are aligned with clinical needs.

The data complexity (multiparametric, multimodal, and high-dimensional) also affects the reliable extraction of physiological and functional information from the MR images. ML-based image analysis, radiomics, and other feature extraction techniques are faced with handling small and/or heterogeneous datasets. While pretrained models or recent foundation models may offer potential solutions for improving performance, they are often overconfident in their predictions, easily deceived and vulnerable to adversarial attacks, are notorious to fine-tune/adapt to the desired application or currently have overwhelming computational demands. Ingenious data fusion and embedding of prior knowledge (e.g., pretrained models) will be key to advancing clinical image analysis. Image-to-image translations between datasets can be supportive in this context, for example for generating synthetic information of missing data entries, or for the restoration of image quality in the form of artifact reduction. Transfer learning and human-in-the-loop learning strategies will also be important for extending analysis to rare diseases with limited available training data. There is significant potential in integrating image acquisition and analysis in an end-to-end fashion, where clinical objectives influence the imaging sequence (k-space trajectory, timings, ...). Similar to image quality control discussed above, approaches could include joint reconstruction, segmentation, registration, and calculation of clinical outcomes.

For successful translation into the clinical routine, these end-to-end solutions need to be as transparent as possible, provide meaningful and verifiable intermediate results, and/or supply objective quality or certainty measures. By developing more generalizable models, more efficient handling of complex and/or scarce data, the field will continue to progress toward clinical adoption.

26.5 Generalization, fairness, and reproducibility

The volume and complexity of data are increasing as MRI technology evolves. In addition, personalized MRI protocols will be increasingly implemented to enhance image information. As ML models are deployed in these diverse clinical settings, ensuring their scalability, fairness, and generalizability will be critical. Chapters 19–23 outline different application-specific ML solutions. Multitask ML models that can handle multiparametric and multimodal datasets in a scalable and efficient manner are needed to support future clinical applications. Across different domains (anatomical regions, MR sequences, patient populations, …) ML models will need to be refined to address the application-specific challenges such as motion correction, artifact mitigation, or objective biomarker extraction, while being as general applicable as possible. An important goal is to achieve real-time, early-detection models that leverage clinical context to improve clinical decision-making. Chapter 16 highlights that, "In MRI scans many different components can cause a domain shift. It is important to remember that compared to other medical imaging systems, such as e.g., Computed Tomography scanners which measure density in absolute units, MRI scanners are not measurement devices." On top of the multitude of application-tailored MRI sequences, this variability in MRI signal underscores the need for robust models that can adapt to different scanners, settings, and patient populations. In addition, variability in MRI protocols across different institutions complicates the development of ML models that generalize well. Chapter 16 states "Generalization has always been at the heart of machine learning, and if we are able to understand why we have problems generalizing, we have a rich toolbox to mitigate the problem." Tackling the challenges of generalization will thus require an understanding of domain shifts and careful design of training data to ensure fair and accurate results across different domains. In this regard, cross-institutional collaboration and large-scale datasets are needed to enhance model performance and generalizability.

However, large annotated datasets are often not available for training these models, particularly in rare conditions or populations. Privacy concerns, data heterogeneity, and lack of standardization across institutions and imaging systems often hinder data sharing and collaboration. While large-scale and open accessible datasets can contribute to a more widespread development and increased generalization, they also pose significant challenges and risks. Data management, including storage, processing time, and data integration of diverse systems, with proper documentation following the FAIR principles [2], is needed which can be time- and cost-intensive. The availability of only a few publicly accessible databases bears the risk of over-engineered solutions. In this context, reproducibility remains also a significant challenge. Inconsistent MRI protocols, lack of standardized datasets, and varying clinical practice hinder the reliable validation of ML models and the development of standardized benchmarks. Benchmarks should reflect clinical outcomes, ensuring that ML models are not only accurate but also clinically relevant. Existing evaluation metrics do not necessarily fully capture clinical relevance or may be over-fitted to the specific benchmarks. Hence, future work must focus on the development of standardized benchmarks and multicenter datasets to increase accessibility to data and to facilitate model evaluation.

The key lies in transparent and open source access. As noted in Chapter 24, "Why open source?" "The answer is a confluence of beneficial incentives, both in terms of using and contributing to open-source software. Open-source software brings communities together. It gives would-be developers the chance to contribute to something and raise their own skill set in the process. It provides software that, defying its early critics, can beat proprietary software in terms of security, transparency, and consistency." Thus, datasets, benchmarks, and ML models should be developed open source to foster collaboration and standardization. As also emphasized in Chapter 25, "[ML] challenges have a tremendous impact on the research community," the reproducibility crisis in AI research [3] underscores the importance of collaboration, open access, and transparency in research and development.

Ethical and legal considerations with regulatory frameworks are equally important as AI in healthcare continues to expand globally. Healthcare companies and research institutions need to collaborate to develop transparent and human-interpretable models to avoid the so-called "black box" nature of ML. As noted in Chapter 18, "How users communicate with the AI model, [...] is [equally ...] important [as] to discuss how the AI model can communicate with the user." Combining expert clinical input with ML models is difficult but necessary to ensure that models are reliable, context-sensitive, and trustworthy in real-world scenarios. Hence, future developments need to consider how to appropriately integrate the human-in-the-loop to best obtain explainable and transparent ML systems. Ultimately this could result in improved efficiency, patient outcomes, and quality-adjusted life years.

By developing more generalizable models and incorporating explainable AI, in an open-source research environment, the field will continue to progress toward clinical adoption. Collaborative efforts to standardize data formats and benchmarks, along with advances in computational efficiency, will be key to ensuring that these technologies are scalable, reproducible, and useful in diverse clinical environments.

26.6 Conclusion

In conclusion, existing challenges and potential directions offer broad opportunities for future research. Moving forward, the primary goal will be to create ML models that not only perform well on isolated datasets but are also robust, generalizable, interpretable, and directly impactful in clinical environments. Innovations in MRI will always challenge existing ML solutions. The combination of data sharing, standardization, and thoughtful benchmarking will help accelerate the transition of ML technologies from research to real-world clinical use. By bridging gaps between clinical expertise and data processing, ML can significantly enhance diagnostic workflows, and ultimately improve patient outcomes across various MRI applications. We would like to give the last concluding words to GPT-o1 by asking it "What would you conclude for the future of machine learning in magnetic resonance imaging?". GPT-o1: "The future of ML in MRI lies in making models more interpretable, generalizable, and clinically relevant. Advances in physics-informed learning, standardized benchmarks, and open collaboration will enhance robustness and trustworthiness. Integrating ML across the imaging pipeline will improve efficiency and personalization, but challenges like data heterogeneity, computational demands, and ethical concerns must be addressed. With transparency and human oversight, ML-driven MRI can revolutionize precision medicine and patient care."-—We can only agree.

References

[1] Introduction. In: Akçakaya M, Doneva M, Prieto C, editors. Magnetic resonance image reconstruction, Advances in magnetic resonance technology and applications. vol. 7. Academic Press; 2022. xxi–xxii. https://doi.org/10.1016/B978-0-12-822726-8.00008-7. https://www.sciencedirect.com/science/article/pii/B9780128227268000087.

[2] Wilkinson MD, Dumontier M, Aalbersberg IJ, Appleton G, Axton M, Baak A, et al. The fair guiding principles for scientific data management and stewardship. Sci Data 2016;3(1):1–9.

[3] Ball P. Is AI leading to a reproducibility crisis in science? Nature 2023;624(7990):22–5.

Index

'*Note:* Page numbers followed by "f" indicate figures.'

C

D

G

H

O

P

T